2019 IEEE Radio Frequency Integrated Circuits Symposium (RFIC 2019)

Boston, Massachusetts, USA
2 – 4 June 2019

IEEE Catalog Number: CFP19MMW-POD
ISBN: 978-1-7281-1702-7

**Copyright © 2019 by the Institute of Electrical and Electronics Engineers, Inc.
All Rights Reserved**

Copyright and Reprint Permissions: Abstracting is permitted with credit to the source. Libraries are permitted to photocopy beyond the limit of U.S. copyright law for private use of patrons those articles in this volume that carry a code at the bottom of the first page, provided the per-copy fee indicated in the code is paid through Copyright Clearance Center, 222 Rosewood Drive, Danvers, MA 01923.

For other copying, reprint or republication permission, write to IEEE Copyrights Manager, IEEE Service Center, 445 Hoes Lane, Piscataway, NJ 08854. All rights reserved.

*** *This is a print representation of what appears in the IEEE Digital Library. Some format issues inherent in the e-media version may also appear in this print version.*

IEEE Catalog Number:	CFP19MMW-POD
ISBN (Print-On-Demand):	978-1-7281-1702-7
ISBN (Online):	978-1-7281-1701-0
ISSN:	1529-2517

Additional Copies of This Publication Are Available From:

Curran Associates, Inc
57 Morehouse Lane
Red Hook, NY 12571 USA
Phone: (845) 758-0400
Fax: (845) 758-2633
E-mail: curran@proceedings.com
Web: www.proceedings.com

TABLE OF CONTENTS

THE DIGITAL FUTURE OF RFICS 1
Greg Henderson

DO THE NETWORKS OF THE FUTURE CARE ABOUT THE MATERIALS OF THE PAST? 2
Michael Peeters

A 1.2-2.8 GHZ TUNABLE LOW-NOISE AMPLIFIER WITH 0.8-1.6 DB NOISE FIGURE 3
Hao Gao ; Zhe Song ; Zhe Chen ; Domine M. W. Leenaerts ; Peter G. M. Baltus

A 28-GHZ CMOS LNA WITH STABILITY-ENHANCED G_m-BOOSTING TECHNIQUE USING
TRANSFORMERS 7
Sunwoo Kong ; Hui-Dong Lee ; Seunghyun Jang ; Jeehoon Park ; Kwang-Seon Kim ; Kwang-Chun Lee

KA-BAND CMOS ABSORPTIVE SP4T SWITCH WITH ONE-THIRD MINIATURIZATION 11
Bosung Suh ; Byung-Wook Min

A COMPACT, HIGH-POWER, 60 GHZ SPDT SWITCH USING SHUNT-SERIES SIGE PIN
DIODES 15
Yunyi Gong ; Jeffrey W. Teng ; John D. Cressler

LOW-COST, HIGH-GAIN ANTENNA MODULE INTEGRATING A CMOS FREQUENCY
MULTIPLIER DRIVER FOR COMMUNICATIONS AT D-BAND 19
*Francesco Foglia Manzillo ; Jose Luis Gonzalez-Jimenez ; Antonio Clemente ; Alexandre Siligaris ; Benjamin
Blampey ; Cedric Dehos*

SCALABLE ANALYTICAL MODEL OF 1.7 THZ CUT-OFF FREQUENCY SCHOTTKY DIODES
INTEGRATED IN 55NM BICMOS TECHNOLOGY 23
*Vincent Gidel ; Fréderic Gianesello ; Pascal Chevalier ; Grégory Avenier ; Nicolas Guitard ; Victor Milon ;
Michel Buczko ; Charles-Alex Legrand ; Cyril Luxey ; Guillaume Ducournau*

EXCELLENT 22FDX HOT-CARRIER RELIABILITY FOR PA APPLICATIONS 27
*T. Chen ; C. Zhang ; W. Arfaoui ; A. Bellaouar ; S. Embabi ; G. Bossu ; M. Siddabathula ; K. W. J. Chew ; S. N.
Ong ; M. Mantravadi ; K. Barnett ; J. Bordelon ; R. Taylor ; S. Janardhanan*

22NM FULLY-DEPLETED SOI HIGH FREQUENCY NOISE MODELING UP TO 90GHZ
ENABLING ULTRA LOW NOISE MILLIMETRE-WAVE LNA DESIGN 31
*L. H. K. Chan ; S. N. Ong ; W. L. Oo ; K. W. J. Chew ; C. Zhang ; A. Bellaouar ; W. H. Chow ; T. Chen ; R. Rassel
; J. S. Wong ; C. K. Lim ; C. W. F. Wan ; J. Kim ; W. H. Seet ; D. Harame*

22NM ULTRA-THIN BODY AND BURIED OXIDE FDSOI RF NOISE PERFORMANCE 35
Ousmane M. Kane ; Luca Lucci ; Pascal Scheiblin ; Sylvie Lepilliet ; François Danneville

A 76-81GHZ FMCW TRANSCEIVER WITH 3-TRANSMIT, 4-RECEIVE PATHS AND 15DBM
OUTPUT POWER FOR AUTOMOTIVE RADARS 39
*Zongming Duan ; Dongfang Pan ; Bowen Wu ; Yan Wang ; Bingbing Liao ; Dong Huang ; Yanhui Wu ; Daiguo
Xu ; Hua Xu ; Wei Lv ; Yuefei Dai ; Pei Li ; Yan Wang ; Fujiang Lin*

RECONFIGURABLE 60-GHZ RADAR TRANSMITTER SOC WITH BROADBAND
FREQUENCY TRIPLER IN 45NM SOI CMOS 43
Wooram Lee ; Tolga Dinc ; Alberto Valdes-Garcia

A 94GHZ 2×2 PHASED-ARRAY FMCW IMAGING RADAR TRANSCEIVER WITH 11DBM
OUTPUT POWER AND 10.5DB NF IN 65NM CMOS 47
Dong Huang ; Li Zhang ; Huabing Zhu ; Boshen Chen ; Yang Tang ; Yan Wang

X/KU-BAND FOUR-CHANNEL TRANSMIT/RECEIVE SIGE PHASED-ARRAY IC 51
Prabir Saha ; Sriram Muralidharan ; Jinzhou Cao ; Ozan Gurbuz ; Christopher Hay

ULTRA-WIDEBAND 8-45 GHZ TRANSMITTER FRONT-END FOR A RECONFIGURABLE
FMCW MIMO RADAR 55
Mantas Sakalas ; Songhui Li ; Niko Joram ; Paulius Sakalas ; Frank Ellinger

A 51.5 - 64.5 GHZ ACTIVE PHASE SHIFTER USING LINEAR PHASE CONTROL TECHNIQUE
WITH 1.4 PHASE RESOLUTION IN 65-NM CMOS 59
Tianjun Wu ; Chenxi Zhao ; Huihua Liu ; Yunqiu Wu ; Yiming Yu ; Kai Kang

DIGITALLY-ASSISTED 27-33 GHZ REFLECTION-TYPE PHASE SHIFTER WITH ENHANCED
ACCURACY AND LOW IL-VARIATION 63
Jingjing Xia ; Mahitab Farouk ; Slim Boumaiza

A 21 TO 30-GHZ MERGED DIGITAL-CONTROLLED HIGH RESOLUTION PHASE SHIFTER-
PROGRAMMABLE GAIN AMPLIFIER WITH ORTHOGONAL PHASE AND GAIN CONTROL
FOR 5-G PHASE ARRAY APPLICATION 67
Wei Zhu ; Wei Lv ; Bingbing Liao ; Yanping Zhu ; Yuefei Dai ; Pei Li ; Lei Zhang ; Yan Wang

A 20 ~ 43 GHZ VGA WITH 21.5 DB GAIN TUNING RANGE AND LOW PHASE VARIATION FOR 5G COMMUNICATIONS IN 65-NM CMOS 71
Tianjun Wu ; Chenxi Zhao ; Huihua Liu ; Yunqiu Wu ; Yiming Yu ; Kai Kang

A 26-GHZ VECTOR MODULATOR IN 130-NM SIGE BICMOS ACHIEVING MONOTONIC 10-B PHASE RESOLUTION WITHOUT CALIBRATION 75
Ilker Kalyoncu ; Abdurrahman Burak ; Mehmet Kaynak ; Yasar Gurbuz

A 20-32GHZ DIGITAL QUADRATURE TRANSMITTER WITH NOTCHED-MATCHING AND MODE-SWITCH TOPOLOGY FOR 5G WIRELESS AND BACKHAUL 79
Huizhen Jenny Qian ; Yiyang Shu ; Jie Zhou ; Xun Luo

A WIDEBAND DIGITAL POLAR TRANSMITTER WITH INTEGRATED CAPACITOR-DAC-BASED CONSTANT-ENVELOPE DIGITAL-TO-PHASE CONVERTER 83
Tong Li ; Liang Xiong ; Yun Yin ; Yangzi Liu ; Hao Min ; Na Yan ; Hongtao Xu

A 5GHZ TO 6GHZ CMOS TRANSMITTER FOR FULL-DUPLEX WIRELESS WITH WIDEBAND DIGITAL CANCELLATION 87
Nimrod Ginzberg ; Dror Regev ; Genadiy Tsodik ; Shimi Shilo ; Doron Ezri ; Emanuel Cohen

A SUB-MW ALL-PASSIVE RF FRONT END WITH IMPLICIT CAPACITIVE STACKING ACHIEVING 13 DB GAIN, 5 DB NF AND +25 DBM OOB-IIP3 91
Vijaya Kumar Purushothaman ; Eric Klumperink ; Berta Trullas Clavera ; Bram Nauta

A 0.3-TO-1.3GHZ MULTI-BRANCH RECEIVER WITH MODULATED MIXER CLOCKS FOR CONCURRENT DUAL-CARRIER RECEPTION AND RAPID COMPRESSIVE-SAMPLING SPECTRUM SCANNING 95
Guoxiang Han ; Tanbir Haque ; Matthew Bajor ; John Wright ; Peter R. Kinget

A 0.5-20 GHZ RF SILICON PHOTONIC RECEIVER WITH 120 DB·HZ$^{2/3}$ SFDR USING BROADBAND DISTRIBUTED IM3 INJECTION LINEARIZATION 99
Navid Hosseinzadeh ; Aditya Jain ; Kang Ning ; Roger Helkey ; James F. Buckwalter

A 65NM CMOS CONTINUOUS-TIME ELECTRO-OPTIC PLL (CT-EOPLL) WITH IMAGE AND HARMONIC SPUR SUPPRESSION FOR LIDAR 103
Ali Binaie ; Sohail Ahasan ; Harish Krishnaswamy

A 6.5-GHZ CRYOGENIC ALL-PASS FILTER CIRCULATOR IN 40-NM CMOS FOR QUANTUM COMPUTING APPLICATIONS 107
Andrea Ruffino ; Yatao Peng ; Fabio Sebastiano ; Masoud Babaie ; Edoardo Charbon

DESIGN CONSIDERATIONS FOR SPIN READOUT AMPLIFIERS IN MONOLITHICALLY INTEGRATED SEMICONDUCTOR QUANTUM PROCESSORS 111
M. J. Gong ; U. Alakusu ; S. Bonen ; M. S. Dadash ; L. Lucci ; H. Jia ; L. E. Gutierrez ; W. T. Chen ; D. R. Daughton ; G. C. Adam ; S. Iordanescu ; M. Pasteanu ; N. Messaoudi ; D. Harame ; A. Müller ; R. R. Mansour ; S. P. Voinigescu

DIRECT DIGITAL SYNTHESIZER WITH 14 GS/S SAMPLING RATE HETEROGENEOUSLY INTEGRATED IN INP HBT AND GAN HEMT ON CMOS 115
Steven Eugene Turner ; Mark E. Stuenkel ; Gary M. Madison ; Justin A. Cartwright ; Richard L. Harwood ; Joseph D. Cali ; Steve A. Chadwick ; Michael Oh ; John T. Matta ; James M. Meredith ; Justin M. Byrd ; Lawrence J. Kushner

A 1 V 54-64 GHZ 4-CHANNEL PHASED-ARRAY RECEIVER IN 45 NM RFSOI WITH 3.6/5.1 DB NF AND -23 DBM IP1DB AT 28/37 MW PER-CHANNEL 119
Hyunchul Chung ; Qian Ma ; Gabriel M. Rebeiz

A FULLY INTEGRATED 60 GHZ 10 GB/S QPSK TRANSCEIVER WITH DIGITAL TRANSMITTER AND T/R SWITCH IN 65NM CMOS 123
Zheng Song ; Jianfu Lin ; Yutian Li ; Jialiang Ye ; Ruichang Ma ; Baoyong Chi

A 60 GHZ POLARIZATION-DUPLEX TX/RX FRONT-END WITH DUAL-POL ANTENNA-IC CO-INTEGRATION IN SIGE BICMOS 127
Yao Liu ; Arun Natarajan

A 180-GHZ SUPER-REGENERATIVE OSCILLATOR WITH UP TO 58 DB GAIN FOR EFFICIENT PHASE RECOVERY 131
Hatem Ghaleb ; Christian Carlowitz ; David Fritsche ; Corrado Carta ; Frank Ellinger

A BROADBAND DIRECT CONVERSION TRANSMITTER/RECEIVER AT D-BAND USING CMOS 22NM FDSOI 135
Ali A. Farid ; Arda Simsek ; Ahmed S. H. Ahmed ; Mark J. W. Rodwell

ENHANCED PASSIVE MIXER-FIRST RECEIVER DRIVING AN IMPEDANCE WITH 40DB/DECADE ROLL-OFF, ACHIEVING +12DBM BLOCKER-P1DB, +33DBM IIP3 AND SUB-2DB NF DEGRADATION FOR A 0DBM BLOCKER 139
Sashank Krishnamurthy ; Ali M. Niknejad

A CODE-DOMAIN RF SIGNAL PROCESSING FRONT-END FOR SIMULTANEOUS TRANSMIT
AND RECEIVE WITH 49.5 DB SELF-INTERFERENCE REJECTION, 12.1 DBM RECEIVE
COMPRESSION, AND 34.3 DBM TRANSMIT COMPRESSION 143

Hussam Alshammary ; Cameron W. Hill ; Ahmed Hamza ; James F. Buckwalter

A CMOS 0.5-2.5GHZ FULL-DUPLEX MIMO RECEIVER WITH SELF-ADAPTIVE AND
POWER-SCALABLE RF/ANALOG WIDEBAND INTERFERENCE CANCELLATION 147

Yuhe Cao ; Jin Zhou

A 0.5-TO-3.5 GHZ SELF-INTERFERENCE-CANCELING RECEIVER FOR IN-BAND FULL-
DUPLEX WIRELESS 151

Ali Ershadi ; Kamran Entesari

A BASEBAND-MATCHING-RESISTOR NOISE-CANCELING RECEIVER ARCHITECTURE TO
INCREASE IN-BAND LINEARITY ACHIEVING 175MHZ TIA BANDWIDTH WITH A 3-STAGE
INVERTER-ONLY OPAMP 155

Anoop Narayan Bhat ; Ronan Van Der Zee ; Salvatore Finocchiaro ; Francesco Dantoni ; Bram Nauta

A 350MV COMPLEMENTARY 4-5 GHZ VCO BASED ON A 4-PORT TRANSFORMER
RESONATOR WITH 195.8DBC/HZ PEAK FOM IN 22NM FDSOI 159

Omar El-Aassar ; Gabriel M. Rebeiz

X-BAND NMOS AND CMOS CROSS-COUPLED DCO'S WITH A "FOLDED" COMMON-MODE
RESONATOR EXHIBITING 188.5 DBC/HZ FOM WITH 29.5% TUNING RANGE IN 16-NM
CMOS FINFET 163

R. Levinger ; D. Ben-Haim ; I. Gertman ; S. Bershansky ; R. Levi ; J. Kadry ; G. Horovitz

A 18.2-29.3 GHZ COLPITTS VCOS BANK WITH -119.5 DBC/HZ PHASE NOISE AT 1 MHZ
OFFSET FOR 5G COMMUNICATIONS 167

F. Quadrelli ; F. Panazzolo ; M. Tiebout ; F. Padovan ; M. Bassi ; A. Bevilacqua

A 9.6 MW LOW-NOISE MILLIMETER-WAVE SUB-SAMPLING PLL WITH A DIVIDER-LESS
SUB-SAMPLING LOCK DETECTOR IN 65 NM CMOS 171

Hao Wang ; Omeed Momeni

A -40-DBC INTEGRATED-PHASE-NOISE 45-GHZ SUB-SAMPLING PLL WITH 3.9-DBM
OUTPUT AND 2.1% DC-TO-RF EFFICIENCY 175

Sangyeop Lee ; Kyoya Takano ; Shinsuke Hara ; Ruibing Dong ; Shuhei Amakawa ; Takeshi Yoshida ; Minoru
Fujishima

A HIGH EFFICIENCY 39GHZ CMOS CASCODE POWER AMPLIFIER FOR 5G
APPLICATIONS 179

Hyun-Chul Park ; Byungjoon Park ; Yunsung Cho ; Jaehong Park ; Jihoon Kim ; Jeong Ho Lee ; Juho Son ; Kyu
Hwan An ; Sung-Gi Yang

A COMPACT E-BAND PA WITH 22.37% PAE 14.29 DBM OUTPUT POWER AND 26 DB POWER
GAIN WITH EFFICIENCY ENHANCEMENT AT POWER BACK-OFF 183

Liang Chen ; Lei Zhang ; Li Zhang ; Yan Wang

AN E-BAND COMPACT POWER AMPLIFIER FOR FUTURE ARRAY-BASED BACKHAUL
NETWORKS IN 22NM FD-SOI 187

Umut Çelik ; Patrick Reynaert

AN E-BAND FULLY-INTEGRATED TRUE POWER DETECTOR IN 28NM CMOS 191

Valdrin Qunaj ; Patrick Reynaert

A COUPLER-BASED DIFFERENTIAL DOHERTY POWER AMPLIFIER WITH BUILT-IN
BALUNS FOR HIGH MM-WAVE LINEAR-YET-EFFICIENT GBIT/S AMPLIFICATIONS 195

Huy Thong Nguyen ; Hua Wang

VSWR ROBUST LINEARIZER TO IMPROVE SWITCH IMD BY >20DB 199

Thomas Meier ; Atif Mehmood ; Jonas Kaps

A BLOCKER-TOLERANT TWO-STAGE HARMONIC-REJECTION RF FRONT-END 203

Faizan Ul Haq ; Mikko Englund ; Yury Antonov ; Kari Stadius ; Marko Kosunen ; Kim B. Östman ; Kimmo Koli ;
Jussi Ryynänen

A LOW NOISE FIGURE 28GHZ LNA IN 22NM FDSOI TECHNOLOGY 207

Chi Zhang ; Frank Zhang ; Shafiullah Syed ; Michael Otto ; Abdellatif Bellaouar

A 1.7-DB MINIMUM NF, 22-32 GHZ LOW-NOISE FEEDBACK AMPLIFIER WITH
MULTISTAGE NOISE MATCHING IN 22-NM SOI-CMOS 211

Bolun Cui ; John R. Long ; David L. Harame

A 112-GS/S 1-TO-4 ADC FRONT-END WITH MORE THAN 35-DBC SFDR AND 28-DB SNDR UP
TO 43-GHZ IN 130-NM SIGE BICMOS 215

X.-Q. Du ; M. Grözing ; A. Uhl ; S. Park ; F. Buchali ; K. Schuh ; S. T. Le ; M. Berroth

A DUAL-28GB/S DIGITAL-ASSISTED DISTRIBUTED DRIVER WITH CDR FOR OPTICAL-
DAC PAM4 MODULATION IN 40NM CMOS 219

Qiwen Liao ; Shang Hu ; Jian He ; Bozhi Yin ; Patrick Yin Chiang ; Jian Liu ; Nan Qi ; Nanjian Wu

A 77DB-SFDR MULTI-PHASE-SAMPLING 16-ELEMENT DIGITAL BEAMFORMER WITH 64 4GS/S 100MHZ-BW CONTINUOUS-TIME BAND-PASS ΔΣ ADCS......223
Rundao Lu ; Sunmin Jang ; Yun Hao ; Michael P. Flynn

A WIDEBAND DIGITALLY CONTROLLABLE RFIC WITH GAIN AND WAVELENGTH TUNABILITY AND BUILT-IN SELF TEST FUNCTIONALITIES FOR OPTICAL TRANSCEIVER MODULES IN FTTX APPLICATIONS......227
Sreekesh Lakshminarayanan ; Harman Malhotra ; David Navara ; Norbert Reiss ; Klaus Hofmann

A COMPACT SINGLE-ENDED DUAL-BAND RECEIVER WITH CROSSTALK AND ISI REDUCTIONS FOR HIGH-DENSITY I/O INTERFACES......231
Jieqiong Du ; Jia Zhou ; X. Shawn Wang ; Chien-Heng Wong ; Huan-Neng Chen ; Chewn-Pu Jou ; Mau-Chung Frank Chang

A 26 DBM 39 GHZ POWER AMPLIFIER WITH 26.6% PAE FOR 5G APPLICATIONS IN 28NM BULK CMOS......235
Kaushik Dasgupta ; Saeid Daneshgar ; Chintan Thakkar ; James Jaussi ; Bryan Casper

A 24-43 GHZ LNA WITH 3.1-3.7 DB NOISE FIGURE AND EMBEDDED 3-POLE ELLIPTIC HIGH-PASS RESPONSE FOR 5G APPLICATIONS IN 22 NM FDSOI......239
Li Gao ; Gabriel M. Rebeiz

A 4-ELEMENT 28 GHZ MILLIMETER-WAVE MIMO ARRAY WITH SINGLE-WIRE INTERFACE USING CODE-DOMAIN MULTIPLEXING IN 65 NM CMOS......243
Manoj Johnson ; Armagan Dascurcu ; Kai Zhan ; Arman Galioglu ; Naresh Adepu ; Sanket Jain ; Harish Krishnaswamy ; Arun Natarajan

A 16-ELEMENT PHASED-ARRAY CMOS TRANSMITTER WITH VARIABLE GAIN CONTROLLED LINEAR POWER AMPLIFIER FOR 5G NEW RADIO......247
Yunsung Cho ; Woojae Lee ; Hyun-Chul Park ; Byungjoon Park ; Jeong Ho Lee ; Jihoon Kim ; Jooseok Lee ; Seokhyeon Kim ; Jaehong Park ; Sangyong Park ; Kyu Hwan An ; Juho Son ; Sung-Gi Yang

A 37-40 GHZ PHASED ARRAY FRONT-END WITH DUAL POLARIZATION FOR 5G MIMO BEAMFORMING APPLICATIONS......251
Ankur Guha Roy ; Ozgur Inac ; Amitoj Singh ; Tsvika Mukatel ; Ohad Brandelstein ; Thomas W. Brown ; Salah Abughazaleh ; Joseph S. Hayden ; Byungho Park ; Greg Bachmanek ; Te-Yu Jason Kao ; Josef Hagn ; Sidharth Dalmia ; Doron Shoham ; Brandon Davis ; Iris Fisher ; Raanan Sover ; Amit Freiman ; Bin Xiao ; Baljit Singh ; Jonathan Jensen

AN 802.11BA 495μW -92.6DBM-SENSITIVITY BLOCKER-TOLERANT WAKE-UP RADIO RECEIVER FULLY INTEGRATED WITH WI-FI TRANSCEIVER......255
Renzhi Liu ; Asma Beevi K. T. ; Richard Dorrance ; Deepak Dasalukunte ; Mario A. Santana Lopez ; Vinod Kristem ; Shahrnaz Azizi ; Minyoung Park ; Brent R. Carlton

A -80.9DBM 450MHZ WAKE-UP RECEIVER WITH CODE-DOMAIN MATCHED FILTERING USING A CONTINUOUS-TIME ANALOG CORRELATOR......259
Vivek Mangal ; Peter R. Kinget

A 4 × 4 × 4-MM³ FULLY INTEGRATED SENSOR-TO-SENSOR RADIO USING CARRIER FREQUENCY INTERLOCKING IF RECEIVER WITH -94 DBM SENSITIVITY......263
Li-Xuan Chuo ; Yejoong Kim ; Nikolaos Chiotellis ; Makoto Yasuda ; Satoru Miyoshi ; Masaru Kawaminami ; Anthony Grbic ; David Wentzloff ; Hun-Seok Kim ; David Blaauw

A 55NM SAW-LESS NB-IOT CMOS TRANSCEIVER IN AN RF-SOC WITH PHASE COHERENT RX AND POLAR MODULATION TX......267
Ps. Tseng ; W. Yang ; Mj. Wu ; Lm. Jin ; Dp. Li ; Ec. Low ; Ch. Hsiao ; Ht. Lin ; Kh. Yang ; Sc. Shen ; Cm. Kuo ; Cl. Heng ; Gk. Dehng

A 1.04 - 4V, DIGITAL-INTENSIVE DUAL-MODE BLE 5.0/IEEE 802.15.4 TRANSCEIVER SOC WITH EXTENDED RANGE IN 28NM CMOS......271
Nam-Seog Kim ; Myoung-Gyun Kim ; Ashutosh Verma ; Gyungseon Seol ; Shinwoong Kim ; Seokwon Lee ; Chilun Lo ; Jaeyeol Han ; Ikkyun Jo ; Chulho Kim ; Chih-Wei Yao ; Jongwoo Lee

A 24.5-43.5GHZ COMPACT RX WITH CALIBRATION-FREE 32-56DB FULL-FREQUENCY INSTANTANEOUSLY WIDEBAND IMAGE REJECTION SUPPORTING MULTI-GB/S 64-QAM/256-QAM FOR MULTI-BAND 5G MASSIVE MIMO......275
Min-Yu Huang ; Taiyun Chi ; Fei Wang ; Sensen Li ; Tzu-Yuan Huang ; Hua Wang

A 39GHZ 64-ELEMENT PHASED-ARRAY CMOS TRANSCEIVER WITH BUILT-IN CALIBRATION FOR LARGE-ARRAY 5G NR......279
Yun Wang ; Rui Wu ; Jian Pang ; Dongwon You ; Ashbir Aviat Fadila ; Rattanan Saengchan ; Xi Fu ; Daiki Matsumoto ; Takeshi Nakamura ; Ryo Kubozoe ; Masaru Kawabuchi ; Bangan Liu ; Haosheng Zhang ; Junjun Qiu ; Hanli Liu ; Wei Deng ; Naoki Oshima ; Keiichi Motoi ; Shinichi Hori ; Kazuaki Kunihiro ; Tomoya Kaneko ; Atsushi Shirane ; Kenichi Okada

A 24.2-30.5GHZ QUAD-CHANNEL RFIC FOR 5G COMMUNICATIONS INCLUDING BUILT-IN TEST EQUIPMENT 283

D. Dal Maistro ; C. Rubino ; M. Caruso ; M. Tiebout ; I. Maksymova ; M. Ilic ; P. Thurner ; M. Zaghi ; K. Mertens ; S. Vehovc ; I. Tsvelykh ; E. Schatzmayr ; M. Druml ; R. Druml ; M. Mueller ; M. Anderwald ; J. Wuertele ; U. Rueddenklau

A HIGHLY LINEAR 28GHZ 16-ELEMENT PHASED-ARRAY RECEIVER WITH WIDE GAIN CONTROL FOR 5G NR APPLICATION 287

Youngchang Yoon ; Kyu Hwan An ; Daehyun Kang ; Kihyun Kim ; Sangho Lee ; Jae Sik Jang ; Donggyu Minn ; Bohee Suh ; Jooseok Lee ; Jihoon Kim ; Meeran Kim ; Jeong Ho Lee ; Sung Tae Choi ; Juho Son ; Sung-Gi Yang

A QUADRATURE CLASS-G COMPLEX-DOMAIN DOHERTY DIGITAL POWER AMPLIFIER 291

Shih-Chang Hung ; Si-Wook Yoo ; Sang-Min Yoo

A FREQUENCY TUNEABLE SWITCHED-CAPACITOR PA IN 65NM CMOS 295

Zhidong Bai ; Ali Azam ; Jeffrey S. Walling

A BROADBAND HIGH-EFFICIENCY SOI-CMOS PA MODULE FOR LTE/LTE-A HANDSET APPLICATIONS 299

A. Serhan ; D. Parat ; P. Reynier ; R. Berro ; R. Mourot ; C. De Ranter ; P. Indirayanti ; M. Borremans ; E. Mercier ; A. Giry

A 27 GHZ ADAPTIVE BIAS VARIABLE GAIN POWER AMPLIFIER AND T/R SWITCH IN 22NM FD-SOI CMOS FOR 5G ANTENNA ARRAYS 303

Christian Elgaard ; Stefan Andersson ; Peter Caputa ; Eric Westesson ; Henrik Sjöland

A 9 DB NOISE FIGURE FULLY INTEGRATED 79 GHZ AUTOMOTIVE RADAR RECEIVER IN 40 NM CMOS TECHNOLOGY 307

Tomotoshi Murakami ; Nobumasa Hasegawa ; Yoshiyuki Utagawa ; Tomoyuki Arai ; Shinji Yamaura

A COMPACT 76-81 GHZ 3TX/4RX TRANSCEIVER FOR FMCW RADAR APPLICATIONS IN 65-NM CMOS TECHNOLOGY 311

Liang Chen ; Lei Zhang ; Weiping Wu ; Li Zhang ; Yan Wang

A FULL-BAND MULTI-STANDARD GLOBAL ANALOG & DIGITAL CAR RADIO SOC WITH A SINGLE FIXED-FREQUENCY PLL 315

Lucien J. Breems ; Jan Van Sinderen ; Tom Fric ; Hans Stoffels ; Franco Fritschij ; Hans Brekelmans ; Hendrik Van Der Ploeg ; Ulrich Moehlmann ; Robert Rutten ; Muhammed Bolatkale ; Shagun Bajoria ; Jan Niehof ; Bert Oude-Essink ; Gerard Lassche

LASER SPECTRAL LINEWIDTH REDUCTION USING AN INTEGRATED POUND-DREVER-HALL STABILIZATION SYSTEM IN 180 NM CMOS SOI 319

Mohamad Hossein Idjadi ; Firooz Aflatouni

22NM FD-SOI TECHNOLOGY WITH BACK-BIASING CAPABILITY OFFERS EXCELLENT PERFORMANCE FOR ENABLING EFFICIENT, ULTRA-LOW POWER ANALOG AND RF/MILLIMETER-WAVE DESIGNS 323

S. N. Ong ; L. H. K. Chan ; K. W. J. Chew ; C. K. Lim ; W. L. Oo ; A. Bellaouar ; C. Zhang ; W. H. Chow ; T. Chen ; R. Rassel ; J. S. Wong ; C. W. F. Wan ; J. Kim ; W. H. Seet ; D. Harame

A LOW POWER FULLY-INTEGRATED 76-81 GHZ ADPLL FOR AUTOMOTIVE RADAR APPLICATIONS WITH 150 MHZ/US FMCW CHIRP RATE AND -95DBC/HZ PHASE NOISE AT 1 MHZ OFFSET IN FDSOI 327

Ahmed R. Fridi ; Chi Zhang ; Abdellatif Bellaouar ; Man Tran

AN 82.2-TO-89.3 GHZ CMOS VCO WITH DC-TO-RF EFFICIENCY OF 14.8% 331

A. Tarkeshdouz ; M. Haghi Kashani ; E. Hadizadeh Hafshejani ; S. Mirabbasi ; E. Afshari

A 62 GHZ TX/RX 2X128-ELEMENT DUAL-POLARIZED DUAL-BEAM WAFER-SCALE PHASED-ARRAY TRANSCEIVER WITH MINIMAL RETICLE-TO-RETICLE STITCHING 335

Umut Kodak ; Bhaskara Rupakula ; Samet Zihir ; Gabriel M. Rebeiz

A 1-4 GHZ 4×4 MIMO RECEIVER WITH 4 RECONFIGURABLE ORTHOGONAL BEAMS FOR ANALOG INTERFERENCE REJECTION 339

Sajad Golabighezelahmad ; Eric Klumperink ; Bram Nauta

Author Index

Plenary Speaker 1

Dr. Greg Henderson
Senior Vice President
Automotive, Communications and
Aerospace & Defense
Analog Devices

The Digital Future of RFICs

Abstract: Through significant advances in RFIC technology that have shrunk form factors and price points, high complexity RF, Microwave, and Millimeter wave solutions for communications and sensing are reaching the point of ubiquity. Large, complex multi-antenna and phased array solutions that previously only government organizations could justify have become the basis of modern wireless communications and automotive radar. Cars include millimeter-wave radar technology as a standard feature and 77GHz radar is playing a critical role in the autonomous vehicle revolution. Wireless bandwidth has grown from a trickle to a torrent and high channel count, multi-antenna systems are the key enabler for 5G, whose impact is predicted to extend beyond enabling that torrent of mobile data to revolutionizing industries as varied as agriculture, automotive, healthcare, and industrial.

To date, most of the advances in RFIC technology have largely been driven by the industry moving to high volume advanced geometry CMOS processes and massive increases in system-on-chip integration of complete antenna-to-bits signal chains. Since these are not the most friendly process technologies for traditional RF and microwave circuit blocks, the advances of tomorrow need new RF signal chain and circuit block architectures that exploit the strengths of advanced CMOS processes, while mitigating the disadvantages. This talk will show how such novel architectures and circuit innovations are enabled through leverage of high-performance digital capabilities, resulting in important performance advances that in fact exceed what could be obtained from traditional "RF friendly" process technologies. The talk will show how digitally-assisted-and-enabled RFICs are enabling the future of wireless sensing and communications with real world examples for applications like 5G and automotive radar.

This talk will review the journey of mmWave technology over the last two decades, and outline the possibilities of a future where multi-functional mmWave circuits are a key differentiator in vertically integrated "Antennas to AI" cognitive systems.

About Dr. Greg Henderson

Dr. Greg Henderson was appointed Senior Vice President of Analog Devices' Automotive, Communications and Aerospace & Defense Group in 2017. Prior to this role, he served as vice president of the RF and Microwave business unit, responsible for the creation and execution of Analog Devices' strategy for its full suite of RF and microwave products and solutions.

Dr. Henderson has served in leadership roles in the microwave, semiconductor, and wireless communications industry for more than 20 years. He joined Analog Devices as part of the company's acquisition of Hittite Microwave Corporation, where he served as Vice President of RF and Microwave business units. From 2009 to 2013, Dr. Henderson served as Hittite's director of broadband products and prior to Hittite, he served as the director of product management, for the Public Safety and Professional Communications Division of Harris Corporation. Prior to Harris Corporation, Dr. Henderson held various management and R&D/ product development positions at TriQuint Semiconductor, IBM, and M/A-COM.

Dr. Henderson earned a B.S. in electrical engineering from Texas Tech University and was granted a Ph.D. in electrical engineering from the Georgia Institute of Technology. He holds seven patents in wireless communications and semiconductor technologies and has published more than 20 conference and journal papers.

Plenary Speaker 2

Dr. Ir. Michael Peeters
Program Director
Connectivity+Humanized Technology
imec

Do the Networks of the Future Care About the Materials of the Past?

Abstract: The traffic in today's networks, 4G, 5G, mobile or otherwise, seems to be following nicely the exponential expectations projected each year. On the one hand, this is driven by and drives further CMOS scaling for the digital processing of information; on the other hand, this has pushed communication channels to use ever wider bandwidths. Unfortunately, not only the individual endpoint throughputs are increasing, but the amount of endpoints and their capabilities is skyrocketing as well. Moreover, capacity as a KPI is being complemented by reliability and latency as use-cases branch out beyond the traditional human-centric communications and entertainment into areas such as industrial automation, AR/VR and autonomous vehicles.

This is creating a perfect storm at the interface of the analog and digital worlds, where traditional scaling does not necessarily buy you performance; physical dimensions are dictated not by atom sizes but by quarter-wavelengths of one kind or another; and speeds seem to all be converging at a point where switching frequencies venture far into the super-100GHz territory. For the first time in history, this is true for chip-to-chip, board-to-board, rack-to-rack, datacenter-to-datacenter, fiber and mobile wireless access systems.

Across the design space, this (finally!) has generated renewed interest into solution spaces that are less obvious, or were considered distinctly niche only a couple of years ago. We take a look at how we can tackle this, not only from an RFIC circuit design space, but also how new network capacity, reliability and latency requirements can drive technology choices for the next 10 years. This includes novel design and integration options for III-V, more exotic telluride and graphene approaches, but also dielectrics, ceramics and nanostructured materials.

About Dr. Ir. Michael Peeters

A passionate leader with a background in both research and strategy, Dr. Ir. Michael Peeters is Program Director Connectivity+Humanized Technology at imec. Michael has been identifying and implementing state-of-the-art technology opportunities in telecommunications through a career that spans two decades.

Both as Head of the Nokia Incubator and the Innovation Portfolio at Nokia, as well as CTO for the Wireless Division at Alcatel-Lucent, his role required him to make sense out of the uncertainty that exists when technological possibilities have to be balanced with business case realities. His team's responsibility: to see beyond the business analysis and help customers envision how emerging technologies and trends, such as 5G and AI, will impact their networks and end-user community.

Prior to his role as CTO for the Wireless Division, he was CTO for the Wireline Division. The team looked beyond the product roadmap and identified what new trends, technologies and tools were on the horizon and determined how those future opportunities fit into the Alcatel-Lucent pipeline. It was also during this period that the business commercialized VDSL2 Vectoring, an idea conceived 7 years earlier while leading the Bell Labs Access Nodes and DSL Technology department.

He has authored more than 100 peer-reviewed publications, many white papers and holds patents in the access and photonics domains. Michael earned a Ph.D. in Applied Physics and Photonics from Vrije Universiteit Brussel as well as a master's degree in Electrotechnical Engineering.

Outside of work, Michael is passionate about cooking and continues to refine the recipe for the perfect lasagna, balanced by bouts of long-distance running to offset the caloric intake inherent with such a quest.

A 1.2-2.8 GHz Tunable Low-noise Amplifier with 0.8-1.6 dB Noise Figure

Hao Gao[#], Zhe Song[#], Zhe Chen[#], Domine M.W.Leenaerts[#*], Peter G.M. Baltus[#]

[#]IC Group, Eindhoven University of Technology, Eindhoven, The Netherlands

[*]NXP Semiconductors, Eindhoven, The Netherlands

h.gao@tue.nl

Abstract—**This paper presents a tunable wideband low-noise amplifier (LNA) covering 1.2 to 2.8 GHz and is realized in a 0.25 μm SiGe:C BiCMOS technology. The LNA covers 80% fractional bandwidth in 16 states using a dual-LC tanks input broadband noise matching technique and a switch capacitor output frequency selection network. The measured minimal noise figure (NF) is 0.8 dB at 1.4 and 1.8 GHz, and the average NF is 1.2 (± 0.4) dB from 1.2 to 2.8 GHz. The best gain is 14 dB at 2 GHz and with ±0.5 dB flatness bandwidth covers from 1.4 GHz to 2.6 GHz. The measured input 1-dB compression point and input IP3 are better than −8 dBm and 3.5 dBm, respectively.**

Keywords— **LNA, tunable, dual-LC tanks, SiGe, Switch capacitor network.**

I. INTRODUCTION

With an increasing demand for high data rate services such as social networking, video streaming, VR and IR, wireless devices are facing the challenge of supporting multiple air interface technologies such as 3G, 4G, and 5G [1][2]. Also, with a growing number of frequency bands used in different geographies around the world, wireless devices are required to operate on multiple frequency bands to facilitate international roaming [3]. Thus, the market for multi-band RF front-end components is growing rapidly [4].

In such a multi-band RF front-end, the low-noise amplifier (LNA) design is one of the biggest challenges. The LNA should provide significant low-noise in all frequency bands, sufficient bandwidth in each mode, wide frequency coverage for multi-band, and small gain variation across all frequency bands. Various solutions are proposed for multi-band systems [5][6][7]. For instance, combining narrowband LNAs and switching among bands [8], which resulting in larger die area, higher cost and power consumption. Or a single broadband LNA which has the advantage on smaller area and power consumption [9], however the sensitivity would suffer from the non-linearity.

In this paper, a tunable LNA is proposed to combine the advantage of narrowband LNA and broadband frequency coverage into a single LNA with low noise performance. A 4-bit digitally controlled LNA is presented with 0.8-1.6 dB NF and $|S_{11}|<$-10 dB over the complete 1.2 to 2.8 GHz band. To achieve both low-noise performance and wideband frequency coverage, a dual-LC tank broadband input noise matching method [10][11] and a switch capacitor output frequency selection network are proposed and implemented.

This paper is organized as follows. In section II, the proposed digitally controlled tunable wideband amplifier is analyzed. The implemented LNA is presented in the Section III. The measurement results are provided in section IV. Conclusions are drawn in Section V.

II. LNA CIRCUIT DESIGN

A. Dual-LC Tank Broadband Input Noise Matching Network

The equivalent small-signal model of the dual-LC tank matching network is shown in Fig.1, which consists of a shunt tank (L_1, C_1), a series tank (L_2, C_2) and transistor input equivalent resistance (R_{EQ}). Both shunt and series tanks resonant at the same frequency ω_C. The input impedance, Z_{IN}, can be expressed as a function of frequency (ω) and R_{EQ} in (1).

$$Z_{IN} = \left[j\omega C_1 \left(1 - \frac{\omega_C^2}{\omega^2}\right) + \frac{1}{j\omega L_2 \left(1 - \frac{\omega_C^2}{\omega^2}\right) + R_{EQ}} \right]^{-1} \quad (1)$$

The Z_{IN} curves are plotted on the Smith chart in Fig.1 with R_{EQ} varying from 32 to 53 Ω. The dashed circle indicates the region for return loss (Γ_{IN}) better than -10 dB. The $|S_{11}|<$-10 dB can be derived by solving ω from (1), as expressed in (2). The source impedance (Z_S) is shown in Fig. 2. The minimal

$$\omega_H - \omega_L = \frac{\sqrt{R_1 R_{EQ} - R_{EQ}^2}}{L_2} \quad (2)$$

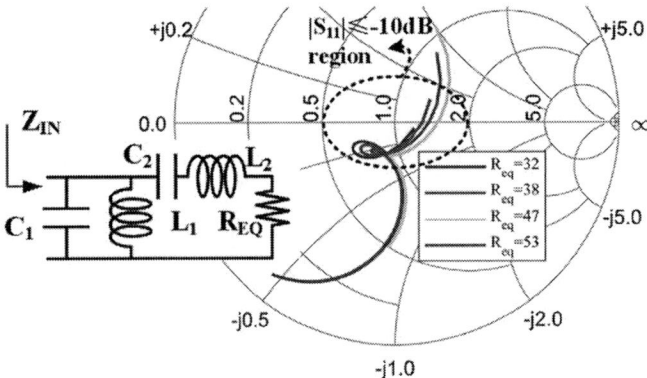

Fig. 1. The simulated Z_{IN} of dual-LC tanks matching network in the Smith chart ($Z_0 = 50$ Ω) for different values of R_{EQ} (32/38/47/53 Ω). Frequency increases from 1 to 3 GHz. The dashed circle indicates the design targeted region of $|S_{11}|<$ −10 dB.

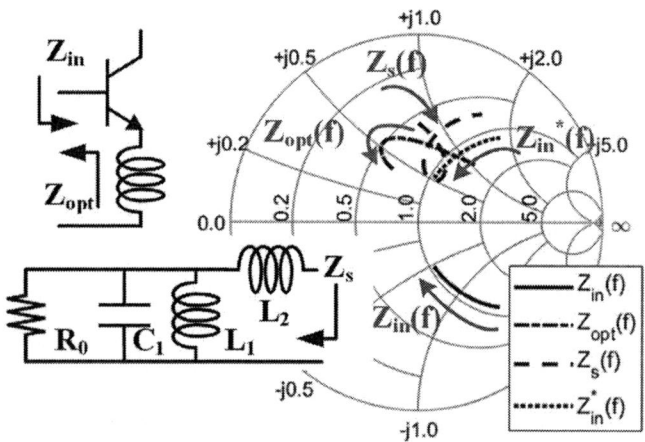

Fig. 2. The simulated input impedance Z_{IN} (with its conjugate matching Z_{IN}^*) and optimal noise impedance Z_{OPT} of inductive degenerated common-emitter transistor and the simulated sourece impedance Z_S of the dual-LC tank matching network in the Smith chart ($Z_0 = 50\ \Omega$). Frequency increases from 1 to 3 GHz.

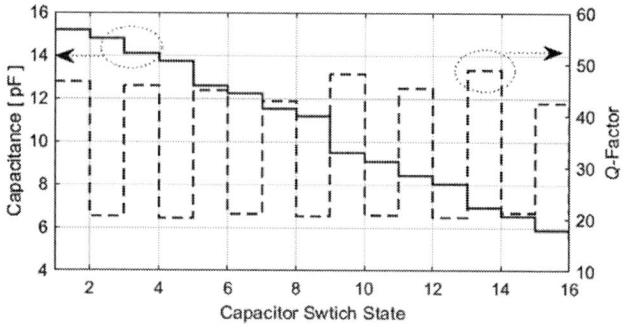

Fig. 3. Simulated effective capacitance and corresponding Q-factor at 2.5 GHz for 16 states.

noise figure (NF_{MIN}) of a transistor is achieved under the noise matching condition of $Z_S = Z_{OPT}$, where Z_{OPT} is the optimal noise impedance of the transistor. If $Z_S = Z_{OPT} = Z_{IN}^*$, then the transistor achieves simultaneous noise and power matching. For an inductive degenerated CE stage (Fig. 2), Z_{OPT} and Z_{IN} are expressed in (3) and (4) respectively.

$$Z_{OPT} \approx \frac{1}{\omega(C_{BE} + C_{BC})}\left[\sqrt{\frac{g_m}{2}(R_E + R_B)} + j\right] \quad (3)$$

$$Z_{IN} \approx \frac{g_m L_E}{C_{BE}} + \frac{1}{j\omega(C_{BE} + C_{BC})} \quad (4)$$

where C_{BE} and C_{BC} are base-emitter and base-collector capacitance; R_E and R_B are emitter and base resistance; g_m is the transconductance and L_E is the degeneration inductance. Z_S, Z_{IN}, Z_{IN}^* and Z_{OPT} are plotted in the Smith chart from 1 to 3 GHz in Fig. 2. By applying the dual-LC tanks matching method, Z_S and Z_{IN} can be matched ($|S_{11}| < -10$ dB) at ω_L, ω_C and ω_H, which means Z_S curve follows Z_{IN}^* curve between ω_L and ω_H. At the same time, Z_S also follow Z_{OPT} for a broadband noise matching.

B. Switch Capacitor Output Frequency Selection Network

Due to the high isolation of the cascode structure, the tunable frequency selection function is achieved through a digitally controlled output switch capacitor network without influence the input noise matching. In a fixed inductive load situation, each frequency shift is determined by (5), the achievable relative bandwidth is determined by (6),

$$\Delta f = \frac{1}{2\pi\sqrt{LC_n}} - \frac{1}{2\pi\sqrt{LC_{n-1}}} \quad (5)$$

$$B_{ratio} = \sqrt{\frac{C_{load_{max}}}{C_{load_{min}}}} \quad (6)$$

where the C_{load_max}, is the maximal capacitance value of the switch capacitor network, the C_{load_min} is the minimal capacitance value of the switch capacitor network and C_n is the

effective capacitance value in state N. The corresponding effective capacitance and the Q factor in each state are shown in Fig. 3. Because the input signal current I_{IN} is almost constant due to the wideband S_{11} matching, the gain flatness is achieved through the output switch capacitor network. In each state, the output amplitude is determined by the Q factor of the load LC tank. Because the inductive load dominated Q factor of the load LC tank, the inductive load can flat gain variation and compensate high-Q capacitive impedance introduced gain drop.

III. CIRCUIT DESIGN

The schematic of the proposed LNA is shown in Fig. 4. The cascode structure (Q_1 and Q_2) is chosen to alleviate the Miller effect and to improve the reverse isolation. The method of dual-LC tank matching network is implemented through C_{ESD}, C_1, L_2 and L_1, C_2, Q_1. The ESD pad capacitance C_{ESD}, the shunt inductor L_2 and the shunt capacitor C_1 form the first tank. The series inductor L_1, the linear capacitor C_2, the input capacitance of transistor Q_1 (C_{BE}) and the degeneration inductor L_3 form the second tank. Transistor Q_1 is biased at a 3.5 mA/μm^2 current density for optimal noise performance. Degeneration inductor L_3 together with C_3 generates the required real part of Z_{in} for power matching and noise matching towards Q_1. L_3 also improves also the linearity.

In this digitally controlled tunable LNA, the input network has a broadband behaviour, while the output network is narrowband and tunable. The input nonlinearity is mostly due

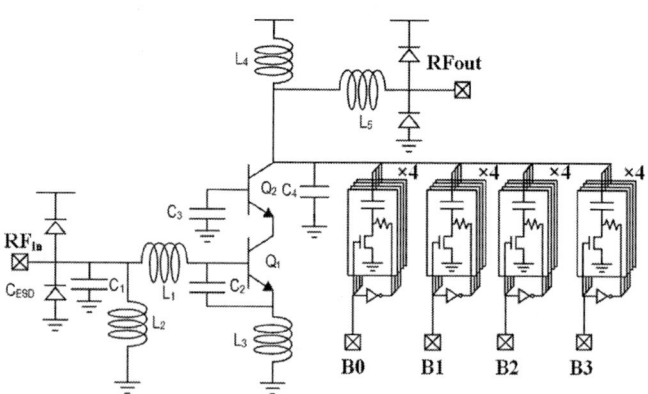

Fig. 4. Simplified schematic of the proposed LNA.

to the bipolar C_{be}, and by adding an extra high-Q capacitor (C_2), this effect can be reduced as shown in Fig 5.

The switch capacitor output frequency selection network is formed by 4 capacitor banks, controlled through B_0 to B_3. Each capacitor bank is distributed into four unit capacitor elements for symmetric parasitic capacitor distribution. Passive structures including inductors, transmission lines and bond pads are all simulated using electromagnetic software.

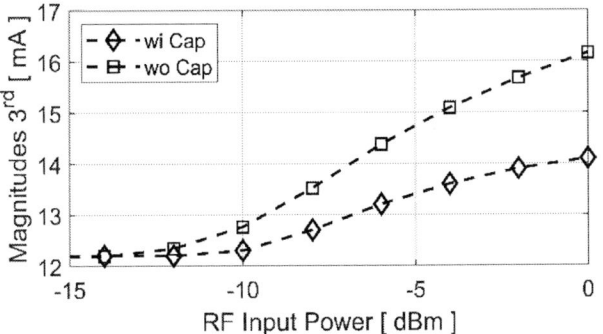

Fig. 5. Simulated 3rd harmonics collector currents of the cascode stages versus the input power.

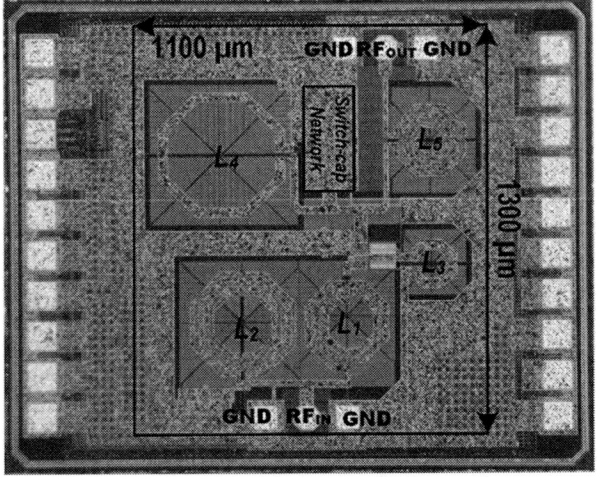

Fig. 6. Die photo.

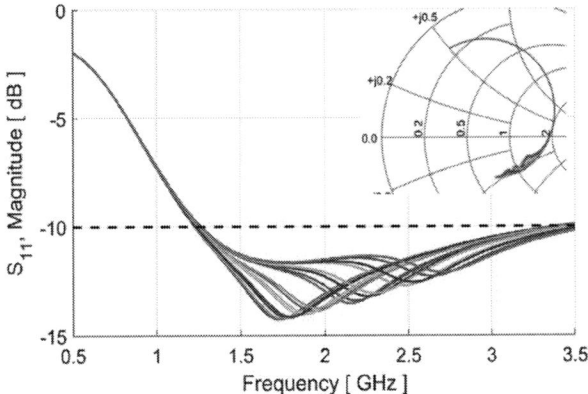

Fig. 7. Measured input return loss, S_{11} as a function of the 16 states.

IV. MEASUREMENT

The LNA is implemented in NXP's 0.25µm SiGe:C BiCMOS technology with a peak f_T/f_{MAX} of 177/216 GHz [12]. The die photo is shown in Fig. 6, the total die area is $1.1 \times 1.3 mm^2$

The measured S_{11} is lower than -10 dB from 1.2 to 3.5 GHz, as shown in Fig.7. The measured S_{22} and S_{21} are shown in Fig. 8 and Fig. 9 for the 16 states. The best power gain is 14 dB at 2 GHz and the ±0.5 dB flatness bandwidth covers 1.4 GHz to 2.6 GHz. Its 3-dB power gain bandwidth covers 1.2 GHz to 2.8 GHz, corresponding to an 80% fractional bandwidth. In each state, the 3-dB bandwidth is varying from 650 MHz to 800 MHz, as shown in Fig. 10, covering the LTE requirements. The measured noise performance is shown in Fig. 11. The measured minimal NF is 0.8 dB at 1.4 and 1.6 GHz and the average NF is 1.2 (± 0.4) dB from 1.2 to 2.8 GHz. The linearity performance is shown in Fig. 12 for input IP3 (IIP_3, two-tone with 10 MHz offset) and input 1-dB compression point (ICP_{1dB}) respectively. The LNA draws 20.8 mA from 2.5 V power supply resulting in 52 mW in total.

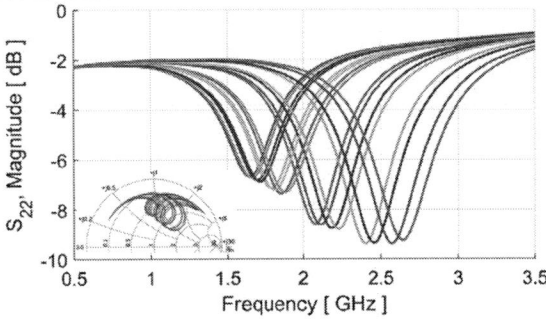

Fig. 8. Measured output return loss, S_{22}, as a function of 16 states.

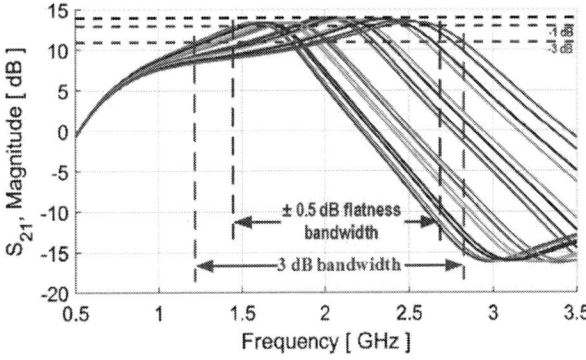

Fig. 9. Measured power gain, S_{21}, as a function of 16 states.

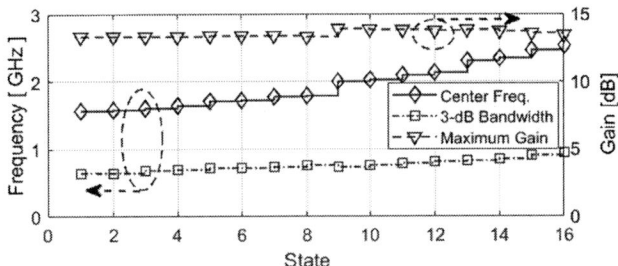

Fig. 10. Measured 16 states centre frequency, gain maximum, and corresponding 3-dB bandwidth.

978-1-7281-1702-7/19 $31.00 © 2019 IEEE

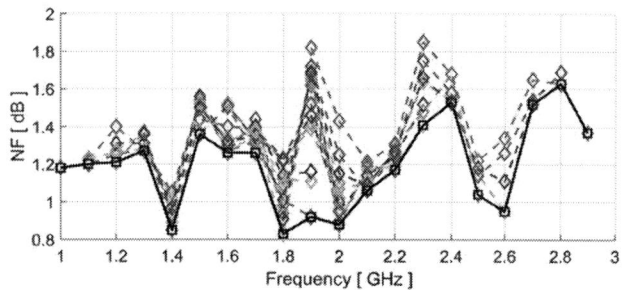

Fig. 11. Measured 16 states noise performances, and the in-band noise performance (black curve).

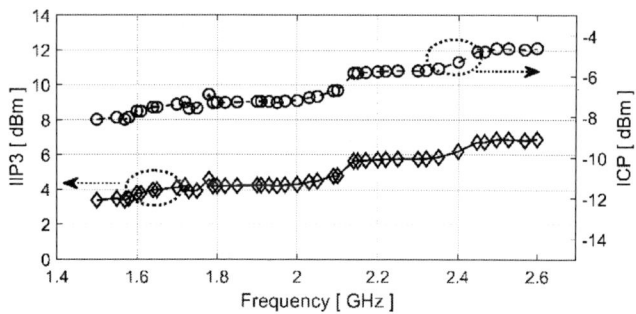

Fig. 12. Measured lineary performance.

Table 1 shows the benchmark with other silicon-based LNAs operation under 3 GHz. The proposed work achieves the best performance in combination of low average noise, wide frequency coverage and sufficient S_{11} bandwidth.

V. CONCLUSION

In this paper, the design of tunable wideband LNA with 80% fractional bandwidth is presented in a 0.25 μm SiGe technology. The input noise matching and output frequency selection network is the most crucial part to achieve a broadband noise, wide frequency coverage, and small bandwidth in each state. Thus, a dual-LC tanks broadband noise matching method and switch capacitor output network are analyzed and applied to this LNA design. The measured *NF* is 0.8-1.6 dB from 1.2 to 2.8 GHz. To our knowledge, this work achieves the best average noise performance with wide frequency coverage and gain flatness. This LNA also enable the RF front-end with low cost single silicon based LNA solution for a under 3 GHz and future 5G communication system.

ACKNOWLEDGMENT

The authors would like to acknowledge the financial contribution of CATRENE project EAST (CAT121) to this work.

REFERENCES

[1] C. Wang et al., "Cellular architecture and key technologies for 5G wireless communication networks," in IEEE Communications Magazine, vol. 52, no. 2, pp. 122-130, February 2014.

[2] F. Boccardi, R. W. Heath, A. Lozano, T. L. Marzetta and P. Popovski, "Five disruptive technology directions for 5G," in IEEE Communications Magazine, vol. 52, no. 2, pp. 74-80, February 2014.

Table 1. Performance summary and comparison.

	This work	JSSC 2015 [13]	TMTT 2016 [14]	MWCL 2017 [15]
Gain (dB)	14	14 - 24	10.1	37.5
3-dB Gain BW (GHz)	1.2 - 2.8 (16 states)	0.2 – 1.6	0.05 - 3	0.3-15
NF (dB)	0.8 - 1.6	3.6	3.4 - 4	1.8-2.2
IIP3 (dBm)	3.5 - 6.8	3	7.1	N/A
OICP (dBm)	4.9 - 8.4	N/A	N/A	10
Power (mW)	52	15.8 – 20.2	28	52
Technology	**0.25 μm SiGe**	65 nm CMOS	0.18 μm CMOS	0.13 μm SiGe

[3] S. Pellerano, A. Mirzaei, C. Hung, J. Craninckx, K. Okada and V. Vidojkovic, "F3: Radio architectures and circuits towards 5G," 2016 IEEE International Solid-State Circuits Conference (ISSCC), San Francisco, CA, 2016, pp. 498-501.

[4] A. Mirzaei, H. Darabi, A. Yazdi, Z. Zhou, E. Chang and P. Suri, "A 65 nm CMOS Quad-Band SAW-Less Receiver SoC for GSM/GPRS/EDGE," in IEEE Journal of Solid-State Circuits, vol. 46, no. 4, pp. 950-964, April 2011.

[5] H. Hashemi and A. Hajimiri, "Concurrent multiband low-noise amplifiers-theory, design, and applications," in IEEE Transactions on Microwave Theory and Techniques, vol. 50, no. 1, pp. 288-301, Jan. 2002.

[6] Run Chen, Hossein Hashemi, "Dual-Carrier Aggregation Receiver With Reconfigurable Front-End RF Signal Conditioning", Solid-State Circuits IEEE Journal of, vol. 50, no. 8, pp. 1874-1888, 2015.

[7] Desheng Ma, Fa Foster Dai, Richard C. Jaeger, J. David Irwin, "An 8 – 18 GHz wideband SiGe BiCMOS low noise amplifier", Microwave Symposium Digest 2009. MTT '09. IEEE MTT-S International, pp. 929-932, 2009

[8] S. Wu and B. Razavi, "A 900-MHz/1.8-GHz CMOS receiver for dual-band applications," in IEEE Journal of Solid-State Circuits, vol. 33, no. 12, pp. 2178-2185, Dec. 1998.

[9] A. Bevilacqua and A. M. Niknejad, "An ultrawideband CMOS low-noise amplifier for 3.1-10.6-GHz wireless receivers," in IEEE Journal of Solid-State Circuits, vol. 39, no. 12, pp. 2259-2268, Dec. 2004.

[10] Z. Chen, H. Gao, D. Leenaerts, D. Milosevic and P. Baltus, "A 29–37 GHz BiCMOS Low-Noise Amplifier with 28.5 dB Peak Gain and 3.1-4.1 dB NF," 2018 IEEE Radio Frequency Integrated Circuits Symposium (RFIC), Philadelphia, PA, 2018, pp. 288-291.

[11] Z. Chen, H. Gao, D. M. W. Leenaerts, D. Milosevic and P. Baltus, "A 16–43 GHz low-noise amplifer with 2.5–4.0 dB noise figure," 2016 IEEE Asian Solid-State Circuits Conference (A-SSCC), Toyama, 2016, pp. 349-352.

[12] Q. Ma, D. M. W. Leenaerts and P. G. M. Baltus, "Silicon-Based True-Time-Delay Phased-Array Front-Ends at Ka-Band," in IEEE Transactions on Microwave Theory and Techniques, vol. 63, no. 9, pp. 2942-2952, Sept. 2015.

[13] J. Zhu, H. Krishnaswamy and P. R. Kinget, "Field-Programmable LNAs With Interferer-Reflecting Loop for Input Linearity Enhancement," in IEEE Journal of Solid-State Circuits, vol. 50, no. 2, pp. 556-572, Feb. 2015.

[14] D. Im and I. Lee, "A High IIP2 Broadband CMOS Low-Noise Amplifier With a Dual-Loop Feedback," in IEEE Transactions on Microwave Theory and Techniques, vol. 64, no. 7, pp. 2068-2079, July 2016.

[15] S. Zeinolabedinzadeh, A. Ç. Ulusoy, M. A. Oakley, N. E. Lourenco and J. D. Cressler, "A 0.3–15 GHz SiGe LNA With >1 THz Gain-Bandwidth Product," in IEEE Microwave and Wireless Components Letters, vol. 27, no. 4, pp. 380-382, April 2017

A 28-GHz CMOS LNA with Stability-Enhanced G_m-Boosting Technique Using Transformers

Sunwoo Kong[1], Hui-Dong Lee, Seunghyun Jang, Jeehoon Park, Kwang-Seon Kim, Kwang-Chun Lee

ETRI (Electronics and Telecommunications Research Institute)

Gajeong-ro 218, Yuseong-gu, Daejeon, South Korea

[1]swkong@etri.re.kr

Abstract — In this paper, we propose a low noise amplifier (LNA) using a g_m-boosting technique with improved stability using transformers in the millimeter-wave (mm-Wave) band. The transformer composed of three inductors improves not only stability, but also gain and low-noise performance of the LNA. The conditions for stability shows that the proposed structure can guarantee good stability over a high frequency range. The chip was fabricated using the TSMC 65-nm CMOS process and it has an active chip area of 0.11 μm^2. The fabricated LNA has a gain of 18.33 dB and a noise figure (NF) of 3.25–4.2 dB. The stability factor μ values are 9.7 and 5.2 at the source and load sides of the LNA, respectively. The 3-dB bandwidth of the LNA is 24.9–32.5 GHz and the chip consumes 17.1-mA current from a 1.2-V supply.

Keywords — LNA, transformer, gm-boosting, stability

I. INTRODUCTION

The bandwidth of the wireless communication system is gradually increasing with the demand of high data rates. Recently, in order to guarantee a wide bandwidth, the operation frequency has started to expand to the millimeter-wave (mm-Wave) band, and 5G cellular network is one of the promising mm-Wave application. A low noise amplifier (LNA) is a component that greatly affects the sensitivity of the system depending on its performance. The performance such as noise figure (NF), gain, bandwidth, and power consumption of the LNA are in a trade-off relationship with each other. In mm-Wave frequency, the stability problem is likely to occur and the performance degradation due to the signal loss is remarkable because the influence of the parasitic component increases. The cascode structure is one of good candidates for improving performance while ensuring stability. The cascode structure can increase the gain and reduce power consumption by stacking common-source (CS) stage and common-gate (CG) stage. Since the cascode structure has high input to output isolation, it ensures good stability.

However, in the mm-Wave band, the influence of the parasitic cap C_{p1} which is at the connection point of the CS and CG satges increases and the signal leaks through it as shown in Fig. 1a. To solve this problem, the inductors connected in series and in parallel are used to compensate for the effect of C_{p1} and C_{p2}, in Fig. 1b and 1c [1][2]. Furthermore, Fig. 1d shows a structure that improves gain and NF of the amplifier while compensating C_{p1} and C_{p2} using a transformer which is composed of L_{SG} and L_P [3]. This structure boosts effective g_m of the CG stage by constructing a negative feedback network using the

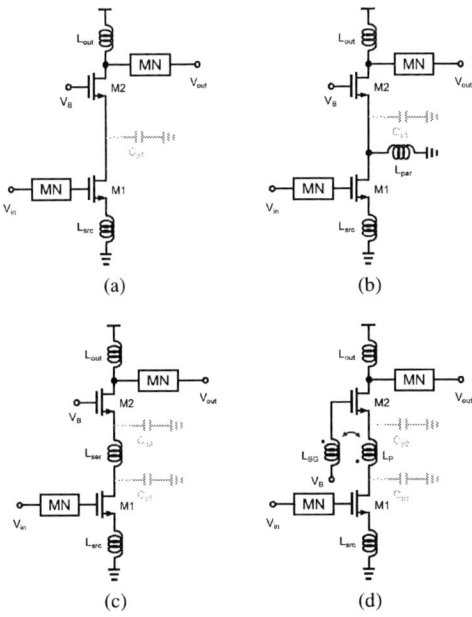

Fig. 1. Stacked LNAs. (a) Cascode. (b) Cascode with a parallel inductor. (c) Cascode with a series inductor. (d) Cascode with a transformer.

transformer, so the conditions for negative feedback must be maintained. Moreover, the cascode with a transformer poses a potential risk of the stability problem due to the inductor connected to the gate of the CG stage [4].

We have proposed an LNA using a g_m-boosting technique with improved stability using transformers in the mm-Wave band. The stability problem of the g_m-boosting technique using transformers was analyzed and new transformers were designed to solve it.

II. CIRCUIT DESIGN

A. Stability-Enhanced G_m-Boosting Technique

Fig. 2a shows a cascode LNA with the proposed stability-enhanced g_m-boosting technique. Transistors M_1 and M_2 constitute CS and CG structures, respectively. Transformers consisting of three inductors, which are L_P, L_{SG}, and L_{SD}, are placed around M_2. L_P is located between the source of M_2 and the drain of M_1, and L_{SG} is connected to the gate of M_2. To enlarge gate-source voltage swing for g_m-boosting, anti-phase relationship due to the magnetic coupling between L_P and L_{SG} is used. L_{SD} is connected to the drain of M_2. The effect of gate-drain capacitance is

978-1-7281-1702-7/19 $31.00 © 2019 IEEE

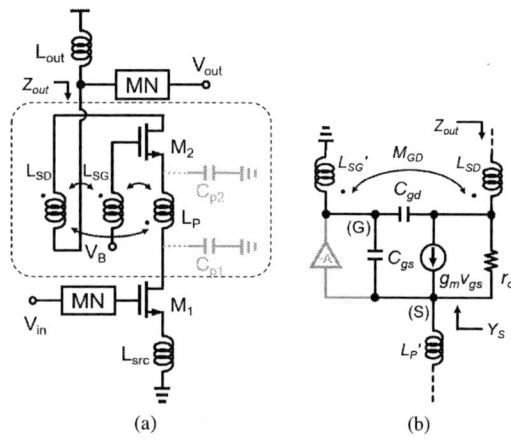

(a)   (b)

Fig. 2. (a) Schematic of the proposed LNA with transformers consisting of L_P, L_{SG}, and L_{SD}. (b) Simplified equivalent model of the circuit in the dotted box around M_2 in Fig. 2a.

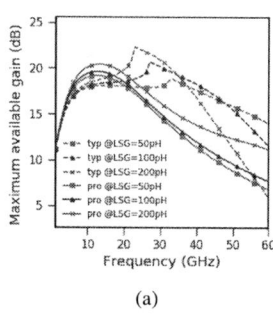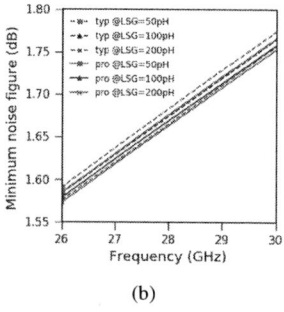

(a)   (b)

Fig. 3. Simulated (a) MAGs and (b) NFs of the proposed structure (Fig. 1d) and the typical cascode structure with a transformer (Fig. 2a) when L_{SG}s are 50 pH, 100 pH, and 200 pH.

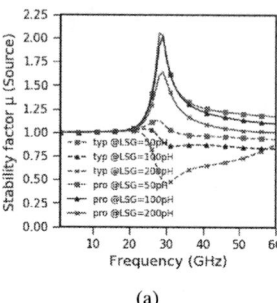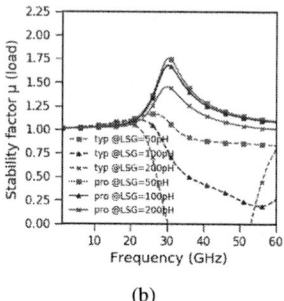

(a)   (b)

Fig. 4. Simulated stability factors (a) μ (source) and (b) μ (load) of the proposed structure (Fig. 1d) and the typical cascode structure with a transformer (Fig. 2a) when L_{SG}s are 50 pH, 100 pH, and 200 pH.

neutralized using the magnetic coupling between L_{SD} and L_{SG} and they make in-phase relationship at the drain and gate of M_2. The transformer composed of L_P and L_{SG} cancels the effects of C_{p1} and C_{p2} and can be represented by L'_P and L'_{SG} in Fig. 2b [3]. L'_{SG} and C_{gs} form a negative feedback factor of A between source and gate. Fig. 2b shows a simplified equivalent model of the circuit in the dotted box around M_2 in Fig. 2a. Without L_{SD}, feedback factor A can be expressed as

$$A = \frac{j\omega L'_{SG}}{1/\omega C_{gs} - \omega L'_{SG}} \quad (1)$$

where C_{gs} is the gate-to-source capacitance of M_2. To maintain negative feedback, $\omega < 1/\sqrt{L'_{SG}C_{gs}}$ must be met. Applying the mutual inductance M_{GD} of the transformer formed by L_{SD} and L'_{SG}, feedback factor becomes

$$A' = \frac{j\omega C_{gs}(j\omega L'_{SG} + \omega_t M_{GD})}{1 + j\omega C_{gs}(j\omega L'_{SG} + \omega_t M_{GD})} \quad (2)$$

where $\omega_t = g_m/C_{gs}$ and g_m are the angular cutoff frequency and trans-conductance of M_2, respectively. The effective g_m of M_2 is boosted as

$$G_m = (1 + A')g_m. \quad (3)$$

Y_S is the admittance seen in (S) and the conductance part is expressed as

$$Re(Y_S) = \frac{(1 - \omega^2(L'_{SG} - M_{GD})C_{gs})g_m}{(1 - \omega^2 L'_{SG}C_{gs})^2 + (\omega M_{GD}g_m)^2}. \quad (4)$$

If $M_{GD} > L'_{SG}$, the proposed structure is stable because $Re(Y_S)$ is always positive. If $M_{GD} < L'_{SG}$, the proposed structure is stable when the following condition is maintained:

$$\omega < \frac{1}{\sqrt{(L'_{SG} - M_{GD})C_{gs}}}. \quad (5)$$

When a transformer composed of L_{SD} and L'_{SG} is applied, the proposed g_m-boosted topology has always stable condition or a wider stable band than when it is not.

Fig. 3a shows that the simulated maximum available gain (MAG)s of the proposed structure (Fig. 2a) and the typical cascode structure with a transformer (Fig. 1d) increase as the value of L_{SG} increases. The two structures commonly have transistor size of 36 μm and coupling factor k of 0.7 for L_P (100 pH) and variable L_{SG}s (50 pH, 100 pH, and 200 pH). For the proposed structure, L_{SD} has a value of 100 pH and the coupling factors between L_{SG} and L_{SD} and between L_P and L_{SD} are 0.7 and 0.6, respectively. Fig. 4a and Fig. 4b show the stability factors (μ) of load and source. The magnitude of μ is a measure of the stability and the larger value of the μ, the higher the stability of 2-port systems. Due to the coupling between L_{SD} and L_{SG}, the proposed structure is stable in all bands with all L_{SG} values. However, the upper band of the stable region of the typical structure decreases as the gate inductor value increases. The minimum NF is shown in Fig. 3b and there is slight NF improvement with the stability-enhanced G_m-boosting technique.

The impedance seen at the output node in Fig. 2b, Z_{out}, is shown on a Smith chart for three types of circuit configurations (OA, OB, and OC) in Fig. 5.

- OA: typical cascode with transformer (without L_{SD})
- OB: typical cascode with transformer (with L_{SD})
- OC: proposed structure (with L_{SD} and M_{GD})

For the three configurations, Z_{out}s are displayed when the values of the gate inductors are varied from 50 pH to 200 pH. In the configuration OA and OB, increasing the value of

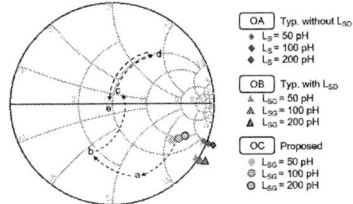

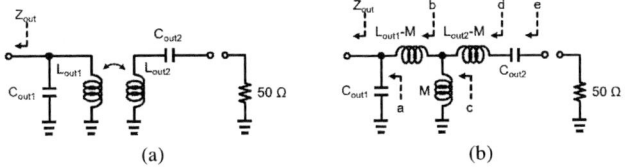

Fig. 5. Output impedance transformation on Smith Chart with different gate inductors. Z_{out} starts from three different points OA, OB, and OC.

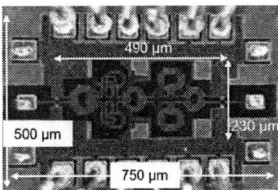

(a)   (b)

Fig. 6. Simplified output matching circuit: (a) output transformer and (b) its equivalent circuit.

L_{SG} causes the amplifier to go into the unstable region. Z_{out} is moved from OA to OB by adding L_{SD}. L_{SD} provides positive reactance term to Z_{out}. Z_{out} shifts from OA to OC as the real impedance and reactance are added by the gate-drain transformer. In this case, stability is guaranteed against the change of L_{SG}. The transformer in Fig. 6a can be used to match Z_{out} to 50 ohms. The transformer performs the function of matching and differential to single ended conversion. Fig. 5 shows the transformed trace of Z_{out} from a to e adding the matching components as an equivalent model of the transformer in Fig. 6b.

B. Circuit Implementation

Fig. 7 shows the overall differential schematic of the proposed LNA. Section II-A explained the single-ended structure of the proposed LNA. In this section, the proposed technique is applied to the differential structure which is robust to noise and effects of bond wires and system integration issues are also considered. The first and second stage uses a CS and cascode structure, respectively, and the source degeneration inductors L_{src1} and L_{src2} are used for input matching. The input matching circuit is composed of a single-ended to differential transformer and the output stage uses a differential to single-ended transformer. The first and the second stages are matched by an inter-stage transformer. The CS stage compensates the gate-to-drain feedback cap using a drain-to-gate transformer [5]. The cascode stage uses the proposed stability-enhanced g_m-boosting technique.

III. MEASUREMENT

The chip was fabricated using TSMC 65-nm CMOS process. Fig. 8 shows a micrograph of the fabricated chip. The chip size is 750 x 500 μm^2 including pads and the active chip area is 490 x 230 μm^2. The performance of the fabricated LNA was investigated by on-probe measurement. At a supply voltage of 1.2 V, 17.1-mA current was consumed. Fig. 9 compares the measured S-parameter of the fabricated

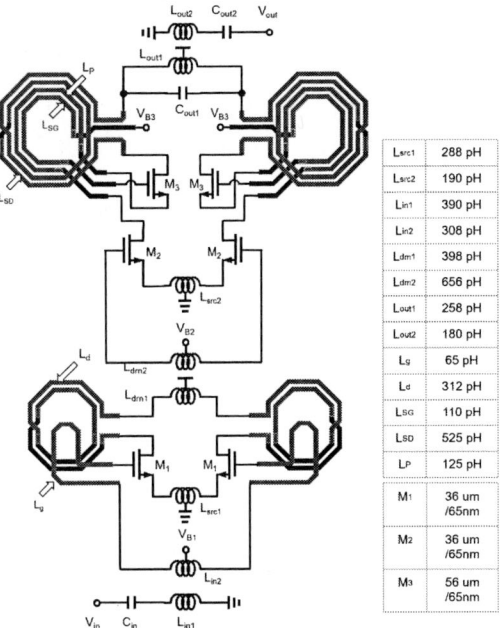

Fig. 7. Schematic of the proposed differential LNA

L_{src1}	288 pH
L_{src2}	190 pH
L_{in1}	390 pH
L_{in2}	308 pH
L_{dm1}	398 pH
L_{dm2}	656 pH
L_{out1}	258 pH
L_{out2}	180 pH
L_g	65 pH
L_d	312 pH
L_{SG}	110 pH
L_{SD}	525 pH
L_P	125 pH
M_1	36 um /65nm
M_2	36 um /65nm
M_3	56 um /65nm

Fig. 8. Micrograph of the chip.

LNA with the simulated S-parameter of it. The measured gain is 18.33 dB and the simulation gain is 21.8 dB at 28 GHz. The 3-dB bandwidth for measurement and simulation are 7.6 GHz (24.9–32.5 GHz) and 5.6 GHz (25.6–31.2 GHz), respectively. The measurement results has lower peak S21 and wider bandwidth than the simulation results because the measured S22 is down-shifted from the simulation result. Fig. 10 compares the measured NF with the simulation results. The both of NFs have similar values because S11 was matched. The measured NF is 3.25 dB at 28 GHz. An Agilent PNA-X N5247B instrument was used for cold-source NF measurements and the wideband power sensor was the Agilent U2011XA instrument. As shown in Fig. 11, the stability factor μ has values of 9.7 and 5.2 on the source and load sides. The peak value of μ is 27 and 20 for the source and load. the fabricated LNA is unconditionally stable because $\mu > 1$ over the entire 20–40 GHz band.

Table 1 summarizes the performance of the proposed LNA and compares it with state of the art results. The LNA has the lowest NF and the smallest active chip area with full differential structure. Also, the gain per power consumption of this work is the second highest in the comparison group. The LNA has a higher stability factor μ than the other work

978-1-7281-1702-7/19 $31.00 © 2019 IEEE

Table 1. Performance summary and comparison with state of the art

	This work	**[6]**	**[2]**	**[7]**
Freq [GHz]	24.9–32.5	26–33	33‡	16–30
Tech.	65 nm	40 nm	28 nm	65 nm
Stages	1 Diff. CS, 1 Diff. Cas.	1 Cas., 2 Diff. CS	2 Cas.	1 Cas.
Gain [dB]	18.33	27.1/18.4	18.6	10.2
NF [dB]	3.25-4.2	3.3-4.4	4.9	3.3-5.7
S12 [dB]	-47	-38†	-45	-
μ (source, load)	9.7, 5.2	-	3, 5†	-
P1dB [dBm]	-24	-21.6/-13.4	-25.5	N/A
Power [mW]	20.5 @1.2 V	31.4/21.5 @1.1 V	9.7 @1.2 V	12.4 @N/A
Act./total area [mm^2]	0.11/0.38	0.26/0.54†	0.23/0.51†	0.13†/0.18

† Graphically estimated. ‡ 3-dB Bandwidth of 4.7 GHz.

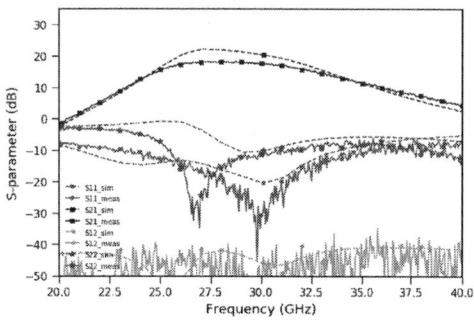

Fig. 9. Comparison of measured and simulated S-parameters of the LNA.

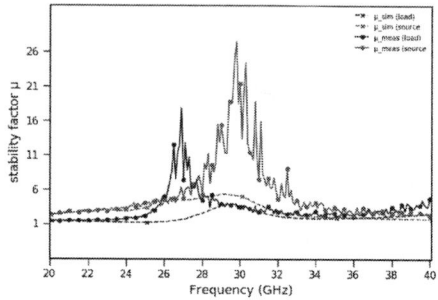

Fig. 11. Comparison of measured and simulated stability factor μ of the LNA.

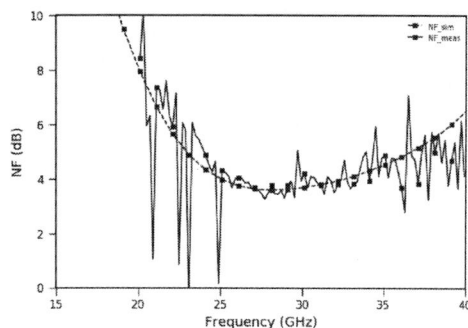

Fig. 10. Comparison of measured and simulated NF of the LNA.

which reported stability factors. It was also confirmed that the LNA had the lowest reverse isolation S12, which is an indirect indicator for good stability.

IV. Conclusion

In this paper, we proposed an LNA using a stability-enhanced g_m-boosting technique with transformers consisting of three inductors in the mm-Wave band. The g_m-boosting technique brought the LNA good gain and low-noise results per power consumption. The comparison of the conditions for stability shows that the proposed structure can guarantee good stability over a wider frequency range. The fabricated LNA has the stability factor μ values of 9.7 and 5.2 at the source and load sides, respectively, and 3.25–4.2-dB NF with 18.33-dB gain at the same time.

Acknowledgment

This work was supported by Institute for Information & communications Technology Promotion (IITP) grant funded by the Korea government (MSIT) (No. 2017-0-00409, Development on millimeter-wave beamforming IC for 5G mobile communication).

References

[1] H. Samavati, H. R. Rategh, and T. H. Lee, "A 5-GHz CMOS Wireless LAN Receiver Front End," *IEEE J. Solid-State Circuits*, vol. 35, no. 5, pp. 765–772, 2000.

[2] M. K. Hedayati, A. Abdipour, R. S. Shirazi, C. Cetintepe, and R. B. Staszewski, "A 33-GHz LNA for 5G Wireless Systems in 28-nm Bulk CMOS," *IEEE Trans. Circuits Syst. II Express Briefs*, vol. 65, no. 10, pp. 1460–1464, 2018.

[3] S. Guo, T. Xi, P. Gui, D. Huang, Y. Fan, and M. Morgan, "A Transformer Feedback Gm-Boosting Technique for Gain Improvement and Noise Reduction in mm-Wave Cascode LNAs," *IEEE Trans. Microw. Theory Tech.*, vol. 64, no. 7, pp. 2080–2090, 2016.

[4] Hsieh-Hung Hsieh and Liang-Hung Lu, "A 40-GHz Low-Noise Amplifier With a Positive-Feedback Network in 0.18-um CMOS," *IEEE Trans. Microw. Theory Tech.*, vol. 57, no. 8, pp. 1895–1902, 2009.

[5] M. P. van der Heijden, L. C. N. de Vreede, and J. N. Burghartz, "On the Design of Unilateral Dual-Loop Feedback Low-Noise Amplifiers With Simultaneous Noise, Impedance, and IIP3 Match," *IEEE J. Solid-State Circuits*, vol. 39, no. 10, pp. 1727–1736, 2004.

[6] M. Elkholy, S. Shakib, J. Dunworth, V. Aparin, and K. Entesari, "A Wideband Variable Gain LNA With High OIP3 for 5G Using 40-nm Bulk CMOS," *IEEE Microw. Wirel. Components Lett.*, vol. 28, no. 1, pp. 64–66, 2018.

[7] P. Qin and Q. Xue, "Compact Wideband LNA With Gain and Input Matching Bandwidth Extensions by Transformer," *IEEE Microw. Wirel. Components Lett.*, vol. 27, no. 7, pp. 657–659, 2017.

Ka-Band CMOS Absorptive SP4T Switch With One-Third Miniaturization

Bosung Suh and Byung-Wook Min

School of Electrical and Electronic Engineering, Yonsei University, Seoul, Korea

bssuh@yonsei.ac.kr, bmin@yonsei.ac.kr

Abstract—An Ka-band absorptive single-pole four-throw (SP4T) switch in 28-nm CMOS process is presented. By capacitive matching and loading method, only four $\lambda/6$ transmission lines (t-line) are used for the proposed switch without additional series switches. The length of the t-line used in the proposed switch is only 1/3 of that of the quarter-wave t-line based absorptive switch. To improve switch performance, low threshold-voltage transistors and source and drain of transistors are biased for power handling capability. The measured insertion loss and isolation are 3.5 dB and 20 dB at 28 GHz. Return losses of on-state and off-state ports are less than -10 dB and -16 dB from 26 GHz to 33 GHz. The measured input 1-dB compression point is more than 15 dBm. The chip area is 0.53 mm^2 and electrical size is $0.0047 \times (\lambda_g)^2$ excluding pads and internal matching circuits that is the smallest electrical size among the millimeter-wave CMOS absorptive SP4T switches.

Keywords—Absorptive switch, phased array, capacitive loading, , power compression point.

I. INTRODUCTION

Millimeter-wave phased-array systems have been considered as a candidate for next generation technology. Beam-forming network is required for different phase shifts to antenna element. Butler matrix is attractive beam-forming network, due to its simplicity and low loss characteristics [1]-[2]. The beam direction can be determined by a switch that allows the signal to pass through a specific input port of Butler matrix. At that time, the other input ports should be matched to keep the isolation of couplers and prevent multiple internal reflections. For these reasons, an absorptive switch is needed for Butler matrix based beam-forming network.

In millimeter wave, absorptive switches are commonly designed by quarter-wavelength transmission lines (t-line) and shunt transistors [3] or PIN-diodes [4] as shown in Fig. 1. Since this design does not use series transistor swithces, there is an advantage of less loss. Also, wide shunt transistors between ground and quarter-wave t-line provide sufficient isolation and return losses. However, large chip area and increased loss due to a number of quarter-wave t-line are drawbacks. There are attempts to avoid using quarter-wave t-lines, such as using bridged-T coil [5] and series-shunt configuration with high-impedance t-line [6]. But both method uses series transistors that can induce additional loss.

In this paper, we propose a miniaturized absorptive SP4T switch by capacitive matching and loading. Four $\lambda/6$ t-lines with shunt switches are used for the proposed switch, while eight $\lambda/4$ lines are used at conventional $\lambda/4$-based absorptive

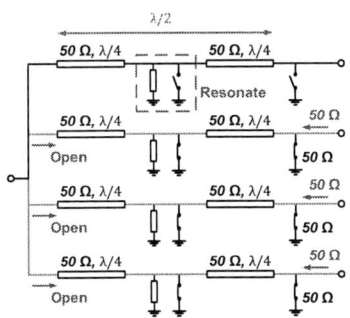

Fig. 1. Conventional $\lambda/4$ transmission line based absorptive SP4T switch.

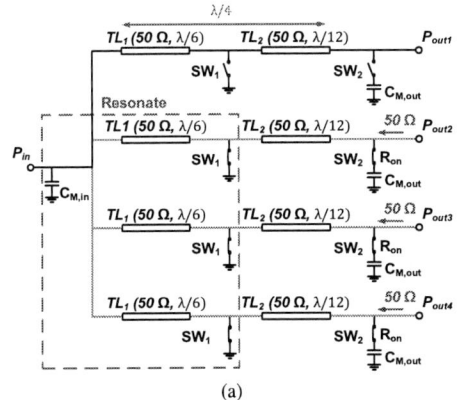

(a)

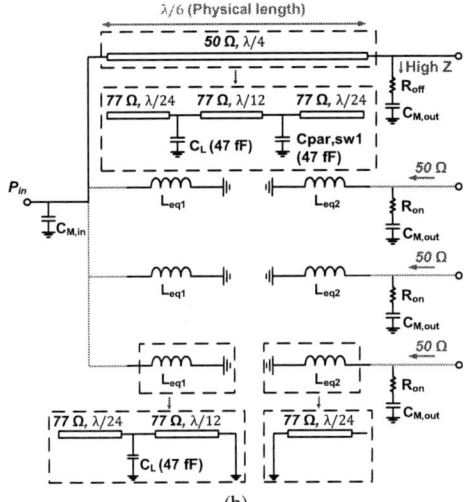

(b)

Fig. 2. (a) A half and (b) one-third miniaturized absorptive switch compared with conventional $\lambda/4$ transmission line based absorptive SP4T switch.

switches [3]-[4]. Also, low loss is achieved, since short t-lines are used and series switches are not used. The designed switch has 3.5 dB insertion loss at 28 GHz and better than 3.9 dB from 26 to 33 GHz. Isolation is better than 20 dB in all frequency bands. The input P1dB is more than 15 dBm and chip area is $1.37 \times 0.39 \ mm^2$ excluding pads and internal matching circuits respectively.

II. DESIGN

To miniaturize absorptive SP4T switch, t-line can be miniaturized as shown in Fig. 2 (a). The electrical length of TL_1 from the input port to the ground switch (SW_1) is $\lambda/6$ and therefore TL_1 is considered as a inductive short stub when the SW_1 is on. The inductance can be resonated and the input can be matched by $C_{M,in}$. Also, TL_2 between SW_1 and SW_2 is used to match the off-state port. The first t-line is set longer, since more short stubs are connected on the input port than the off-state port. For impedance matching at off-state port, capacitors ($C_{M,out}$) are added between the switch (SW_2) and ground. Therefore, an absorptive SP4T switch is one-half miniaturized through capacitive matching by $C_{M,in}$ for short stubs connected at input port. At that time, bandwidth is reduced by about 1.2 GHz as shown in Fig. 3. If miniaturized further, bandwidth is further reduced due to reduction of equivalent inductance of short stubs.

For further miniaturizing the TL_1 and TL_2, capacitive loading is used as shown in Fig. 2 (b). Quarter-wave t-line is composed of high-impedance t-lines, metal-oxide-metal (MOM) capacitor (C_L) and parasitic capacitance of SW_1 as shown in Fig. 2(b). By setting the impedance of t-line and capacitances to 77 Ω and 44 fF, respectively, the physical length is reduced by two-third compared with the structure in Fig.2 (a), while keeping the electrical length same at 28 GHz. Thus, total physical length is reduced to one-third compared with conventional switch shown in Fig. 1. More high impedance t-line and more capacitances can be used for further miniaturization, but this degrades insertion loss due to increase of the loss from high-impedance t-line. Since, impedance of the off-state SW_2 is quite high, return loss of the on-path output port does not get much worse. Equivalent inductance of L_{eq1} composed of high impedance t-lines and MOM capacitor is higher than that of 50-Ω 60° short stub in Fig. 2(a) and therefore the bandwidth can be wider in Fig. 2(b) as shown in Fig. 3.

The schematic of proposed absorptive SP4T switch is shown in Fig. 4. The widths of T_1 and T_2 are 110 μm and 75 μm, respectively. 20-KΩ resistors are used at gate and body of transistors to prevent signal leakage and gate oxide breakdown. T-lines are designed as microstrip line with side shields or grounded coplanar waveguide (CPW).

Isolation of the proposed switch is determined by the switch transistor T_1. For sufficient isolation, the T_1 should have low on-resistance and therefore, low threshold-voltage transistors whose RC time constant is lower than that of general NFET transistors are used. The shunt NFET transistors limit the power handling capability. To overcome the problem,

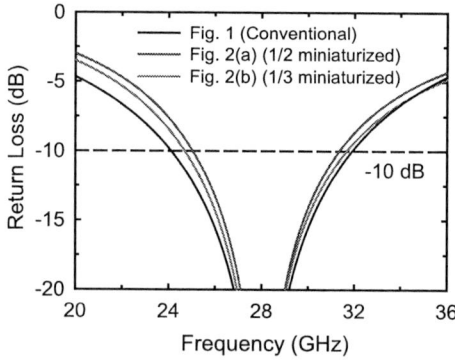

Fig. 3. Return loss of absorptive SP4T switches shown in Fig. 1 and Fig. 2

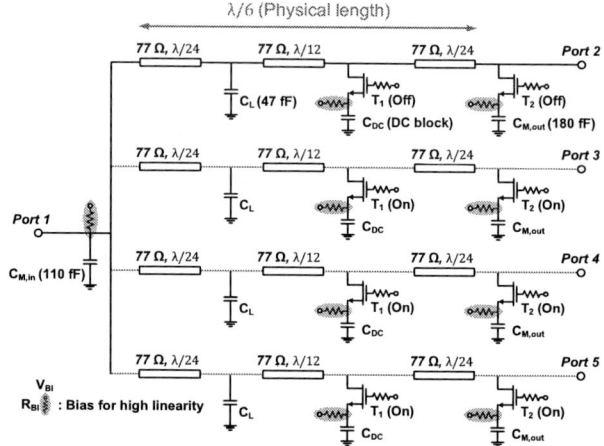

Fig. 4. Schematic of the proposed absorptive SP4T switch

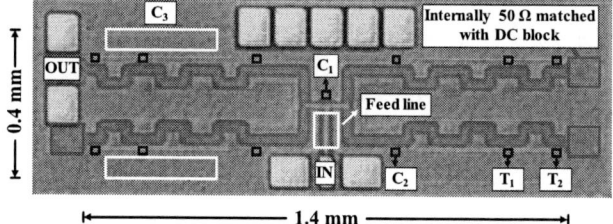

Fig. 5. Photograph of miniaturized absorptive SP4T switch.

source and drain are biased to 0.3 V and the DC block capacitors, C_{DC} are required [7]. By source and drain bias, IP1dB is improved about 6 dB in simulation.

III. MEASUREMENT

The absorptive SP4T switch was fabricated in Samsung 28-nm CMOS process. The chip photograph of proposed switch is shown in Fig. 5. The feed line is designed with 50-Ω grounded CPW. The output ports without pad are internally matched by 50-Ω with large DC block capacitors. The overall chip area is $1.37 \times 0.39 \ mm^2$ excluding internal matching circuits and pads. GSG pads are located on input and one of the output ports. All t-lines and interconnecting lines were simulated by full-wave electromagnetic (EM) software

(Sonnet). The absorptive SP4T switch was measured by on-wafer probing with Agilent E8364A network analyzer.

Fig. 6 shows the measured and simulated S-parameters of the on-state absorptive SP4T switch. The measured insertion loss is 3.5 dB at 28 GHz. Return losses are less than −10 dB from 26 GHz to 33 GHz. The measured and simulated results of isolation and return loss of the off-state port are shown in Fig. 7. The isolation and return loss of the off-state port are 20 dB and −21 dB, respectively and better than 20 dB and −16 dB from 26 GHz to 33 GHz, respectively. Isolation is limited by performance of ground switch (T_1), whose on-resistance is 5 Ω.

Power measurements are conducted at 28 GHz. High input power is generated by Anritsu MG3694C with high power option and output-power is measured by spectrum analyzer. Because of the output power limit of the signal analyzer, input power was varied up to 15 dBm. Power handling capability of the proposed switch is shown in Fig 8. When source and drain are not biased, the absorptive SP4T switch has input P1dB of 11 dBm. When the source and drain are biased, the measured input P1dB is higher than 15 dBm.

The performances of millimeter-wave CMOS absorptive SP4T switches are compared in Table 1. The proposed switch has the smallest electrical size compared with the reported millimeter-wave absorptive switches. Also, low insertion loss is achieved due to the absence of series switches and reduction of t-lines.

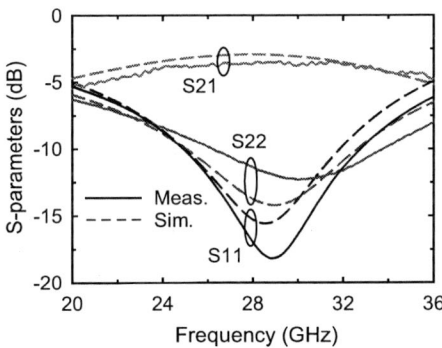

Fig. 6. Measured and simulated S-parameters of on-state absorptive SP4T switch.

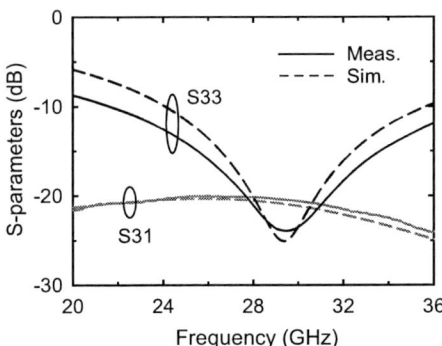

Fig. 7. Measured and simulated S-parameters of off-state absorptive SP4T switch.

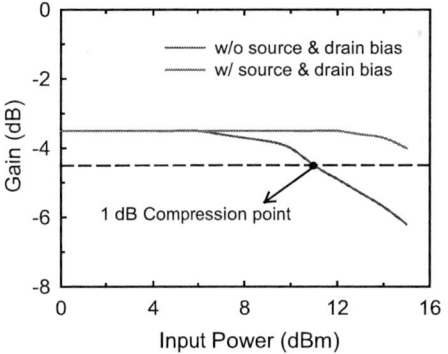

Fig. 8. Measured P1dB of the proposed switch

Table 1. Performance comparison with the previous millimeter-wave CMOS absorptive SP4T switches

	[3]	[6]	This work
Topology	4×λ/4 t-lines	Series-shunt	4×λ/6 t-lines
Process	130 nm CMOS	40 nm CMOS	28 nm CMOS
Frequency (GHz)	57−63	57−66	26−33
Insertion loss (dB)	4.5@60 GHz	4.1@60 GHz	3.5@28 GHz
Isolation (dB)	>31	>18.2	>20
Return losses of the on-state ports (dB)	<−13	<−9.3	<−10
Return loss of the off-state port (dB)	<−14	<−16.4	<−16
IP1dB (dBm)	N/A	N/A	>15
Chip size (mm^2)	1.4×0.4*	0.62×0.44*	1.37×0.39# 1.56×0.51*
Electrical size	0.0224×$(\lambda_g)^2$ @60 GHz*	0.0109×$(\lambda_g)^2$ @60 GHz*	0.0047×$(\lambda_g)^2$ @28 GHz#

*:Including pads #:Excluding pads

IV. CONCLUSION

This paper presented a Ka-band miniaturized absorptive SP4T switch with low insertion loss. By using capacitive matching and loading, the length of t-line is used by one-third compared with previous quarter-wave t-line based switches. To improve switch performance, low threshold-voltage transistors are used whose RC time constant is lower than general transistors. Also, source and drain bias method is used for power handling capability improvement. The insertion loss is 3.5 dB at 28 GHz, whereas isolation is better than 20 dB. Return losses of the on-state ports are less than −10 dB from 26 GHz to 33 GHz. Return loss of the off-state port is better than −16 dB in the same frequency range. Input P1dB is better than 15 dBm. The chip area with and without pads and internal matching circuits are 1.56 × 0.51 mm^2 and 1.37 × 0.39 mm^2, respectively. Electrical sizes are 0.15λ_g × 0.048λ_g and 0.13λ_g × 0.036λ_g, respectively.

ACKNOWLEDGMENT

This work was supported by Yonsei-Samsung Research Center (YSSRC) and Space Core Technology Program through the NRF of Korea funded by the MSIT (NRF-2017M1A3A3A02016255). CAD softwares was supported by IDEC, and measurement equipments was supported by Rohde and Schwarz.

REFERENCES

[1] C.-C. Chang, R.-H. Lee, and T.-Y. Shih, "Design of a beam switching/steering Butler matrix for phased array system, *IEEETrans. Antennas Propag.*, vol. 58, no. 2, pp. 367–374, Feb. 2010.

[2] B. Cetinoneri, Y. A. Atesal, and G. M. Rebeiz, "An 8×8 Butler matrix in 0.13 μm CMOS for 5-6-GHz multibeam applications, *IEEETrans. Microw. Theory Tech.*, vol. 59, no. 2, pp. 295–301, Feb. 2011.

[3] W. Choi, K. Park, Y. Kim, K. Kim, and Y. Kwon, "A V-band switched beam-forming antenna module using absorptive switch integrated with 4×4 Butler Matrix in 0.13 μm CMOS, *IEEE.Trans. Trans. Microw. Theory Tech.*, vol. 58, no. 12, pp. 4052–4059, Dec. 2010.

[4] J. G. Yang and K. Yang, "High-linearity K-band absorptive-type MMIC switch using GaN PIN-diodes, *IEEE Microw. Wireless Compon. Lett.*, vol. 23, no. 1, pp. 37–39, Jan. 2013.

[5] W.-T. Fang, C.-H. Chen, and Y.-S. Lin, "2.4-GHz absorptive MMIC switch for switched beamformer application, *IEEE.Trans. Trans. Microw. Theory Tech.*, vol. 65, no. 10, pp. 3950–3961, Dec. 2017.

[6] D. Lin, K. Kao and K. Lin, "A 40-nm CMOS V-band single-pole quadruple-throw absorptive switch for phased-array applications, in *Proc. IEEE Asia-Pacific Microw. Conf. (APMC)*, Nov. 2017, pp. 268-271.

[7] B. Suh and B.-W. Min, "DC−X-band high-power SOI CMOS T/R switch, *Electron. Lett.*, vol. 52, no. 11, pp. 937–939, May. 2016.

A Compact, High-Power, 60 GHz SPDT Switch Using Shunt-Series SiGe PIN Diodes

Yunyi Gong, Jeffrey W. Teng, and John D. Cressler

School of Electrical and Computer Engineering

Georgia Tech, 777 Atlantic Drive, NW, Atlanta, GA 30332-0250 USA

Abstract—This work describes the design of a compact, 60 GHz, SPDT switch implemented using PIN diodes in a 130 nm SiGe BiCMOS technology. The SPDT RF switch employs a novel shunt-series topology with a resistive biasing scheme to "self-reverse-bias" the off-state shunt diode, thereby improving power handling capability of the SPDT. A coupled inductor matching network is used to minimize the switch size. The proposed design achieves a minimum insertion loss of 2.0 dB, more than 26 dB of isolation, and input-referred P1dB (at 60 GHz) of 22 dBm, with a 0.20 × 0.33 mm² footprint.

Keywords—SPDT switch, PIN diode, SiGe, 60 GHz.

I. INTRODUCTION

Millimeter-wave RF transmit/receive (T/R) single-pole double-throw (SPDT) switches are important elements in modern front-ends for pulsed radar and wireless communication applications. Low insertion loss is required for millimeter-wave T/R switch designs to ensure high transmitter efficiency and modest receiver noise figure. High power-handling capability is also a critical criterion for SPDT design, in order to prevent signal distortion at high transmitted output power levels. Additionally, high isolation is also an important design concern for preventing undesired signal leakage among the signal ports.

High-performance silicon-based millimeter-wave SPDT switches have been successfully demonstrated with bulk CMOS, CMOS on SOI, SiGe HBT, and PIN diodes using various circuit topologies [1]-[7]. A series-shunt SPDT topology was implemented using CMOS SOI and SiGe PIN diodes in [4] and [6], respectively. At millimeter-wave frequencies, a quarter-wave(λ/4)-shunt SPDT is also a popular topology, since it improves insertion loss by avoiding the series device on the signal path, and the physical layout separation between the transmitter (Tx) and receiver (Rx) paths helps to improve isolation. Such a topology has been demonstrated using CMOS, SiGe HBTs and PIN diodes in [1], [5], and [7]. To avoid the use of bulky λ/4 transmission lines, and to achieve a compact SPDT design, researchers have also proposed designs with lumped-element matching networks [8] and compact transformers [2], [3], and [5].

In general, using CMOS in RF switch design is advantageous over PIN diodes and SiGe HBTs in terms of power consumption and layout area. On the other hand, PIN diodes have higher R_{ON}/C_{OFF} ratio than CMOS and SiGe HBTs, which is beneficial for high performance RF switch design [6], [7]. Moreover, PIN diode switches can achieve better power handling capabilities than CMOS and SiGe HBT switches at a similar technology node, due their relatively high turn-on

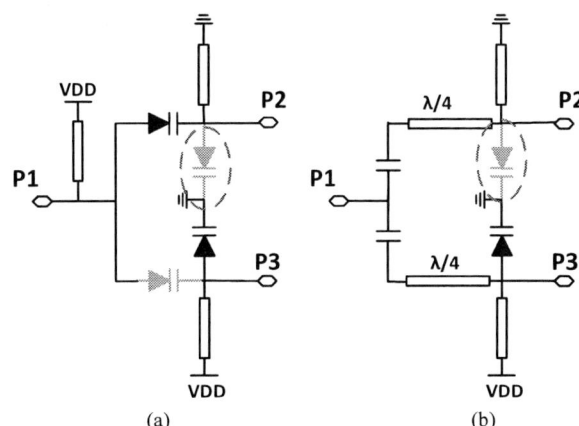

(a)   (b)

Fig. 1. Simplified schematic of (a) series-shunt and (b) λ/4-shunt PIN diode switches with P1 as the antenna port, P2 as the thru port, and P3 as the isolation port. The off-state shunt diodes, circled in the schematics, are usually the limiting factor of the SPDT designs.

voltage and breakdown voltage.

In practice, however, biasing PIN diode switches to fully take advantage of their power handling capability is not straight forward. The most common way to bias SiGe HBTs and PIN diodes in switch designs is to use transmission line stubs to present high impedance, or to resonate with off-capacitance of the devices at the operating frequencies while presenting a low impedance at DC [5]-[7], as illustrated in the simplified typical series-shunt and λ/4-shunt PIN diode SPDTs shown in Fig. 1. Despite short physical length at millimeter-wave frequencies, the biasing transmission lines still inevitably occupy a significant portion of the total footprint of the switch design, even with meandering.

Biasing PIN diodes and SiGe HBTs with resistors, on the other hand, is a more desirable alternative compared to biasing using transmission lines, in term of chip area. However, using resistive biasing requires higher bias voltage and power consumption since, unlike CMOS devices, PIN diodes and SiGe HBTs require significant DC bias current to be turned on. The choice of bias resistance value also requires extra attention, since a bias resistance that is too low will lead to RF signal leakage and thus higher insertion loss, while a bias resistance that is too high may cause voltage clipping at high input power level and thus lead to signal distortion.

Furthermore, it is well understood that, in typical series-shunt (Fig. 1(a)) and λ/4-shunt (Fig. 1(b)) PIN diode switches, the off-state PIN diode on the thru path usually limits the P1dB of the switch design, since large voltage swing across it at high input power level can turn the diode on during the positive

978-1-7281-1702-7/19 $31.00 © 2019 IEEE

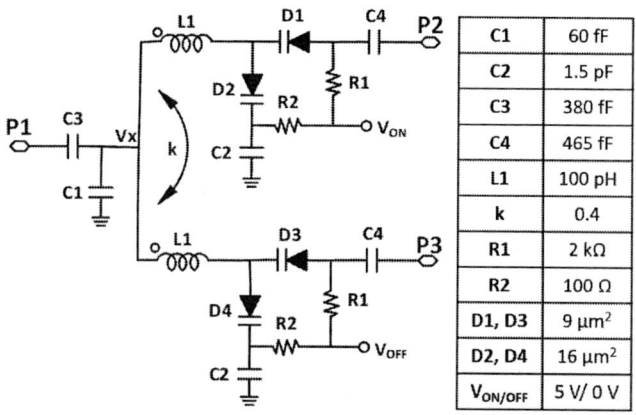

C1	60 fF
C2	1.5 pF
C3	380 fF
C4	465 fF
L1	100 pH
k	0.4
R1	2 kΩ
R2	100 Ω
D1, D3	9 μm²
D2, D4	16 μm²
$V_{ON/OFF}$	5 V / 0 V

Fig. 2. Schematic, passive component values and bias condition of the proposed PIN diode SPDT switch design.

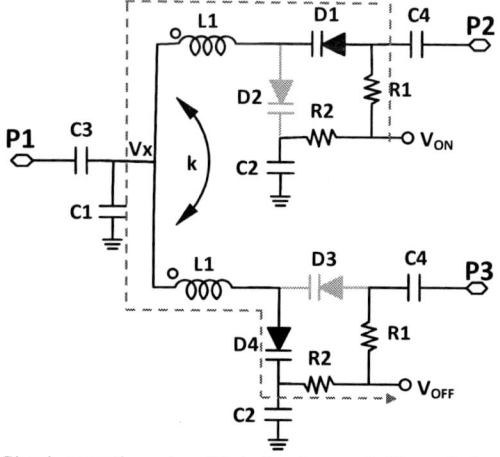

Fig. 3. Circuit operation when P2 is the thru port. The red dotted arrow indicates the DC current path.

signal cycle, causing distortion. To fully utilize the maximum turn-on-to-breakdown voltage swing, Song *et al.* in [7] proposed a negative bias voltage for the off-state shunt diode. The use of negative bias on the off-state device to improve P1dB also applies to SPDT switch using SiGe HBTs, as shown in [5], where a 5 dB increase in P1dB is observed when a -0.8 V is applied to the base of the off-state SiGe HBT. Though effective, the use of a secondary negative supply voltage to obtain high P1dB leads to higher system integration complexity and is thus undesirable.

To avoid the use of space-consuming bias transmission lines and to increase the power handling capability of the PIN diode switch without using negative supply, this paper proposes a novel, resistively biased, shunt-series PIN diode SPDT switch.

II. CIRCUIT DESIGN

The schematic, passive component values, and bias conditions of the proposed SPDT switch topology are shown in Fig. 2. This novel switch topology is constructed using two identical branches, with shunt diodes D2 and D4 placed before series diodes D1 and D3. The matching network consists of a shunt capacitor C1 and two coupled inductors of value L1. Large bypass capacitors (C2) are used to present good RF

ground to diode D2 and D4. Capacitors C3 and C4 are employed as DC blocks.

As shown in Fig. 3, when routing a signal between ports P1 and P2, 5 V on V_{ON} is applied with the V_{OFF} node grounded, and D1 on the thru path and D4 on the isolation path are thus turned on. DC bias current is sourced from node V_{ON}, flows through R1, D1, D4 and R2, and is sunk at node V_{OFF}. From this current flow path, a potential difference is established between V_{ON} and V_X (shown in Fig. 3), which equals the IR drop across R1 plus the voltage drop across D1. The off-state diode D2 is reverse-biased by the voltage difference between V_{ON} and V_X. This helps to prevent D2 from turning on when a large voltage swings across its cathode and anode at high power levels, and hence improves the P1dB of this SPDT design.

The sizes of the diodes are first determined considering the design tradeoff between R_{ON}, R_{OFF}, and C_{OFF} of the devices [7]. Given a supply voltage and desired diode ON current, the sum of R1 and R2 could be readily estimated. The voltage across D2 can be effectively engineered by changing the resistance ratio between R1 and R2. To obtain a large reverse-bias voltage across D2 and to prevent RF signal leakage through R1, a large R1 value of 2 kΩ is chosen for this design. A relatively small resistance value of 100 Ω is used for R2 to minimize its voltage headroom consumption. However, the impedance of R2 is still much greater than that of C2 (1.8 Ω at 60 GHz), preventing undesired RF signal coupling into the bias network.

The matching network of the proposed design uses a similar topology to that presented in [8], except instead of using two individual lumped inductors, coupled inductors are used. The small signal equivalent circuit of configuration in Fig. 3 is as shown in Fig. 4(a). For simplicity, the DC blocking capacitor C3 and C4, RF shorting capacitor C2, and bias resistors R1 and R2 are neglected. Assuming the R_{ON} of the diodes is small, the equivalent circuit of the thru path between P1 and P2 can be further simplified to as shown in Fig. 4(b). The use of coupled inductors reduces the total area required for the matching network, since it reduces the amount of self-inductance needed for matching and occupies one inductor footprint instead of two. The coupling factor also provides another degree of freedom for matching network optimization.

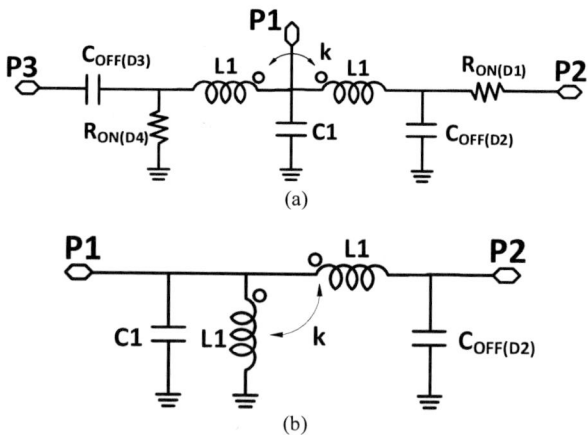

Fig. 4. (a) Simplified small signal equivalent circuit of Fig. 3, and (b) further simplification of (a).

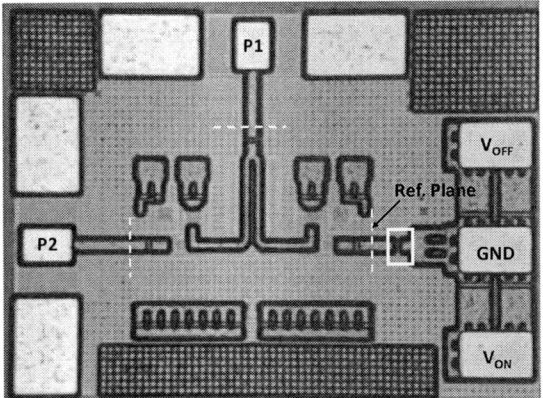

Fig. 5. Chip microphotograph of the proposed PIN SPDT. P3 is terminated with on-die 50 Ω resistor to facilitate two-port measurements. The active chip area is 0.20 × 0.33 mm².

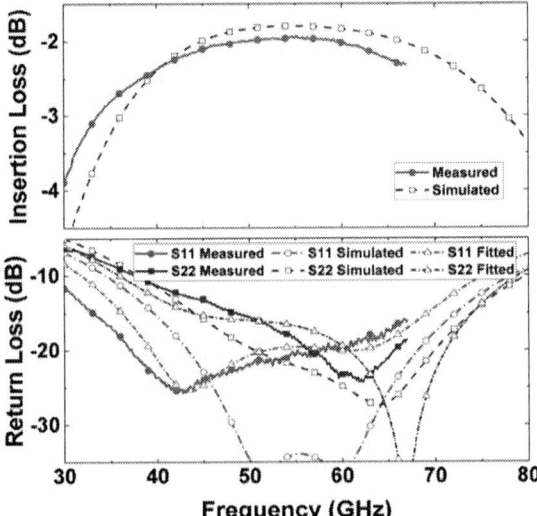

Fig. 6. Simulated and measured insertion loss and return loss.

III. MEASUREMENT RESULTS

The SPDT switch was designed and fabricated using the 130 nm GlobalFoundries 8HP SiGe BiCMOS process (f_T/ f_{MAX} = 200/ 265 GHz) using the 7AM BEOL option. The 8HP process features an integrated PIN diode optimized for millimeter-wave applications, as described in [9].

To facilitate two-port on-wafer measurements, two versions of the SPDT with different on-die port terminations were fabricated. Fig. 5 shows the chip microphotograph of the version where P3 is 50 Ω terminated with an on-die resistor. Another version has the antenna port P1 terminated to characterize the Tx to Rx path isolation. The active chip area is 0.20×0.33 mm². The circuit consumes 8.5 mW of DC power under 5.0 V bias.

A Keysight E8361C PNA was used for on-wafer S-parameter measurements from 30 GHz to 67 GHz. A standard line-reflect-match (LRM) calibration was performed with a pair of 150 μm GSG RF co-axial probes. The measurement was de-embedded to the reference plane (shown in Fig. 5) using a tensor-based multi-line TRL calibration method [10].

The simulated and measured insertion loss and return loss are plotted in Fig. 6. To ensure good simulation and measurement agreement, a full circuit simulation including all passive components enclosed by the reference plane (as shown in Fig. 5) was performed at the end of the design process using the commercial Integrand EMX full-wave 3D solver. Fig. 6. shows good agreement between the simulated and measured S_{21} of the SPDT. The simulated input and output return loss could be reasonably fitted to the measurement by merely adding two single L-C sections of 5 fF shunt capacitor and a 6 pH series inductor before the two shunt PIN diodes. These values are low enough to be within the errors of the EM simulations and the diode model. The measured minimum insertion loss of the switch is 2.0 dB, and the insertion loss remains less than 2.5 dB from 38 to beyond 67 GHz, showing very wide-band performance. The measured S_{11} and S_{22} are less than -10 dB from 37.8 GHz to beyond 67 GHz. The measured and simulated isolation are plotted in Fig. 7. The measured isolation between Rx/Tx to the antenna port (S_{31}) is greater than 23 dB up to 67 GHz, and the measured isolation between the Tx and Rx port (S_{32}) is greater than 26 dB up to 67 GHz.

The input-referred P1dB of the SPDT switch is measured on-wafer using a pair of 100 μm pitch WR-15 waveguide probes at 60 GHz. The signal source setup consists of an OML S15MS source module, a Millitech 53-70 GHz power amplifier, and a Quinstar mechanical attenuator. The input and output power are measured using a 10 dB WR-15 directional coupler and a pair of Keysight V8486 series waveguide power sensors.

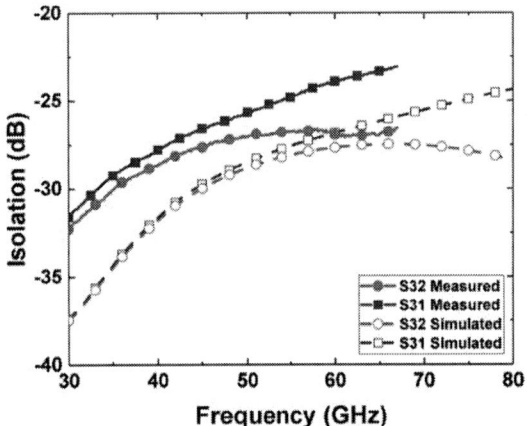

Fig. 7. Simulated and measured switch isolation.

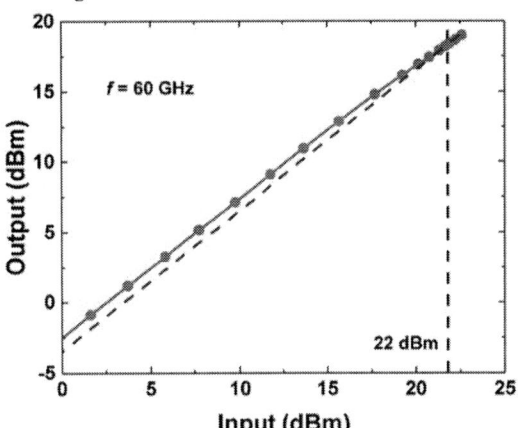

Fig. 8. P1dB measurement result at 60 GHz.

978-1-7281-1702-7/19 $31.00 © 2019 IEEE

Table 1. Performance Comparison with SOA Millimeter-wave SPDT Switches

Reference	[1]	[2]	[3]	[4]	[5]	[6]	[7]	[8]	This Work
Technology	90 nm CMOS	90 nm CMOS	65 nm CMOS	45 nm CMOS SOI	90 nm SiGe	130 nm SiGe	90 nm SiGe	130 nm CMOS	**130 nm SiGe**
Device	NMOS	NMOS	NMOS	NMOS	Rev. Sat. HBT	PIN	PIN	NMOS	**PIN**
Topology	λ/4-shunt	Transformer based	Lumped 4-way comb.	Series-shunt	λ/4-shunt	Series-shunt	λ/4-shunt	Matching network	**Shunt-series**
Freq. (GHz)	50-70	50	58-85	DC-60	73-110+	50-78	73-133	57-66	**38-67+**
Min. IL (dB)	1.5	3.4	1.8	2.5 @ 60 GHz	1.1	2.0	1.4[b]	1.7	**2.0**
RL (dB)	>8	>10	>10	>10	>10	>12	>10	>10	**>10**
ISO (dB)	25-30	13.7	22-30	>25	22	25-35	19-22	>21.1	**>23**
P1dB (dBm)	13.5	13	10	7.1	17/ 22[a]	-	>+24[a, c]	13.8	**22**
P_{DC} (mW)	-	-	-	-	5.9	16.8	10.2	-	**8.5**
Core Area (mm²)	0.022	0.18	0.015	0.040	0.213	0.11	0.14	0.02	**0.066**

[a] Negative bias used
[b] Pad loss not de-embedded
[c] Measurement limited by available source power

The loss of the isolators and waveguide components were carefully measured and accounted for, but the pad loss is not de-embedded from this measurement. The measured large signal results are shown in Fig. 8. The measured input-referred P1dB of the proposed SPDT is 22 dBm at 60 GHz.

Table 1 compares the performance of the proposed shunt-series PIN diode SPDT switch with other state-of-the-art millimeter-wave switches. The proposed SPDT switch design shows good bandwidth performance with the coupled inductor matching network. A high P1dB is achieved with the proposed shunt-series topology without the need for a secondary negative bias level. Though the addition of series diodes incurs some additional loss for the SPDT, the insertion loss of the design remains competitive with the performance of other state-of-the-art designs.

IV. CONCLUSION

This paper demonstrates a 60 GHz PIN diode SPDT switch with a novel shunt-series topology. The resistive bias network of the proposed does not require any transmission line stubs, which leads to a reduced chip area. The off-state shunt PIN diode on the thru path is reverse-biased by the voltage drop from the series diode on the thru path and its bias resistor. This biasing scheme prevents the shunt diode from turning on by the large voltage swing across its anode and cathode at input high power levels. It thus leads to an improved P1dB, compared to conventional series-shunt and λ/4-shunt SPDT topologies, without the need for an additional negative power supply. Though the use of series diode on the signal path incurs some additional insertion loss compared to topologies using only shunt devices, the overall insertion loss of the proposed design remains competitive with the state-of-the-art millimeter-wave SPDT designs. The use of series diode also helps to ensure a good port-to-port isolation. In addition to the use of a resistive bias network, matching network with coupled inductors are also employed to further reduce the overall footprint of the design.

ACKNOWLEDGMENT

The authors are grateful to GlobalFoundries SiGe team (especially D. Harame, A. Joseph, N. Cahoun, and V. Jain) for their support and contributions. We would also like to acknowledge the contributions and help from the members of the SiGe Devices and Circuits Group and the GEMS group at Georgia Tech.

REFERENCES

[1] M. Uzunkol and G. Rebeiz, "A Low-Loss 50–70 GHz SPDT Switch in 90 nm CMOS," in *IEEE Journal of Solid-State Circuits*, vol. 45, no. 10, pp. 2003-2007, Oct. 2010.

[2] E. Adabi and A. M. Niknejad, "A mm-wave transformer based transmit/receive switch in 90nm CMOS technology," *2009 European Microwave Conference (EuMC)*, Rome, 2009, pp. 389-392.

[3] R. Shu and Q. J. Gu, "A Transformer-Based V-Band SPDT Switch," in *IEEE Microwave and Wireless Components Letters*, vol. 27, no. 3, pp. 278-280, March 2017

[4] M. Parlak and J. F. Buckwalter, "A 2.5-dB Insertion Loss, DC-60 GHz CMOS SPDT Switch in 45-nm SOI," *2011 IEEE Compound Semiconductor Integrated Circuit Symposium (CSICS)*, Waikoloa, HI, 2011, pp. 1-4.

[5] R. L. Schmid *et al.*, "On the Analysis and Design of Low-Loss Single-Pole Double-Throw W-Band Switches Utilizing Saturated SiGe HBTs," in *IEEE Transactions on Microwave Theory and Techniques*, vol. 62, no. 11, pp. 2755-2767, Nov. 2014.

[6] K Lam *et al.*, "Wideband millimeter wave pin diode SPDT switch using IBM 0.13μm SiGe technology," *2007 European Microwave Integrated Circuit Conference*, Munich, 2007, pp. 108-111.

[7] P. Song *et al.*, "A high-power, low-loss W-Band SPDT switch using SiGe PIN diodes," *2014 IEEE Radio Frequency Integrated Circuits Symposium*, Tampa, FL, 2014, pp. 195-198.

[8] J. He, Y. Xiong and Y. P. Zhang, "Analysis and Design of 60-GHz SPDT Switch in 130-nm CMOS," in *IEEE Transactions on Microwave Theory and Techniques*, vol. 60, no. 10, pp. 3113-3119, Oct. 2012.

[9] B. A. Orner *et al.*, "p-i-n Diodes for Monolithic Millimeter Wave BiCMOS Applications," *2006 International SiGe Technology and Device Meeting*, Princeton, NJ, 2006, pp. 1-2.

[10] Y. Rolain *et al.*, "A Tensor-Based Extension for the Multi-Line TRL Calibration," in *IEEE Transactions on Microwave Theory and Techniques*, vol. 64, no. 7, pp. 2121-2128, July 2016.

Low-cost, High-Gain Antenna Module Integrating a CMOS Frequency Multiplier Driver for Communications at D-band

Francesco Foglia Manzillo, José Luis Gonzalez-Jimenez, Antonio Clemente, Alexandre Siligaris, Benjamin Blampey, Cedric Dehos

CEA-Leti and Univ. Grenoble Alpes, Minatec Campus, 38054 Grenoble, France

{francesco.fogliamanzillo, joseluis.gonzalezjimenez}@cea.fr

Abstract—**This paper presents a compact D-band antenna module for ultrafast short-range communications. The system comprises a planar lens fed by an antenna-in-package embedding an integrated 28-nm bulk CMOS frequency multiplier. Both lens and primary source are fabricated using low-cost printed circuit board (PCB) technology. The module achieves a peak gain of 25 dBi and works between 114 GHz and 138 GHz with a gain drop lower than 3 dB. The proposed solution paves the way for low-loss and low-profile wireless systems entirely integrated in planar multilayer substrate modules.**

Keywords— **millimeter-wave antennas, systems-in-package, D-band, beyond-5G, advanced flat lens array.**

I. INTRODUCTION

The exploitation of the mm-wave spectrum beyond 100 GHz is currently foreseen to pursue the increase of data rate in future mobile communications. The D-band (110-170 GHz) is suitable to attain data rates greater than 10 Gbit/s over relatively short links [1], [2]. To this end, antennas achieving gain values of 25 dBi or higher and operating fractional bandwidths (BWs) of about 20% are necessary.

Most state-of-the-art D-band wireless modules rely on dielectric lens antennas [3]-[5] to meet such challenging requirements. However, high-gain lenses are oversized, bulky and have non-planar shapes. At (sub)mm-wave frequencies, they are generally fabricated using technologies, such as silicon micromachining and additive manufacturing, other than that employed for the primary source. Additional steps are thus required to assembly lens and source and to guarantee the stability of the overall system.

Novel cost-effective antenna concepts ensuring compactness and reduced thickness are clearly needed to further push the integration and enhance the efficiency of millimetre-wave front-ends, in view of the massive increase of the number of wireless systems expected in the next decade.

Flat lenses, often referred to as transmitarray antennas [6]-[9], represent an attractive solution to achieve high gain and reduced insertion loss employing relatively short focal distances. They can be realized using planar fabrication processes, such as PCB technology. However, due to the limited manufacturing resolution, only a few transmitarrays in PCB technology [7] have been presented in the literature beyond 100 GHz and achieve degraded performance.

This paper proposes a novel antenna module composed of a high-gain fixed-beam transmitarray and a compact in-package antenna feed. Both subsystems are fabricated using standard PCB technology. To the best of authors' knowledge, this is the first transmitarray antenna system fabricated in PCB technology achieving a -3-dB gain bandwidth of 20% at D-band. A 28-nm CMOS integrated circuit (IC) is co-integrated on the primary source. It realizes an eightfold frequency multiplication of the input signal (from K-band to D-band) and provides an output power greater than -10 dBm over the operating band.

The proposed design accounts for the challenging constraints of packaging and IC assembly and demonstrates the feasibility of such a compact, wideband, high-gain antenna module using a low-cost PCB manufacturing process.

II. ANTENNA MODULE: DESIGN AND OPTIMIZATION

A. Focal source

The primary source of the module is an antenna-in-package (AiP) co-integrating a 2×2 array of aperture-coupled patches and the frequency multiplier. The stack-up is shown in Fig. 1. It comprises two substrate layers (Rogers RT/Duroid 6002, $\varepsilon_r = 2.94$, tan $\delta = 1.2 \times 10^{-3}$) bonded by an adhesive film (Rogers 4450F, $\varepsilon_r = 3.52$, tan $\delta = 4 \times 10^{-3}$). The total thickness of the AiP is only 515 µm. The minimum feature size of the proposed AiP design is 75 µm, which makes it suitable for standard PCB manufacturing. A square array of four patches was designed to achieve a -10-dB beamwidth of about 80° in both principal planes at 140 GHz. The inter-element spacing along both x- and y-axis is of 1.25 mm, i.e. $0.58 \cdot \lambda_0$, where λ_0 is the free-space wavelength at 140 GHz. The array is located at the center of the AiP. The distance between the radiating patches and the ground ($0.16 \lambda_0$) was selected to enhance the antenna gain bandwidth. A rectangular cavity, realized with via-fences (see Fig. 1), surrounds the array to mitigate the impact of surface-waves [5]. The cavity is shorted to the ground of the patches (on M2) and to the bottom ground of the antenna board (on M1). The simulated gain of the array at 130 GHz is 8.5 dBi and the -3-dB gain bandwidth spans from 110 GHz to 145 GHz.

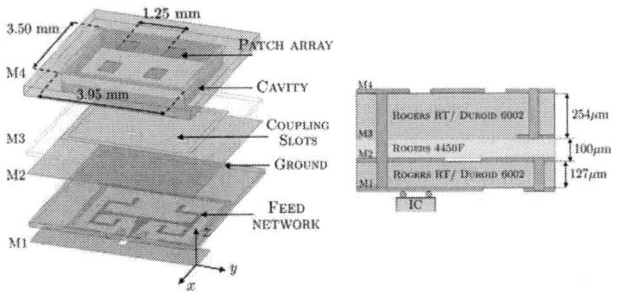

Fig. 1. Exploded view and stack-up of the 2×2 patch array in the AiP.

978-1-7281-1702-7/19 $31.00 © 2019 IEEE

B. Transmitarray

The flat discrete lens has been optimized using an *ad-hoc* tool based on ray tracing, considering the simulated radiation patterns of the primary source and of the lens elements (unit cells). It consists of 40×40 square unit cells (UCs). The overall electrical size is $20\lambda_0 \times 20\lambda_0$.

Most D-band transmitarrays in the open literature rely on complex stack-ups and technologies. For instance, the design in [8] comprises six low-temperature co-fired ceramic (LTCC) tapes and nine metal layers. Conversely, the proposed transmitarray includes only 3 metal layers and two dielectric substrates. The stack-up is similar to that shown in Fig. 1 and employs the same materials. The minimum conductor spacing and via diameter in the design are 75 µm and 100 µm, respectively. The planar lens was synthetized using eight different UCs (*i.e.* 3-bit phase quantization as in [6], [8]), realized as either aperture-coupled or probe-fed patch antennas. The use of both coupling mechanisms increases the degrees of freedom in the design and makes it possible to achieve phase errors lower than 20° and insertion losses lower than 1.5 dB over a large fractional bandwidth (21.3%). The 3-bit quantization determines a directivity loss of only 0.3 dB. Figure 2 shows the distribution of the cells on the lens, optimized at 126 GHz, as well as the fabricated prototype. The final distance between the AiP source and the lens is relatively short: 32 mm, corresponding to a focal-to-diameter (*F/D*) ratio of 0.75. State-of-the-art designs use much larger *F/D* values, e.g. 1.87 in [8], to cover similar bandwidths.

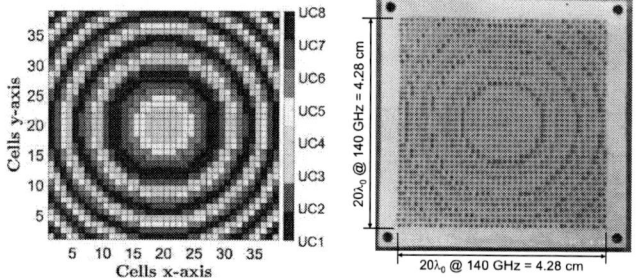

Fig. 2. Distribution of the UCs on the planar lens and picture of the prototype.

C. Antenna driver IC

An IC driving the antenna module is integrated on the AiP for validating the packaging approach and characterizing the antenna. D-band on-wafer probes, with a pitch of 50 µm, could have been used to test the AiP. However, such a narrow pitch cannot be attained using standard PCB technology. Therefore, the driver circuit is used to up-convert an input K-band signal to D-band and feed the input line of the AiP. The circuit is shown in Fig. 3. It is designed and fabricated in 28-nm bulk CMOS technology. It relies on three stages of self-mixing multiplication. Two versions of the circuit have been fabricated. The first one, featuring C4 µbumps with a pitch of 150 µm, was mounted on the feedline of the source antenna (see Fig. 4). The second one includes output pads with a pitch of 50 µm and was realized for on-wafer probing. The characterization of the latter version made it possible to de-embed the effect of the driving circuit from the antenna measurements. The electronc

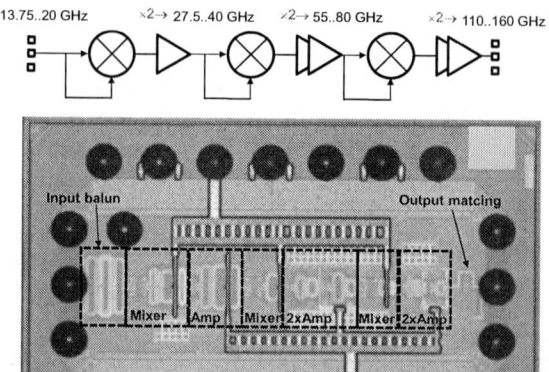

Fig. 3. Antenna driver IC: block diagram and layout (flip-chip version).

schematic of the self-mixing multipliers is shown in Fig. 5. All NMOS devices use a modified layout version of the PDK low-Vth NMOS native transistor for improving f_{max} parameter. Some transistors were fabricated and measured in a separate die. They achieve a peak $f_{max} = 202$ GHz for current densities of 0.05 mA/µm. The active core of all three mixers is equal, with $M_{1,2,3,4,5,6}$ of 36 µm width and minimum length. An input balun is used to generate the differential signal for the first mixer. Each multiplication stage is coupled to the next one using transformers. Each transformer has a central coupling region and input and output series inductors (see Fig. 5) that were tuned during the design optimization. They are implemented using the two topmost thick metal layers available in the 28-nm process. A central tap in the primary winding is used to connect the power supply by means of an ALUCAP feedline. The sizes of the transformer and inductors are modified in each stage according to the operating frequency. The amplifiers between each mixing stage are conventional common source pseudo-differential stages with 70 µm width and minimum length

Fig. 4. Antenna driver IC mounted on the back side of the focal source.

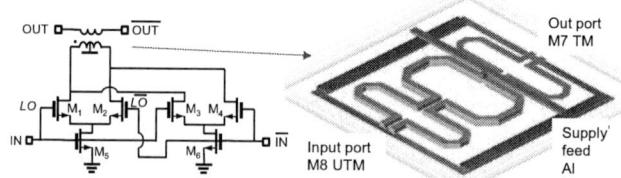

Fig. 5. Self-mixing stage: schematic and transformer layout.

transistors. A single amplifier is used at the first stage and two cascaded amplifiers are employed at the second and last stages. The center frequency of each of the two stages is tuned differently to enlarge the overall BW. The output of the last amplifier is matched to 50 Ohms at D-band using a series-parallel transmission line network, sized differently in the flip-chip and the on-wafer probing version of the circuit to take into account the different output pads and C4 µbumps parasitics.

III. EXPERIMENTAL RESULTS

The ×8 frequency multiplier IC used as antenna driver has been characterized separately using a semi-automatic on-wafer probing bench and a D-band VDI mm-wave down-converter head. The latter is connected through a WR-6.5 waveguide and probes to the output of the IC. The input is provided using a conventional mm-wave signal generator at K-band. The frequency and output power characteristics of the circuit are shown in Fig. 6, after de-embedding the input and output test fixtures. Simulation results are also reported for comparison. The measurements show that the IC provides a nearly constant output signal from 110 GHz to 140 GHz, and that it operates in saturation for an input power greater than 0 dBm. Figure 7 (left) shows the output reflection coefficient of the IC, demonstrating a broadband matching. These measurements are used to determine the input power level that feeds the antenna. The impact of the pads and the C4 bumps, which in the antenna module are different from those used in the measured circuit, is also taken into account in the calculation of the delivered power.

The full antenna system is then characterized in an anechoic chamber. The board containing the in-package source antenna and the frequency multiplier is assembled with the transmitarray using plastic spacers, as shown in Fig. 7(right). The system size is $20\lambda_0 \times 20\lambda_0 \times 15\lambda_0$. The module was excited by a signal generator injecting a power of 13 dBm in the 13.75-

21.25-GHz band. The receiver (Rx) includes a 20-dBi standard gain horn and the D-band down-converter connected to a spectrum analyzer (same setup used for the characterization of the IC). Figure 8 (right) shows measured and simulated far-field radiation patterns at 130 GHz (H-plane cut). Numerical results and measurements are in excellent agreement, proving the reliability of the PCB technology and of the proposed design. The measured -3-dB beamwidths in the principal planes are about 3.2°. The cross-polarization is less than -24 dB. Figure 9 (left) shows the equivalent isotropic radiated power (EIRP) as a function of output frequency, showing a peak level of 15 dBm and a 3-dB BW ranging from 114 GHz to 138 GHz. Considering that the driver IC provides a nearly constant power at the antenna feed in this frequency range, the estimated gain of the antenna module reaches 25 dBi, including the insertion losses of the interconnect between the IC and the antenna. Note that the BW of the module, in terms of EIRP, is limited by the drop of the IC output power beyond 140 GHz.

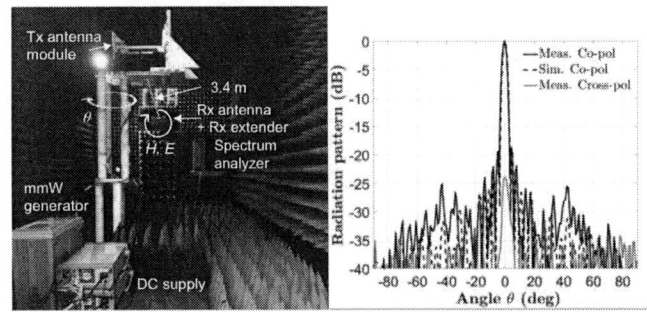

Fig. 8. Far-field characterization setup (left). Simulated and measured radiation pattern (H-plane cut) of the antenna module at 130 GHz (right).

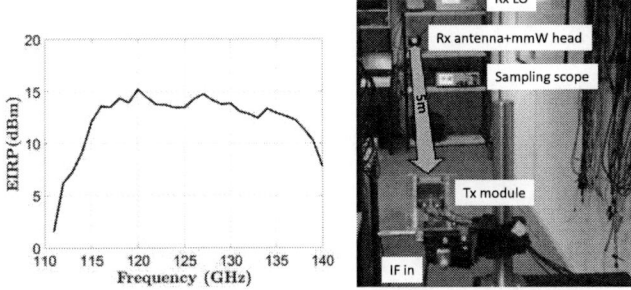

Fig. 9. Left: measured EIRP of the antenna module fed by the frequency multiplier IC. Right: *Tx-Rx* link setup.

IV. DISCUSSION AND COMPARISON WITH THE STATE OF THE ART

The main goal of this work was to prove the feasibility of a broadband, high-gain compact antenna system integrated with the electronic transmitter using a low-cost packaging technology. The antenna driver IC was realized as a frequency multiplier to characterize the antenna in the same environment as when connected to a D-band transceiver IC. The IC driving the antenna is mounted on the backside of the in-package source antenna. The measurements account for the parasitic effects due to the interconnections.

The main limitation imposed by the PCB technology is the minimum track width and spacing of 75 µm, but fabrication

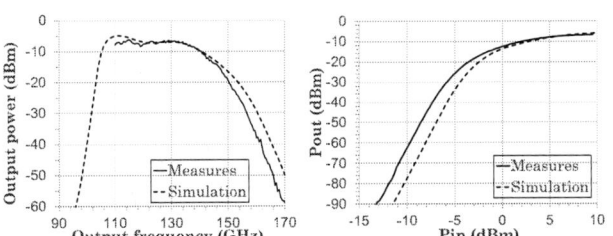

Fig. 6. Frequency response (left) and output power at D-band as a function of input power at K-band (right) of the frequency multiplier IC.

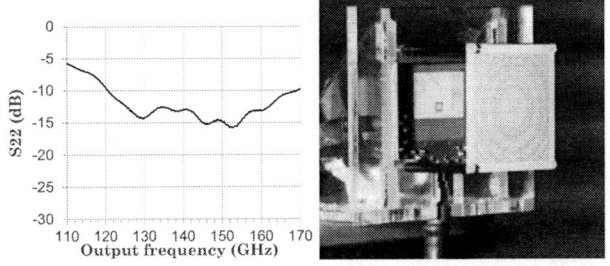

Fig. 7. Output matching characteristic of the driver IC (left). Antenna module, comprising the focal source AiP and the transmitarray, under test in the anechoic chamber (right).

tolerances have also an important impact. For example, an up-shift (about 5 GHz) of the frequency response of the transmitarray was found in a separate measurement, using a 10-dBi feeding horn. As confirmed by full-wave simulations, this discrepancy can be attributed to the observed over-etching of the lens elements: about 30 μm in the worst case, corresponding to ~6% of the average patch size.

Table I shows a comparison of the implemented D-band antenna with similar mm-wave system. With respect to D-band modules [5], [8], the proposed design achieves a performance similar to [8] using a much simpler fabrication process. Moreover, it offers a more compact solution than [5] which relies instead on a dielectric dome lens. Even though a more complex PCB process was employed in [5] for the primary source, enabling a minimum track width of 50 μm, our design exhibits similar gain and bandwidth. Finally, as opposed to all other works reported in Table 1, the values of gain and bandwidth of our prototype account for the loss due to the interconnection between the driving IC and the antenna.

Table 1. Comparison of the proposed antenna with the state of the art

Ref.	[5]	[7]	[8]	[9]	This work
Freq. (GHz)	140	250	140	83.5	**130**
Primary source	PCB 4 layers	Horn antenna	Horn antenna	Slot antenna	**PCB 4 layers**
Source gain (dBi)	7.8	15	20	≈6	**≈8.4****
Technology	3-D printed plastic	PCB 2 layers	LTCC 7 layers	PCB 3 layers	**PCB 3 layers**
Diameter	$19\lambda_0$	$14\lambda_0$	$20\lambda_0$	$25\lambda_0$	**$20\lambda_0$**
F/D	1.06	0.30	1.87	0.32	**0.75**
Peak gain (dBi)	28	28.8	33.5	33.9	**25**
-3dB EIRP BW	20%	12%	24%	n.a.	**18%***

*limited by the BW of the driver IC. **Simulated value.

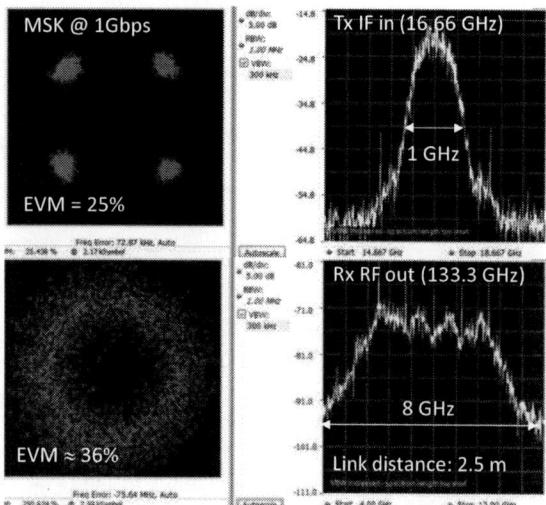

Fig. 10. Input and output constellation and modulation spectra for a 2.5m-link (see Fig. 9) using the proposed antenna module as Tx unit.

A final validation of the antenna module is done using it as transmitter (Tx) in a point-to-point link (see Fig. 9). The Rx consists of a similar 33-dBi transmitarray fed by a 10-dBi horn, connected to a D-band down-converter. The transmitarray is based on the same cells used for the lens of the Tx module but was optimized for the feeding horn. A 1-Gbps MSK modulated signal generated by an AWG and centered at 16.66 GHz is applied at the input of Tx module. The frequency multiplier IC produces a large spectral regrowth due to its inherent non-linear behavior. As a consequence, the phase information is lost, hindering the demodulation at the Rx. Nevertheless, the EVM of the received signal can be estimated in the radial direction from the dispersion of the Rx constellation trajectory. The latter is well preserved, as shown in Fig. 10. The link was validated up to 5 m as shown in Fig. 9. EVM was degraded by the Tx from 25% at the input to 36% at 2.5 m and 46% at 5 m. Note that the antenna and driving system could transmit up to three 8-GHz channels, as the one shown in Fig. 10.

V. CONCLUSION

This paper demonstrates a compact D-band antenna-in-package in low-cost PCB technology achieving a -3-dB EIRP BW of 24 GHz (114-138 GHz) and a peak gain of 25 dBi. The antenna consists of a primary source comprising four patch antennas and a $20\lambda_0 \times 20\lambda_0$ flat discrete lens. The antenna performance is measured taking into account the insertion losses due to the connection to the driving circuit. The latter is a 28 nm CMOS ×8 frequency multiplier IC flip-chipped on the back-side of the primary source. It provides a nearly constant output power of -10 dBm from 110 to 145 GHz. The antenna module is used to demonstrate a MSK modulated 5 m link at D-band on a single channel of 8 GHz of bandwidth. This is, to the best of our knowledge, the first planar high-gain D-band antenna module entirely built using standard PCB technology.

REFERENCES

[1] A. Hirata et al., "120-GHz-band wireless link technologies for outdoor 10-Gbit/s data transmission," IEEE Trans. Microw. Theory Techn., vol. 60, no. 3, pp. 881–895, Mar. 2012.

[2] C. Wang et al., "A 10-Gbit/s wireless communication link using 16-QAM modulation in 140-GHz band," IEEE Trans. Microw. Theory Techn., vol. 61, no. 7, pp. 2737–2746, July 2013.

[3] A. J. Alazemi, H. Yang, and G. M. Rebeiz, "Double bow-tie slot antennas for wideband millimeter-wave and terahertz applications," IEEE Trans. THz Sci. Technol., vol. 6, no. 5, pp. 682–689, Sept 2016.

[4] B. Gottel et al., "Miniaturized 122 GHz short range radar sensor with antenna-inpackage (aip) and dielectric lens," in Proc. 8th Eur. Conf. Antennas Propag., pp. 709–713, The Hague, Netherlands, April 2014.

[5] A. Bisognin et al., "Ball grid array module with integrated shaped lens for 5G backhaul/fronthaul communications in F-band," IEEE Trans. Antennas Propag., vol. 65, no. 12, pp. 6380–6394, Dec. 2017.

[6] C. Jouanlanne et al., "Wideband linearly polarized transmitarray antenna for 60 GHz backhauling," in IEEE Trans. Antennas Propag., vol. 65, no. 3, pp. 1440-1445, March 2017..

[7] H. Yi, S. W. Qu, and C. H. Chan, "Low-cost two-layer terahertz transmit array," Electron. Lett., vol. 53, no. 12, pp. 789–791, 2017.

[8] Z. W. Miao, Z. C. Hao, G. Q. Luo, L. Gao, J. Wang, X. Wang, and W. Hong, "140 GHz high-gain LTCC-integrated transmit-array antenna using a wideband SIW aperture-coupling phase delay structure," IEEE Trans. Antennas Propag., vol. 66, no. 1, pp. 182–190, Jan. 2018.

[9] Y. Kasahara et al., "Low-profile transmitarray antenna with single slot source and metasurface in 80-GHz band," in Proc. IEEE Int. Antennas Propag. Symp., July 2018.

Scalable Analytical Model of 1.7 THz Cut-off Frequency Schottky Diodes Integrated in 55nm BiCMOS Technology

Vincent Gidel[1][2][3], Fréderic Gianesello[1], Pascal Chevalier[1], Grégory Avenier[1], Nicolas Guitard[1], Victor Milon[1], Michel Buczko[1], Charles-Alex Legrand[1], Cyril Luxey[2] and Guillaume Ducournau[3]

[1]STMicroelectronics, France
[2]Polytech'Lab, Univ. Nice Sophia A., France
[3]IEMN, UMR CNRS 8520, France
vincent.gidel@st.com

Abstract—In this paper, an innovative Schottky diode architecture is proposed and implemented in 55 nm BiCMOS technology. A State-of-the-art 1.7 THz cut-off frequency is measured and an analytical scalable model is proposed and experimentally validated paving the way for further performance improvement. In addition, this analytical model can be integrated in a Design Kit library in order to enable sub-THz Schottky diode-based circuit designs in advanced BiCMOS.

Keywords— Schottky diode, modeling, THz.

I. INTRODUCTION

Our connected society asks a never ending consumption for more data, which requests a continuously growing demand for higher data rates therefore driving the carrier frequencies to be extended up-to the millimeter wave (mmW) [1] and even terahertz (THz) range. Recently, a new standard (802.15.3d-2017) has been established around 300 GHz by the IEEE, and data rates up-to 100 Gb/s has already been demonstrated in [2].

From the transmitter side, such high data rates have been achieved with III-V technologies using photonics-based techniques leveraging their intrinsic broadband and linear performances. But as it was the case some years ago for the previous developments at mmW frequencies, silicon is today starting to seriously compete with those photonics-based transmitters leveraging Silicon Photonics developments [3].

Consequently, the question is now to determine the feasibility of high performance sub-THz receivers in silicon technology. At such high frequencies, mixer first receiver is the architecture of choice. Available Local Oscillator (LO) power being a constraint, subharmonic mixer is a natural choice and state-of-the-art Sub-Harmonics Mixer (SHM) are today implemented in III-V using Schottky diodes [4]. The availability of high-performance Schottky diodes in advanced CMOS or BiCMOS technology could then make feasible the development of a full silicon-based chipset solution enabling higher integration and low-cost solution.

In this paper, we present in section II an innovative Schottky diode architecture implemented in 55 nm BiCMOS process. In section III, we propose an analytical model for this Schottky diode which is validated by experimental characterizations in section IV.

II. INNOVATIVE SCHOTTKY DIODE ARCHITECTURE IN 55 NM BiCMOS TECHNOLOGY

A. Schottky diodes state-of-the-art

The first ever integrated Schottky diode has been developed in III-V technology using whiskered architecture [5]. The evolution of the fabrication processes allowed the development of planar architecture in order to facilitate integration inside a chip. The progress of silicon technologies to address high frequency applications enables to support planar Schottky diode architecture. In CMOS technologies, lateral architecture with Shallow Trench Insulation (STI) has been proposed first [6], [7], [8] and then polysilicon spacer architecture has emerged in order to reduce achievable series resistance of the diode [9], [10]. The major limitation of CMOS structure is linked to parasitic substrate capacitance which reduces achievable performances in antiparallel configuration [11]. In BiCMOS technology, leveraging the buried collector module dedicated to bipolar transistor has enabled to develop an improved resistive Schottky diode [7]. However, in advanced BiCMOS technology, the collector is not buried deeply enough and a deeper layer, requiring a specific implantation, is needed to avoid an increase of the parasitic capacitance [12].

B. Proposed Innovative Schottky Diode Architecture

Taking advantage of those state-of-the-art results, we propose here a new Schottky diode architecture (Fig. 1) leveraging both the advantages of CMOS architecture and specific bipolar process steps available in BiCMOS technology.

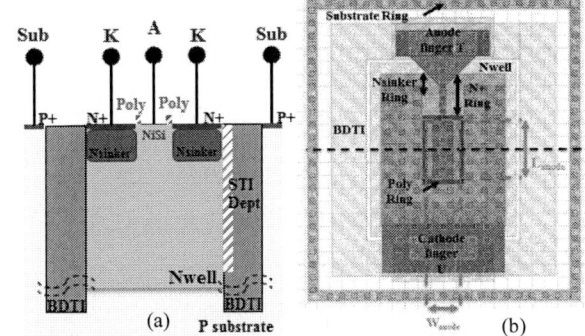

Fig. 1. Innovative BiCMOS 55nm Schottky Diode: (a) Side-cut view; (b) Top view

The proposed architecture of the diode takes advantage of the polysilicon spacer developed in CMOS technology combined with the Nsinker implantation (in order to reduce series resistance) and deep trenches module [13] of bipolar transistor (in order to reduce the substrate parasitics).

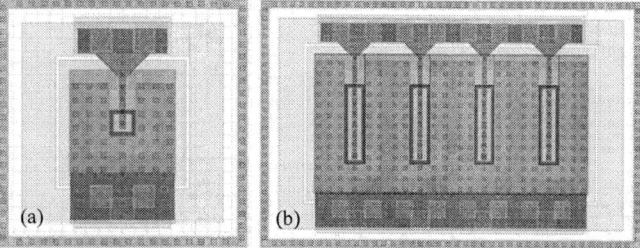

Fig. 2. Schottky diode layout with: (a) 1 Anode fingers; (b) 4 Anode fingers

As depicted in Fig. 1, the metal-semiconductor junction has been integrated at the surface on the Nwell with NiSi silicide deposit (inside the polysilicon ring). The diffusion current flows inside the Nwell between the anode and the cathode. Moreover, the anode-to-cathode isolation is achieved with a polysilicon spacer in order to reduce the current resistive path. Nsinker implant enables to further reduce the series resistance value. Then, the cathode-substrate insulation is improved using deep trenches BiCMOS Deep Trenches Insulation (BDTI) module.

C. Measured 55 nm BiCMOS Schottky diode performances

The main parameters which sets the performance of the Schottky diode are the series resistance R_s and the zero-bias junction capacitance C_{j0}. For some applications, it could be useful to consider the barrier height Φ_b and the ideality factor η as well. The cut-off frequency of the diode is given by Eq. 1:

$$F_c = \frac{1}{2\pi \times R_s \times C_{j0}} \qquad (1)$$

Table 1. Comparison of Schottky Diode performances

Ref.	A (μm^2)	R_s (Ω)	C_{j0} (fF)	Φ_b (V)	η
[7]	3.68	57	0.43	N.R.	N.R.
[6]	1.63	10	7	0.52	1.3
[9]	2.3	15	8.2	0.38	1.34
[10]	0.42	210	3	N.R.	N.R.
[8]	4	13	14.3	N.R.	N.R.
[13]	4	5.4	21	N.R.	N.R
This work	2	39.9	2.34	0.65	1.16

Table 1 summarizes extracted parameters from the measured I-V curve of one of our typical achieved diode in ST 55 nm BiCMOS technology (This work). The junction capacitance C_{j0} has been de-embedded thanks to Technology Computer-Aided Design (TCAD) simulation (to assess doping) coupled with S_{ij} parameter measurements in order to dissociate the capacitance

linked to metallic access C_{BE} (Back-End capacitance) from the junction capacitance C_{j0}. An extracted cut-off frequency of 1.7 THz is obtained which represents a state-of-the-art value (Fig. 3).

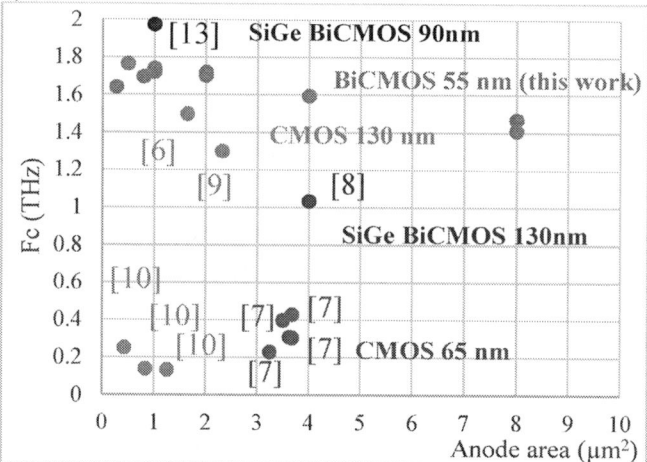

Fig. 3. Schottky diodes cut-off frequency achieved with 55 nm BiCMOS process (this work) presented with state-of-the-art other diodes.

III. SCALABLE ANALYTICAL MODEL DEVELOPMENT

A. Proposed equivalent lumped model

In the aim to enable future circuit designs and identify possible improvements for the proposed Schottky diodes, it seems meaningful to elaborate an analytical model describing key electrical figure in function of device key technological and layout parameters. We propose to use the circuit model presented Fig. 4 an assuming the three following hypotheses:

- The current circulation between the anode and the cathode fingers will be modelled by an average path due to the distributed nature of the conductive path.
- The inductive contribution of the anode and the cathode fingers will be neglected due to their small lengths and their interdigitated configuration.
- The doping value N_d inside the Nwell layer given by the TCAD report simulation is reliable and accurate.

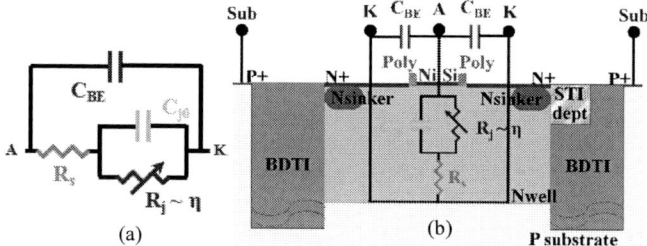

Fig. 4. (a) Schottky diode circuit model related to (b) side cut-view of the diode

B. Series resistance modelling strategy

The series resistance R_s was estimated thanks to the existing Design Rules Manual (DRM) of the BiCMOS 55 nm technology and the Schottky diodes layout drawing parameters. The R_s model is given by the following equations:

$$R_s = R_{anode} // R_{silicide} + R_{nwell} + R_{n+} + R_{cathode} \qquad (2)$$

$$R_{anode} = \frac{L_{access} + \frac{L_{anode}}{2}}{(\frac{1}{R_{\square\_m1}} + \frac{4}{R_{\square\_mx}}) \times W_{m1\_anode} \times N_{fing}} \quad (3)$$

$$R_{silicide} = \frac{R_{\square\_silicide} \times W_{anode}}{2 \times L_{anode}} \quad (4)$$

$$R_{nwell} = \frac{R_{\square\_nwell} \times (L_{poly} + W_{anode})}{L_{anode} \times N_{fing}} \quad (5)$$

$$R_{n+} = \frac{R_{\square\_n+} \times W_{cathode}}{2 \times L_{anode}} \quad (6)$$

$$R_{cathode} = \frac{1}{\frac{W_{m1\_cathode}}{R_{\square\_m1} \times (L_{access} + \frac{L_{anode}}{2})} + \frac{N_{contact} \times N_{fing}}{R_{contact}}} \quad (7)$$

R_s equation (2) models the resistive path seen by the current in working configuration. This equation includes the weight of the back-end stack metallization of the anode (equation (3)) and cathode (equation (7)). Moreover, the resistive weight of the current in the doping layers have been splintered in equations for Nwell (equation (5)) and N+ (equation (6)). Finally, we have chosen to include the silicide resistive value for the Schottky anode contact (equation (4)) and to neglect the contribution of the contacts and vias. Fig. 5 shows the equivalent current path of the series resistance inside the cross-section of the diode with some width and length used to determine each resistive contribution due to the series resistance R_s.

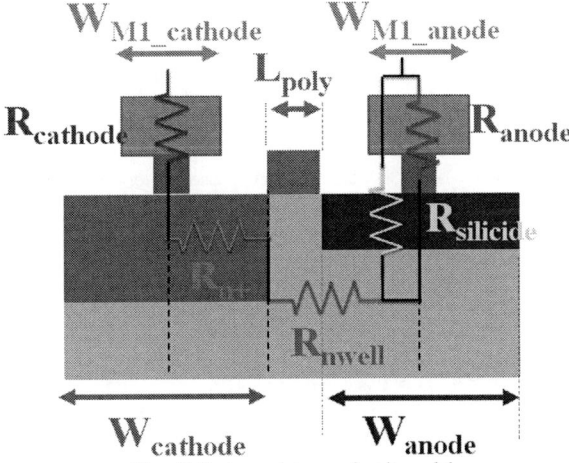

Fig. 5. Series resistance circuit model

C. Parasitic capacitance modelling strategy

The total parasitic capacitance C_{tot} is composed of the sum of the junction capacitance C_j and the back-end capacitance C_{BE} estimated using the DRM of the 55 nm BiCMOS technology and the Schottky diodes layout drawing parameters. The C_{tot} model is given by the following equations:

$$C_{tot} = C_j + C_{BE} \quad (8)$$

$$C_{BE} = C_{acces} + C_{A\_K} + C_{anode\_N+} \quad (9)$$

$$C_j = L_{anode} \times W_{anode} \times N_{fing} \times \frac{\sqrt{q \times \varepsilon_{si} \times N_d}}{\sqrt{2\Phi_b} \times \left(1 - \frac{V_d}{\Phi_b}\right)^{\gamma = 0,5}} \quad (10)$$

$$C_{acces} = \frac{\varepsilon_0 \times \varepsilon_r \times (W_{anode} + 2 \times L_{poly} + W_{cathode}) \times h_{Mx}}{L_{anode} + 2 \times L_{poly} + 2 \times L_{N+OD} + L_{N+\_to\_BDTI}} \times N_{finger} \quad (11)$$

$$C_{A\_K} = \frac{\varepsilon_0 \times \varepsilon_r \times L_{anode} \times h_{Mx} \times 2 \times N_{finger}}{\frac{W_{anode}}{2} + L_{poly} + W_{Co\_to\_poly} - WK_{Co\_to\_m1} - \frac{W_{Co}}{2} - WA_{Co\_to\_m1}} \quad (12)$$

$$C_{anode\_N+} = \frac{\varepsilon_0 \times \varepsilon_r \times L_{N+OD} \times W_{m1\_anode}}{h_{Mx\_to\_N+}} \times N_{finger} \quad (13)$$

The total parasitic capacitance C_{tot} (equation (8)) has two major contributors which are the junction capacitance C_{j0} (equation (10)) and the back-end capacitance C_{BE} (equation (9)). C_{j0} induces Schottky area, doping values, Schottky barrier and silicon material dependency. C_{BE} can be split in three different capacitances. The anode and the cathode fingers are very close and induce two capacitances given by the equations (11) and (12). The last equation (13) is due to the charges flowing under the anode finger in the N+ ring implanted for the cathode layer. Fig. 6 shows some width and thickness used to determine the total parasitic capacitance C_{tot} and highlight the fact that the anode-to-cathode capacitance is due to the metal (only first metal layer) finger stack anode and cathode accesses. Moreover, this cross section underlines that junction capacitance C_j is defined by the anode area NiSi/Nwell doping interface dimensions.

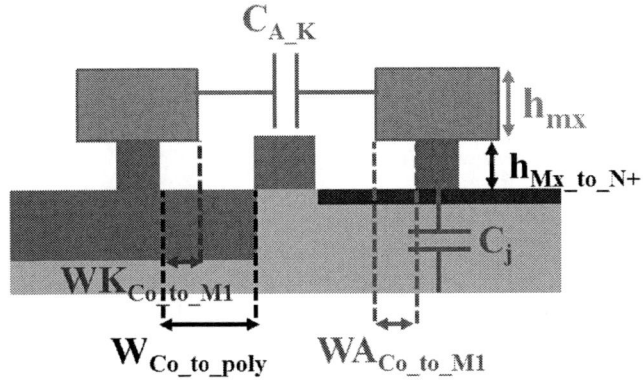

Fig. 6. Total capacitance circuit model

IV. SCALABLE MODEL EXPERIMENTAL VALIDATION

In this part, we will bring to highlight the agreement reached between the proposed model and experimental data for the series resistance (Fig. 7) and the parasitic capacitance (Fig. 8)

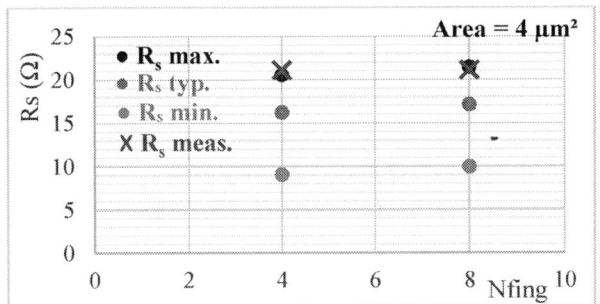

Fig. 7. Series resistance analytical model versus extracted measurement parameters.

Concerning the measured series resistance, as illustrated in Fig. 7, a good agreement with the predicted value from the analytical model is obtained using worst-case process parameters. This worst case is induced by the variation of the square resistance of the Nell which is established in the DRM for several Nwell geometries (more or less aggressive). Consequently, it is reasonable to assume that an acceptable

978-1-7281-1702-7/19 $31.00 © 2019 IEEE

model accuracy can be achieved using some calibration on corners lot which take account of process dispersion as it is done for all other models integrated in ST Process Desing Kit (PDK).

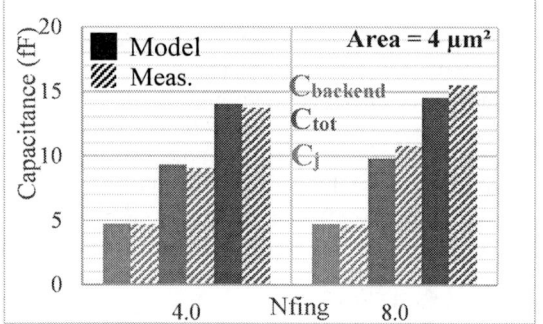

Fig. 8. Parasitic capacitance analytical model versus extracted measurement parameters

Dealing with parasitic capacitance, the total capacitance C_{tot} has been extracted from S_{ij} parameters and the junction capacitance has been extracted using III.A hypothesis. While using previous hypothesis, the extracted and modeled C_{j0} values are by definition equivalent, we can see in Fig. 8 that predicted and extracted parasitic capacitor related to back-end-of-line are a good agreement. This confirms the validity of the chosen hypothesis and the accuracy of the proposed methodology.

Finally, the development of the analytical model of the proposed Schottky diodes has enabled to identify which part of the device is limiting the achievable performances. Fig. 9 shows that the major resistive contributor is the path in the Nwell layer and consequently this is where we have to put some efforts at the process or the layout level for further improvement of the performances.

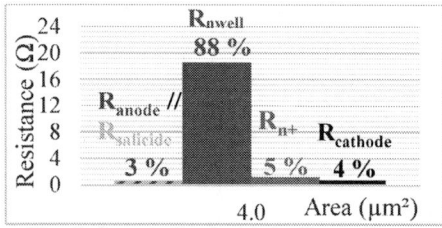

Fig. 9. Impact of each resistive R_s contributors

The total capacitance C_{tot} is mainly dominated by the back-end-of-line capacitance C_{A_K}. Some improvement could be performed on the metal access spacing to reduce this anode-to-cathode parasitic capacitance. It is though more complicated to optimize the junction capacitance C_j without having to introduce any specific implantation steps and so adding costs.

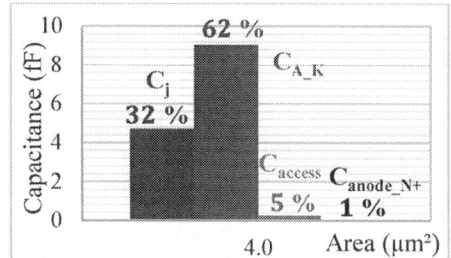

Fig. 10. Impact on the total capacitance C_{tot} of each parasitic capacitance

V. CONCLUSION

State-of-the-art Schottky diodes with cut-off frequency of ~1.7 THz have been achieved using an innovative BiCMOS diode architecture combining aggressive 55 nm poly-gate spacer and bipolar dedicated implantation to reduce parasitic resistance. In addition, we proposed an analytical model demonstrating a reasonable agreement with experimental measurements validating its future integration in a design kit library. Those results pave the way for the design of high performance subharmonic mixers in BiCMOS technology in order to develop highly integrated and low-cost receivers targeting communication links at 300 GHz leveraging IEEE standard 802.15.3d-2017.

REFERENCES

[1] Cisco, "Mobile VNI Forecast," 09 February 2017. [Online]. Available: https://www.cisco.com/c/en/us/solutions/service-provider/visual-networking-index-vni/vni-infographic.html.

[2] VK Chinni and al., "Single-channel 100 Gbit/s transmission using III–V UTC-PDs for future IEEE 802.15. 3d wireless links in the 300 GHz band," *IET Electronics Letters*, vol. 55, no. 10, pp. 638-640, 17 May 2018.

[3] E. Lacombe and al., "300 GHz OOK Transmitter Integrated in Advanced Silicon Photonics Technology and Achieving 20 Gb/s," in *IEEE Radio Frequency Integrated Circuits Symposium (RFIC)*, Philadelphia, PA,USA, 10-12 June 2018.

[4] Paul Richard Wilkinson, "Development of 664GHz Sub-harmonic Mixers," Thesis Report, April 2014. [Online]. Available: http://etheses.whiterose.ac.uk/8079/1/Thesis%201.31.pdf.

[5] T. Crowe et al, "GaAs Schottky Barrier Diodes for Space Based Applications at Submillimeter Wavelengths," in *International Symposium on Space Terahertz Technology (ISSTT)*, Oxford and Didcot, UK, 23-25 March 2010.

[6] S. Sankaran and al., "Schottky Barrier Diodes for Millimeter Wave Detection in a Foundry CMOS Process," *IEEE Electron Device Letters*, vol. 26, no. 7, pp. 492-492, July 2005.

[7] M. K. Matters-Kammerer and al., "RF Characterization of Schottky Diodes in 65-nm CMOS," *IEEE Transactions on Electron Device*, vol. 57, no. 5, pp. 1063-1068, May 2010.

[8] U. R. Pferiffer and al, "Schottky barrier diode circuits in silicon for future millimeter-wave and terahertz applications," *IEEE Transactions on Microwave Theory and Techniques (TMTT)*, vol. 56, no. 2, pp. 364-371, february 2008.

[9] Junyu Shi and al., "A CMOS Schottky Barrier Diode with the Four-Sided Cathode," in *International Conference on Integrated Circuits and Microsystems (ICICM)* , Chengdu, China, 23-25 November 2016.

[10] Suna Kim and al., "A New Resistance Model for a Schottky Barrier Diode in CMOS Including N-well Thickness Effect," *Journal of Semiconductor Technology and Science*, vol. 13, no. 4, pp. 381-386, August 2013.

[11] E. Seok and al, "Progress and Challenges Towards Terahertz CMOS Integrated Circuits," *IEEE Journal of Solid-State Circuits*, vol. 45, no. 8, pp. 1554-1564, August 2010.

[12] F. Stein and al, "Advanced Extraction Procedure for Parasitic Collector Series Resistance Contributions in High-Speed BiCMOS Technologies," in *BiCMOS Circuits and Technology Meeting (BCTM)*, Bordeaux, France, September 30 to October 3.

[13] V. Jain and al., "Schottky Barrier Diodes in 90nm SiGe BiCMOS," in *IEEE BiCMOS Circuits and Technology Meeting (BCTM)*, Bordeaux, France, September 30 to October 3.

Excellent 22FDX Hot-Carrier Reliability for PA Applications

T. Chen[#1], C. Zhang[#], W. Arfaoui[^], A. Bellaouar[#], S. Embabi[#], G. Bossu[^], M. Siddabathula[^],
K.W.J. Chew[*], S.N.Ong[*], M. Mantravadi[#], K. Barnett[#], J. Bordelon[#], R. Taylor[#], S. Janardhanan[#]

[#]GLOBALFOUNDRIES, USA
[*]GLOBALFOUNDRIES, Singapore
[^]GLOBALFOUNDRIES, Germany
[1]tianbing.chen@globalfoundries.com

Abstract—**This work shows the excellent HCI (hot-carrier injection) reliability that 22FDX demonstrates for mmWave PA applications. The underlying device physics to explain this performance are also shown. Due to the fact that fully depleted SOI (FDSOI) eliminates the lateral bipolar device, the MOSFETs in 22FDX® technology have an increased BVDSs when compared to a device in a partially depleted SOI (PDSOI) technology. A 2-stack PA is presented that demonstrates excellent reliability against all HCI stress. The device aging model is built based on the device stress data specific for PA applications. RelXpert is used to simulate device aging based on the model and suggests excellent PA reliability even under the worst mismatch condition.**

Keywords—**mmWave power amplifier, low noise amplifier, switch, SOI, FDSOI, aging, HCI, reliability, load pull, 22nm, RelXpert, 22FDX®.**

I. INTRODUCTION

The deployment of 5G networks demands mmWave technologies that are capable of high data rate, low latency, and low power product designs. The 22FDX platform is well-suited for mmWave applications because: (1) It provides transistor with 350+GHz fT and 400+GHz fmax, (2) its ultralow parasitic capacitance provides a natural benefit for broadband mmWave design, and (3) the low capacitance of FDSOI allows for the FETs to be stacked. All of these are important for high output power (Pout) and high power added efficiency (PAE) design of power amplifiers. Furthermore, 22FDX allows for switch (FEM) integration with transceivers [1-2]. The high density (>5M gates/mm2), high performance, and available ultra-low power DSP will also help to drive the next wave of mmWave systems with the lowest power consumption, the lowest system footprint, and the lowest system cost. The excellent RF and digital capabilities of 22FDX have made it possible to integrate all digital, analog, and mixed-signal components for a system on chip (SoC). SoCs consume much less power, take up much less area, and often offer a lower system cost than multiple-chip designs with the similar functionality.

At mmWave frequencies, the PA is one of the most challenging design components in the front-end. Simultaneously designing a PA with high PAE, high 1dB compression (P1dB) and sufficient reliability has become a daunting task for even every seasoned PA designers. On one hand we need the improved technology scaling to provide devices with ever-higher fT and fmax for mmWave design, but on the other hand the shorter channel length and hence the higher electric field makes the HCI more significant. In the past

HCI has been found to be responsible for the performance degradations of CMOS PAs designed with bulk CMOS technologies [3-4]. In this work, we will demonstrate the excellent HCI reliability of a 28GHz PA designed with 22FDX. The device physics that results in this are also discussed.

II. DEVICE TECHNOLOGY & PHYSICS

In PA application the device bias follows the load line: from high V_{GS} and low V_{DS} bias to low V_{GS} and high V_{DS} bias conditions. The main reliability concern at high V_{GS} low V_{DS} bias condition is the electro-migration (EM) current limit and self-heating. The HCI effect is usually the main concern for the low V_{GS} and high V_{DS} conditions. In this section we focus on the difference in HCI between FDSOI and PDSOI (also bulk) CMOS technologies. The simplified device cross section comparison is shown in Fig. 1. For bulk/PDSOI, the drain, source, and the p channel region form the lateral bipolar. For an FDSOI technology such as 22FDX, the SOI thickness is as thin as 6nm, which is 10x thinner than typical PDSOI thickness. The channel becomes fully depleted at a very small or even zero V_D. This completely eliminates the lateral bipolar device. HCI is a phenomenon in electron device that an electron or hole gains sufficient kinetic energy to overcome a potential barrier necessary to create an interface state. This typically results in Vt and Idsat degradation in transistors. The larger V_D, hence the larger electrical field in channel, will lead to more HCI degradation. The HCI situation however becomes different in a fully-depleted FDSOI transistor. The maximum electric field in the depletion region is decided by the channel doping concentration [5]. The FDSOI technology is more immune to HCI effect as compared to PDSOI or bulk technology.

Fig. 2 shows the typical $I_D V_D$ family curves of a 2um PDSOI device. When the device is at "on" condition ($V_{GS}>Vt$), it does not breakdown up to 5V V_{DS}. However, in non-conducting mode ($V_{GS}<Vt$, the bottom 3 curves) the device enters breakdown mode at $V_{DS} <5V$. At conducting mode ($V_{GS}>Vt$) the channel is on and the base node of the lateral bipolar is approximately in a fixed bias state. So the breakdown voltage is close to BV_{CBS}. At non-conduction mode ($V_{GS}<Vt$) the base node of the lateral device is floating, and the breakdown voltage is close to BV_{CEO}. The $I_D V_D$ family curves of a 24nm 22FDX SLVT device are shown in Fig. 3. As can be seen in the curves the breakdown voltages are actually slightly higher for the non-conducting mode of operations than for the conducting mode. This suggests that with the elimination of the

978-1-7281-1702-7/19 $31.00 © 2019 IEEE

lateral bipolar device in FDSOI, the transistor is more robust at low V_{GS}, high V_{DS} conditions. We will investigate the impact of this improvement on PA reliability in section IV and V.

III. PA PERFORMANCE

To investigate the performance and reliability of FDSOI for mmWAve design, a 28GHz pseudo differential 2-stack PA was designed with transformers at both the input and output for proper impedance matching. To achieve a balance between the performance and reliability the 24nm Lg devices were used in the design. The PA S-parameter characterization and the power sweep tests were done on a 2-port 50 Ohm system. The PA ruggedness tests were done on a load-pull system with 4:1 VSWR.

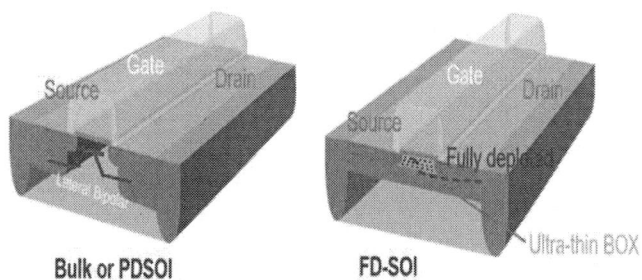

Bulk or PDSOI   **FD-SOI**

Fig. 1 Simplified cross-section comparison between Bulk/PDSOI and FDSOI MOSFET

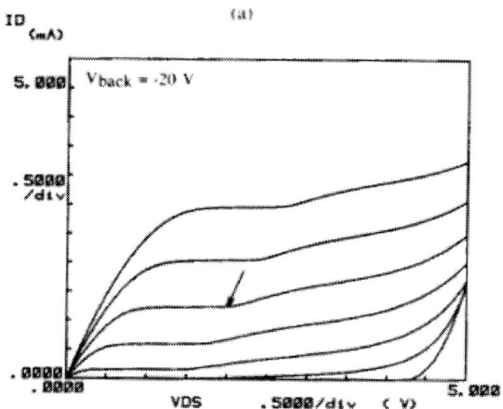

Fig. 2 $I_D V_D$ behaviour for a PDSOI MOSFET[6]. V_{GS} increase from 0 to 3V in step of 0.5V; W/L = 50um/2um.

The measured S-parameter data and the power sweep data were shown in Fig. 5a and Fig. 5b, respectively. The Iddq is at 16mA for 1.4V Vdd. Also, the gain is 12.4dB, P1dB is 16dBm, and the peak PAE is 41%. To assess the PA overdrive capability, the power sweeps were done at Vdd of 1.25V, 1.4V, 1.6V, 1.8V, and 2V, respectively. The overdrive data were shown in Fig. 6. As expected both PAE and P1dB improves at increased Vdd. The s-parameter and power sweep performance remain the same as those shown in Fig. 5. This is a good indication of PA robustness.

IV. DEVICE RELIABILITY

Proper accurate device aging models secure product success from the earliest steps of RF/mmWave circuit design. During

the entire design flow they enable to target best Performance-Power-Area (PPA) trade-offs, while keeping the IC reliable and pass state-of-the-art RF silicon IP validation standards. This is especially important as 22FDX platform is a serious CMOS technology candidate to support the emergence and growth of the mmWave technologies bringing attractive options while featuring good PPA for logic periphery, in a field which has long been dominated by III-V technologies. To insure high efficiency stacking for PA and switch in 22FDX, the designer pushed the devices to their limits and thus applies high voltage/current stress which raised a lot of concern of the devices reliability especially for the HCI as it is the major reliability concern of deep-submicron technology [7].

Extensive DC stress campaigns were carried on at device level to evaluate the HCI reliability under PA operating condition voltages and to build an accurate HCI model able to cover a wide range of $V_{GS}/V_{DS}/V_{BS}$ to help designer on their PA reliability simulation. In order to mimic the operation voltages seen by the transistors during the PA lifetime, the stress conditions were extended beyond the upper usage boundaries of the technology and beyond the standard HCI stress range. Stress conditions with very low V_{GS} and very high V_{DS} were applied, as well as extensive wide range back bias conditions to support this unique feature of FDSOI technology.

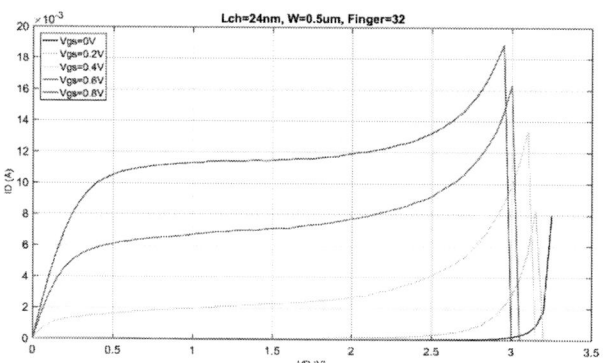

Fig. 3 $I_D V_D$ behaviour of a FDSOI MOSFET. The drop of I_D reference the point where the transistor hit current compliance

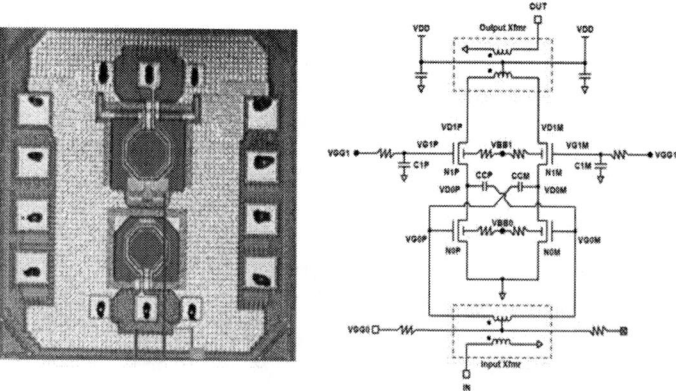

Fig. 4 Layout and schematic a two-stack PA.

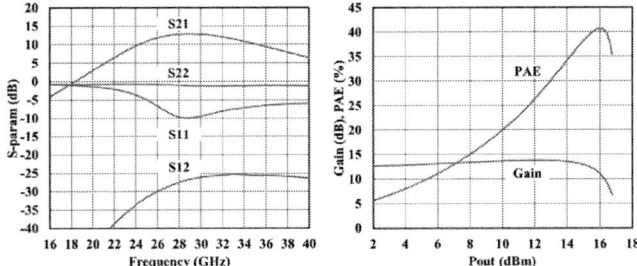

Fig. 5 Small-signal and power sweep data for the 2-stack PA.

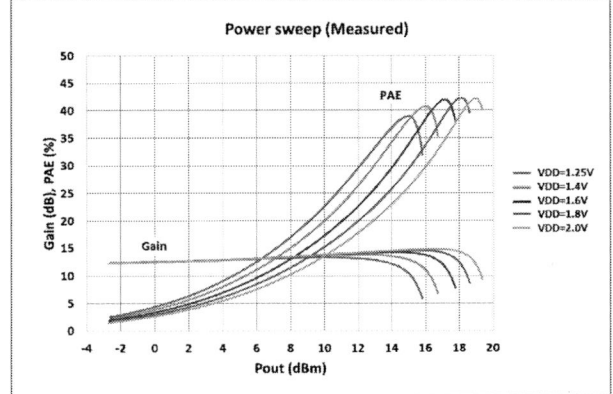

Fig. 6 Overdrive of the 2-stack PA during the power sweep characterization.

22FDX Core N-channel FETs processed in Globalfoundries with short gate-length L= 20nm is shown here to support this RF design. Fig.7 shows the Nfet saturation drain current Idsat drift as a function of the gate voltage for 3 back bias conditions (0V, 1V and 2V) with a fixed drain voltage V_{DS}= 1.9V captured at a fixed stress time. HCI degradation increases for high Vb due to the increased stress current under Forward Back Bias. The model follows well the measured degradation and reproduces the back bias effect.

In 22FDX technology, the threshold voltage Vth is linearly proportional to the back bias in FBB (Forward Back Bias) and RBB (Reverse Back Bias) [8]. To take this behaviour into account, the back bias dependency was included on the gate voltage term $V_{GS}'=V_{GS}+B(Vb-Vbref)$ as shown in Fig. 8. An unified HCI device aging model is developed accordingly and will be published.

V. PA RUGGEDNESS & RELIABILITY

To test the ruggedness of the PA, the Pout was set at ~P1dB (fixed Pin). Then the load tuner was set to a VSWR of 4:1 and its phase was swept from 0 to 350 degrees in step of 10 degrees with the hold time of 15 seconds at each phase setting. Fig. 9 shows the measured load circle on a smith chart during the ruggedness test. This procedure is repeated through 3 cycles for total stress time of 27 minutes (15*36*3=1620 seconds=27 minutes). The power sweep at 28GHz were done both before and after the ruggedness test. The pre- and post-stress power sweep results were compared afterwards and are shown in Fig. 10. It can be seen from Fig. 10 that there has been very minimal

performance change after the ruggedness test. As a result, the PA is viewed to have passed the 4:1 VSWR ruggedness test.

The voltage swing at the output of the PA varies as the load is changed. To investigate the PA reliability, the HCI degradation is simulated at the worst case VSWR (mag, phase) conditions where the worst case voltage swing is found (V_{DS}_max). The procedure to find the worst wave form is as follows: (1)the input power if first fixed at IP1dB, (2)set gamma_mag=0.6 (VSWR=4:1), (3)the gamma phase is swept from 0 to 350 degree in step of 10, (4) the full V_{DS} and V_{DG} wave forms during the Gamma phase sweep is recorded in Fig. 11. Both the V_{DS}_max and V_{DG}_max are found to be approximately 2.4V. The worst case V_{DS} and V_{GS} wave forms for the cascoded devices are shown in Fig. 12. The V_{DS}_max and V_{GS}_max for the common source devices are found to be 1.36V and 0.58V, respectively. The V_{DS}_max and V_{GS}_max for the common gate devices are found to be 2.40V and 0.68V, respectively.

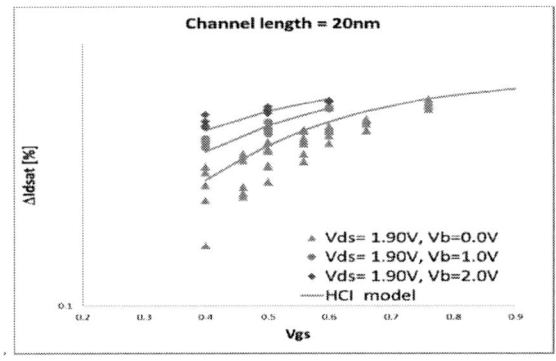

Fig. 7 Idsat degradation of thin gate-oxide Core FDSOI NFET under HCI with fixed high V_{DS} (V_{DS}= 1.90V) and various back bias for high to low VG stress.

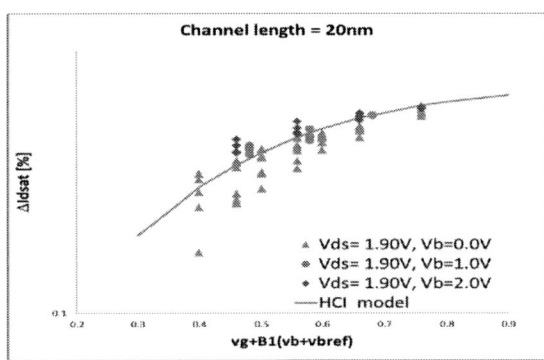

Fig. 8 Idsat degradation of thin gate-oxide Core FDSOI NFET under HCI with fixed high V_{DS} (V_D= 1.90V) and various back bias as a function of V_{GS}=B(Vb-Vbref)

RelXpert simulation results in Table 1 shows expected BTI and HCI degradation after operating the PA under the worst wave form conditions for 10 years. The HCI and BTI degradation are reported only for the top device. The BTI degradation is only 0.06 mV for the Vt shift, and the HCI degradation has only leads to 0.02% Idsat change. This clearly suggests the excellent 22FDX device robustness for mmWave PA applications.

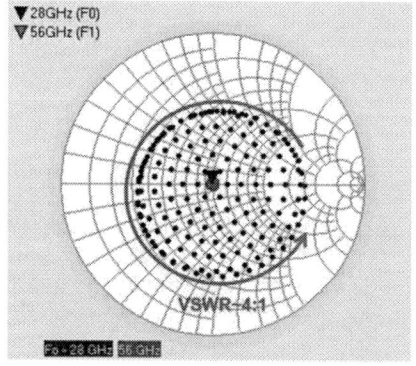

Fig. 9 Smith chart used for ruggedness test

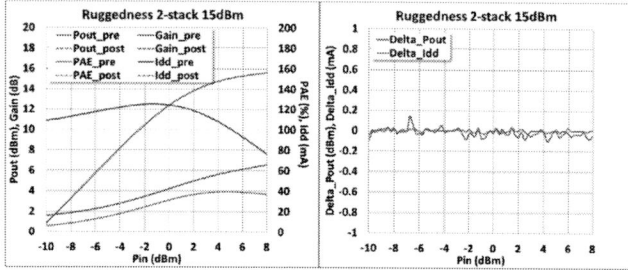

Fig.10 Power sweep data pre- and post-ruggedness test

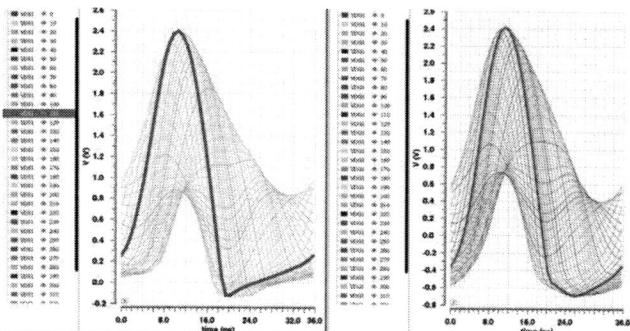

Fig.11 VDS and Vdg swing at VDD=1.4V (VSWR=4:1)

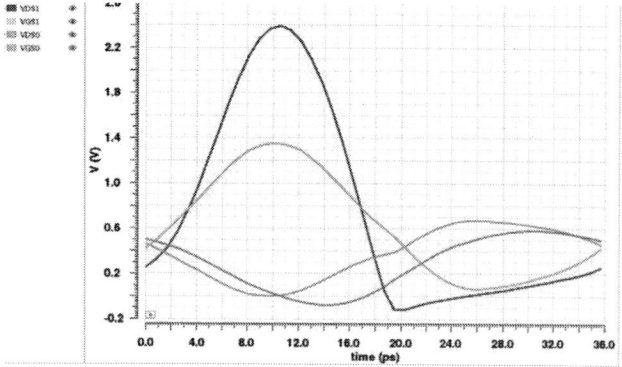

Fig.12 V_{DS} and V_{GS} swing at V_{DD}=1.4V (worst case)

To further evaluate the long-term PA reliability, the device were subjected to a 12-hour on-wafer stress. During the on wafer stress, the phase of the load tuner was rotated through 80 cycles as compared to only 3 circles for the earlier ruggedness test. The total stress time is 80*9=720 minutes=12 hours. The

power sweep characterizations were done before and after the on-wafer stress. The power sweep data shows very little change after the 12-hour on-wafer stress, as shown in Fig. 13. The dc bias current drops by ~1% over time and is likely caused by the probe contact resistance change. To avoid this contact probe related degradation, further in-package long-term aging stress testing will be conducted. Previously reported results have shown that HCI induced significantly more PA performance degradation in both CMOS RF (<6GHz) PA [3] and mmWave PA [4] designs.

Table 1. RelXpert simulation at worst case V_{DS} condition

VDD (V)	OP1dB (dBm)	Gamma (deg)	Vds_max (V)	BTI Degrd. (mV)	HCI Degrd. (%)
1.4	15.9	110	2.40	0.06	0.02

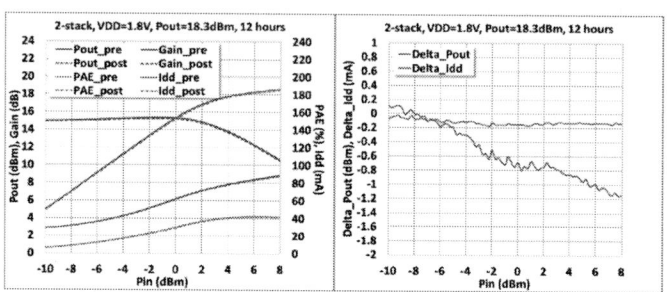

Fig. 13 2-stack PA on-wafer 12-hour stress

VI. CONCLUSION

The HCI effect in 22nm FDSOI transistors is compared to that in bulk/FDSOI technology. HCI degradation is mitigated in FDSOI at low V_{GS} high and V_{DS} conditions due to the elimination of the parasitic lateral bipolar device under the active transistor. A 28GHz PA with 41% PAE and 16dB P1dB has shown superior PA reliability against HCI effect as compared to previous results reported in bulk and PDSOI technologies. The excellent PA reliability correlates well to the device physics analysis on the FDSOI and PDSOI transistors.

REFERENCES

[1] R. Carter, et. al., "22nm FDSOI Technology for Emerging Mobile, Internet-of-things, and RF Applications," IEDM Technical Digest, pp. 27-30, 2016

[2] S.N. Ong, et. al, "A 22nm FDSOI Technology Optimized for RF/mmWave Applications," RFIC Technical Digest, pp. 72-75, 2018.

[3] E. Xiao, "Hot Carrier Effect on CMOS RF Amplifiers," Proc. IEEE 43 Annual International Reliability Physics Symposium, pp. 680-681, Apr. 2005.

[4] T. Quemerais et al., "Hot-Carrier Stress Effect on a CMOS 65-nm 60-GHz One-Stage Power Amplifier", IEEE Elect. Dev. Let., vol. 31, no. 9, Sept. 2010.

[5] S.M. Sze, Semiconductor Devices--Physics and Technology. Wiley , Toronto, 1985

[6] A. Chauhan and A. Prakash, "Analysis of Floating Body Effects in SOI Transistors", Intl. Research J of Eng. & Tech. V.03, pp.1123-1131, 2016

[7] S. Rauch and G. La Rosa, "The energy driven paradigm of NMOSFEThot carrier effects", IEEE International Reliability Physics Symposium IRPS, Proceedings.43rd Annual., vol. 5, no. 4, pp. 708–709, 2005.

[8] J.P. Noel, O.Thomas, et al, "Multi-Vt UTBB FDSOI Device Architectures for Low-Power CMOS Circuit", IEEE Electron Devices, Vol. 58, pp.2473 – 2482, Aug. 2011.

22nm Fully-Depleted SOI High Frequency Noise Modeling up to 90GHz Enabling Ultra Low Noise Millimetre-Wave LNA Design

L.H.K. Chan[#1], S.N. Ong[#], W. L. Oo[#], K.W.J. Chew[#], C. Zhang[^], A. Bellaouar[^], W.H. Chow[#],
T. Chen[^], R. Rassel[^], J.S. Wong[#], C.K. Lim[#], C.W.F. Wan[#], J. Kim[#], W.H. Seet[#], D. Harame[%]

[#] GLOBALFOUNDRIES Singapore
[^] GLOBALFOUNDRIES USA
[%]now with Research Foundation State University Polytechnic at Albany, USA
[1]lyehock.chan@globalfoundries.com

Abstract—This paper reports the high frequency (HF) noise characterized performance and modeling on 22nm FD-SOI technology transistor (GLOBALFOUNDRIES' 22FDX® technology) from 2 GHz to the maximum E-band millimetre-Wave (mmWave) frequency of 90 GHz. The measurement was performed using the Focus Microwaves noise system with different customised setups and optimised for each discrete frequency bands. The data measured from each frequency bands were subsequently combined to produce the noise spectrum covering from 2 GHz to 90 GHz, with high accuracy, good continuity and excellent correlation to the compact model. The 22FDX® technology transistor demonstrated very low mmWave noise figure, which is favourable for RF and mmWave applications such as LNA.

Keywords— E-band, mmWave, noise, RF, 22FDX®, FDSOI.

I. INTRODUCTION

The FD-SOI technology has the advantage of offering very low-power consumption yet achieving superb RF and mmWave performance with extremely high cut-off frequency, fT and maximum oscillation frequency, fMAX [1]. A promising high frequency noise performance is essential to the emerging mmWave transceiver design. The mmWave noise characterization on CMOS has been discussed up to 67 GHz in [2]. The noise characterization on an InP HEMT for a narrow frequency range of 79-94 GHz frequency band has been reported in [3]. This paper presents the high frequency (HF) noise characterization and modeling capability for a wide frequency band from 2 GHz up to 90 GHz on GLOBALFOUNDRIES' 22FDX® technology. The characterization was performed using the Focus Microwaves noise system with different setups customised and optimised for each discrete frequency bands, which will be described in Section II. In Section IV, the HF noise figure will be benchmarked against other advanced CMOS technology nodes. A 28GHz low noise amplifier (LNA) achieving ultra-low noise figure will be demonstrated in Section V.

II. ON-WAFER CHARACTERIZATION

In this work, the device under test (DUT) structures, as shown in Fig. 1, were designed using a two-port GSG pad frame with common source-bulk configuration, where the probe pads and metal interconnects have been carefully optimized to ease the de-embedding without impacting the actual noise performance of the device.

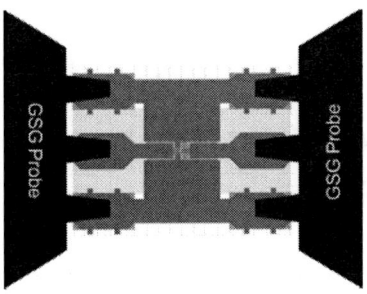

Fig. 1. Layout view of 2-port transistor, with GSG probes.

Table 1. Multi-band Measurement Setup.

No.	Frequency bands	Setup
1	0.7-26 GHz	Tuner: iCCMT-26070
2	10-50 GHz	Tuner: iCCMT-67100
3	50-67 GHz	Tuner: iCCMT-67100 with 47GHz LO Mixer
4	60-77 GHz	Tuner: iCCMT-90600 with 53GHz LO Mixer
5	78-90 GHz	Tuner: iCCMT-90600 with 68GHz LO Mixer

A. Multi-band Measurement Setup

DC current and S-parameters were characterized using the CASCADE RF probe station together with Keysight PNA and B1500 DC parametric analyzer. The noise measurements were carried out on a series of cascaded setups, each dedicated to a band or sub-band of frequencies starting from 0.7-26GHz, followed by 10-50GHz, 50-67GHz, 60-77GHz and 78-90GHz, as summarized in Table 1, with diagrams of each frequency band as illustrated in Fig. 2 (a) to (e) respectively. In each setup, HF noise measurements were performed using Keysight N5247A PNA-X (Nonlinear Vector Network Analyzer) with varying source admittances generated by the Focus Microwaves electromechanical tuners. The entire system setup is confined in an electromagnetic shielded room with the aim to insulate the system setup from external interferences.

In addition to the PNA-X noise receiver, external LNAs are connected to the DUT output to boost the receiver sensitivity and minimizing measurement uncertainties. Five LNAs with optimized gain and noise performances have been used for full coverage of the entire noise measurement frequency range (0.7-26GHz, 10-50GHz, 50-67GHz, 60-

978-1-7281-1702-7/19 $31.00 © 2019 IEEE

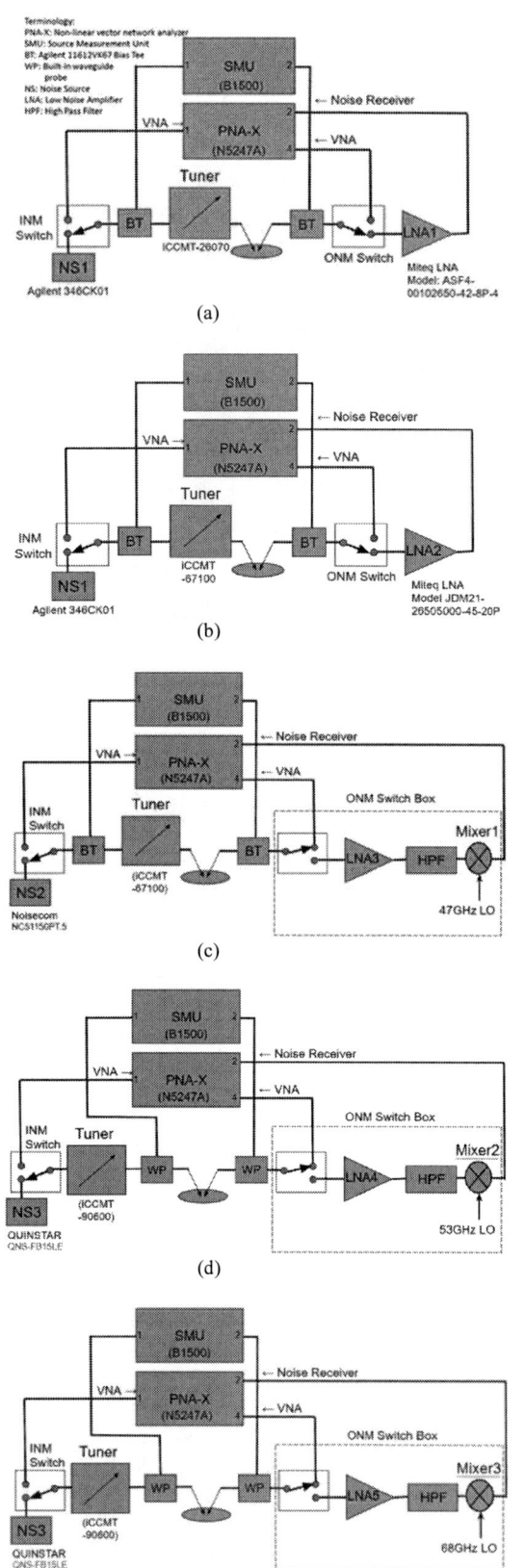

Fig. 2. Diagram of the multi-band HF noise measurement setup (a) 0.7-26 GHz, (b) 10-50 GHz, (c) 50-67 GHz, (d) 60-77 GHz, (e) 78-90 GHz.

77GHz and 78-90GHz). For frequencies above 50GHz, the LNAs are connected in conjunction with down-converters of 47GHz, 57GHz and 75GHz for down-conversion of HF noise signal to match the operation frequency range of the 50GHz PNA-X noise receiver. Meanwhile, the DC biasing conditions of the transistor at input and output terminals are provided by Agilent B1500 Source Measurement Unit (SMU). DC bias is connected to the built-in bias tee of the waveguide probes, i.e., Cascade i90-T-GSG-50-BT for setup 2(d) and (e).

B. Noise Calibration, De-embedding & Extraction

Calibration is essential to both the S-parameters and noise characterizations in order to shift the measurement reference plane to the RF probe tips. In this work, the Load-Reflect-Reflect-Match (LRRM) calibration [4] was performed, prior to the S-parameters measurements. The "Cold Source" technique described in [5], has been adopted for noise figure measurement. In the sub-band frequencies of 0.7-26 GHz and 10-50 GHz, the noise figure calibration was performed using the LRRM method. For the higher frequency sub-band, such as 50-67 GHz, 60-77GHz and 78-90 GHz, the noise figure calibration was carried out using the multi-line Thru-Reflect-Line (TRL) technique [6]. Upon extracting the S-parameters and noise parameters, OPEN-SHORT de-embedding [7] was performed to remove unwanted parasitic from the pad and metal interconnects to obtain the de-embedded noise parameters, which consists of minimum noise figure NF_{min}, equivalent noise resistance R_n, and optimum reflection coefficient Γ_{opt}. Thereafter, the drain and the gate noise power spectral density (S_{id}, S_{ig}) were extracted [8] from the de-embedded measured noise data.

III. HIGH FREQUENCY NOISE MODELING

A. Model

The high frequency noise model was developed based on the BSIM-IMG compact core model. A geometry dependent empirical model has been developed and integrated on to the noise fitting parameter in BSIM-IMG, with the idea to improve the noise model scalability across a wider range of device geometry. Sub-circuit model consists of extrinsic parasitic resistance and diode has been built, to calculate the extrinsic parasitic effects up to M1 reference plane, as illustrated in Fig. 3. In this paper, the gate resistance model includes both the vertical and horizontal poly-PC resistor components [9], as well as M1 metal resistance at the gate. All these resistors generated resistance thermal noise S_{i_R} which is $4kTR$ [10], and eventually being collected at drain terminal contributing to the total drain current noise S_{id}.

B. Validation

HF noise figure-of-merit (FoM) versus frequency at different gate biases, of a NFET device, is shown in Fig. 4, with (a) drain current noise S_{id}, (b) normalized noise resistance R_n, (c) minimum noise figure NF_{min}, and (d) noise figure terminated at 50Ω source resistance NF_{50}. Symbols are measured data, while solid lines are model. It can be seen that excellent continuity in noise spectrum across different bands of measured data has been achieved. The drain current noise S_{id} demonstrated white noise characteristic, while NF_{min} is proportional to frequency, up to 90GHz. Good agreement between data and model is achieved up to the frequency of 90 GHz.

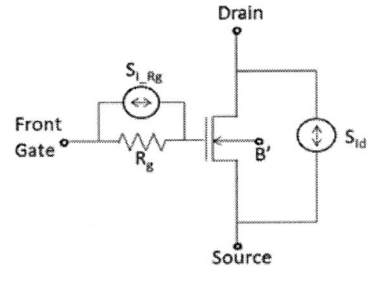

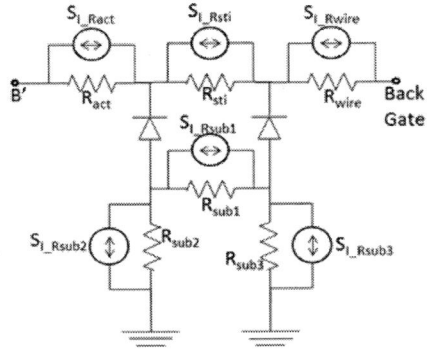

Label	Parasitic Component
Rg	Gate resistance
Ract	Well resistance under active area
Rsti	Well resistance under STI
Rwire	Wiring resistance
Rsub	3R Substrate resistance network

Fig. 3. RF and HF noise sub-circuit model

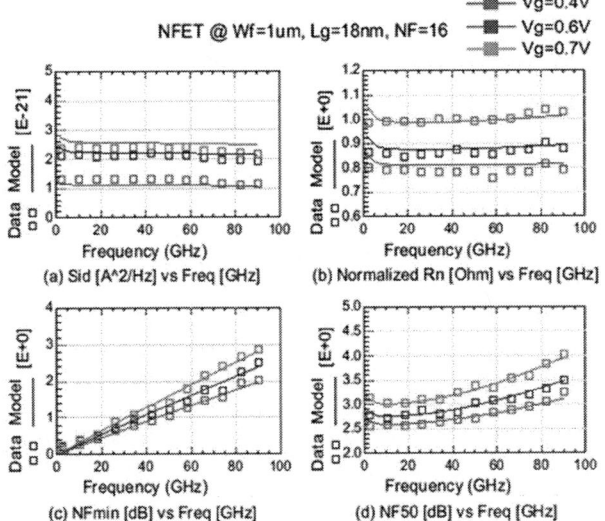

Fig. 4. HF noise FoM of NFET, with (a) Sid-Freq, (b) Rn -Freq, (c) NFmin-Freq, and (d) NF50 -Freq.

IV. HF NOISE BENCHMARKING

The comparison of HF noise FoM to several advanced CMOS technologies is shown in Table 2. FD-SOI technology is a fully dielectric isolation process, with absence of channel and pocket doping, causing lower noise coupling and improving the gain. Moreover, 22FDX® technology has low gate parasitic and resistance, and thus exhibiting the lowest minimum noise figure, NF_{min}, at different frequencies among the advances technologies. This paper also revealed that 22FDX® technology produced extremely low NF_{min} of 1.99 dB, achieved at 90 GHz.

Table 2. HF Noise FoM comparison across technologies

	NFmin (dB)
22nm FD-SOI (This work)	0.18 @ 6GHz, 0.63 @ 26GHz, 1.06 @ 50GHz, 1.99 @90GHz
14nm FinFET [11]	0.89 @26GHz
22nm FinFET [12]	<1.7 @50GHz
28nm HiKMG Bulk CMOS [13]	0.4 @6GHz, 2.7 @26GHz
28nm PolySiON Bulk CMOS [13]	0.37 @6GHz, 2.08@26GHz

V. MMWAVE LOW NOISE AMPLIFIER

A 28GHz LNA has been fabricated using the 22FDX® technology process, and has demonstrated the exceptionally low noise figure. The lowest NF achieved is 1.4 dB and 1.46 dB at 26 GHz and 28 GHz, respectively, with only 13 mW of power dissipation.

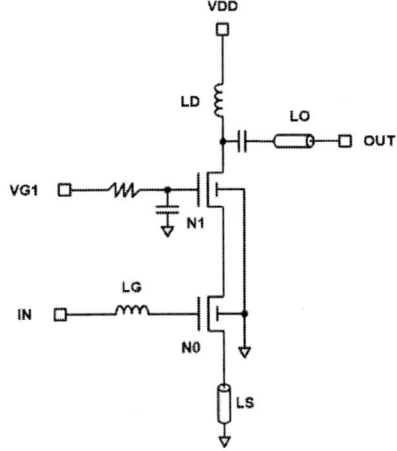

Fig. 5. 28GHz LNA circuit schematic.

The basic cascoded common-source with inductive degeneration topology was used in the 28GHz LNA design. The schematic is shown in Fig. 5. The main noise contributors in the LNA circuit includes R_g and the thermal noise from the common source device, as well as the series resistance (or Q) of the series gate matching inductor Lg. This is a result of the low noise characteristic of the core active device, together with the thick copper metal option offered by the 22FDX® technology process, where both low device noise and high Q for the inductor component can be achieved at the same time. As a result, very low noise figure has been achieved. Both simulated and measured NF results on multiple sites across the wafer are shown in Fig. 6. Excellent model-to-hardware correlation trend has been established with small discrepancy which could be due to the passive component simulation accuracy. Multiple sites NF measured data also shows very tight process variation across frequency.

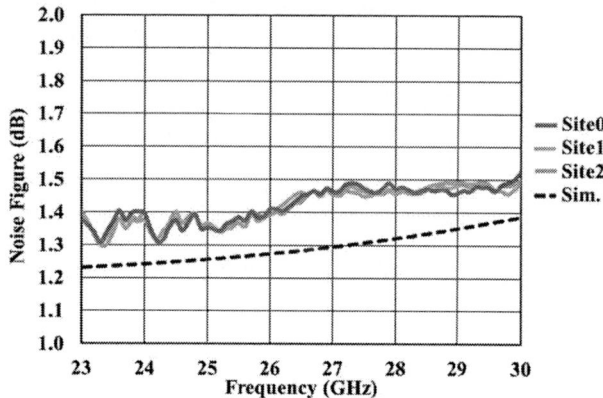

Fig. 6. Measured and simulated *NF* for a 28GHz LNA.

VI. CONCLUSION

HF noise characterization and modeling up to E-band mm-wave frequency based on 22FDX® technology has been described. The noise spectrum covering the frequency from 2 GHz to 90 GHz, were obtained by combining the measured results from different measurement setups dedicated to different frequency bands. The noise frequency spectrum also shows excellent correlation to the result predicted by model. A 28GHz LNA achieved the lowest reported noise figure of 1.4dB, demonstrating the promising high frequency noise performance for mmWave applications based on 22FDX® technology.

ACKNOWLEDGEMENT

The authors are grateful to Focus Microwaves teams for their great support in measurement setup in this work.

REFERENCES

[1] S. N. Ong, *et. al.,* "A 22nm FDSOI Technology Optimized for RF/mmWave Applications," *2018 IEEE Radio Frequency Integrated Circuits Symposium (RFIC)*, Philadelphia, PA, June 2018, pp. 72-75.

[2] X. S. Loo, *et. al.,* "MM-wave noise characterization of 40nm CMOS transistor for up to 67 GHz," *2014 IEEE Radio Frequency Integrated Circuits Symposium (RFIC)*, Tampa, FL, June 2014, pp. 191-194.

[3] T. Vaha-Heikkila, M. Lahdes, M. Kantanen, and J. Tuovinen, "On-wafer noise-parameter measurements at W-band," *IEEE Trans. Microw. Theory Techn.*, vol. 51, no. 6, pp. 1621–1628, Jun. 2003.

[4] A. Davidson, K. Jones, and E. Strid, "LRM and LRRM calibrations with automatic determination of load inductance," in *ARFTG Microw. Meas. Conf. Dig.*, Nov. 1990, vol. 36, pp. 57–63.

[5] Adamian V.,Uhlir A., "A novel procedure for receiver noise characterization", *IEEE Trans. Instrumentation and measurement*, vol. 22, no. 2, pp. 181- 183, 1973.

[6] R.B. Marks, "A multiline method of network analyzer calibration," *IEEE Trans. Microwave Theory Tech.*, vol. MTT-39, no. 7, pp. 1205-1215, July 1991.

[7] A. Issaoun, X. Yong Zhong, S. Jinglin, J. Brinkhoff, and L. Fujiang, "On the deembedding issue of CMOS multigigahertz measurements," *IEEE Trans. Microw. Theory Techn.*, vol. 55, no. 9, pp. 1813–1823, Sep. 2007.

[8] C. H. Chen, M. J. Deen, M. Matloubian, and Y. Cheng, "Extraction of the induced gate noise, channel thermal noise and their correlation in sub-micron MOSFETs from RF noise measurements," in *Proc. Int. Microelectron. Test Structures Conf.*, Kobe, Japan, 2001, pp. 131–135.

[9] C. Schwan, *et. al.,* "CMOS RF performance gain by gate resistance optimization," *2016 IEEE Radio Frequency Integrated Circuits Symposium (RFIC)*, San Francisco, CA, May 2016, pp. 114-117.

[10] H. Nyquist, "Thermal agitation of electric charge in conductors," *Phys. Rev.*, vol. 32, pp. 110–114, 1928.

[11] J. Singh et al., "14nm FinFET Technology for Analog and RF Applications," in Proc. 2017 VLSI Technology Digest.

[12] B. Sell, et. al, "22FFL: A High Performance and ultra Low Power FinFET Technology for Mobile and RF Applications," *IEEE International Electro Devices Meeting (IEDM)*, December 2017.

[13] K.W.J Chew, et al "RF Performance of 28nm PolySiON and HKMG CMOS Devices" *2015 IEEE Radio Frequency Integrated Circuits Symposium (RFIC)*, Phoenix, AZ, May 2015, pp. 43-46.

22nm Ultra-Thin Body and Buried Oxide FDSOI RF Noise Performance

Ousmane M. Kane[#1], Luca Lucci[$], Pascal Scheiblin[$], Sylvie Lepilliet[*], François Danneville[*]

[#]CEA Leti, Lille University, France

[$]CEA Leti, France

[*]CNRS, Université Lille, ISEN, Université Valenciennes, UMR 8520 - IEMN, Lille, France

[1]Ousmane.kane@cea.fr

Abstract—**The drastic downscaling of the transistor size along with advances in material sciences allowed the development of low power CMOS technologies with competitive RF figure of merits suitable for millimeter applications. In this context, this paper presents the RF and noise characterization (up to 110 GHz) of an advanced 22 nm UTBB FDSOI technology developed by Globalfoundries. In addition to the excellent DC performance, the technology presents promising RF characteristics. Indeed, a maximum transconductance of 1.78 S/mm and a F_{max} of 435 GHz are achieved. The technology also offers a state-of-the-art minimum noise figure (NF_{min}) of 0.45 dB at 20 GHz (with an associated Gain of 13 dB) for a drain current of 185 mA/mm.**

Keywords— **CMOS, FDSOI, noise measurement, millimeter wave.**

I. INTRODUCTION

Increasing demand in wireless telecommunication combined with low power consumption for internet-of-things (IoT) applications require the development of new electronic devices. In this context, a CMOS technology of interest is the 22 nm ultra-thin body and back-oxide (UTBB) fully-depleted silicon-on-insulator (FDSOI) technology from Globalfoundries (22FDX). Fig. 1.a shows an illustrative cross section for a NMOS.

Work reported in [1-3] already showed competitive RF performances for this technology in millimeter wave (mmW) applications, including IoT, 5G, and radar. In [1-3], values for the transit frequency (F_t) and the maximum oscillation frequency (F_{max}) were already extracted. Though such figure-of-merit were reported, this paper focuses not only on the radio frequency performance, but also on broadband noise one.

This paper is organized as follows. The second section presents the device geometry for the best selected NMOS transistor, and its DC and RF (F_t, F_{max}) performance. The third section focuses on the extraction of the small signal equivalent circuit (SSEC) as well as the extraction of the drain noise temperature. In section four, a presentation of the noise performance and a validation of this extraction in mmW range is shown. This paper ends with a benchmarking of the CMOS 22 nm UTBB FDSOI with the state-of-the-art.

II. DC AND RF PERFORMANCE

Many NMOS transistors have been measured having different unit finger widths varying between 0.3 μm and 1.07 μm. Also devices with relaxed poly pitch (Fig. 1.b) were measured; this has a considerable impact on F_t because of a lower C_{gg} capacitance and a higher transconductance, resulting from enhanced stress response [4]. Each NMOS device was accompanied by a dedicated open and short structure for de-embedding at lower metal reference plane and all of these transistors were characterized. But, a selected device, whose dimensions are reported in Table 1, was chosen for this paper.

Table 1. Device geometry

Gate length (Lg) [nm]	Unit gate finger width (Wf) [μm]	Number of gate fingers (Nf)	Total gate width (Wtot) [μm]
18	0.3	192	57.60

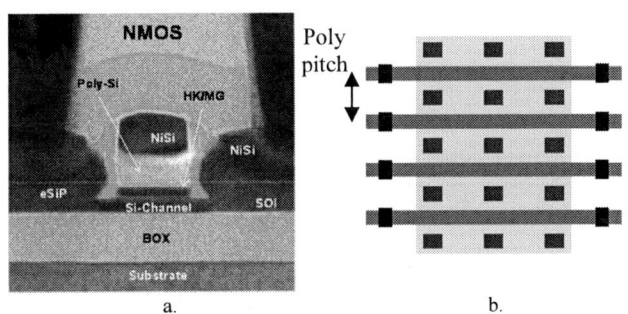

Fig. 1. (a) Typical cross section of a 22 nm FDSOI [2] (b) FET layout schematic showing gate poly pitch.

The drain current and the transconductance as a function of gate voltage are shown in Fig. 2. A maximum DC normalized drain current of 787 mA/mm and a maximum DC transconductance of 1.78 S/mm are obtained.

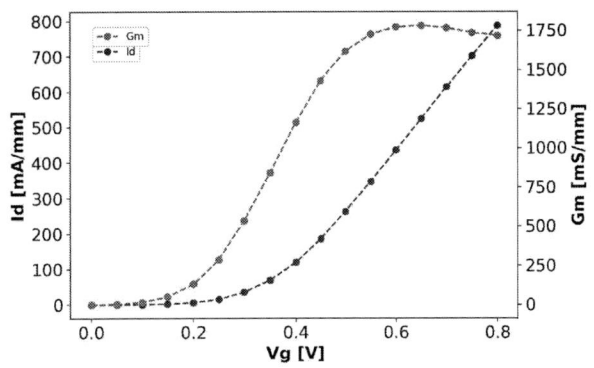

Fig. 2. Drain current (Id) and transconductance (Gm) as a function of gate voltage (Vg) for a drain-source voltage Vds=0.8 V.

978-1-7281-1702-7/19 $31.00 © 2019 IEEE

For the RF characterization, S parameters are measured up to 50 GHz using Keysight network analyser PNAX N5245B. A short-open-load-reciprocal thru calibration is performed to bring the reference plane to the probe tips. For the de-embedding of the pad and metallic interconnections contribution, a Pad-Short-Open (PSO) de-embedding [6] is performed, a simple Open-Short (OS) de-embedding not being appropriate for our types of device structures.

As it is shown in Fig. 3, peak F_t/F_{max} of 332/435 GHz are extracted from current gain (H21) and mason gain (U) at a drain current of I_d=347mA/mm (this current density approximately corresponds to the maximum of the DC transconductance).

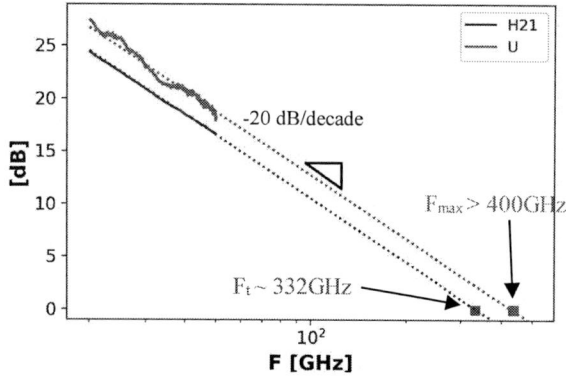

Fig. 3. Current gain (H21) and Mason gain (U) at I_d=347mA/mm

III. RF SSEC AND NOISE MODEL EXTRACTION

The small signal equivalent circuit (SSEC) of the transistor, which includes the HF noise sources in chain representation, is given in Fig. 4. The F50 method [15] was used to extract Tout; once Tout is determined, the four noise parameters (NF$_{min}$, R$_n$, Γ_{opt}) of the transistor are known. Note that an accurate SSEC is required for this method (we shall return to this point in section III.B). The noise figure was measured with the PNAX N5245B, which has an integrated noise source at its port 2. And the noise calibration was done using the Keysight power meter U8487A.

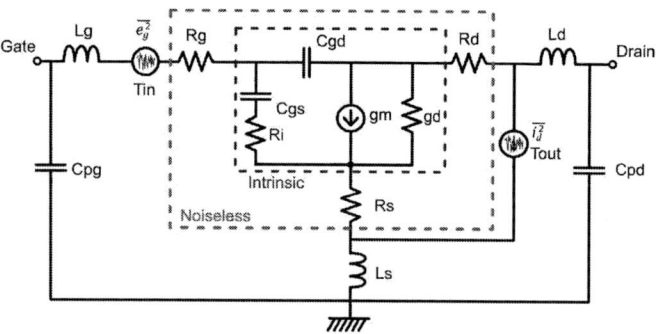

Fig. 4. MOSFET SSEC including noise sources.

For noise temperature extraction Pospiezalski's noise model [11-12] was used. The thermal noise of the device is modeled by two uncorrelated noise sources: an input voltage noise source ($\overline{e_g^2}$) and an output current noise source ($\overline{i_d^2}$). $\overline{e_g^2}$ is

proportional to an input equivalent noise temperature T$_{in}$ and $\overline{i_d^2}$ is proportional to an output equivalent noise temperature T$_{out}$. T$_{in}$ is usually set to room temperature. In order to extract T$_{out}$ through 50 Ω noise figure data (NF$_{50}$), the lumped equivalent circuit of the pads and metallic interconnections surrounding the device, which are modeled using the PSO de-embedding procedure, must be added to the small signal model (represented by DUT in Fig. 5, which shows the model that was used). The values of the different elements are given in Table 2.

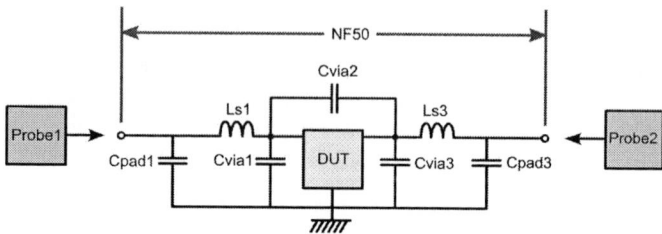

Fig. 5. Equivalent model of the pads and interconnections (shows the reference plane for NF$_{50}$ measurement).

Table 2. Impedance and admittance values of the pads and interconnections

Cpad1	Cpad3	Cvias1	Cvias2	Cvias3	Ls1	Ls3
fF					pH	
44	41.6	29	19.7	34.5	27	30

A. RF SSEC Extraction

The SSEC was extracted using the methodology described in [5] and [7]. The series resistances (R$_g$, R$_s$, R$_d$) are extracted first at cold bias (V$_{ds}$=0V). The values of the extrinsic and intrinsic elements are given in Table 3.

Table 3. Extrinsic and intrinsic parameters values of the transistor

R$_g$	R$_d$	R$_s$	L$_g$	L$_d$	L$_s$	C$_{pd}$	C$_{pg}$
Ω			pH			fF	
5.78	1.38	1.02	0.5	0.5	1.31	9.5	5.6

C$_{gd}$	C$_{gs}$	g$_m$	g$_d$	R$_i$
fF		mS		Ω
14.52	38.81	121	7	4

The voltage gain of the transistor (g$_m$/g$_d$) is equal to 17, signature of an excellent channel control. Values for the residual capacitances after de-embedding (C$_{pg}$ and C$_{pd}$) are due to the multiplicity of cells for the RF device, which is equal to 6 (because of the large number of fingers). As observed in Fig. 6, C$_{pg}$ and C$_{pd}$ are almost flat over the entire frequency range using a PSO de-embedding.

The SSEC with the pads and metallic interconnection models has been implemented on Agilent Advanced Design Systems (ADS) software and S parameters were simulated up to 50 GHz. Fig. 7 presents a comparison of the S parameters between measurement and ADS simulation. A good agreement is obtained up to 50 GHz.

978-1-7281-1702-7/19 $31.00 © 2019 IEEE

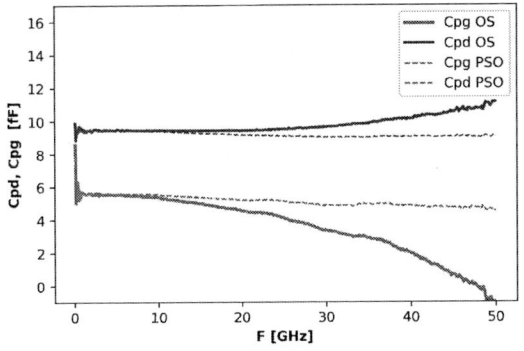

Fig. 6. Cpd and Cpg versus frequency after a Pad-Short-Open (PSO) deembedding and after an Open-Short (OS) deembedding.

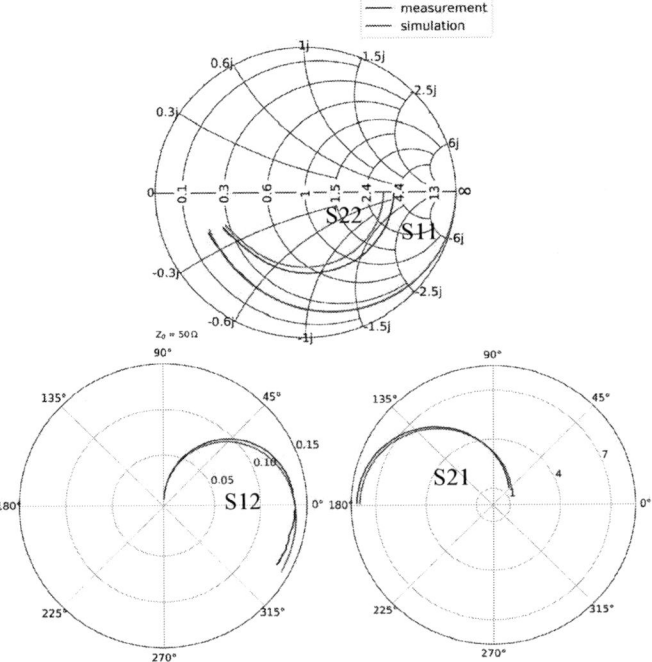

Fig. 7. S parameters comparison between measurement and ADS simulation

B. Tout Extraction

After the validation of the SSEC, the two noise sources ($\overline{e_g^2}$ and $\overline{i_d^2}$) were added to the circuit with Tin set to room temperature. In order to determine T_{out}, the NF_{50} measurement

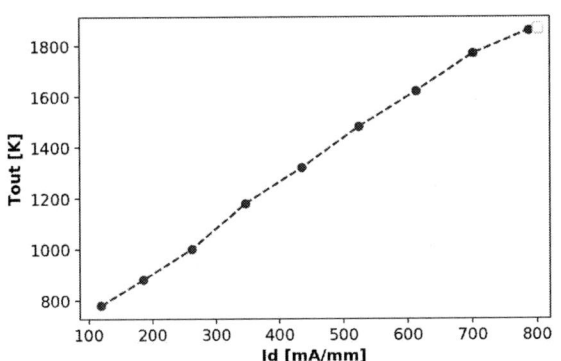

Fig. 8. T_{out} as a function of I_d

curve is used to tune Tout until the simulated NF_{50} (using ADS) reproduces NF_{50} measured data (in shape and magnitude). This process is repeated and validated for different drain currents. It is important to point out that the NF_{50} measurement reference planes are at the probe tips (Fig. 5). The resulting Tout as function of Id is shown in Fig. 8. T_{out} increases monotonically from 780 K to 1860 K, when Id increases from 100 to 800 mA/mm. These values are comparable results extracted in Bulk technologies [8].

IV. NOISE PERFORMANCE, ROBUSTNESS OF THE EXTRACTION

The extraction of T_{out} allows to extract the four noise parameters. In Fig. 9, the minimum noise figure (NF_{min}) and the associated gain (G_{assoc}) are reported as a function of drain current @ 20 GHz, while the equivalent noise resistance (R_n) is reported in Fig. 10 and the optimum source admittance (Γ_{opt}) is reported in Fig. 11. The lowest values for NF_{min}/R_n, 0.45 dB/17 Ω are obtained for Id=185mA/mm. For this current, the associated gain is almost at its maximum (13 dB).

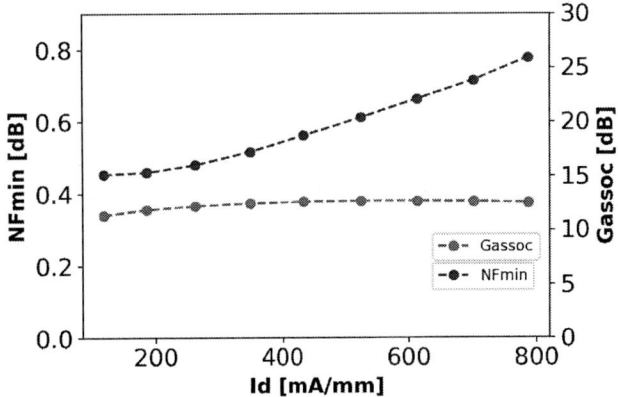

Fig. 9. Minimum noise figure and associated gain versus Id @ 20 GHz

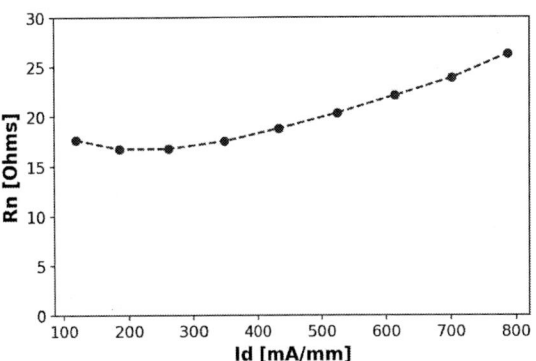

Fig. 10. Equivalent resistance versus Id @ 20 GHz

In order to check the validity of the extraction, NF_{50} measurements were performed in W band. The measurement was done with the instruments presented in [14]. Fig. 12 shows that the simulated NF_{50} data (using ADS) in W band compares well with measured data, which validates the robustness of our noise extraction.

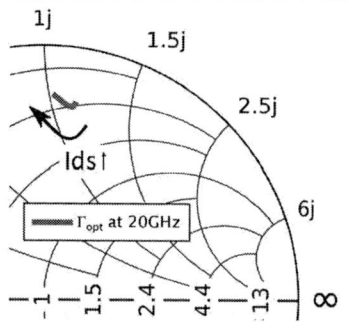

Fig. 11. Γ_{opt} versus Id @ 20 GHz

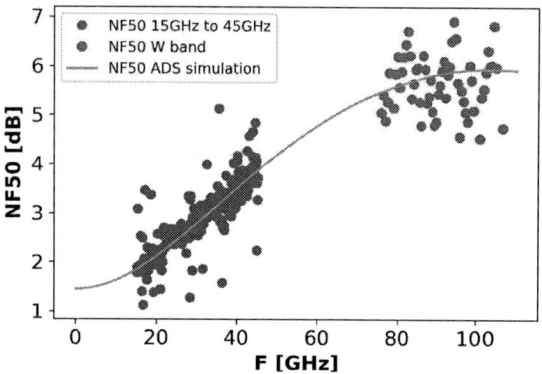

Fig. 12. NF50 measurement (dots) versus simulation (line) up to W band @ Id=347mA/mm

V. COMPARISON WITH STATE OF THE ART

A benchmarking with previously reported CMOS technology nodes is presented on Fig. 13. It is observed that the technology performs very well when NF_{min} is compared to data reported in [8] and [13]. It is to be mentioned that the data from [13] features a very high equivalent noise resistance (250 Ω), as compared to roughly 20 Ω for this technology; indeed, in [13], the geometry of the investigated RF NMOS is not well balanced (with a total gate width of 8 μm). Furthermore, [16] (not depicted in Fig. 13), according to the authors' knowledge, was the only previous report of a UTBB FDSOI noise characterization and it presented very high Rn and NF_{min} compared to our work but on a different technology.

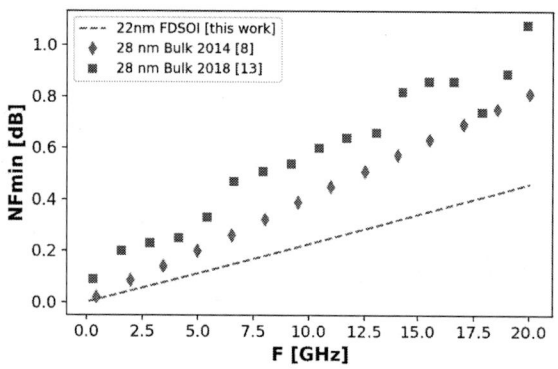

Fig. 13. Benchmarking with previous CMOS

VI. CONCLUSION

In this paper, a high frequency noise characterization of the 22 nm FDSOI technology, preceded by its SSEC extraction, was presented and validated up to 110 GHz. The device achieves a F_t/F_{max} of 332/435 GHz, and very good noise properties with an NF_{min}/R_n of 0.45dB/17Ω @ 20 GHz for a drain current of 185 mA/mm, and an associated gain of almost 13 dB.

REFERENCES

[1] S.N.Ong, et al, "A 22nm FDSOI Technology Optimized for RF/mmWave Applications," *in IEEE Radio Frequency Integrated Circuits Symposium (RFIC)*, Philadelphia, Pennsylvania, USA, 2018, pp 72-75.

[2] R.Carter, et al., "22nm FDSOI Technology for Emerging Mobile, Internet-of-Things, and RF Applications," in *IEDM Technical Digest*, 2016.

[3] J. Watts et al, "RF-pFET in Fully Depleted SOI demonstsrates 420 GHz Ft," *in IEEE Radio Frequency Integrated Circuits Symposium (RFIC)*, Honolulu, Hawaii, USA, 2017, pp 84-87.

[4] Lee Sungjae, et al., "Record RF performance of 45-nm SOI CMOS Technology," in IEEE Int. Electron Devices Meeting, Washington, DC, USA, Dec. 2007, pp. 255-258.

[5] A. Bracale, et al., "A new approach for SOI devices small signal parameters extraction," *Analog Integrated Circuits and Signal Processing*, vol. 25, pp. 157-169, 2000.

[6] R.Torres-Torres, et al., "Analytical model and parameter extraction to account for the pad parasitic in RF-CMOS," *IEEE Transactions on Electronic Devices*, 2005, pp. 1335-1342.

[7] N.Waldhoff, et al, "Improved characterization methodology for MOSFET up to 220 GHz," *IEEE Transactions on Microwave Theory and Techniques*, vol. 57, no. 5, pp. 1237-1243, 2009.

[8] F.Danneville, et al, "RF broadband investigation in High-k/Metal gate 28nm CMOS bulk transistor," *International Journal of Numerical Modelling*, vol. 27, no. 5-6, pp. 736-747, 2014.

[9] P. VanDerVoorn, et al, "A 32-nm Low Power RF CMOS SOC Technology Featuring High-k/Metal Gate," in *IEEE VLSI Technology*, pp. 137-138, Joune 2010.

[10] C. H. Jan, et al, "RF CMOS Technology Scaling in High-k/Metal Gate Era for RF SoC (System-on-Chip) Applications," in *IEEE Int. Electron Devices Meeting*, December 2010, pp. 27.2.1 – 27.2.4.

[11] G. Dambrine, et al, "High-frequency four noise parameters of silicon-on-insulator-based technology MOSFET for the design of low-noise RF integrated circuits," *IEEE Transactions on Electronic Devices*, vol 46, no. 8, pp. 1733-1741, August 1999.

[12] M. W. Pospieszalski, "Modeling of Noise Parameters of MESFET's and MODFET's and Their Frequency and Temperature Dependence," *IEEE Transactions on Microwave Theory and Techniques,* vol. MTT-37, pp. 1340-1350, September 1989.

[13] H. Zhang, et al, "Extraction of Drain Current Thermal Noise in a 28 nm High-k/Metal Gate RF CMOS Technology," *IEEE Transactions on Electronic Devices,* vol 65, no. 6, pp. 2393-2399, June 2018.

[14] H. Zhang, et al, "Extraction of Drain Current Thermal Noise in a 28 nm High-k/Metal Gate RF CMOS Technology," *IEEE Transactions on Electronic Devices,* vol 65, no. 6, pp. 2393-2399, June 2018.

[15] N.Waldhoff, et al, "Small Signal and Noise Equivalent Circuit for CMOS 65 nm up to 110 GHz," in *European microwave conference*, pp. 321-324, October 2008.

[16] P. Kushwaha, et al, "Characterization of RF Noise in UTBB FD-SOI MOSFET", *IEEE Journal of the Electron Devices Society*, vol. 4 no. 6, pp. 379-386, November 2016.

A 76-81GHz FMCW Transceiver with 3-Transmit, 4-Receive Paths and 15dBm Output Power for Automotive Radars

Zongming Duan[1], Dongfang Pan[1], Bowen Wu[2], Yan Wang[2], Bingbing Liao[2], Dong Huang[3], Yanhui Wu[4], Daiguo Xu[4], Hua Xu[4], Wei Lv[2], Yuefei Dai[2], Pei Li[2], Yan Wang[3], Fujiang Lin[1]

[1]University of Science and Technology of China, China
[2]APELAM lab, East China Research Institute of Electronic Engineering, China
[3]Tsinghua University, China
[4]STAIC lab, the 24th Research institute of China Electronics Technology Group Corporation, China

Abstract—A fully-integrated 76-81GHz FMCW transceiver in 65-CMOS is presented for automotive radar applications. The transceiver consists of 3-transmiter, 4-receiver, FMCW synthesizer and ADC with decimation filters. The transmitter, with a TX chain including a push-push doubler, a 3-stage power amplifier employing neutralization technique and 2 power-combined topologies, shows an output power of 15dBm. The receiver, with an LNA employing a 3-stage neutralized common source (CS) with inductive degeneration, achieves a noise figure of 15dB, an input P-1dB of -20dBm, and an ENOB of 10.2-b for ADC. The double tuning modulated PLL provides FMCW waveform with -81dBc/Hz phase noise at 1MHz offset and chirp rate of 50MHz/us. The power consumption of the entire transceiver is 1.06W.

Keywords— Automotive radar, CMOS transceiver, transmitter, receiver, FMCW PLL, power amplifier.

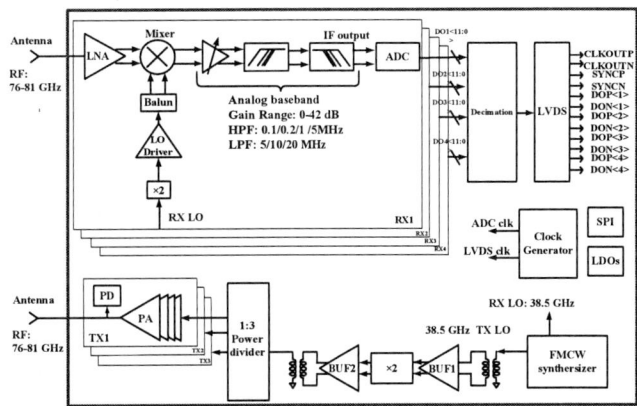
Fig. 1 Block diagram of the proposed FMCW transceiver.

I. INTRODUCTION

Millimeter-wave (mm-wave) radars, thanks to their inherent detecting capability and robustness to light and environmental conditions, has now became a key role sensor for advanced driver-assistance systems in vehicles [1]-[2]. Multimode radars which support long-range-radar (LRR), middle-range radar (MRR) and short-range radar (SRR) at the same time can allow the same hardware to support different functions. The radar transceiver must have multi-channel in the transmitter (TX), to support diverse transmit antennas and achieve high output power to enable long-range sensing especially. The receiver (RX) needs configurable capabilities to adopt different applications and multi-channel to enable accurate angle-of-arrival measurement. In addition, for low-cost requirements, it is desirable to achieve a fully-integrated radar with RF transceivers, analog and digital modules in a single chip, and preferably fabricated in CMOS technology [3]-[5].

This paper presents a fully-integrated 76-81GHz CMOS frequency-modulated continuous wave (FMCW) transceiver. With integration of 3-TX, 4-RX, FMCW synthesizer, internal ADCs, decimation filters and LVDS interface (Fig. 1), the transceiver shows relatively high-integration compared with some pervious works [5]-[8]. Output power of TX is enhanced by employing a 3-stage power amplifier and 2-way power-combined topologies, achieving 15dBm typically. With 3-channel TX and configurable capabilities including gain, bandwidth (BW) in the RX, the transceiver can be configured as different applications. The integrated FMCW synthesizer

with double tuning technique provides various radar waveforms including triangular and sawtooth frequency-modulated waves. The chip is fabricated in 65-nm CMOS technology, with a power consumption of 1.06W under 1V and 3.3V supply. Measured and analysed results show state-of-art performance compared with that of up to date works.

II. SYSTEM ARCHITECTURE

The chip block diagram is shown in Fig. 1. This transceiver includes a 4-channel RX path with an LO chain, a 3-channel TX path, a 38.5GHz FMCW synthesizer, 12-bit ADCs along with 1/2/4/8 decimation filters and LVDS interface. For angular resolution and blind spot monitoring of radar systems, multi-channel solutions are required, while consistency between each channel, delay and isolation are significant issues in system-level design. In this paper, 4-RX and 3-TX paths are integrated in a single chip, similarly to [3], [6] and [7]. As a trade-off of reduction for TX and LO chains design and PLL design difficulties, FMCW synthesizer is set to half of transmitter frequencies ranging from 38~40.5GHz. Triangular and sawtooth frequency-modulated waves can be generator from the synthesizer with reconfigurable bandwidth (BW) and period. One output of the synthesizer is to drive the TX chains, firstly doubling through a push-push doubler, then splitting by a 1:3 power divider, and finally amplified to an output of 15dBm typically by the power amplifiers. The other output of the synthesizer is to drive the RX chains which are similarly to the

978-1-7281-1702-7/19 $31.00 © 2019 IEEE

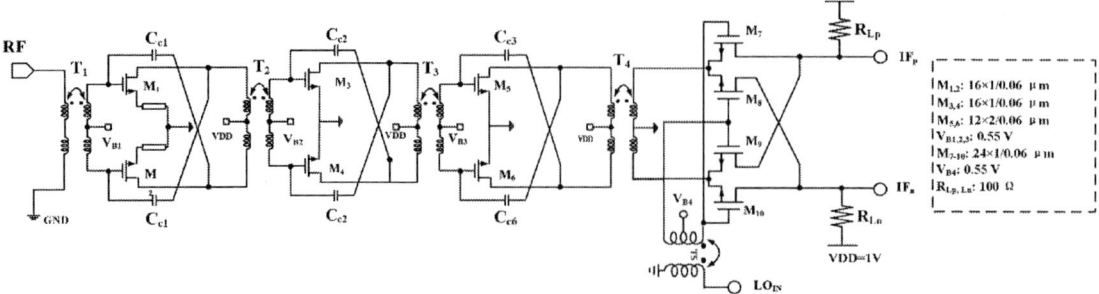

Fig. 2 Schematic of mm-wave front of the proposed receiver.

TX chains. The RX chain consists of 3 parts including mm-wave front-end, analog baseband (ABB), and ADC with decimation filters and LVDS interface which is also included in [3]-[4] while not included in [5]-[8]. For different range detection and multimode applications, some RX parameters, such as the gain, BW and decimation coefficient, can be reconfigurable through SPI. A clock generator PLL with the same reference as FMCW synthesizer is adopted for ADC clock and overall system synchronization.

III. CIRCUIT DESIGN

A. Receiver chain

The block diagram of RX chain is shown in Fig. 1. A mm-wave front-end, an ABB, an ADC, decimation filters and a LVDS interface compose the RX chain. The schematic of mm-wave front-end is shown in Fig. 2. It consists of a three-stage differential LNA and an active mixer. The LNA employs a three-stage neutralized common source (CS) with inductive degeneration for high linearity. The input balun provides NF matching and ESD protection. In the first stage of LNA, the differential pair ($M_{1,2}$) bias is optimized to achieve a minimal NFmin with an optimal current density, which is 0.15mA/μm in this process. The value of the neutralized capacitors in each stage is chosen by optimizing the maximum available gain (MAG) and stability simultaneously. In this design, the optimal stability factor (Kf) and MAG are achieved with 6fF of C_{c1}, 6fF of C_{c2}, and 10fF of C_{c3} respectively. The active mixer consists of quad switches made of transistors M_7-M_{10}. Benefit of removing the G_m pair in the conventional mixer, which requires large current for high input P-1dB, the power consumption and the temperature dependence of the mixer are reduced. The transformer T4 is used to improve the linearity of the mixer by using the secondary coil as the source degeneration inductor of the mixer. This allows for the mixer to be powered by a 1.0V supply and helps to reduce the power consumption as well.

The ABB consists of a 2rd order high-pass filter and 5th order low-pass filter, which inserted in the analog baseband to filter out unwanted signals and avoid DC-offset. 42dB with 6dB step receiver tuning gain range is provided by the three-stage PGA. Supply voltage (2.5V) of analog baseband block is provided by on-chip LDO regulator. The last stage buffer drives the internal 12-bit ADCs connected to the internal decimators before sending the digital data through the LVDS interface. The 12-bit ADC with a 50MHz sampling clock is adopted by SAR architecture. The following multistage filters with 1/2/4/8

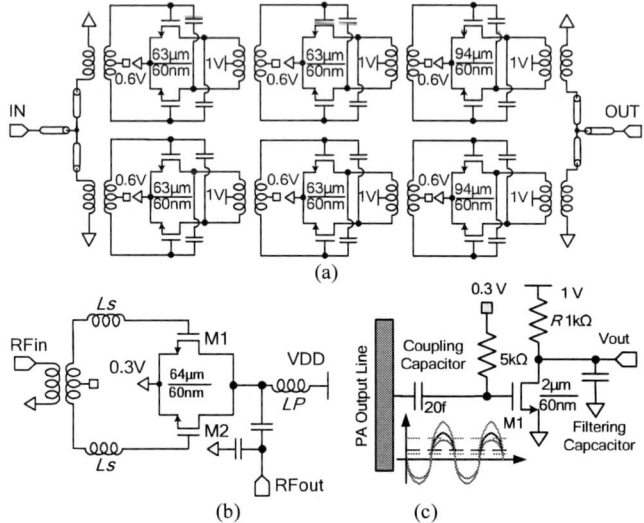

Fig. 3 Schematic of (a) PA, (b) push-push doubler and (c) power detector.

decimations coefficients, can also achieve offset error adjustment. Measurement results show that an ENOB of 10.2, a DNL of +0.48/-0.85 LSB and an INL of +1.5/-1.5 LSB can be achieved.

B. Transmitter

The TX chain consists of a single-to-differential balun, a 38GHz buffer, a 38GHz doubler, a 77GHz buffer, a 1:3 power divider and output PAs, as shown in Fig. 1. Firstly, the 38~40.5GHz chirp signal is amplified by the 38GHz buffer and injected into the doubler to generate 76-81GHz signal. Then the 76-81GHz signal is delivered to 3 PAs through a 1:3 power divider. Differential topology, neutralization technique and on-chip transformers are used to achieve the buffers and amplifiers to obtain great gain and output power and a compact layout. In order to achieve a large output power, PA with 2-way power-combined topology is also used, which obtains a maximum output power of 15dBm typically, as shown in Fig. 3 (a). The 38GHz doubler is achieved by a push-push common source topology, as shown in Fig. 3 (b), which helps to achieve a large conversion gain. The 1:3 power divider applies two stage topology and three Wilkinson power dividers to obtain high isolation among the three PAs. In order to detect the large output power on-chip, a NMOS transistor biased at subthreshold is used as detector is also used at the output of the PA, as shown in Fig. 3 (c), which has a detection range from 0dBm to 15dBm with a 6-bit resolution ADC. Measurement

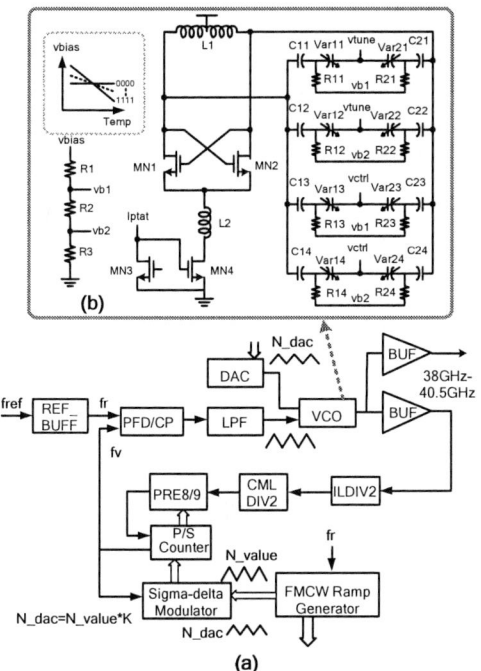

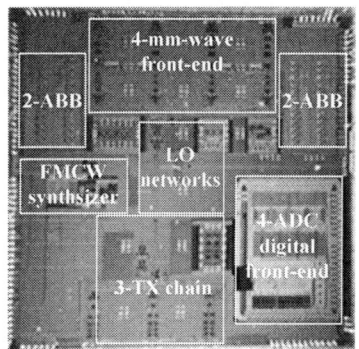

Fig. 4 (a) Block diagram of the FMCW PLL (b) Schematic of 38.5~40.5GHz VCO.

results show that an output of 15dBm can be achieved from 76-81GHz band due to the proposed design technique.

C. FMCW synthesizer

The block diagram of the FMCW synthesizer is shown in Fig. 4(a). Schematic of wideband on-chip VCO ranging from 38 to 40.5GHz is also show in Fig. 4(b). To improve detection resolution for short range detection, VCO should cover wide frequency range. For K_v linearization consideration, more capacitors are added to LC tank. Optimized size of cross-coupled NMOS transistors and inductor L_1 are obtained by simulation carefully. The VCO with two tuning ports, one is from loop filter and the other is from DAC (as show in Fig. 4(a)), achieves linearity and chirp rate improvements. To compensate the impact of temperature variation, a frequency temperature calibration circuit is proposed as shown in Fig. 4(b). To improve phase noise performance, the noise from V_{bias} generated by temperature calibration circuit is filtered by RC network. Simulation result shows that a phase noise of -97.5dBc/Hz @1MHz offset can be achieved.

To reduce noise from charge pump, the pulse width of charging and discharging is designed to a narrow level, 400ps typically. The feedback path of the synthesizer consists of divide-by-2 injection-locked frequency divider (ILFD), current mode logic divide-by-2 frequency divider (CML), dual-mode /8/9 prescaler, P/S counter and delta-sigma modulator. The total currents of the synthesizer are about 38mA in 1V supply and 13mA for 3.3V supply.

IV. MEASUREMENT

The proposed FMCW transceiver is fabricated in commercial 65-nm CMOS process. Fig. 5 shows the die

photograph. The chip size is 5.8mm×5.8mm with pads including. The total power consumption is approximate 1.06W under 1V and 3.3V supplies. On-wafer measurements are performed by a Keysight's 110GHz measurement system.

Fig. 6 shows the measured results of the RX. In Fig. 6 (a), the gain with 20-60dB and the NF with 14-18dB, are reached for the mm-wave front-end with ABB at 78GHz, in 3-bit control with 6dB gain step. S11 (RF port) is lower than -10dB over 72-84GHz band. As shown in Fig. 6(b) and Fig. 6(c), the ADC digital front-end achieves a SNDR of 65dB and a SFDR of 85dB, while a DNL of +0.48/-0.85 LSB and an INL of +1.5/-1.5 LSB.

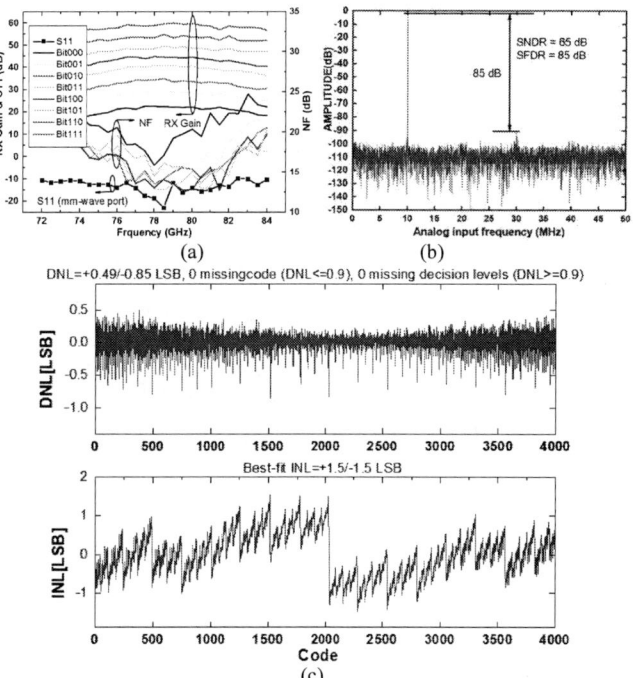

Fig. 5 Die photograph of the proposed transceiver.

Fig. 6 Measured results of (a) mm-wave front-end with ABB, (b) AC performance and (c) DC performance of ADC digital front-end.

Fig. 7 (a) shows the measured results of the TX. The output power of the TX is more than 15dBm in 76-77GHz band which is essential for long range radar, and more than 13dBm in 79GHz band in 25°C. The power shows 2-3dB reduction at 125°C, while 1-2dB improvement at -45°C. The measured S22 (the mm-wave output port) is lower than -10dB in 76-81GHz band, over the temperature of -45~+125°C.

978-1-7281-1702-7/19 $31.00 © 2019 IEEE

Table 1. Performance summary and comparison with state-of-the art 77GHz radar transceivers

	This work	ISSCC'18 [3]	JSSC'17 [4]	JSSC'16 [5]	RFIC'12 [6] [7]	EuMC'14 [8]
Technology	**65-nm CMOS**	45-nm CMOS	28-nm CMOS	65-nm CMOS	SiGe HBT (chipset)	65-nm CMOS (chipset)
Integration block, key feature	**3TX/4RX /FMCW/ADC**	3TX/4RX /FMCW/ADC	2TX/2RX /PMCW/ADC	1TX/2RX /FMCW	3TX/4RX	2TX/4RX
Output power	**15 dBm typ.**	10.8dBm	8.5dBm	13dBm	11.7dBm	11dBm
Noise figure	**13-20dB**	18dB	12dB	5dB	14.5dB	12dB
Input P-1dB	**-20 dBm**	-22dBm	--	--	-3dBm	--
IF BW	**LPF: 5/10/20 MHz HPF; 0.1/0.2/1/5MHz**	15MHz	1GHz	LPF: 0.5/1/25MHz HPF: 0.01MHz	--	LPF: 200 KHz HPF: 1 KHz
RX ADC	**10.2 ENOB**	10.5ENOB	6 ENOB	w/o	w/o	w/o
Phase noise @1MHz	**-81.3dBc/Hz**	-91dBc/Hz	-85dBc/Hz	-87dBc/Hz	-96dBc/Hz	-84dBc/Hz
Ramp BW	**500MHz**	4GHz	5GHz (PMCW)		3GHz	5GHz
Ramp rate	**50MHz/us**	100MHz/us	0 (PMCW)	1.96Gbps	--	240MHz/us
Temperature	**-45 to +125℃**	-40 to +125℃	+27℃	+27 ℃	-40 to +125 ℃	-40 to +90℃
Area	**33.64mm²**	22mm²	7.9mm²	4.64mm²	18mm²	33.6mm²
Power cons.	**1.06W**	3.5W	1 W	0.343W	3.15W	1.1W

Fig. 7(b) and(c) show the measured results of the FMCW synthesizer. The phase noise measured at the output of CML divider is about -93.3dBc/Hz @1MHz offset, which shows a phase noise performance of -81.3dBc/Hz @1MHz offset in 76-81GHz band. Fig. 7(c) shows the measured frequency vs. time of ramps in triangular and sawtooth modulation respectively.

The performances of proposed FMCW transceiver and recent works are summarized in Table 1. The proposed transceiver shows a relatively high-integration level and

demonstrates some performance improvements as well such as output power, wide range operation temperature.

V. CONCLUSION

This paper presents a 76-81GHz CMOS FMCW transceiver for automotive radar applications. With integration of 3-TX, 4-RX, FMCW synthesizer, internal ADCs, decimation filters and LVDS interface in a single chip, the transceiver shows relatively high-integration level compared with state-of-the works. The measured TX shows the output power of 15dBm, with a chirp ramp rate of 50MHz/us and phase noise of -81.3 dBc/Hz @1MHz. The measured RX shows a NF of 15dB and an ENOB of 10.2, with gain, BW reconfigurable additionally. The transceiver exhibits state of the art performance with low fabrication cost for 76-81GHz band automotive radar applications.

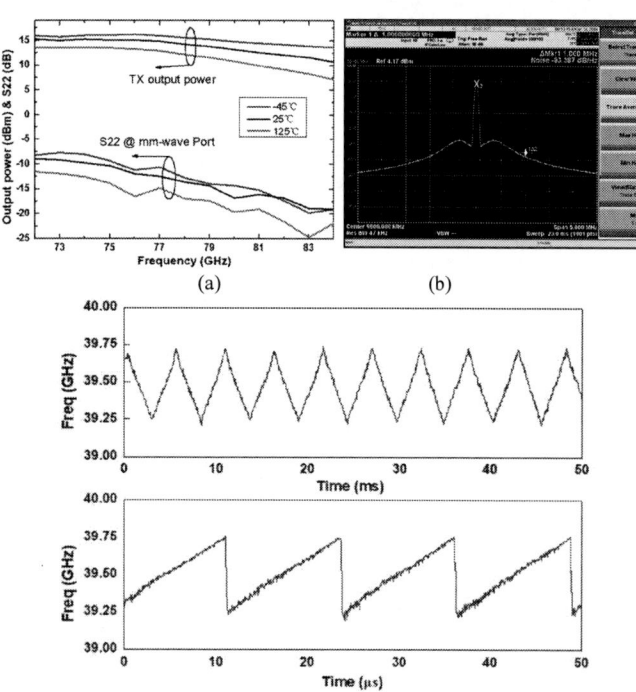

Fig. 7 Measured results of (a) TX output power and S22 (b) synthesizer phase noise at /4 divider (9.6GHz) (c) frequency vs. time of ramps in triangular and sawtooth modulation.

REFERENCES

[1] Y. Li, M. Hung, S. Huang, J. Lee, "A Fully Integrated 77GHz FMCW Radar System in 65nm CMOS", ISSCC Dig. Tech. Papers, pp. 216-217, Feb. 2010.

[2] J. Hatch et al., "Millimeter-wave technology for automotive radar sensors in the 77 GHz frequency band," *IEEE Trans. Microw. Theory Techn.*, vol. 60, no. 3, pp. 845–860, Mar. 2012.

[3] B. P. Ginsburg et al., "A multimode 76-to-81ghz automotive radar transceiver with autonomous monitoring," in 2018 IEEE International Solid - State Circuits Conference - (ISSCC), Feb 2018, pp. 158–160.

[4] D. Guermandi et al., "A 79-GHz 2x2 MIMO PMCW Radar SoC in 28-nm CMOS," *IEEE J. of Solid-State Circuits*, vol. 52, no. 10, pp.2613-2626, Oct. 2017.

[5] H. Jia et al., "A 77 GHz Frequency Doubling Two-Path Phased-Array FMCW Transceiver for Automotive Radar," *IEEE J. of Solid-State Circuits*, vol. 51, no. 10, pp. 2299-2311, Oct. 2016.

[6] H. Knapp et al., "A 77GHz 4-Channel Automotive Radar Transceiver in SiGe", Proc. RFIC, pp. 233-236, June 2008.

[7] C. Wagner, J. Boak, et al. "A 77GHz Automotive Radar Receiver in a Wafer Level Package", Proc. RFIC, pp. 511-514, June 2012.

[8] T. Shimura, et al., "Multi-Channel Low-Noise Receiver and Transmitter for 76-81 GHz Automotive Radar Systems in 65 nm CMOS," European Microwave Conf., pp. 596-599, 2014.

Reconfigurable 60-GHz Radar Transmitter SoC with Broadband Frequency Tripler in 45nm SOI CMOS

Wooram Lee[1], Tolga Dinc, Alberto Valdes-Garcia

IBM T. J. Watson Research Center, Yorktown Heights, NY, USA

[1]leewoora@us.ibm.com

Abstract — **A reconfigurable 60-GHz radar transmitter with a broadband frequency tripler is proposed to support CW/FMCW, pulse, and PMCW radar waveforms from a single front-end. The proposed IC consists of a wide-band frequency tripler, a two-stage driver, two power mixers with baseband circuitry and serial I/O circuitry. The IC measurements in CW mode operation show an output power of 12.8 dBm (average) and 14.7 dBm (peak) from 54 GHz to 67 GHz with harmonic suppression greater than 27 dB. Pulse and PMCW mode operations are also demonstrated to generate short pulses with the minimum pulse width of 25 ps corresponding to 40 GHz signal bandwidth and 10-Gb/s PRBS modulated signals, respectively. Fabricated in a 45-nm CMOS SOI process, the IC consumes 0.51 W and occupies an active area of 1.95 mm^2 excluding pads.**

Keywords — **CMOS, multi-modal radar, FMCW radar, 60 GHz, pulse radar, PMCW radar, wideband frequency tripler.**

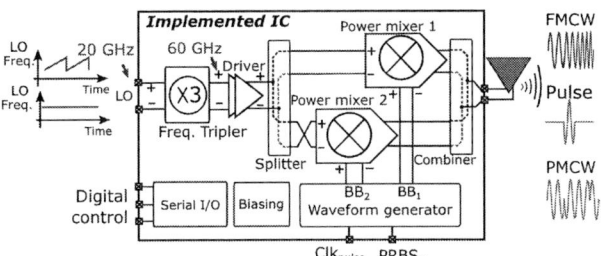

Fig. 1. Proposed reconfigurable radar transmitter architecture.

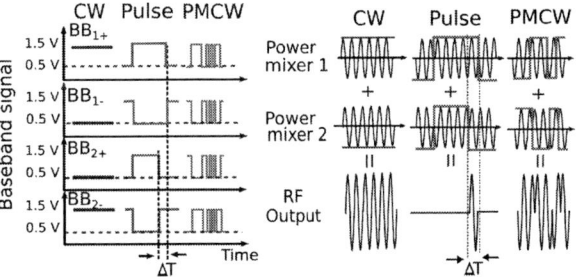

Fig. 2. Baseband signals to modulate the two power mixers for generating different radar waveforms.

I. INTRODUCTION

Millimeter wave (mmWave) radars realized using Si-based ICs have enabled the exploration of an increasing number of applications, such as automotive, drone navigation, gesture recognition, and remote biometrics monitoring. Different usage scenarios require different waveforms and varying performance trade-offs between detection range (available SNR) and resolution (signal bandwidth). For instance, short-pulse radar is suitable for medical applications requiring high spatial resolution within a short range[1]. By contrast, phase-modulated continuous-wave (PMCW)[2] or frequency-modulated continuous-wave (FMCW) radar[3], [4] is optimal for automotive applications requiring a long detection range and moderate resolution. Low cost and reduced form factor are key enablers for the mass adoption of mmWave sensors, so a single-chip multi-mode radar solution is desirable[5]. In this paper, we propose a reconfigurable broadband V-band radar transmitter (TX) capable of generating (1) continuous wave (CW/FMCW), (2) pulse, and (3) PMCW radar waveforms from a single CMOS front end, providing adaptability for different applications and support for a multimodal radar system.

II. RECONFIGURABLE TRANSMITTER ARCHITECTURE

Fig. 1 shows the proposed reconfigurable transmitter architecture. A broadband frequency tripler up-converts an LO input in the range of 17–22 GHz to 51 GHz–66 GHz. The tripler is followed by a two-stage amplifier and a pair of power mixers. The power mixers are configured/driven by baseband (BB) signals generated from a waveform generator and their combined outputs form the final output of the radar TX. The waveform generator is key to the configurability of the TX. Fig. 2 shows the BB signals and the waveform generation principle corresponding to each output radar waveform type. In CW mode, the power mixers operate as power amplifiers with DC BB signals. The BB signal polarities between the two power mixers are opposite for constructive combining as the LO input polarity is swapped for the 2nd power mixer. When coupled to a wideband PLL which generates an output frequency of 17–22 GHz with frequency chirp generation capability, e.g. [6], this mode enables the creation of FMCW radar signals from 51 GHz to 66 GHz. For pulse mode operation, two clock signals with aligned rising edges but different falling edges by ΔT are applied to the power mixers. The mixer's outputs are added constructively only when the two BB signals are not overlapped (added destructively otherwise), thus generating a sharp pulse with the programmable pulse width down to 25 ps at the output. A key advantage of this approach is that pulse generation occurs at the combiner, therefore the bandwidth of the front-end circuitry does not directly limit the minimum pulse width, enabling the maximum pulse signal bandwidth of 40 GHz. In PMCW mode, the two power mixers are fed with complementary code waveforms to support binary phase

978-1-7281-1702-7/19 $31.00 © 2019 IEEE

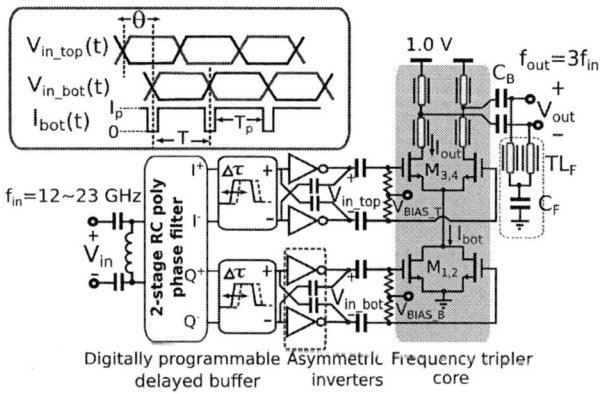

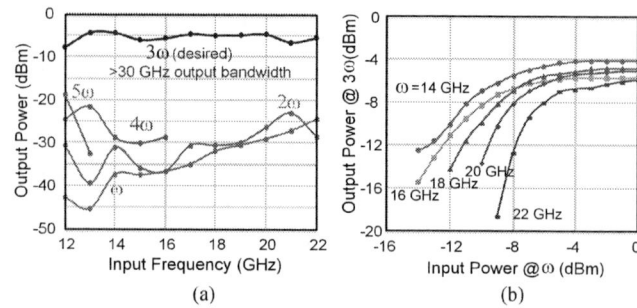

Fig. 3. Schematic of frequency tripler with >30-GHz output frequency range.

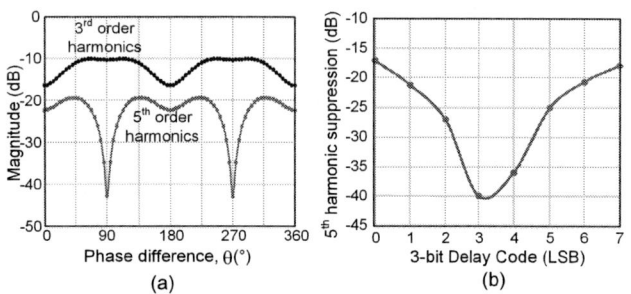

Fig. 4. (a) Calculated output current magnitude at $3\omega_{in}$ and $5\omega_{in}$ with respect to phase difference (b) measured $5\omega_{in}$ suppression across 3-bit fine delay tuning code for an input frequency of 13 GHz.

Fig. 5. Measured frequency tripler output using a separate breakout circuitry

modulation. Since only the last stage of the transmitter is directly modulated with the baseband signal, the front end can support code data rates exceeding 10 Gb/s.

III. CIRCUIT-LEVEL IMPLEMENTATION IN 45-NM SOI

A. Wide-band Frequency Tripler Design

Fig. 3 shows the proposed frequency tripler with input I/Q generation and programmable delay preamplifiers. The frequency tripler core consists of a push-push frequency doubling differential pair $M_{1,2}$ and an up-conversion differential pair $M_{3,4}$ in a current-reuse topology. Assuming (1) an ideal square-wave switching of $M_{3,4}$ and (2) a rectangular waveform of the combined drain current of $M_{1,2}$, $I_{bot}(t)$, with a duty ratio of $\alpha = T_p/T$, the output current $I_{out}(t)$ for an input frequency of ω_{in} is written as

$$I_{out}(t) = \frac{4}{\pi} I_{bot}(t) \sum_{m=1} \frac{\sin\left((2m-1)\omega_{in}t\right)}{2m-1} \quad (1a)$$

$$I_{bot}(t) = \sum_{n=0} I_n \cos(2n(\omega_{in}t + \theta)), \quad (1b)$$

where $I_0 = \alpha I_p$, $I_n = I_p/(n\pi)\sin(2n\pi\alpha)$ for $n > 1$, and θ is the phase difference between driving voltages of $M_{1,2}$ and $M_{3,4}$ as shown in the inset of Fig. 3. Eq. (1) indicates that the kth harmonic frequency component of the output current results from the combination of different mixing processes which meets $k = |2m \pm 2n - 1|$ as a function of θ. To find out the optimum θ for the maximum

conversion gain, Fig. 4(a) shows the calculated 3rd and 5th harmonic frequency components of the output current across θ for $m, n \leq 3$. $\theta = 90°$ provides the maximum conversion gain at $3\omega_{in}$ and the minimum conversion gain at $5\omega_{in}$ which is the dominant unwanted harmonic for low input frequencies. To drive $M_{1,2}$ and $M_{3,4}$ with an accurate 90° difference, a two-stage RC poly phase filter generates broadband quadrature signals using staggered tuning over a bandwidth >10 GHz, and the programmable delay buffers, based on current-starved inverters, provide fine phase control of ~0.3 ps resolution over ~2 ps tuning range with 3-bit digital control. Fig. 4(b) shows the measured 5ω harmonic suppression over 3-bit delay code of the input buffer (realized in a separate breakout), demonstrating the effect of fine I/Q timing control. Asymmetric inverters following the I/Q timing circuitry drive $M_{1,2}$ with the optimum bias voltage and duty ratio α for maximum $2\omega_{in}$ generation in I_{bot}. An output filtering network consisting of C_B, TL_F, and C_F is designed to suppress ω_{in} and $4\omega_{in}$ in which C_B and TL_F form a high pass filter to suppress ω_{in}, and TL_F and C_F form a notch filter using a series resonance to suppress $4\omega_{in}$. Fig. 5 shows the measurement of the frequency tripler in a separate breakout. Output bandwidth is greater than 30 GHz with a peak output power of -4.4 dBm; harmonic rejection up to $4\omega_{in}$ is better than -20 dB over an input frequency range of 14 GHz to 20 GHz. The harmonic suppression is further improved through the front-end circuitry that follows the frequency tripler. Fig. 5(b) shows the measured output power at $3\omega_{in}$ with respect to input power for different frequencies. The maximum conversion gain is higher than 0 dB for input frequencies up to 20 GHz, and the saturation output power is around -5 ± 0.9 dBm. The measured power consumption of the frequency tripler ranges from 33 mW to 39 mW from 1 V power supply for input frequencies from 12 GHz to 22 GHz. The proposed frequency tripler presents a wider output bandwidth and higher conversion gain with a comparable power dissipation compared to a prior wide-band V-band frequency tripler with 18-GHz bandwidth and -5.2-dB conversion gain[7].

B. TX Front-end and Waveform Generator Design

Fig. 6 shows the front-end and waveform generator schematic. The two power mixers are driven by two-stage

978-1-7281-1702-7/19 $31.00 © 2019 IEEE

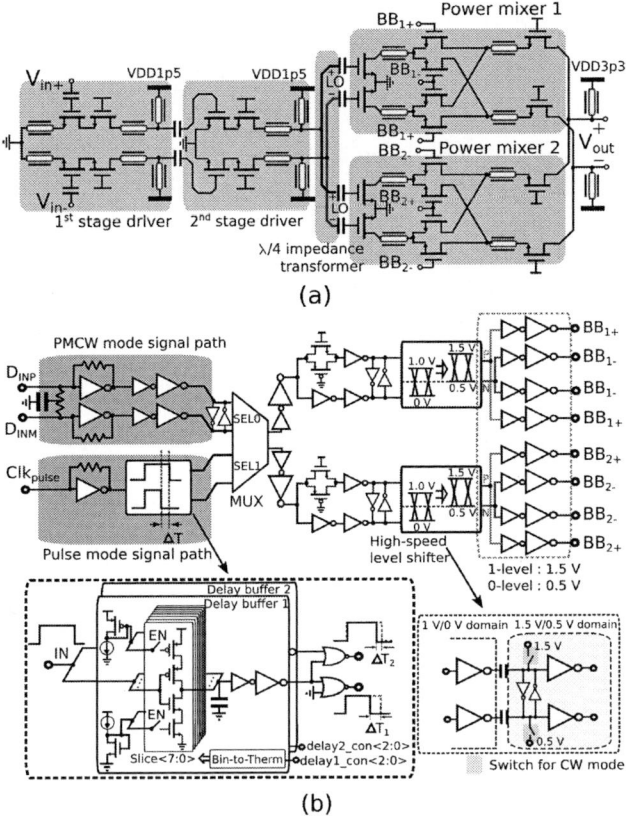

(a)

(b)

Fig. 6. Schematic of (a) front-end and (b) waveform generator.

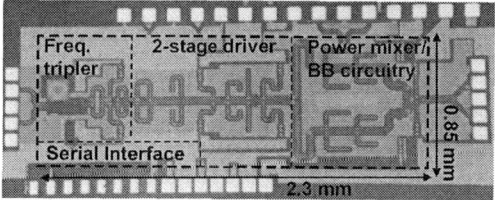

Fig. 7. Chip photograph.

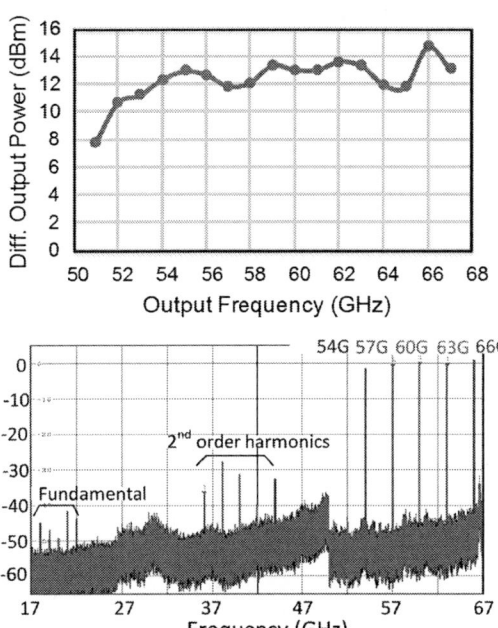

Fig. 8. Measured differential output power across frequency and frequency spectrum at the single-ended output.

differential cascode amplifiers through $\lambda/4$ impedance transformers. The 1st stage driver has a degeneration inductor for a broadband input matching to provide a consistent load impedance to the frequency tripler over a wide frequency range. The power mixer is similar to a differential Gilbert mixer with cascode transistors on the top. The quad-switching devices are modulated with BB signals from the waveform generator as shown in Fig. 2. The simulated front-end gain in CW mode is 27.7 dB at 60 GHz with 3-dB bandwidth of 17 GHz. The waveform generator has separate signal paths for pulse and PMCW modes, selected by a MUX. The pulse mode input receives a clock signal CLK_{pulse} which determines the pulse repetition frequency. The clock signal is applied to a programmable pulse generation block after resistor-feedback inverters. The pulse generation circuitry consists of two programmable delay blocks followed by NOR gates to generate two outputs with aligned rising edge but time-shifted falling edges. The falling edge difference $\Delta T = \Delta T_1 - \Delta T_2$ determines the pulse width at the RF output. Each delay block is independently controlled with 3-bit slices, covering a 140-ps range with $\sim$23-ps resolution. After the MUX selects one of the two input paths, the signal is converted into a differential signal and level-shifted from the 0/1.0V to the 0.5/1.5V domain to properly drive the power mixer's quad-switching devices. For CW mode, the switches inside the level shifters are turned on to provide the proper DC level to the BB output.

IV. MEASUREMENT RESULT

The IC was fabricated in 45-nm SOI CMOS process with an active chip area of 2.3 mm×0.85 mm excluding pads; a die photo is shown in Fig. 7. All measurements have been performed on-wafer at 25 °C. Fig. 8 shows the CW mode operation. The measured average output power (measured using a U8488A power sensor with 0 dBm tripler input power) from 54 GHz to 67 GHz was 12.8 dBm, with peak power of 14.7 dBm at 66 GHz. The power variation from 59 GHz to 63 GHz is less than $\sim$0.3 dB, which can support 4-GHz FMCW radar generation. The frequency spectrum (measured using a single-ended output) also demonstrates harmonic rejection greater than 27 dB over a wide bandwidth; this result could be further improved if measured differentially, considering that dominant harmonic component at the single-ended output is the second order harmonic frequency. The pulse and PMCW mode operations were measured at a single-ended output in the time and frequency domains using a 70-GHz BW DSA8300 Digital Serial Analyzer and FSU67 spectrum analyzer. Fig. 9 shows results from pulse mode operation with CLK_{pulse} of 400 MHz and output carrier frequency of 66 GHz for different ΔT settings of the delay blocks in the waveform generator. The measured pulse width at half maximum ranges from 25

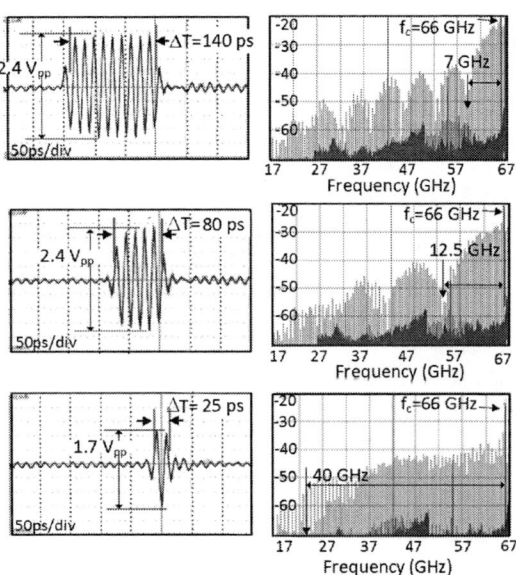

Fig. 9. Measured pulse mode output for different pulse width settings in time and frequency domains.

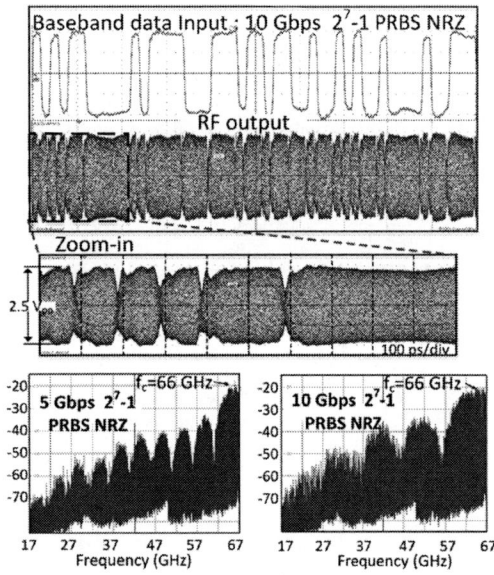

Fig. 10. Measured PMCW mode output in time and frequency domains.

ps to 140 ps, which corresponds to signal bandwidth from 7 GHz to 40 GHz. Note that 40-GHz BW can be translated into a 3.75 mm spatial resolution. The peak output power differed by less than 1 dB compared to the CW mode until the pulse width is reduced to <45 ps. The PMCW mode of operation was also measured with a 10-Gbps $2^7 - 1$ PRBS baseband signal in the time and frequency domains, demonstrating a wide signal bandwidth as shown in Fig. 10. The average power consumption of the IC is 0.51 W (0.36 W for power mixers, 0.1 W for pre-drivers, and 0.05 W for frequency tripler and BB/digital circuitry). Table 1 summarizes the performance of the proposed design compared to prior single-mode and multi-modal radar transmitter ICs.

Table 1. Performance summary and comparison

	This Work	[5]	[1]	[4]	[2]
Supported Waveforms	**Pulse, FMCW/CW, Code (PMCW)**	Pulse, FMCW/CW	Pulse	FMCW	Code (PMCW)
Frequency	**60GHz**	24/26GHz	90GHz	77GHz	79GHz
Pulse Width	**25ps-140ps**	500ps-1ns	25ps-220ps	N/A	N/A
FMCW BW	**4GHz[a]**	0.32GHz	N/A	4GHz	N/A
CW Output Power	**12.8dBm(ave.) 14.7dBm(peak)**	-0.56dBm	17.2dBm (PA Psat)	10.8dBm[b]	8.5dBm[c]
Code Rate	**10Gb/s**	N/A	N/A	N/A	2Gb/s
Integration level	**Freq. tripler, waveform gen., TX front-end**	VCO, freq divider, waveform gen., driver	VCO, PA, antenna, waveform gen.	PLL, 3TX,4RX, Baseband, A2D, MCU	PLL, 2TX, 2RX, A2D, digital core
Chip Area	**1.95mm²**	3.64mm²	1.2mm²	22mm²	7.9mm²
Technology	**45nm SOI CMOS**	130nm CMOS	130nm SiGe	45nm CMOS	28nm CMOS
Power Cons.	**0.51W**	0.14W	0.74W	3.5W	1W

[a] A chirp bandwidth of 4 GHz for FMCW radar is supported with <0.3-dB output power variation, but an external PLL with chirp generation capability is required.
[b] Across -40°C to 125°C   [c] Including flip-chip assembly and module loss

V. CONCLUSION

This work has presented a wide-bandwidth reconfigurable mm-Wave radar transmitter integrated with a frequency tripler in CMOS. The proposed transmitter architecture enables a single front-end to generate FMCW/CW, pulse, and PMCW radar signals, thereby covering a wide variety of applications. The novel frequency tripler design technique has been integrated for a wide-bandwidth LO generation with high harmonic suppression.

ACKNOWLEDGEMENT

This research was developed with funding from the Defense Advanced Research Projects Agency (DARPA). The views, opinions and/or findings expressed are those of the authors should not be interpreted as representing the official views or policies of the Department of Defense or the U.S. Government. The authors thank M. Ferriss, B. Sadhu, M. Yeck, and H. Liu for technical discussions and support and Daniel Friedman for management support.

REFERENCES

[1] A. Arbabian et al., "A 90 GHz hybrid switching pulsed-transmitter for medical imaging," *IEEE J. Solid-State Circuits*, vol. 45, no.12, pp. 2667–2680, Dec. 2010.

[2] D. Guermandi et al., "A 79-GHz 2×2 MIMO PMCW radar SoC in 28-nm CMOS," *IEEE J. Solid-State Circuits*, vol. 52, no.10, pp. 2613–2624, Oct. 2017.

[3] Y. Li et al., "A fully integrated 77GHz FMCW radar system in 65nm CMOS," in *IEEE Int. Solid-State Circuit Conf. Dig. Tech. Papers*, pp. 216–218, February 2010.

[4] B.P. Ginsburg et al., "A multimode 76-to-81GHz automotive radar transceiver with autonomous monitoring," in *IEEE Int. Solid-State Circuit Conf. Dig. Tech. Papers*, pp. 158–160, February 2018.

[5] G. Pyo et al., " K-band single-path dual-mode CMOS transmitter for FMCW/UWB radar," *IEEE Microw. Compon. Lett.*, vol. 26, no. 10, pp. 858–860, Oct. 2016.

[6] M. Ferriss et al., "A 12-to-26 GHz fractional-N PLL with dual continuous tuning LC-D/VCOs," in *IEEE Int. Solid-State Circuit Conf. Dig. Tech. Papers*, pp. 196–198, February 2016.

[7] M. Chou et al., " A 60-GHz CMOS frequency tripler with broadband performance," *IEEE Microw. Compon. Lett.*, vol. 27, no. 3, pp. 281–283, Mar. 2017.

2019 IEEE Radio Frequency Integrated Circuits Symposium

A 94GHz 2×2 Phased-Array FMCW Imaging Radar Transceiver with 11dBm Output Power and 10.5dB NF in 65nm CMOS

Dong Huang[1], Li Zhang[1], Huabing Zhu[2], Boshen Chen[2], Yang Tang[2] and Yan Wang[1]

[1]Institute of Microelectronics, Tsinghua University, Beijing, 100084, China

[2]Microsystem & Terahertz Research Center, China Academy of Engineering Physics, Chengdu, China

Abstract—This paper presents a 94GHz 2×2 phased-array frequency-modulated continuous wave (FMCW) imaging radar transceiver. The transceiver consists of two receiving paths, two transmitting paths and a local oscillator (LO) generation path. In order to improve gain and noise performance, multi-paralleled transistors with small number-of-fingers are used to design power amplifier (PA) and low noise amplifier (LNA). A 24GHz voltage-controlled oscillator (VCO) is used to generate wideband FMCW signals, and two differential push-push cascode doublers are used to act as a quadrupler, which has a low power and high stability. Phase shifter (PS) with constant insertion loss (IL) compensation is used to achieve constant gain among different phase states. The LO path has a continuous frequency tuning range of 14GHz. The transmitter has a maximum output power of 11 dBm and 2.8dB power flatness over 14GHz bandwidth. The receiver has a conversion gain of 20dB and a noise figure (NF) of 10.5dB.

Keywords—CMOS, millimeter-wave, 94GHz, imaging radar, transceiver, phased array.

I. INTRODUCTION

Recently, millimeter-wave (mm-wave) imaging radar obtains an ever-increasing attention [1-4]. Mm-wave imaging radar is harmless to human body and has a great resolution, which shows a great potential on security check, material analysis, biomedical detection and other short-range applications [1-4]. CMOS technique is an excellent way to fabricate phased-array mm-wave imaging radar circuits and systems with large volume and low cost. However, there are still a few challenges of CMOS technique when higher mm-wave frequency imaging radar is designed, for example, a *W*-band imaging radar designed with a typical 65nm CMOS technique. Firstly, it is difficult to obtain effective gain and good noise property because of performance degradation of CMOS technique with frequency approaching to f_t and f_{max} [4, 5]. Secondly, when designing an FMCW radar, a wideband and low power LO generation path requires, which leads to difficulties considering the limited f_t and f_{max}. Thirdly, design of mm-wave phase shifter (PS) with low gain variation among different phase shifting states is still challenging.

In this paper, we designed a 94GHz 2×2 phased-array transceiver for FMCW imaging radar, shown in Fig. 1. The transceiver consists of two receiving paths, two transmitting paths and an LO generation path. The transmitter and receiver share the same LO and RF phase shifting topology is adopted. In order to improve gain and noise performance, multi-paralleled transistors with small number-of-fingers are used to design PA and LNA. A 24GHz VCO is used to generate wideband FMCW signal, and two cascaded differential push-

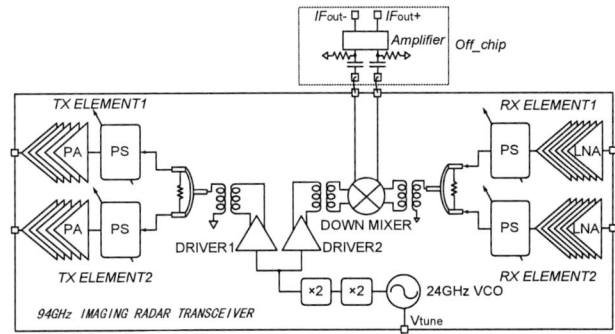

Fig. 1 Block diagram of the 2×2 phased-array FMCW imaging radar transceiver.

push cascode doublers are used to act as a quadrupler, which has a low power and high stability. PS with constant IL compensation technique [6] is used to achieve constant gain among different phase states.

Based on the above techniques, the 94GHz phased-array transceiver has a continuous frequency tuning range of 14GHz, a maximum transmitter output power of 11dBm, and a receiver NF of 10.5dB and conversion gain of 20dB.

II. CIRCUITS IMPLEMENTATION

A. Amplifiers with Multi-Paralleled Transistors with Small Number-of-Fingers

Because of the limited f_{max}, it is difficult to achieve a substantial gain for each stage using 65nm CMOS technique to design 94GHz amplifiers [4, 5], which results in more gain stages and power consumption. Therefore, it is important to improve the gain of CMOS amplifying stages. f_{max} is a crucial parameter to calculate the achievable gain of CMOS technique and f_{max} of a CMOS transistor is shown in (1), in which R_i, R_{load}, C_{gs} and g_m are transistor's gate resistor, load resistor gate-source capacitor and transconductance, respectively.

$$f_{max} = \frac{g_m}{4\pi C_{gs}} \sqrt{\frac{R_{load}}{R_i}} \qquad (1)$$

Therefore, decreasing R_i promotes f_{max}. The noise of an amplifier is also influenced by the gate resistor R_i and a smaller R_i leads to a better noise performance.

The relationship between the number-of-fingers of a transistor and R_i and f_{max} is analysed in [7], and an optimized number-of-fingers provides a small R_i and a maximum f_{max}. However, the analysis only considered a single transistor with different number-of-fingers and only aimed at providing a

978-1-7281-1702-7/19 $31.00 © 2019 IEEE

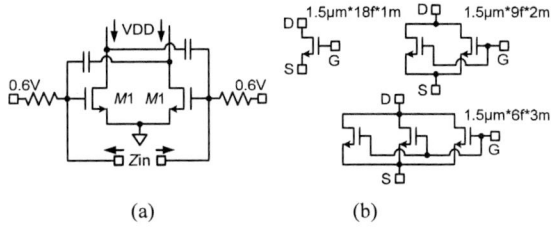

(a)   (b)

Fig. 2 (a) Differential transistors with neutralization technique. (b) Transistors M_i in (a) with same total width but different finger numbers and multipliers.

Table 1. R_i, C_{gs} @94GHz and f_{max} and G_{max} of transistors with same total width but different number-of-fingers and multipliers

	R_i (Ω)	C_{gs} (fF)	f_{max} (GHz)	G_{max} (dB)
1.5μm*18f*1m	12	33.1	181.6	8.4
1.5μm*9f*2m	9.68	34.3	195.1	9.57
1.5μm*6f*3m	9	35.3	196.6	10.14

considerable gain but didn't include power and noise performance, which is not enough when transistors with large total width must be used to provide large output power and low noise performance. We therefore analyse and simulate the gate resistor R_i and f_{max} of transistors with same total width but different number-of-fingers and multipliers.

Differential topology with neutralization technique is used in amplifier design, and input impedance (Z_{in}) of the transistor pairs is simulated, shown in Fig. 2(a). We simulated transistors with 1.5μm*18f (finger numbers)*1m (multipliers), 1.5μm*9f *2m and 1.5μm*6f *3m, shown in Fig. 2(b). R_i and C_{gs} of the transistors are shown in table 1. f_{max} and maximum achievable gain (G_{max}) are also obtained. It can be seen that multi-paralleled transistors with small number-of-fingers have smaller R_i and C_{gs} hardly changes. f_{max} and G_{max} of transistors with small number-of-fingers but large multipliers are also greater than transistors with large number-of-fingers and small multipliers. It can also be seen that the G_{max} of transistors with 1.5μm*6f *3m is 1.7dB greater than the G_{max} of transistors with 1.5μm*18f *1m.

A six stage LNA and five stage PA are designed with multi-paralleled transistors with a finger number of 6 and a width-per-finger of 1.5μm. The schematics are shown in Fig. 3. Fig. 4 shows the layouts of two differential transistor pairs with width of 1.5μm*6f*3m and 1.5μm*6f *9m in Fig. 3. The simulated performance of the LNA is shown in Fig. 5. In order to verify the performance of multi-paralleled transistors with small number-of-fingers, another LNA with transistors with a total width of 31.5μm, a finger number of 18 and a multiplier of 1 is also simulated. The former LNA achieves a gain of 40dB and an NF of 5.7dB. The latter LNA achieves a gain of 30dB and an NF of 7.2dB. Therefore, the simulated results show multi-paralleled transistors with small number-of-fingers can increase gain and decrease noise greatly. The large gain of the LNA decreases the influence of the IL of PS and 1/f noise of the down mixer. The PA achieves a simulated gain of 34dB and a maximum output power of 11.3 dBm.

B. Design of Wideband and Low Power LO

In FMCW radar, the resolution is mainly dependent on the bandwidth of the continuous modulated frequency. In order to achieve wide bandwidth, a wideband voltage-controlled oscil-

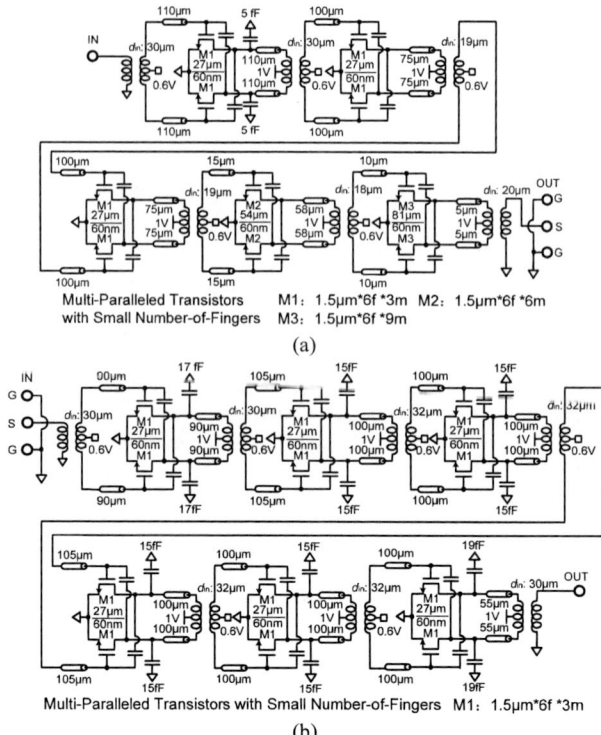

Multi-Paralleled Transistors with Small Number-of-Fingers   M1: 1.5μm*6f *3m   M2: 1.5μm*6f *6m   M3: 1.5μm*6f *9m

(a)

Multi-Paralleled Transistors with Small Number-of-Fingers   M1: 1.5μm*6f *3m

(b)

Fig. 3 Schematic of the (a) PA and (b) LNA with multi-paralleled transistors with small number-of-fingers.

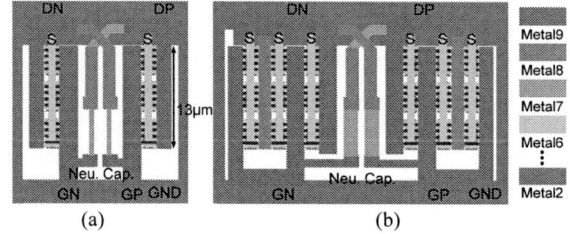

(a)   (b)

Fig. 4 Layouts of differential transistor pairs with total width of (a) 1.5μm*6f *3m and (b) 1.5μm*6f *9m.

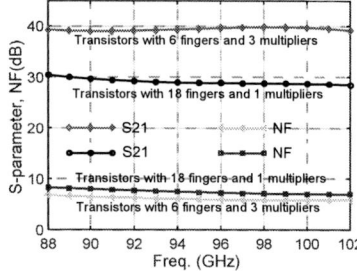

Fig. 5 Simulated gain and NF of LNAs with multi-paralleled transistors with small number-of-fingers and a single transistor with large number-of-fingers.

lator is designed to generate wideband signal [8], shown in Fig. 6. The center frequency is chosen as 24GHz and is quadrupled to 94GHz. A center frequency of VCO of 24GHz reduces the influence of parasitic capacitors of cross-coupled nMOS pair and the buffer, and therefore increases the value of varactors to enlarge the oscillating frequency range. A 50Ω series tail resistor is placed at the source of the cross-coupled nMOS pair to decrease phase noise [1]. The VCO has a simulated tuning range of 4GHz and a relative range of 16.7%. The simulated

2019 IEEE Radio Frequency Integrated Circuits Symposium

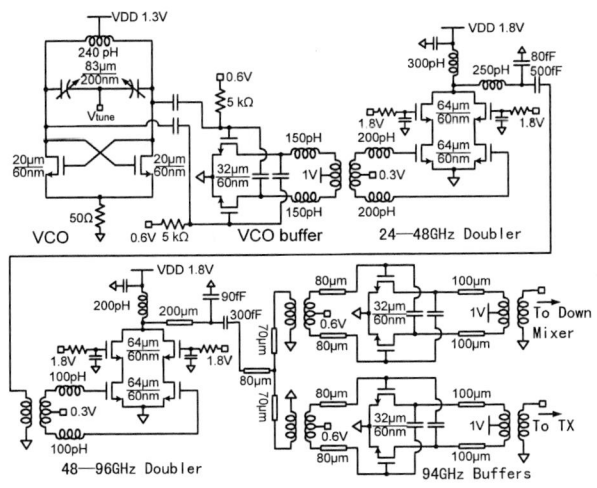

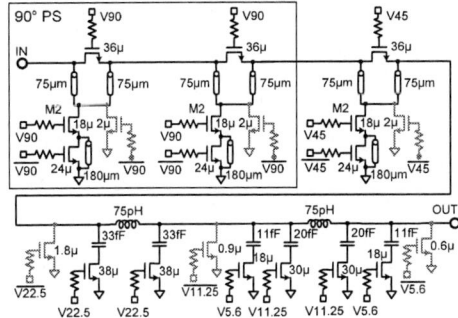

Fig. 6 Schematic of LO path.

Fig. 7 Schematic of 5-bit passive PS with constant IL compensation.

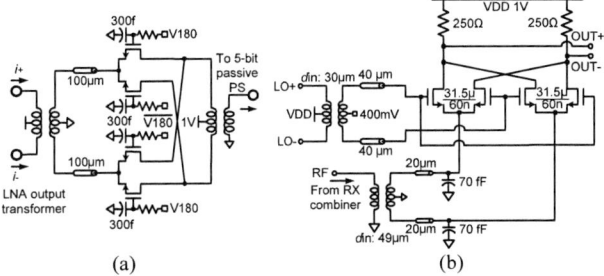

(a)   (b)

Fig. 8 Schematic of (a) the 180° active PS and (b) the down mixer.

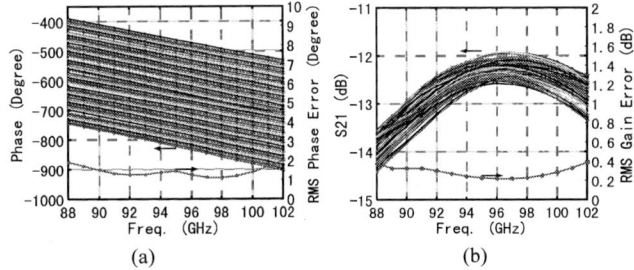

(a)   (b)

Fig. 9 Simulated (a) phase and (b) S_{21} of the 6-bit phase shifter.

phase noise is -108dBc at 1MHz offset from 24GHz. The FMCW signal is generated by tuning the control voltage V_{tune} in Fig. 6 and the nonlinearity between the frequency and V_{tune} can be calibrated by digital control [8].

Doublers with push-push cascode topology are used to decrease power and increase stability compared with common source topology. Decoupled capacitors are used at the gate of cascode transistors to improve gain and 1.8V V_{DD} and gate bias voltage of cascode transistors are used to improve gain. Transient simulation is done and the results show that the voltage swing at every point of the circuits is within 2V. The LO path consumes a total power of 113mW when it works.

C. Phase Shifters with Constant IL Compensation and Mixer

In this design, a 94GHz 360° 6-bit PS is designed. The 180° PS is achieved with active topology and the 90° to 5.6° PS are achieved with passive topology.

The 90° PS is designed with two cascaded 45° high-pass/direct-pass PSs, and the 22.5°, 11.25° and 5.6° PSs are achieved with *C-L-C* type, shown in Fig. 7. The gain varies with different phase states. In order to achieve constant gain among different phase states, constant IL compensation technique is used [6]. Each single-bit PS has an extra nMOS transistor and the transistor is used as an attenuator to connect the signal path to ground, shown in Fig .7 with red color. The gate of the transistor is controlled by the negative voltage of the phase control signal. When a single-bit PS is controlled not to shift phase, the transistor is closed. When the single-bit PS is controlled to shift phase, the transistor is on and attenuates the insertion loss to the non-phase-shifting-state. Constant IL compensation is also used in *C-L-C* topology and it can be derived that IL compensation won't induce extra phase variation. The 180° PS adopts active current steering topology, shown in Fig .8(a). Simulated phase and gain of the 6-bit PS are shown in Fig. 9. Root-mean-square (RMS) phase error is about 2°. The average gain of the PS is about -13dB. Because of the use of the IL compensation technique, the gain variation among different phase states is within 1dB over 14GHz bandwidth.

Fig. 8(b) depicts the down mixer with a Gilbert topology.

Both RF and LO signals are injected to the Gilbert unit through a transformer. Simulated voltage gain and 1dB input compression point (IP1dB) of the mixer are 0dB and 0dBm at 94GHz respectively, with a dc power consumption of 4mW.

III. MEASURED RESULTS

The chip photograph is shown in Fig. 10, with an area of 3.06mm × 1.38mm. The dc power consumption is 519mW.

Simulated and measured output power of transmitting paths are shown in Figure. 11. By tuning the V_{tune} of the VCO, the output power covers 14-GHz bandwidth with a maximum value of 11dBm. The output power variation over 14-GHz bandwidth is 2.8dB. The power difference between two paths is within 0.3dB.

According to link calculation, the IF signal of the output of the mixer is smaller than 5MHz and greater than 50kHz. Therefore, off-chip high-pass *C-R* filters with a cut-off frequency of 10kHz to eliminate the dc offset voltage and attenuate 1/f noise are used and an off-chip amplifier is also used to provide a proper gain and drive load, shown in Fig. 1. Fig. 12 shows the conversion gain (CG) and NF of two receiving paths. The measured maximum gain is 20dB excluding off-chip amplifier and 4dB smaller than simulated gain, which may be caused by inaccuracy of PDK model.

978-1-7281-1702-7/19 $31.00 © 2019 IEEE   49

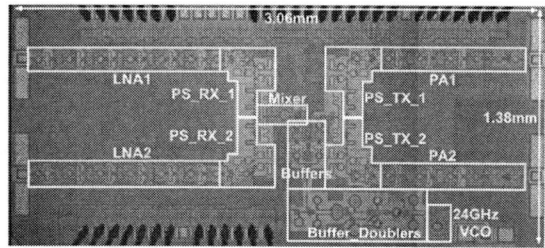

Fig. 10 Chip photograph of 2 × 2 phased-array FMCW imaging radar transceiver.

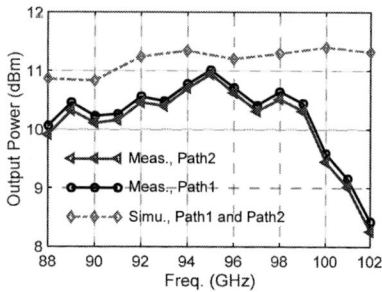

Fig. 11 Simulated and measured output power vs. frequency of transmitting paths.

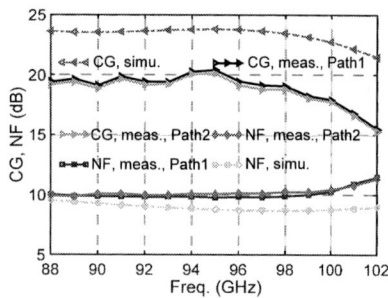

Fig. 12 Simulated and measured CG and NF vs. frequency of receiving paths.

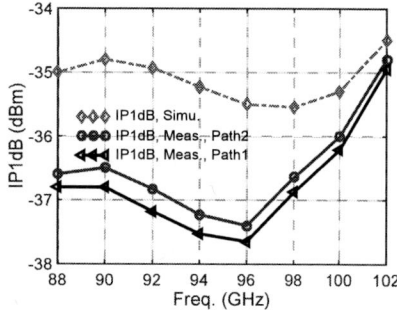

Fig. 13 Simulated and measured IP_{1dB} vs. frequency of receiving paths.

20dB maximum gain demonstrates that multi-paralleled transistors with small number-of-fingers promotes gain greatly. 3dB bandwidth is from 88GHz to 100GHz. NF is measured through an external 24GHz LO signal instead of the VCO and at an IF frequency of 3MHz. Measured NF (SSB) is 10.5dB and 1dB greater than simulated results. NF deteriorates with frequency beyond 100GHz because of the decrease of gain. Measured IP1dB is about -37dBm, shown in Fig. 13.

Table 2 shows a comparison of this work with other the-state-of-arts and this work achieves a maximum output power and continuous frequency tuning range.

Table 2. Performance Comparison with The-State-Of-Arts

Ref	[1]	[2]	[3]	[4]	This work
Tech.	130nm SiGe	120nm SiGe	130nm SiGe	65nm CMOS	65nm CMOS
Channels	4RX, 4TX	16TX, 32RX	4RX, 4TX	4RX, 4TX	2RX, 2TX
Freq. (GHz)	89-95	88-96	90-102	87.8-98.9	88-102
Max. Pout (dBm)	6.4	3.2	6.5	8	11
RX NF (dB)	12.5	8.2	16.5	10.5	10.5
RX CG (dB)	25–38	30	7	26-40	20(excluding baseband)
PS resolution(°)	9	11.25	-	2.8	5.6
TX power per ch.(mW)	135	181	147	150	175*
RX power per ch.[mW]	130	138	101	90	148*
Level of Integration	TX+RX+ mixer+VCO +divider	TX+mixer +RX+PLL	TX+RX + mixer+LO buffer	TX +RX +LO	TX+RX+ Mixer+LO

$*P_{channel}=(1/2)(P_{TX\,or\,RX}+P_{LO})$

IV. CONCLUSION

A 94GHz 2 × 2 phased-array FMCW imaging radar transceiver is presented in this paper. Multi-paralleled transistors with small number-of-fingers are used to design PA and LNA to improve gain and noise performance. LO path is designed with wide bandwidth and low power. PS with constant IL compensation is used to achieve constant gain among different phase states. The transceiver has a continuous frequency tuning range of 14GHz and the transmitter has a maximum output power of 11dBm and 2.8dB power flatness. The receiver has a CG of 20dB and an NF of 10.5dB.

ACKNOWLEDGMENT

This work is supported by National Natural Science Foundation of China under grants 61574084 and National Major Research Program 2016YFA0201903.

REFERENCES

[1] A. Townley, et al., "A 94-GHz 4TX–4RX phased-array FMCW radar transceiver with antenna-in-package," IEEE J. Solid-State Circuits, vol. 52, no.5, pp. 1245-1259, May 2017.

[2] A. Natarajan, et al., "W-band dual-polarization phased-array transceiver front-end in SiGe BiCMOS," IEEE Trans. Microw. Theory Techn., vol. 63, no. 6, pp. 1989–2002, Jun. 2015.

[3] M. Jahn, et al., "A four-channel 94-GHz SiGe-based digital beamforming FMCW radar," IEEE Trans. Microw. Theory Techn., vol. 60, no. 3, pp. 861-869, Mar. 2012.

[4] P. Peng, et al., "A 94 GHz 3D image radar engine with 4TX/4RX beamforming scan technique in 65 nm CMOS technology," IEEE J. Solid-State Circuits, vol. 50, no.3, pp. 656-668, Mar. 2015.

[5] K. Tsai, et al., "A W-band Power Amplifier in 65-nm CMOS with 27GHz bandwidth and 14.8dBm saturated output power," in Proc. IEEE RFIC Symp., Jun. 2012, pp. 69–72.

[6] D. Huang, et al., "A 60-GHz 360° 5-bit phase shifter with constant IL compensation followed by a normal amplifier with ±1 dB gain variation and 0.6 dBm OP-1dB," IEEE Trans. Circuits Syst. II, Exp. Briefs, vol. 64, no. 12, pp. 1437-1441, Nov. 2017.

[7] C. Ko, et al., "A 210-GHz amplifier in 40-nm digital CMOS technology," IEEE Trans. Microw. Theory Techn., vol. 61, no. 6, pp. 2438–2446, Jun. 2013.

[8] A. Mostajeran, et al., A 170-GHz fully integrated single-chip FMCW imaging radar with 3-D imaging capability. IEEE J. Solid-State Circuits, vol. 52, no. 10, pp. 2721-2734, Oct. 2017.

X/Ku-Band Four-Channel Transmit/Receive SiGe Phased-Array IC

Prabir Saha[#1], Sriram Muralidharan[#], Jinzhou Cao[#], Ozan Gurbuz[#], Christopher Hay[$]

[#]North-West Labs, Analog Devices Inc., USA

[$]Colorado Springs, Analog Devices Inc., USA

[1]prabir.saha@analog.com

Abstract—This paper presents a phased-array core chip integrating four transmit/receive channels and supporting circuitry on a single chip working over 8–16 GHz, covering X and Ku frequency bands. Each RX and TX channel has a VGA and a phase shifter featuring precise amplitude and phase control for RF beamforming. The phase-compensated VGA provides 16 dB of gain control range in 0.5 dB steps while limiting the corresponding phase variation to 3°. The phase shifter, based on the principles of a vector modulator, provides 360° phase control range with resolution better than 2.8°, rms phase and gain error less than 2.8° and 0.3 dB respectively. Each RX channel has more than 10 dB of gain, and the noise figure for maximum gain condition is better than 10 dB. In receive mode, input P1dB is -16 dBm per channel and in transmit mode, output P1dB is greater than 9 dBm over 8–16 GHz. Each RX and TX channel draws 65 mA and 85 mA respectively from a 3.3 V supply. The chip, fabricated in a 0.18 μm SiGe BiCMOS technology, occupies 4mm X 4mm and was put in a 7mm x 7mm LGA package.

Keywords—Phased array, Core-chip, Radar, AESA, VGA, Phase shifter, Beamforming.

I. INTRODUCTION

Electronically steered antenna arrays have the ability to steer a directed beam much faster than the mechanical counterparts. They can also alleviate problems due to interference by controlling placement of nulls in the radiation pattern. Despite being very well-suited for a wide range of applications including different kinds of radars and communication systems [1], use of phased arrays beyond the defense sector is rather limited due to high cost, which is primarily driven by the use of discrete components in existing solutions.

In this paper, we present a highly-integrated four-channel phased array core-chip with accurate and near-orthogonal control of phase and gain over a wide range of frequencies (8–16 GHz). The chip was implemented in a 0.18 μm SiGe BiCMOS technology. A simplified block diagram of the phased array core-chip is shown in Fig. 1. Four transmit/receive channels, each with independent gain and phase control and a 4:1 Wilkinson power combiner/splitter provide the core functionality necessary for RF beamforming. Each channel also has a low-noise amplifier (LNA) and a medium power amplifier (MPA) to provide low noise figure (NF) and high output power for the receive and transmit path respectively. For applications requiring lower NF or higher output power, higher-performance GaAs or GaN based transmit/receive (T/R) modules can be used in conjunction with the SiGe core-chip to achieve required performance. In addition to the core functionality, several other useful features

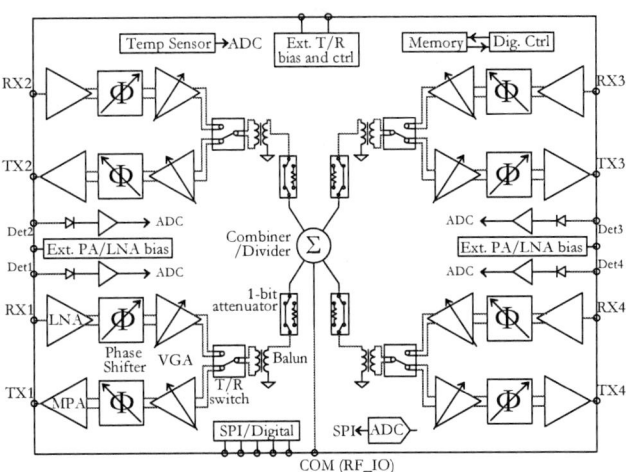

Fig. 1. Simplified block diagram of the four-channel phased array IC.

were integrated on this chip, including bias and control for external T/R modules if needed, power detectors, temperature sensors, ADC for digital readout from the sensors, memory for storing beam positions etc. Digital control of various aspects of the chip including gain/phase control and bias of the different sub-blocks are provided through a 4-wire SPI interface. A dedicated T/R control is also included for faster switching between transmit and receive modes. The level of integration and performance will make it easier and cheaper to build, calibrate and maintain phased arrays.

II. DESIGN DETAILS

Design details of the various sub-blocks inside each channel are discussed in the following sub-sections.

A. LNA and MPA

The LNA (Fig. 2) and the MPA (Fig. 3) are cascode amplifiers with a resistive feedback to widen the bandwidth of operation. On-chip transformer baluns were used at the input of the LNA and at the output of the MPA for single-ended to differential conversion. The balun, along with shunt capacitors, was designed to improve impedance matching over the wide bandwidth. The power amplifier has two stages to provide more gain and output power.

B. Phase Shifter

The phase shifter, shown in Fig. 4 is based on the principle of vector modulator. The input signal is split into two orthogonal components, referred to as in-phase (I) and quadrature (Q), using the quadrature filter [2]. The I and Q

978-1-7281-1702-7/19 $31.00 © 2019 IEEE

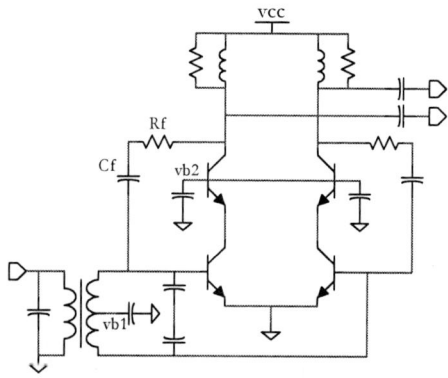

Fig. 2. Schematic of the low noise amplifier with input balun.

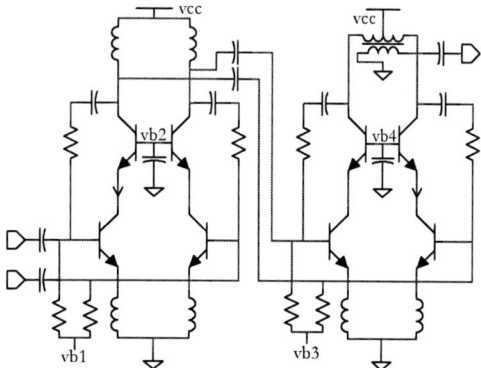

Fig. 3. Schematic of the medium power amplifier.

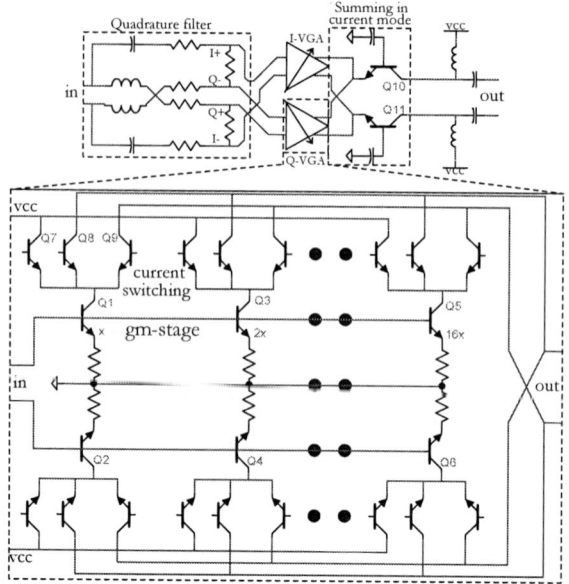

Fig. 4. Simplified schematic of the vector modulator based phase shifter and the VGA used in it.

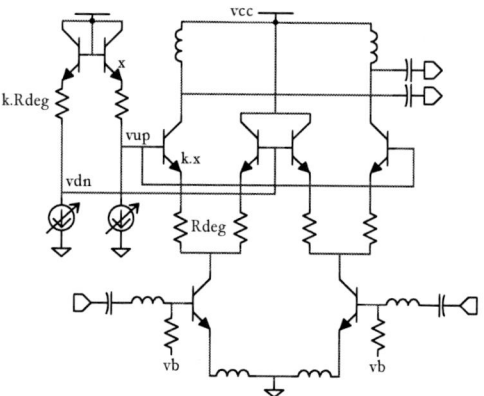

Fig. 5. Simplified schematic of the phase-compensated VGA.

signals are then scaled using two bi-phase VGAs and summed at the output to produce a phase shifted version of the input signal. By changing the gains of the VGAs, the phase of the output signal can be changed.

The VGA consists of 5 binary-weighted sections in parallel. Inside each section, the current from the gm-stage either flows to the output or is switched to the supply by digital control. The digital control also allows switching the sign of the output current, which enables operation in all four phase quadrants. For each VGA, there are 5 bits for changing the amplitude from zero to full-scale and 1 bit controls the sign. In this VGA architecture, all the active transistors in the signal path operate at a constant current density, irrespective of the gain setting. This translates to more uniform performance over various phase settings as compared to the traditional approach where the bias current is scaled to change the I and Q weights. This approach also allows the phase shifter to maintain constant phase shift up to about P1dB, whereas the current-scaled implementations suffer from significant AM/PM distortion for certain phase settings when the bias current gets small.

The output of the two VGAs are summed in current mode and fed to a common-base stage followed by an LC matching network at the output. The common-base stage helps to reduce parasitic capacitance at the output, which is important for wideband operation. The low input impedance of the common-base stage also helps to reduce inaccuracies

due to varying VGA output impedance with changing gain.

C. Phase Compensated VGA

The VGA, shown in Fig. 5 is based on a current-steering topology. Voltages at the base of the differential pairs (vup and vdn) control the fraction of current from the g_m-stage flowing at the outputs. The current-steering differential pairs are degenerated using resistors (R_{deg}) to reduce phase variation with changing current density. The underlying principle is same as described in [3], where the g_m transistor was degenerated to achieve phase invariance. One significant advantage of applying the phase compensation in the cascode pairs as opposed to the g_m transistor is the fact that it does not affect the gain control range. The input impedance also remains constant for all gain settings in this topology as the current density of the input g_m transistor does not change.

The VGA together with the phase shifter provides precise and near-orthogonal control of phase and amplitude.

2019 IEEE Radio Frequency Integrated Circuits Symposium

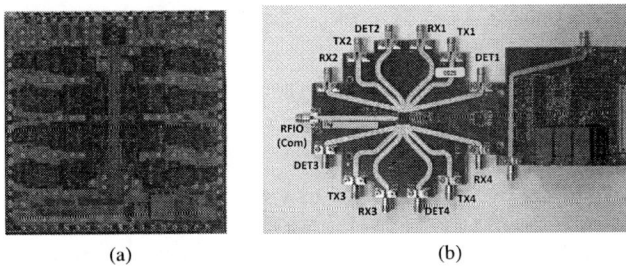

Fig. 6. Pictures of (a) the die, (b) the circuit board with the packaged IC.

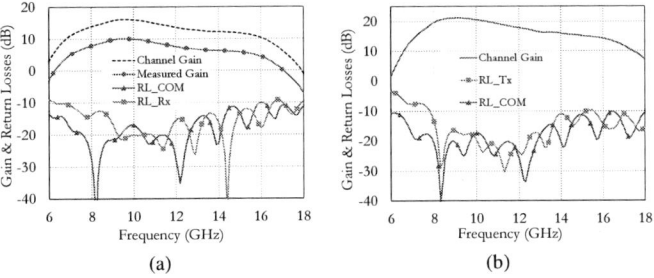

Fig. 7. Gain for a single channel in: (a) receive and (b) transmit modes. In receive mode, exciting one receive channel and measuring at the common port results in the "Measured Gain" plot. Actual gain of the channel is 6 dB higher when all four channels are coherently excited.

III. MEASUREMENT RESULTS

The chip was fabricated in a 0.18 μm SiGe BiCMOS technology. A die photo is shown in Fig. 6a. The chip was put in a 7 mm X 7 mm LGA package and mounted on a circuit board (Fig. 6b) to perform the measurements. Rogers 4350 was used in the top layer of the board for the high-frequency traces.

Gain and return losses in the receive and transmit modes are shown in Fig. 7a and Fig. 7b. Average gain in RX and TX modes are 12 and 17 dB respectively. As shown in Fig. 8, the VGA, controlled by a 7-bit word, has a gain range of 23 dB; although, only the top 16 dB of that range will normally be used. The 1-bit attenuator provides an additional 16 dB of gain control. The phase variation corresponding to the normal gain range of the VGA is less than 3° (Fig. 9).

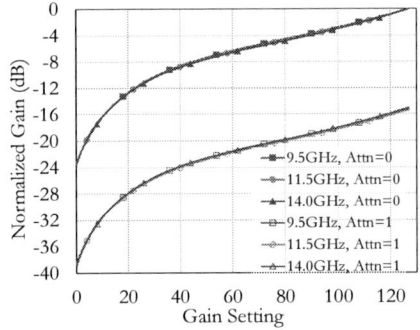

Fig. 8. Gain control in RX mode using the VGA and the 1-bit attenuator. Gain numbers are normalized with respect to the highest gain and plotted against the decimal equivalent of the 7-bit gain control word. Combined gain control range of 31 dB in 0.5 dB steps.

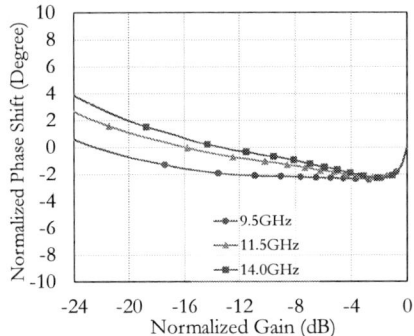

Fig. 9. Phase varies by less than 3 degree for the top 16 dB of the gain control range of the VGA.

The phase control of the channels were characterized from 0-360° in 2.8125° steps. Relative phase shift for all phase states, referenced to the 0° state, are plotted in Fig. 10. No cross-over was detected over the frequency range of 8–16 GHz. For these measurements, The I-Q phase imbalance was measured and the phase shifter settings were adjusted accordingly. Other than that, no other calibration was performed.

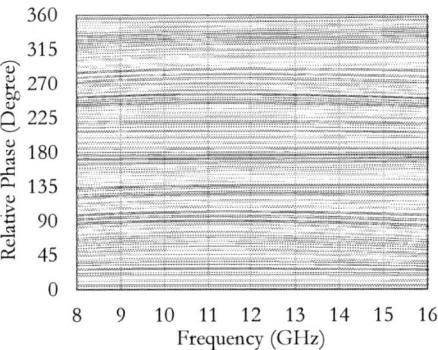

Fig. 10. Relative phase shift vs. frequency over 0° to 360° in 2.8125° steps. 0° was used as the reference for this plot. No cross-over was detected over 8 to 16 GHz.

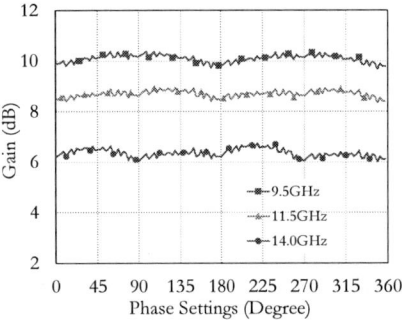

Fig. 11. Variation in gain at different frequencies for all phase settings. Peak to peak gain variation is about 0.5 dB for 0-360 degree phase change; corresponding rms gain error is less than 0.3 dB.

Gain variation with change in phase, as shown in Fig. 11, was limited to ±0.3 dB. The rms phase and gain errors over phase states for the receive channel are shown in Fig.

978-1-7281-1702-7/19 $31.00 © 2019 IEEE   53

12. Over 8 to 16 GHz, the rms phase error was below 2° for most of the frequency range, climbing up to 2.8° at 16 GHz. Gain and phase control performance for the transmit channel was similar to the receive channel and is not shown here for brevity.

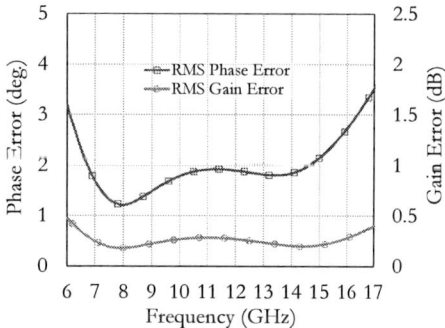

Fig. 12. RMS phase and gain errors in receive mode over all phase states for 7-bit precision vs frequency. RMS phase error stays below 2° for most of the frequency range and RMS gain error is less than 0.3 dB.

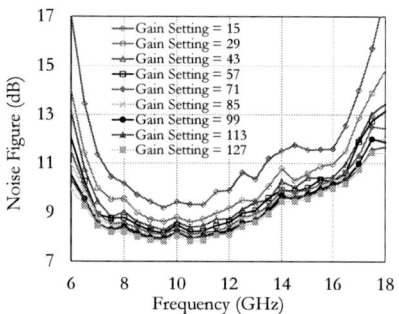

Fig. 13. Noise figure of the receive channel.

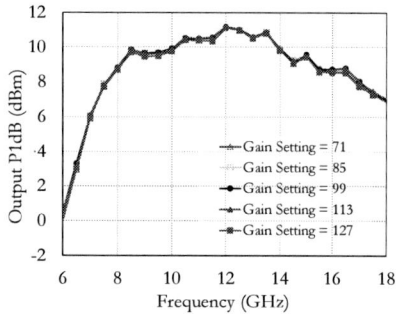

Fig. 14. Output P1dB of the transmit channel.

Measured noise figure for the receive channel was below 10 dB (Fig. 13) including package loss. In a large array, the effective noise performance of the array will improve significantly over that of a single channel. Output P1dB of the transmit channel was measured to be greater than 8.5 dBm over 8-16 GHz (Fig. 14) and saturated power was more than 13 dBm.

IV. CONCLUSION

A wideband (8–16 GHz) four-channel phased array IC implemented in a SiGe BiCMOS technology was presented.

Table 1. Key performance metric of the phased array core-chip and comparison with prior work.

Parameter	[4]	[5]	[6]	[7]	This Work
Technology	0.18 μm SiGe BiCMOS	0.13 μm CMOS	0.13 μm CMOS	0.13 μm SiGe BiCMOS	0.18 μm SiGe BiCMOS
Frequency (GHz)	6–18	7.9-9.6	8.5–10.5	9–11	8–16
Gain RX/TX (dB)	10–24.5/–	–/11.5	3.5/3.5	25/22	10–16/15–21
Gain range /step (dB)	–	5/3-bit	31/1	–	31/0.5
Phase resolution	22.5°	22.5°	5.6°	11°	2,8°
Phase Error RMS	<5.7°	5°	4.3°	3.8°	<2.8°
Gain Error RMS (dB)	0.9	0.5	0.8	1.2	0.3
RX NF (dB)	4.2–11	–	7.5	3	<10
TX OP1dB (dBm)	–	8.8	6.5	28	>8.5
RX/TX chan. Power (mW)	58/–	195/195	154/154	352/4128	215/280

Excellent phase and gain accuracy over a wide range of frequencies and near-orthogonal control are distinguishing features of this core-chip. In addition to the core functionality, various useful features were integrated on the chip which will make building phased arrays easier and cheaper and hopefully result in more widespread adoption.

ACKNOWLEDGMENT

We would like to thank Sue Nuthman, Ann Carbonari and Rod Dobler for layout and Ron Simonson, Quy Tran for characterization.

REFERENCES

[1] R. Mailloux, *Phased Array Antenna Handbook*, ser. Antennas and Propagation Library. Artech House, 2005.

[2] S. Y. Kim, D. Kang, K. Koh, and G. M. Rebeiz, "An Improved Wideband All-Pass I/Q Network for Millimeter-Wave Phase Shifters," *IEEE Transactions on Microwave Theory and Techniques*, vol. 60, no. 11, pp. 3431–3439, Nov 2012.

[3] B. Sadhu, J. F. Bulzacchelli, and A. Valdes-Garcia, "A 28GHz SiGe BiCMOS Phase Invariant VGA," in *2016 IEEE Radio Frequency Integrated Circuits Symposium (RFIC)*, May 2016, pp. 150–153.

[4] K. Koh and G. M. Rebeiz, "An X- and Ku-Band 8-Element Phased-Array Receiver in 0.18-μm SiGe BiCMOS Technology," *IEEE Journal of Solid-State Circuits*, vol. 43, no. 6, pp. 1360–1371, June 2008.

[5] D. Shin, C. Kim, D. Kang, and G. M. Rebeiz, "A High-Power Packaged Four-Element *X*-Band Phased-Array Transmitter in 0.13-μmCMOS for Radar and Communication Systems," *IEEE Transactions on Microwave Theory and Techniques*, vol. 61, no. 8, pp. 3060–3071, Aug 2013.

[6] S. Sim, L. Jeon, and J. Kim, "A Compact X-Band Bi-Directional Phased-Array T/R Chipset in 0.13μmCMOS Technology," *IEEE Transactions on Microwave Theory and Techniques*, vol. 61, no. 1, pp. 562–569, Jan 2013.

[7] C. Liu, Q. Li, Y. Li, X. Deng, X. Li, H. Liu, and Y. Xiong, "A Fully Integrated X-Band Phased-Array Transceiver in 0.13-μmSiGe BiCMOS Technology," *IEEE Transactions on Microwave Theory and Techniques*, vol. 64, no. 2, pp. 575–584, Feb 2016.

Ultra-Wideband 8-45 GHz Transmitter Front-End for a Reconfigurable FMCW MIMO Radar

Mantas Sakalas[#], Songhui Li[#], Niko Joram[#], Paulius Sakalas[*], Frank Ellinger[#]

[#]Chair for Circuit Design and Network Theory, Technische Universität Dresden, Germany
[*]Chair for Electron Devices and Integrated Circuits, Technische Universität Dresden, Germany
mantas.sakalas@tu-dresden.de

Abstract—This paper presents an ultra-wideband 8-45 GHz FMCW MIMO radar transmitter (Tx) front-end, designed in a 130 nm SiGe BiCMOS technology. Various design techniques were applied to achieve the ultra-wide bandwidth, high linearity and the output power, which is at the edge for the given technology. The measured output power at 1 dB compression point was $15-20$ dBm in a frequency range of $8-45$ GHz and 3 dB output power flatness of 17-20 dBm was achieved for a $9-41$ GHz bandwidth. The compact design features a total active IC area of 2.72 mm² and a peak overall system PAE of 9 %. The total DC power consumption was 1.45 W.

Keywords— BiCMOS integrated circuits, MIMO radar, silicon germanium, transmitters, ultra wideband technology.

I. INTRODUCTION

An ability to operate a radar in a vast range of frequencies gives several major advantages. An operator can define the frequency with respect to the standards, licensed at a specific location, improve the detection accuracy by simultaneously utilizing different frequency bands [1, 2], or even to take use of frequency hopping [3, 4]. However, designing a wideband radar front-end hardware is challenging, since the design steps are followed by multiple performance trade-offs.

A number of publications dealing with wideband radar transmitters (Tx), front-ends and components can be found [5]–[8]. However, each of these publications deals with a limitation in terms of bandwidth, transmit power, integration and technology cost.

Addressing these issues, this work focuses on designing an FMCW MIMO imaging radar Tx front-end that covers an ultra-wide bandwidth, delivers cutting edge output power for the technology, has a good degree of integration and a low technology cost.

II. DESIGN TECHNIQUE

The Tx front-end system consists of 4 major ultra-wideband components, namely an active balun, a frequency multiplier, a driver amplifier and a power amplifier (Fig. 1).

A frequency multiplication is used to achieve the required bandwidth and to overcome the oscillator pulling problem. The

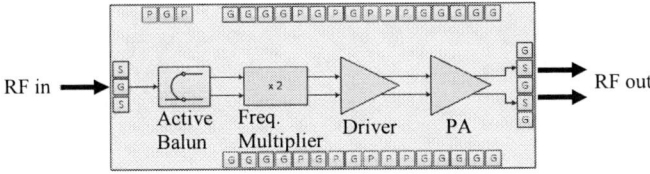

Fig. 1. The top-level schematic and brief description of the proposed MIMO radar Tx front-end.

multiplication degrades the phase noise by 6 dB, however at lower frequencies higher quality factor of the LC-tank and hence a lower phase noise of the voltage controlled oscillator (VCO) can be achieved.

A. Active Balun

On the input side, the Tx front-end receives single ended chirp signals, generated by an N-fractional PLL, consisting of broadband VCO cores, such as [9]. The power level of the chirp signal is 0 dBm and the frequency range is $4-22.5$ GHz.

For driving the frequency multiplier, the active balun needs to deliver a very flat output power in the range from $0-1$ dBm. This is because the frequency multiplier uses a 2nd harmonic component to multiply the frequency. Consequently, increasing the input power level by 1 dB, raises the output power by approximately 2 dB. Furthermore, the single ended input Tx front-end requires a good broadband matching to mitigate the mismatch and performance degradation introduced by the bond-wire, when bonded to a PCB.

To address the above issues, a topology of a single stage, distributed active balun was chosen (Fig. 2).

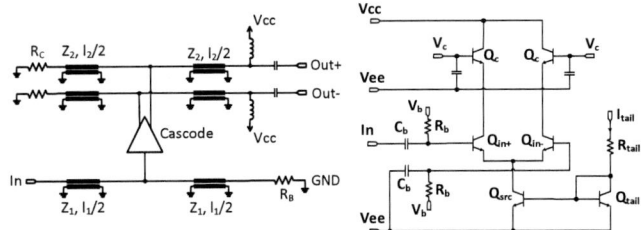

Fig. 2. Single stage distributed active balun. Overall circuit view (left) and the cascode stage (right).

The core of the circuit consists of a differential pair that has one side connected to the input and the other side connected to the ground. Both connections are implemented through bias-tees to enable the transistor biasing. For maintaining the flat gain response, a cascode configuration is used. The broadband impedance matching is implemented by using a distributed technique. The input transistor (Q_{in+}) is matched by two transmission lines (TLs), each of length $l_1/2$ and characteristic wave impedance Z_1. With the termination resistor R_B, the overall structure forms an ultra-wideband 50 Ω TL. The same approach is used to implement the output matching, except the differential output TLs are designed to form an overall impedance equal to the input of the following multiplier circuit, rather than 50 Ω. The circuit is biased for class A operation to enable the maximum output voltage swing and hence to

978-1-7281-1702-7/19 $31.00 © 2019 IEEE

increase the linearity. It consumes 30 mA from a 3 V power source.

B. Frequency Multiplier

The frequency multiplier receives a differential, 0 dBm, $4 - 22.5$ GHz input signal. The required output frequency is $8 - 45$ GHz along with best possible conversion gain flatness.

Conventional frequency multiplier topologies employ either differential pairs or gilbert cell mixers with the RF signal partially fed to the LO ports [10, 11]. However, for large input signals the power consumption of such active topologies is relatively high and the common emitter/source, (or common collector/drain) input stages make the broadband impedance matching complicated. To overcome these issues, a quasi-passive, balanced frequency multiplier was designed (Fig. 3).

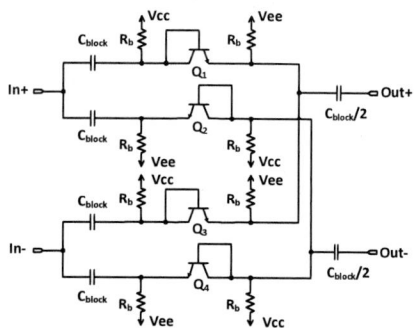

Fig. 3. The proposed quasi-passive frequency multiplier topology.

This multiplier makes use of diode connected bipolar devices to extract the 2nd harmonic frequency component. The bias voltage supply is implemented by using the resistors (R_b) and the blocking capacitors (C_{block}). The cross-coupled output structure serves for suppressing the fundamental wave and the odd harmonic signals (Fig. 4).

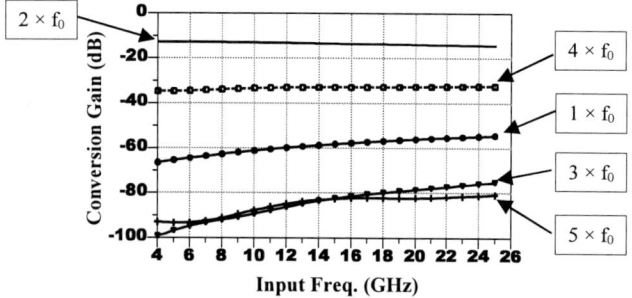

Fig. 4. Simulated gain response of the proposed frequency multiplier. The input drive level was 0 dBm. Results based on electromagnetic layout simulation.

The diode connected bipolar devices were chosen instead of the diodes due to the more accurate model provided within the process development kit (PDK). The conversion gain flatness was 1.25 dB and the input matching was better than -10 dB for the targeted frequency and impedance domain. This circuit consumes a total of 6 mA from a 3 V power source.

C. Driver Amplifier

The task of the driver amplifier is to compensate for the frequency multiplier loss, to reject the common mode signal, coming from the frequency multiplier and to provide sufficient output power in order to drive the power amplifier (PA). The required gain is 14 dB, since the power level at the output of the frequency multiplier is -12 dBm and the power amplifier requires a 2 dBm input drive. Furthermore, the delivered power needs to be flat vs. frequency and the harmonic components provided by the frequency multiplier need to be suppressed.

The strongest harmonic component passed from the multiplier originates at $4 \times f_0$ with respect to its input frequency. This can also be seen as the 2nd harmonic for the driver amplifier stage. Hence, in order to minimize the impact of such even harmonic frequency components, a symmetric, fully differential driver topology was implemented (Fig. 5 a, b).

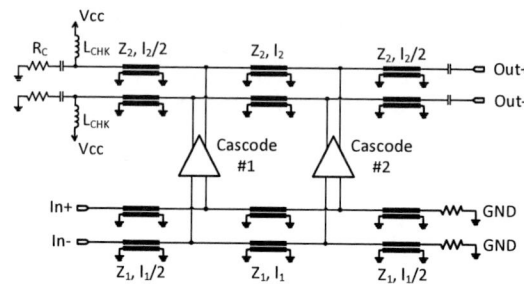

Fig. 5a. Fully differential, two stage driver amplifier. Overall circuit view.

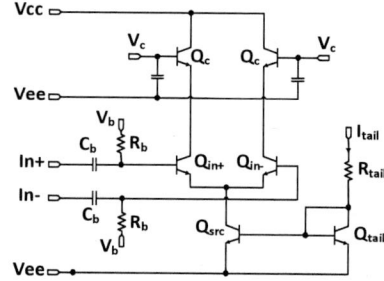

Fig. 5b. Fully differential, two stage driver amplifier. Cascode stage.

The DC power is supplied via the choke inductors L_{CHK} designed to have a self-resonance frequency just above 45 GHz. These inductors also serve as low pass filters, grounding the undesired harmonic frequencies. The symmetry of the circuit was optimized by means of electromagnetic (EM) simulations, the power output of the 2nd harmonic (4th multiplier harmonic) was between $45 - 60$ dB below the fundamental wave at the input drive level of $P_{IN} = -12$ dBm. The designed driver amplifier also demonstrated flat gain response of $12 - 14$ dB within the $8 - 45$ GHz frequency band (Fig. 6).

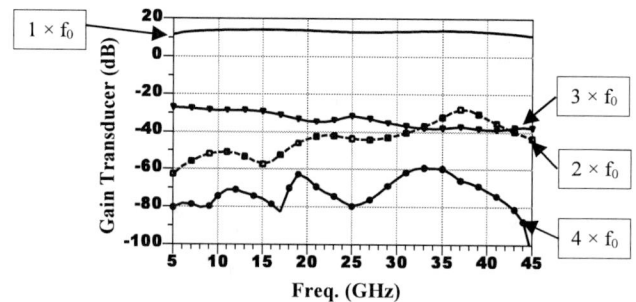

Fig. 6. Simulated gain response of the driver amplifier. The input drive level was -12 dBm. Results based on electromagnetic layout simulation.

The output referred 1 dB compression point ($P_{OUT, 1dB}$) was reached at the output power of 7.2 dBm, the circuit consumes 41 mA from a 3 V power source. Folded layout and patterned ground structure was used to minimize the IC size.

D. Power Amplifier

The core specifications for the PA were to achieve the maximum possible output power, as close to 20 dBm as possible, and to operate at an ultra-wideband frequency range of at least $10 - 40$ GHz.

A distributed differential PA topology was employed (Fig. 7a, b), featuring 4 cascode stages with stacked bipolar devices to increase the headroom for the output voltage swing. The DC power supply was implemented using $\lambda/4$ DC feeds. A total of 4 DC feeds were used to achieve a nearly uniform supply power distribution. The wavelength (λ) was chosen with respect to the middle frequency of 25 GHz.

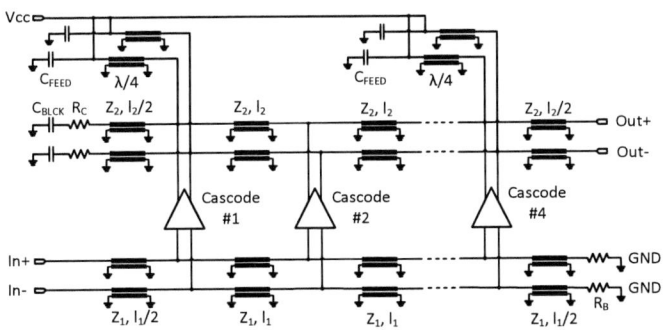

Fig. 7a. The distributed, differential 4 stage PA. Overall circuit view.

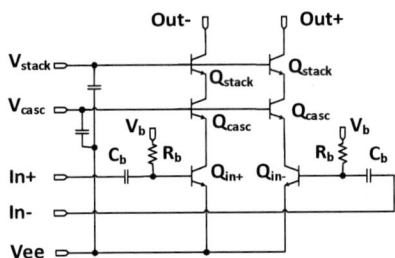

Fig. 7b. The distributed, differential 4 stage PA. Cascode stage.

With respect to [7], [12] the total optimum output power of a distributed PA can be expressed by

$$P_{OUT} = \left(\frac{1}{1 + R_{OPT} / R_C} + (N - 1) \right) P_{MAX}. \quad (1)$$

Here P_{MAX} describes the maximum output power of each cascode stage, N is the number of cascode stages, R_C is the collector line termination resistor and R_{OPT} is the optimum load impedance, described as the ratio of the transistor knee voltage (V_K) subtracted from the breakdown voltage (V_{CEO}) and divided by the maximum collector current ($I_{C,MAX}$):

$$R_{OPT} = \frac{V_{CEO} - V_K}{I_{C,MAX}} \quad (2)$$

It is obvious that for $R_C \rightarrow \infty$ the maximum output power is achieved. However this condition strongly degrades the output matching. Considering the above case (Fig. 7a), R_{OPT} amounts

to approximately 220 Ω and the PA has N = 4 stages, hence the optimum output power for $R_C = 50$ Ω and $R_C \rightarrow \infty$ would result in: $P_{OUT, 50 \Omega} = 3.18 \times P_{MAX}$ and $P_{OUT, \infty} = 4 \times P_{MAX}$ respectively. The simulated output power and the return loss (RL) vs. R_C of the PA circuit is shown below (Fig. 8).

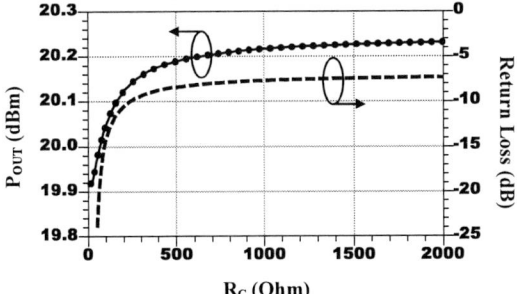

Fig. 8. Simulated P_{OUT} (circles) and RL (dashed) vs. R_C at 25 GHz frequency.

A compromise was met by setting $R_C = 300$ Ω, which gave an optimum output power of $P_{OUT, 300 \Omega} = 3.6 \times P_{MAX}$ and a good output matching of RL < -8 dB.

III. FABRICATION AND MEASUREMENT RESULTS

The above described MIMO radar Tx front-end was fabricated in 130 nm IHP SG13S SiGe BiCMOS technology, featuring f_T/f_{MAX} of 340 and 250 GHz respectively. The layout was partially folded and patterned TL ground was used to make the design more compact (Fig. 9).

The total fabricated IC size is 3.84 mm^2, whereas the active area is 2.72 mm^2.

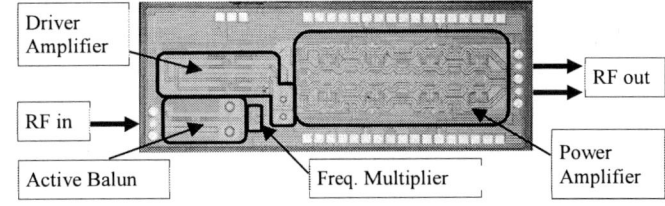

Fig. 9. Fabricated Tx front-end IC photograph and brief description.

Measurements of key performance parameters of the Tx front-end are presented below. Fig. 10 shows the results vs. the output frequency ($2 \times f_0$), Fig. 11 shows the output power and power added efficiency (PAE) vs. the input drive level and Fig. 12 presents the measured phase noise at 1 MHz offset frequency.

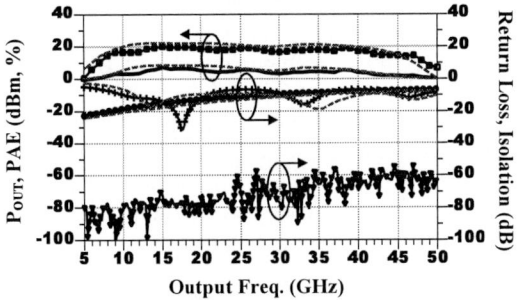

Fig. 10. Measured P_{OUT} (squares) and PAE (solid) at 1 dB compression point, measured Input (circles) and output (crosses) Return Loss and measured output to input isolation (triangles). The corresponding simulation results are in given dashed blue.

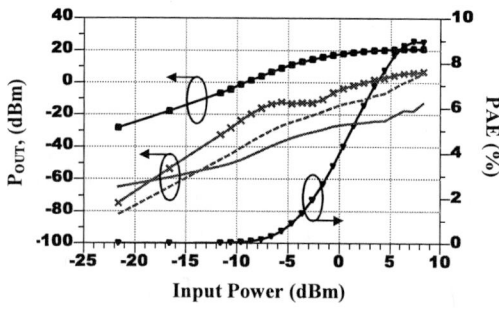

Fig. 11. Measured P_{OUT} (squares) and PAE (triangles) vs. the input drive level at fundamental output frequency ($2 \times f_0 = 10$ GHz). The unwanted power components are given in blue: $1 \times f_0$ (solid), $3 \times f_0$ (dashed) and $4 \times f_0$ (crosses).

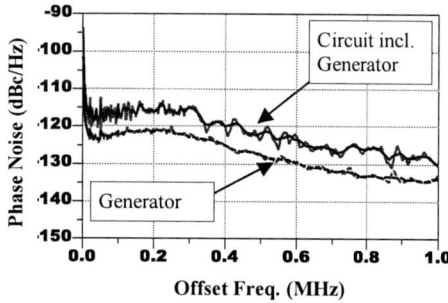

Fig. 12. Measured Tx front-end contribution to the phase noise at input drive level of 0 dBm and fundamental output frequency of ($2 \times f_0 = 25$ GHz). The measurements performed with (black) and without (blue) averaging.

The measured output power at the 1 dB compression point is between $17 - 20$ dBm for $9 - 41$ GHz and between $15 - 20$ dBm for the $8 - 45$ GHz frequency range. The input and output matching are better than -11 dB and -7 dB respectively, the overall peak PAE of the front-end was measured to 9 % and the output to input isolation was better than -60 dB. The suppression of the unwanted input frequency components was better than 21 dBc for $4 \times f_0$, 31.5 dBc for $3 \times f_0$ and 44 dBc for $1 \times f_0$. The measured Tx front-end contribution to the phase noise is approximately 6 dB/Hz, which results from frequency multiplication. The total DC power consumption was 1.45 W, where the active balun, the frequency multiplier and the driver amplifier together consumed 78 mA from a 3 V DC source and the power amplifier drew 270 mA from a 4.5 V DC source.

IV. CONCLUSION

An ultra-wideband FMCW MIMO radar Tx front-end was designed. A distributed, differential design technique was employed to achieve high bandwidth and to suppress the unwanted even harmonic components, transistor stacking and increasing the collector line termination resistance were used to increase the linearity and the output power, and layout folding was employed to reduce the IC size. Up to 20 dBm output power was reached, which is at the edge of what has been reported for the given IC technology. Furthermore the achieved $8 - 45$ GHz bandwidth proves the feasibility of the design.

This Tx front-end will be used as an enabler solution for designing a very high performance MIMO radar receiver.

ACKNOWLEDGMENT

The research leading to these results has received funding from the German Federal Ministry of Education and Research (BMBF) within the frame of project SIGMA 5G, from the European Union's Horizon 2020 Research and Innovation Program under project RANGER with Grant Agreement n°[700478], under project TARANTO with Grant Agreement n°[737454] and under DFG project SCHR695/12.

REFERENCES

[1] A. Patyuchenko et al., "Highly integrated dual-band digital beamforming Synthetic Aperture Radar," in *2015 European Radar Conference (EuRAD)*, 2015, pp. 1–4.

[2] Y. Kwag, I. Woo, H. Kwak, and Y. Jung, "Multi-mode SDR radar platform for small air-vehicle Drone detection," in *2016 CIE International Conference on Radar (RADAR)*, 2016, pp. 1–4.

[3] A. Srinivas, S. Badrinath, and V. U. Reddy, "Frequency-hopping code optimization for MIMO radar using the hit-matrix formalism," in *2010 IEEE Radar Conference*, 2010, pp. 631–636.

[4] A. R. Hunt, "Use of a Frequency-Hopping Radar for Imaging and Motion Detection Through Walls," *IEEE Transactions on Geoscience and Remote Sensing*, vol. 47, no. 5, pp. 1402–1408, May 2009.

[5] S. Lee, S. Sim, and S. Hong, "A CMOS Ultra-wideband radar transmitter with pulsed oscillator," in *2010 IEEE Radio Frequency Integrated Circuits Symposium*, 2010, pp. 509–512.

[6] A. Shoykhetbrod, A. Hommes, and N. Pohl, "A scanning FMCW-radar system for the detection of fast moving objects," in *2014 International Radar Conference*, 2014, pp. 1–5.

[7] B. Sewiolo, G. Fischer, and R. Weigel, "A 12-GHz High-Efficiency Tapered Traveling-Wave Power Amplifier With Novel Power Matched Cascode Gain Cells Using SiGe HBT Transistors," *IEEE Transactions on Microwave Theory and Techniques*, vol. 57, no. 10, pp. 2329–2336, Oct. 2009.

[8] D. Palombini, A. Bentini, M. Palomba, S. Dibello, and E. Limiti, "Design methodology for distributed power amplifier in Software-Defined Radio applications," in *2013 European Microwave Integrated Circuit Conference*, 2013, pp. 512–515.

[9] T. Drechsel, N. Joram, and F. Ellinger, "A 6.5 to 15.1 GHz ultra-wideband SiGe LC VCO with 80 % continuous tuning range," in *2017 European Conference on Circuit Theory and Design (ECCTD)*, 2017, pp. 1–4.

[10] S. Vehring and G. Boeck, "Truly Balanced K-Band Push-Push Frequency Doubler," in *2018 IEEE Radio Frequency Integrated Circuits Symposium (RFIC)*, 2018, pp. 348–351.

[11] S. Yuan and H. Schumacher, "90–140 GHz frequency octupler in Si/SiGe BiCMOS using a novel bootstrapped doubler topology," in *2014 9th European Microwave Integrated Circuit Conference*, 2014, pp. 158–161.

[12] M. Campovecchio, B. L. Bras, M. Lajugie, and J. Obregon, "Optimum design of distributed power-FET amplifiers. Application to a 2-18 GHz MMIC module exhibiting improved power performances," in *Proceedings of 1994 IEEE Microwave and Millimeter-Wave Monolithic Circuits Symposium*, 1994, pp. 125–128.

A 51.5 - 64.5 GHz Active Phase Shifter Using Linear Phase Control Technique With 1.4° Phase resolution in 65-nm CMOS

Tianjun Wu, Chenxi Zhao, Huihua Liu, Yunqiu Wu, Yiming Yu, Kai Kang*

University of Electronic Science and Technology of China, China

*kangkai@uestc.edu.cn

Abstract—This paper presents a V-band active phase shifter using proposed linear phase control technique with 1.4° phase step in 65-nm CMOS technology. Different from conventional active phase shifter, the linear phase control technique achieves a linear relationship between output phase and control signals. It makes the control of output phase more accurate and greatly improves the phase resolution. Furthermore, a current-reuse technique is used to improve gain in mm-wave frequency. The measurement results show that the measured 3-dB bandwidth of 51.5 ~ 64.5 GHz is achieved. The measured RMS phase error of 8-bit phase resolution is 0.5° ~ 1.2° in 3-dB bandwidth and the measured RMS gain variation error is 0.17 ~ 0.25 dB. The measured input-referred P1dB at maximum gain phase states is -5 dBm. The chip consumes 16.9 mA from 1.2 V voltage supply and the core area of the phase shifter is 720 um × 560 um.

Keywords—phased-array, millimeter wave integrated circuit, active phase shifter, CMOS, V-band.

I. INTRODUCTION

Due to the features of beam forming and fast electronic beam scanning, phased-array technique is widely used in millimeter wave wireless systems, including 5G [2] and V-band high-data-rate short range wireless communication systems [1]. As the key element in phased arrays, phase shifter limits the beam steering resolution, beam pointing accuracy and beam steering range of phased arrays [6].

Generally, phase shifters have two types, passive and active ones. The reflective loading [3] and switched-filter [4] architectures are usually implemented in passive phase shifter. However, the reflective loading architecture features narrow bandwidth and low phase resolution [3], while the switched-filter type phase shifter suffers from large chip size and high insertion loss [4]. Compared to passive phase shifter, active phase shifter has better performances on chip size, insertion loss and phase resolution [6][8][9], and is popular in phased arrays [1][2].

However, the phase resolution of active phase shifter is greatly restricted by the gain control accuracy of variable gain amplifier (VGA). In active phase shifter, VGA is usually implemented with current steering [6] or tail current controlled Gilbert structure [7], and the gain of VGA is controlled by a current DAC [6][7]. As shown in Figure. 1(b), the g_m or gain of those VGAs are proportional to the square root of the output current of DAC. Because of the non-linearity, it is difficult to accurately set the gain of VGA, resulting in bad phase resolution.

In order to alleviate this problem, an active phase shifter with translinear VGA is proposed (Fig. 1(a)). Based on the translinear VGA (Fig. 1(c)), a linear phase control technique is

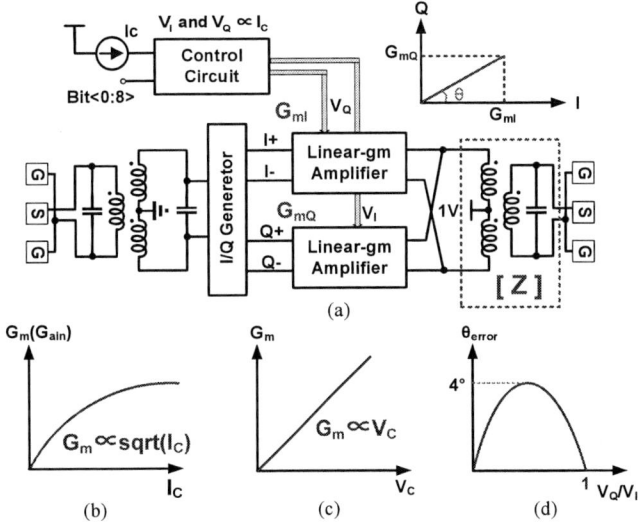

Fig. 1. (a) Block diagram of the proposed active phase shifter; (b) The transconductance/gain curve of conventional VGA versus control current; (c) The transconductance versus control voltage of the translinear VGA in the proposed phase shifter; (d) The theoretic phase error of the proposed phase shifter.

proposed to simplify the control logic and improve phase resolution extremely. The bit width of the control signal is 9-bit and the phase shifter achieves 8-bit phase resolution with measured RMS phase error of 0.5° ~ 1.2° and RMS gain error of 0.17 ~ 0.25 dB across 51.5 ~ 64.5 GHz.

This paper is organized as follows. Section II introduces the principle of the proposed linear phase control technique. Section III presents the circuit design of the phase shifter. The measurement results are shown in Section IV. Section V shows the conclusions of the phase shifter.

II. LINEAR PHASE CONTROL TECHNIQUE

The block diagram of the proposed phase shifter is shown in Fig. 1(a). G_{mI} and G_{mQ} denote the transconductance of the translinear VGA in I- and Q-path respectively, while [Z] denotes the impedance matrix of output matching network. V_I and V_Q are the control voltage of VGA in I- and Q-path, and are proportional to the external control current I_C.

The output phase (θ) of the active phase shifter is given by

$$\theta = arctan \frac{G_{mQ}Z_{21}}{G_{mI}Z_{21}} = arctan \frac{V_Q}{V_I} \qquad (1)$$

where Z_{21} denotes the transfer impedance of output matching network. When $0 < V_Q / V_I < 1$, which means $0 < \theta < \pi/4$, equation (1) can be simplified as equation (2).

$$\theta = arctan\frac{V_Q}{V_I} \approx \frac{\pi}{4}\frac{V_Q}{V_I} \ . \tag{2}$$

Equation (2) indicates that the output phase of the proposed phase shifter is proportional to V_Q/V_I. Hence, an active phase shifter with linear output phase characteristic is achieved. When $1 < V_Q / V_I < +\infty$, which means $\pi/4 < \theta < \pi/2$, equation (1) can be transferred into

$$\theta = \frac{\pi}{2} - arc\,cot\frac{V_Q}{V_I} = \frac{\pi}{2} - arctan\frac{V_I}{V_Q}$$

$$\approx \frac{\pi}{2} - \frac{\pi}{4}\frac{V_I}{V_Q} \ . \tag{3}$$

Equations (2) and (3) imply that a pair of complementary output phases are achieved by exchanging the control voltage of VGA, simplifying the complexity of control circuit greatly. The linear approximation from equation (1) to (2) and (3) induces phase error inevitably. The phase error is defined as $\theta_{error} = \pi V_Q/(4V_I) - arctan(V_Q/V_I)$, and is shown in Fig. 1(d). Fortunately, θ_{error} can be easily compensated because of the linear phase characteristic. The method for compensation will be introduced in Section III.

III. CIRCUIT IMPLEMENT

A. I/Q Generator

Fig. 2(a) shows the schematic of the two-stage Hybrid I/Q generator. The transformer-based Hybrid I/Q generator was proposed in [5]. However, its output ports (Couple and Thru port) locate at opposite side in layout, resulting in asymmetric route and phase error. In order to alleviate this problem, a Hybrid layout with output ports (Tr and Cp) locating at one side is proposed (Fig. 2(a)). Single-stage Hybrid has wide phase bandwidth but narrow amplitude bandwidth. To expand the bandwidth of amplitude balance, a two-stage Hybrid is implemented with six single Hybrids. The single Hybrid is achieved with broadside coupled microstrip line to guarantee a complete ground plane. As shown in Fig. 5, the stage two are arranged as a circle to maintain layout symmetry to minimize phase and amplitude mismatches. As results, the simulated maximum phase and amplitude error among Q+, Q-, I- and I- is below 2.5° and 0.4 dB across 51 ~ 66 GHz. The insertion loss is about 4.3 dB.

B. Translinear VGA

As shown in Fig. 2(b), the translinear VGA consists of a common gate amplifier ($M_1 \sim M_2$) and a common source variable gain transconductance amplifier ($M_3 \sim M_6$) with current-reuse technique. When the VGA is controlled by a small voltage V_c, the transconductance of the VGA is given by

$$G_m = G_{cg}(g_{m3,6} - g_{m4,5})$$

$$\approx \sqrt{2}G_{cg} \bullet \mu_n C_{ox}(\frac{W}{L})_{3,4,5,6}V_c \tag{4}$$

where G_{cg}, $g_{m3,6}$ and $g_{m4,5}$ denote the gain of the common gate amplifier, the transconductance of M_3 and M_6, and transconductance of M_4 and M_5, respectively. According to equation (4), the transconductance of the VGA is proportional

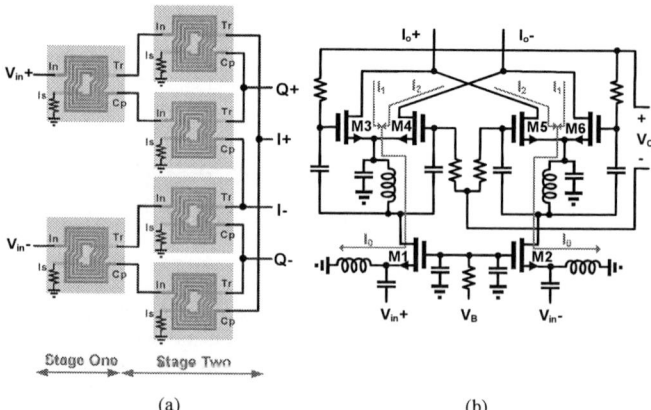

(a)   (b)

Fig. 2. (a) The two-stage Hybrid I/Q generator; (b) The schematic of the proposed current-reuse translinear VGA.

to V_c, demonstrating a translinear characteristic. The insertion phase of the VGA can be inverted with a negative V_c, which is helpful to make the phase shifter cover full 360°. The maximum control voltage V_c is limited by equation (5).

$$V_c < \sqrt{\frac{I_0}{\mu_n C_{ox}(W/L)_{3,4,5,6}}} \tag{5}$$

where I_0 denotes the bias current of common gate amplifier. If V_c does not satisfy equation (5), the gain of the VGA will become saturated. The maximum transconductance of the proposed VGA is given by (6) and the minimum value is zero ($V_c = 0$). Theoretically, the VGA has an infinite gain tuning range.

$$G_{m,max} = \sqrt{2\mu_n C_{ox}(W/L)_{3,4,5,6}I_0} \ . \tag{6}$$

C. Control Circuit

As shown in Fig. 3, the control circuit consists of a 6-bit DAC, a gain variation compensation circuit and three double-pole-double-throw (DPDT) switches. V_1 and V_2 are generated by the 6-bit DAC and are used as the gain control voltage of VGA. As shown in Fig. 1(d), the uniform distribution of V_Q/V_I between 0 and 1 results in a phase error up to 4°. In order to compensate the phase error, the output of the 6-bit DAC is designed with slightly non-linearity, which is shown in Fig. 4(a). The simulated output phase errors controlled with uniform DAC and the proposed DAC are compared in Fig. 4(b). As shown in Fig. 4(b), the simulated output phase errors controlled by the proposed DAC are smaller and its maximum value is < 1.3°. The signal Bit6 decides V_Q/V_I equals to V_1/V_2 or V_2/V_1, making the phase shifter cover 90° phase range. Signals Bit7 and Bit8 reverse the polarity of V_I and V_Q. According to equation (4), the insertion phase of VGA in I- and Q-path are also reversed. Therefore, signals Bit7 and Bit8 make the phase shifter cover full 360°.

The bias voltages (V_{b1}, V_{b2}) of the DAC are generated by the gain variation compensation circuit. V_{b1} and V_{b2} are tuned by signals Bit2 ~ Bit5 to partly counteract the gain variation among different phase states. The gain variation compensation circuit is biased by an external current I_c. As shown in Fig. 3, V_1 and V_2 are proportional to I_c. According to equation (4),

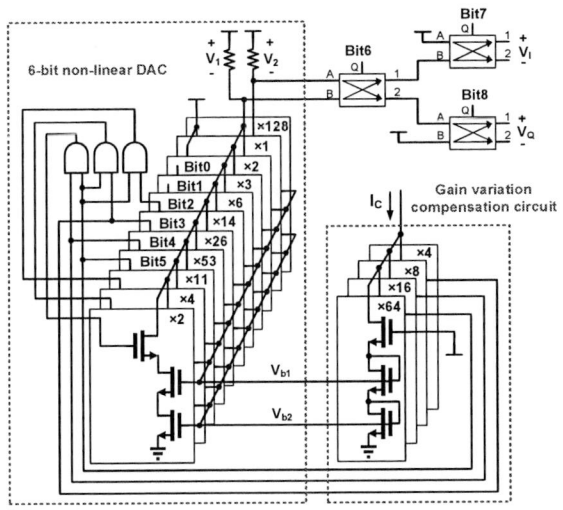

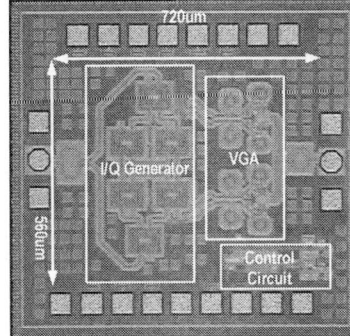

Fig. 5. Chip photo of the 60-GHz active phase shifter.

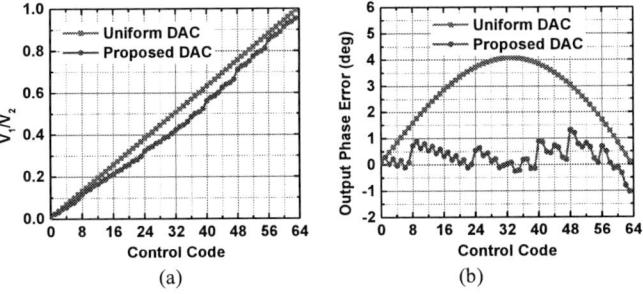

Fig. 3. The schematic of the control circuit.

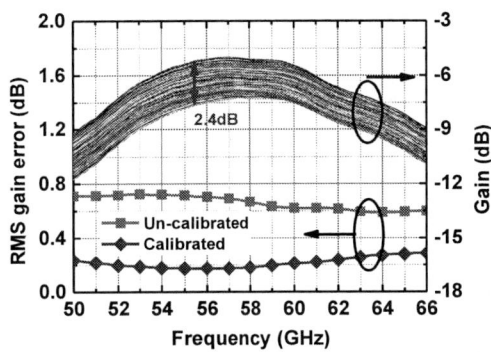

Fig. 6. Measured gains of 512 states and rms gain variation error of the phase shifter.

Fig. 4. (a) The phase control signal (V_1/V_2) produced by uniform DAC and proposed DAC; (b) The simulated output phase errors of the phase shifter controlled with uniform DAC and proposed DAC.

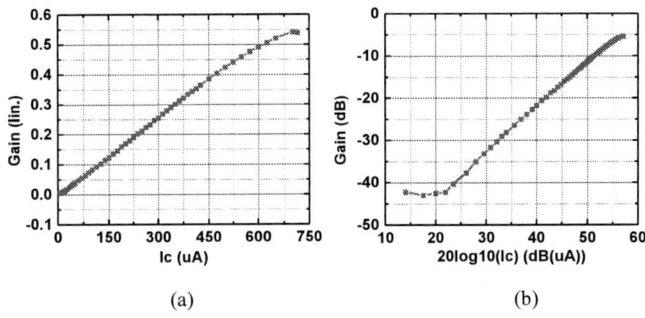

Fig. 7. The measurement results of the translinear VGA with Bit0 ~ Bit8 fixing to VDD (@57 GHz): (a) The measured amplitude of proposed phase shifter versus Ic. (b) The logarithmic form of measured gain of proposed phase shifter versus Ic.

the gain of the phase shifter can be adjusted by the current I_c. Therefore, the proposed phase shifter has variable gain characteristic.

IV. MEASUREMENT RESULTS

The chip is fabricated in 65-nm CMOS Technology, with core area of 720 um × 560 um (Fig. 5). The measurement was done on-chip with Agilent vector network analyzer N5247A. The control signals, supply and GND are wire bonded to a PCB.

The 9-bit digital control signals (Bit0 ~ Bit8) generate 512 different insertion losses and phase states. Fig. 6 shows the measured 512 different insertion losses and rms gain variation error when the external gain control current (I_c) is 650 uA. The peak gain is -5 dB with 3-dB bandwidth of 51.5 ~ 64.5 GHz. The maximum gain variation is 2.4 dB with rms gain variation error of 0.6 ~ 0.7 dB in 3-dB bandwidth. The gain variations calibrated by the external current I_c with 25 uA step is also measured. With the gain calibration, the rms gain variation error is 0.17 ~ 0.25 dB in 3-dB bandwidth (Fig. 6). The gain linearity of translinear VGA is measured at 57 GHz by fixing Bit0 ~ Bit8 to VDD and tuning the external bias current I_c (Fig. 7). The measured results show that the translinear VGA has great gain linearity. The measured rms phase error of 8-bit (signal Bit0 is fixed to GND) is 2.6° ~ 3.6°

(Fig. 8) in 3-dB bandwidth. In order to get lower phase error, 256 phase states are selected from the measured 512 states. The rms phase error of the selected 256 phase states is 0.5° ~ 1.2° in 3-dB bandwidth, achieving a 8-bit phase resolution. The measured S_{11} and S_{22} of the 512 states are < -11 dB in 3-dB bandwidth (Fig. 9). The chip consumes 16.9 mA from 1.2 V supply. The measured input-referred P1dB is -5 dBm when the phase shifter works at maximum state at 60 GHz. The FOM that combines phase resolution, conversion gain, 3-dB bandwidth, RMS gain and phase error [6] is calculated. The comparisons with other prior state-of-art phase shifters are summarized in Table 1.

978-1-7281-1702-7/19 $31.00 © 2019 IEEE

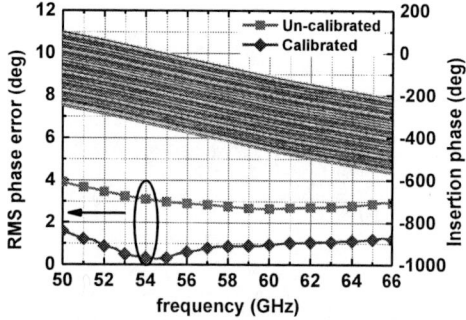

Fig. 8. Measurement insertion phase of 256 states and rms phase error of the phase shifter.

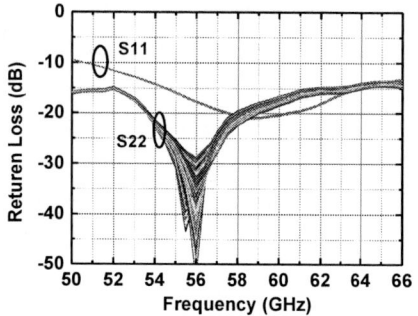

Fig. 9. Measurement port return loss of 512 states of the phase shifter.

V. CONCLUSION

This paper proposes a linear phase control technique that builds linear relationship between output phase and control signals of the phase shifter. The accuracy of phase control is extremely improved and the complexity of the control logic is simplified greatly. The bit width of the control signal is only 9-bit and a 8-bit phase resolution is achieved. The measured RMS phase and gain error of 8-bit phase resolution are 0.5° ~ 1.2° and 0.17 ~ 0.25 dB in 51.5 GHz ~ 64.5 GHz. The active phase shifter is very suitable for IEEE 802.11ad and IEEE 802.11ay standards phased-array systems because of its great performances. Because the gain of the proposed phase shifter is controlled by I_c, while the output phase is decided by V_Q/V_I

only, the proposed control technique is the fist method that succeeds in separating the gain control from phase control in active phase shifter. With the proposed technique, attenuator and VGA can be left out from phased arrays and a compact phased-array structure is achieved.

ACKNOWLEDGMENT

This work is supported by National Natural Science Foundation of China (Grant No. 61771115, 61874020, 61804024).

REFERENCES

[1] T. Sowlati et al., "A 60GHz 144-element phased-array transceiver with 51dBm maximum EIRP and ±60° beam steering for backhaul application," ISSCC, pp. 66-68, Feb. 2018.

[2] K. Kibaroglu, M. Sayginer and G. M. Rebeiz, "A Low-Cost Scalable 32-Element 28-GHz Phased Array Transceiver for 5G Communication Links Based on a 2×2 Beamformer Flip-Chip Unit Cell," IEEE JSSC, vol. 53, pp. 1260-1274, May 2018.

[3] R. Garg and A. S. Natarajan, "A 28-GHz Low-Power Phased-Array Receiver Front-End With 360° RTPS Phase Shift Range," in IEEE Transactions on Microwave Theory and Techniques, vol. 65, no. 11, pp. 4703-4714, Nov. 2017.

[4] W. Li, Y. Chiang, J. Tsai, H. Yang, J. Cheng and T. Huang, "60-GHz 5-bit Phase Shifter With Integrated VGA Phase-Error Compensation," in IEEE Transactions on Microwave Theory and Techniques, vol. 61, no. 3, pp. 1224-1235, March 2013.

[5] J. S. Park and H. Wang, "A Transformer-Based Poly-Phase Network for Ultra-Broadband Quadrature Signal Generation," in IEEE Transactions on Microwave Theory and Techniques, vol. 63, no. 12, pp. 4444-4457, Dec. 2015.

[6] B. Wang, H. Gao, M. K. Matters-Kammerer and P. G. M. Baltus, "A 60 GHz 360° Phase Shifter with 2.7° Phase Resolution and 1.4° RMS Phase Error in a 40-nm CMOS Technology," 2018 IEEE Radio Frequency Integrated Circuits Symposium (RFIC), Philadelphia, PA, 2018, pp. 144-147.

[7] X. Quan et al., "A 52－57 GHz 6-Bit Phase Shifter With Hybrid of Passive and Active Structures," in IEEE Microwave and Wireless Components Letters, vol. 28, no. 3, pp. 236-238, March 2018.

[8] D. Pepe and D. Zito, "Two mm-Wave Vector Modulator Active Phase Shifters With Novel IQ Generator in 28 nm FDSOI CMOS," in IEEE Journal of Solid-State Circuits, vol. 52, no. 2, pp. 344-356, Feb. 2017.

[9] Y. Yu et al., "A 60-GHz 19.8-mW Current-Reuse Active Phase Shifter With Tunable Current-Splitting Technique in 90-nm CMOS," in IEEE Transactions on Microwave Theory and Techniques, vol. 64, no. 5, pp. 1572-1584, May 2016.

Table 1. Performance comparison summary

	This Work	RFIC2018 [6]	MWCL2018[7]	JSSC2017[8]	MTT2016[9]
Process	65nm CMOS	40nm CMOS	40nm CMOS	28nm CMOS	90nm CMOS
Method	Vector modulator	Vector modulator	Hybrid	Vector modulator	Vector modulator
3-dB Frequency Band	51.5~64.5GHz	58~62GHz	52-57GHz	78.8~92.8GHz	57~64GHz
Phase Resolution	1.4°	2.7°	5.6°	22.5°	22.5°
RMS P.E. in B_{3dB}	0.5~1.2°	1.4~2.7°	2.8~3.76°	8.4~9.4°	2.3~7.6°
RMS G.E. in B_{3dB}	<0.25dB	0.13~0.2dB	2.0~2.23dB	1.5~2dB	0.75~1.6dB
Peak Gain	-5dB	-0.4dB	-9dB	2.3dB	1.1dB
P_{dc}	20.3mW	38mW	14.3mW	21.6mW	19.8mW
Chip Area	0.72×0.56mm²	1.5×0.73mm²	0.72×460mm²	0.12mm²	0.94×0.66mm²
20lg(FOM)	68.8	55.4	41.6	52.9	46.5

$$FOM = \frac{f_0(GHz) \times Gain(lin.) \times B_{3dB}(GHz) \times Re\,solution(bits)}{RmsPhaseErrorInB_{3dB}(°) \times RmsGainErrorInB_{3dB}(lin.)} [6]$$

Digitally-Assisted 27-33 GHz Reflection-Type Phase Shifter with Enhanced Accuracy and Low IL-Variation

Jingjing Xia, Mahitab Farouk and Slim Boumaiza

Emerging Radio Systems Group, University of Waterloo, Waterloo, Ontario, N2L 3G1, Canada

j4xia, meladwy, sboumaiza@uwaterloo.ca

Abstract— **This paper presents a new millimeter-wave 360° digitally-assisted reflection-type phase shifter (DA-RTPS). It is composed of a cascade of three compact, low-loss and fully-differential transformer-based hybrid couplers with two of their ports terminated by switch-controlled capacitor banks. Each stage is configured to produce a phase change of up to 60°. An additional phase inverting stage is used to attain 360° of phase shift. The proposed DA-RTPS arrangement enabled low insertion loss (IL) variation and enhanced phase accuracy compared to single-stage RTPS-based typologies. A proof-of-concept prototype, implemented using 45nm silicon-on-insulator CMOS technology, demonstrated a 1-dB RF bandwidth spanning from 27 to 33 GHz, a low root-mean-square phase error of 0.3^o, and an IL of 6.8 dB$\pm$0.25dB, while covering 360° of phase shift at 5.6^o resolution. Furthermore, it maintained a group delay below $\pm$12ps and an input 1dB compression point >7.3 dBm at all phase-shift settings between 27 and 33 GHz.**

Keywords— **millimeter-wave, passive phase shifter, RF beamforming, SOI.**

I. INTRODUCTION

Fifth-generation (5G) technology is expected to connect billions of devices and enable ultra broadband mobile communication with low latency and improved robustness. To meet these unprecedented demands, 5G technology will tap into millimeter-wave (mmwave) frequency bands to benefit from the gigahertz of available spectrum. It will also involve large-scale phased-array-based radio hardware to minimize the interference and maximize the spatial multiplexing. These two requirements, among others, pose significant challenges to the design of 5G radio hardware for attaining satisfactory RF performance while minimizing power consumption and cost.

Among the various building blocks of 5G radio hardware, the RF phase shifter (PS) is a key component for the successful realization of viable large-scale phased-array systems. The PS needs to exhibit low insertion loss (IL) and high phase accuracy, while minimizing the group delay and IL-variation versus phase shift settings and frequency. In the literature one can distinguish two major types of PS: active and passive [1]–[7]. Passive PSs have several advantages that make them highly attractive to large-scale antenna array realization. These include zero power consumption, bi-directionality, and improved linearity. These advantages have spurred growing research interest in passive PSs, including switched-type [1], loaded transmission line [2] and reflection-type (RTPS) [3]–[6] configurations. In this work, an RTPS was chosen since they typically combine compact size, moderate IL and high resolution phase shift; very relevant attributes for the

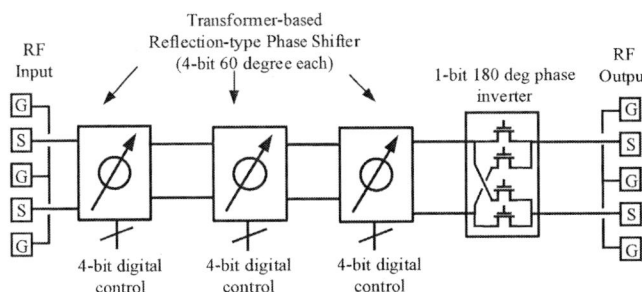

Fig. 1. Block diagram of proposed DA-RTPS.

realization of viable mmwave large-scale phased-array-based 5G radio hardware.

The RF performance of an RTPS is directly linked to the characteristic and the topology of its two major building blocks, namely, the hybrid coupler and the tunable reflective load. Several circuit arrangements have been suggested to maximize the phase tuning range (ideally to reach 360°) of an RTPS. For example, authors in [3] suggested a cascade of three RTPS units that used varactor-based reflective loads to attain a phase tuning range of 180°. Yet, this arrangement yielded an IL-variation of $\pm$2dB versus phase setting. Alternatively, single-stage RTPS arrangements that incorporate more complicated tunable reflective loads-including LC- [6], CLC- [5] and transformer-based [4] networks-allow for extended phase tuning range and reduced size and IL-variation. However, the limited quality factor of on-silicon passive components used to implement the underlying tunable reflective loads engendered relatively high IL (7.7 to 10 dB). In addition, the underlying reflective load circuits often limit the achievable RF bandwidth.

This paper opts for a multi-stage RTPS architecture that aims to realize a compact 360° phase shift tuning range while minimizing the IL and its variation and widening the RF bandwidth. This architecture takes advantage of the compactness and low-loss of fully-differential transformer-based hybrid couplers and the very low IL-variation across the tuning range of a newly devised switch-controlled (SC) capacitor reflective load to create a three-stage digitally-assisted (DA) RTPS with 180° phase tuning range. This tuning range can be extended to 360° using a simple phase-inverting stage.

978-1-7281-1702-7/19 $31.00 © 2019 IEEE

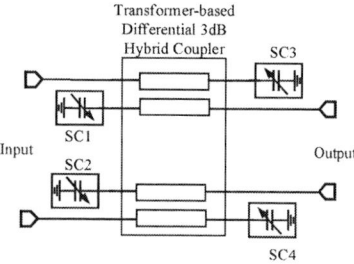

Fig. 2. Block diagram of single-stage 60° 4-bit DA-RTPS cell.

II. PROPOSED DIGITALLY-ASSISTED RTPS

Fig. 1 shows the block diagram of the proposed DA-RTPS. It includes three RTPS stages of 60° followed by a 1-bit 180° phase inverting stage used to attain a total phase tuning range of 360°. As shown in Fig. 2, each stage consists of a differential pair of hybrid couplers terminated by reflective loads at their transmission and coupling ports. Each reflective load uses carefully designed 4-bit SC capacitor banks (SC1 to SC4 in Fig. 2) to minimize their IL and size while ensuring a tuning range equal to 60° in each stage.

Assuming for now that the hybrid coupler's IL (IL_{hybrid}) is the only source of nonideality (i.e. loss) in the single-stage RTPS of Fig. 2, its overall IL (IL_{RTPS}) and phase shift ($\Delta\Phi$) can be expressed as follows

$$\Delta\Phi(^o) = \angle\Gamma_L - 90^o \qquad (1)$$

$$IL_{RTPS}(dB) = 2IL_{hybrid}(dB) + |\Gamma_L|(dB) \qquad (2)$$

where Γ_L denotes the reflection coefficient of the reflective loads (SC1 - SC4). Based on (1) and (2), to improve the performance of a single-stage RTPS, it is important to i) minimize IL_{hybrid} as it counts twice in the overall IL_{RTPS} in (2), and ii) maximize $|\Gamma_L|$ (minimize the losses) of the reflective load, and iii) minimize the variation of the $|\Gamma_L|$ with the phase shift.

As previously mentioned, the design of the reflective load in an RTPS needs to maximize the phase tuning range and minimize the IL and its variation with the phase setting. A single-stage RTPS with sophisticated reflective load can yield a large phase shift tuning range, but can potentially yield high values for IL and its variation versus phase shift. Alternatively, a single-stage RTPS with a tunable-capacitance based reflective load would allow for low IL_{hybrid} as well as high $|\Gamma_L|$ and potentially low IL-variation versus phase setting. Yet, theoretically, to obtain a phase-tuning range of 180°, a simple tunable-capacitance based reflective load with $C_{max}/C_{min} = \infty$ (where C_{max} and C_{min} denote the maximum and minimum capacitance of the reflective load) would be needed. However, this ratio is not practically conceivable and only a finite value of C_{max}/C_{min} is realizable; consequently, yielding a phase shift lower than 180°. Furthermore, varactor-based tunable capacitances would require bipolar tuning and typically exhibit relatively large IL-variation versus tuning value [3]. In this work, a new

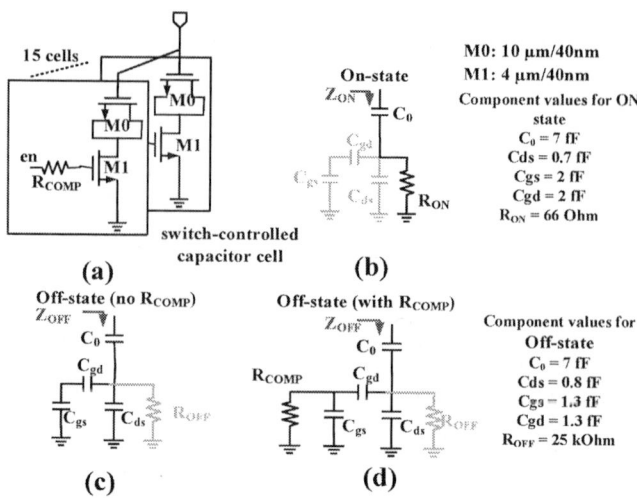

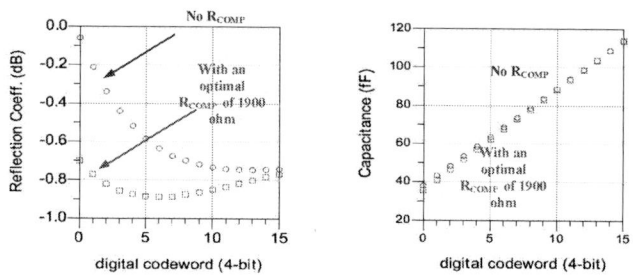

Fig. 3. (a) Schematic and equivalent small-signal at (b) ON-state (c) OFF-state without R_{COMP} and (d) OFF-state with R_{COMP}.

Fig. 4. Post-layout simulation results at 30 GHz comparing proposed SC cell with and without R_{COMP}.

digitally-controlled SC capacitance bank was chosen to realize the tunable capacitance.

Fig. 3(a) depicts the schematic of the proposed SC reflective load. It is composed of 15 identical cells controlled using a 4-bit digital code word. Each cell is implemented using a MOSFET capacitor (MO) of $\sim$7 fF, and includes a transistor, M1, that serves as an enabling/disabling switch. Fig. 3(b) and (c) show the equivalent small-signal circuit of a unit cell when the switch is turned ON and OFF, where Z_{ON} and Z_{OFF} denote the respective input impedances. During the ON state, the imaginary and real parts of Z_{ON} are determined by the MOSCAP and R_{ON} of the switch, respectively. During the OFF state, Z_{OFF} is dominated by the parasitic capacitance of M1 and includes a negligible resistive component. The difference between Z_{OFF} and Z_{ON} of the 15 unit cells in the reflective load engenders a significant variation of $|\Gamma_L|$ with the code word and consequently leads to a large variation of IL in the RTPS with the phase setting. To alleviate this problem, a compensation resistor, R_{COMP}, is added to the gate of the transistors (M1 in each cell) to minimize the IL-variation with phase setting as shown in Fig. 3(d). The resistor's value is chosen so that it minimizes the IL-variation with the phase setting while meeting the following constraints: $R_{ON} << R_{COMP} << R_{OFF}$ (R_{ON} and R_{OFF} designate

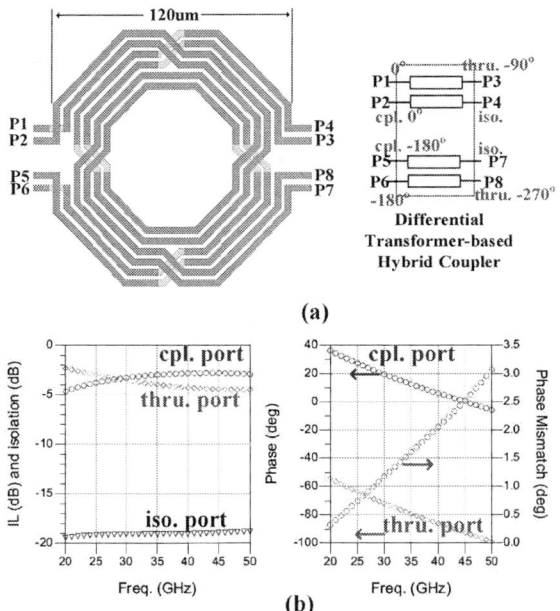

(a)

(b)

Fig. 5. (a) Layout and (b) EM-simulated performances of the differential transformer-coupled hybrid coupler.

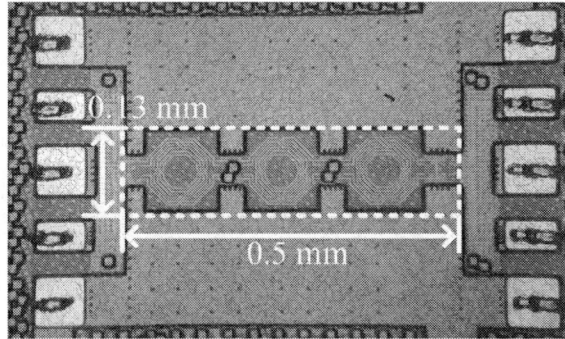

Fig. 6. Microphotograph of fabricated DA-RTPS prototype.

the resistances of the switch in its ON and OFF states). Fig. 4 shows the magnitude of the reflection coefficient (in dB) as a function of the code word for $R_{COMP} = \infty$ and $R_{COMP} = 1900$ ohm, obtained using post-layout simulation results. It confirms the minimization of $|\Gamma_L|$ over the code word range even as the capacitance is maintained over the tuning range $C_{max}/C_{min} = 113fF/35fF = 3.1$. Given that this capacitance range would allow a phase tuning of about 60°, three RTPS stages (RTPS circuitry shown in Fig. 2) would be required to achieve 180° of phase shift.

To minimize the overall IL of the three-stage RTPS, it is imperative to reduce IL_{hybrid}. To accomplish this, a transformer-based coupler was chosen. This transformer takes advantage of multiple thick-metal layers of the 45nm SOI-CMOS process to minimize its IL as well as size. Based on the transformer reported in [8], Fig. 5(a) shows the layout of the transformer-based differential hybrid coupler used to realize the proposed DA-RTPS. With a size of 120 by 120 μm, the electromagnetic (EM)-based optimization of the line width, spacing, coil dimension and metallization choice yield a low IL of 0.45 dB and a phase mismatch of $<1^o$ at 29 GHz (see Fig. 5(b)).

III. MEASUREMENT RESULTS

The proposed DA-RTPS was fabricated in GlobaFoundries' 45nm SOI-CMOS process; a microphoto of the fabricated prototype is shown in Fig. 6. A vector network analyzer (Keysight N5247A) was used to measure the small- and large-signal performances of the DA-RTPS. Furthermore, digital code words were fed to the DA-RTPS chip using an SPI-interface device (NI USB-8452).

The proposed DA-RTPS has a total of 13 control bits (i.e., 4-bit for each of the three 60° RTPSs and a 1-bit for the phase inverter). In order to deduce the optimal code word, a pre-characterization step which sweeps all code word combinations was conducted. As shown in Fig. 7(a), the corresponding phase shifts and IL obtained at 29 GHz are plotted. These results confirmed that the proposed DA-RTPS achieved 1) a maximum phase shift of 360° and 2) an excellent IL-variation of ± 0.4 dB, regardless of the code word setting. Based on the pre-characterization results, a look-up-table (LUT) with 64 entries was constructed using the optimal code words needed to achieve phase shifts over 360° in a step of 5.6^o while maximizing the phase accuracy and minimizing the IL-variation. The DA-RTPS performance at 29 GHz with these optimal code words are highlighted in Fig. 7(a), and demonstrate an excellent IL-variation of ± 0.1dB and a low RMS phase error of only 0.3^o. Furthermore, Fig. 7(b) shows measured small-signal performances (ILs, input and output return losses) corresponding to the 64 LUT entries. According to Fig. 7(b), the proposed DA-RTPS has an average IL of -6.8 dB at 29 GHz and maintained within 1 dB of magnitude variation between 27 - 33GHz (20% bandwidth). The measured input and output return losses were below -13.5 dB and -9 dB in this bandwidth, respectively. Lastly, the measured group delays are shown in Fig. 7(c). A relatively constant group delay (± 12 ps between 27 - 33 GHz) was achieved over 360° of phase shifts.

Fig. 8 summarizes the measured IL-variations and RMS phase errors versus frequency. It is of note that at each frequency point the optimal code words were re-calculated. According to Fig. 8, the IL-variation and RMS phase errors were maintained within ± 0.25dB and 0.3^o for the targeted band of 27 to 33 GHz. Furthermore, the proposed DA-RTPS demonstrated the potential to achieve even wider bandwidth. It maintained a good IL-variation ($< \pm 0.75$ dB) and low phase errors ($< 1.1^o$) from 24 - 40 GHz. Lastly, large-signal measurement results revealed that the input 1dB compression point was maintained above 7.3 dBm at all of the 360° phase shift settings between 27 - 33GHz.

Table I summarizes the measured performance of the fabricated DA-RTPS at its centre frequency and compares to recently reported passive PSs at their best-reported

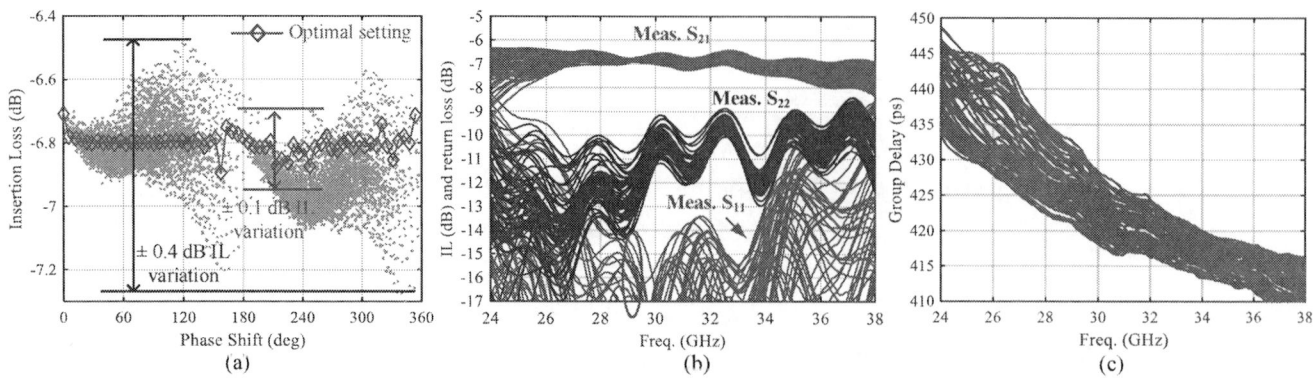

Fig. 7. Performance of proposed DA-RFPS: (a) measured phase shifts versus IL using all code words (blue) and selected code word (red) that achieve optimal performances; (b) measured S-parameters using the optimal settings; and (c) measured group delays using the optimal settings.

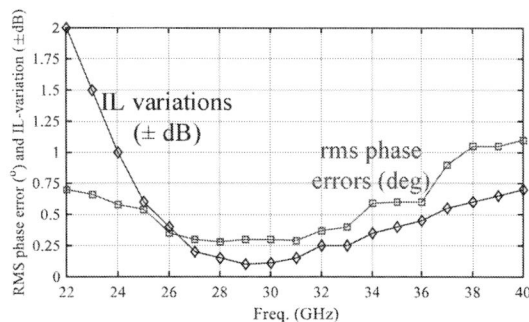

Fig. 8. Measured IL-variations and RMS phase errors for proposed DA-RTPS. Code words are optimized at each frequency point.

Table 1. Comparison of recent passive phase shifters.

	This work	[5]	[6]	[4]	[2]
CMOS Tech.	**45 nm SOI**	65 nm	65 nm	130nm SiGe	130nm SiGe
Topology	**DA-RTPS**	RTPS	RTPS	RTPS	DA TLine
Freq. (GHz)	**29**	28	29	62	28
Avg. IL (dB)	**-6.8**	-7.7	-8.3	-10	-9.3
IL-Variation (dB)	**±0.1**	±0.3	±0.2	±0.35	±0.25
RMS phase error (°)	**0.3**[(1)]	0.3[(1)]	n.a.	n.a.	0.6[(2)]
Phase resolution	**5.6°**	11.25°	22.5°[(3)]	n.a.	4.75°
Area (mm²)	**0.065**	0.16	0.076	0.16	0.18
Control Signal	**Digital**	Analog	Analog	Analog	Digital
FoM1[(4)]	**52.1**	45	42.3	35.5	19.9
FoM2[(5)]	**253**	185.8	n.a.	n.a.	93.3

[(1)] over 360° phase shift [(2)] over 180° phase shift [(3)] estimated from fig.

[(4)] $FoM1 = \frac{total\ phase\ shift\ (deg)}{max\ IL\ (dB)}$ [4] [(5)] $FoM2 = \frac{f(GHz) \times ave.\ IL\ (lin) \times Resolution\ (bits)}{RMS\ Phase\ error\ (deg) \times |IL-variation|(lin)}$ based on [7]

Table 2. IL-variation and RMS phase errors obtained on multiple chips at 29 GHz using the optimal code word from one chip (ref chip).

	Ref Chip	Chip 1	Chip 2	Chip 3	Chip 4	Chip 5	Chip 6	Chip 7
IL Variation (dB)	±0.1	±0.12	±0.15	±0.08	±0.1	±0.1	±0.1	±0.09
RMS Phase Error (°)	0.3	0.58	0.41	0.43	0.5	0.39	0.44	0.46

frequency point. According to Table I, the proposed DA-RTPS demonstrated the best RMS phase errors (0.3°), the lowest IL -6.8±0.1dB and occupied the smallest area. Two figures-of-merit (FoM) were defined for the purpose of comparison: FoM1 uses the definition in [4] and FoM2 takes the operating frequency and PS accuracy into consideration based on [7]. In both cases the proposed DA-RTPS demonstrated the best FoM.

In order to assess process variation on the proposed DA-RTPS, 7 additional fabricated samples were measured using the same code words, which were calculated from the pre-characterization results of one reference chip at 29 GHz. According to Table II, these 7 chips maintained similar IL-variation and RMS phase accuracy compared to the reference. This demonstrated the potential of the proposed DA-RTPS for attaining high accuracy without requiring pre-characterization on every chip.

IV. CONCLUSION

This paper presents the design of a new DA-RTPS for the 28 GHz band. These results confirm the validity of the proposed DA-RTPS topology in attaining high accuracy and low IL and IL-variation while maintaining a compact size. Furthermore, the digitally-assisted design minimizes the complexity required to integrate the RTPS into a large-scale 5G antenna system.

REFERENCES

[1] M. Elkholy, *el. al.*, "Low-loss highly linear integrated passive phase shifters for 5G front ends on bulk CMOS," *IEEE TMTT*, vol. 66, no. 10, pp. 4563–4575, Oct 2018.

[2] Y. Tousi and A. Valdes-Garcia, "A Ka-band digitally-controlled phase shifter with sub-degree phase precision," in *Proc. RFIC*, May 2016, pp. 356–359.

[3] M. Tabesh, A. Arbabian, and A. Niknejad, "60GHz low-loss compact phase shifters using a transformer-based hybrid in 65nm CMOS," in *2011 Proc. CICC*, Sep. 2011, pp. 1–4.

[4] T. Li and H. Wang, "A millimeter-wave fully integrated passive reflection-type phase shifter with transformer-based multi-resonance loads for 360° phase shifting," *IEEE TCAS-I*, vol. 65, no. 4, pp. 1406–1419, April 2018.

[5] R. Garg and A. S. Natarajan, "A 28-GHz low-power phased-array receiver front-end with 360° RTPS phase shift range," *IEEE TMTT*, vol. 65, no. 11, pp. 4703–4714, Nov 2017.

[6] P. Gu and D. Zhao, "Ka-band cmos 360° reflective-type phase shifter with ±0.2 dB insertion loss variation using triple-resonating load and dual-voltage control techniques," in *Proc. RFIC*, June 2018, pp. 140–143.

[7] D. Pepe and D. Zito, "Two mm-wave vector modulator active phase shifters with novel IQ generator in 28 nm FDSOI CMOS," *IEEE JSSC*, vol. 52, no. 2, pp. 344–356, Feb 2017.

[8] J. S. Park and H. Wang, "A transformer-based poly-phase network for ultra-broadband quadrature signal generation," *IEEE TMTT*, vol. 63, no. 12, pp. 4444–4457, Dec 2015.

2019 IEEE Radio Frequency Integrated Circuits Symposium

A 21 to 30-GHz Merged Digital-Controlled High Resolution Phase Shifter-Programmable Gain Amplifier with Orthogonal Phase and Gain Control for 5-G Phase Array Application

Wei Zhu[#1], Wei Lv[*$], Bingbing Liao[*$], Yanping Zhu[*$], Yuefei Dai[*$], Pei Li[*$], Lei Zhang[#], Yan Wang[#]

[#]Institute of Microelectronics, Tsinghua University, Beijing 100084, China

[*]East China Research Institute of Electronic Engineering, Hefei 230088, China

[$]Anhui Province Engineering Laboratory for Antennas and Microwave, China

Abstract—**This paper presents a 21 to 30-GHz merged passive vector sum phase shifter (PS) and programmable gain amplifier (PGA) with sub-degree phase resolution for 5-G phased array application. In PS, a transformer-based full-differential high-order resonant coupler functions as a quadrature generator (QG) with a novel layout strategy to achieve broad bandwidth, low loss, and accurate quadrature phase within only one inductor-footprint. Compared to the conventional transformer-based fourth order resonant coupler, the proposed resonant coupler with higher order features much wider quadrature bandwidth. Two phase invariant 6-bit binary-weighted arrays of vector modulators scale the quadrature signals to achieve the desired high resolution vector phase interpolation. In PGA, a phase invariant and dB-linear gain is achieved by adopting a "fractional-bit-based" PGA design. The chip prototype is fabricated in a 65-nm CMOS process, this implementation achieves 43% fractional BW$_{-3dB}$ (20 to 31-GHz). The phase control operates with 0.8° steps while maintaining a minimum RMS phase error of 0.42°, demonstrating the best phase accuracy when compared to state-of-the-art mm-wave PSs.**

Keywords— **Phase shifter, programmable gain amplifier, ultra-compact, high-order, quadrature generator, phased array.**

I. INTRODUCTION

Emerging 5-G mm-wave networks require large array-size beamforming systems to improve the link budget and system capacity. Fig. 1 illustrates an example of All-RF phased array transceiver front-end module (FEM). The phase shifter (PS) is one of the most critical blocks in phased array systems, since its phase resolution and phase shifting range dominate the beam-forming. The phase resolution directly translates to beam steering resolution (Fig. 1), thus a high resolution PS is highly recommended. Phase-invariant programmable gain amplifier (PGA) with accurate dB-linear gain control and high resolution is another critical block required at each element to compensate for gain/loss variations between radiating elements and reduce side lobes through amplitude tapering [1]. From a system perspective, orthogonal beam steering, sidelobe suppression functions and accurate dB-linear gain control are essential to ease the calibration of phase array. To achieve orthogonal beam steering control and sidelobe suppression, the phase and amplitude control in each element should be independent [1].

Conventional passive PSs like reflection-type PSs (RTPS) [2] or switch-type PSs (STPS) [3] naturally feature high linearity but also rely on higher loss and larger die area for

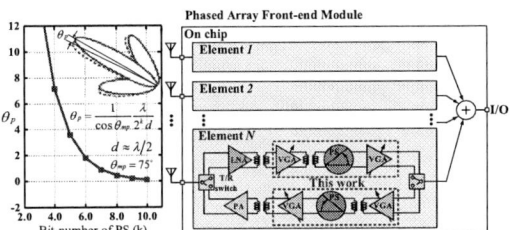

Fig. 1. Typical block diagram of an N-element All-RF phased array transceiver (right) and the effect of PS bit-number (i.e. resolution) on beam steering step-size (left), where d represents the element spacing, θ_{mp} represents the phase gradient, and θ_p represents the beam steering resolution.

higher resolution. Active vector modulator-based PSs [4] typically suffer from low linearity and narrow bandwidth. To addresses all the aforementioned issues, a transformer-based full-differential ultra-compact high-order resonant coupler is proposed in this work to function as a quadrature generator (QG) to achieve broad bandwidth, low loss, and accurate quadrature phase within only one inductor-footprint. Two phase invariant 6-bit binary-weighted arrays of vector modulators (VMs) scale the quadrature signals to achieve the desired high resolution vector phase interpolation, independent phase control and high linearity (Fig. 2).

To achieve independent and accurate dB-linear gain control, The PGA utilizes a novel robust "fractional-bit-based" method and achieves mathematically zero phase variant and accurate dB-linear gain across a wide tuning range with low power consumption and wide bandwidth (Fig. 2).

II. VECTOR MODULATOR-BASED PASSIVE PS

A. A Fold Transformer-Based Full Differential Ultra-Compact High-order QG

The QG is a critical block in a VM-based PS, as its performance largely governs the phase interpolation quality. Compared to RC-CR poly-phase filter and Lange Coupler, the transformer-based QG (Fig. 3(a)) is preferred in mm-wave applications due to the low insertion loss and compact layout. The bandwidth of the conventional transformer-based QG is mainly limited by the I/Q magnitude and phase mismatches. In [2], multiple transformer-based QG are cascaded to form a high-order poly-phase network to substantially extend the quadrature signal generation bandwidth with the cost of larger die area and greater passive loss, which is unacceptable for

978-1-7281-1702-7/19 $31.00 © 2019 IEEE

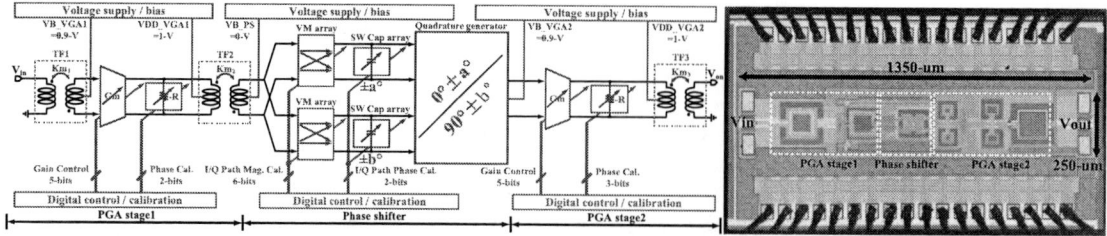

Fig. 2. Block diagram and die photo of the proposed merged PS-PGA, core area: 0.34-mm².

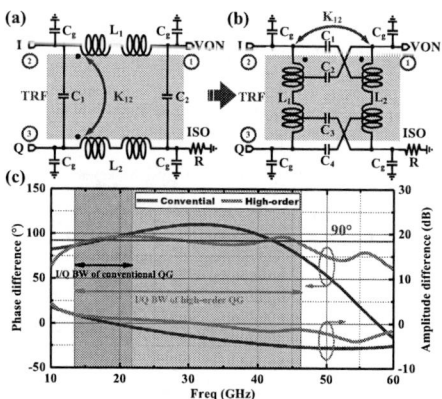

Fig. 3. (a) The circuit schematic of the conventional transformer-based QG, where C_g represents the grounded capacitor. (b) The single-ended equivalent form of the proposed transformer-based high-order resonant coupler. (c) Comparison of the I/Q bandwidth of conventional and proposed high-order QG.

large array-size beamforming systems. To address the challenge, we propose a folded transformer-based high-order poly-phase resonant coupler to suppress I/Q magnitude and phase mismatches and achieve ultra-broadband operation, the simplified single-ended equivalent form is shown in Fig. 3(b). Two additional capacitors are introduced and all the capacitors (C_{1to4}) are placed between a primary/secondary port and part of the secondary/primary inductor to increase the order of the resonant coupler. With higher order, more pairs of pole-zero poles are introduced, making the 3-dB quadrature relationship ($S_{21}/S_{31}=j$) of the resonant coupler been interpolated into more roots, so that the phase and magnitude errors are substantially suppressed and the quadrature fractional bandwidth (phase error < 2°, magnitude error < 0.5-dB) is greatly expanded from 49.3% (13 to 21.5-GHz) to 112.6% (13 to 46.5-GHz) (Fig. 3(c)). In addition, with the re-arrangement of the port order, the resonant coupler can also function as an impedance converter for the front and rear stages with easier dc feed, and the added capacitors decrease the required inductance and leads to further size reduction.

The single-ended transformer based quadrature generation scheme can be extended to a fully differential scheme with constructive magnetic coupling for further substantially size reduction. Two identical singled-ended high-order resonant coupler-based QG can be placed side-by-side for full differential operation (Fig. 4(a)). With a differential mode operation, the current directions in the two transformers are contrary, making it feasible to fold the two transformers into one structure within a single inductor footprint (Fig. 4(b)). The

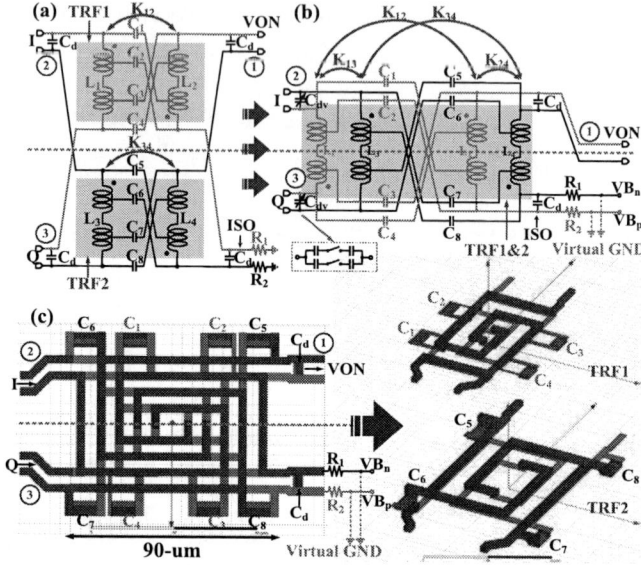

Fig. 4. (a) The circuit schematic of the transformer-based full differential high-order QG, where $C_d = C_g/2$. (b) The circuit schematic of the fold transformer-based full differential high-order QG. (c) The physical layout implement of the fold transformer-based full differential high-order QG.

folding operation introduces additional magnetic coupling (K_{13} and K_{24}) and ensures magnetic coupling enhancement of the two single-ended resonant couplers, the additional magnetic coupling leads to further reduction of required inductance and die area. The shunt capacitors at I and Q ports C_{dv} are implemented with switched capacitor arrays for quadrature calibration. The physical layout implementation of the proposed fold transformer-based full differential high-order QG is shown in Figure 4 (c). The coils of TRF1 (L1 and L2) occupy the second and fourth windings, while the coils of TRF2 (L3 and L4) occupy the first, third and part of the fifth windings. The coil lengths of TRF1 and TRF2 are kept the same to ensure the same inductance. It is worth noting that the core area of the proposed QG is only 90-μm by 70-μm, achieving size reduction of over 100 times compared to a 28-GHz λ/4 Lange Coupler. The simulated behavior of the proposed QG is shown in Fig.3, demonstrating ultra-wide quadrature bandwidth with 0.9-dB passive loss at 28-GHz.

B. Passive Vector Modulator-Based PS Design

The architecture of the proposed PS is shown in Fig. 2. Two phase invariant VM arrays scale the quadrature signals to achieve the desired high resolution vector phase interpolation. Each VM array is implemented as phase variant 6-bit binary-

978-1-7281-1702-7/19 $31.00 © 2019 IEEE

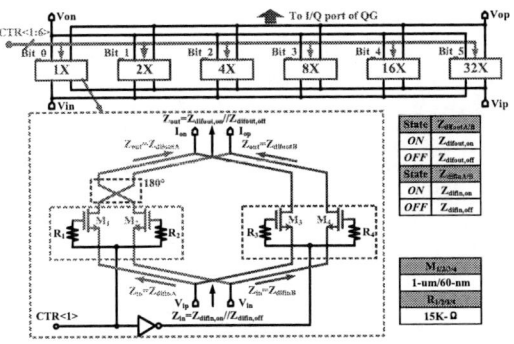

Fig. 5. The circuit schematic of the 6-bit binary weighted VM array. The Bit_N VM cell is used to generate normalized "2^N" or "-2^N" weighting.

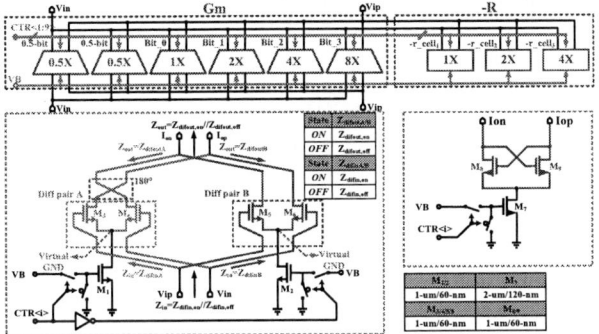

Fig. 6. The circuit schematic of the 6-bit binary weighted gm-cell array and negative resistance array (Gm and –R in Fig. 2).

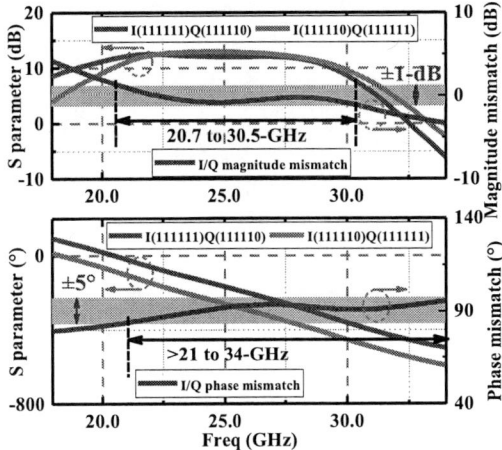

Fig. 7. The I/Q magnitude (up) and phase (bottom) difference of the merged PS-PGA.

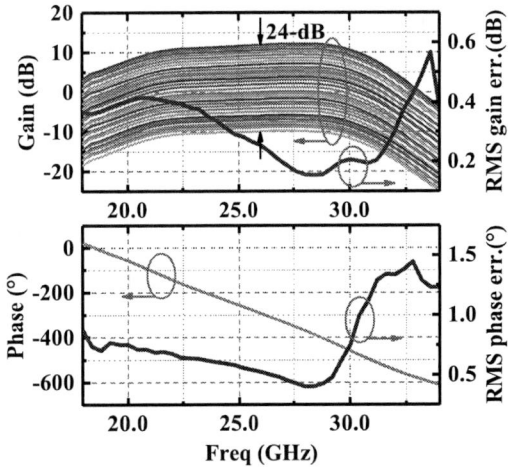

Fig. 8. Measured gain response and RMS gain error (up). Measured phase response and RMS phase error (bottom).

weighted VM cells connected in parallel (Fig. 5). The VM cells are implemented as polarity selectors with unchanged input and output impedance to ensure phase invariance and thus achieve high phase resolution for PS. The VM array can achieve normalized AC current for all the odd integers within range of -63 to +63 ($\pm 2^0 \pm 2^1 \pm 2^2 \pm 2^3 \pm 2^4 \pm 2^5$) with the resolution of 2. Summation of the weighted currents from I/Q paths is achieved by connecting two VM arrays to the proposed QG. The VM array naturally shows superior linearity performance than active amplifiers and the low input impedance of VM array further reduces the passive loss of QG compared to active common-source amplifying stages.

III. "FRACTIONAL-BIT-BASED" PHASE-INVARIANT PGA

The architecture of the "fractional-bit-based" PGA is shown in Fig. 2. The binary weighted cells in the PGA are implemented as polarity selectors with digital controlled differential common-source amplifiers (Fig.6). The complementary control signal determines the polarity of the output currents with unchanged input and output impedance. Constant input and output impedance guarantees unchanged parasitic capacitive loadings for a bit-setting-independent and phase invariant frequency response. As the integer-bit (0/1/2/3-bit) cell is used to generate normalized "2^N" or "-2^N" weighting, the PGA can achieve normalized AC output currents for all the odd integers within range of -15 to +15 ($\pm 2^0 \pm 2^1 \pm 2^2 \pm 2^3$) with the resolution of 2. To improve the gain resolution and obtain accurate dB-linear gain, two fractional-bits (0.5-bit) are introduced to ensure the normalized gain of

the PGA can be set between –16 to +16 ($\pm(0.5 \pm 0.5) \pm 2^0 \pm 2^1 \pm 2^2 \pm 2^3$) to achieve dB-linear gain with the resolution of 1. It is noteworthy that this scaling method is independent of transistor modeling, and thus robust to modeling inaccuracy and PVT variations.

IV. MEASUREMENT RESULT

The chip is fabricated in a 65-nm CMOS process while occupies 1350-μm by 250-μm die area (Fig. 2) and consumes 12-mA from 1-V supply voltage. We perform all characterizations using an N5247A network analyser. The I/Q gain and phase difference of the proposed high-order QG was indirectly measured by turning on only the I-path and Q-path VM arrays of PS (*ON* word = '111111/000000', *OFF* word = '111110/000001') as shown in Fig. 7, The measured quadrature frequency of the merged PS-PGA is wider than 37% (21 to 30.5-GHz), demonstrating the quadrature bandwidth of the proposed high-order QG is much wider than 37%. The higher-order effect is also evident in Fig. 7. Fig. 8 shows the measured gain response of the merged PS-PGA, which achieves 43% fractional BW$_{-3dB}$ (20 to 31-GHz) with 12.2-dB

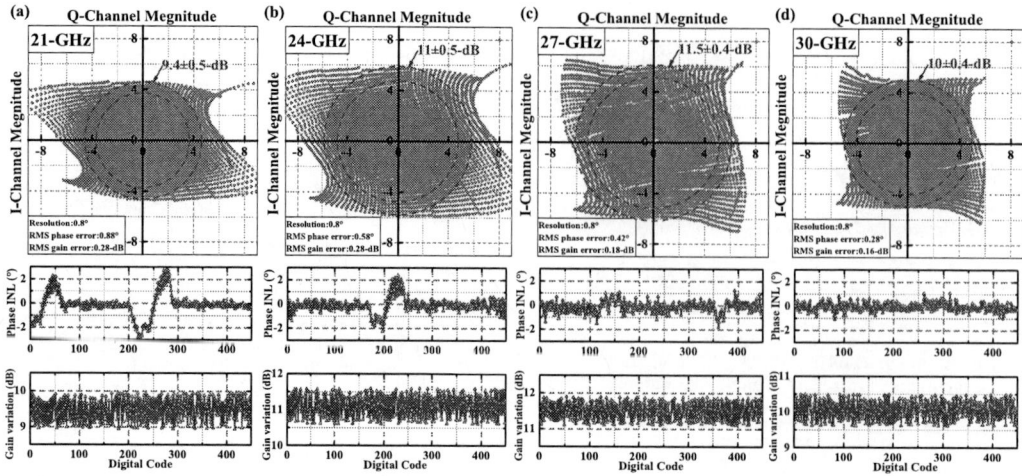

Fig. 9. Measured constellation points on the I/Q channel coordinates showing four phase interpolation examples with 0.8° step and 0.4/0.5-dB magnitude variations at (a) 21-GHz, (b) 24-GHz, (c) 27-GHz and (d) 30-GHz. The RMS phase errors are 0.88°, 0.58°, 0.42°and 28°, respectively. The RMS gain errors are 0.28, 0.28, 0.18 and 0.16-dB, respectively (up). Measured phase Integral nonlinearity (INL) error and gain variation vs. digital code (middle and bottom)

Table 1. Performance comparison of state-of-the art

	[1]	[3]	[4]	[5]	**This Work**
Type	Passive		Active		Passive +PGA
Tech.	130-nm SiGe	40-nm CMOS	SiGe	65-nm CMOS	65-nm CMOS
Freq (GHz)	28	22-36	20.5-26.5	27-29	21-30
Range (deg)	190	360	360	360	360
Phase res.(°)	91	45	5	5.625	0.8
RMS phase err. (deg)	0.6	<12.8	0.55-21	0.54	0.28-0.88
Gain var.(dB)	0.25	0.5	0.75	0.47	0.4-0.5
RMS gain err. (dB)	N/A	<0.6	>0.5	0.13	0.16
Gain (dB)	-9.3	-5.6	0.51	-3.1	12.2
Area (mm2)	0.18	0.132	0.12	0.32	0.052
Power (mW)	0	0	10	25.2	12

¹Estimate from figures   ²Only core area of PS

maximum gain and 24-dB gain range with 0.75-dB gain step. The RMS gain error is less than 0.4-dB from 18 to 32-GHz with the minimum value of 0.18-dB at 28-GHz, demonstrating high accuracy of dB-linear gain. The RMS phase error is plotted in Fig. 8 (bottom), and the value is less than 1° from 18 to 30-GHz, demonstrating that the "fractional-bit-based" PGA features independent, accurate gain control.

Fig. 9 shows the measured constellation points on the I/Q channel coordinates with 0.8° phase step within 0.4/0.5-dB amplitude variation at 21/24/27/30-GHz. The results demonstrate that thanks to the proposed high-order QG the merged PS-PGA achieve phase and magnitude bandwidth of wider than 37%. Table I compares this work with state-of-the-art, this work demonstrates the best phase accuracy with highest resolution and wide bandwidth.

V. CONCLUSION

This paper demonstrates a novel passive VM-based PS merged with two "fractional-bit-based" PGA stages. The PS employs a folded transformer-based high-order full-differential QG to substantially extend the quadrature bandwidth and reduce the die area. The PGA utilizes a novel robust "fractional-bit-based" method and achieves mathematically zero phase variant and accurate dB-linear gain across a wide tuning range. The phase control operates with 0.8° steps while maintaining a minimum RMS phase error of 0.42°, and the gain control operates with 0.75-dB steps while maintaining a minimum RMS gain error of 0.18-dB.

ACKNOWLEDGMENT

This work is supported by National Natural Science Foundation of China under grants 61574084 and National Major Research Program 2016YFA0201903. (Corresponding author: Yan Wang.)

REFERENCES

[1] B. Sadhu et al., "A 28-GHz 32-Element TRX Phased-Array IC With Concurrent Dual-Polarized Operation and Orthogonal Phase and Gain Control for 5G Communications," in IEEE Journal of Solid-State Circuits, vol. 52, no. 12, pp. 3373-3391, Dec. 2017.

[2] J. S. Park and H. Wang, "A Transformer-Based Poly-Phase Network for Ultra-Broadband Quadrature Signal Generation," in IEEE Transactions on Microwave Theory and Techniques, vol. 63, no. 12, pp. 4444-4457, Dec. 2015.

[3] M. Elkholy, S. Shakib, J. Dunworth, V. Aparin and K. Entesari, "Low-Loss Highly Linear Integrated Passive Phase Shifters for 5G Front Ends on Bulk CMOS," in IEEE Transactions on Microwave Theory and Techniques, vol. 66, no. 10, pp. 4563-4575, Oct. 2018.

[4] J. S. Park and H. Wang, "A K-band 5-bit digital linear phase rotator with folded transformer based ultra-compact quadrature generation," 2014 IEEE Radio Frequency Integrated Circuits Symposium, Tampa, FL, 2014, pp. 75-78.

[5] J. Pang, R. Kubozoe, Z. Li, M. Kawabuchi and K. Okada, "A 28GHz CMOS Phase Shifter Supporting 11.2Gb/s in 256QAM with an RMS Gain Error of 0.13dB for 5G Mobile Network," 2018 48th European Microwave Conference (EuMC), Madrid, 2018, pp. 807-810.

978-1-7281-1702-7/19 $31.00 © 2019 IEEE

2019 IEEE Radio Frequency Integrated Circuits Symposium

A 20 ~ 43 GHz VGA with 21.5 dB Gain Tuning Range and Low Phase Variation for 5G Communications in 65-nm CMOS

Tianjun Wu, Chenxi Zhao, Huihua Liu, Yunqiu Wu, Yiming Yu, Kai Kang*

University of Electronic Science and Technology of China, China

*kangkai@uestc.edu.cn

Abstract—This paper presents a broadband variable gain amplifier (VGA) with low phase variation in 65 nm CMOS technology. The mechanism of phase variation in CMOS VGA is analyzed. According to the analysis, the feedforward paths formed by parasitic capacitive and inductive couplings are one of the main factors that result in phase variation in CMOS VGA. In order to achieve low phase variation, a parasitic capacitor elimination technique is proposed to remove the feedforward path formed by gate-drain parasitic capacitor. Besides, an isolation enhancement layout technique is proposed to minimize the inductive coupling and parasitic capacitors in the layout of the variable gain stage. As results, the measured phase variation is 0.2° ~ 2° in 18 GHz ~ 37 GHz and 0.2° ~ 5.4° in 18 GHz ~ 45 GHz when the gain variation range is 21.5 dB. The measured peak gain is 14.5 dB with 3-dB bandwidth of 20 GHz ~ 43 GHz. The gain ripple in 3-dB bandwidth is < 2.5 dB. The noise figure and input-referred P1dB are 5.5 dB and -16.5 dBm in the maximum gain state, respectively. The chip consumes 30.8 mW from 1.1 V voltage supply and the core area is 370 um × 930 um.

Keywords—VGA, phase-invariant, CMOS, phased arrays, millimeter-wave, broadband.

I. INTRODUCTION

Phased-array technique is a key method to improve link robustness and effective isotropic radiated power (EIRP) in mm-wave wireless systems, including fifth-generation (5G) and IEEE 802.11ad/ay standards communication systems. In phased-array systems, sidelobe suppression is achieved by tuning the gain of VGA with gain tapering function [1]. VGA is also widely used in active phase shifter [3]. Therefore, VGA plays an important role in phased arrays.

The insertion phase of VGA should keep invariant over gain settings to maintain beam pointing direction [1]. Some advanced phase invariance techniques have been reported in SiGe technology [1][4][5][7]. In [1] and [5], the phase invariance technique based on local feedback is proposed. With the local feedback technique, a VGA with 18 dB gain tuning range and 5° phase variation is presented [5]. Works [4] and [7] use poles and zeros compensation technique to achieve 2° phase variation and 22 dB gain tuning range. However, those techniques are based on SiGe technology and cannot be applied in CMOS technology because of the different work mechanisms of BJT and MOS transistor.

In this paper, the mechanism of phase variation in CMOS VGA is analyzed, and a simple but effective phase invariance technique based on parasitic capacitor elimination and isolation enhancement layout is proposed. The proposed VGA achieves 21.5 dB gain tuning range and < 2° phase variation across 18 ~ 37 GHz.

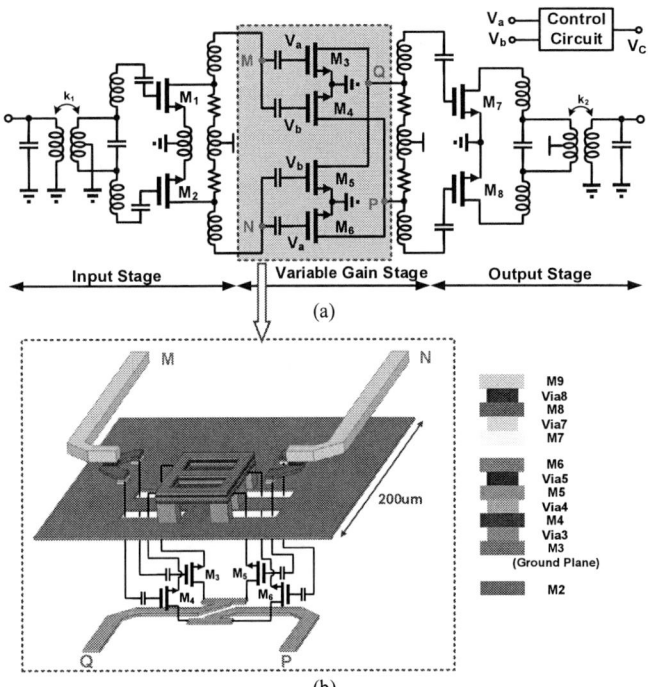

Fig. 1. (a) The schematic of the proposed VGA based on parasitic capacitor elimination technique; (b) The isolation-enhanced layout of the variable gain stage in VGA.

This paper is organized as follows. Section II analyzes the mechanism causing phase variation in CMOS VGA and describes the circuit and principle of the proposed phase-invariant technique. Section III shows the measurement results. Finally, the design and measurement conclusions are drawn in Section IV.

II. VGA TOPOLOGY AND PHASE-INVARIANT TECHNIQUE

Fig. 1(a) shows the schematic of the proposed VGA. The VGA includes three cascaded amplifiers made up of $M_1 \sim M_8$. The variable gain stage consists of $M_3 \sim M_6$, which have the same size. In Fig. 1(a), V_a denotes the bias voltage of M_3 and M_6, while V_b denotes the bias voltage of M_4 and M_5. V_a and V_b are generated by an external control voltage V_c.

A. Phase Variation Analysis of CMOS VGA

Common source (CS) amplifier is widely used in microwave and mm-wave bands because of its simple structure and high gain. Fig. 2 shows the schematic of a CS amplifier and its small signal equivalent circuit with parasitic

978-1-7281-1702-7/19 $31.00 © 2019 IEEE

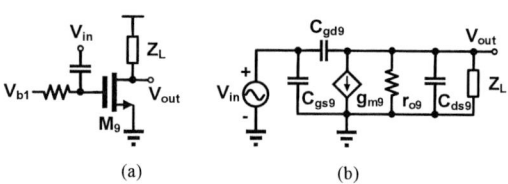

Fig. 2. Schematic of a CS amplifier and its small signal equivalent circuit.

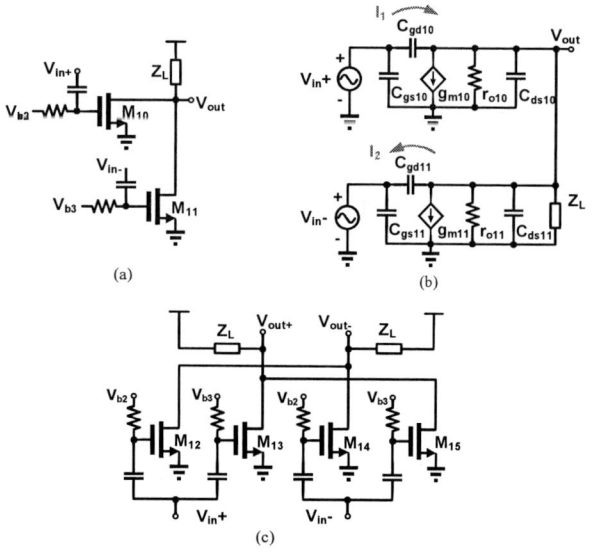

Fig. 3. Simplified schematic of the proposed variable gain amplifier (a) and its small signal equivalent circuit (b); (c) The full schematic of the proposed variable gain stage amplifier.

elements. The voltage transfer function is given by

$$\frac{V_{out}}{V_{in}} = \frac{g_{m9} - j\omega C_{gd9}}{-(\frac{1}{Z_L} + \frac{1}{r_{o9}} + j\omega C_{gd9} + j\omega C_{ds9})} \quad (1)$$

where g_{m9}, C_{gd9}, C_{ds9} and r_{o9} denote the transconductance, gate-drain parasitic capacitor, drain-source parasitic capacitor, and channel impedance of M_9, respectively. When the CS amplifier is used as VGA, the gain variation (ΔG) between two different gain states is defined based on (1) and written as

$$\Delta G = \frac{g_{m9,1} - j\omega C_{gd9}}{g_{m9,2} - j\omega C_{gd9}} \quad (2)$$

where $g_{m9,1}$ and $g_{m9,2}$ denote the transconductances of M_9 in different gain states, respectively. ΔG is a complex number and its amplitude and phase vary with frequency because of the non-zero C_{gd9}. Therefore, equation (2) reveals that the phase variation of a CS VGA is caused by the gate-drain parasitic capacitor.

B. Parasitic Capacitor Elimination Technique

According to equation (2), the feedforward path formed by C_{gd9} causes the phase variation and makes gain variation vary with frequency. In order to alleviate those problems, a VGA topology based on CS amplifier is proposed (Fig. 3(a)). M_{10} and M_{11} have the same size to guarantee that the gate-drain parasitic capacitors C_{gd10} and C_{gd11} have the same values. As

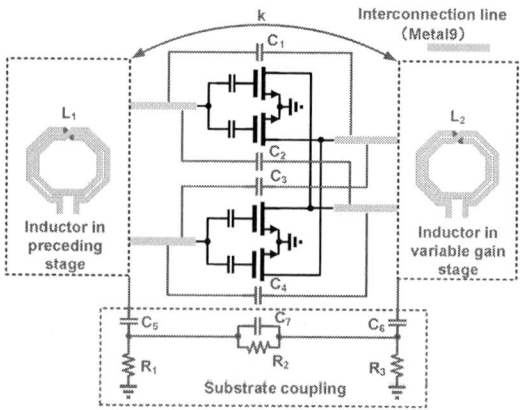

Fig. 4. Parasitic capacitors formed between interconnect lines, and inductive coupling through substrate and air.

shown in Fig. 3(b), the VGA is driven by a pair of differential signal V_{in+} and V_{in-} to make the currents flowing through C_{gd10} and C_{gd11} (I_1, I_2) have the same amplitude but opposite phase. The currents I_1 and I_2 are counteracted with each other at output node. Therefore, the feedforward paths formed by C_{gd10} and C_{gd11} are neutralized and have no impact on the phase and gain variation of the proposed VGA. The voltage transfer function of the proposed topology is given by

$$\frac{V_{out}}{V_{in+} - V_{in-}} = \frac{g_{m10} - g_{m11}}{-(\frac{1}{Z_L} + \frac{2}{r_{o10,11}} + 2j\omega C_{gd10,11} + 2j\omega C_{ds10,11})}. \quad (3)$$

The gain variation between two different gain states is defined based on (3) and given by

$$\Delta G = \frac{g_{m10,1} - g_{m11,1}}{g_{m10,2} - g_{m11,2}} \quad (4)$$

where $g_{m10,1}$, and $g_{m10,2}$ denote the transconductances of M_{10} in two different gain states, while $g_{m11,1}$ and $g_{m11,2}$ denote the transconductances of M_{11} in the same way. Equation (4) indicates that the VGA has no phase variation and its gain variation is independent on frequency. The gain variation range of the proposed VGA can be infinity theoretically because the denominator of equation (4) can achieve zero. Figure. 3(c) shows the full differential topology of the proposed variable gain stage amplifier.

C. Isolation-enhanced Layout Technique

As predicted by equation (2), the feedforward path formed by C_{gd} causes the phase variation in CMOS VGA. If there is no feedforward path, as calculated in equation (3) and (4), the VGA will have no phase variation.

However, there are many parasitic capacitors in layout that induced by interconnection lines. As shown in Fig. 4, parasitic capacitors ($C_1 \sim C_4$) are generated among the interconnection lines at the input and output nodes of the MOS transistors. Because those metal lines are very close and occupy large areas, the parasitic capacitors $C_1 \sim C_4$ cannot be ignored. The inductors in preceding stage and variable gain stage also present coupling. As shown in Fig. 4, the magnetic coupling (k) between the inductors L_1 and L_2 is inevitable because of the compact layout, where L_1 and L_2 denote the inductors in

978-1-7281-1702-7/19 $31.00 © 2019 IEEE   72

preceding stage and variable gain stage respectively. Due to the low resistivity of CMOS technology, the substrate provides another coupling path (formed by $C_5 \sim C_7$ and $R_1 \sim R_3$) for inductors. All the feedforward paths formed by capacitive coupling, magnetic coupling and substrate coupling in layout have high-pass frequency characteristic. Hence, the couplings increase with frequency and make the phase variation become worse with frequency increasing.

In order to minimize the coupling caused by layout, the layout of the variable gain stage is implemented with isolation enhancement technique. As shown in Fig. 1(b), the interconnection lines at the input nodes (M and N) of the variable gain stage are located in Metal9 and Metal8, while the others at the output nodes (P and Q) are placed in Metal2. Metal9 and Metal8 are isolated from Metal2 by ground plane (Metal3). Therefore, the parasitic capacitors ($C_1 \sim C_4$) and capacitive coupling are minimized. The magnetic coupling and substrate coupling are alleviated by increasing the distance between L_1 and L_2 to 200 um.

D. Gain Control Circuit

The gain control circuit of the variable gain stage is shown in Fig. 5 [2]. The two bias voltages V_a and V_b are generated from an external control voltage V_c. According to equation (3), the voltage transfer function of variable gain stage is given by

$$G_{var} = \frac{g_{m16-17} V_c \sqrt{n\mu_n C_{ox} (W/L)_{3-6}}}{-(1/Z_L + 2/r_o + 2j\omega C_{gd} + 2j\omega C_{ds})\sqrt{I_{SS}}} \quad (5)$$

where C_{gd}, C_{ds}, r_o and Z_L denote the gate-drain capacitor, drain-source capacitor, channel impedance of M_{3-6} and differential load impedance of the variable gain stage measured at P and Q nodes, respectively. From equation (5), the voltage transfer function of the variable gain stage is proportional to the control voltage V_c.

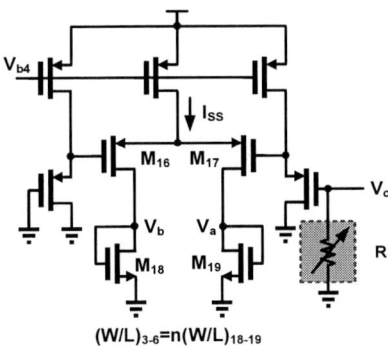

(W/L)$_{3-6}$=n(W/L)$_{18-19}$

Fig. 5. The schematic of control voltages generator.

III. MEASUREMENT RESULTS

The chip is fabricated in 65 nm CMOS technology with core area of 370 um × 930 um (Fig. 6). For measurement, the chip is mounted on a printed circuit board (PCB) of FR-4 using an electrically conductive silver epoxy. All bias feeding pads and ground pads are wire bonded to the PCB with gold wires. Measurements were conducted on chip by placing ground signal-ground (GSG) probes at the input and output

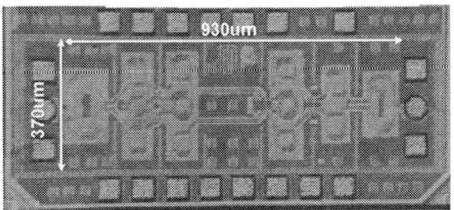

Fig. 6. Chip photo of the proposed VGA.

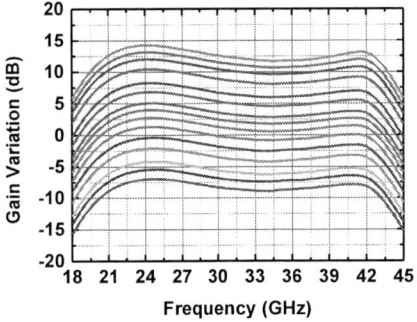

Fig. 7. Measured gain curves for various gain settings.

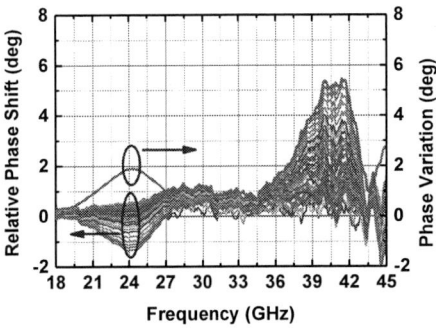

Fig. 8. Measured relative phase shift and maximum phase variation under different frequency when the gain variation is 21.5 dB.

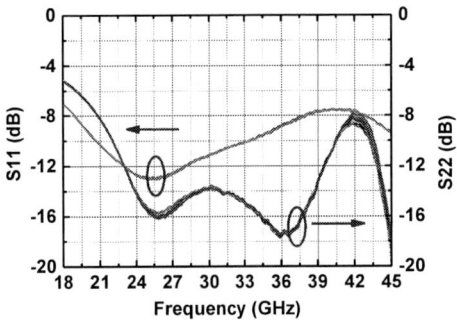

Fig.9. Measured input and output port return loss for various gain settings.

pads.

The s-parameter of the VGA was measured using vector network analyzer N5247A. As shown in Fig. 7, the measured gain variation is 21.5 dB across 18 ~ 45 GHz and the 3-dB bandwidth is 20 ~ 43 GHz with 2.5 dB gain fluctuation. The peak gain is 14.5 dB at 23.5 GHz. The phase variation is 0.2° ~ 2° across 18 ~ 37 GHz (shown in Fig. 8). The maximum phase variation in 3-dB bandwidth is 5.4°. The VGA provides

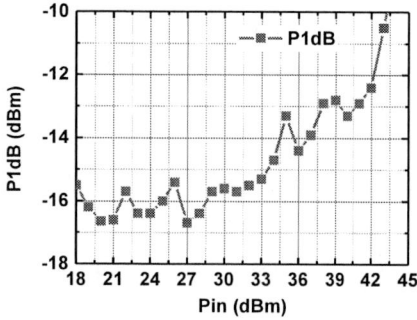

Fig. 10. Measured input-referred P1dB as the VGA in the maximum gain state.

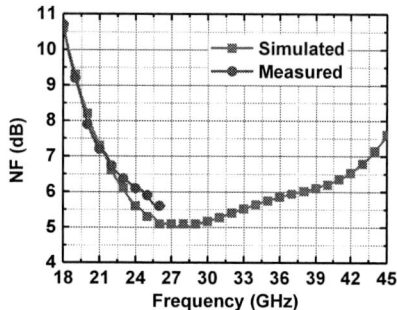

Fig. 11. Measured and simulated noise figure when VGA is at the maximum gain state.

input and output return loss of < -7 dB in 3-dB bandwidth for various gain states (Fig. 9). The linearity of the maximum gain state is measured. As shown in Fig. 10, the input-referred P1dB is -16.5 ~ -10.5 dBm in 3-dB bandwidth. The variation in P1dB results from the gain unflatness in each stage of the VGA. The noise figure is measured with noise figure analyzer N8975A, which has a maximum measurement frequency of 26.5 GHz. The measured noise figure is shown in Fig. 11. It can be observed that the VGA achieves measured noise figure of 5.5 ~ 8.2 dB in 3-dB bandwidth, which is consistent with simulation result. The measured dc current of the VGA is 28 mA under 1.1 V voltage supply. The performances are summarized in Table 1, along with the comparison with other state-of-art VGAs.

IV. CONCLUSION

In this paper, the mechanism that causes phase variation in CMOS VGA is analyzed, and the phase-invariant technique

based on parasitic capacitor elimination and isolation-enhanced layout is proposed. The feedforward path formed by gate-drain parasitic capacitor is neutralized by the parasitic capacitor elimination technique, and the capacitive coupling and inductive coupling in the layout of the variable gain stage are minimized by the isolation enhancement layout technique. The proposed CMOS VGA achieves a measured gain tuning range of 21.5 dB with 3-dB bandwidth of 20 ~ 43 GHz. In 26.5 ~ 35.5 GHz, the measured phase variation is < 1.2°. In 18 ~ 37 GHz, the measured phase variation is < 2°. The measured maximum phase variation across 3-dB bandwidth is 5.4°. Benefited from the wide 3-dB bandwidth and low phase variation, the VGA is very suitable for K- and Ka-band phased-array applications.

ACKNOWLEDGMENT

This work is supported by National Natural Science Foundation of China (Grant No. 61771115, 61874020, 61804024).

REFERENCES

[1] B. Sadhu et al., "A 28-GHz 32-Element TRX Phased-Array IC With Concurrent Dual-Polarized Operation and Orthogonal Phase and Gain Control for 5G Communications," in IEEE Journal of Solid-State Circuits, vol. 52, no. 12, pp. 3373-3391, Dec. 2017.

[2] S. Ray and M. M. Hella, "A 10 Gb/s Inductorless AGC Amplifier With 40 dB Linear Variable Gain Control in 0.13 um CMOS," in IEEE Journal of Solid-State Circuits, vol. 51, no. 2, pp. 440-456, Feb. 2016.

[3] B. Wang, H. Gao, M. K. Matters-Kammerer and P. G. M. Baltus, "A 60 GHz 360° Phase Shifter with 2.7° Phase Resolution and 1.4° RMS Phase Error in a 40-nm CMOS Technology," 2018 IEEE Radio Frequency Integrated Circuits Symposium (RFIC), Philadelphia, PA, 2018, pp. 144-147.

[4] F. Padovan, M. Tiebout, A. Neviani and A. Bevilacqua, "A 12 GHz 22 dB-Gain-Control SiGe Bipolar VGA With 2° Phase-Shift Variation," in IEEE Journal of Solid-State Circuits, vol. 51, no. 7, pp. 1525-1536, July 2016.

[5] B. Sadhu, J. F. Bulzacchelli and A. Valdes-Garcia, "A 28GHz SiGe BiCMOS phase invariant VGA," 2016 IEEE Radio Frequency Integrated Circuits Symposium (RFIC), San Francisco, CA, 2016, pp. 150-153.

[6] Y. Yi, D. Zhao and X. You, "A Ka-band CMOS Digital-Controlled Phase-Invariant Variable Gain Amplifier with 4-bit Tuning Range and 0.5-dB Resolution," 2018 IEEE Radio Frequency Integrated Circuits Symposium (RFIC), Philadelphia, PA, 2018, pp. 152-155.

[7] F. Padovan, M. Tiebout, A. Neviani and A. Bevilacqua, "A 15.5–39GHz BiCMOS VGA with phase shift compensation for 5G mobile communication transceivers," ESSCIRC Conference 2016: 42nd European Solid-State Circuits Conference, Lausanne, 2016, pp. 363-366.

Table 1. Performance comparison summary

	This Work	JSSC2016[4]	RFIC2016[5]	RFIC2018[6]	ESSCIRC2016[7]
Technology	**65nm CMOS**	SiGe	SiGe	65nm CMOS	SiGe
Frequency	**20~43GHz**	10~14.4GHz	28GHz	27~42GHz	15.5~30GHz
Peak Gain	**14.5dB**	13dB	22dB	9.6dB	17dB
ΔG	**21.5dB**	22dB	18dB	7.5dB	23dB
Phase Variation	**0.6°~2° #**	1.2°~2°	<5°	0.7°~3.5° *	1.5°~3°
NF	**5.5~8.2dB ***	5.1~6.8dB	-	-	3.6~9dB
P_{dc}	**30.8mW**	83mW	35mW	15.6mW	104mW
Chip Area	**0.34mm^2**	0.7mm^2	-	0.08mm^2	0.33mm^2

\# < 37GHz   *RMS phase error   ★ 20 GHz ~ 26.5GHz

A 26-GHz Vector Modulator in 130-nm SiGe BiCMOS Achieving Monotonic 10-b Phase Resolution Without Calibration

Ilker Kalyoncu[#1], Abdurrahman Burak[#], Mehmet Kaynak[$], Yasar Gurbuz[#]

[#]Faculty of Engineering and Natural Sciences, Sabanci University, Turkey
[$]IHP – Leibniz-Institut fur innovative Mikroelektronik, Germany
[1]ikalyoncu@sabanciuniv.edu

Abstract—This paper presents a high-resolution (10-b) vevtor-modulator (VM) phase shifter (PS) in 130-nm SiGe BiCMOS targeting 5G applications at 26 GHz. It employs a Gilbert-cell RF core, the tail current of which is controlled by an 8-b low-power current-steering DAC and 2-b I/Q sign switches. The DAC includes an on-chip PTAT current reference with process compensation capabilities. A 2-stage RC polyphase filter (PPF) is used to generate the quadrature signals. Without any calibration or correction of PS control signals, the measured results demonstrate completely monotonic 2^{10} phase states, covering the full 0-360° range without any dead zones or overlapping phase states. The worst case (maximum) phase difference between any adjacent states is 0.65°. The VM exhibits an average insertion loss of 0.5 dB at 26 GHz with a 3-dB BW of 8 GHz, an rms amplitude error of 0.2 dB, IP_{1dB} of 2 dBm, and 23 mW dc power dissipation. Potential applications are in RF beamforming and RF self-interference cancellation.

Keywords — Vector modulator, phase shifter, SiGe, 5G

I. INTRODUCTION

Next-generation (5G) mobile wireless communications have challenging requirements such as peak data rate up to 20 Gbps and ultra-low user plane latency down to 1 ms. Several promising technologies such as mm-waves, small cells, phased arrays, full-duplex etc. have been investigated both by academia and industry in recent years to achieve these requirements [1].

SiGe technology has been the focus of research on 5G transceivers, due to its low cost, high integration capability, low noise and moderate output power performance, in comparison to CMOS and III-V technologies. Recently, several works on SiGe-based multi-channel RF beamforming 5G phased arrays have been reported [2]–[4].

Vector summation is a promising phase shifter topology for mm-wave 5G applications as it provides the highest bit resolution in a compact area. Despite the fact that many studies can be found in the recent literature on vector modulators [3]–[6], no work has been reported yet on high resolution vector modulators for 26 GHz 5G applications.

High resolution phase shifters are demanded to achieve precise beam steering in phased arrays, which will be crucial in overcoming increased path loss of mm-waves. For instance, 0.1° beam-steering resolution was reported in [7] using LO phase shifting. High resolution phase control is also critical in improving self-interference cancellation performance of full-duplex front-ends, widening the cancellation bandwidth, and even enabling hybrid technologies that combine full-duplex operation and beamforming [8].

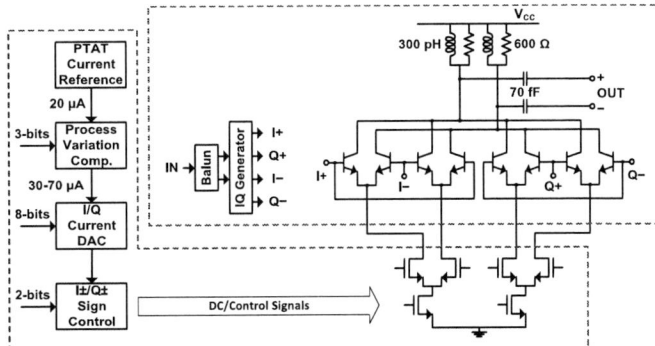

Fig. 1. The architecture of the process-compensated high-resolution (10-b) vector modulator phase shifter. Red box: RF part, blue box: dc/control part.

In this paper, we present a high resolution (10-b) vector modulator (VM) in 130 nm SiGe BiCMOS process for 5G applications at 26 GHz. The design employs a Gilbert-cell type core, an 8-b on-chip current-steering DAC to control the tail currents, a 2-b I/Q sign switches, an on-chip PTAT current reference, and a process compensation circuitry. The VM monotonically generates all the phases in 0-360° range without any dead zones, with a maximum 0.65° phase difference between any adjacent phase states. It can be configured as a state-of-the-art 8-b phase shifter with rms phase and amplitude errors of 0.2° and 0.2 dB, respectively.

II. CIRCUIT DESIGN

The design of the vector modulator can be divided into RF circuitry and dc/control circuitry, as shown in Fig. 1. The RF part includes a balun, quadrature generator and Gilbert-cell type vector modulator; and the dc design part includes a PTAT current reference, process compensation circuitry, current steering DAC, and cascode tail current source.

A. Transformer Balun

In order to achieve high linearity, the vector modulator uses a transformer balun instead of an active balun. At 26 GHz, transformer baluns are considerably small compared to transmission line based baluns. The design can be seen in Fig. 2. It converts a single-ended 50 Ω to differential 100 Ω. The 200 pH primary coil was realized by a single turn top-metal 2 trace of width 12-μm and inner diameter of 80-μm. The 440 pH secondary coil was realized by a two-turn top-metal 1 trace of width 10-μm and inner diameter of 43-μm. This

978-1-7281-1702-7/19 $31.00 © 2019 IEEE

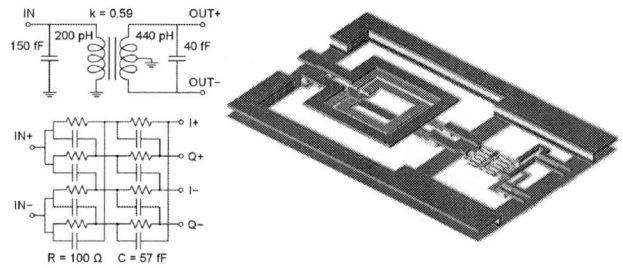

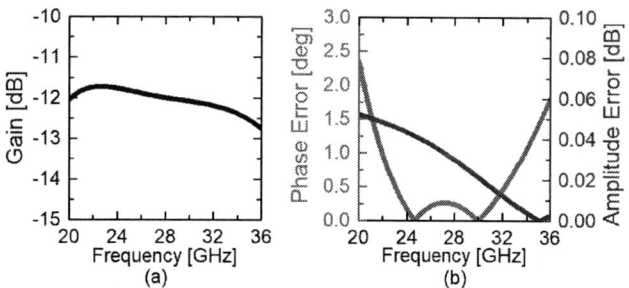

Fig. 2. The schematic and 3D layout view of the transformer balun and 2-stage RC polyphase filter.

Fig. 3. The simulated (a) gain and (b) phase/amplitude imbalance of the cascaded transformer balun and 2-stage RC polyphase filter.

geometry produces a moderate magnetic coupling coefficient of $k = 0.59$ to enable a wideband matching performance. It was optimized to maximize the available power gain of the transformer. The 150 fF and 40 fF MIM capacitors tune out the coil inductances and perform the matching. Two 80 fF capacitors are used in series to ensure symmetry. The overall balun has a simulated insertion loss of 1.5 dB at 26 GHz.

B. Quadrature Generator

The balun is followed by a 2-stage RC polyphase filter. Its advantages are good amplitude and phase balance between its outputs, high linearity, and very compact area; and its main disadvantage is insertion loss. A quadrature all-pass filter (QAPF) could have been used instead, providing significantly less insertion loss, but we preferred phase/amplitude balance advantage of the 2-stage PPF over the insertion loss advantage of QAPF. We used the constant amplitude configuration of the 2-stage RC PPF as shown in Fig. 2, where $R = 100\ \Omega$ and $C = 57$ fF. The actual layout implementation is slightly different than the schematic drawing to ensure symmetry and utilizes dummy resistors and capacitors for better device matching.

Fig. 3 shows the EM simulation results of the cascaded balun and quadrature generator. Its insertion loss is 12 dB at 26 GHz and input return loss is better than 10 dB from 20 to 36 GHz. By tuning the line lengths after the 2-stage PPF, the phase error was adjusted to exhibit two zero-crossings that were placed on separate sides of 26 GHz, as can be seen in Fig. 3(b), to maintain a good phase balance between the outputs even under process variations. The phase error is less than 1° in 22-34 GHz band and the amplitude error is negligible.

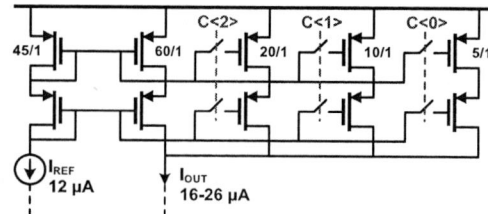

Fig. 4. Schematic view of the PMOS cascode current mirror with 3-b process compensation capability.

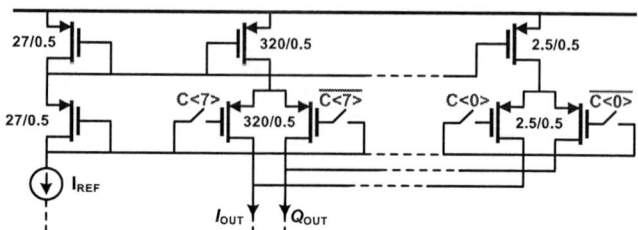

Fig. 5. Schematic view of the 8-b, PMOS, binary-weighted, current-steering DAC.

C. Vector Modulator Core

Differential I and Q signals of the PPF are fed to the core of the vector modulator as shown in Fig. 1. The HBT sizes are $8 \times 0.48\ \mu$m and they are biased for a maximum tail current of 7.5 mA. The HBT sizes are chosen to provide a trade-off between power consumption and RF output power. The 300 pH shunt inductor and 70 fF series capacitor form the output matching network for a 100 Ω differential load, and a 600 Ω resistor is used to widen the output matching bandwidth. This type of modulator has the benefit of a constant power consumption that is independent of its setting and input drive power, unlike the current steering type modulators whose power consumption changes as a function of the phase setting. The tail current of the HBTs are provided by NMOS transistors. The bottom NMOS acts as a current source and the upper ones act as a sign switch for I and Q signal paths.

D. DC Bias and Control Circuit

The performance of a high-resolution vector modulator strongly depends on its dc bias and control voltages. In this work, to tolerate process and temperature variations, dc bias and control voltages are generated by on-chip circuit blocks: a PTAT current reference followed by a 3-b process compensating PMOS current mirror, an 8-b current-steering DAC, and a cascode current mirror with 2-b sign switches.

The on-chip PTAT current reference has a nominal output current of 12 μA at room temperature. It draws 720 μA from a 2.5 V supply for a power consumption of 1.8 mW. To compensate for process variations, a PMOS cascode current mirror with 3-b control was used, as seen in Fig. 4. The output current of this stage is in the range of 16-26 μA, i.e. a 21 μA nominal current with $\pm 25\%$ tunable range.

The critical part of the dc/control circuitry is the current steering DAC. It is based on cascode PMOS current steering topology, as shown in Fig. 5. The sizes of PMOS transistors

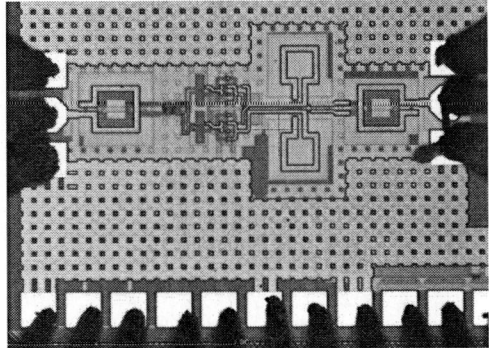

Fig. 6. Die photo.

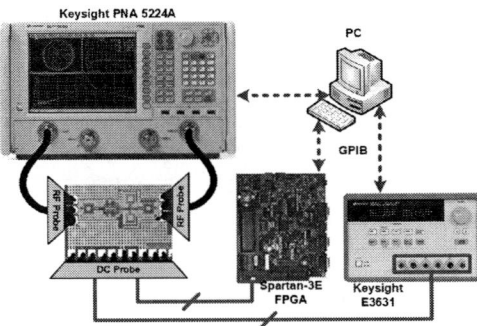

Fig. 7. Measurement setup of the vector modulator.

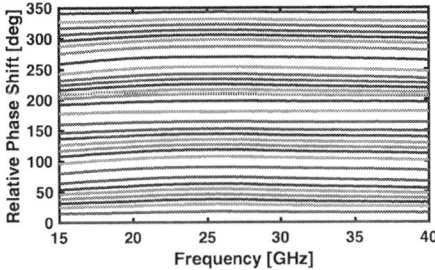

Fig. 8. Measured relative phase shifts of the VM for different states. Only 32 major states are shown for clarity.

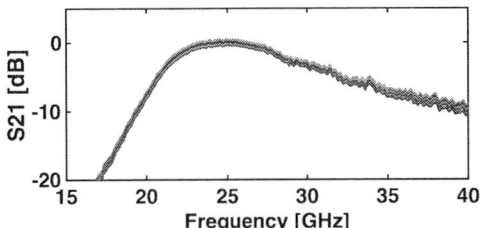

Fig. 9. Measured gain of the VM for all states.

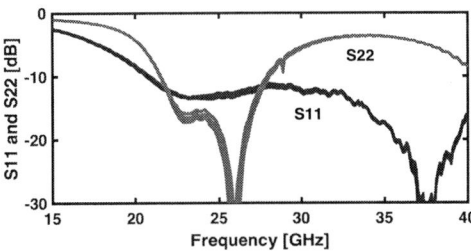

Fig. 10. Measured input and output matching for all phase states.

in each DAC cell are binary weighted between $W/L = 2.5\mu\text{m}/0.5\mu\text{m}$ and $320\mu\text{m}/0.5\mu\text{m}$. The long channel devices ($L = 0.5$ μm) help increase the linearity of the DAC. The maximum output current of the DAC is 500 μA.

The final part of the dc/control circuitry is NMOS cascode current mirrors to provide the tail currents of the Gilbert-cell type vector modulator, as shown in Fig. 1. The common-source transistors are sized $10/0.5$ μm and $150/0.5$ μm, and the common-gate transistors are sized $6/0.13$ μm and $90/0.13$ μm, to provide a maximum tail current of 500 μA $\times 15 = 7.5$ mA. The common gate transistors are controlled via 2-b NMOS sign switches.

III. MEASUREMENT RESULTS AND DISCUSSION

The PS is fabricated in IHP SG13S BiCMOS technology. The process offers HBTs with an f_T/f_max of $250/340$ GHz, CE breakdown voltage of 1.6 V, 5 thin and 2 thick top metal layers, and 1.2 V logic and 3.3 V I/O CMOS. The chip micrograph is shown in Fig. 6. It includes input/output baluns for single-ended measurements. The IC also features a custom-designed SPI for the 10-b phase and 3-b process compensation controls. All digital pads include ESD protection circuitry. The chip area is $0.8 \times 0.6 = 0.48$ mm², excluding the pads; and the core area is $0.4 \times 0.6 = 0.24$ mm², excluding the baluns.

The pad-to-pad S-parameters are measured with a PNA N5224A network analyzer and an RF probe station, using the setup shown in Fig. 7. 100-um GSG Z-probes are used for the

input/output, and the supply/digital controls are provided by a GGB dc probe.

Fig. 8 shows the relative insertion phase across different phase settings. Here, only the 5-MSB control is swept to display the phase shifts for brevity and clarity. These results are obtained without any phase calibration. These phase shifts do not correspond to a 5-b phase shifter, since the DAC in this work was implemented in a binary fashion. Nevertheless, if these phase states are treated as the states of a 5-b phase shifter, the VM achives 4-5° rms phase error without any calibration. The other 5-LSB control can easily be used to improve the performance the phase shifter. For instance, we were able to synthesize an 8-b phase shifter with 0.2° rms phase error, which is a state-of-the-art performance.

Fig. 9 shows the measured insertion gain across different phase settings. After deembedding the combined 3 dB loss of the input/output baluns that were separately measured, the average loss is around 0.5 dB, with an rms gain error of 0.2 dB. Input/output of the PS is well-matched to 50 Ω, as seen in Fig. 10, which is predominantly determined by the baluns. The measured group delay is around 40 ps at 26 GHz center frequency.

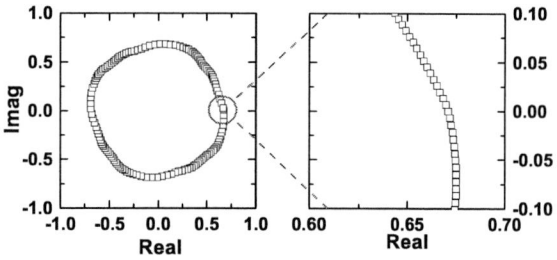

Fig. 11. Measured constellation of S21 at 26 GHz. (Left) One out of eights states is shown for brevity. (Right) Zoomed in version showing all the states.

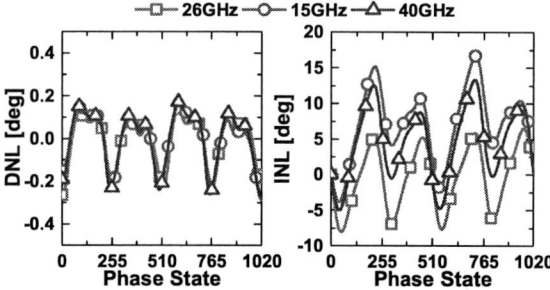

Fig. 12. Measured INL and DNL for the vector modulator.

Our aim in this work was to obtain perfectly monotonic phase states. We measured all the $2^{10} - 4$ phase states, and observed perfectly monotonic phase shifts, without any dead zones from 0 to 360°. (−4 is due to the four redundant phase states in $I/Q_{\pm}$ transitions). Fig. 11 shows the S21 constellation, and its zoomed-in version. The measured worst case (largest) phase difference between any adjacent states is 0.65°. As would be expected and can be seen in the figure, these largest gaps occur around the quadrature reference vectors.

Fig. 12 shows the DNL and INL of the VM for each 1020 control code, at three different operating frequencies, treating the VM as a digital-to-phase conterter. For almost all control codes, the measured DNL is between ±0.2° and its worst case value is −0.3°, meaning the phase states are completely monotonic. The best INL performance is obtained at the center frequency (between 5 and −8°) and it degrades further for higher/lower frequencies. This INL performance was expected, though, since the DAC cells were binary weighted. The target application of this work was adaptive self-interference cancellation in full-duplex radios, which requires good DNL but not INL.

The vector modulator core draws 7.5 mA from a 2.5 V supply. The total power consumption is 23 mW including the dc biasing and control circuitry.

For completeness, Table 1 compares state-of-the-art silicon VMs at 26-28 GHz. Two versions are included in the table for this work. The 5-b version represents the uncalibrated PS performance obtained by sweeping only the 5-MSBs, without using the other 5-b control. The 8-b version represents the PS that was synthesized after measuring all the phase states. The presented work achieves the highest resolution and linearity

Table 1. State-of-The-Art Silicon-Based Vector-Modulators at 26-28 GHz

Reference	This W.	[3]	[4]	[5]	[6]
Technology	130nm SiGe	180nm SiGe	130nm SiGe	130nm CMOS	180nm CMOS
Phase Resolution	8-b/5-b	6-b	4-b	6-b	4-b
Phase Err. [°rms]	0.2/4	3.4	5.4	2.6	1.5
Gain [dB]	−0.5	1	10.5	−5	−15
Gain Err. [dB rms]	0.2	0.5	0.6	0.31	0.45
NF [dB]	17	12	7	18	15
IP$_{1dB}$ [dBm]	2	−7	−17	−10	12
P$_{DC}$ [mW]	23	-	136	27	0
Chip Area [mm^2]	0.45	-	0.3	0.3	0.31

(as [6] is fully passive), lowest rms phase/gain error, and comparable performance in other aspects.

IV. CONCLUSION

A 26-GHz high-resolution (10-b) vector modulator in 130-nm SiGe have been presented. High resolution is obtained by an on-chip current steering DAC and process compensation circuitry. High linearity is achieved via a transformer balun and 2-stage RC PPF. The modulator achieved, without any calibration, completely monotonic phase states in 0-360° range without any dead zones or overlapping phase states, with a maximum 0.65° phase difference between any adjacent states. It has an average 0.5 dB insertion loss, 0.2 dB rms gain error, 2 dBm IP$_{1dB}$, and 23 mW dc power consumption. These state-of-the-art measured results are promising for 5G phased array transceivers and self-interference canceling circuits.

ACKNOWLEDGMENT

This work was supported by Aselsan doctorate scholarship program. Authors would like to thank IHP Microelectronics for IC fabrication.

REFERENCES

[1] S. Talwar *et al.*, "Enabling technologies and architectures for 5G wireless," in *IEEE MTT-S Int. Microw. Symp. Dig.*, pp. 1-4, June 2014.

[2] B. Sadhu *et al.*, "A 28-GHz 32-element TRX phased-array IC with concurrent dual-polarized operation and orthogonal phase and gain control for 5G communications," *IEEE J. Solid-State Circuits*, vol. 52, no. 12, pp. 3373-3391, Dec. 2017.

[3] K. Kibaroglu, M. Sayginer and G. M. Rebeiz, "An ultra low-cost 32-element 28 GHz phased-array transceiver with 41 dBm EIRP and 1.01.6 Gbps 16-QAM link at 300 meters," in *IEEE RFIC Symp. Dig. Papers*, June 2017, pp. 73-76.

[4] Y.-S. Yeh, E. Balboni and B. Floyd, "A 28-GHz phased-array transceiver with series-fed dual-vector distributed beamforming" in *IEEE RFIC Symp. Dig. Papers*, June 2017, pp. 65-68.

[5] F. Akbar and A. Mortazawi, "A frequency tunable 360° analog CMOS phase shifter with an adjustable amplitude," *IEEE Trans. Circuits Syst. II, Exp. Briefs*, vol. 64, no. 12, pp. 1427-1431, Dec. 2017.

[6] C.-W. Wang, H.-S. Wu, and C.-K. C. Tzuang, "CMOS passive phase shifter with group-delay deviation of 6.3 ps at K-band," *IEEE Trans. Microw. Theory Tech.*, vol. 59, no. 7, pp. 1778-1786, July 2011.

[7] J. Pang *et al.*, "A 28GHz CMOS Phased-Array Transceiver Featuring Gain Invariance Based on LO Phase Shifting Architecture with 0.1-Degree Beam-Steering Resolution for 5G New Radio," *IEEE Radio Freq. Integ. Circuits Symp. (RFIC)*, 2018, pp. 56-59.

[8] M. B. Dastjerdi, N. Reiskarimian, T. Chen, G. Zussman and H. Krishnaswamy, "Full duplex circulator-receiver phased array employing self-interference cancellation via beamforming," *IEEE RFIC Symp. Dig. Papers*, June 2018, pp. 108-111.

A 20–32GHz Digital Quadrature Transmitter with Notched-Matching and Mode-Switch Topology for 5G Wireless and Backhaul

Huizhen Jenny Qian, Yiyang Shu, Jie Zhou, and Xun Luo

Center for Integrated Circuits, UESTC, Chengdu 611731, China

Abstract — In this paper, a mm-wave wideband digital quadrature transmitter with improved efficiency for the 5G wireless and backhaul communication is presented. A novel synthesized notched-matching network is proposed to decrease the impedance mismatch from the interconnection of power digital-to-analog converter (power-DAC) cells operating at mm-wave bands. The performance is further optimized at the low- and high-bands by mode-switch of the quadrature signal generator, interstage matching of sign map, and power-DAC output-matching, simultaneously. Based on the mechanisms mentioned above, a 2×10-bit digital quadrature transmitter operating at 20–32GHz is implemented and fabricated using a conventional 28-nm CMOS technology, which exhibits the saturated output power of 19.02dBm, 34.4% maximum drain efficiency, and maximum system efficiency of 22.1%, respectively. Such mm-wave transmitter can support 64QAM modulation signals with 3Gb/s data-rate, −28.9dB EVM, 9.96dBm output power, and −33.6dBc ACPR.

Keywords — CMOS, digital quadrature transmitter, mode-switch, notched-matching, power-DAC, wideband.

I. INTRODUCTION

Transmitters with the high output power and efficiency are demanded for gigabits data-rate applications, including the 5G wireless and point-to-point backhaul systems. In the traditional analog transmitters, the digital-to-analog converters (DACs) are essential to convert the I/Q digital baseband signal to an analog type. However, commercial DACs with gigabits data-rate typically consumes watt-level dc power. Thus, the total efficiency of analog transmitters is decreased significantly considering the power consumption of high data-rate DACs. Fully digital quadrature transmitter [1] directly converts the digital baseband signal to the radio frequency (RF) signal, which has multiple functions of DAC, frequency up-conversion, and power amplification. Such transmitters can operate with a low voltage supply in deep submicron CMOS technology to achieve the relatively higher output power and efficiency, which exhibit better compatibility with digital baseband and higher integration level as well. For applications in mm-wave with gigabits data-rate, the digital quadrature transmitter with power-DAC is potentially competitive to improve the system efficiency and output power. Recently, some mm-wave power-DACs are presented [2-3]. However, due to the parasitics of complex interconnections in power-DAC, energy efficiency digital quadrature transmitter with high resolution and wideband operation in mm-wave is still challenging.

In this paper, a 20–32GHz 2×10-bit digital quadrature transmitter with the synthesized notched-matching network is

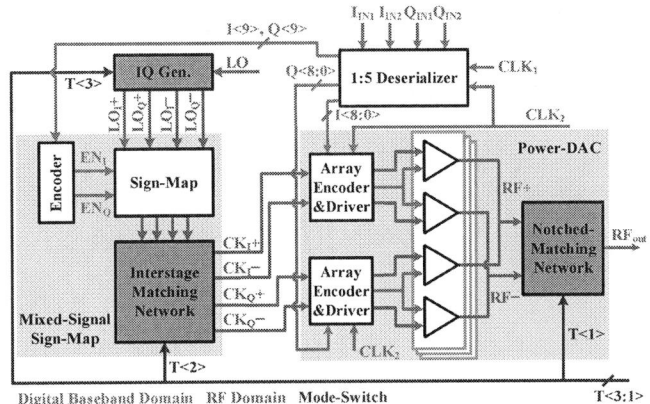

Fig. 1. Block diagram of the proposed digital quadrature transmitter.

proposed, which can suppress the impedance mismatch due to the parasitics of power-DAC cells interconnection within a wideband. To further enhance the operation bandwidth and efficiency of the mm-wave transmitter, a mode-switch topology is introduced. The proposed wideband digital transmitter in 28-nm CMOS achieves the state-of-the-art power efficiency and data-rate.

II. DIGITAL QUADRATURE TRANSMITTER DESIGN

A. Transmitter Architecture

As the simplified block diagram shown in Fig. 1, the proposed 2×10-bit digital quadrature transmitter is composed of the quadrature signal generator, mixed-signal sign-map, mm-wave quadrature power-DAC with notched-matching network, and deserializer. The quadrature signal generator converts the input LO signal to quadrature signals (i.e., LO_I+, LO_Q+, LO_I-, LO_Q-). The 2×10-bit base-band IQ data (i.e., $I<9:0>$ and $Q<9:0>$) are restored by a high data-rate deserializer from the serial inputs (i.e., I_{IN1}, I_{IN2}, Q_{IN1}, and Q_{IN2}). Two sampling clocks are required for the deserializer, with $CLK_1 = 5\times CLK_2$. The quadrant selection is realized by the mixed-signal sign-map block. The sign bits of $I<9>$ and $Q<9>$ are converted to enable signals of the sign-map. Such sign-map with a tunable interstage matching network amplifies the output ac current of CK_I+, CK_Q+, CK_I-, CK_Q-, and provides dc bias for the inputs of the power-DAC. The power-DAC with the array encoder, drivers, switch-cells array, and notched-matching network directly converts the digital baseband to amplified RF output. Meanwhile, the IQ generator, interstage matching of the sign-map, and output notched-matching are reconfigurable with switched capacitors

978-1-7281-1702-7/19 $31.00 © 2019 IEEE

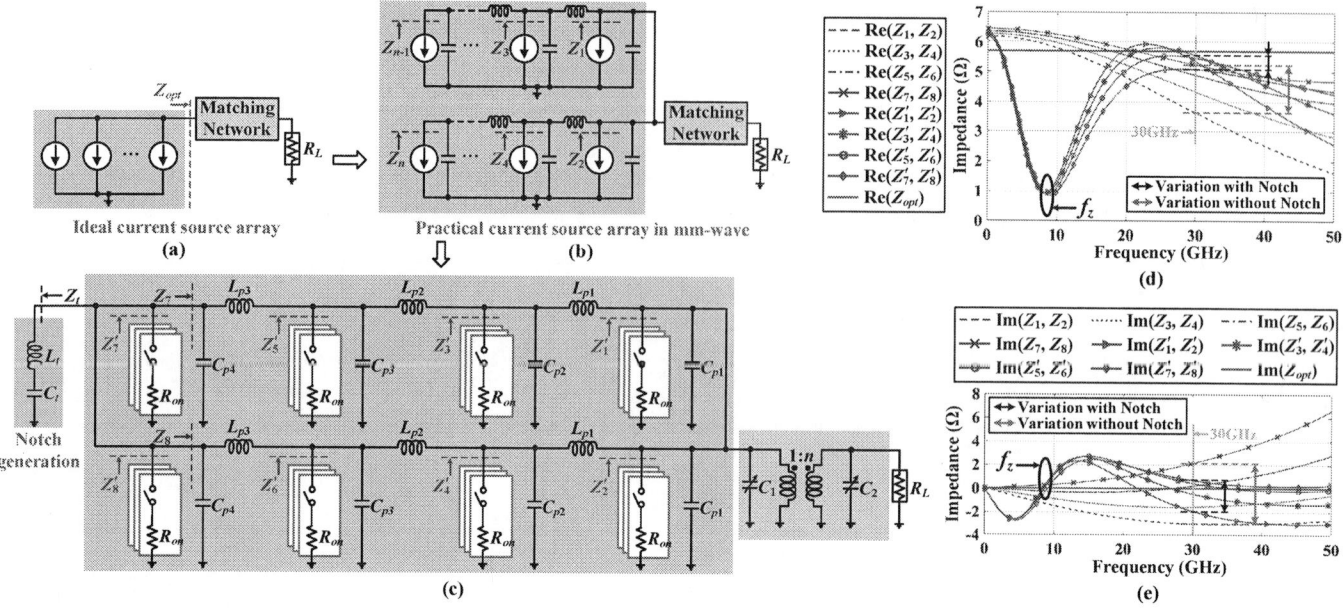

Fig. 2. Model of (a) ideal power-DAC, (b) practical power-DAC in mm-wave, (c) power-DAC with notch generation, simulated (d) real and (e) imaginary parts of the sub-array's output impedance in the three models.

to achieve a mode-switch operation, which can enhance the wideband performance of the proposed transmitter.

B. Notched-Matching Network for Power-DAC

As depicted in Fig. 2(a), the ideal RF power-DAC is normally consisted of the current source array with matching network. However, at mm-wave, the parasitic inductance and capacitance from interconnections of unit cells in the current source array have notable impact on the performance of the power-DAC, as shown in Fig. 2(b). Note that, the actual impedances (i.e., Z_i, where i = 1, 2, ..., 8) in Fig. 2(b) seen by the current sources are varied compared to the ideal impedance of Z_{opt} in Fig. 2(a). Therefore, to decrease the impact of the interconnection parasitics, a resonant tank with L_t and C_t connected in parallel with the current source array is introduced, as depicted in Fig. 2(c). Such resonant tank generates a notch with the frequency of f_z, which is lower than the power-DAC operation band. Here, the mm-wave current source array in Fig. 2(c) is distributed as 8 sub-arrays with load impedances of Z_i'. To investigate the characteristics of sub-arrays, the Z_7' or Z_8' is derived as follow:

$$Z_7' = Z_8' = Z_7 || Z_8 || Z_t$$
$$= X \sqrt{\frac{A^2 + B^2}{A^2 + (B+X)^2}} e^{j(\frac{\pi}{2} + tan^{-1}\frac{B}{A} - tan^{-1}\frac{B+X}{A})} \quad (1)$$

where the X, A, and B are determined by

$$Z_t = jX = j(\omega L_t - \frac{1}{\omega C_t}) \quad (2)$$

$$Z_7 || Z_8 = Z_7/2 = A + jB \quad (3)$$

here Z_7 and Z_8 are the input impedances as labeled in Fig. 2(c). From (1), it can be concluded that the resonant tank with

an impedance of Z_t introduces the same resonant frequency of f_z for the power-DAC, once the $im(Z_7', Z_8') = 0$.

Note that, the following two issues cause the performance degradation of the mm-wave power-DAC: 1) The impedance mismatch of Z_i compared to Z_{opt} in the target operation band (i.e., 20–32GHz) is significant; 2) The load impedance variation of different sub-arrays increases at a higher operation frequency, especially for mm-wave. The real/imaginary parts of the sub-arrays' load impedance (i.e., Z_i' and Z_i) with/without the notch generation are compared with the ideal impedance Z_{opt} in Fig. 2(d) and Fig. 2(e), respectively. Particularly, the ranges of sub-array's load impedance variations at 30GHz are labeled, which is decreased with the notched-matching. Meanwhile, with such matching network, the average impedance mismatch compared to Z_{opt} is suppressed from 0.77+j1.75Ω to 0.15+j0.87Ω at 30GHz. Therefore, the efficiency and output power of the mm-wave power-DAC can be improved and optimized with the proposed notched-matching network.

C. Power-DAC Array Implementation

The detailed layout-configuration of the power-DAC array including the unit cells, interconnection, and notched-matching network is shown in Fig. 3. The 2×9-bit IQ data are segmented into 7-bit MSB with thermometer code and 2-bit LSB with binary code, respectively. The switch cells with the RF phase of I+ and Q+ are placed in pairs. Meanwhile, the switch cells with the opposite phase of I– and Q– are located in adjacent rows of I+ and Q+ to enhance the common-mode suppression. The MSB array is arranged by 16 rows with 8 MSB cells in each row, except the last row with 7 MSB cells and 1 dummy cell. Note that, the switch unit cell is comprised of cascode circuits with its own driving circuits.

978-1-7281-1702-7/19 $31.00 © 2019 IEEE

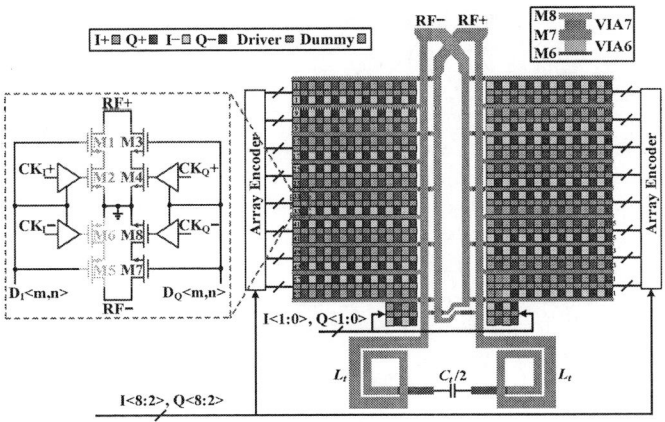

Fig. 3. Layout configuration of the power-DAC array with unit-cells interconnection and notched-matching network.

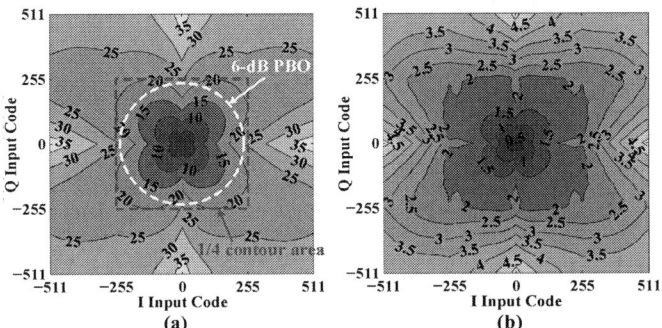

Fig. 4. Simulated (a) DE (%) contour and (b) DE improvement (%) comparing to power-DAC without notched-matching.

The IQ data ($D_I<m,n>$, $D_Q<m,n>$) are synchronized with CLK_2 in digital control blocks. The same data is shared for the gate inputs of M1 and M5 with a differential operation. The RF quadrature signals are shaped to the quasi square-wave by drivers. Here, the drivers are enabled by D_I, D_Q as well. Therefore, both transistors in the cascode circuit are switched on/off under the control of D_I, D_Q. This topology improves the isolation between the LO signal and RF output. The simulated LO leakage is improved by 35dB comparing to the topologies with only one transistor (M1 or M2) enabled by base-band data. To simplify the digital control routings, the array encoders are distributed besides the MSB switch array, which can further decrease the circuit size and interconnection parasitics of the power-DAC.

The notch generation circuits with two inductors and a capacitor are located on the bottom of the power-DAC array, while the output matching network are arranged on the top of the power-DAC array. The simplified layout of the power-DAC array interconnection with notched-matching network is depicted in Fig. 3. Since the power-DAC core operates with the 50% duty cycle waveform, the output power summing by I and Q paths suffers from efficiency loss with overlapped drain current and voltage of the core transistors in 25% cycle. Therefore, the simulated drain efficiency (DE) achieves peak of 37.9% with I = 511 and Q = 0 (only one

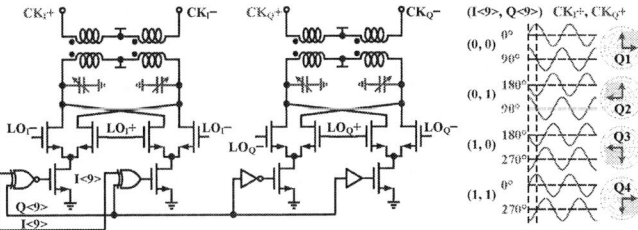

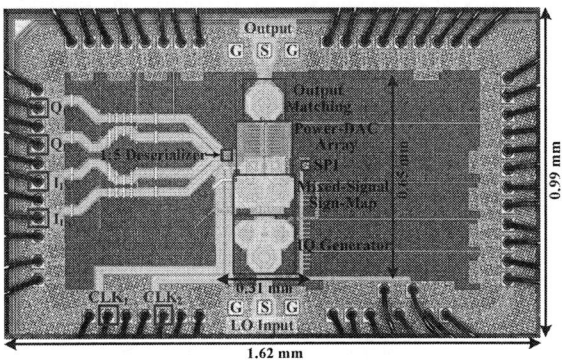

Fig. 5. Schematic of the mixed-signal sign-map and its behavior model.

path turned-on). At saturated power (I=511, Q=511), DE is degraded to 27.5%. As shown in Fig. 4(a), DE is larger than 20% with more than 3/4 contour area, which indicates a relatively high average efficiency. Besides, at the 6-dB power back-off (PBO), the average DE is 19.8%. With the notched matching network, DE is improved by maximum 5.2% (I = 511, Q = 0), as depicted in Fig. 4(b).

D. Mixed-Signal Sign-Map

The mixed-signal sign-map adopts two Gilbert-like circuits and transformer-based matching network, as shown in Fig. 5. The sign bits (I<9>, Q<9>) are converted to enable signals by the logic circuits, which could control the output polarity of CK_I+, CK_Q+, CK_I-, CK_Q-. The quadrant of the transmitter output signal is determined accordingly. Switched capacitors are used to achieve the optimized dual-mode interstage matching in a wideband. The power consumption of the sign-map circuits is 22–31mW within the whole operation band.

III. MEASUREMENT AND COMPARISION

Finally, a 2×10-bit digital quadrature transmitter is implemented and fabricated in a conventional 28-nm CMOS technology. The total chip size is 1.6mm² with core size of 0.2mm², as shown in Fig. 6. The chip is wire-bonded to a 4-layer PCB for measurement. The LO input and transmitter output signals are feed and measured with ground-signal-ground (GSG) probes without any off-chip matching network. 4 serial inputs (I_{IN1}, I_{IN2}, Q_{IN1}, Q_{IN2}), CLK_1, and CLK_2 are generated by signal generators. The mode control signals are sent by a SPI adapter to the chip. With the CW signal, the saturated output power, peak DE of

Fig. 6. Chip microphotograph.

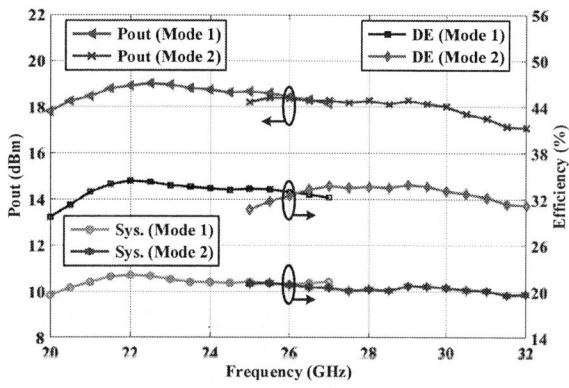

Fig. 7. Measured peak output power, peak drain efficiency, and peak system efficiency.

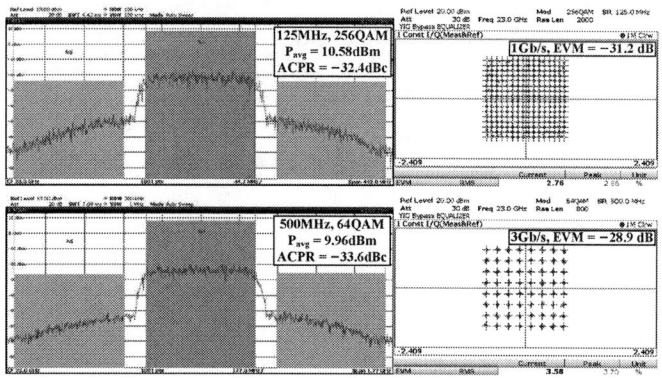

Fig. 8. Measured transmitter output spectrum and constellation at 23GHz.

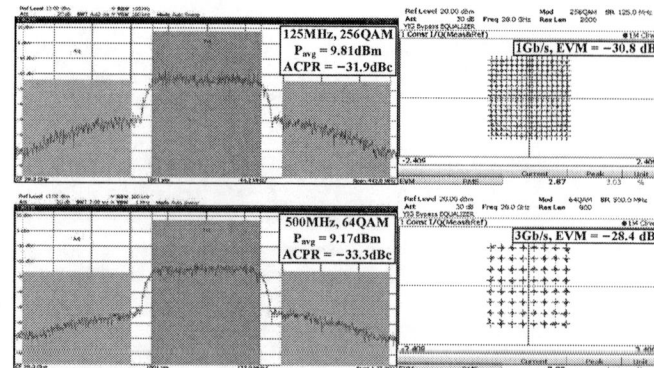

Fig. 9. Measured transmitter output spectrum and constellation at 28GHz.

Table 1. Comparision with state-of-the-art works

	This Work	[4]	[5]	[2]	[3]
Tech.	28-nm CMOS	130-nm SiGe	65-nm CMOS	45-nm SOI CMOS	45-nm SOI CMOS
Architecture	2×10-bit IQ power-DAC	2-stage differential linear	2-stage linear PA	2×6-bit IQ power-DAC	2×8-bit IQ power-DAC
Supply (V)	1	1.9	1.1	4.5–6	4
Freq. (GHz)	20–23	19–29.5	27–31	20–32	42–47
Psat (dBm)	19.02	17.5	15.6	19.9	21.3
Peak PA η	34.4% DE	43.5% PAE	41% PAE	15.6% PAE	-
System η	22.1%	-	-	10.3%	16%
Mod.	64QAM 3Gb/s 9.96dBm	64QAM 6Gb/s 10.7dBm	64QAM 2.04Gb/s 9.8dBm	64QAM 12Gb/s	QPSK 1.25Gb/s
EVM (dB)	−28.9	−27.6	−26.4	−28	−25.2
ACPR (dBc)	−33.6	-	−30	-	-
Area (mm²)	1.6 (0.2*)	0.29*	0.24*	2.4	1.15

*Core size.

the power-DAC, and peak system efficiency of the transmitter versus frequency in dual-mode are plotted in Fig. 7. It achieves the maximum output power of 19.02dBm at 22.5GHz with DE of 24.5% (I = 511, Q = 511), peak DE of 34.4% at 22GHz (I = 511, Q = 0), and peak system efficiency of 22.1% (I = 511, Q = 0) at 22GHz, respectively. The average DE at 6-dB PBO (22GHz) is 17.9%. 500MSym/s 64QAM and 125MSym/s 256QAM modulation signals are measured with 4 and 16 upsampling rate, respectively. The baseband data are deserialized and resampled with 2GHz CLK_2. To improve the transmitter linearity, 2-dimensional DPD is adopted. The digital compensation for the phase/amplitude variation in wideband is further added to improve the EVM of wideband modulation signals. At 23 and 28GHz, the transmitter exhibits EVM $\leq$ −28.4dB, output power $\geq$9.17dBm, ACPR $\leq$ −31.9dBc, respectively (Fig. 8 and Fig. 9).

As shown in Table I, the proposed mm-wave digital quadrature transmitter is compared with state-of-the-arts. It demonstrates 3Gb/s data-rate with good ACPR, and better efficiency compared to the power-DACs [2], [3]. It exhibits comparable PA efficiency with conventional analog power amplifiers[4], [5]. Note that, to have a fair comparison with proposed power-DAC transmitter, a dual-channel high data-rate DAC ($\geq$2GHz CLK) with 1W dc power should be included for the system efficiency estimation. Thus, the whole system efficiency of the analog PA and DAC is <5%, which

is much lower than the proposed digital transmitter.

IV. CONCLUSION

In this paper, a prototype of mm-wave 2×10-bit digital quadrature transmitter is proposed. The impedance mismatch caused by the interconnection of DAC cells is decreased by a notched-matching network. The operation band is further extended with the mode-switching topology. The fabricated mm-wave transmitter exhibits a competitive performance, including the system efficiency and 3Gb/s data-rate with −33.3dBc ACPR, which is attractive for the 5G wireless and backhaul communication.

REFERENCES

[1] J. Keyzer, et al., "Digital generation of RF signals for wireless communications with band-pass delta-sigma modulation," in *IEEE MTT-S Int.Microw. Symp.*, pp. 2127–2130, Jun. 2001.

[2] S. Shopov, et al., "Ultra-broadband I/Q RF-DAC transmitters," *IEEE Trans. Microw. Theory Techn.*, vol. 65, no. 12, pp. 5411–5421, Dec. 2017.

[3] A. Agah, et al., "A 42 to 47-GHz, 8-bit I/Q digital-to-RF converter with 21-dBm Psat and 16% PAE in 45-nm SOI CMOS," in *Proc. of IEEE RFIC*, pp. 249–252, Jun. 2013.

[4] T. Li, et al., "A continuous-mode harmonically tuned 19-to-29.5GHz ultra-linear PA supporting 18Gb/s at 18.4% modulation PAE and 43.5% peak PAE," in *IEEE ISSCC Dig. Tech. Papers*, pp. 410–411, Feb. 2018.

[5] S. Ali, et al., "A 28GHz 41%-PAE linear CMOS power amplifier using a transformer-based AM-PM distortion-correction technique for 5G phased arrays," in *IEEE ISSCC Dig. Tech. Papers*, pp. 406–407, Feb. 2018.

A Wideband Digital Polar Transmitter with Integrated Capacitor-DAC-Based Constant-Envelope Digital-to-Phase Converter

Tong Li, Liang Xiong, Yun Yin, Yangzi Liu, Hao Min, Na Yan, Hongtao Xu

State Key Laboratory of ASIC and System, Fudan University, Shanghai, China

yiny@fudan.edu.cn

Abstract—This paper presents a wideband digital polar transmitter (DPTX) with integrated capacitor-DAC-based constant-envelope digital-to-phase converter (DPC). The switched-capacitor DAC topology is adopted to improve the linearity. The harmonic rejection and cell-reused techniques are employed to reject the $3^{rd}/5^{th}$-order harmonics and reduce the power consumption, respectively. The measurement results of DPC demonstrate the maximum INL and DNL of 1 and 0.8 degrees with the power consumption of 12.7mW@1.5GHz. The DPTX obtains the peak output power of 20.1dBm with 23.7% system efficiency and wideband frequency coverage over 1.2-2.5GHz with only 0.7dB power variation. When amplifying a 10MHz 64QAM LTE signal at 1.5GHz, it achieves -28.6dB EVM and 15.2% system efficiency at 15.0dBm average output power.

Keywords—CMOS, wideband, polar transmitter, switched-capacitor, digital-to-phase converter (DPC), harmonic rejection, cell-reused.

I. INTRODUCTION

In the face of high peak-to-average power ratio (PAPR) OFDM communication standards, digital polar or outphasing transmitters (TXs) have drawn extensive attention due to their high efficiency [1]-[3]. Digital-to-phase converter (DPC) is the key block in digital polar or outphasing TXs. Recently, many architectures have been explored to implement the phase modulator, such as ADPLL-based two-point modulator [1], delay cell-based digital-to-time converter (DTC) [3]-[4], phase interpolator (PI)-based DPC [2][5], etc. Delay cells offer high linearity and decent time resolution, but it has relatively narrowband frequency range due to frequency dependent time-to-phase conversion, undesired supply modulation and high jitter performance; besides, power-hungry delay mismatch calibration to overcome the PVT variations is required which increases the design complexity. PI-based DPCs usually have wide coverage of the full 2π range but with limited phase resolution due to the amplitude-to-phase nonlinearity; although the nonlinear code mapping can be used to improve the resolution, it's at the cost of complex non-uniform sizing of unit cells.

In our work, a capacitor-DAC (C-DAC)-based all-digital polar TX with equivalent 11-bit constant-envelope 8-phase-reused PI-based phase modulator is proposed, which supports wideband frequency range and achieves high phase resolution by introducing the normalized CORDIC algorithm while maintaining good system efficiency without the need of extra calibration engines.

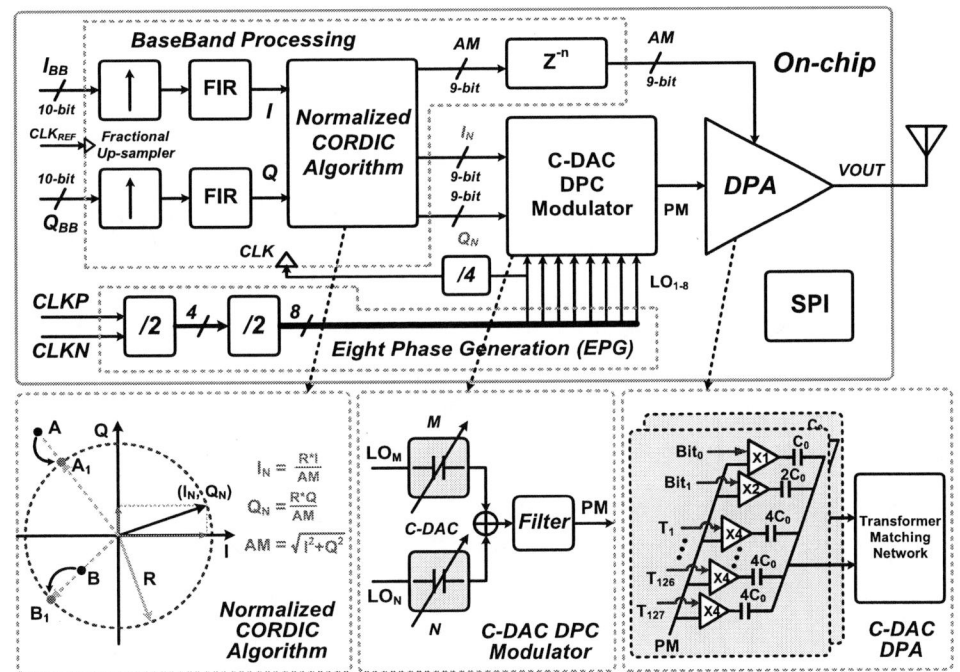

Fig. 1. Architecture of proposed all-digital polar transmitter.

978-1-7281-1702-7/19 $31.00 © 2019 IEEE

II. SYSTEM ARCHITECTURE

Fig. 1 shows the architecture of proposed all-digital polar transmitter. The highly-integrated TX consists of the baseband processing, an 8-phase generator (EPG), a C-DAC-based DPC modulator, a digital PA (DPA) [6-7] and a serial peripheral interface (SPI) for mode reconfiguration. The 10-bit I_{BB}/Q_{BB} signals are first fed into a fractional up-sampler to synchronize with the sampling clock $f_{LO}/4$ that is divided from the LO signal, and then filtered by FIR filters to remove images far away. The up-sampled I/Q signals are converted to the 9-bit AM signal and 9-bit normalized I_N/Q_N signals for phase modulation.

Compared with the traditional polar operation, this proposed CORDIC algorithm normalizes the input I/Q signals onto a constant-envelope circle while reserving the original phase information, which effectively overcomes the PM bandwidth expansion issue and avoids the AM-PM nonlinearity due to the amplitude variations of input OFDM signals. In the design, a PI-based C-DAC DPC is realized to perform the function of phase modulation that will finally be fed into digital PA (DPA) as the PM signal. Here, the phase modulation is conducted by properly weighting and summing the switched-capacitor cells driven by two adjacent fundamental LO phases, where 8-phase LOs are adopted to perform the $3^{rd}/5^{th}$-order LO harmonic rejections while increasing the amplitude of PM signal. Besides, coarse/fine-tuning delay cells are added to adjust the AM-PM delay synchronization. Finally, a 9-bit switched-capacitor DPA which is decoded into 7-bit unary array and 2-bit binary array is employed to recombine the AM and PM signals while providing power amplification.

III. PROPOSED DIGITAL-TO-PHASE CONVERTER

A. Harmonic Rejection and Cell-Reused Technique

Fig. 2 illustrates the principles of proposed 8-phase-reused and harmonic rejection (HR) techniques in the C-DAC DPC modulator using fundamental phase diagrams. First, the I/Q plane (4-phase) can be split into 8 regions by the fundamental 8-phase LOs (LO1-8), and the relationships between 4-phase and 8-phase vector combinations are: $N1=I_N-Q_N$, $N2=\sqrt{2}Q_N$. With the 8-phase-reused technique, the required total number of capacitor cells is $2R/\sqrt{2+\sqrt{2}}$ which is less than the traditional 4-phase IQ-reuse ($\sqrt{2}R$), thus contributing to reduced power consumption and increased PM amplitude. In order to remove the $3^{rd}/5^{th}$-order harmonics, the 8-phase HR is performed with three pairs of vectors which are separated by 45° and vector combined with the gain ratio of 1: $\sqrt{2}$: 1. For the 3rd-order HR, the phase differences among consecutive vectors are tripled and the three vectors are cancelled out with each other completely. The same phenomenon will also happen to the 5th-order HR. Finally, a wideband low-pass filter (LPF) formed by 2nd-order RC circuit is utilized to remove the higher-order harmonics and further improve the phase resolution.

B. Circuit Implementation

Fig. 3 presents the detailed schematic of proposed C-DAC DPC, which is composed of an EPG, a phase selector, a PI-

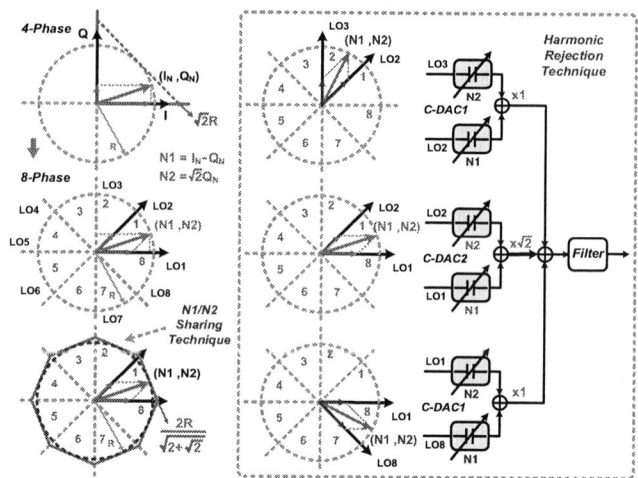

Fig. 2. Operation principles of proposed 8-phase-reused and harmonic rejection (HR) techniques in the C-DAC DPC modulator using fundamental phase diagrams.

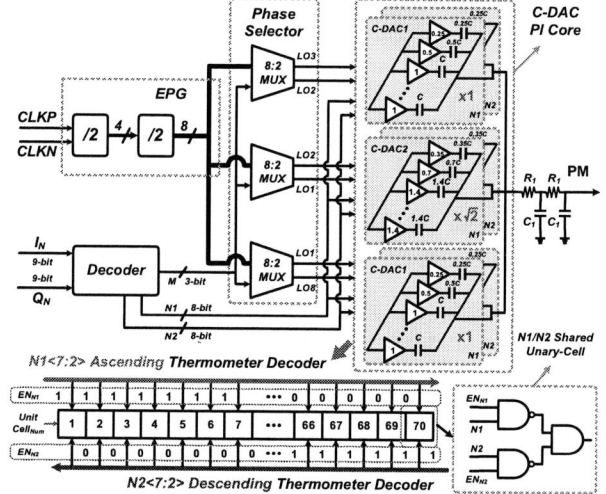

Fig. 3. Detailed schematic of proposed C-DAC DPC modulator.

based C-DAC core and a 2nd-order LPF. First, the number of bits of I_N/Q_N signal is determined by top architecture simulation and verification. A total 9-bit signed I_N/Q_N signals are chosen and decoded into 8-bit unsigned N1/N2 signals with a 3-bit M signal for 8-phase LO selection. Then, the fundamental vectors of the three consecutive regions are selected by three 8:2 multiplexers (MUXes) controlled by the M signal. By adopting the 8-bit switched-capacitor array, the PI-based C-DAC DPC actually realizes the equivalent 11-bit PM resolution with the 3-bit M signal. In each C-DAC array, the 8-bit capacitors are divided into hybrid 6-bit unary cells and 2-bit binary cells so as to obtain better resolution while reducing area overhead. Besides, each unary cell is split into 2 sub-cells which are separately switched to VDD and GND at off-state. This is called the DC averaging technique to stabilize the output DC voltage and avoid undesired AMPM distortion. For the $3^{rd}/5^{th}$-order HRs, the capacitor ratio of the three C-DAC paths is 1: $\sqrt{2}$: 1. Moreover, a total of 70 shared unary cells is employed

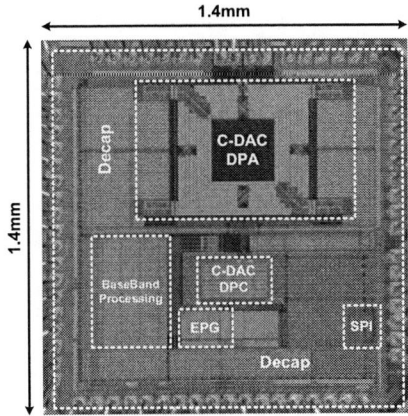

Fig. 4. Die micrograph.

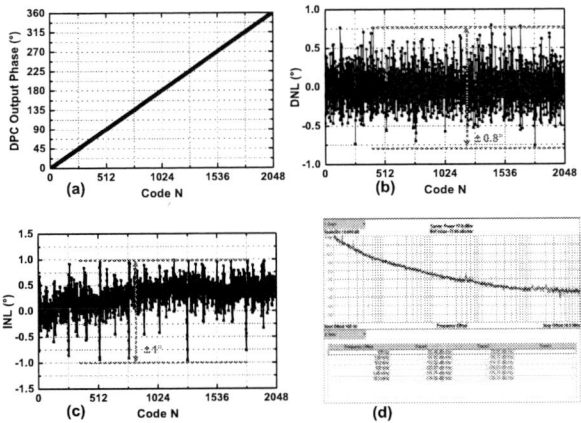

Fig. 5. DPC measurement results at 1.5GHz: (a) transfer function, (b) DNL, (c) INL, (d) phase noise.

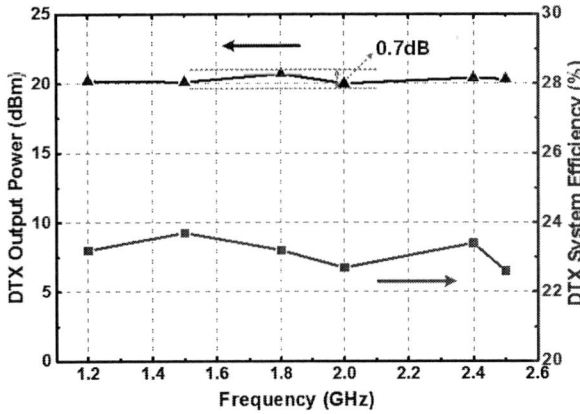

Fig. 6. Measured DTX continuous-wave results of output power, system efficiency and frequency response.

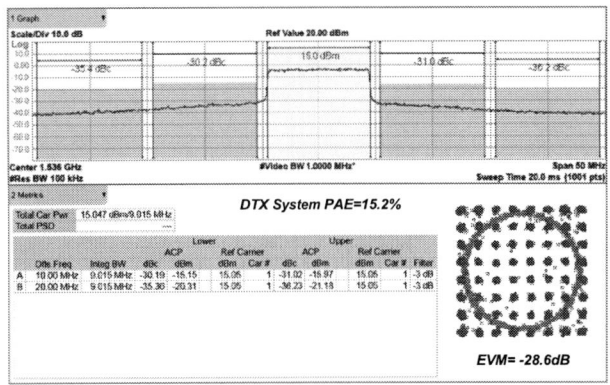

Fig. 7. Measured DTX output spectrum and constellation of a 10MHz 64QAM LTE signal.

in each C-DAC array to express both 6-bit N1<7:2> and 6-bit N2<7:2>, which are decoded into ascending thermometer coding and descending thermometer coding, respectively. Here, each reused unary cell consists of two "NAND-gates" and one "AND-gate" to accomplish the N1/N2 8-phase-resue function.

IV. MEASUREMENT RESULTS

The proposed all-digital polar TX has been implemented in 40nm CMOS, which occupies an area of 1.4×1.4 mm² including all decoupling capacitors and ESD I/O pads, as shown in Fig. 4. The chip is powered by a 1.1V supply, and the power consumption of all drivers, logic blocks, baseband processing, DPC and DPA is included in the calculation of TX system efficiency.

The measured results of the C-DAC DPC modulator are presented in Fig. 5. The DPC power consumption is only 12.7mW at 1.5GHz which contains the EPC, phase selector and C-DAC PI core. Its transfer function of output phase versus the input code N shows the coverage of the full 2π range. As shown in Fig. 5(b)-(c), the measured DNL/INL is ±0.8°/±1° with 0.157°RMS phase error at 1.5GHz, which is enough to support most communication standards. The measured DPC close-in and far-out phase noise performances [Fig. 5(d)] are -125.3 dBc/Hz @ 100kHz frequency offset and -134.3dBc/Hz @

10MHz offset, respectively.

Moreover, the DPTX realizes wideband frequency coverage over 1.2-2.5GHz with only 0.7dB power variation while maintaining good efficiency performance, where it achieves 20.2dBm output power with 23.2% system efficiency at 1.2GHz, 20.1dBm output power with 23.7% system efficiency at 1.5GHz and 20.3dBm output power with 22.6% system efficiency at 2.5GHz, as shown in Fig. 6.

In modulation tests, the 10MHz 64QAM LTE signal is utilized to verify the DTX performance. Without the need of any digital pre-distortion, the measured DTX output spectrum and constellation are shown in Fig. 7. With -28.6dB EVM, the DTX achieves the average output power of 15.0dBm with 15.2% system efficiency. The measured upper and lower ACLRs are -31.0dBc and -30.2dBc, respectively.

Table 1 and Table 2 summarize the measured performance and makes a comparison with the state-of-the-art DPCs and DTXs. By adopting the normalized CORDIC algorithm and 8-phase-reused HR techniques, the proposed C-DAC DPC achieves the 0.157-degree phase resolution with compact die size. Moreover, by utilizing this DPC modulator, this C-DAC-based DTX achieves wideband frequency coverage with superior performance of system efficiency, thus well-fitting multi-band multi-standard wireless communications.

Table 1. DPC Performance Summary and Comparison with Prior Works

	Architecture	Freq. (GHz)	Resolution	DNL/INL	Power (mW)	Voltage (V)	Area (mm²)	CMOS Process
JSSC 2016 [4]	Digitally-controlled Edge Interpolation	2.0	11-bit	0.22°/1.08° @2GHz	19.8	1.1	0.009	28nm
JSSC 2013 [5]	Resistor-based Interpolation	0.1-1.5	8-bit	0.73°/±1.87° @0.5GHz	4.3 @1.5GHz	1.2	0.06	65nm
This work (DPC)	**PI-based 8-phase-reused C-DAC**	**1.2-2.5**	**11-bit**	**±0.8°/±1° @1.5GHz**	**12.7 @1.5GHz**	**1.1**	**0.045**	**40nm**

Table 2. DTX Performance Summary and Comparison with Prior Works

	Freq. (GHz)	PA Peak Pout (dBm)	PA Peak PAE (%)	Modulation Signal	EVM (dB)	Pavg (dBm)	TX System Efficiency (%)	DPD	Voltage (V)	CMOS Process
ISSCC 2016 [1]	0.75-0.93	8	45	1/2MHz 802.11ah	-27.1	0	14	No	1	40nm
JSSC 2013 [2]	2.2	23.3	43 (DE)	20MHz 802.11g	-28	16.8	19.3	Yes	1.2	65nm
ISSCC 2017 [3]	2.4	19	32	20MHz 802.11g	-25	14	19.5†	Yes	1	14nm
This work (DTX)	**1.2-2.5**	**20.1**	**31.3**	**10MHz LTE**	**-28.6**	**15.0**	**15.2**	**No**	**1.1**	**40nm**

†Estimated from Fig. 6 in [3].

V. CONCLUSION

A wideband DPTX with integrated capacitor-DAC-based constant-envelope DPC is presented in this work. With the 3rd/5th-order harmonic rejection and 8-phase-reused techniques, the proposed C-DAC DPC modulator achieves the DNL/INL of ±0.8°/±1° and 0.157° RMS phase error with the power consumption of 12.7mW at 1.5GHz. Moreover, the DPTX obtains the 20.1dBm peak power with 23.7% system efficiency and wideband frequency coverage over 1.2-2.5GHz with only 0.7dB power variation. When amplifying a 10MHz 64QAM LTE signal at 1.5GHz, it achieves -28.6dB EVM and 15.2% system efficiency at 15.0dBm average output power without the need of any digital pre-distortions.

ACKNOWLEDGMENT

This work was supported by the National Science and Technology Major Projects of China under Grant 2018ZX03001005-004, the National Natural Science Foundation of China under Grant 61874153, and the Research Project Fund of State Key Laboratory of ASIC and System, Fudan University under Grant 2018MS007.

REFERENCES

[1] A. Bo, et al., "A 1.3nJ/b IEEE 802.11ah fully digital polar transmitter for IoE applications," IEEE ISSCC 2016, pp. 440-441, Feb. 2016.

[2] L. Ye, et al., "Design considerations for a direct digitally modulated WLAN transmitter with integrated phase path and dynamic impedance modulation," in Solid-State Circuits, IEEE Journal of, vol. 48, no. 12, pp. 3160-3177, Dec 2013.

[3] P. Madoglio, et al., "A 2.4GHz WLAN digital polar transmitter with synthesized digital-to-time converter in 14nm trigate/FinFET technology for IoT and wearable applications," IEEE ISSCC 2017, pp. 226-227, Feb. 2017.

[4] S. Sievert, et al., "A 2GHz 244fs resolution 1.2ps peak-INL edge interpolator-based digital-to-time converter in 28nm CMOS," in Solid-State Circuits, IEEE Journal of, vol. 51, no. 12, pp. 2992-3004, Dec 2016.

[5] M.-S. Chen, et al., "A 0.1-1.5GHz 8-bit inverter-based digital-to-phase converter using harmonic rejection," in Solid-State Circuits, IEEE Journal of, vol. 48, no. 11, pp. 2681-2692, Nov 2013.

[6] Vorapipat V, Levy C, Asbeck P. "A wideband voltage mode Doherty power amplifier," IEEE RFIC Symp. 2016, pp. 266-269, May. 2016.

[7] Liang Xiong, et al., "A broadband switched-transformer digital power amplifer for deep back-off efficiency enhancement," IEEE ISSCC 2019, pp. 76-77, Feb. 2019.

A 5GHz to 6GHz CMOS Transmitter for Full-Duplex Wireless with Wideband Digital Cancellation

Nimrod Ginzberg[#], Dror Regev[+], Genadiy Tsodik[+], Shimi Shilo[+], Doron Ezri[+], Emanuel Cohen[#]

[#]Faculty of Electrical Engineering, Technion, Israel
[+]Toga Networks, a Huawei Company, Israel

Abstract — **This paper presents a quadrature balanced transmitter, assisted by a digital equalization and predistortion self-interference cancellation technique for Full-Duplex wireless applications. An analysis of design trade-offs between low receiver (Rx) loss and high transmitter (Tx) efficiency is laid out and demonstrated on a 5GHz to 6GHz class AB power amplifier implemented in 180nm CMOS. Wideband cancellation of >48dB and >57dB at 20dBm and 10dBm Tx output power, respectively, is measured in CW. Cancellation of >30dB for an actual 160MHz 802.11ac OFDM packet around a carrier frequency of 5.2GHz together with EVM of -33dB is demonstrated. Measured Tx power added efficiency (PAE) for concurrent Tx-SIC operation is 35% and 6.4% at peak and RMS (10dB backoff) power, respectively. Rx loss is lower than 1.6dB at RMS Tx power within the signal frequency band.**

Keywords — **power amplifier, full-duplex, output impedance, self-interference cancellation.**

I. Introduction

Full-Duplex (FD) wireless radio communication is of high interest in recent research due to its theoretical potential to double spectral efficiency. Recent FD transceiver works proposed different approaches for primary passive Tx-Rx isolation, followed by various secondary RF or analog self-interference cancellation (SIC) filters. Reported passive and active multi-tap delay-line or N-path filters [1]-[3] were connected between the PA output and the LNA Input, thereby imposing substantial Tx and Rx losses and SNR degradation. A digital SIC approach with minimal primary isolation is proposed in [4], where a replica of the Tx leakage is injected from a cancelling DAC at the LNA input. A quadrature balanced power amplifier (QBPA) architecture proposed in [5] offers a competitive passive primary isolation and does not impose Tx power loss, however with narrow bandwidth (BW) SIC capabilities. The QBPA topology also benefits from good port matching and immunity to antenna VSWR variations [6].

In this paper we propose an efficient and wideband QBPA RF front-end (RFFE) transmitter configuration, along with a digital quadrature equalization and predistortion (DPD) SIC technique. Digital channel estimation and SIC signal synthesis is performed at each subcarrier frequency of the modulated Wi-Fi signal, hence facilitating the phase and amplitude equalization and eliminating the BW limitations present in [1]-[5]. The SIC signal is introduced at the PA input and transmitted through the same amplification path as the Tx signal, therefore Tx efficiency is preserved. The Tx noise floor elevation can be digitally subtracted from the Rx due the correlated deterministic nature of the quantization and nonlinear errors [4].

The high S_{22} of the Tx required for this topology [5] is obtained in this work by designing the PAs to introduce high output impedance [7]. An analysis of the design trade-off between high S_{22} and high PAE is conducted, according to which efficient high output impedance class AB PAs are implemented in TSMC 180nm CMOS and employed in the Tx path of the proposed transmitter.

II. Theory of Operation

A. Transmit Path

The transmitter's operation principle is shown in Fig. 1. The Tx signal is split in the digital and fed into PA1 and PA2 from two separate DACs at 90° phase offset. The Tx-Rx leakage is determined by the antenna reflection coefficient and the S_{22} of the PAs, as well as by mismatches between the PAs and phase and amplitude imbalance of the quadrature hybrid. Cancellation of this leakage is achieved by applying two digitally synthesized quadrature SIC signals at the PAs' inputs, which add up at the Rx port and introduce similar amplitude and opposite phase as the antenna reflection, as well other Tx leakages arriving at the LNA input.

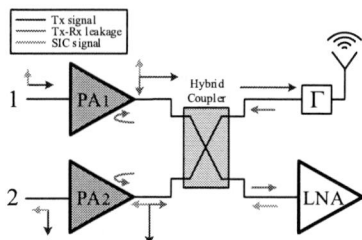

Fig. 1. Operation principle

Let $v_{in1} = \frac{A(\omega)}{\sqrt{2}}e^{j\omega t}$ and $v_{in2} = \frac{A(\omega)}{\sqrt{2}}e^{j(\omega t-\pi/2)}$ be two Tx signals that are input to PA1 and PA2, respectively, where $A(\omega)$ is a complex amplitude and ω is angular frequency. The signal at the antenna can be expressed as $v_{ANT} = A(\omega)h_{Tx}(\omega)$, where $h_{Tx}(\omega)$ is the transfer function of the transmit path, which equals roughly to $S_{21}(\omega)$ of the PAs, assuming identical amplifiers. The signal at the Rx is given by $v_{Rx,leak} = A(\omega)h_{leak}(\omega)$, where $h_{leak}(\omega)$ represents the transfer function of the signal that is reflected from the antenna and reconstructs at the Rx. For clarity, if assuming a single reflection and identical PAs, $h_{leak}(\omega)$ can be written as:

$$h_{leak}(\omega) = S_{21}(\omega)\Gamma(\omega)S_{22}(\omega), \quad (1)$$

where $\Gamma(\omega)$ is the frequency dependent antenna reflection coefficient and $S_{21}(\omega)$ and $S_{22}(\omega)$ are the forward gain and output return loss of the PAs, respectively.

To eliminate the signal leaking to the Rx, two SIC signals $v_{sic1} = \frac{B(\omega)}{\sqrt{2}}e^{j(\omega t + \pi/2)}$ and $v_{sic2} = \frac{B(\omega)}{\sqrt{2}}e^{j(\omega t - \pi)}$ are injected to PA1 and PA2, respectively. The $90°$ phase shift of the SIC signals with respect to the Tx signals results in their reconstruction at the Rx port. The SIC signal at Rx can be expressed as $v_{Rx,sic} = B(\omega)h_{SIC}(\omega)$, where $h_{SIC}(\omega)$ is the input-to-Rx transfer function of the SIC signals. Therefore, the total signal at the Rx is now given by:

$$v_{Rx,tot} = A(\omega)h_{leak}(\omega) + B(\omega)h_{SIC}(\omega). \quad (2)$$

To obtain cancellation we demand that $v_{Rx,tot} = 0$, from which the complex amplitude of the SIC signals $B(\omega)$ can be calculated:

$$B(\omega) = h_{SIC}^{-1}(\omega)\big[-A(\omega)h_{leak}(\omega)\big]. \quad (3)$$

If assuming, for simplicity, a single antenna reflection and identical PAs, then $h_{SIC}(\omega) \approx S_{21}(\omega)$ and $B(\omega) \approx -A(\omega)\Gamma(\omega)S_{22}(\omega)$. The input signals in this case are as follows:

$$
\begin{aligned}
v_{in,1} &= \frac{A(\omega)}{\sqrt{2}}e^{j\omega t}\Big(1 - \Gamma(\omega)S_{22}(\omega)e^{j\pi/2}\Big), \\
v_{in,2} &= \frac{A(\omega)}{\sqrt{2}}e^{j(\omega t - \pi/2)}\Big(1 - \Gamma(\omega)S_{22}(\omega)e^{-j\pi/2}\Big).
\end{aligned}
\quad (4)
$$

$B(\omega)$ is computed using a MATLAB emulation of a compliant receiver that decodes and estimates the transfer function at each subcarrier frequency.

It is noted from (4) that the magnitude ratio between the Tx and SIC signals is roughly $|\Gamma(\omega)S_{22}(\omega)|$. For typical values of antennas' reflection coefficients and PAs' return losses, this ratio is in the order of 15-20dB, which eliminates the need to sacrifice Tx power and efficiency in favor of the SIC operation.

B. Rx Channel

Rx signals that enter the system from the antenna split when passing through the hybrid coupler, reflect at the output of the PAs and reconstruct at the LNA input after having been multiplied by the S_{22} of the PAs. This path is similar to that of the Tx signal reflecting from the antenna. Therefore the SNR degradation of Rx signals depends on the reflectivity of the PAs, which are designed to introduce high S_{22}.

III. HIGH S_{22} PA DESIGN CONSIDERATIONS

For low Rx loss, the design guideline was that PAs introduce high output reflection coefficient while maintaining high PAE and sufficient output power for Wi-Fi applications. In order to calculate the S_{22} seen from the output, we consider the properties of a passive and lossless two-port matching network S (Fig. 2a). Such a network fulfills $S^T \cdot S = 1$, which can be written as follows:

$$\begin{cases} |S_{11}|^2 + |S_{12}|^2 = 1, \\ |S_{21}|^2 + |S_{22}|^2 = 1. \end{cases} \quad (5)$$

As reciprocity dictates $S_{12} = S_{21}$, we get that $|S_{11}| = |S_{22}|$. Consequently, $|S_{22}|$ seen from the output can be expressed

using the PA's output impedance, Z_{PA}, and the optimum load, Z_{opt}, as follows:

$$|S_{22}| = \left| \frac{Z_{out} - Z_{opt}^*}{Z_{out} + Z_{opt}} \right|. \quad (6)$$

High S_{22} can be achieved through maximization of Z_{out}. For high drain resistance to begin with, a Cascode topology was chosen. Together with the process parameters of the relatively long-channel TSMC 180nm technology, the PA exhibits $S_{22} \approx -2.5dB$ for optimal PAE (Fig. 2b).

S_{22} can be increased by trading-off PAE through nonoptimal selection of Z_{opt}, denoted here as Z'_{opt}. This trade-off is represented graphically in (Fig. 2b). Equal S_{22} circles are plotted on the Smith chart according to (6), together with equal power and PAE contours. By decreasing Z_{opt} towards the 50Ω load, the $S_{22} = -1.1dB$ circle intersects with a PAE circle of 43%. Assuming an LC matching network with a quality factor $Q = 15$, the vicinity of Z'_{opt} to 50Ω implicates a 0.4dB lower insertion loss and wider BW compared to Z_{opt}. Therefore, the Tx penalty resulting from the 1.4dB improvement in Rx loss is only 0.5dB.

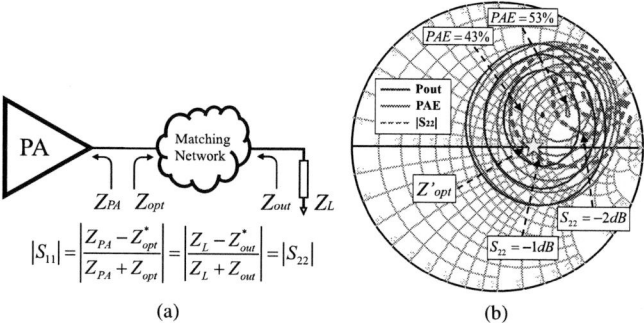

(a)　　　　　　(b)

Fig. 2. (a) S_{22} calculation, (b) PAE and S_{22} circles intersection

The PA was designed as a class AB Cascode (Fig. 3) to operate around 5.5GHz with a BW of 1GHz, to comply with IEEE 802.11ac standard. The current mirror used for biasing also provides a low impedance node at the bias branch, which is essential for stability. Input and output impedance matching were implemented using L-C networks. A low Q RC bypass network is placed between the supply and ground nodes to introduce low impedance over a wide frequency range [8]. ESD protection diodes were placed at all nodes interfacing external signals.

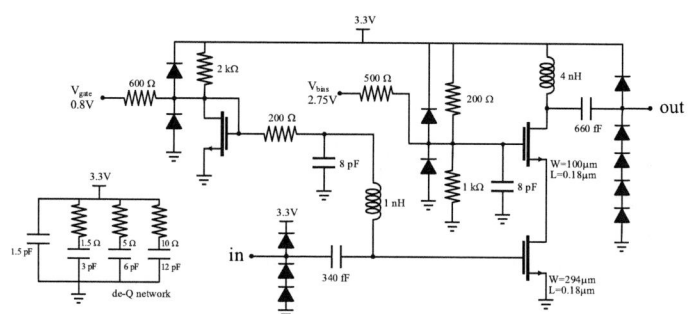

Fig. 3. PA Schematic

IV. MEASUREMENTS

A. Standalone High S_{22} PA

The standalone PA measurements were performed on-die in a probe-station. The PA delivers saturated power (P_{sat}) of 17dBm from a 3.3V supply at 5.5GHz with a BW of nearly 1GHz, and achieves maximum power gain of 16.5dB and 38% peak PAE (Fig. 4). The PA's stability factor is >1 at all frequencies. The die micrograph is shown in Fig.5. The total die size is $600\mu m \times 850\mu m$.

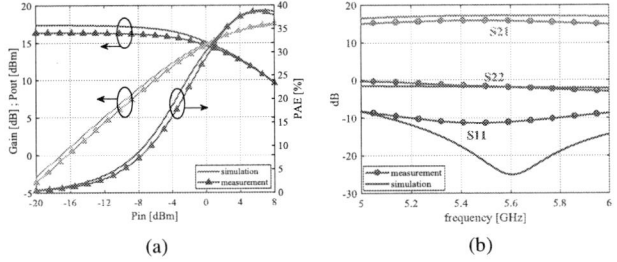

(a)   (b)

Fig. 4. Measured and simulated standalone PA: (a) Gain, PAE and output power, (b) frequency response.

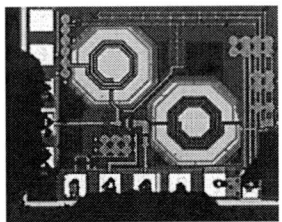

Fig. 5. PA Micrograph. The die size of one PA is $600\mu m$ x $850\mu m$.

B. Cancellation Measurement

The PA was mounted on a Rogers 4350B PCB that was designed and fabricated for the system's measurements. The measurement setup is shown in Fig. 6. The Tx and SIC signals are generated in MATLAB, and uploaded to a four channel 8-bit Keysight 8195A arbitrary waveform generator, from which they are fed into hybrid power combiners that drive the respective QBPA input. A 90° Krytar hybrid coupler with insertion of 0.3dB at 5.2GHz is used for quadrature combining of the two power amplifiers. The Tx port is connected to an actual Wi-Fi antenna through a 10dB coupler. The coupled Tx port, as well as the Rx port, are connected to a Keysight DSOS604A VSA.

1) CW Signals

The system was measured using CW signals at saturation (20dBm) and RMS (10dB backoff) power levels, and SIC equalization was performed per subcarrier frequency within a 160MHz BW around 5.2GHz. Tx-Rx isolation of >48dB and >57 dB is measured at saturation and backoff, respectively. The limited cancellation at P_{sat} is due to the clipping of the SIC signal, which cannot gain sufficient magnitude to suppress the leakage. Average measured PAE for concurrent operation of Tx and SIC at P_{sat} and backoff is 35% and 6.4%, respectively (Fig. 7). Interestingly, the SIC operation slightly

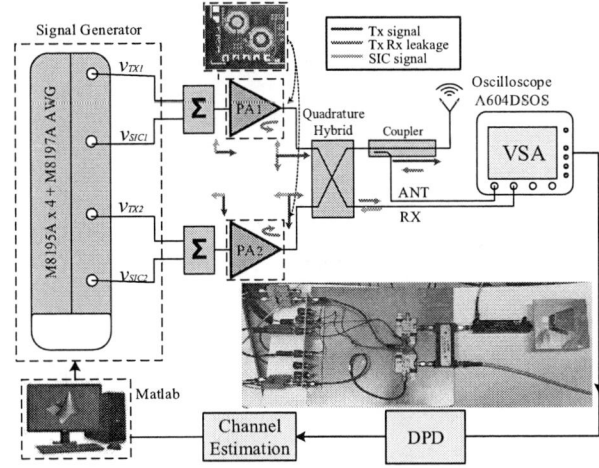

Fig. 6. Measurement setup. The reflected signal from the antenna (red) reconstructs at the LNA input and generates interference. A SIC signal (green) with 90° phase shift with respect to the Tx signal (blue) is injected to the input to appear at the LNA with similar magnitude and opposite phase as the Tx-Rx leakage.

improves PAE. This occurs since the SIC signal opposes the reflection from the antenna and compensates for imbalances in the Tx path.

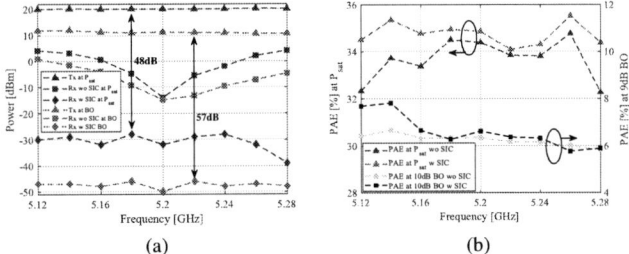

(a)   (b)

Fig. 7. (a) Measured Tx-Rx isolation for CW signals. (b) Measured PAE at peak and 10dB backoff power levels with and without SIC. The SIC operation slightly improves PAE both by equalizing the Tx paths and by opposing the antenna reflection.

2) OFDM Signals

Wideband cancellation of modulated signals was demonstrated using a 802.11ac VHT160 MSC7 packet transmitted around 5.2GHz carrier frequency. A piecewise DPD algorithm [9] was used for EVM correction. Then both the Tx and SIC signals were multiplied by the DPD coefficients and fed into the system. Finally, a channel estimation process was performed to calculate the channel's frequency response and shape the SIC signal accordingly.

The measurement results are summarized in Fig. 8. A primary isolation of 10-25dB is measured within the signal BW, complemented by additional 5-20dB of leakage suppression obtained by the SIC signal. The leakage is suppressed down to the noise floor of the measurement setup, and a total cancellation of 30dB on average over the entire 160MHz BW is measured. The cancellation level achieved in this measurement is limited only by the noise floor of the measurement setup, as can be inferred from the isolation achieved in CW. The DPD corrects TX linearity by 7.6dB.

The SIC signal does not contaminate the Tx and EVM level is maintained for concurrent Tx-SIC operation (Fig. 9).

In Fig. 8 it is observed that Tx-Rx isolation level without DPD and SIC (purple) is similar inband and out-of-band. This implies that linear as well as intermodulation distortion terms are suppressed at the Rx by the quadrature hybrid topology. In addition, note that the SIC without DPD (black) cancels also out-of-band leakage, although the SIC is an inband signal.

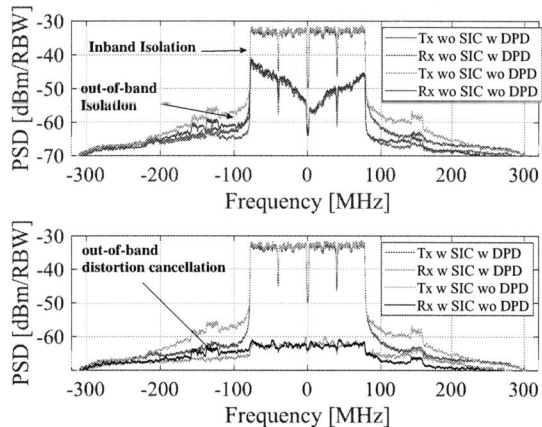

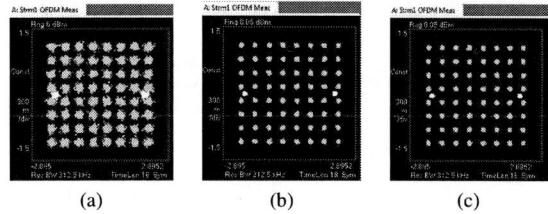

Fig. 8. Measured spectrum of an 802.11ac VHT160 signal at 10dBm RMS output power. SIC operation further suppresses out-of-band distortion on top of the DPD. Intermodulation distortion terms as well as linear terms are canceled out at the Rx.

Fig. 9. Measured Tx EVM of a VHT160 signal at 10dBm RMS output power: (a) w SIC wo DPD, EVM = -25.4dB. (b) wo SIC w DPD, EVM = -33dB. (c) w SIC w DPD, EVM = -33dB. SIC operation does not affect Tx linearity.

C. Rx Loss

Rx loss consists of the PA's S_{22} and the insertion loss of the hybrid coupler. At RMS power, S_{22} is lower than 1.6dB around 5.2GHz, and increases with output power up to 3.4dB at P_{sat} (Fig. 10a). However, since the Tx processes high PAPR OFDM signals, peak power events are statistically infrequent and Rx insertion loss remains the RMS value of 1.6dB. The amplitude and phase cross-modulation from Tx to Rx (Fig. 10) may require a 2D DPD correction at Rx [9].

Table 1 summarizes this work compared to prior published works on FD systems.

V. CONCLUSION

In this paper we proposed a FD transmitter architecture and a digital SIC technique achieving wideband cancellation over a 160MHz BW around a center frequency of 5.2GHz with no additional power for canceller operation and with improved Tx PAE under antenna VSWR variations. An efficient high S_{22} class AB PA was designed and fabricated in 180nm CMOS

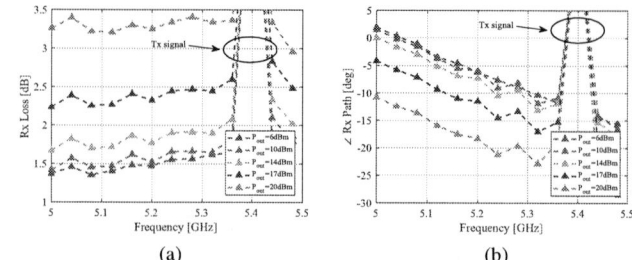

Fig. 10. Measured (a) insertion loss and (b) phase of the Rx path. At RMS Tx power, Rx loss is lower than 1.6dB around the 5.2GHz operating frequency.

Table 1. COMPARISON TABLE

	[1]	[2]	[3]	[4]	This Work
SIC Architecture	N-path G_m-C Filter	Passive 16-tap Delay-Line FIR Filter	Active 5-tap Delay-Line FIR Filter	Cancelling DAC	QBPA+Digital Equalization
Frequency (GHz)	0.8-1.4	2.41-2.49	1.7-2.2	1.0-1.8	5-6
Primary Isolation (dB)	34[1]	15[2]	[30-35][2]	3	[10-25][3]
Cancellation BW (MHz)	27	80	42	20	>160
Tx-Rx Isolation (dB)	54	60	55	50	57[4]/48[5]
Maximum Tx Output Power (dBm)	N/A	20	25	12.6	20
Peak Tx PAE	N/A	N/A	32%	N/A	35%
Tx EVM (dB)	N/A	N/A	-25.8	N/A	-33
Canceller Power Consumption (mW)	44-91	>100[6]	11.5	60	0
Tx Loss	N/A	4[6]	0.5[6]	0.94[7]	0.3[8]
Rx NF Degradation (dB)	0.9-1.8	4[6]	1.6	1.1[9]	1.6[4]/3.4[5]
Integration	External PA[1]	No	External circulator	Full	External hybrid and LNA
Technology	65nm	N/A	40nm	65nm	180nm[8]

[1] Antenna pair [2] Circulator [3] Quadrature Balance [4] @10dB backoff from P_{sat} [5] @Maximum output power [6] Estimated [7] Calculated for 12Ω series resistance [8] External hybrid coupler [9] Measured @2dBm FDD

and employed in the Tx path to minimize Rx loss. Over 48dB and 57dB isolation is shown in CW at peak and at 10dB backoff power, respectively, and 30dB on average for a VHT160 signal. The SIC signal does not contaminate the Tx, and correct EVM level is maintained for concurrent Tx-SIC operation. An analysis of the trade-off between Tx PAE and low Rx loss is conducted and shown to yield useful design guidelines for the Tx path in FD transceivers.

REFERENCES

[1] J. Zhou et al, "Receiver with >20MHz bandwidth self-interference cancellation suitable for FDD, co-existence and full-duplex applications", *ISSCC* 2015.

[2] D. Bharadia et al, "Full Duplex Radios", *SigComm* 2013.

[3] T. Zhang et al, "A 1.7-to-2.2GHz Full Duplex Transceiver System with >50dB Self-Interference Cancellation over 42MHz Bandwidth", *ISSCC* 2017.

[4] L. Calderin et al, "Analysis and Design of Integrated Active Cancellation Transceiver for Frequency Division Duplex Systems", *Journal of Solid-State Circuits*, vol. 52, pp. 2038 - 2054, June 2017.

[5] D. Regev et al, "Modified re-configurable quadrature balanced power amplifiers for half and full duplex RF front ends", *WMCS* 2018.

[6] G. Berretta et al, "A balanced cdma2000 SiGe HBT load insensitive power amplifier", *IEEE Radio and Wireless Symposium*, April 2006.

[7] P. Aflaki et al, "Broadband class-E power amplifier with high cold output impedance suitable for load modulated dual branch amplifiers", *LAMC* 2016.

[8] D. Chowdhury et al, "A Fully Integrated Dual-Mode Highly Linear 2.4GHz CMOS Power Amplifier for 4G WiMax Applications", in *IEEE Journal of Solid-State Circuits*, vol. 44, no. 12, December 2009.

[9] Tomer Gidoni et al, "Digital predistortion on concurrent noncontiguous transmitters using 2D piecewise vector decomposition", *COMCAS*, 2017.

978-1-7281-1702-7/19 $31.00 © 2019 IEEE

A Sub-mW All-Passive RF Front End with Implicit Capacitive Stacking Achieving 13 dB Gain, 5 dB NF and +25 dBm OOB-IIP3

Vijaya Kumar Purushothaman[1], Eric Klumperink[2], Berta Trullas Clavera, Bram Nauta[3]

IC Design group, University of Twente, Enschede, The Netherlands

[1]v.k.purushothaman@utwente.nl, [2]e.a.m.klumperink@utwente.nl, [3]b.nauta@utwente.nl

Abstract—This paper presents a sub-mW mixer-first RF front-end that exploits a novel capacitive stacking technique in an altered bottom-plate N-path filter/mixer to achieve passive voltage gain and high-linearity at low noise figure. Capacitive stacking is realized implicitly by reading out the voltage from the bottom-plate of N-path capacitors instead of their top-plate, which provides a 2x gain at the read-out capacitors. Additional passive voltage gain is achieved using impedance upconversion while improving the out-of-band linearity performance of small switches. With no other active circuitry, only clock generation circuits determine the total power consumption of this RF front-end. A prototype is fabricated in GF22 nm FDSOI technology. Operating at f_{LO}=1 GHz, the prototype achieves a voltage gain of 13 dB, 5 dB Noise Figure and +25/+66 dBm Out-of-band IIP3/IIP2 at 160 MHz offset while consuming only 600 μW of power from a 0.8 V supply.

Keywords — Passive mixer, N-path filter, mixer-first receiver, bottom-plate mixing, capacitive stacking, high linearity, low-power, RF front-ends.

I. INTRODUCTION

Massive deployment of wireless sensor nodes and scarce available radio spectrum make the receivers in the sensor networks susceptible to interference problems. On top of the strict limits on cost and power consumption, the interference robustness is becoming a major concern for these radios. RF front-ends adopt techniques such as N-path filters/mixers, in which large switches and capacitors are used, to achieve high out-of-band linearity (> 30 dBm) at the cost of power in the clock drivers [1]–[4]. Such front-ends exhibit low conversion loss and high input impedance at tunable switching frequency. Often, the N-path filters/mixers are cascaded with active blocks to provide impedance matching at RF input and IF gain at the desired operating frequency [1],[3]. These active blocks consume a lot of power to achieve low noise figure and high linearity [3],[4].

In this paper, we present a low power and highly linear all-passive RF front-end employing an altered bottom-plate N-path mixer and a passive matching network. The architecture, implementation and performance of the novel RF front-end is discussed in detail in the rest of the paper.

II. ARCHITECTURE

Bottom-plate N-path filters achieve higher linearity compared to conventional top-plate N-path filters [4]. As shown in Fig.1a and Fig.1b, modulation of switch resistance due to input-dependent V_{GS} is reduced in the bottom-plate configuration by grounding the V_S node of the switch. Hence, it results in higher linearity. However, when the switch is

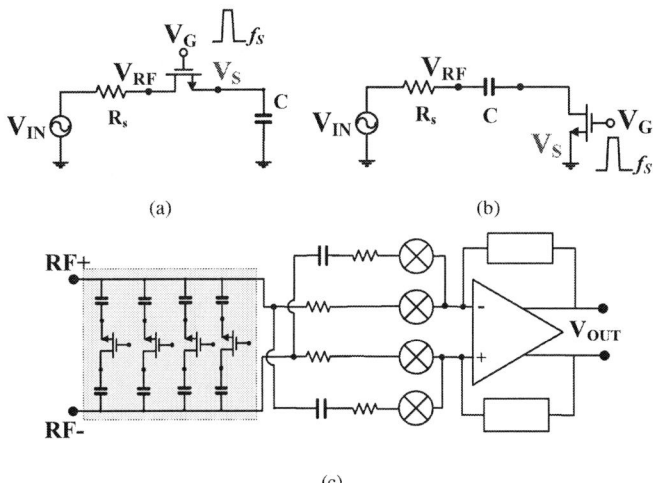

Fig. 1. (a) Top-plate N-path filter, (b) Bottom-plate N-path filter and, (c) Differential bottom-plate N-path and cross-coupled switch-RC front-end [4]

open, the capacitor becomes floating and stores the down-converted baseband voltage across its terminals (without being referenced to ground). As such, it complicates the possibility of extracting the baseband signal from the capacitor. Hence in [4], the bottom-plate configuration is simply used as bandpass filter and the RF voltage from the top-plate of the N-path capacitors is down-converted using a cross-coupled switch-RC network (Fig.1c). It achieves an out-of-band IIP3 of 44 dBm, which is 15-20 dB higher than the top-plate configuration.

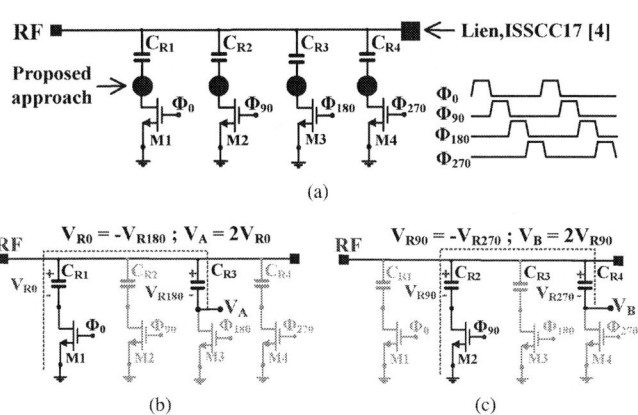

Fig. 2. (a) Proposed down-conversion technique in bottom-plate N-path filtering; Capacitive stacking (b) when M1 is conducting and (c) when M2 is conducting

Unlike [4], we propose to sense the voltage from the bottom-plate of the N-path capacitors in this work (Fig.2a). The resultant behavior can be explained as follows.

Consider a 4-path single-ended bottom-plate N-path filter in Fig.2a. Suppose that RF sinusoidal input f_{in}, is down-converted to baseband by clocks, ϕ_{0-270}, switching at frequency f_{LO}. Let the resulting down-converted baseband voltage across the mixing capacitors C_{R1}-C_{R4} be V_{R0}, V_{R90}, V_{R180} and V_{R270} respectively. For $f_{in} \approx f_{LO}$, the baseband voltages V_{R0} and V_{R180} are in anti-phase since the clock phases (ϕ_0, ϕ_{180}) driving their corresponding switches (M1,M3) are 180 degrees phase-shifted. Similarly V_{R90} and V_{R270} are in anti-phase. In simpler terms, the in-band instantaneous voltages can be expressed as,

$$V_{R0} = -V_{R180}$$
$$V_{R90} = -V_{R270} \tag{1}$$

When switch M1 is conducting, the bottom-plate of the capacitor C_{R1} is connected to ground and other capacitors remain floating. When observed from node A in Fig. 2b, the capacitors C_{R3} and C_{R1} are connected in series to ground. In other words, capacitor C_{R1} and C_{R3} appears to be stacked upon each other. Therefore the resultant voltage at node A,

$$V_A = -V_{R180} + V_{R0} = 2 \times V_{R0} \tag{2}$$

Note that from node A the bottom-plate of the Capacitor C_{R3}, V_{R180} is seen inverted and hence the negative sign. Thus, a voltage gain of 2 is achieved for baseband signals by simply tapping them from the node A. Likewise, when switch M2 is closed, $V_B = 2 \times V_{R90}$ can be achieved by tapping the signal from node B, at the bottom-plate of the Capacitor C_{R4}, as shown in Fig. 2c. Similar voltage gain can be achieved by reading out the signal from the bottom-plate of capacitors C_{R2} and C_{R4} when switches M2 and M4 are closed respectively.

In a fully-differential implementation of the bottom-plate N-path filters, the balanced RF inputs RF+ and RF- are in anti-phase. Therefore, when a switch is closed, a zero voltage point occurs in the middle of the inversion region in the channel of the switching transistor. The zero-voltage point behaves similar to ground in the single-ended implementation and can be used for differential capacitor stacking. The implicit capacitor stacking in differential bottom-plate N-path filters can be realized as shown in Fig. 3. Here, the additional read-out capacitors C_{B1} and C_{B2} are used to read-out the $2\times BB$ signal from the bottom-plate of capacitors C_{R3} and C_{R7} when M1 is closed (Fig.3a). Additionally, they can be re-used to read-out the signal from the bottom-plate of capacitors C_{R5} and C_{R1} when M3 is closed (Fig.3b). Such capacitor re-use is possible due to differential RF inputs and 180 degrees phase shifted clocks driving the switches. Since the read-out capacitors share the charges with two N-path capacitors at different time instants, an additional order of infinite impulse response filtering is achieved in this scenario.

Further, to achieve the desired input matching and more voltage gain, we propose to use passive matching network instead of power-hungry active blocks [2],[5],[6]. The loss

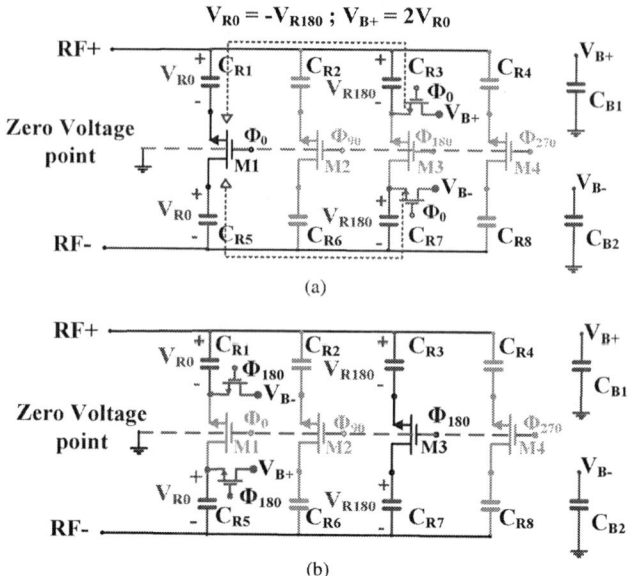

Fig. 3. (a) Capacitive stacking in the fully-differential proposed RF front-end and (b) read-out capacitors reuse.

due to the parasitic substrate capacitance of N-path capacitors and charge sharing behavior of read-out capacitors together determine the impedance to be matched and the achievable voltage gain and bandwidth.

III. CIRCUIT IMPLEMENTATION

The circuit schematic of the proposed RF front-end is shown in Fig. 4. The core of the RF front-end is composed of fully differential bottom-plate N-path filters implemented together with the proposed read-out technique. RF mixing and IF read-out capacitors are chosen to be 6.4 pF and 4 pF respectively to achieve an IF bandwidth of 15 MHz and desired impedance matching at 1 GHz.

Passive mixer switches are sized to provide a differential resistance of 30 Ω. This is a trade-off between power consumption of the clock drivers and maximum achievable linearity. As shown in Fig. 4, additional switches are used to set the common-mode bias voltage, V_C of the mixer and sampling switches via an external supply.

All the switches are driven by 4-phase non-overlapping 25% duty-cycle clocks provided by divide-by-2 circuit. GF22 nm FDSOI CMOS process uses SiGe channel in PMOS transistors to achieve a driving capability similar to that of NMOS transistors. Hence the input capacitance of the digital circuits are significantly reduced. When compared to other CMOS processes with Si channel in MOSFETs, where PMOS transistors are often 2-3x larger than NMOS for similar driving capability, almost 2x higher speed and lower power consumption is achieved in the digital implementation.

IV. MEASUREMENT RESULTS

An experimental prototype is fabricated in 22 nm FDSOI CMOS process and a QFN40 package is used. The chip photo is shown in Fig. 5a and the active area is 0.65×0.35 mm². To

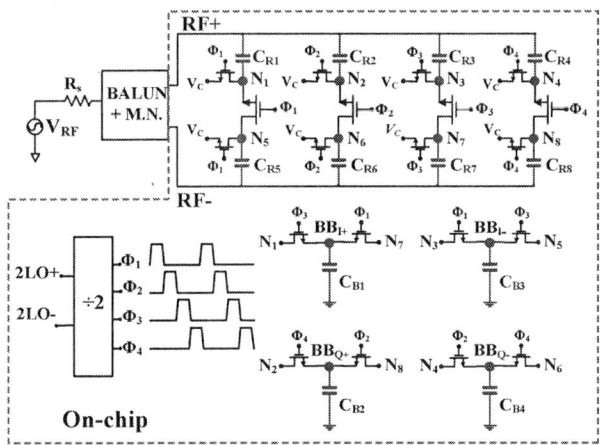

Fig. 4. Complete architecture of the implemented RF front-end

be able to experiment with multiple impedance up-conversion ratios, off-chip transformers are used as matching network and as balun simultaneously. For all the reported measurements below, a Minicircuits TC4-14X+ transformer with an impedance up-conversion ratio of $1 : 4$ is used. The prototype operates in the frequency range of $0.6 - 1.3\,\text{GHz}$ (Fig. 5b). At a switching frequency of $1\,\text{GHz}$, the dividers and clock drivers in the front-end consume $600\,\mu\text{W}$ of power from a $0.8\,\text{V}$ supply.

Figure 5b and Figure 5c show that the front-end achieves a small-signal conversion gain of $13 - 14\,\text{dB}$, an IF bandwidth of $16\,\text{MHz}$, and a minimum double-sided NF of $5 - 6\,\text{dB}$. The parasitic substrate capacitance of the RF capacitors provides low-pass filtering at the RF terminals. Such passive filtering limits the bandwidth and attenuates the input signal. Together with limited transformer bandwidth, it degrades the NF of the front-end by $\sim 3\,\text{dB}$ at higher RF frequency [4]. It also shifts the S11 minima to left in accordance with the LPF phase response [1].

The measured IIP3 and IIP2 performance versus frequency offset of the interferers from the f_{LO} is shown in Fig. 6a. The offset frequency of the two-tone interferers are swept while keeping the resulting IM3 or IM2 product at a baseband frequency of $5\,\text{MHz}$. With no active circuit elements, the proposed RF front-end achieves an in-band IIP3 of $+10\,\text{dBm}$. As the frequency offset increases, IIP3/IIP2 increases due to N-path filtering. The front-end achieves a maximum out-of-band IIP3 of $+25\,\text{dBm}$ and IIP2 of $+66\,\text{dBm}$ at an offset of $160\,\text{MHz}$ (10x BW).

Large signal performance of the RF front-end is evaluated by measuring the in-band $1\,\text{dB}$ compression point (CP1dB) and out-of-band blocker $1\,\text{dB}$ compression point (B1dB). Figure 6a shows the B1dB compression point versus blocker frequency offset normalized to bandwidth. For a blocker located at $80\,\text{MHz}$ offset (5x BW), the proposed work achieves $-1\,\text{dBm}$ B1dB compression point. As shown in Fig. 6b, the front-end achieves a large in-band compression point (CP1dB) of $-7.5\,\text{dBm}$ thanks to no active circuitry.

As shown in Fig. 6c, LO leakage at the RF port was measured for 4 different samples. We see that the proposed RF front-end achieves $< -70\,\text{dBm}$ LO leakage across the operating LO frequency range. At $f_{LO} = 1\,\text{GHz}$, it exhibits $< -80\,\text{dBm}$ LO leakage.

The performance summary of the proposed RF front-end and the comparison with other state-of-the-art mixer-first front-ends is shown in Table 1. From the table, it is evident that this work achieves comparable out-of-band linearity performance while consuming 10x - 20x lower power than several high-performance mixer-first front-ends.

Strikingly, when compared to other sub-mW RF front-ends in Table 2, the proposed work shows $\sim 20\,\text{dB}$ improvement in the out-of-band IIP3 performance while exhibiting low noise figure. Additional baseband amplification is necessary to adopt this architecture in low-power RX. And, it would degrade the linearity performance of the proposed front-end. However, it should be noted that at out-of-band frequencies, this RF front-end provides more than $20\,\text{dB}$ attenuation. Such attenuation minimizes the linearity degradation of the RX and facilitates competitive performance. Finally, additional attenuation can be achieved by using high-Q LC matching network instead of transformers.

V. CONCLUSION

This paper reports a proof of concept of implicit capacitive stacking in a bottom-plate N-path filter. Such capacitive stacking is achieved by reading-out the down-converted signal from the bottom-plate of N-path capacitors. It results in an inherent passive voltage conversion gain of $6\,\text{dB}$ and facilitates low noise figure at the cost of additional capacitor area. Further, a passive matching network is employed to realise impedance matching, to achieve more voltage gain and high linearity with small mixer switches.

ACKNOWLEDGMENT

The authors would like to thank Global Foundries for supporting Chip Fabrication, Gerard Wienk for CAD assistance and Henk de Vries for measurement setup.

REFERENCES

[1] C. Andrews and A. C. Molnar, "A Passive Mixer-First Receiver With Digitally Controlled and Widely Tunable RF Interface," *IEEE Journal of Solid-State Circuits*, vol. 45, no. 12, pp. 2696–2708, Dec 2010.

[2] A. Ghaffari, E. A. M. Klumperink, and B. Nauta, "A differential 4-path highly linear widely tunable on-chip band-pass filter," in *2010 IEEE Radio Frequency Integrated Circuits Symposium*, May 2010, pp. 299–302.

[3] C. Wu, Y. Wang, B. Nikolic, and C. Hull, "A passive-mixer-first receiver with LO leakage suppression, 2.6dB NF, gt;15dBm wide-band IIP3, 66dB IRR supporting non-contiguous carrier aggregation," in *2015 IEEE Radio Frequency Integrated Circuits Symposium (RFIC)*, May 2015, pp. 155–158.

[4] Y. Lien, E. Klumperink, B. Tenbroek, J. Strange, and B. Nauta, "A high-linearity CMOS receiver achieving +44 dBm IIP3 and +13 dBm B1dB for SAW-less LTE radio," in *2017 IEEE International Solid-State Circuits Conference (ISSCC)*, Feb 2017, pp. 412–413.

[5] A. Selvakumar, M. Zargham, and A. Liscidini, "Sub-mW Current Re-Use Receiver Front-End for Wireless Sensor Network Applications," *IEEE Journal of Solid-State Circuits*, vol. 50, no. 12, pp. 2965–2974, 2015.

[6] S. Krishnamurthy, F. Maksimovic, and A. M. Niknejad, "580 μW 2.2-2.4 GHz Receiver with +3.3 dBm Out-of-Band IIP3 for IoT Applications," in *ESSCIRC 2018 - IEEE 44th European Solid State Circuits Conference (ESSCIRC)*, Sep. 2018, pp. 106–109.

2019 IEEE Radio Frequency Integrated Circuits Symposium

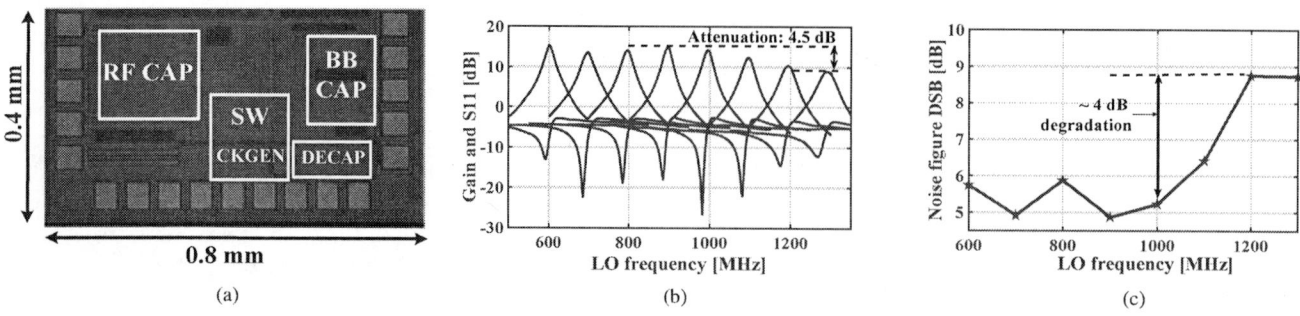

(a)   (b)   (c)

Fig 5   (a) Die Micrograph and Small signal performance: (b) Conversion gain and S11 vs LO frequency; (c) DSB Noise figure vs LO frequency

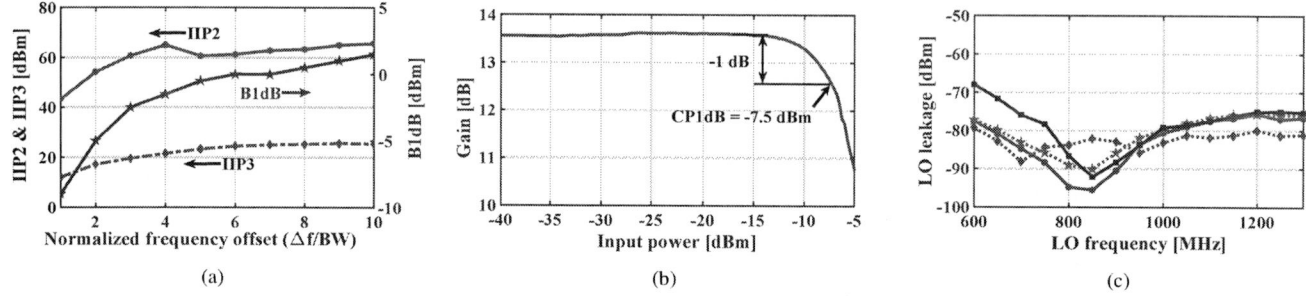

(a)   (b)   (c)

Fig. 6.   (a) Linearity performance: IIP3, IIP2 and B1dB (IF BW = 16 MHz); (b) Inband 1 dB Compression point and (c) LO leakage at RF port (measured for 4 different samples)

Table 1.   Result summary and comparison with high-performance mixer-first receivers.

Features	Andrews JSSC10	Nejdel RFIC15	Lin ISSCC15	Westerveld RFIC16	Lien RFIC17	This Work
Technology	65 nm	65 nm	65 nm	65 nm	45 nm SOI	22 nm FDSOI
Frequency [GHz]	0.1 - 2.4	2 - 3	0.1 - 1.5	0.03 - 0.3	0.2 - 8	0.6 - 1.3
Power [mW]	37 - 70	27 - 75	11@1.5 GHz	46.1	50 + 30 mW/ GHz	0.6@1 GHz
Gain [dB]	40 - 70	7.5	38	21-36	21	13 - 14
IF BW [MHz]	10	10	2	2 - 40	10	16
DSB-NF [dB]	3 - 5	2.5 - 4.5	2.9	6	2.3 - 5.4	5 - 6 ($\leq$1 GHz)
OOB IIP3[dBm @ Δf/BW]	25 @ 10	26 @ 10	13 @ 15	41 @ 20	39 @ 8	25 @ 10
OOB IIP2[dBm @ Δf/BW]	56 @ 10	65 @ 10	47 @ 15	90 @ 20	88 @ 8	66 @ 10
LO leakage [dBm]	-65	-60	N.A.	N.A.	-65	<-70
Supply [V]	1.2/2.5	1.2	0.7/1.2	1.2	1.2	0.8
Active Area [mm^2]	0.75	0.23	0.028	0.8	0.8	0.23

Table 2.   Comparison with low-power RF front-ends.

Features	Bryant RFIC12	Lin ISSCC14	Selvakumar JSSC15	Lee TMTT18	Krishnamurthy ESSCIRC18	This Work
Technology	65 nm	65 nm	130 nm	28 nm	28 nm	22 nm FDSOI
Frequency [GHz]	2.45	0.43 - 0.96	2.4	2.4	2.4	0.6 - 1.3
Power [mW]	0.4	1.15	0.6	0.64	0.58	0.6@1 GHz
Gain [dB]	27.5	50	55.5	50	19	13 - 14
BW [MHz]	N.A.	N.A.	2	1	3.6	16
DSB-NF [dB]	9	8.1	15.1	6.5	11.9	5 - 6 ($\leq$1 GHz)
OOB IIP3[dBm @ Δf/BW]	-21 @ N.A.	-20.5 @ N.A.	-15.8 @ 2.5	0.9 @ 10	3.3 @ 13.9	25 @ 10
Supply [V]	0.8	0.5	0.8	0.8	1	0.8
Active Area [mm^2]	0.24	0.2	0.25	0.25	N.A.	0.23

978-1-7281-1702-7/19 $31.00 © 2019 IEEE   94

A 0.3-to-1.3GHz Multi-Branch Receiver with Modulated Mixer Clocks for Concurrent Dual-Carrier Reception and Rapid Compressive-Sampling Spectrum Scanning

Guoxiang Han, Tanbir Haque, Matthew Bajor, John Wright, and Peter R. Kinget

Dept. of Electrical Engineering, Columbia University, New York, NY 10027, USA

Abstract— **A flexible RF receiver is introduced that uniquely uses CW-modulated clocks for the down-conversion mixers in its mixer-first and low-noise transconductance branches, thereby enabling tuned matching and reception concurrently at two RF carriers. Turning off the modulation reverts the receiver back to single-carrier operation, whereas using PN sequences to modulate the mixer clocks enables rapid, wideband compressive-sampling spectrum scanning. All three functions are accomplished within a single unified architecture. A prototype of the multi-branch modulated-mixer-clock receiver was developed in 65nm CMOS and operates from 0.3 to 1.3GHz. For single-carrier reception, the receiver delivers 15MHz RF bandwidth, 42dB conversion gain, 3.3dB NF, +3.3dBm B1dB, and +12.2dBm OB-IIP3. Concurrent dual-carrier reception at 500MHz and 900MHz offers -8.4dBm B1dB and <6dB NF. In rapid CS spectrum scanning mode, the receiver achieves 66dB dynamic range with -75dBm sensitivity over a 630MHz RF span within 0.71us and consumes 18.5nJ per detected signal.**

I. INTRODUCTION

Ever-increasing wireless throughput demands drive the need to perform concurrent signal reception from multiple carriers that are hundreds of MHz apart. Different receiver architectures have been proposed to address this challenge. The frequency-translational quadrature-hybrid receiver [1] can aggregate two carriers with a low NF by introducing an off-chip hybrid coupler. Its broadband impedance matching offers no selectivity at RF and thus limits out-of-band (OB) linearity. The gain-boosted N-path filter (NPF) receiver [2] aggregates two carriers and offers RF selectivity by placing two NPF filters in feedback. However, the active feedback limits both in-band (IB) and OB linearity. The FTNC receiver [3] offers a tuned RF impedance and good OB linearity by translating the baseband impedance to RF with its mixer-first branch. Its input impedance is 50Ω inside the desired band but low outside, making the parallel combination of multiple FTNC receivers impractical for multi-carrier reception.

In this paper, we explore how modulating the driving clocks of the down-conversion mixers offers unique abilities to a flexible multi-branch modulated-mixer-clock (MMC) receiver. It leverages several advantages of aforementioned receivers and additionally provides a rapid spectrum scanning feature [4]. With a continuous-wave (CW) clock modulation, the lowpass baseband impedance in the mixer-first branches (Fig. 1) is translated as a tuned response at F_{RF1} and F_{RF2}, offering RF selectivity, good OB linearity, and concurrent reception from these two RF carriers. Two LNTA branches are incorporated into the system to offer noise cancellation

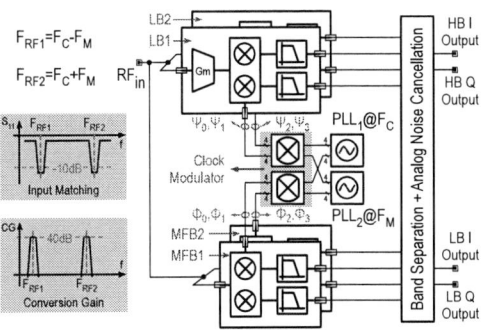

Fig. 1. Diagram of the proposed multi-branch modulated-mixer-clock receiver

for better noise performance. With a pseudo-noise (PN) clock modulation, a widespread conversion gain profile is realized that enables the rapid compressive-sampling (CS) spectrum scanning feature with a superior combination of detection sensitivity and energy consumption. Without any clock modulation, the receiver is a single-carrier receiver and delivers excellent B1dB and OB-IIP$_3$.

II. MULTI-BRANCH MMC RECEIVER ARCHITECTURE

Four pairs of tri-level-modulated mixer clocks φ_0 to φ_3 are derived from two 4-phase 25% non-overlapping clocks at F_C and F_M with a digital clock modulator by either flipping, or not flipping, the polarity of the input clocks, or holding both outputs low (Figs. 1 and 2). When flipping the RF clock at a clock rate of F_M, the F_C tone in the RF clock is moved to two separate tones at $F_C \pm F_M$. When one modulated clock is low, one of the other three modulated outputs is switching. While the F_C and F_M clocks do not need to be synchronized, the four clocks φ_0 to φ_3 are still guaranteed to be non-overlapping in the time domain and they have linearly-independent spectra.

Applying φ_0 to φ_3 to the mixer-first branches (MFB$_1$ and MFB$_2$ in Fig. 1) translates the lowpass baseband impedance to F_{RF1} and F_{RF2} as dual-band tuned input matching. Meanwhile, the mixers concurrently down-convert the RF input signals at F_{RF1} and F_{RF2} to linearly independent baseband outputs. Simple addition and subtraction separate the I/Q components for each RF carriers, as shown in Fig. 2. Short clock pulses can occur that are potentially too short to turn on the mixer switches; however, these pulses carry little energy and mainly contribute to higher-order inter-mixing components that have significantly lower power levels and do not affect the main operation at $F_C \pm F_M$.

978-1-7281-1702-7/19 $31.00 © 2019 IEEE

2019 IEEE Radio Frequency Integrated Circuits Symposium

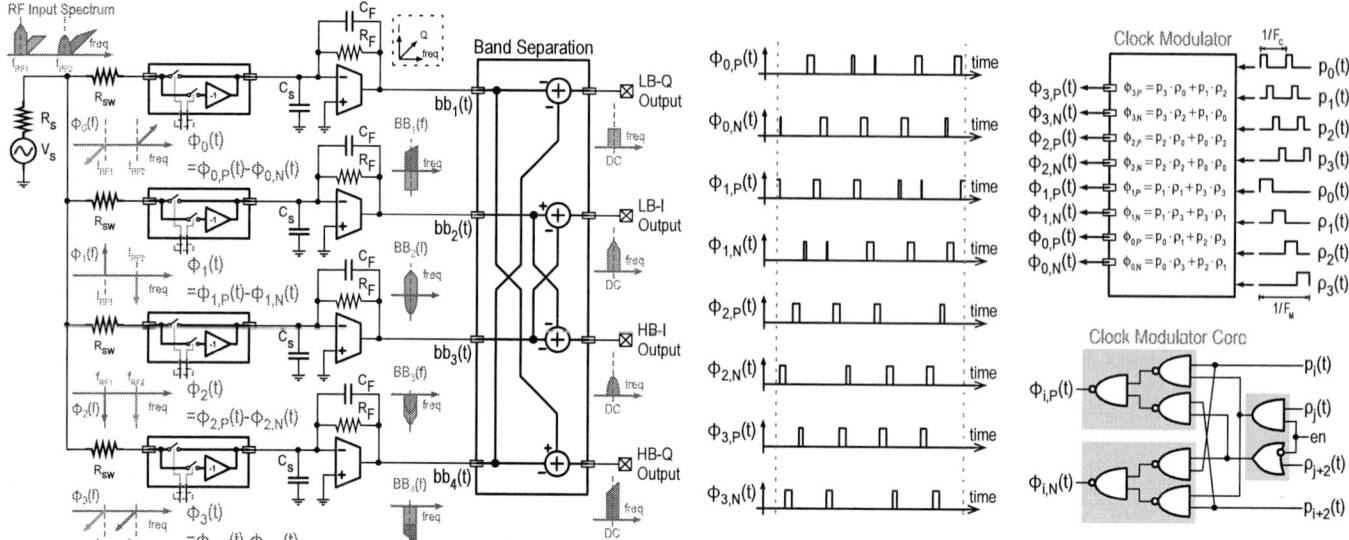

Fig. 2. Conceptual diagram for the operation of the mixer-clock modulation and the two mixer-first branches MFB$_1$ and MFB$_2$. A single-ended version is shown, but the circuit implementation is fully differential. Tri-level modulations of an RF clock (at F$_C$ = (F$_{RF1}$ + F$_{RF2}$)/2) and a modulation clock (at F$_M$ = (F$_{RF2}$ − F$_{RF1}$)/2) results in concurrent input matching at F$_{RF1}$ and F$_{RF2}$ and concurrent reception from these two frequencies.

The addition of two LNTA branches (LB$_1$ and LB$_2$ in Fig. 1) driven by the modulated clocks Ψ_0 to Ψ_3 allows for the noise cancellation [3] and improves system's NF. These two groups of non-overlapping modulated clocks $\Psi_{0,1}$ and $\Psi_{2,3}$ are generated with 50% 4-phase F$_M$ clocks modulated with the 25% 4-phase RF clocks at F$_C$.

III. MULTI-BRANCH MMC RECEIVER IMPLEMENTATION

The fully-differential MMC receiver prototype shown in Fig. 3 with bias and control circuitry has been implemented on 1.8mm^2 in 65nm CMOS and operates from a 1.2V supply. Each LNTA branch comprises a cascoded common-source (C-S) LNTA, differential passive mixers, and two-stage Miller-compensated TIAs. The cascoded C-S LNTA shown in Fig. 4a provides 88mS transconductance and consumes 13mA. The passive mixers use transmission gates. The baseband TIAs are implemented with 4-bit programmable feedback resistance and 5-bit feedback capacitance for gain and bandwidth control.

Each mixer-first branch consists of RF switches, baseband folded-cascode TIAs, and Cherry-Hooper voltage buffers. The RF switches are placed in a floating-body configuration for small OFF capacitance and biased at a 0.2V source voltage for low ON resistance. The Cherry-Hooper voltage buffers in Fig. 4b have 6-bit programmable degeneration resistance for linearity control and 2-bit programmable feedback resistance for gain control. The baseband TIAs in Fig. 4c and the voltage buffers together consume 4.92mA per branch.

The clock path contains clock dividers, non-overlapping clock generators, a PRBS waveform synthesizer, and two clock modulators. The modulator cores in Fig. 4d are based on custom high-speed NAND and NOR gates. The PRBS synthesizer can generate shift-register-based and LFSR-based PN sequences with various lengths. The LNTA and the mixer-first branches can operate independently with different clock sources or can be driven synchronously with the same source.

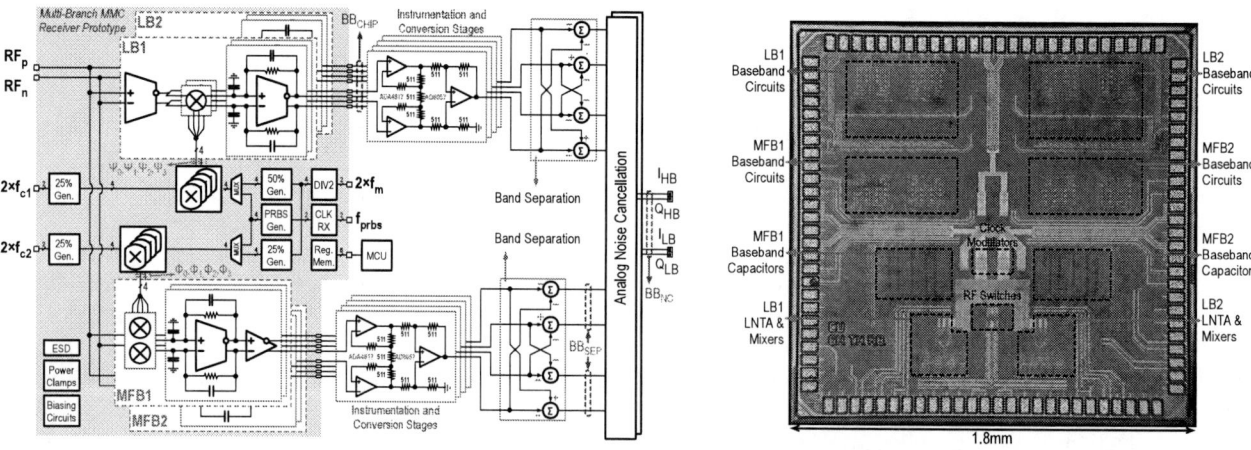

Fig. 3. The multi-branch modulated-mixer-clock receiver system architecture and the 65nm CMOS die micrograph. The fully-differential receiver prototype occupies an active area of 1.8mm^2 and operates from a 1.2V power supply.

978-1-7281-1702-7/19 $31.00 © 2019 IEEE

2019 IEEE Radio Frequency Integrated Circuits Symposium

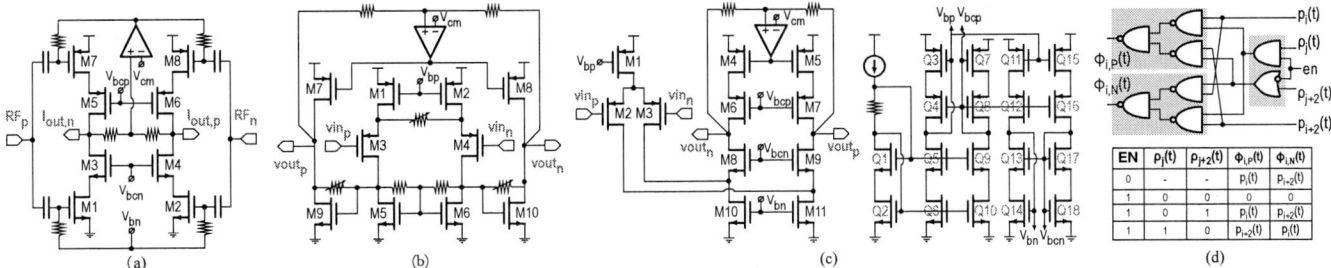

Fig. 4. Key circuit blocks of the MMC receiver, including (a) cascoded common-source LNTAs, (b) Cherry-Hooper voltage amplifiers, (c) folded-cascode TIA OTA core with biasing circuitry, and (d) clock modulator cores and its truth table.

IV. EXPERIMENTAL RESULTS

The tuned input impedance matching in the MMC receiver offers OB linearity improvements over a broadband matched receiver (like [1]). Fig. 5 shows the swept-B1dB and the input matching, S_{11}, measurements. To provide a baseline, the performance with broadband matching is first measured, where the RF switches in the mixer-first branches are disabled and the RF input is resistively terminated. Then, the mixer-first termination is employed and the MMC receiver now has a tuned impedance matching, resulting in a reduction of the voltage swing from the OB blocking signal at the RF input. Therefore, the OB linearity is improved by up to 10dB in the single-carrier reception mode and by more than 5dB in the concurrent dual-carrier reception mode. We estimate the routing and switch resistance in our prototype to be around 13Ω, resulting in a -3dB and -5dB S_{11} floor in both modes. Better OB matching and further improved linearity are expected with better layout for less routing resistance and advanced technologies for less switch resistance.

In the single-carrier reception mode, only the LB_1 branch and the MFB_1 branch are active and the clock modulation is disabled. The RF clock can be swept between 300MHz and 1300MHz; the conversion gain can be programmable from 25dB to 46.7dB, and the baseband bandwidth can be set from 7MHz to 33MHz. The receiver has 42dB conversion gain at 700MHz with a 15MHz RF bandwidth and +3.3dBm B1dB for a 350MHz offset (Fig. 6a). The measured OB-IIP3 is +12.2dBm with tones at 175MHz and 349MHz offsets. The measured NF is 3.3dB at 700MHz after cancellation.

For the concurrent dual-carrier reception, both LNTA and mixer-first branches are active and the clock modulation is enabled. The modulation clock can be set from 100MHz to 300MHz, thus supporting concurrent dual-carrier reception with a carrier separation from 200MHz to 600MHz. The conversion gain to the chip baseband outputs (BB_{CHIP} in Fig. 3) decreases by 3dB since the clock modulation bifurcates the RF clock to F_{RF1} and F_{RF2}. Concurrent down-conversion is shown with different modulation clock rates in Fig. 6b. For

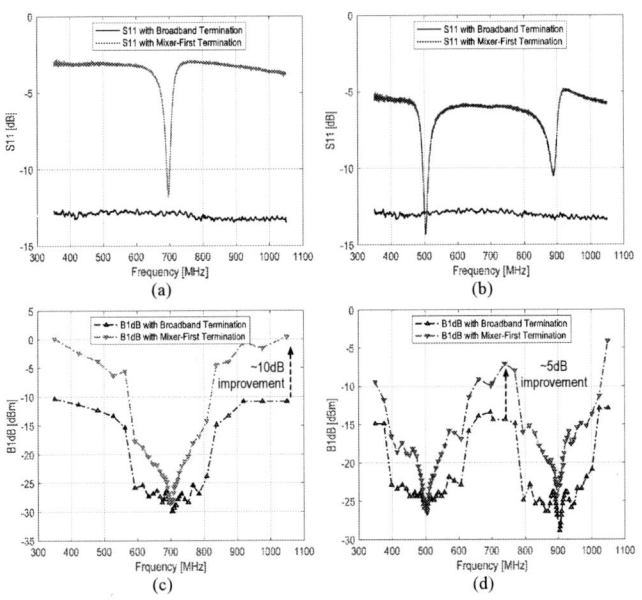

Fig. 5. A comparison of return loss [(a), (b)] and swept-B1dB performance [(c), (d)] of the MMC receiver for a broadband resistive termination versus a tuned mixer-first termination in the single-carrier reception mode at 700MHz and in the concurrent dual-carrier reception mode at 700MHz ± 200MHz.

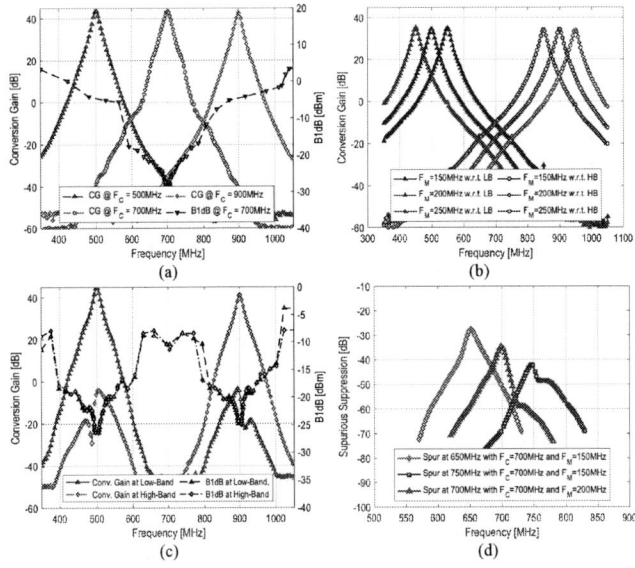

Fig. 6. Measured receiver performance: (a) conversion gain and B1dB in the single-carrier reception mode, (b) demonstration of concurrent dual-carrier reception at BB_{NC} with varying F_M, (c) conversion gain and B1dB at BB_{NC} from 500MHz and 900MHz in the concurrent dual-carrier reception, and (d) spurious suppression at BB_{NC} in the concurrent dual-carrier reception mode.

978-1-7281-1702-7/19 $31.00 © 2019 IEEE

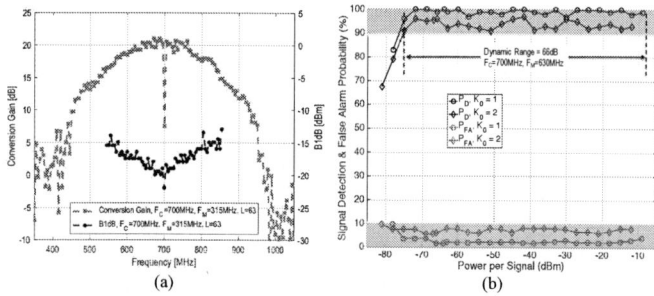

Fig. 7. Measurements for rapid CS spectrum scanning: (a) conversion gain and B1dB profiles with F_M = 315MHz and length-63 sequences, (b) P_D and P_{FA} vs. input power level for detecting one and two ($K_0 = 1, 2$) input signals.

Table 1. Multi-branch MMC receiver performance summary and comparison with other concurrent dual-carrier receiver and CS spectrum scanner

Metric	This Work	[1]	[4]	[2]	[3]
Design Type	MMC	FTQH	DRF2IC	GB-NPF	FTNC
RF Frequency [MHz]	300 ~ 1300	600 ~ 2200	600 ~ 3000	300 ~ 1400	100 ~ 3300
Max. Gain [dB]	46.7	48	41.5	38.5	N/R
BB Bandwidth [MHz]	7 ~ 33	6 ~ 35	20	1	3
Number of Clock Phases	4	8	4	4	4
Clock Power [mW/GHz]	12.0	10.5	N/R	N/R	8.0
CMOS Technology	65nm	65nm	65nm	65nm	28nm
Active Area [mm2]	1.8	1.1	0.56	0.31	5.2
Single-Carrier Reception Mode					
Noise Figure [dB]	2.7 ~ 4.4	0.9 ~ 1.8	3.6	3.4 ~ 4.9	1.8 ~ 3.8
IB IIP3 [dBm]	12.1	12.6	11.0	26.0	N/R
OB-IIP3 [dBm]	+12.2	+8	+4	+7.5	+11.5
B1dB-CP [dBm]	+3.3	-10	-2	-11.8	-2.5
Analog Power [mW]	39.4	96.0	46.5	25.5	36.0
Concurrent Dual-Carrier Reception Mode					
Noise Figure [dB]	4.6 ~ 5.9	1.3 ~ 3.2	N/A	3.4 ~ 4.9	N/A
B1dB-CP [dBm]	-8.4	-10		-11.8	
Analog Power [mW]	78.8	96.0		25.5	
Rapid CS Spectrum Scanning Mode					
Detection Range [MHz]	385 ~ 1015		635 ~ 2840		
Number of Branches	4		4		
Scan Time [us]	0.71		1.2		
Enery per Scan [nJ]	36.9	N/A	132	N/A	N/A
Enery per Detected Signal [nJ]	18.5		22.0		
Sensitity [dBm]	-75		-71		
Dynamic Range[dB]	66		66		
N/R = Not Reported				N/A = Not Applicable	

dual-carrier reception at 500MHz and 900MHz, the measured mid-band B1dB is better than -10dBm. After band separation, the conversion gain increases by 3dB and is the same as the conversion gain under the single-carrier reception mode. The rejection between two bands is 46dB after standard calibration for gain and delay mismatches. The measured NF at BB_{NC} (Fig. 3) is 4.6dB at 500MHz and 5.9dB at 900MHz, respectively. Spurious responses exist in the system due to the higher-order inter-mixing components in the modulated clocks. But they have significantly lower gain, and >30dB spurious suppression is observed in Fig. 6d. More suppression is expected with higher-order clocking systems [3].

In the rapid CS spectrum scanning mode, only the mixer-first branches are active. The mixer-clock modulation is enabled and the RF clock F_C at 700MHz is modulated with a maximal-length PRBS sequence of length L=63 clocked at F_M = 630MHz, resulting in a widespread conversion gain as shown in Fig. 7a. The RF input span extends from ($F_C - F_M/2$) to ($F_C + F_M/2$) and signals from all L channels are simultaneously down-converted to baseband. With CS DSP, the spectral locations of a few strong signals within the input span can be rapidly found. The measured sensitivity for a signal detection probability P_D > 90% and a false alarm probability P_{FA} < 10% is -75dBm and the dynamic range is 66dB over a 630MHz bandwidth with 75 samples per measurement. More signals can be detected with more sequences [4].

V. COMPARISON TO THE STATE OF THE ART

The performance of the multi-branch MMC receiver is summarized and compared in Table 1. In the single-carrier reception mode, the MMC receiver delivers excellent B1dB and OB-IIP$_3$ thanks to the tuned impedance matching and the up-front filtering. The IB-IIP$_3$ is slightly better than [1] but it is 14dB better than [2]. In the concurrent dual-carrier reception mode, the MMC receiver delivers excellent B1dB. Though [2] consumes less analog power, the MMC receiver offers at least 6x more bandwidth. In the rapid CS spectrum scanning mode, the MMC receiver delivers superior detection sensitivity and energy consumption when compared to other published CS spectrum scanner [4].

VI. CONCLUSIONS

The multi-branch MMC receiver presented in this paper offers a single unified architecture for high-performance single-carrier and concurrent dual-carrier reception, and rapid CS spectrum scanning by modulating the mixer clocks. With a CW clock modulation, the receiver realizes dual-band tuned RF input matching for excellent OB linearity and enables concurrent dual-carrier reception. With a PN clock modulation, the receiver widely spreads the conversion gain profile and can detect a few strong signals within a wide RF span. These features render the proposed multi-branch MMC receiver a promising front-end solution for great reconfigurabilities and tunabilities.

ACKNOWLEDGMENT

This work was supported by NSF Award 1733857.

REFERENCES

[1] J. Zhu and P. R. Kinget, "9.3 a very-low-noise frequency-translational quadrature-hybrid receiver for carrier aggregation," in *2016 IEEE International Solid-State Circuits Conference (ISSCC)*, Jan 2016, pp. 168–169.

[2] A. Agrawal and A. Natarajan, "An interferer-tolerant cmos code-domain receiver based on n-path filters," *IEEE Journal of Solid-State Circuits*, vol. 53, no. 5, pp. 1387–1397, May 2018.

[3] D. Murphy, H. Darabi, and H. Xu, "3.6 a noise-cancelling receiver with enhanced resilience to harmonic blockers," in *2014 IEEE International Solid-State Circuits Conference Digest of Technical Papers (ISSCC)*, Feb 2014, pp. 68–69.

[4] T. Haque, M. Bajor, Y. Zhang, J. Zhu, Z. A. Jacobs, R. B. Kettlewell, J. Wright, and P. R. Kinget, "A reconfigurable architecture using a flexible lo modulator to unify high-sensitivity signal reception and compressed-sampling wideband signal detection," *IEEE Journal of Solid-State Circuits*, vol. 53, no. 6, pp. 1577–1591, June 2018.

A 0.5-20 GHz RF Silicon Photonic Receiver with 120 dB·Hz$^{2/3}$ SFDR using Broadband Distributed IM3 Injection Linearization

Navid Hosseinzadeh[1], Aditya Jain[2], Kang Ning[1], Roger Helkey[2], James F. Buckwalter[1]

[1]Department of Electrical and Computer Engineering, University of California - Santa Barbara, Santa Barbara, CA 93106, USA
[2]Institute for Energy Efficiency, University of California - Santa Barbara, Santa Barbara, CA 93106, USA
navid@ece.ucsb.edu, buckwalter@ece.ucsb.edu,

Abstract — **Radio-over-fiber (RoF) supports microwave and millimeter-wave communication with remote antenna heads. However, RoF links suffer from low spur-free dynamic range (SFDR) due to the low gain and high nonlinearity of silicon photonic (SiP) Mach-Zehnder modulators (MZM). This work demonstrates the first distributed silicon-germanium (SiGe) HBT LNA co-designed for linearization of a broadband SiP-based RoF link. The SiGe LNA features a distributed LNA intermodulation (IM) injection scheme that is inherently wideband, improving IIP3 over a 10 GHz range. The assembled SiGe LNA and SiP MZM prototype demonstrates an SFDR as high as 120dB·Hz$^{2/3}$ at 9 GHz, a 19 dB improvement over previous SiP RoF links.**

Keywords — **Microwave photonics, Radio over fiber, silicon photonics, linearization, Mach-Zehnder modulator, distributed LNA, SFDR, analog predistortion, RF photonics.**

I. INTRODUCTION

Radio-over-fiber (RoF) links are useful in environments where antenna remoting improves signal coverage; typically in situations with high RF path loss or blockage. Furthermore, RoF links are capable of capturing signals over large instantaneous frequency range and transporting signals with low-loss over optical fiber. RoF links are typically characterized by spur-free dynamic range (SFDR), defined by $\frac{2}{3}(IIP3 - MDS)$, where MDS is the minimum detectable signal assuming 1-Hz BW and IIP3 is the 3rd-order input intercept point. Typically, RoF links require an SFDR exceeding 120 dB·Hz$^{2/3}$, compatible with CMOS RF receivers [1], [2].

Fig. 1 illustrates a conventional RoF link consisting of an LNA followed by an electro-optical (E/O) conversion using a Mach-Zehnder modulator (MZM), transport over fiber, and opto-electrical (O/E) conversion back to RF through a photodetector. The IIP3 of the RoF link is generally limited by the V_π of the MZM, the voltage required to change the optical carrier phase by 180°. Unfortunately, higher V_π also reduces the link gain and increases the noise figure (NF).

Different material systems offer fundamental limits on the $V_\pi L$ product of the MZM, where L is the length of the MZM. For instance, InP MZMs typically exhibit 4 V-mm, while SiP MZMs exhibit 15 V-mm due to the weaker change in the index of refraction induced by plasma dispersion. This suggests that the SiP modulator has lower gain and higher NF. Prior work illustrates that the SiP MZM has an SFDR of around 110 dB·Hz$^{2/3}$ at 1 GHz that reduces at

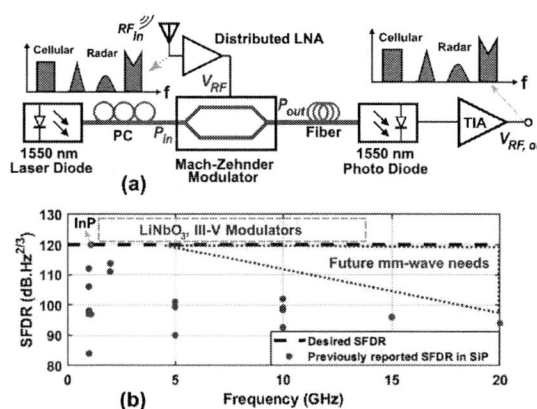

Fig. 1. Block diagram of an RoF link with a SiGe LNA driving an MZM (top). Recently published SFDR results for SiP MZMs in RoF links. (bottom)

higher frequency as shown in Fig. 1(b). Therefore, a 20-dB SFDR gap exists between current SiP MZMs and the 120 dB·Hz$^{2/3}$ target, particularly in microwave applications. While SiP devices are inferior, SiP processes offer many attractive features including the integration of the RoF receiver in a single process with electronics for larger scale integration of an RF system. Prior optical linearization techniques include feedforward modulators and ring-assisted MZMs. While these approaches have demonstrated a 112 dB·Hz$^{2/3}$ SFDR at 1 GHz and 94 dB·Hz$^{2/3}$ SFDR at 20 GHz for a LiNbO3 MZM [3], these approaches are typically demonstrated with discrete components and not in monolithic IC processes. Here, electronic linearization is proposed to compensate the SiP MZM and and reduce the SFDR performance gap between SiP MZM and more expensive LiNbO$_3$ or InP MZMs.

This work reports the first SiP MZM co-designed with a broadband SiGe LNA to linearize the RoF link. Prior work has demonstrated high-speed distributed SiGe drivers for optical communication [4] but the design here focuses on high SFDR. A record 120 dB·Hz$^{2/3}$ SFDR at 9 GHz and a response above 110 dB·Hz$^{2/3}$ to 18 GHz is demonstrated. Three innovations are presented: 1) a distributed LNA that covers 0.5-20 GHz, 2) segmentation of the MZM into four sections, and 3) distributed injection of the 3rd-order intermodulation products (IM3) to linearize the cascade of the SiGe LNA and MZM. The measured SiGe/SiP RoF indicates that the improvement in IIP3 exists over more than 10 GHz, demonstrating broadband

978-1-7281-1702-7/19 $31.00 © 2019 IEEE

linearization.

II. DISTRIBUTED IM3 INJECTION FOR BROADBAND LINEARIZATION

The optical transfer function of the MZM is

$$\frac{P_{OUT}}{P_{IN}} = \frac{e^{-\alpha L}}{2}(1 + \cos(\frac{\pi v_{RF}}{V_\pi} + \frac{\pi}{2})) \qquad (1)$$

where α is the optical attenuation in the waveguide. Consequently, the IIP3 of the MZM is $IIP3_V = \frac{2\sqrt{2}v_\pi}{\pi}$. Along with V_π, other factors in SiP processes impacting the IIP3 include the refraction index nonlinearity and C-V variation [5]. For a given V_π, SiP MZMs will be relatively longer (< 4 mm) than III-V MZMs incurring significant frequency-dependent traveling wave (TW) losses. To overcome these losses, the TW-MZM is instead broken into several segments. The segmented MZM (S-MZM) mitigates high-frequency roll-off in microwave bands.

A four-segment MZM is shown in Fig. 2, where each segment includes a differential p-i-n optical phase shifter to confine the optical field in the waveguide and linearize the MZM response through a combination of plasma dispersion and Kerr effect modulation [6]. While the S-MZM improves the 3-dB bandwidth compared to a TW-MZM, it does not improve the SFDR beyond 110 dB·Hz$^{2/3}$ at 1 GHz. To increase the SFDR further demands an active linearization approach to improve IIP3 that does not penalize NF.

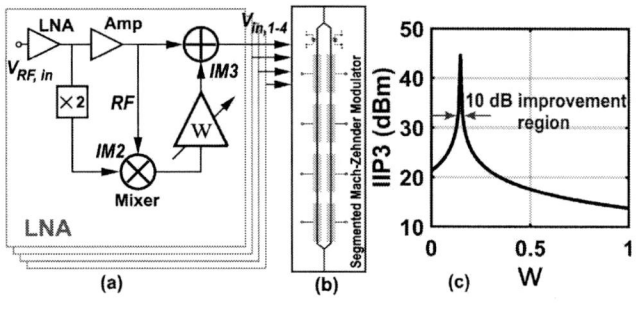

Fig. 2. Conceptual illustration of the broadband linearization scheme with block diagram of distributed IM3 injection (a), segmented MZM (b), and (c) theoretical improvement in IIP3.

Segmenting the MZM allows a distributed mechanism to linearize the optical response in (1) through an analog predistortion. The proposed block diagram of the distributed, analog predistortion LNA is shown in Fig. 2. The RF signal is tapped off from a TW line at each segment and amplified to set the NF. The signal is then be split between two paths. In one path, the RF signal is amplified while through an auxiliary path the RF signal is squared and low-pass filtered to generate a low-frequency 2nd-order intermodulation (IM2) component. Next, the IM2 component is mixed with the original RF signal to generate an IM3 component that can be weighted and added to the original RF signal to cancel the IM3 components generated by the MZM. By adjusting the IM3 weight (w), the inherent MZM IIP3 can be increased substantially as

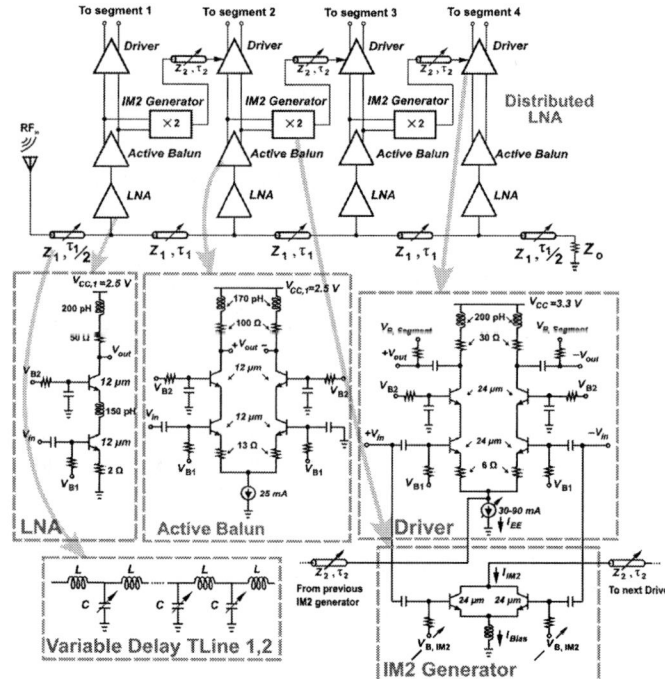

Fig. 3. Circuit schematic illustrating the specific components including the single-ended LNA, wideband active balun, highly-linear differential driver, and IM2 generator

illustrated in Fig. 2(c). The 4-segment MZM is shown in 2(b) and length of each segment is 800 μm.

The circuit schematic of the four-stage SiGe distributed IM3-injection low noise amplifier (LNA) is illustrated in Fig. 3 and is co-designed specifically for the SiP S-MZM stages. The wideband LNA is a resistively degenerated cascode stage with inductive peaking to extend the bandwidth. Resistive degeneration is used for broadband frequency response. Next, an active balun produces differential RF signals with less than 1 dB amplitude and under 5° of phase imbalance up to 25 GHz. A linear, high-voltage differential cascode driver amplifies the RF signal up to 2-Vpp to drive the p-i-n phase shifter segments in the S-MZM.

The IM3 injection circuitry taps the differential signals and uses an inductively-degenerated frequency doubler to square and filter the IM2 products. The weighting and mixing function are produced by mirroring the IM2 current into the bias current of the high-voltage differential driver. Ignoring the emitter degeneration (R_E) of the driver stage, the OIP3 of the linearized LNA is

$$OIP3 \simeq \frac{2R_L(I_{EE} + I_{BIAS})}{\sqrt{|\frac{3I_{BIAS}}{I_{EE}+I_{BIAS}} - 1|}} \qquad (2)$$

where I_{BIAS} is the DC current through the IM2 generator, I_{EE} is the DC current of the differential driver stage, R_L is the output load resistance. Consequently, the OIP3 is tuned through I_{BIAS} and is optimized to cancel the MZM nonlinearity. Here, the bias point for all IM2 generators and drivers were all the same. The average simulated OIP3 of the driver is about 41 dBm, an increase of more than 10 dB in

2019 IEEE Radio Frequency Integrated Circuits Symposium

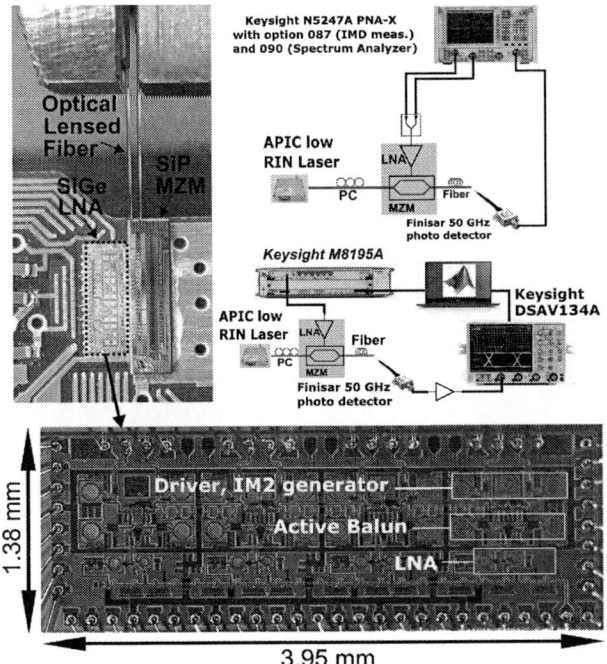

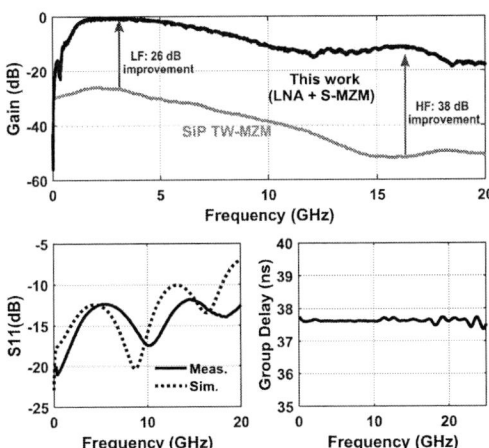

Fig. 5. Measured RoF link gain and group delay from S-parameters.

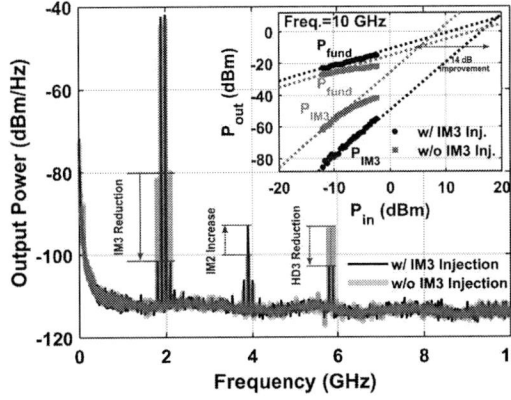

Fig. 4. Micrograph of the assembled chips (top left), measurement setups for linearity and QAM measurements (top right), and micrograph of the distributed LNA chip fabricated in 130 nm SiGe BiCMOS technology (bottom).

Fig. 6. Measured power spectrum for two tones at 1.9 and 2 GHz at input of the driver, and input power sweep at 10 GHz with and without IM3 injection.

the absence of IM3 injection. To match the optical traveling wave velocity in the MZM with the RF traveling wave, the transmission line delay is adjusted with a varactor. The distributed design allows differential delay control between the RF input traveling wave line and the IM2 generation between stages.

III. MEASUREMENTS

The PCB assembly is illustrated in Fig. 4 and shows the SiGe distributed LNA and SiP MZM chips. The driver chip area is 3.95 x 1.38 mm^2 and SiP S-MZM area is 5.5 x 0.65 mm^2, respectively. Measurement setups are also shown in Fig. 4. Measurement setups are also shown in Fig. 4. A 1550-nm low-RIN laser drives the MZM. Measured loss for optical edge couplers is about 3-dB and total optical loss of MZM is less than 10-dB, including light coupling loss.

Fig. 5 plots the measured S-parameters for the RoF link with an external high-linearity photodiode. The link gain is plotted in comparison to a driverless Si-based TW-MZM in the same technology. The active driver improves the gain by 26 dB at 3 GHz and substantially improves gain above 15 GHz relative to the TW-MZM. The simulated and measured input return loss is better than 10 dB to more than 20 GHz. The measured group delay also indicates low variation up to 24 GHz.

Fig. 6 plots the measured output power spectrum for two tones at 1.9 and 2 GHz. The IM3 components decrease by more than 23 dB while an IM2 term at 3.9 GHz increases by 8 dB. Additionally, the HD3 components are reduced by 10 dB. The inset figure shows the input power sweep at 10 GHz

and indicates the IIP3 improvement of 14 dB. The weight of the IM3 component is controlled by bias current of IM2 generator and the the maximum linearity improvement point is found by tuning.

The wideband IIP3 is shown in Fig. 7 with and without IM3 injection at two different bias points. Over the entire 1-20 GHz range, the IIP3 is increased relative to the LNA without IM3 injection by at most 17 dB. The bias on the IM3 injection can be adjusted to achieve higher IIP3 improvement above and below 10 GHz. Second order nonlinearty is important in systems with more than an octave bandwidth. Therefore, the IIP2 is measured before and after IM3 injection. Notably, IIP2 does not substantially change over the frequency range with IM3 injection. However, the results indicate an IIP2 improvement of more than 20 dB at 5 GHz.

Fig. 8 plots the measured NF compared to the NF of a Si-based TW-MZM. The LNA improves the NF by more than 20 dB with a minimum 13.6 dB NF at 3 GHz. From the measured IIP3 and NF, the SFDR is 120 dB·Hz$^{2/3}$ at 9 GHz, a 19 dB improvement over earlier work and 22 dB improvement at 17 GHz. The power consumption of the driver is 1.70 W (425 mW per stage) without IM3 injection and increases by

978-1-7281-1702-7/19 $31.00 © 2019 IEEE   101

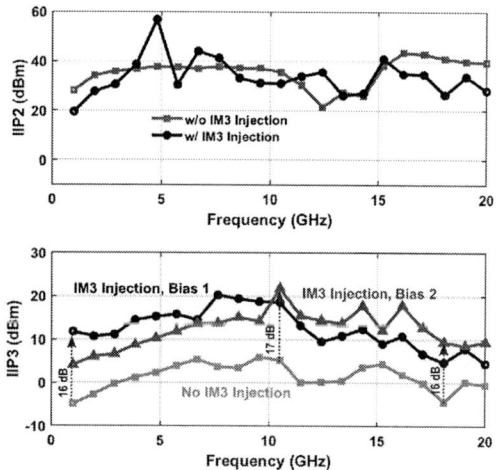

Fig. 7. Measured IIP2 versus frequency before and after IM3 injection (top), and measured IIP3 versus frequency with and without IM3 injection at two bias points (bottom).

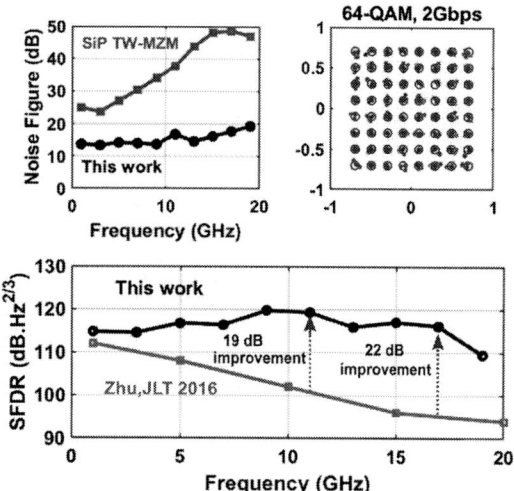

Fig. 8. Measured NF of the RoF link compared to the NF of a Si-based TW-MZM (top left), Measured 64-QAM constellation at 2 Gbps (top right), and calculated SFDR from measurement data and comparison with [3] (bottom)

60 mW with the use of IM3 injection to provide the 10-20 dB IIP3 improvement. The RoF link is also tested with a 2-Gbps 64-QAM modulated signal at carrier frequency of 11 GHz achieving EVM of 6.5% without employing any digital predistortion.

This work is compared to earlier reported linearized RoF links in Table 1. The presented work demonstrates the best SFDR over the highest frequency range in Silicon and on par with LiNbO3 modulators. Additionally, this is the first reported work based on an integrated electronic/photonic solution with small size and weight.

IV. CONCLUSION

This work reports a SiP based integrated RoF link with high SFDR of up to 120 dB·Hz$^{2/3}$ and operating up to 20 GHz. The first co-designed analog predistortion SiGe LNA

Table 1. Comparison to Reported Linearized MZMs

Param.\Ref.	[3]	[7]	[8]	[9]	This Work
Frequency (GHz)	1-20	12	1-18	2-20	**0.5-20**
SFDR ($dB \cdot Hz^{2/3}$)	94-111	116	124	114	**109-120**
IIP3 (dBm)	2	18	42	2	**22**
Technology	LiNbO3	LiNbO3	LiNbO3	LiNbO3	Si-SiGe
Technique	Discrete Predistorter	Dual Parallel MZMs	Dual Series MZMs	-	IM3 Injection

and segmented SiP MZM with a new approach to cancel IM3 distortion generated by the SiP MZM is reported. This work demonstrates that SiP MZMs are capable of meeting high-performance criteria for RoF links and meets or surpasses commercial LiNbO3 and InP MZM performance.

ACKNOWLEDGMENT

This material is based on research conducted under AIM Photonics and sponsored by Air Force Research Laboratory via agreement number FA8650-15-2-5220. Special thanks are given to Erman Timurdogan and Zhan Su of Analog Photonics for providing the p-i-n phase shifter building block and GlobalFoundries for access to the 8HP process technology. The authors would also like to thank Prof. Schow for helpful discussions, and Prof. Bowers and Prof. Rodwell for access to measurement equipment. Finally, we acknowledge Integrand for access to EMX software.

REFERENCES

[1] R. W. Ridgway, C. L. Dohrman, and J. A. Conway, "Microwave Photonics Programs at DARPA," *Journal of Lightwave Technology*, vol. 32, no. 20, pp. 3428–3439, Oct 2014.

[2] N. Hosseinzadeh, A. Jain, R. Helkey, and J. F. Buckwalter, "RF Silicon Photonics for Wideband, High Dynamic Range Microwave and Millimeter-wave Signal Processing," in *2018 IEEE SiRF*, Jan 2018, pp. 41–44.

[3] R. Zhu, X. Zhang, D. Shen, and Y. Zhang, "Ultra Broadband Predistortion Circuit for Radio-over-Fiber Transmission Systems," *Journal of Lightwave Technology*, vol. 34, no. 22, pp. 5137–5145, Nov 2016.

[4] R. J. A. Baker, J. Hoffman, P. Schvan, and S. P. Voinigescu, "SiGe BiCMOS Linear Modulator Drivers with 4.8-Vpp Differential Output Swing for 120-GBaud Applications," in *2017 IEEE RFIC*, June 2017, pp. 260–263.

[5] N. Hosseinzadeh, A. Jain, R. Helkey, and J. Buckwalter, "Sources of RF Intermodulation Distortion in Silicon Photonic Modulators," in *2018 IEEE AVFOP*, Nov 2018, pp. 1–2.

[6] E. Timurdogan, C. V. Poulton, M. Byrd, and M. Watts, "Electric field-induced second-order nonlinear optical effects in silicon waveguides," *Nature Photonics*, vol. 11, no. 3, p. 200, 2017.

[7] W. Jiang, Q. Tan, W. Qin, D. Liang, X. Li, H. Ma, and Z. Zhu, "A Linearization Analog Photonic Link With High Third-Order Intermodulation Distortion Suppression Based on Dual-Parallel MachZehnder Modulator," *IEEE Photonics Journal*, vol. 7, no. 3, pp. 1–8, June 2015.

[8] A. Karim and J. Devenport, "High Dynamic Range Microwave Photonic Links for RF Signal Transport and RF-IF Conversion," *Journal of Lightwave Technology*, vol. 26, no. 15, pp. 2718–2724, Aug 2008.

[9] A. Nikolov, D. Guenther, W. Liu, R. Cendejas, and R. Dutt, "Advancements and Challenges for Photonic Components For Avionic Interconnects," in *2013 IEEE AVFOP*, Oct 2013, pp. 5–6.

A 65nm CMOS Continuous-Time Electro-Optic PLL (CT-EOPLL) with Image and Harmonic Spur Suppression for LIDAR

Ali Binaie[#], Sohail Ahasan[#], Harish Krishnaswamy[#]

[#]Department of Electrical Engineering, Columbia University, New York, NY, 10027, USA

Abstract — An integrated continuous-time electro-optic phase-locked loop (CT-EOPLL) is presented that features image and harmonic spur suppression, and is used in a frequency-modulated continuous-wave (FMCW) LIDAR. The proposed EOPLL has its loop bandwidth equal to its reference frequency, which enables it to relax the trade-off between chirp bandwidth and Mach-Zehnder (MZ) delay and consequently reduce the area and loss associated with the silicon-photonic delay implementation by 10×. Image and harmonic spurs are rejected through single-sideband (SSB) and harmonic-reject (HR) mixing techniques. This EO-PLL is integrated in 65nm CMOS technology, suppresses the highest spur by more than 25dB, and is used in a LIDAR system that can detect an object at ranges exceeding 3.3 meters with an RMS depth precision of 558μm at 2m distance and 9.4mm depth resolution.

Keywords — electro-optic phase-locked loop (EO-PLL), frequency-modulated continuous wave (FMCW) LIDAR, single side-band (SSB) mixing, harmonic-reject mixing (HRM)

I. INTRODUCTION

Recently, considerable effort has been devoted to the research and development of LIDAR (Light Detection and Ranging) systems due to their broad range of applications [1]. Among the different flavors of LIDAR, FMCW (Frequency Modulated Continuous Wave) LIDAR is of high interest as it relaxes the timing precision required when compared with pulsed LIDAR [2]. One of the fundamental elements in FMCW LIDAR is the light source which in turn is difficult to implement as the frequency of the laser has to be modulated accurately based on an input modulation waveform. However, the tuning characteristics of lasers are typically very nonlinear and sensitive to changes in the ambient temperature. Fig. 1 depicts a LIDAR based on an open-loop FMCW laser modulation architecture. A laser is directly frequency modulated with an appropriate electrical current injected to the laser input. In practice, the nonlinearity associated with the laser's transfer function causes not only a bias shift in the beat frequency (error in the accuracy of ranging) but also a broadening of the beat tone (lower precision) (Fig. 1).

The FMCW modulation can be linearized and stabilized by placing the laser within an electro-optic PLL. Previously, FMCW EO-PLLs have been reported using discrete-time (DT-EOPLL) and continuous-time (CT-EOPLL) architectures, shown in Fig. 2a and 2b respectively. In either case, using a specific delay (τ_{MZI}) in one arm of a Mach-Zehnder Interferometer (MZI), a beat tone with frequency f_{beat} is generated due to the interference of the signals with and without the delay at the output of the photodiode (PD). As f_{beat} is the product of chirp slope (α) and τ_{MZI}, locking

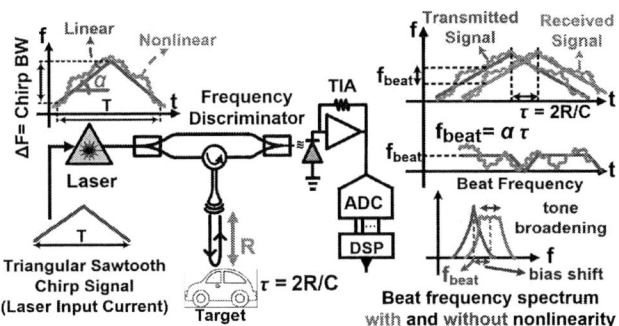

Fig. 1. Open-loop FMCW chirp LIDAR suffers from laser nonlinearity and drift, which cause beat tone broadening and bias shift in the beat frequency.

it to an electrical reference frequency (f_{Ref}) guarantees that α remains constant even with the laser's nonlinearity and frequency drift.

Although the digital-intensive DT-EOPLL [2] (Fig. 2a) seems promising for applications like micro-imaging in which a chirp bandwidth (ΔF) of more than 100GHz is needed to ensure micron-scale depth resolution, it poses fundamental problems for other applications like automotive LIDAR where cm-scale range resolution and consequently ΔF of a few GHz are required. In a DT-(EO)PLL, f_{Ref} must be at least 10 times higher than the loop bandwidth, B, for loop stability due to the Gardner limit (practically 20x from spur suppression considerations [3]):

$$f_{Ref} \geq 10 \times B. \tag{1}$$

Furthermore, as the chirp changes its polarity every T/2 second, the PLL settling time (1/B) has to ensure that the loop locks quickly after each change which results in

$$\frac{1}{B} \leq 0.1 \times \frac{T}{2}. \tag{2}$$

Also, according to Fig. 1, we have .

$$\alpha = \frac{f_{beat}}{\tau_{MZI}} = \frac{\Delta F}{T/2}. \tag{3}$$

Considering (1), (2) and (3) and $f_{Ref} = f_{beat}$, it is simple to show that $\alpha \times \tau_{MZI} \geq 100/T/2$ which finally results in

$$\Delta F \times \tau_{MZI} \geq 100. \tag{4}$$

This equation results in the need for a substantial τ_{MZI} in applications where the chirp bandwidth is a few GHz. For instance, ΔF=16GHz dictates a minimum τ_{MZI} of ~6nS which translates to a 60cm long optical delay that would result in a 120dB loss in a silicon photonic implementation!

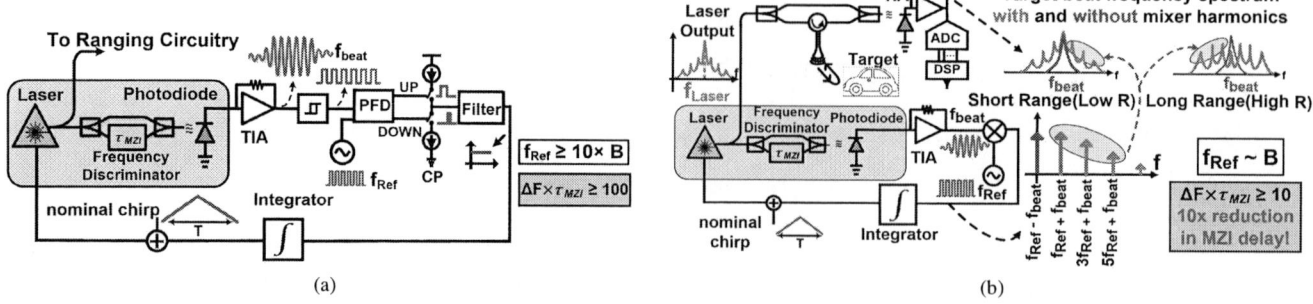

Fig. 2. FMCW chirp LIDAR electro-optic PLL using : (a) a discrete-time architecture, and (b) a continuous-time architecture.

On the other hand, a CT-EOPLL (Fig. 2b) does not suffer from the large optical delay problem since its phase detector (PD) can be realized with either a mixer or an XOR gate without any discrete-time sampling. Therefore, theoretically, B of such a structure can be as high as f_{Ref}, and therefore the optical delay requirement reduces by 10×. The downside of increasing the loop bandwidth is the reduction in spur filtering. A square-wave mixer that mixes the feedback signal with f_{Ref} will produce spurs at $f_{beat} + (2n-1) \times f_{Ref}$, where n is an integer greater than 1. These spurs produce spurs in the laser output signal, which in turn produce unwanted tones around the beat signal at f_{beat} in the LIDAR receiver. These unwanted tones will be interpreted by the LIDAR as shadow objects. The situation gets worse at high ranges, since the power of the spurs can exceed the power of the desired beat signal itself. In [4], in a discrete-component-based CT-EOPLL, an effort has been made to eliminate the spurious tones using a replica mixer in a feed-forward structure. Such a feed-forward cancellation approach requires calibration across temperature and other environmental drifts.

In this work, a 65nm CMOS CT-EOPLL is presented that surpasses Gardner's Limit with a loop bandwidth equal to the reference frequency, while utilizing single-sideband (SSB) and harmonic-reject (HR) mixing to suppress spurs by 25dB without the need for calibration. In conjunction with a silicon-photonic implementation of the frequency-discriminating MZI, a LIDAR system with 16GHz chirp bandwidth is demonstrated that allows us to detect an object at ranges exceeding 3.3 meters with an RMS depth precision of 558μm at 2m distance and depth resolution of 9.4mm.

II. Wideband and Spur-Less EOPLL Design

Fig. 3 shows our proposed architecture, which uses concepts borrowed from the software-defined radio community for image and harmonic rejection. At the output of the TIA, we generate I and Q signals from the beat frequency using, for instance, a polyphase filter, and then feed those to an SSB and HR mixer. Therefore, under lock, the mixer output is free from the image spur ($f_{Ref} + f_{beat}$) because of SSB mixing, and third ($3f_{Ref} \pm f_{beat}$) and fifth ($5f_{Ref} \pm f_{beat}$) harmonic spurs because of 8-phase HR

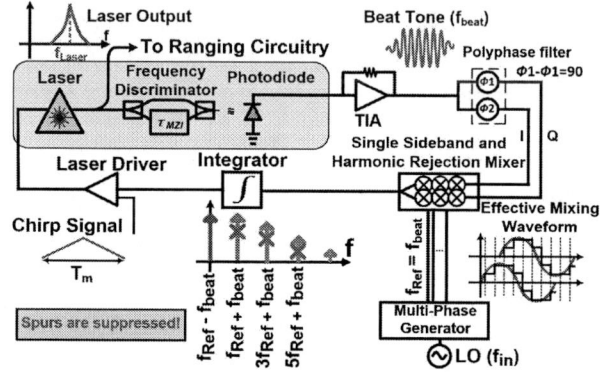

Fig. 3. Proposed CT-EOPLL using SSB and HR mixer for spur suppression.

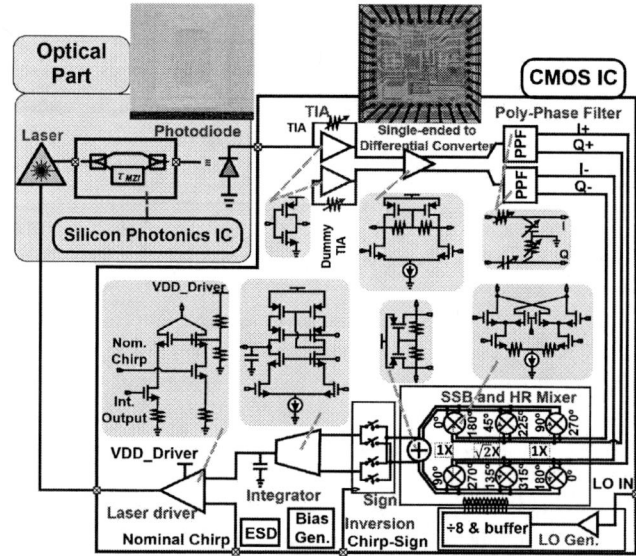

Fig. 4. Proposed EO-PLL block and circuit diagrams, and the electrical and optical chip photographs.

mixing. As in a typical FMCW EO-PLL, the mixer output is then fed into an integrator, which cancels out the differential behavior of the MZI, and then to a laser driver, which provides the laser with both the required nominal current and the correction current provided from the loop.

Fig. 4 shows the EO-PLL block and circuit diagrams as well as the chip photographs of the CMOS and

978-1-7281-1702-7/19 $31.00 © 2019 IEEE

silicon-photonic chips. The silicon-photonic optical integrated circuit (OIC) includes the MZI with a 600ps delay with near 2dB/cm loss which results in a total loss of 20dB (including the coupling losses). The electrical current signal at the output of the PD is brought on to the 65nm CMOS IC and is amplified to a voltage by a trans-impedance amplifier (TIA). The 1GHz bandwidth TIA is implemented as an inverter amplifier with variable resistive-feedback that allows us to adjust its gain from 250Ω to 8kΩ. For the 1kΩ case, the input-referred current noise of the TIA is 19pA/$\sqrt{Hz}$ at 20KHz and lower than that for higher frequencies. A dummy TIA is included to generate the bias for the single-ended to differential converter. After single-ended to differential conversion, a differential tunable polyphase filter (PPF) is used to generate phase shifts of +45° and -45°. These signals are mixed in six doubly-balanced Gilbert-style mixers with six differential LOs with different phases including -45°, 0° and 45° in the I path and 45°, 90° and 135° in the Q path for SSB and HR mixing. These different LO phases are generated inside the chip using a Johnson counter from an input signal at $8 \times f_{ref}$. The outputs of the mixers are connected to the same load to add all signals in current. The weighting of (1, $\sqrt{2}$, 1) in the mixers for the purpose of harmonic rejection is accomplished using device and current scaling as well as scaling of the degeneration resistors which are used for increasing the linearity. The product of the mixing, which is at $f_{Ref} - f_{beat}$, is fed into the gm-C integrator. The gm-C integrator is a high-swing telescopic operational transconductance amplifier (OTA) which achieves a dominant pole of a few tens of hertz by using long channel length for the transistors in conjunction with an off-chip capacitance. Before the integrator, sign inversion switches are used to adjust the loop sign during the up-chirp and down-chirp periods. Finally, the laser driver provides the laser with the appropriate input current including the nominal modulation signal and the correction signal from the loop. High voltage devices as well as transistor stacking are used in this block to support the high voltage levels required.

Along the loop, the gain from the input of the laser to the output of the single ended to differential stage is 103.82 V/A, and the gain of the TIA is adjusted to see a voltage swing of 100mV at the input of PPF. The gain of the next stages including the PPF and the SSB and HR mixers is 8/π. The gain of the integrator and the driver are 504.34Hz and 150mS respectively. Therefore, the loop gain of the whole system is equal to its loop bandwidth and its f_{Ref}, which is 20KHz.

III. MEASUREMENT RESULTS

The electrical chip is interfaced with an Alcatel DFB 1546nm laser with 2MHz line-width that was coupled to the OIC which in turn was coupled to the photodiode. The laser tuning BW is 250KHz, which limits the loop BW (and hence, reference frequency) to around 20KHz to ensure that the loop does not suffer from phase margin degradation. From (2), the chirp repetition rate is chosen as 1KHz. The availability of a laser with fast tuning capability, as in [2], would

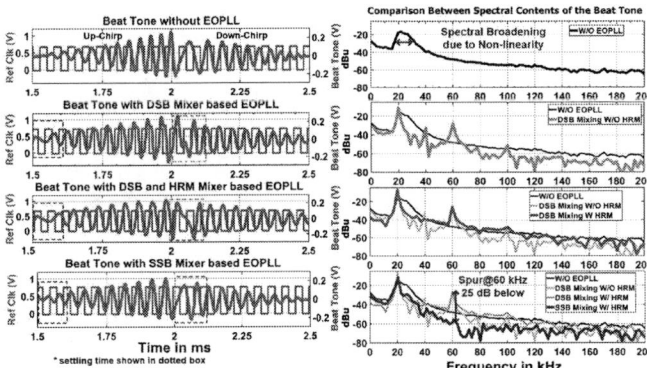

Fig. 5. Measured EO-PLL beat frequency for four cases – without the EOPLL, with a DSB mixer-based EOPLL, with DSB and HRM mixer-based EOPLL, and with SSB and HRM mixer-based EOPLL.

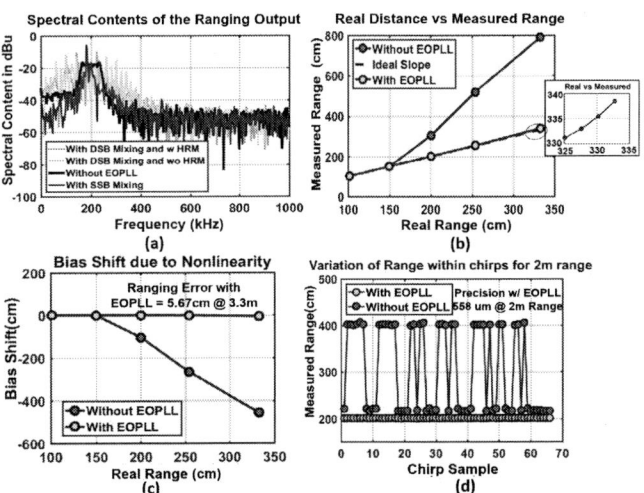

Fig. 6. (a) The spectrum of the beat signals from the ranging experiment. (b),(c) and (d) The actual distance and the measured range have been plotted for two cases - with and without the EOPLL. The significant broadening of the beat signal without the EOPLL results in the substantial error beyond 1.5m, while the experiment with the proposed EOPLL engaged shows a maximum ranging error of 5.67cm at 3.3m range. The ranging precision is 558um calculated for 2m.

enable microsecond repetition rates. The chirp BW is 16GHz, which leads to an MZI delay of 600ps in conjunction with the reference frequency mentioned earlier. The time-domain and frequency-domain results of the beat tone at the output of the EO-PLL TIA for different configurations of the LIDAR system (including open loop and various closed loop configurations) are represented in Fig. 5. Bias shift and a significant broadening of the beat tone are evident in the open loop configuration. Both problems can be eliminated by closing the loop using one mixer (without SSB and HR mixing), yet spur levels are prohibitively high. Enabling SSB mixing can remove the first spur ($f_{Ref} + f_{beat} = 40$KHz at the mixer output, which shows up at 60KHz in the beat signal) by more than 25dB. Finally, enabling HR mixing leads to elimination of third and fifth harmonics of mixing and a clean spectrum. The total power consumption of the CMOS chip is 13mw and the power consumption of the input buffer for the

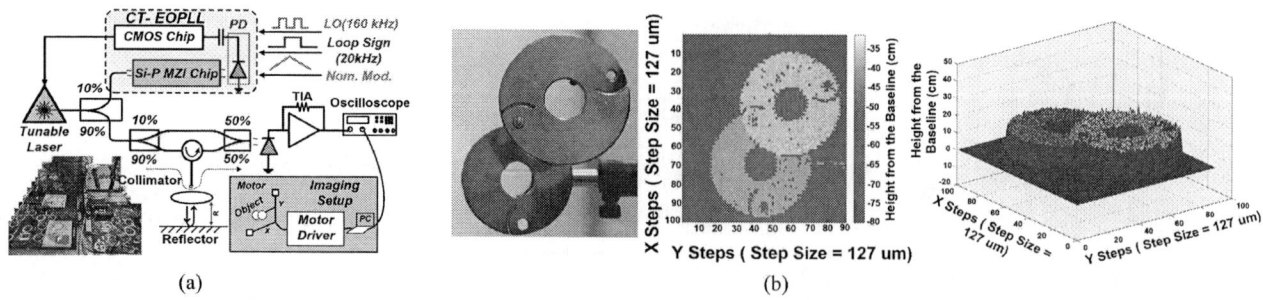

Fig. 7. (a) Ranging and imaging setup. (b) 2D and 3D images of a real object using the proposed EOPLL with an XY resolution of $127\mu m$.

Table 1. Comparison with state-of-the-art works.

	ISSCC 2016 [2]	Opt. Exp. 2009 [5]	CLEO 2018 [4]	This Work
Architecture	DT-EOPLL	CT-EOPLL	CT-EOPLL	CT-EOPLL
Technology	180nm CMOS, Si-Photonics	Discrete	Discrete	65nm CMOS, Si-Photonics
Wavelength, nm	1550	1539	N/A	1546
Chirp BW $(\Delta F), GHz$	122	100	16.67**	16**
Loop BW$(B)/f_{Ref}$	0.07	0.14	1	1
$\Delta F \times \tau_{MZI}$	40.26	2860	10	10
Range, m	1.4	N/A*	N/A*	>3.3***
Transmitted Power, dBm	10	N/A*	N/A*	5
Precision, μm	8	150	N/A	550
Spur Suppression, dB	N/A	N/A	17	25
Range Resolution, mm	N/A	1.5	N/A	9.4

* Ranging was not done.   **BW dictated by the application.   ***Enabled by spur suppression.

laser is 150mw.

Fig. 6 shows the result of ranging experiments in which the received beat tone (Fig. 6a) is analyzed to extract the range. The setup is shown in Fig. 7a - two 90/10 couplers are used to divide the laser power between the EO-PLL(10%), the reference path (9%), and the transmit path(81%). A circulator is used to simultaneously transmit the laser power to free space (through a collimator) and receive the signal after reflection from an object. A 50/50 coupler, PD and TIA are used to mix the received signal and the reference laser power and convert it to a voltage which is eventually analyzed in a digitizer. With the EO-PLL on but without spur suppression mechanisms, the spurs in the received beat signal are incredibly severe. As shown in the Figs. 6b and 6c, in the open loop configuration, the substantial errors in the estimated range result from the broadening of the beat tone as well as the bias shift. With the EO-PLL in lock and spur suppression engaged, the maximum ranging error is 5.67cm at 3.3m distance, and the ranging precision is $558\mu m$ at 2m distance (Fig. 6d). The maximum range is limited by laboratory space restrictions, and not by the spectral purity of the received beat signal.

Fig. 7 shows the results of an imaging experiment in which the object is moved in the lateral direction using a stepper motor. 2D and 3D images of the object are extracted from the received signal (Fig. 7b). This experiment illustrates the ability of the this LIDAR to achieve cm-scale range resolution. Finally, the performance of the proposed system is summarized and compared to prior art in Table 1. The proposed EO-PLL achieves an order-of-magnitude

improvement in $\Delta F \times \tau_{MZI}$ compared to [2] and [5]. Compared to [4], it achieves superior spur suppression while realizing an integrated implementation.

IV. CONCLUSION

The CT-EOPLL presented in this paper uses a wide bandwidth architecture, with a loop bandwidth equal to the reference frequency, thus enabling the integration of the MZI delay for GHz-range LIDAR bandwidths by reducing the area consumption and loss of the optical delay by $10\times$. The associated increase in spur levels is mitigated using SSB and HR mixing. These different mechanisms for spur suppression enable this LIDAR to achieve superior range than the prior art. The ranges currently measured are purely limited by laboratory space restrictions, and not by the spectral purity of the received beat signal.

ACKNOWLEDGMENT

The first two authors are equal contributors in this work.

REFERENCES

[1] R. Fatemi et al., "A low power PWM optical phased array transmitter with 16° field-of-view and 0.8° beamwidth," in RFIC, 2018, pp. 28–31.

[2] B. Behroozpour et al., "Chip-scale electro-optical 3D FMCW lidar with $8\mu m$ ranging precision," in ISSCC, 2016, pp. 214–216.

[3] L. Kong and B. Razavi, "A 2.4GHz 4mw inductorless RF synthesizer," in ISSCC, 2015, pp. 1–3.

[4] J. Sharma et al., "Continuous-time electro-optic PLL with decimated optical delay/loss and spur cancellation for LIDAR," in Conference on Lasers and Electro-Optics, 2018, p. JTu2A.81.

[5] N. Satyan et al., "Precise control of broadband frequency chirps using optoelectronic feedback," Opt. Express, vol. 17, no. 18, pp. 15 991–15 999, Aug 2009.

A 6.5-GHz Cryogenic All-Pass Filter Circulator in 40-nm CMOS for Quantum Computing Applications

Andrea Ruffino[#1], Yatao Peng[#], Fabio Sebastiano[$], Masoud Babaie[$], Edoardo Charbon[#2]

[#]École Polytechnique Fédérale de Lausanne, Switzerland
[$]Delft University of Technology, The Netherlands
[1]andrea.ruffino@epfl.ch, [2]edoardo.charbon@epfl.ch

Abstract — Cryogenic solid-state quantum processors require classical control and readout electronics; to achieve compactness and scalability, cryogenic integrated circuits have been recently proposed for this goal. Circulators are widely used in readout circuits, however they are typically discrete bulky devices, thus preventing miniaturization. To address this issue, we propose a fully integrated 40-nm CMOS 6.5-GHz circulator operating from 300 K to 4.2 K. At 300 K, it achieves a 2.2-dB insertion loss, an 18-dB isolation, and a 2.4-dB noise figure over the 1-dB bandwidth from 5.6 GHz to 7.4 GHz, with a core power of only 2.5 mW. This improves to 2.1 mW core power at 4.2 K, while showing 1.3-dB insertion loss and 17-dB isolation over the 1-dB bandwidth from 5.8 GHz to 7.6 GHz. The circuit achieves a record-low core power and a 1.6× wider fractional bandwidth than the state-of-the-art, thus allowing its use for multiple channels in power-constrained cryogenic refrigerators. These advances are enabled by a fully-passive architecture based on LC all-pass filters, allowing the use of a lower clock frequency than in prior art.

Keywords — Cryo-CMOS, circulator, qubit, spin qubit, superconducting qubit, qubit readout, quantum computing.

I. INTRODUCTION

Quantum computing is a promising solution to the ever-increasing demand for computational power. State-of-the-art solid-state quantum processors, such as those based on spin quantum bits (qubits) [1] or on superconducting qubits [2], operate at deep-cryogenic temperature while the complex RF setup required for their control is implemented by off-the-shelf instrumentation at room temperature. Although this approach is feasible for the few qubits (<100) available today, it will become unpractical to wire room-temperature electronics to the thousands of qubits required in practical quantum computers. CMOS circuits operating directly at cryogenic temperatures (cryo-CMOS) can pave the way for co-integration of qubits and classical control, so as to achieve compact and scalable systems in the near future [3].

Cryogenic circulators are commonly used in qubit readout systems, as shown in Fig. 1, however they are currently implemented with bulky ferrite devices, thus preventing system integration. Integrated circulators operating at room temperature have been recently proposed for full-duplex radio transceivers, exploiting time-varying circuits to achieve the circulator's non-reciprocal response. CMOS implementations of circulators have been demonstrated in the K-band, based on switched transmission lines [4], and in the GSM band, based on N-path filters [5], however neither approach is well suited for the 5-8 GHz band required for superconducting qubits [2].

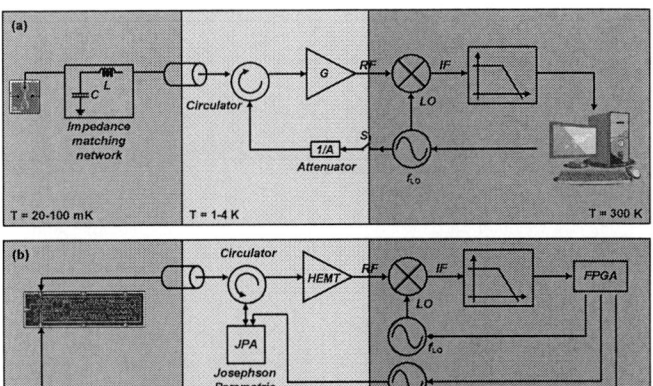

Fig. 1. Circulators used in readout circuits for (a) spin qubit processors [1] and (b) superconducting qubit processors [2].

Switched transmission lines require $\lambda/4$ transmission lines at the switching frequency (i.e., 1/3 of the 6.5-GHz operating frequency), thus resulting in a significant penalty on the circulator's die area and insertion loss. N-path filters require non-overlapping clock phases at the 6.5-GHz operating frequency, thus resulting in excessive power dissipated by the clock drivers. Minimizing power consumption is essential in cryogenic applications because of the very limited cooling power available in existing cryostats. Substantial power savings can be achieved by re-using the same electronics for multiple qubit channels via frequency multiplexing, thus requiring bandwidth maximization.

To address these issues, we propose the first fully integrated CMOS circulator operating at cryogenic temperatures. By exploiting a novel architecture based on all-pass filters, we avoid both large-area transmission lines and power-hungry high-frequency clock drivers. This results in a passive circulator operating down to 4.2 K with large bandwidth, while dissipating only 2.1 mW. Target applications involve superconducting qubits (hence the operating frequency band), but do not exclude spin qubit experiments at 5-8 GHz.

II. DESIGN AND ARCHITECTURE

Existing non-magnetic CMOS circulators are formed by a loop with two reciprocal branches, providing 90° reciprocal phase shifts, and a non-reciprocal branch, the core component, creating a non-reciprocal 0°/180° phase shift depending on the signal direction, as shown in Fig. 2a.

978-1-7281-1702-7/19 $31.00 © 2019 IEEE

2019 IEEE Radio Frequency Integrated Circuits Symposium

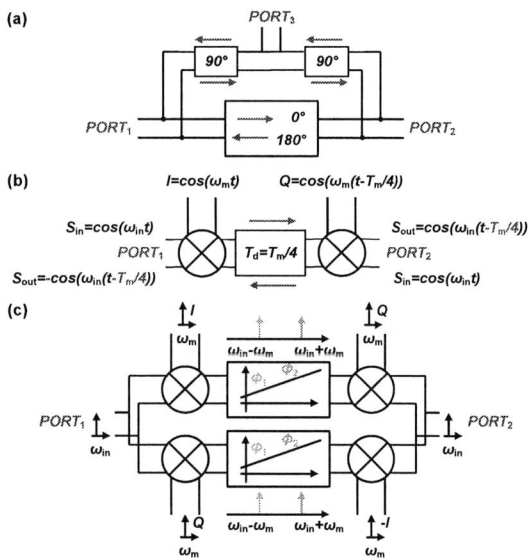

Fig. 2. Block diagram of (a) the circulator, including (b) the time domain analysis and (c) the frequency domain analysis of the non-reciprocal branch.

Both reciprocal and non-reciprocal branches create the required phase shifts using passive filters, which are implemented as CLC low-pass filter T sections [5], or lumped-element Bragg-limited transmission line sections [4].

The non-reciprocal branch, shown in Fig. 2b, is realized by inserting a filter between two I/Q mixers switching at a frequency ω_m, while the incoming signal travels at frequency ω_{in}, with $\omega_m < \omega_{in}$ to enable low-power operation. In order to have a lossless transmission, the internal filter needs to produce a time delay T_d equivalent to $T_m/4$, where T_m is the clock (LO) period [4]. In the forward direction, this yields a transmission with a delay of $T_m/4$, while in the reverse direction this gives a delay of $T_m/4$ and a sign flip. Time-variance introduced by mixers results in non-reciprocity.

In order to reduce the circulator power consumption, ω_m must be reduced (or its period T_m increased). This means that the time delay introduced by the internal filter needs to be larger. If this is implemented with transmission lines, then the corresponding line needs to be longer, which requires more equivalent LC sections. This would result in a larger insertion loss and area, or, if a smaller number of sections is used, reduced bandwidth.

If, however, one considers the non-reciprocal branch shown in Fig. 2c in the phase-frequency domain, a signal travelling at ω_{in} is I/Q mixed with the clock at ω_m, thus generating two mixing products at $\omega_L = \omega_{in} - \omega_m$ and $\omega_H = \omega_{in} + \omega_m$, with corresponding phase shifts $\phi_1 = (\omega_{in} - \omega_m) T_d$ and $\phi_2 = (\omega_{in} + \omega_m) T_d$, respectively. Those components mix with Q/I clock signals and appear again at ω_{in} at the other port of the non-reciprocal branch [4]. If ϕ is the phase shift between the set of mixers, the conditions for non-reciprocal operation can be expressed as:

$$
\begin{cases}
\phi = -\pi/2 \\
\phi_1 - \phi_2 = 2\phi \\
\phi_1 = (\omega_{in} - \omega_m) T_d = \pi/2.
\end{cases}
$$

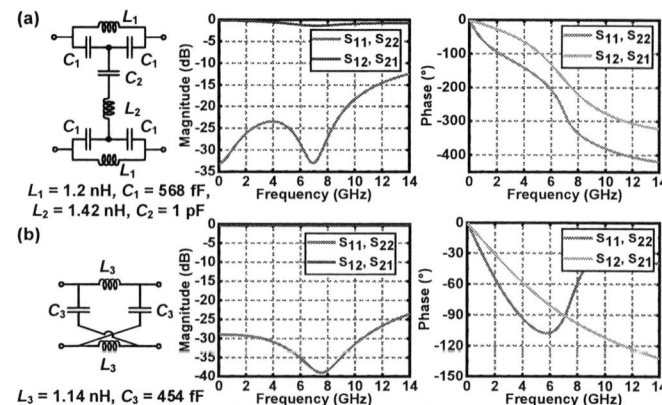

Fig. 3. Schematic and transfer function post-layout simulation of (a) a second order bridged-T all-pass filter and (b) a first order lattice all-pass filter.

The first equation sets the choice to have 90° phase shift between clock signals, to yield a differential and I/Q clocking scheme and mitigate the effect of duty cycle mismatch [4].

The second equation describes the core functionality of the circulator and establishes that the phase shifts of ω_L and ω_H components need to be 180° apart. If ω_m is reduced to minimize power consumption, the ω_L and ω_H components get closer to one another, requiring a sharply dispersive filter.

Instead of employing first-order transmission line sections, we propose to use a second-order filter, namely a bridged-T LC all-pass filter, shown in Fig. 3a. In this case, the center frequency of the circulator can be placed at the center of the all-pass filter, where the phase shift is −180°, while ω_L and ω_H are placed where the phase shift of the filter is −90° and −270° respectively. In such second-order all-pass filters, thanks to their sharp phase response, the second non-reciprocity condition is satisfied at frequencies closer to the center frequency. Consequently, a modulation index $\omega_{in}/\omega_m = 5$ can be achieved without transmission line trade-offs.

Such an approach allows to reduce clock frequency to minimize power consumption, while achieving area compactness, lower insertion loss and noise figure, since fewer inductors are in series with the signal path. The use of a second-order all-pass filter also allows to increase the circulator bandwidth thanks to the highly linear (as well as steep) phase response around the center frequency, capable of maintaining the required phase relationship for an extended frequency band. This is beneficial for frequency multiplexing, to accommodate multiple qubits on the same readout chain.

Finally, the third equation sets the insertion phase, which is chosen to be 90°, in order to eliminate the additional 45° equivalent sections in [4], so as to further reduce the count of passives, with benefits on area and complexity.

We also propose to implement the reciprocal branches as LC all-pass filters, but in this case first-order lattice filters, shown in Fig. 3b, suffice to create 90° phase shifts.

The non-reciprocal branch is thus resulting in an overall 0°/180° non-reciprocal phase shift depending on the signal direction, and with the action of reciprocal branches, this yields a final constructive/destructive interference at all ports. The entire circulator is shown in Fig. 4.

978-1-7281-1702-7/19 $31.00 © 2019 IEEE

2019 IEEE Radio Frequency Integrated Circuits Symposium

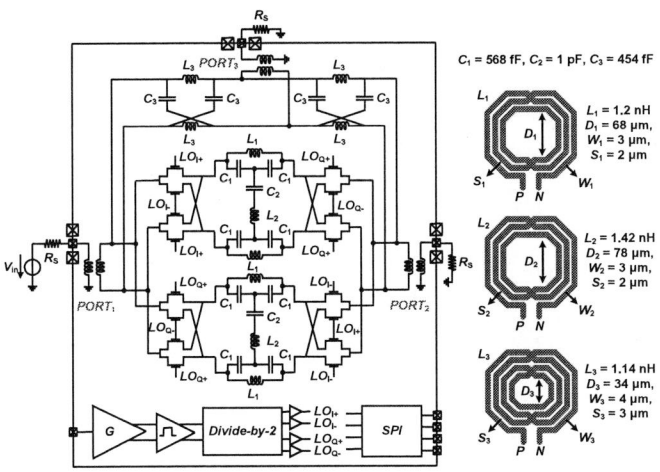

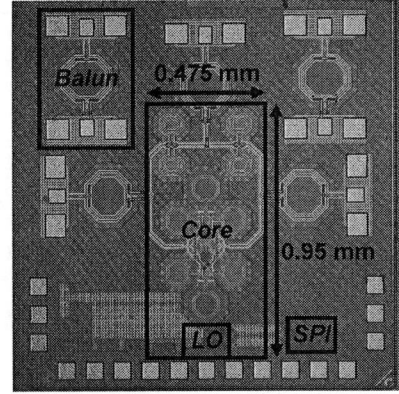

Fig. 4. Block diagram and circuit schematic of the designed circulator. Circulator core and auxiliary structures are shown.

Fig. 5. Circulator chip micrograph, showing the circulator core, transformer baluns, and auxiliary structures.

III. CHIP IMPLEMENTATION

The proposed circulator was fabricated in 40-nm CMOS with an ultra-thick top metal layer, and the chip micrograph is shown in Fig. 5. Thanks to the all-pass filter implementation, the circulator occupies a core area of only $0.45\,\text{mm}^2$. The LC filters have been realized with multi-turn spiral inductors and interdigitated MOM capacitors, for compactness and integration. Mixers have been realized as nMOS-only passive differential mixers with $W/L = 50\,\mu\text{m}/40\,\text{nm}$, optimized to trade off R_{ON} resistance, causing additional insertion loss, and parasitic capacitance C_p, producing unwanted phase shifts, which have been compensated for by filter design.

The system has been designed to operate at a center frequency of 6.5 GHz, with a clock signal at 1.3 GHz, thus yielding a modulation index m = 6.5 GHz/1.3 GHz = 5.

Transformer baluns have been used to convert on-chip differential signals to single-ended signals for RF probing on GSG pads and have been de-embedded from measurements.

The on-chip non-overlapping clock generator outputs the four required I/Q differential signals from a single external sinusoidal input. An active single-ended to differential amplifier at the input, cascaded by further gain stages, is used to clip the signal to a square wave, which is then fed to a C^2MOS latch-based divider-by-2, finally reinforced by proper buffers and phase aligners. An on-chip SPI allows tuning of the amplifier bias and port impedance during testing.

IV. MEASUREMENT RESULTS AND DISCUSSION

The fabricated prototype has been bonded to a PCB, while the GSG pads at the three ports have been probed with a Lake Shore CPX probe station at 300 K and 4.2 K.

The DC power consumption of the circulator core (including frequency divider and buffers driving the mixers) is 2.3 mA from a 1.1 V power supply, while the auxiliary on-chip clock generation (amplifiers and phase aligners) consumes an additional 9 mA from a 1.1 V supply, when the internal clock signal is 1.3 GHz (the external LO signal is 2.6 GHz). These values reduce to 1.9 mA and 7.7 mA respectively when the circuit is cooled down to 4.2 K under the same conditions.

The measured S-parameters at 300 K, after calibration and balun de-embedding, are reported in Fig. 6. The circuit operates over a 1-dB insertion loss-isolation bandwidth of 5.6 GHz - 7.4 GHz, thus yielding a 28% fractional bandwidth, with a 2.2-dB insertion loss and an 18-dB isolation. Impedance matching at all ports is always below −10 dB. These results are in good agreement with post-layout simulations.

Fig. 7 shows the measured S-parameters at 4.2 K. The 1-dB operational bandwidth is 5.8 GHz - 7.6 GHz, the minimum insertion loss is reduced to 1.3 dB, while the isolation becomes 17 dB. This improvement at cryogenic temperatures can be explained by an increase in quality factor of passives, in particular inductors, due to the reduction of substrate losses thanks to carrier freeze-out, and by a reduction of R_{ON} resistance of the mixer switches in series with the signal path, due to larger carrier mobility [3].

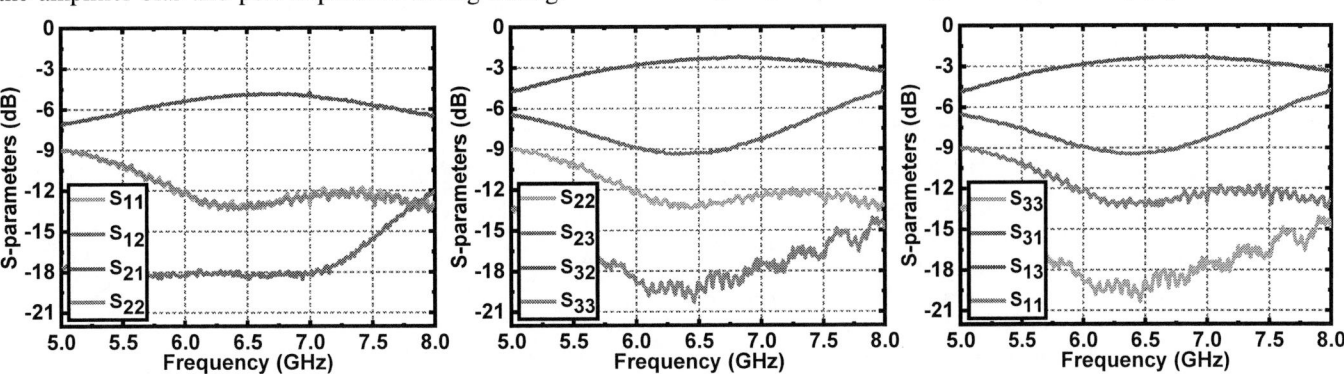

Fig. 6. Circulator S-parameter measurements at 300 K.

978-1-7281-1702-7/19 $31.00 © 2019 IEEE   109

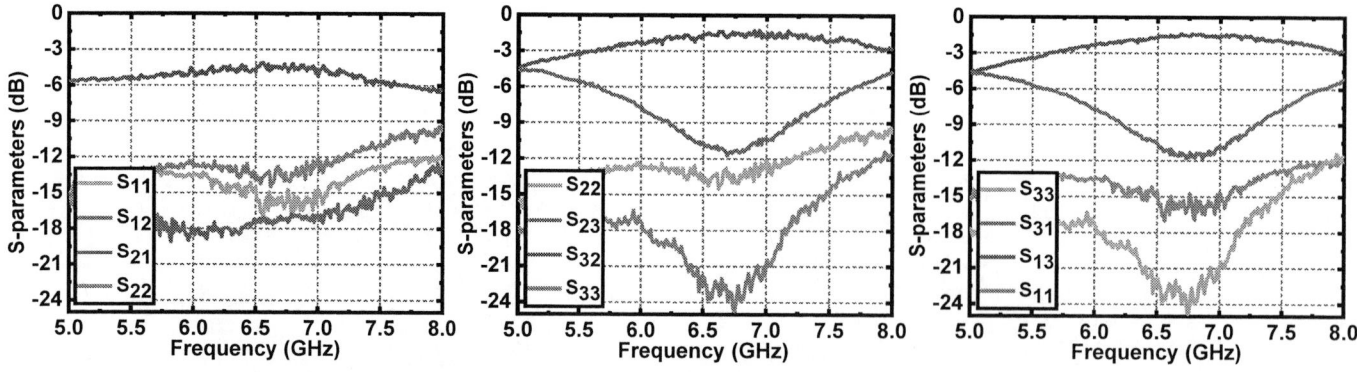

Fig. 7. Circulator S-parameter measurements at 4.2 K.

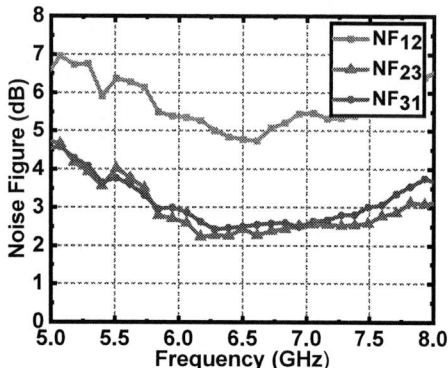

Fig. 8. Circulator noise figure measurements at 300 K.

Table 1. Comparison table with state-of-the-art circulators.

	This work	[5]	[4]	[6]	[7]
Technology	40-nm CMOS	65-nm CMOS	45-nm SOI	180-nm SOI	Ferrite
Working temperature (K)	4.2-300	300	300	300	0.02-300
Architecture	All-pass	N-path	T-line	T-line	Discrete
Frequency (GHz)	5.6-7.4	0.61-0.97[1]	22.7-27.3	0.86-1.08	4-8
Modulation index	5	1	3	3	N.A.
Insertion loss (dB)	2.2	1.7	3.2	2.1	0.4
Isolation (dB)	18	>20	18.5	>25	18
Fractional bandwidth[2] (%)	28	4.3	18	17	66
Noise figure (dB)	2.4-3.4	4.3	3.3-4.4	2.9-3.1	0.4
IIP3 (dBm)	>+17.5	+27.5	+20.1	+50	N.R.
Core area (mm^2)	0.45	25	2.16	16.5	1575
Power consumption (mW)	2.5/12.4[3]	59	78.4	170	0
Normalized power P_{DC}/f_0 (mW/GHz)	1.9	75	3.1	175	0

[1]Range of center frequency tunability, [2]Isolation and 1-dB insertion loss bandwidth, [3]Core (divider and mixer buffers) power and overall power consumption respectively.

Noise measurements of the circulator at 300 K are presented in Fig. 8 and show good agreement with insertion loss results, with a minimum noise figure of 2.4 dB.

The measured circulator IIP3 is higher than +17.5 dBm in all directions of circulation at both 4.2 K and 300 K, which is sufficient for quantum computing applications, where the received signal is well below the non-linear region (−90 dBm).

In the target quantum computing application, the proposed circulator is expected to be integrated in a full receiver [3] operating at 4.2 K, so that the clock signal would be generated by the on-chip PLL, thus minimizing noise. Clock feedthrough could be a concern in such application, given the very low-noise requirements and the low signal power levels (−90 dBm). While coupling from clock generation circuitry through the substrate is highly reduced at 4.2 K thanks to carrier freeze-out, mixer feedthrough may need to be mitigated in the future by a further reduction of the clock frequency or by other techniques, to limit the amplitude of clock harmonics in the circulator band.

Measured results are summarized and compared to state-of-the-art CMOS and commercial circulators in Table 1.

V. CONCLUSION

A 40-nm CMOS circulator has been designed to operate at both 300 K and 4.2 K for quantum computing applications.

Thanks to the realization of a large modulation index, enabled by the exploitation of a second-order bridged-T all-pass filter with sharp but linear phase response, power consumption, area and fractional bandwidth are significantly better than prior art.

A record-low power consumption of 2.5 mW and a state-of-the-art fractional bandwidth of 28% are achieved in only 0.45 mm^2 active area. This enables miniaturization and multiplexing of qubit frequencies in power-constrained cryogenic refrigerators. Furthermore, the circulator yields a minimum 2.2-dB insertion loss and 2.4-dB noise figure with 18-dB isolation in the band 5.6 GHz - 7.4 GHz.

REFERENCES

[1] M. F. Gonzalez-Zalba *et al.*, "Probing the limits of gate-based charge sensing," *Nature Communications*, vol. 6, p. 6084, Jan 2015.

[2] D. Ristè *et al.*, "Deterministic entanglement of superconducting qubits by parity measurement and feedback," *Nature*, vol. 502, p. 350, Oct 2013.

[3] B. Patra *et al.*, "Cryo-CMOS Circuits and Systems for Quantum Computing Applications," *IEEE JSSC*, vol. 53, no. 1, pp. 309–321, Jan 2018.

[4] T. Dinc *et al.*, "A Millimeter-Wave Non-Magnetic Passive SOI CMOS Circulator Based on Spatio-Temporal Conductivity Modulation," *IEEE JSSC*, vol. 52, no. 12, pp. 3276–3292, Dec 2017.

[5] N. Reiskarimian *et al.*, "A CMOS Passive LPTV Non-Magnetic Circulator and Its Application in a Full-Duplex Receiver," *IEEE JSSC*, vol. 52, no. 5, pp. 1358–1372, May 2017.

[6] A. Nagulu *et al.*, "Fully-Integrated Non-Magnetic 180nm SOI Circulator with >1W P1dB, >+50dBm IIP3 and High Isolation Across 1.85 VSWR," in *2018 IEEE RFIC*, June 2018, pp. 104–107.

[7] QuinStar Technology, Inc. website, http://www.quinstar.com.

Design Considerations for Spin Readout Amplifiers in Monolithically Integrated Semiconductor Quantum Processors

M. J. Gong[1], U. Alakusu[1], S. Bonen[1], M. S. Dadash[1], L. Lucci[2], H. Jia[3], L. E. Gutierrez[3],
W. T. Chen[3], D. R. Daughton[4], G.C. Adam[5], S. Iordănescu[5], M. Păşteanu[5], N. Messaoudi[6]
D. Harame[2], A. Müller[5], R. R. Mansour[3], S. P. Voinigescu[1]

[1]ECE, University of Toronto, Canada
[2]was with GLOBALFOUNDRIES, Germany
[3]ECE, University of Waterloo, Canada
[4]Lake Shore Cryotronics, USA
[5]IMT, Bucharest, Romania
[6]Keysight Technologies, Canada

mecca.gong@utoronto.ca, alakusuu@ece.utoronto.ca, bonensha@ece.utoronto.ca

Abstract— **The high frequency performance of all active and passive devices in a production 22nm FDSOI CMOS technology was measured up to 40 GHz over temperature down to 3.3 Kelvin, targeting applications in cryogenic and quantum computing ICs. It was found that the quality factor of the passives and the f_T and f_{MAX} of both p- and n-MOSFETs improved at 3.3 K. More importantly for circuit design, the peak-f_T and peak-f_{MAX} current densities, and the MOM capacitor and polysilicon resistor values show no variance with temperature. This information and the measured I-V characteristics of electron and hole single- and double-quantum dot structures, measured at 2 K and representative of qubits, were used to design monolithically integrated double quantum dots with readout transimpedance amplifiers output matched to 50 Ω. Transimpedance gain, S_{21}, and bandwidth of 108 dBΩ, 19 dB, and 7.5 GHz, respectively, were measured at 300 K with only 4.5 mW power consumption and S_{22} < -10 dB up to 60 GHz.**

Keywords—**cryogenics, electron-spin, hole-spin, qubit, semiconductor quantum dot, silicon germanium, transimpedance amplifier**

I. INTRODUCTION

One of the biggest hurdles to realizing physical quantum computers has been the development of a reliable hardware platform capable of integrating billions of identical qubits monolithically. The current approach is for qubit array chips to be separately packaged and operated at a base temperature below 100 mK, with their microwave control and readout electronics located on a separate die at 1-4 K and accessed via a large number of 50Ω coaxial cables [1]. This multi-chip approach has historically been motivated by several factors, despite the interconnect and fidelity challenges which arise from it [2], [3]. More specifically, it has been a challenge to accommodate high yield, high f_T/f_{MAX}, transistors and microwave and analog-mixed-signal (AMS) circuits in the experimental laboratory technologies in which the silicon qubits are manufactured [4]. Additionally, state-of-the-art cryogenic cooling systems have also been limited in the dissipated power they can remove while maintaining a certain temperature: a few mW at 100 mK and ~2 W at 4 K [1]. More recently however, there has been mounting evidence that Si-based electron- and hole-spin qubits need not be limited to

sub 1K temperatures [5] and that they are realizable in production CMOS technology, where they can be integrated with classical readout and control electronics [7]. In this paper, we describe for the first time the cryogenic high-frequency performance of all the active and passive components of a commercial 22nm FDSOI CMOS technology [6] and investigate the design considerations and design methodology for monolithically integrated quantum dots (QDs) with readout circuitry achieving record gain and bandwidth.

II. CRYOGENIC TECHNOLOGY CHARACTERIZATION

A full set of MOSFETs, 4μm x 7.5μm metal-oxide-metal (MOM) caps, 100Ω and 200Ω poly resistors, and the associated 6μm x 400μm de-embedding transmission line structures were designed and fabricated in two variants of the 22nm FDSOI CMOS process [6] with thin metal and thick metal BEOL, respectively. A microphotograph of the test chip is reproduced in Fig. 1. All 3.3K MOSFET measurements shown are with floating back gates due to limitations of the cryogenic setup.

Figs. 2-4 compile the measured transconductance, f_T, f_{MAX}, gate, R_G, and source, R_S, resistances for fully metallized, single-gate contact *n*- and *p*-channel MOSFETs with 40 gate fingers, 20nm gate length, 1x source/drain contact pitch, and gate finger widths of 430 nm and 590 nm, at 3.3 K and 300 K.

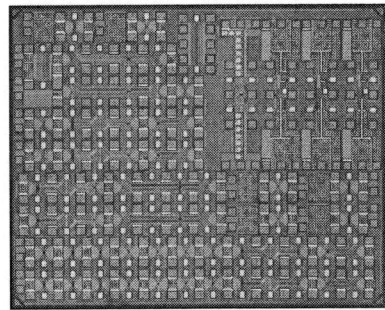

Fig. 1: Die microphotograph with chip dimensions of 2.5mm x 2.0mm. Included on die are 1x, 2x, and 3x pitch transistor test structures, passive components, quantum dot structures, and double quantum dot with readout circuits.

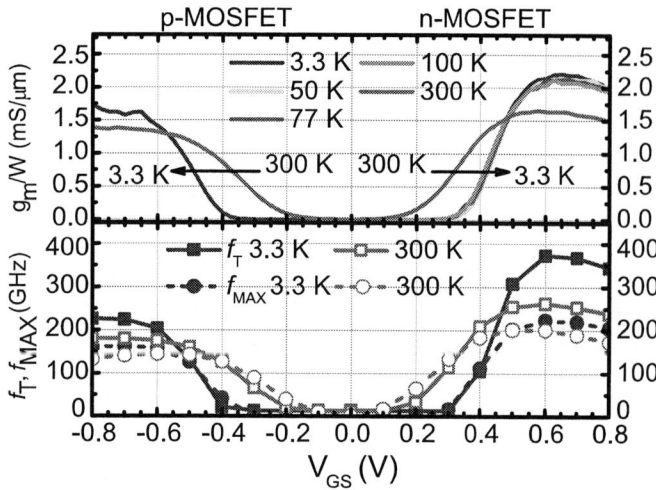

Fig. 2: Measurements for 40x20nmx590nm MOSFETs ($V_{DS}=\pm0.8$V) a) normalized transconductance b) f_T/f_{MAX} vs. V_{GS}

All transistor S-parameter measurements include parasitics of the wiring stack up to the top metal. The latter was designed to satisfy electromigration rules up to 110 °C. However, at 2-4 K, electromigration is not a problem and the wiring stack could be redesigned for significantly reduced parasitics, resulting in better circuit performance than reported here. Compared to 300 K, at 3.3 K the peak transconductance increases by 34% and 25% for the *n*- and *p*-MOSFET, respectively. Peak f_T improves by 42% (to 373 GHz) and 25% (to 226 GHz) for *n*-/*p*-MOSFETs, while peak f_{MAX} improves by about 11% to 223 GHz and 163 GHz, respectively. The peak-f_T, peak-f_{MAX} current densities remain nearly constant across temperature, simplifying the design of circuits that must operate over a wide temperature range, even in the absence of transistor models valid at 2-4 K.

Finally, Fig. 5 shows the measured characteristics of the transmission line, MOM capacitor, and poly-Si resistor at 3.3 K and 300 K. The 50Ω transmission line retains its characteristic impedance from room temperature (as evidenced by its measured S_{11}) and has lower loss at 3.3 K. The MOM capacitor Q improves at lower temperatures while the capacitance does not change. The poly-Si resistor retains the same resistance across temperature, which is again very important in designing transimpedance amplifiers (TIAs) whose output impedance does not change across temperatures.

III. READOUT AMPLIFIER DESIGN

The fabricated qubit structures consist of minimum-size Si *n*-MOSFETs and SiGe *p*-MOSFETs and cascodes. QDs are formed in the thin (<10 nm) undoped semiconductor film below each top gate, which can be biased to control the confinement energies of each QD while the back gate formed in the Si substrate below the buried oxide layer controls the amount of coupling between the QDs in the cascode, which acts as a double quantum dot (DQD) [7]. The QD structures are expected to behave as electron- and hole-spin qubits when a DC magnetic field is applied [4]. Fig. 6 compares measured I-V transfer characteristics of electron- and hole-spin single quantum-dots at 2 K and 300 K. At 2 K and low V_{DS} bias

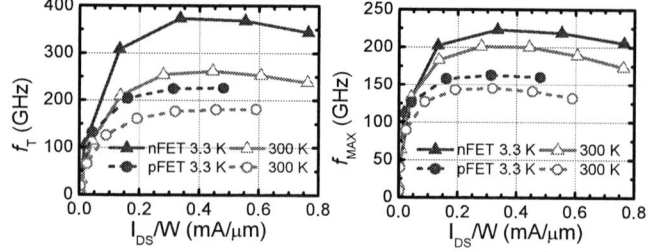

Fig. 3: Measured f_T and f_{MAX} vs. current density for 40x20nmx590nm MOSFETs ($V_{DS}=\pm0.8$V)

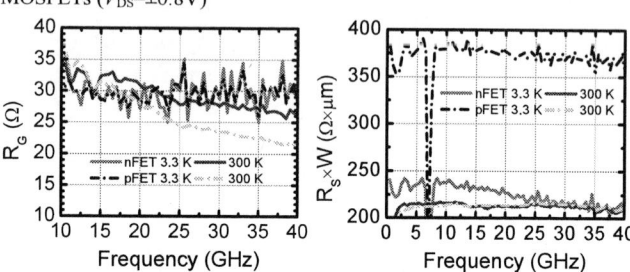

Fig. 4: Measured R_G, R_S for 40x20nmx430nm MOSFETs

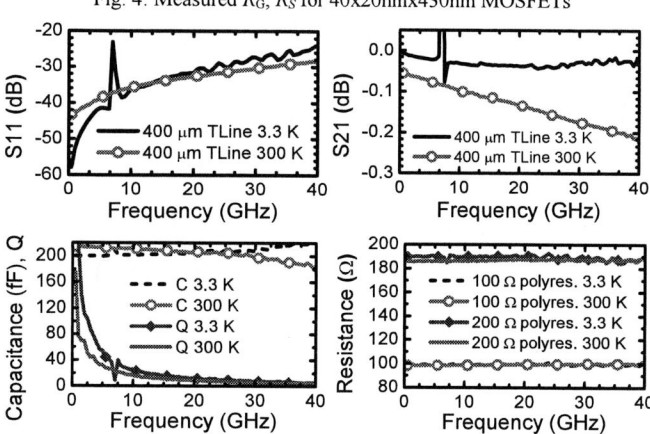

Fig. 5: Measured passives at 3.3 K and 300 K. a) 400µm long transmission line S_{11}, b) 400µm long transmission line S_{21}, c) 200fF MOM capacitor, and d) 100Ω and 200Ω poly resistors.

(<50mV), current oscillations are observed in the sub-threshold region, representative of electron/hole tunnelling events through the discrete energy levels of the QD. The TIA must detect the fast electron or hole charge transfer events and amplify the resulting small tunneling current at the first peak, on the order of 10 pA to 10 nA, to a voltage swing of at least a few mV, which can be easily processed by off-chip test equipment or FPGAs. This requires a low-noise, high-bandwidth readout amplifier with a transimpedance gain of 100-140 dBΩ, capable of driving 50 Ω without significantly loading the minimum-size QDs which have less than 60aF output capacitance. The circuit in Fig. 7 shows a DQD-with-TIA schematic, where the DQD is representative of a coupled-spin qubit when a DC magnetic field is applied. Both the TIA and the DQD are optimally biased, allowing qubit V_{DS} control through the V_{source} terminal and through the V_{DD} of the TIA. Figs. 8-9 show the measured output characteristics of the *n*- and *p*-MOSFET QDs when V_{GS} is set at the locations of the first peak and valley, respectively, as measured at 2 K in Fig. 6. For high fidelity, a QD with a large peak-to-valley current

978-1-7281-1702-7/19 $31.00 © 2019 IEEE

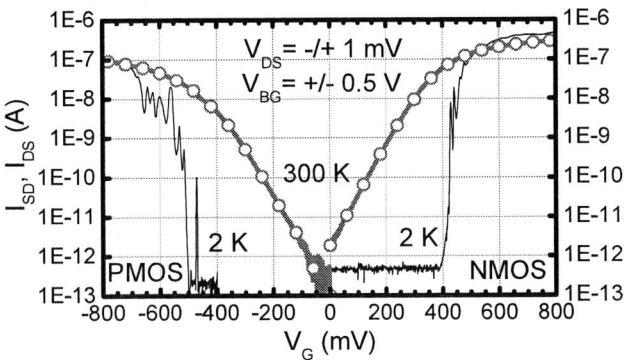

Fig. 6: Transfer characteristics at 2 K and 300 K for a 1x20nmx80nm *n*-MOSFET and *p*-MOSFET QD

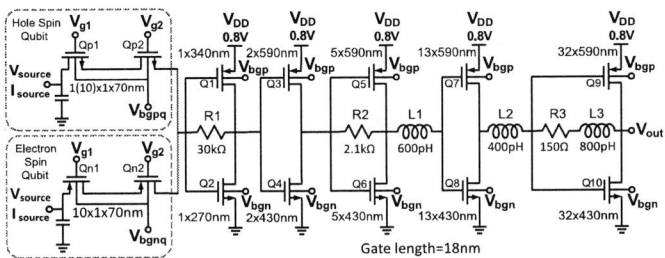

Fig. 7: Schematic diagram of integrated DQD structure and readout circuit

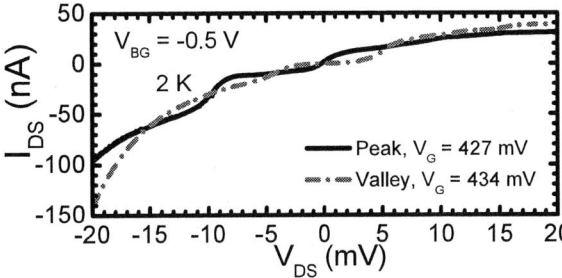

Fig. 8: Measured output characteristics at 2 K for the 1x20nmx80nm *n*-MOSFET QD from Fig. 6 when V_{GS} is set at the first peak and valley of the transfer characteristics in Fig. 6

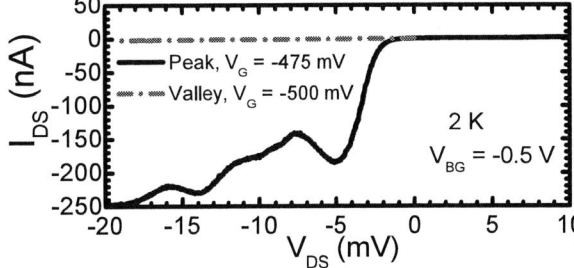

Fig. 9: Measured output characteristics at 2 K for the 1x20nmx80nm *p*-MOSFET QD from Fig. 6 when V_{GS} is set at the first peak and valley of the transfer characteristics in Fig. 6

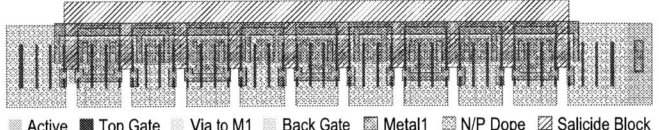

Fig. 10: Layout of integrated qubit structure with ten 1x18nmx70nm cascodes connected in parallel (10x)

ratio, like the *p*-MOSFET in this 22nm FDSOI process, is desirable. In the *n*-MOSFET case, because of smaller source/drain barriers in the QD, the peaks and valley are observable in the transfer characteristics (Fig. 6) but reduce to plateaus in the output characteristics, Fig.8. In both cases, these characteristics show evidence of Coulomb blockade where the electrons/holes selectively tunnel out of the QD [7]. In the readout phase, precise V_{DS} values in the 0 mV to +/-5 mV range, are needed to scan for the first peak and valley in the output characteristics to ensure that the resulting drain tunneling current is from a single electron/hole.

To study the behaviour of both electron- and hole-spin qubits, 3 types of DQD structures were integrated with the same readout TIA: (i) a single finger *p*-MOSFET cascode (1x-DQD hole-spin qubit), (ii) a structure with ten *p*-MOSFET cascodes connected in parallel (10x-DQD hole-spin qubit) and, (iii) a structure with ten *n*-MOSFET cascodes connected in parallel (10x-DQD electron-spin qubit). All QD and TIA MOSFETs have a physical gate length of 18 nm. The QD gate finger width is 70 nm. Fig. 10 shows the layout for the QDs with 10 gate fingers connected in parallel. The readout amplifier consists of 3 cascaded CMOS-inverter TIA stages with CMOS inverters placed in-between to maximize gain. The interstage fanout is below 3 to maximize bandwidth. The MOSFETs and feedback resistor in the output stage were sized for 50Ω matching. To reduce noise, a large 30kΩ feedback resistor was used in the first TIA stage. To further minimize the noise and maximize gain, all MOSFETs are biased at the peak-f_{MAX} current density (J_{pfMAX} = 0.25 mA/μm) by adjusting the backgate voltages, V_{bgn} and V_{bgp}. The simulated transimpedance gain, bandwidth, and equivalent input noise current density at 2 GHz changes from 104 dBΩ, 12 GHz, 0.83 pA/√Hz to 112 dBΩ, 11 GHz, 0.19 pA/√Hz,

respectively as the temperature decreases from 300 K to 12 K, the lowest at which the circuit simulator still works. The bandwidth of the TIA, 12 GHz, is designed to cover DC to 4x Rabi frequency when the DQD gate is excited with a mm-wave signal of up to 20 mV$_{pp}$ in the 60-220 GHz range [7].

IV. READOUT AMPLIFIER MEASUREMENTS

Fig. 11 compiles the measured S_{21} and S_{22} of all three DQD-with-TIA circuits at 300 K. Although, because of their small size and g_m, the standalone DQDs have an S_{21} of -20 dB or less, the 10x *n*-MOS DQD with TIA achieved a maximum S_{21} of 18.9 dB and a 3dB bandwidth of 7.5 GHz. The equivalent 10x *p*-MOS DQD with TIA readout circuit has an S_{21} of 13.5 dB and 7.5GHz bandwidth, while the 1x *p*-MOS DQD circuit reached a maximum S_{21} of 8.89 dB with a 3dB bandwidth of 8.5 GHz. In all 3 versions, S_{22} is better than -10 dB up to 60 GHz. As in the transistor measurements in Fig. 3 and in our previous 3-stage readout TIA [7], it is expected that the gain and bandwidth will improve at 2 K, while the output will remain matched. Because a standalone 5-stage TIA was not fabricated in this run, the transimpedance gain, Z_{21}, of the readout amplifier itself was obtained by applying a variable DC current, I_{source}, through the V_{source} terminal and measuring the output voltage of the TIA, V_{out}, and its derivative, as reproduced in Fig. 12. The current was swept from -40 nA to +40 nA, covering the range of the DQD's tunnelling current.

978-1-7281-1702-7/19 $31.00 © 2019 IEEE

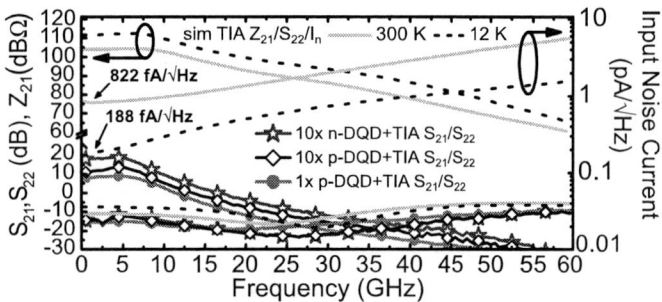

Fig. 11: Measured S-parameters of the integrated DQD-with-TIA circuits at 300 K, and simulated Z_{21}, S_{22}, and equivalent input noise current, I_n, of the TIA at 300 K and 12 K

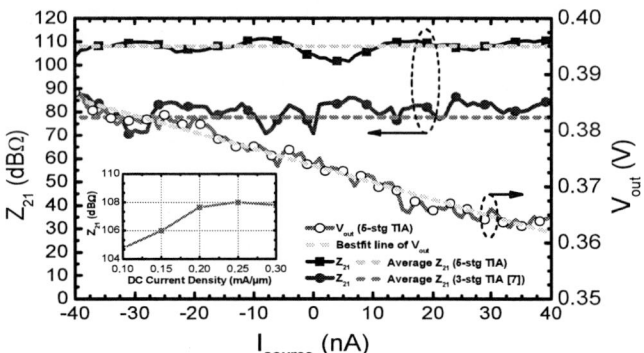

Fig. 12: Measured readout amplifier at 300 K showing Z_{21} from 1x p-MOS DQD plus amplifier circuits with inset showing Z_{21} as a function of TIA MOSFET drain current density

The peak Z_{21} is 108 dBΩ (251 kΩ) at a current density of 0.25 mA/μm, corresponding to J_{pfMAX}, as illustrated in the inset of Fig. 12. The output spectrum of the 1x p-DQD with TIA was measured with variable-amplitude sinusoidal signals in the 1-8GHz range applied to the gate of the DQD. Fig. 13 shows that even at -110dBm output power, the 4GHz sinusoidal signal is clearly visible above the noise floor. Based on the 251 kΩ TIA gain, this corresponds to $3pA_{rms}$ current at the input of the TIA. Table 1 compares the performance of the DQD with TIA to other state-of-the-art qubit readout amplifiers intended to operate at cryogenic temperatures.

V. CONCLUSION

The measured DC transfer and output characteristics of electron and hole single and double quantum dots at cryogenic temperatures were taken into consideration in the design of a readout TIA which was monolithically integrated with single and multi-finger DQD qubit structures in a production 22nm FDSOI CMOS process. Unlike in other applications, this TIA was optimized to read qubit currents in the range of 10 pA and 10 nA with ultra-low input capacitance to avoid overloading the qubits and to maximize the spin-readout bandwidth. Transistor measurements show that most of the performance and V_t variation occur between 300 K and 100 K. The high frequency performance improves at 3.3 K compared to 300 K while the peak f_T and f_{MAX} current densities remained unchanged. Combined with the fact that the quality factor of the passive components also improves, while their values remain practically unchanged at cryogenic temperatures, this makes it possible to reliably design all the readout and control

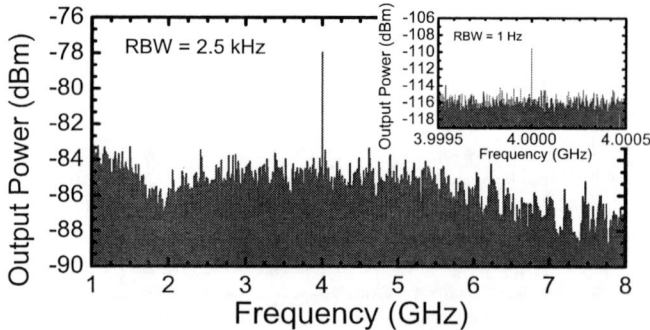

Fig. 13: Measured output spectrum of the 1x p-DQD with TIA at 300 K showing a -78dBm output signal (corresponding to $120pA_{rms}$ TIA input current) at 4 GHz and, in the inset, a -110dBm output signal (corresponding to $3pA_{rms}$ TIA input current) also at 4 GHz on a zoomed-in scale

Table 1: Comparison of state-of-the-art qubit readout amplifiers

Parameter	10x n-DQD with TIA	TIA only [7]	LNA [8]
Technology	22nm FDSOI	22nm FDSOI	160nm Bulk
S_{21} (dB)	19 @ 300 K	15/20 @ 300/2 K	57 @ 4 K
BW (GHz)	7.5 @ 300 K	4/5 @ 300/2 K	0.5 @ 4 K
$Z_{21,TIA}$ (dBΩ)	108/112 @ 300/12* K	78/80 @ 300/2 K	N/A
P_{DC} (mW)	4.5 @ 300 K	3.1 @ 300/2 K	45.9 @ 4 K

*simulated

electronics in a FDSOI CMOS monolithic quantum processor using standard design kit models, calibrated down to 100 K. The 4.5mW TIA power consumption allows for monolithic integration of up to 440 qubits with individual readout amplifiers within a power budget of 2 W. This dissipated power can be removed with state-of-the-art cryostats at 4 K.

ACKNOWLEDGEMENT

The authors would like to thank Dr. Nigel Cave for discussions and GlobalFoundries for chip donation and fabrication, Lake Shore Cryotronics and Keysight for testing support, CMC and Jaro Pristupa for CAD tools and CAD support, and Integrand for the EMX software.

REFERENCES

[1] E. Charbon et al., "Cryo-CMOS for quantum computing," in IEDM Tech. Dig., San Francisco, CA, USA, Dec. 2016, pp. 13.5.1–13.5.4, doi: 10.1109/IEDM.2016.7838410.

[2] S. K. Moore and A. Nordrum, "Intel's New Path to Quantum Computing," IEEE Spectrum: Technology, Engineering, and Science News, 08-Jun-2018.

[3] J. P. G. van Dijk et al., (Mar. 2018). "The impact of classical control electronics on qubit fidelity," Available: https://arxiv.org/abs/1803.06176

[4] R. Maurand et al., "A CMOS silicon spin qubit," in Nature Communications, vol. 7, no. 13575, Nov. 2016, doi: 10.1038/ncomms13575

[5] L. Petit et al., "Spin Lifetime and Charge Noise in Hot Silicon Quantum Dot Qubits," Phys. Rev. Lett., vol. 121, no. 7, Aug. 2018, doi: 10.1103/PhysRevLett.121.076801

[6] R. Carter et al., "22nm FDSOI technology for emerging mobile, Internet-of-Things, and RF applications," in IEDM Tech. Dig., San Francisco, CA, USA, Dec. 2016, pp. 2.2.1–2.2.4

[7] S. Bonen et al., "Cryogenic Characterization of 22nm FDSOI CMOS Technology for Quantum Computing ICs," IEEE Electron Device Letters, Vol. 40, No. 1, Jan. 2019, pp. 127-130, doi: 10.1109/LED.2018.2880303

[8] B. Patra et al. "Cryo-CMOS Circuits and Systems for Quantum Computing Applications," IEEE JSSC, Vol. 53, no. 1, Sep. 2017, pp.309-321, doi: 10.1109/JSSC.2017.2737549

Direct Digital Synthesizer with 14 GS/s Sampling Rate Heterogeneously Integrated in InP HBT and GaN HEMT on CMOS

Steven Eugene Turner[1], Mark E. Stuenkel, Gary M. Madison, Justin A. Cartwright,
Richard L. Harwood, Joseph D. Cali, Steve A. Chadwick, Michael Oh, John T. Matta,
James M. Meredith, Justin M. Byrd, Lawrence J. Kushner

BAE Systems, USA

[1]steven.e.turner@ieee.org

Abstract— **A 14 GS/s direct digital synthesizer (DDS) heterogeneously integrated with InP and GaN on CMOS is presented. The DDS includes over 6 million 45 nm CMOS FETs, 2151 InP HBTs, 2 GaN HEMTs, and 9930 heterogeneous interconnects, making it the most complex heterogeneously integrated mixed-signal circuit reported to date. By heterogeneously integrating multiple technologies, a high output power of 6.9 dBm is achieved while maintaining better than 37 dBc Nyquist SFDR and 8.7 W power consumption – performance currently unachievable with state-of-the-art single-technology approaches.**

Keywords— **CMOS integrated circuits, bipolar integrated circuits, heterogeneous integrated circuits, mixed analog digital integrated circuits, direct digital synthesizer, InP, GaN.**

I. INTRODUCTION

Heterogeneous integration of diverse technologies allows designers to choose the optimal devices for each sub-circuit of a system, be they CMOS devices for high-density digital circuits or III-V devices for high-speed circuits. This is a growing area of research [1], but to date, most of the reported heterogeneously integrated circuits have been limited to two technologies. Recently reported examples include InP on CMOS THAs/SHAs [2], [3], a GaN on CMOS TX [4], and an InP HEMT/HBT downconverter [5]. The only reported three-technology heterogeneous integrated design is an InP and GaN on CMOS VCO/amplifier [6].

Complex, mixed-signal circuits, such as direct digital synthesizers (DDS), are particularly well suited to take advantage of the diverse technologies offered by heterogeneous integration. Traditional DDS designs make performance tradeoffs due to the limitations of single process technologies. For example, InP-only DDS circuits are limited to narrow accumulator bit-widths due to power consumption limitations. Likewise, SiGe FETs are restricted to larger nodes, SiGe HBTs have lower breakdown voltages than GaN HEMTs, and SiGe HBTs have lower early voltage than InP HBTs.

This paper reports a 14 GS/s, high output-power DDS (Fig. 1a) implemented in the three-technology Diverse Accessible Heterogeneous Integration (DAHI) process [7]. The DAHI process enables new circuit topologies presented in this paper by intimately integrating diverse devices via gold micro-pillar heterogeneous interconnects (HICs) with as small as a 15 µm pitch. The HICs provide electrical, thermal, and mechanical connections from the facedown InP and faceup GaN 'chiplets' to the CMOS base chip (Fig. 1b). To the

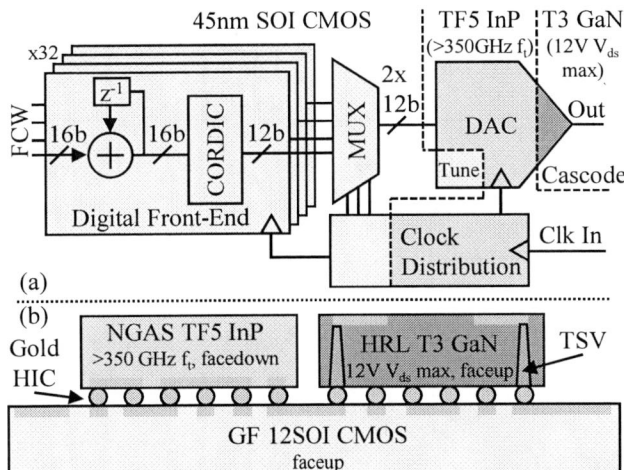

Fig. 1. (a) Direct digital synthesizer using three heterogeneous technologies – CMOS for digital, InP for high-speed mixed-signal, and GaN for high output swing. (b) Heterogeneous stack-up with 45nm CMOS as the base technology, and both InP and GaN chiplets integrated on top with gold micro-pillar heterogeneous interconnects (HICs).

authors' knowledge, this is the most complex three-technology heterogeneously integrated mixed-signal circuit reported, with over 6 million CMOS FETs, 2151 InP HBTs, 2 GaN HEMTs, and 9930 heterogeneous interconnects (HICs). This heterogeneous DDS leverages all three technologies to optimize performance.

II. DDS DESIGN IN HETEROGENEOUS PROCESS

The heterogeneous DDS design partitions all three technologies available in the DAHI process [7] to maximize performance. CMOS is suited for high-density and low power digital circuitry, particularly compared to InP-only implementations of DDS circuits. The CMOS base chip uses GlobalFoundries' 45 nm 12SOI process for the parallelized digital front-end circuitry (accumulator, phase-to-sine converter), MUX, tuning/calibration circuitry, and 7 GHz clock distribution. The >350 GHz f_T NGAS TF5 InP HBTs [7] are ideal for the high-speed mixed-signal operation inherent in DAC current switches and the high-frequency portions of the clock generation circuitry. The 220 GHz f_T, 12 V maximum V_{ds} HRL T3 GaN HEMTs [8] operate beyond the breakdown voltage of either the CMOS FETs or InP HBTs, allowing them to support high-voltage swing as an output cascode.

978-1-7281-1702-7/19 $31.00 © 2019 IEEE

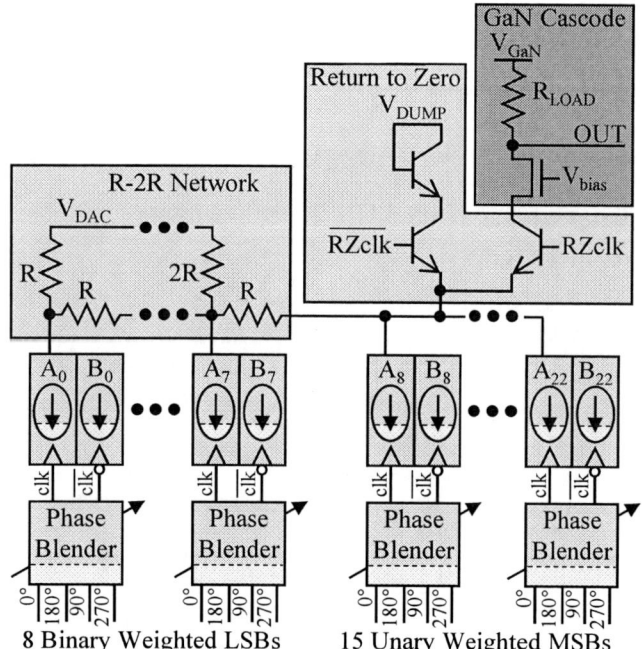

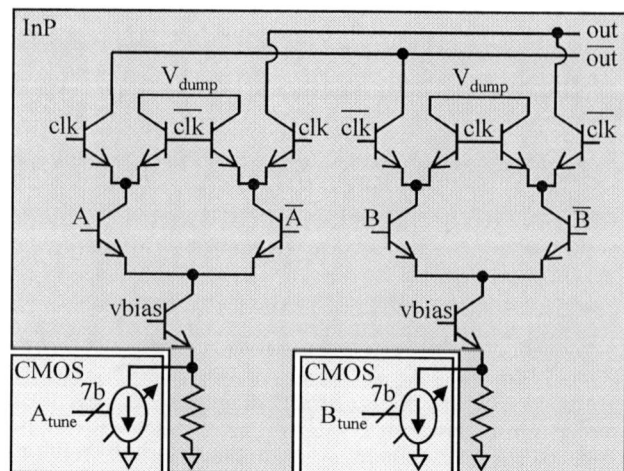

Fig. 2. Return-to-zero DAC circuit leveraging the optimal features of three technologies – CMOS for digitally controlled current tuning, InP for high-speed switching, and GaN for high voltage output. The circuit implementation is differential, but the diagram is single-ended for simplicity.

Fig. 3. Interleaved DAC current switch shunts current of the inactive side to a "dump" rail to keep the devices biased in an active state. The circuit also uses CMOS digital current tuning to compensate for mismatch.

A. CMOS Digital Front-End

The accumulator is 16 bits wide, allowing for output frequency steps of $f_{clk}/2^{16}$. The CORDIC [9], [10], converts the 16-bit phase word to a 12-bit amplitude word to drive the DAC. Digitally (i.e. with an ideal DAC), the CORDIC phase to amplitude conversion has a worst case spurious-free dynamic range (SFDR) >78 dBc, so it does not limit maximum SFDR in this design. Due to power consumption concerns, prior InP and GaAs DDS circuits use 5-bit to 8-bit sine-weighted DAC phase-to-amplitude conversion that limits the maximum achievable SFDR. The parallelized front-end logic (accumulator and CORDIC) runs at 1/32 of the sample rate, leveraging dense CMOS logic not possible in InP. Also, unlike an InP-only design, CMOS design tools enable synthesis of the digital front-end with an RTL/place and route design flow, easing the burden on the designer versus full custom logic design and layout.

B. High-Speed CMOS MUX

The 16:1 MUX up-converts the reduced-rate, parallelized output of the digital front-end to the full DAC clock rate using a mixture of traditional CMOS logic and true single-phase clocked (TSPC) logic. The 1.0 V CMOS supply drastically reduces MUX power compared to InP. The first 2:1 stage, implemented with an RTL/place and route flow, converts 437.5 MS/s data to 875 MS/s. The next 2:1 stage is hand-routed with standard cell gates to convert the data to 1.75 GS/s. In the final 4:1 stage, custom TSPC logic converts the data to the full 7 GS/s data rate feeding the interleaved DAC. The final conversion to 14 GHz occurs in the interleaving DAC circuitry. The physical design of the DAC imposes challenges

on the MUX design. Since the InP devices are large compared to CMOS devices, the DAC layout forces the MUX layout to be spread out over nearly 3 mm. High-speed (7 GHz) clock distribution in the MUX is carefully hand-routed with matched length lines and includes clock synchronization circuitry.

C. Three-Technology Heterogeneous DAC

The interleaved DAC (Fig. 2) has a segmented architecture with 8 binary-weighted bits and 15 unary-weighted (thermometer encoded) bits, equivalent to a 12-bit binary-weighted DAC. The DAC uses an R-2R ladder for the binary portions, with an external GaN cascode supporting high-voltage output swing. The DAC interleaving built into the InP current switch (Fig. 3) allows the A/B DACs to be active on opposite clock phases. Each DAC switch has tunable current sources implemented with CMOS current steering DACs. The current tuning range of ±10% compensates for process variation and mismatch. Signal routing from the InP current switches to the summing junction/R-2R ladder is physically long, so these signals are routed as transmission lines in InP, and modelled with a multi-port s-parameter network. The lines are length matched as much as practical, and differences in timing are calibrated by clock tuning circuitry.

Clock distribution of four phases of 7 GHz in the CMOS process reduces area, saves power (~10x versus InP), and reduces jitter compared to InP circuitry. The clock distribution circuitry buffers, filters, level-shifts, and distributes the four clock phases to 22 InP DAC phase blenders. The phase blenders independently tune the clock of each current-steering switch over a 70° range with 0.14° steps.

Physically, the CMOS DAC clock distribution circuitry resides on the base chip, *under* the DAC summing network located on the InP DAC chiplet. This is a unique circuit topology enabled by this heterogeneous process. A return to zero circuit (RZ) driven by a 14 GHz clock follows the DAC. The RZ circuitry minimizes spurs from code-dependent non-linearity by only exposing the A/B DACs to the output when their data switching is settled, and outputting zero while the A/B DACs transition.

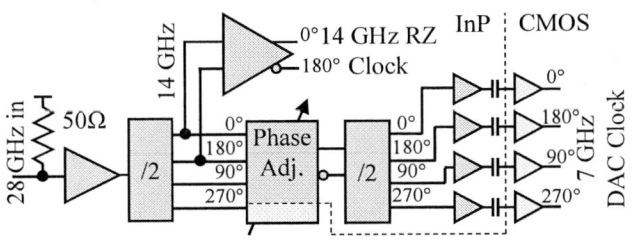

Fig. 4. Heterogeneous global clock generation using InP for the high frequency portions and CMOS to save power on the lower frequency portion.

D. CMOS and InP Global Clock Generation

The global clock circuit (Fig. 4) generates the 14 GHz clock for the DAC RZ switch and the 7 GHz clock for the DAC current switches. The clock chiplet receives a 28 GHz clock and divides it by two to feed the DAC RZ switches with 14 GHz. A tunable phase shifter adjusts the relative timing between the interleaved DAC clock and the RZ clock. This guarantees a settled DAC output prior to RZ assertion. Finally, a divide-by-two circuit generates four 7 GHz DAC clock phases from the phase-shifted 14 GHz clock. Placing this circuitry on a separate InP chiplet from the DAC provides isolation between the clock and the DAC.

III. FABRICATION AND MEASUREMENT RESULTS

The fabricated heterogeneous DDS (Fig. 5) consists of an 8 mm x 6 mm 45 nm CMOS base chip with over 6 million FETs, two facedown InP chiplets (a 2.75 mm x 3.5 mm DAC chiplet with 1865 HBTs and 389 signal HICs, and a 1.75 mm x 1.25 mm clock generation chiplet with 286 HBTs and 16 signal HICs), and a faceup GaN chiplet (1.75 mm x 2.00 mm with 2 HEMTs and 5 signal HICs). In addition to the signal HICs, the chiplets use numerous redundant signal, supply, and thermal HICs: 7290 on the InP DAC chiplet, 1901 on the InP clock distribution chiplet, and 739 on the GaN cascode chiplet. The CMOS base chip is attached to a heatsink and wirebonded to an evaluation board for measurement. Cable and connector

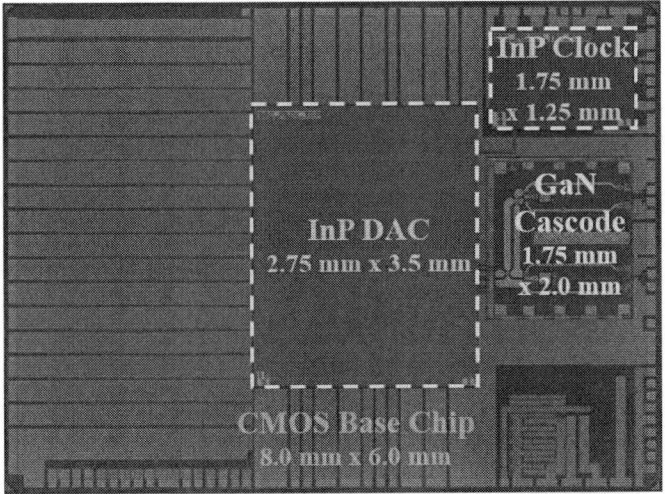

Fig. 5. Microphotograph of the heterogeneously integrated DDS test circuit. The CMOS base chip is 8 mm by 6 mm. There are two InP chiplets integrated facedown, a 2.75 mm by 3.5 mm DAC chiplet in the middle and a 1.75 mm by 1.25 mm clock distribution chiplet in the top right corner. The 1.75 mm by 2 mm faceup-integrated GaN cascode chiplet is on the middle right side.

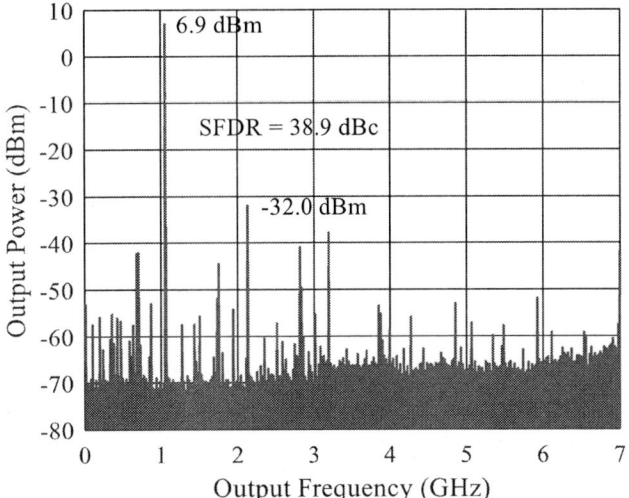

Fig. 6. DDS Output spectrum for FCW=5000 (1.068115 GHz), SFDR=38.4 dBc, showing high SFDR and output power.

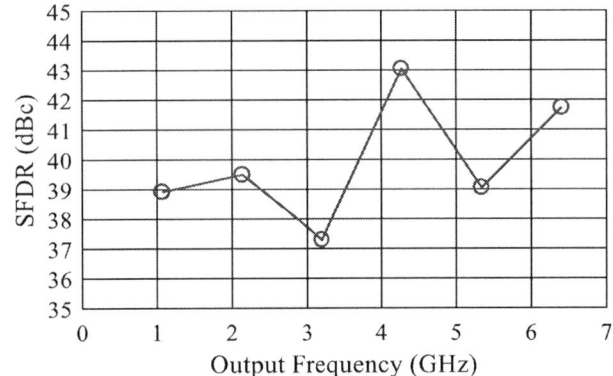

Fig. 7. Full-Nyquist SFDR versus output frequency. The worst-case output is 37.3 dBc at 3.204346 GHz.

losses are de-embedded from the presented measured results. The DDS operates up to a maximum clock rate of 14 GHz, and it can synthesize outputs up to 7 GHz in 32,768 steps of 213.623 kHz. The 8.7 W power consumption breakdown is 850 mW for the digital front-end, 110 mw for the MUX, 515 mW for the clock distribution, 6925 mW for the DAC, and 300 mW for the GaN cascode.

For the DDS output spectrum with frequency control word (FCW)=5000 (Fig. 6), the fundamental output power is 6.9 dBm at 1.068 GHz. The largest spur is -32.0 dBm, occurring at the 2nd harmonic of 2.13 GHz. Measurement of full-Nyquist SFDR versus output frequency for 14 GS/s sample rate (Fig. 7) shows a worst-case SFDR of 37.3 dBc at 3.2 GHz output frequency. For most cases, the highest spur is the 2nd harmonic. The results only include manual calibration of RZ to DAC clock phase. Magnitude and timing calibration of individual DAC current switches is not included due to non-functioning calibration sensors.

Table 1 shows a comparison to other >10 GS/s DDS circuits. Heterogeneous integration allows this design to have a larger accumulator bit-width and higher SFDR than all but the SiGe DDS in [13]. However, the superior SFDR of [13] comes at the expense of over 2x higher power consumption

Table 1. Comparison of direct digital synthesizers with >10 GS/s sample rate. This work has much higher output power and compares favorably in other metrics.

	This work	[11]	[12]	[13]	[14]	[15]	[16]	[17]
Sample Rate (GS/s)	14	10	12	12	13	16.8	24	32
Nyquist SFDR (dBc)	37.6	23.2	22	55	26.7	20	30.7	21.6
Power Consumption (W)	8.7	2.4	1.9	19.9	5.42	0.488	19.8	9.45
Output Power (dBm)	6.9	-22	-6.7	1	-10	-12	-24	-8
Accumulator bit-width	16	8	9	72	8	8	12	8
DAC bit-width	12	5	8	12	7	6	7.5	7
Phase to sine converter	CORDIC	Sine-weighted DAC	Sine-weighted DAC	Not specified	Binary Logic	Diff-pair (tanh)	ROM	Sine-weighted DAC
Technology	Heterogeneous InP and GaN on SOI CMOS	GaAs HBT	SiGe	SiGe	InP	SiGe	InP	InP

compared to this design. The low-power SiGe design [15] uses an interesting triangle-to-sine wave conversion leveraging the tanh characteristic of HBT differential pair. This allows it to have much lower power consumption, but at the expense of low SFDR. The highest sample rate InP designs suffer from either high power consumption [16] or low SFDR [17]. Compared to other reported DDS designs, this work has much larger output power (6.9 dBm), enabled by the heterogeneously integrated GaN HEMT cascode devices with large maximum V_{ds}.

IV. Conclusion

The presented DDS contains over 6 million CMOS FETs, 2151 InP HBTs, 2 GaN HEMTs, and 410 signal HICs (9930 HICs total), making it the most complex three-technology heterogeneously integrated mixed-signal circuit to date. Multiple technologies enable use of optimal devices for each section of the design – CMOS FETs for the large bit-width accumulator, high SFDR CORDIC phase-to-amplitude converter, MUX, and tuning control; InP HBTs for the high-speed DAC; and GaN HEMTs for the high-swing output. With a 14 GS/s sample rate, 37.3 dBc worst case SFDR, 8.7 W power consumption, and 6.9 dBm output power, it has much higher output power than other reported >10 GHz DDS designs, while comparing favorably in other metrics.

Acknowledgment

The authors wish to acknowledge the support of Dr. Daniel Green of DARPA/MTO, Dr. James Wilson of ARL, and the rest of the team under Contract No. W911QX-13-C-0050. The authors would also like to thank the DAHI process team at NGAS. This research was developed with funding from the Defense Advanced Research Projects Agency (DARPA). The views, opinions, and/or findings expressed are those of the authors and should not be interpreted as representing the official views or policies of the Department of Defense or the U.S. Government. Distribution Statement A: Approved for Public Release, Distribution Unlimited.

References

[1] D. S. Green, C. L. Dohrman, J. Demmin, Y. Zheng, and T.-H. Chang, "A Revolution on the Horizon from DARPA: Heterogeneous Integration for Revolutionary Microwave/Millimeter-Wave Circuits at DARPA: Progress and Future Directions," *IEEE Microwave Magazine*, vol. 18, no. 2, pp. 44-59, Feb. 2017.

[2] K. N. Madsen, *et al.*, "A high-linearity, 30 GS/s track-and-hold amplifier and time interleaved sample-and-hold in an InP-on-CMOS process," *IEEE J. Solid-State Circuits*, vol. 50, no. 11, pp. 2692–2702, Nov. 2015.

[3] S.-K. Kim, *et al.*, "A 30 GSample/s InP/CMOS Sample-Hold Amplifier with Active Droop Correction," in *IEEE MTT-S Int. Microwave Symp. Dig.*, May 2016.

[4] M. LaRue, *et al.*, "A Fully-Integrated S/C Band Transmitter in 45nm CMOS/ 0.2μm GaN Heterogeneous Technology," in *IEEE CSICS Dig.*, Oct. 2017.

[5] V. Radisic, D. W. Scott, K. K. Loi, C. Monier, R. Lai, and A. Gutierrez-Aitken, "Heterogeneously Integrated W-Band Downconverter," *IEEE Microwave and Wireless Components Letts.*, vol. 27, no. 8, pp. 739-741, July 2017.

[6] Y.-C. Wu, M. Watanabe, and T. LaRocca, "InP HBT/GaN HEMT/Si CMOS Heterogeneous Integrated Q-Band VCO-Amplifier Chain," in *IEEE RFIC Dig.*, Jun. 2015, pp. 39-42.

[7] A. Gutierrez-Aitken, *et al.*, "A Meeting of Materials: Integrating Diverse Semiconductor Technologies for Improved Performance at Lower Cost," *IEEE Microwave Magazine*, vol. 18, no. 2, pp. 60-73, Feb. 2017.

[8] S. D. Burnham, *et al.*, "Reliability Characteristics and Mechanisms of HRL's T3 GaN Technology," *IEEE Trans. Semicond. Manuf.*, vol. 30, no. 4, pp. 480-485, Nov. 2017.

[9] J. E. Volder, "The CORDIC trigonometric computing technique," *IRE Transactions on Electronic Computers*, no. 3, pp. 330–334, 1959.

[10] J. Qin, J. D. Cali, B. F. Dutton, G. J. Starr, F. F. Dai, and C. E. Stroud, "Selective Spectrum Analysis for Analog Measurements," *IEEE Trans. on Ind. Electron.*, vol. 58, no. 10, pp. 4960–4971, Oct. 2011.

[11] G. Chen, D. Wu, Z. Jin, and X. Liu, "A 10GHz 8-bit Direct Digital Synthesizer Implemented in GaAs HBT Technology," in *IEEE RFIC Dig.*, May 2010, pp. 425-428.

[12] X. Yu, F. F. Dai, J. D. Irwin, and R. C. Jaeger, "A 12 GHz 1.9 W Direct Digital Synthesizer MMIC Implemented in 0.18 μm SiGe BiCMOS Technology," *IEEE J. Solid-State Circuits*, vol. 43, no. 6, pp. 1384–1393, Jun. 2008.

[13] F. Van de Sande, *et al.*, "A 7.2 GSa/s, 14 Bit or 12 GSa/s, 12 Bit Signal Generator on a Chip in a 165 GHz f_T BiCMOS Process," *IEEE J. Solid-State Circuits*, vol. 47, no. 4, pp. 1003–1012, Apr. 2012.

[14] S. E. Turner and D. E. Kotecki, "Direct Digital Synthesizer with ROM-Less Architecture at 13-GHz Clock Frequency in InP DHBT Technology," *IEEE Microwave and Wireless Components Letts.*, vol. 16, no. 5, pp. 296-298, May 2006.

[15] B. Laemmle, C. Wagner, H. Knapp, H. Jaeger, L. Maurer, and R. Weigel, "A Differential Pair-Based Direct Digital Synthesizer MMIC with 16.8-GHz Clock and 488-mW Power Consumption," *IEEE Trans. Microw. Theory Tech.*, vol. 58, no. 5, pp. 1375–1383, May 2010.

[16] S. E. Turner, R. T. Chan, and J. T. Feng, "ROM-Based Direct Digital Synthesizer at 24 GHz Clock Frequency in InP DHBT Technology," *IEEE Microwave and Wireless Components Letts.*, vol. 18, no. 8, pp. 566-568, Aug. 2008.

[17] S. E. Turner and D. E. Kotecki, "Direct Digital Synthesizer with Sine-Weighted DAC at 32-GHz Clock Frequency in InP DHBT Technology," *IEEE J. Solid-State Circuits*, vol. 41, no. 10, pp. 2284–2290, Oct. 2006.

A 1 V 54-64 GHz 4-Channel Phased-Array Receiver in 45 nm RFSOI with 3.6/5.1 dB NF and -23 dBm IP1dB at 28/37 mW Per-Channel

Hyunchul Chung, Qian Ma and Gabriel M. Rebeiz,
University of California, San Diego, La Jolla, CA, USA
hcchung@ucsd.edu, qian.rfic@gmail.com, rebeiz@ece.ucsd.edu

Abstract— This paper presents a low-power, low-noise, high-linearity 4-channel phased-array receiver in 45 nm RFSOI process. An architecture employing an active low-noise balun, a 180° active phase-shifter, and 11/22/45/90° passive phase-shifters results in optimal performance between noise figure and power consumption. The phased-array front-end channel consumes 28 mW from a 1 V supply, with a measured gain and NF of 14 dB and 3.6 dB at 59 GHz, respectively (NF 3.6–4 dB at 57–64 GHz). The four front-end receive channels are followed by a high-linearity down-conversion mixer and an IF amplifier. The chip is flipped and placed on a low-cost printed circuit board (PCB) with matching network for connectorized measurement. The measured electronic gain of the phased array receiver is 20–21 dB with a 3-dB bandwidth of 54-64 GHz and NF of 5.1–5.4 dB at 57–64 GHz, with a system IP1dB of -23 dBm at 150 mW dc power. To author's knowledge, this work achieves the lowest NF and highest dynamic range with <40 mW Pdc/channel at 60 GHz band and enables the construction of a large arrays (256–1024 elements) with low power consumption.

Keywords— CMOS SOI, electronic gain, front-end, low-power, millimeter-wave, phased-array, receiver, 60 GHz.

I. INTRODUCTION

The mm-wave frequency band has seen a lot of interest in the past few years, due to growing demand of data usage with multi-Gbps rates. In particular, the interest in the unlicensed ISM band (57-64 GHz) for 60-GHz wireless applications, has led to significant improvement in mm-wave phased-arrays system. However, due to high propagation loss as well as the limited output power per element at 60 GHz, a large array (256 elements) is needed for long-distance links and high EIRP [1]. On the receive side, it is critical to result in low noise figure at low dc power, as a 3 dB reduction in NF will reduce the element count by half, and any dc power reduction per channel is multiplied by 256 at the system level. Therefore, building a low noise low-power phased-array receiver becomes an important goal to meet the cost, size, power consumption requirements for practical use.

In this work, a low-noise, low-power 54-64 GHz 4-channel phased-array receiver is demonstrated using a 45 nm RFSOI process. The channel is built using active and passive phase shifters with low power and high linearity. Input single-ended amplifiers and active baluns are used to reduce the noise figure penalty of passive baluns. An on-chip mixer is also employed to allow for high IF gain and for scalability to large arrays. Finally, a chip-scale package is used for low-loss transition to the PCB, together with transitions having high channel-to-channel isolation. The result is a start-of-the-art packaged chip operating at 54-64 GHz with excellent performance.

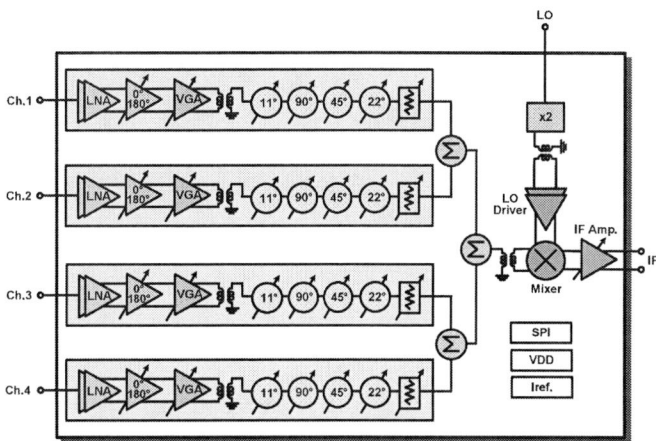

Fig. 1. Block diagram of the low-power 54-64 GHz, 4-channel phased array receiver.

II. PHASED ARRAY RECEIVER DESIGN

A. Technology

The phased-array receiver is designed in the GlobalFoundires 45-nm RFSOI process with option 18 for the back-end-of-line (BEOL) options. Option 18 provides thick metals (LD=4.1 μm, OB=3 μm, OA=3 μm), resulting in high inductor Q with a simulated Q of 30-35 at 60 GHz. The High-Vt NFET is selected and provides an NF_{min} of 0.8 dB at 60 GHz and an associated f_T of 180-200 GHz (referenced to the top metal) at a bias current of J=0.15–0.2 mA/μm. A double-gate contact is employed to reduce the gate resistance (∼ 4 Ω) and a relaxed pitch layout is chosen to reduce the parasitic capacitance [2]. A coplanar waveguide (CPW) 50-Ω transmission line has measured loss of 0.8 dB/mm at 60 GHz.

B. Front-End Design

The phased-array chip is composed of a 4-channel front-end followed by a down-conversion mixer and an IF amplifier, for scalability to large arrays (Fig. 1). The active stages are implemented in a differential configuration (except the first stage) so as to minimize the effect of bump inductance from VDD and GND for stability, and to greatly reduce the channel to channel coupling. In addition, single-ended switched-LC phase shifters are used to reduce the insertion loss with a compact area. The cascode LNA/balun is biased at J=0.15 μm/mA for low-power and low NF (Fig. 2a), and results in a simulated gain of 13-14 dB with 2.8 dB NF at

978-1-7281-1702-7/19 $31.00 © 2019 IEEE

2019 IEEE Radio Frequency Integrated Circuits Symposium

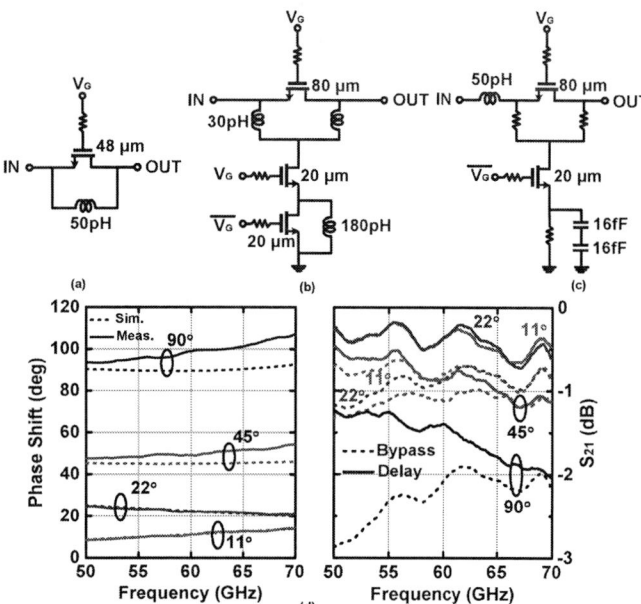

Fig. 2. Schematic of (a) low-noise amplifier (LNA)/balun, (b) active 180° phase shifter with 3.5 dB attenuator, and (c) variable gain amplifier.

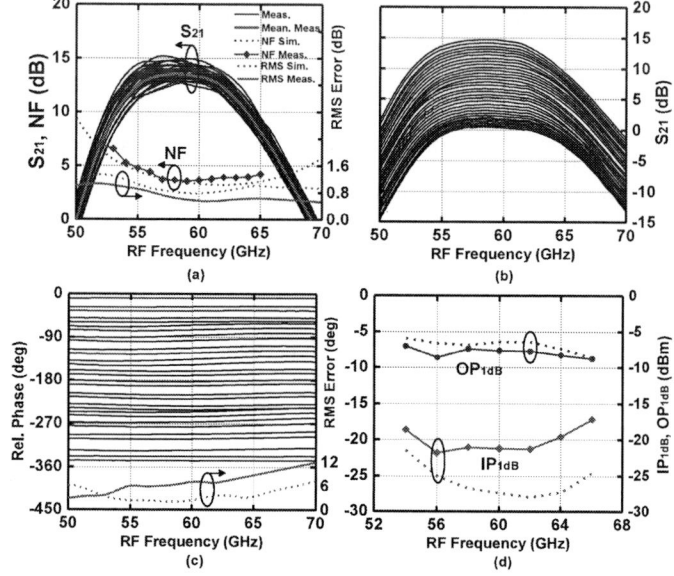

Fig. 3. (a) 11° phase shifter, (b) 22° phase shifter, (c) 3 dB attenuator, (d) measured phase shift and S_{21} of 11°, 22°, 45°and 90° phase shifters.

Fig. 4. Measured (a) S_{21} and NF, (b) 5-bit gain states, (c) 5-bit phase states, and (d) P_{1dB} of a single-channel front-end.

60-64 GHz. The simulated IP1dB of the LNA/balun is -19 dBm with 6 mW P_{DC}. An active 180° phase shifter is chosen over a high-loss switched-LC design, with a simulated gain of 7 dB and an amplitude and phase imbalance of <0.2 dB and <2.5° between the 0° and 180° states. In addition, a 1-bit 3.5 dB gain control is designed for improved linearity using another differential pair with a series resistance between the g_m cell and the switching quad.

The variable gain amplifier is based on current steering with 3-bit 0.5-dB gain steps (Fig. 2c). It has a simulated gain of 7 dB with a NF of 4.5–5 dB. Due to low voltage operation, a tail inductor (j75Ω @ 60 GHz) is used to provide >10 dB common-mode rejection (CMRR). The differential VGA output is converted to a single-ended signal using a passive balun (0.8 dB loss), providing additional 20 dB CMRR.

Fig. 3a presents the 11° and 22° phase shifters and an additional 20-µm transistor is used in the 22° cell to provide a constant phase shift over a wide frequency [3]. An identical topology is utilized for the 45° cell, and the 90° cell is built by

cascading two 45° units. The measured phase shift of the 11°, 22° and 45° phase shifters agree well with simulations with at most a 5° error in the 45° cell. Due to cascaded design of the 90° cell, it shows 10° more phase shift than simulations. This is due to an under-estimated inductance in the delay path, and has been resolved in subsequent chips (not shown for brevity). The measured insertion loss of the 11° to 90° cells is 0.6±0.2 dB to 1.8±0.2 dB (Fig. 3d). The 4 phase-shifter cells have an excellent impedance match (S_{11},S_{22} <-15 dB, not shown), and this results in low RMS gain error. A 6-dB attenuator based on a switched-R circuit is also implemented for added gain control with a measured loss in the bypass/attenuation modes of 1/7 dB, and $S_{nn} < -20$ dB.

The active and passive blocks are then optimized together using inter-stage matching networks to widen the channel bandwidth (55-65 GHz). A stand-alone 1-channel phased-array front-end breakout is measured using GSG probes and probe-tip calibration (Fig. 4). The measured average gain is 14 dB with a 3-dB bandwidth of 53–64 GHz with a gain variation of ±1 dB for 5-bit/32 phase states,

978-1-7281-1702-7/19 $31.00 © 2019 IEEE   120

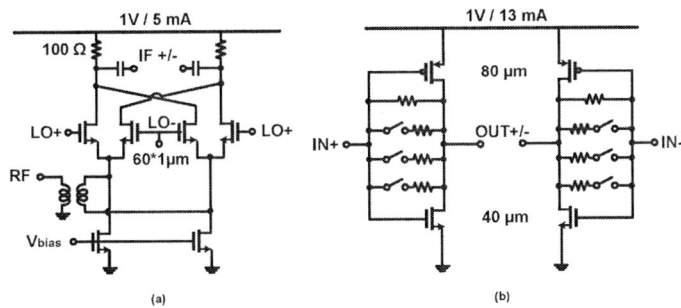

Fig. 5. Schematic of (a) high-linearity down-conversion mixer and (b) IF amplifier.

resulting in an RMS gain error of 0.7–0.8 dB (Fig. 4a). The measured minimum NF is 3.6 dB at 59 GHz (simulated 3.3 dB) and with a NF < 4 dB at 57–64 GHz. The measured gain versus 5-bit control provides 14-dB gain control (Fig. 4b). The measured RMS phase error is 6° at 60 GHz, a bit higher than simulation, due to the phase error in 90° cell. The measured IP_{1dB}/OP_{1dB} is -21 dBm/-8 dBm (Fig. 4d). The single-channel consumes 28 mW including bias circuitry. As a stand-alone phased-array channel, this is the lowest noise figure to-date achieved at 60 GHz using SiGe or CMOS, and with very low power consumption and high linearity, and is due to the judicious choice between the active and passive stages.

C. Down-conversion mixer and IF amplifier

The four phased-array channels are summed using a 4:1 Wilkinson combiner and fed into a down-conversion mixer. The simulated ohmic loss of the 4:1 Wilkinson network is 1 dB at 60 GHz with > 20 dB isolation between the ports. Note that a single-ended network is used for reduced loss.

The mixer must be able to handle 0 dBm to maintain the system linearity when all 4 channels are active, and therefore the g_m stage is removed and the RF signal is fed directly into the LO quads in a current-mode configuration (Fig. 5a). The down-converted signal is fed into an inverter-based IF amplifier with 3-bit switching resistors for 15 dB gain control (Fig. 5b) and is biased at 13 mA for high linearity. The measured power conversion gain and NF of the mixer+IF amplifier on a breakout circuit is 7 dB and 13 dB at 59 GHz, respectively, with a 3-dB RF bandwidth of 33–66 GHz (±1dB agreement with simulations). The IP_{1dB} of the mixer+IF amplifier is -3 dBm at the maximum gain state of 7 dB (limited by the IF amplifier, OP1dB of +2.5 dBm). A two-stage LO driver is implemented together with a frequency doubler to lower the input LO frequency to 25–30 GHz (not shown for brevity). The total power consumption of down-conversion mixer, IF amplifier, and LO path is 38 mW. This is relatively high, but is needed to maintain the linearity of the 4-channel receiver.

III. MEASUREMENTS

The 4-channel phased-array receiver chip is designed in GlobalFoundries 45 nm RFSOI with C4 bumps and with a pitch of 250 μm (Fig. 6a, 2.6 mm×2.3 mm). A row of ground

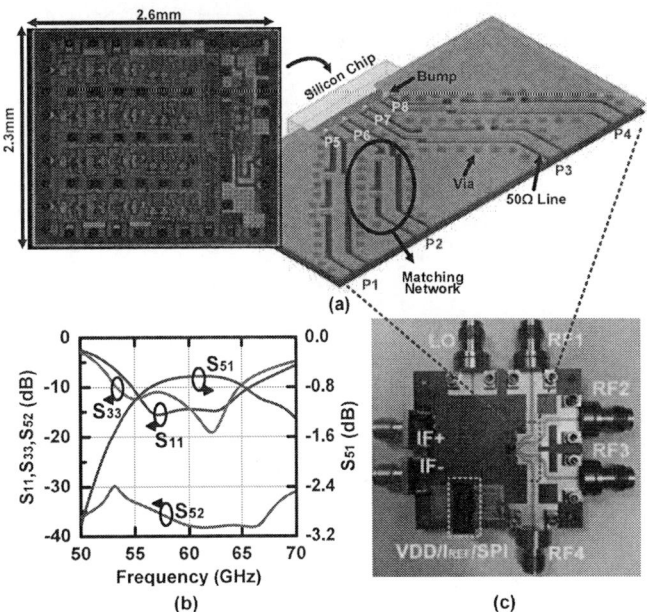

Fig. 6. (a) Microphotograph of 4-channel receiver phased array on PCB with matching network (3D view), (b) simulated bump transition (S_{51}), return loss (S_{11}, S_{22}) and forward isolation (S_{52}), and (c) connectorized test board.

vias is placed between the channels and around the input GSG pads to improve the isolation knowing that the first-stage LNA is single-ended. The chip consumes 150 mW (4x28+38 mW).

The chip is flipped on a low-cost printed circuit board (PCB) with 5 mil Tachyon 100G laminate (ϵ_r=3.0) on top of 40-mil FR-4 for mechanical stability (Fig. 6c). The flip-chip transition is modeled using Ansoft HFSS, and a simulated loss (S_{51}) with a wideband matching network of 0.6 dB is achieved at 60 GHz with a return loss <-10 dB at 53–65 GHz (Fig. 6b). The simulated forward coupled power between two adjacent channels, for example from P2 to P5 (S_{52}) is < -38 dB at 60 GHz.

The phased-array receiver gain is measured at an IF frequency of 1 GHz. The reference planes were calibrated to the 50 Ω ports on the PCB, and loss from the connectors and long T-lines is de-embedded. The electronic gain is derived from the measured gain with one RF channel activated, and +6 dB is added to the measured S_{21} to take into effect the added Wilkinson loss since the other channels are not energized. Fig. 7a shows an average electronic gain of 20–21 dB with a 3-dB bandwidth of 54–64 GHz (32-phase states, same phase response as Fig. 4). A total of 29-dB gain control range is achieved over 5(RF)+3(IF)-bit gain control (Fig. 7b). The measured RMS gain and phase error are 0.5–1 dB, 3–7.5° respectively. The measured IF bandwidth is 0.4-6 GHz. Channel-to-channel coupling measurements show < -30 dB coupling, and prove that the single-ended single-stage cascode design together with an active balun is an excellent choice for low-noise-figure receivers at 60 GHz. This topology does not result in any coupling or stability concerns, while at the same time, provides a very low noise front-end.

Fig. 8 presents the system IP1dB and NF analysis for

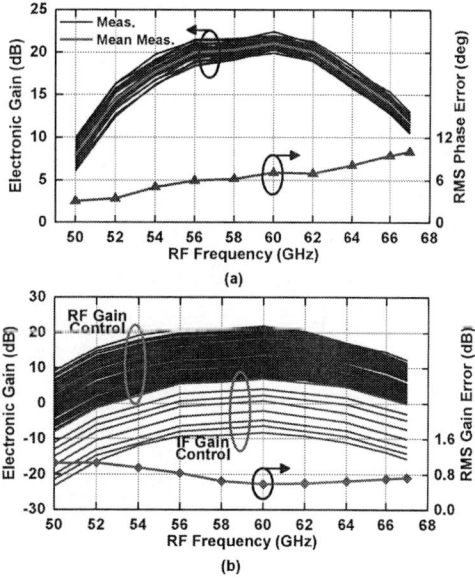

Fig. 7. Measured (a) electronic gain of phased-array receiver and RMS phase error, and (b) electronic gain of phased-array receiver with RF/IF gain control and RMS gain error (for the 32 phase shifter states).

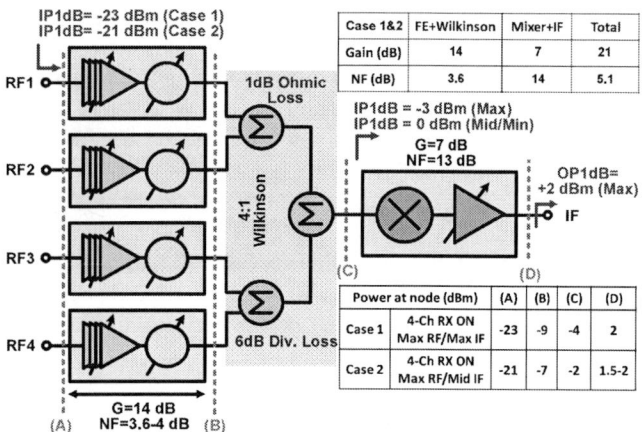

Fig. 8. IP1dB and NF simulations of the 4-channel phased-array receiver.

several different gains, and with all 4-channels activated. At the maximum RF/IF gain state, an input power of -23 dBm at node A results in -4 dBm at node C. Since the mixer IP1dB is -3 dBm (limited by the IF amplifier OP1dB), the system IP1dB is -23 dBm and is limited by both the front-end and the mixer. When a medium IF gain state is used (Case 2), the mixer IP1dB becomes 0 dBm, allowing the system IP1dB to be -21 dBm, and is nearly limited by the front-end. The system NF is calculated using the measured NF of a stand-alone front-end and mixer+IF amplifier, and is 5.1 dB at 59 GHz.

The mixer+IF receiver consume 38 mW, but a lower linearity design (not implemented here) consumes 18 mW. Therefore, the power per channel can be reduced to 32.5 mW/channel instead of 37.5 mW/channel if an IP1dB/channel of -32 dBm is acceptable.

Table 1 summarizes the state-of-the-art 60 GHz CMOS phased array receivers. This work achieves the lowest NF with

Table 1. Performance Comparison of 60 GHz CMOS Front-End Receiver

Ref.	[4]	[5]	[6]	**This work**
Tech.	65nm CMOS	45nm CMOS	45nm CMOS SOI	**45nm CMOS SOI**
Integration	FE+IF	FE	FE+IF	**FE+IF**
Element #	4	8	4	**4**
Freq. (GHz)	57–66	53–65	45–66	**54–64**
Gain (dB)	21^a	16^b	26^a	**20^a**
NF (dB)	5.8-8.4	4.3-6.3	5.5-9	**3.6-4^b 5.1-5.4^a**
IP1dB (dBm)	-12.5^c	N/A	-27^e	**-23**
Area (mm^2)	2.6	7.2	4	**6**
RMS Error (dB)	±1.1^d	<0.5	<2	**0.6–0.8**
RMS Error (deg)	±0.6^d	<5	<5	**3–8**
P_{DC}/ch. (mW)	55	76	30	**28^b/37.5^a**
P_{DC} Total (mW)	320	613	N/A	**150**

^{a} FE+IF ^{b} FE only ^{c} lowest gain ^{d} Phase, Gain error ^{e} Single-channel only

<30-40 mW P_{DC}/ch and a 3-dB BW of 10 GHz. To our knowledge, this is the highest dynamic range receiver achieved to-date at 60 GHz.

IV. CONCLUSION

A low-power, low-noise-figure 4-channel front-end and phased array receiver chip with an integrated down-converter is presented. A design employing single-ended LNAs with active baluns, active and passive phase shifters, down-conversion mixers, and chip-scale packaging results in a chip with ultra-low-noise figure, low coupling between the channels, low power consumption, and is ideal for very large element phased-arrays.

ACKNOWLEDGMENT

This work was supported by the National Institute of Standards and Technology (NIST). The authors thank HRL for help in assembly.

REFERENCES

[1] S. Zihir, O. D. Gurbuz, A. Kar-Roy, S. Raman and G. M. Rebeiz, "60-GHz 64-and 256-elements wafer-scale phased-array transmitters using full-reticle and subreticle stitching techniques," *IEEE Transactions on Microwave Theory and Techniques*, vol. 64, no. 12, pp. 4701-4719, Dec. 2016.

[2] O. Inac, M. Uzunkol and G. M. Rebeiz, "45-nm CMOS SOI technology characterization for millimeter-wave applications," *IEEE Transactions on microwave theory and techniques*, vol. 62, no. 6, pp. 1301-1311, June 2014.

[3] U. Kodak and G. M. Rebeiz, "A 42mW 26-28 GHz phased-array receive channel with 12 dB gain, 4 dB NF and 0 dBm IIP3 in 45nm CMOS SOI," in *IEEE Radio Frequency Integrated Circuits Symposium(RFIC)*, May. 2016, pp. 348–351.

[4] L. Wu, H. F. Leung, A. Li and H. C. Luong, "A 4-element 60 GHz CMOS phased-array receiver with beamforming calibration," *IEEE Transactions on Circuits and System I: Regular Papers*, vol. 64, no. 3, pp. 642-652, March. 2017.

[5] S. Drago *et. al.*, "A 60GHz wideband low noise eight-element phased array RX front-end for beam steering communication applications in 45nm CMOS," in *IEEE Radio Frequency Integrated Circuits Symposium(RFIC)*, June. 2012, pp. 435–438.

[6] S. Kundu and J. Paramesh, "A compact, supply-voltage scalable 45-66 GHz baseband-combining CMOS phased-array receiver," *IEEE Journal of Solid-State Circuits*, vol. 50, no. 2, pp. 527-542, Feb. 2015.

A Fully Integrated 60 GHz 10 Gb/s QPSK Transceiver with Digital Transmitter and T/R Switch in 65nm CMOS

Zheng Song, Jianfu Lin, Yutian Li, Jialiang Ye, Ruichang Ma and Baoyong Chi

Institute of Microelectronics, Tsinghua University, Beijing, 100084, China

Abstract—A fully integrated 60 GHz 10 Gb/s QPSK transceiver (TRX) with digital transmitter and T/R switch in 65nm CMOS is presented. The TRX consists of a direct-conversion receiver, a digital transmitter with on-chip QPSK modulator and a quadrature local-oscillation (LO) signals generation network with 20 GHz integer-N phase-locked loop (PLL) frequency synthesizer. A T/R switch is also integrated to interface with the antenna. RF bandwidth of the TRX is expanded to ~10GHz by using the magnetically coupled resonator based matching network. The QPSK modulation is directly realized in the I/Q digital power amplifier, which simplifies the transmitter complexity and reduces the power consumption. A 20 GHz integer-N PLL and a quadrature injection-locked frequency tripler (QILFT) are integrated on-chip to generate 60 GHz quadrature LO signals. The QILFT utilizes in-phase coupling technique to improve the LO I/Q matching and phase noise performance. The receiver achieves 7.1 dB noise figure and 25-47 dB dynamic gain range. The LO phase noise measured at the transmitter output is -93 dBc/Hz at 1-MHz offset from 60 GHz carrier. The measured error vector magnitude (EVM) of the transmitter is -23.9 dB for 10 Gb/s QPSK signals at 7 dBm output power. The EVM of the Over-the-Air (OTA) modulation-demodulation system is -16.3 dB for 10 Gb/s QPSK signals.

Keywords— CMOS, millimeter wave, 60 GHz, transceiver, wideband, IQ digital power amplifier, quadrature injection-locked frequency tripler.

I. INTRODUCTION

The millimeter-wave (mm-wave) band, especially the unlicensed spectrum around the 60 GHz (57 to 66 GHz), has drawn more attention from both the academic and industry since the growing demands for short-distance high data rate wireless communication applications [1].

Great challenges exist during the implementation of the 60 GHz transceiver (TRX). Firstly, owing to rich spectrum resource around the 60 GHz, a wideband TRX with up to 10 GHz link bandwidth is needed to support high data rate (~tens Gbps) with simple modulation scheme, which requires the broadband design of mm-wave RF front-end circuits such as low-noise amplifier (LNA) and power amplifier (PA). Secondly, it is still challenging to generate quadrature local oscillation (LO) signals with low phase-noise and low phase mismatching in the mm-wave band. Thirdly, the conventional transmitter consists of multiple cascaded circuit stages, which significantly limits the achievable link bandwidth. High sampling rate DAC and high-speed digital processing further increase the design complexity. Some recently reported transmitters realize the on-chip direct QPSK modulator to simplify the complexity [2]. However, the QPSK modulator

only realizes the first up-conversion, and further up-conversion as well as power amplification are needed, which also limits the link bandwidth.

In this paper, a fully integrated 60 GHz 10 Gb/s QPSK TRX with digital transmitter and on-chip T/R switch in 65nm CMOS is presented. Wide link bandwidth is achieved by using magnetically coupled resonator based matching network. The quadrature injection-locked frequency tripler (QILFT) with in-phase coupling technique improves the LO phase noise and I/Q mismatching performance. The QPSK modulation is directly realized in the digital power amplifier (DPA), which simplifies the design complexity and reduces the power consumption. Due to the above techniques, the prototype achieves superior performance.

II. TRANSCEIVER ARCHITECTURE AND CIRCUIT DESIGN

Fig. 1 shows the block diagram of the proposed fully integrated transceiver, consisting of a direct-conversion receiver, a digital transmitter with on-chip QPSK modulation and a quadrature LO generation network. An on-chip λ/4 transmission line based T/R switch is also integrated to interface with the antenna. The receiver adopts a quadrature direct down-conversion architecture. Direct quadrature modulation at the I/Q digital PA is first proposed in mm-wave band, which acts as on-chip QPSK modulator at the same time. A frequency tripling scheme is employed in the LO generation network. A 20 GHz integer-N phase locked loop (PLL) and a quadrature injection-locked frequency tripler are integrated on-chip to generate 60 GHz quadrature LO signals. To simplify the measurement, the baseband data could come from either on-chip PRBS7 generator or off-chip. The data rate of on-chip PRBS generator can be configured to 2.5Gb/s, 5Gb/s or 10Gb/s.

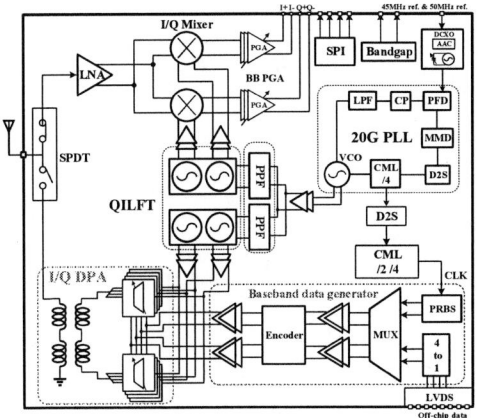

Fig. 1. The block diagram of the proposed transceiver.

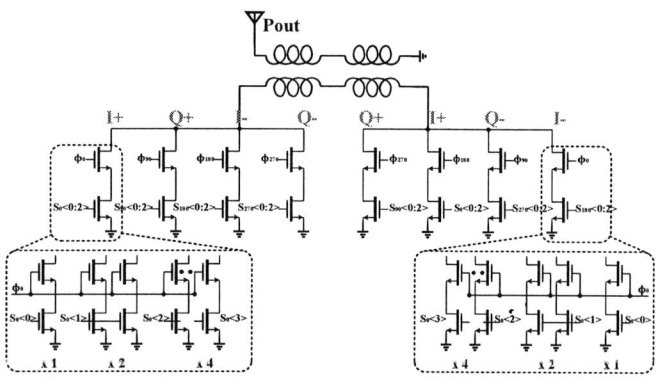

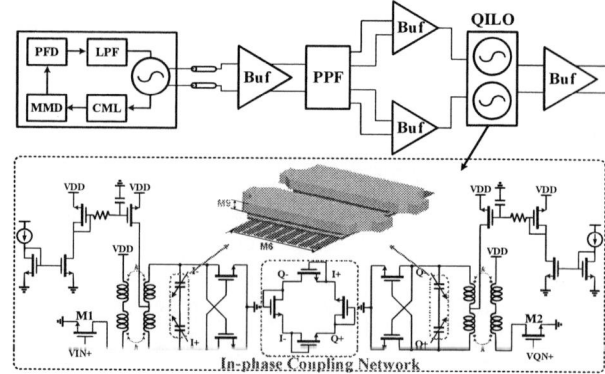

Fig. 2. Schematic of the digital power amplifier.

Fig. 4. Schematic of the LO distribution network.

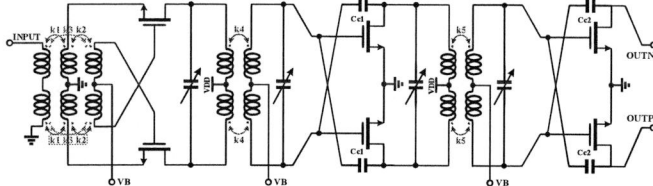

Fig. 3. Schematic of the LNA.

B. Receiver Design

The receiver consists of an LNA, a double balanced Gilbert Mixer and an intermediate frequency (IF) programmable gain amplifier (PGA). The LNA, the Mixer and the PGA should be carefully designed to achieve wide link band. As shown in Fig. 3, the LNA adopts the magnetically coupled resonator (MCR) based matching network to extend the bandwidth. The input three-winding transformer realizes single-ended to differential conversion as well as broadband input matching [4]. The same technique is also used in the design of the DPA to extend the bandwidth. The neutralization capacitors are added to improve the gain, reverse isolation and stability. Four DiCAD transmission lines are inserted to realize the frequency tuning and overcome the process variation. The current-bleeding double balanced Gilbert Mixer has to trade-off between the bandwidth and the gain. The wideband PGA is realized as two flipped voltage follower (FVF) based amplifier stage, where an additional zero is introduced to extend the bandwidth in the FVF cell.

A. QPSK Transmitter

The transmitter features the direct quadrature modulation, which not only simplifies the design complexity but also reduces the power consumption and expands the link bandwidth (since only one stage remains in the TX link). In principle, the QPSK modulated signal can be split into in-phase and quadrature-phase signals. According to the QPSK constellation diagram, each symbol can be presented by the vector summation of the in-phase signal and the quadrature-phase signal. Based on this concept, the QPSK modulation can be realized by controlling the amplitude of the 4-phase LO signals with the baseband data at the DPA [3]. The schematic of the proposed I/Q DPA with the embedded phase modulation and amplitude control is shown in Fig. 2. It consists of two paths, with the output combination at the balun-based wideband network. In each path, four amplifier cores are driven by the 4-phase differential LO signals ($\Phi_0, \Phi_{90}, \Phi_{180}, \Phi_{270}$). Limited by the parasitic capacitance, each amplifier core is constrained to 7 (3-bit) unit cells, which provides 17 dB dynamic output power range theoretically. The cascode transistors are driven by the LO signals while the bottom transistors are switched by the amplitude control words. Since the input of the DPA is 4-phase differential LO signals rather than wideband modulated signals, the drive stage is easy to design. The I/Q DPA achieves the QPSK modulation while also amplifying the signal. It can be noted that the bandwidth of the transmitter is only limited by the DPA output matching network. The output balun converts the combined output from differential to single-ended while achieving wideband output matching.

C. The LO Distribution Network Design

The 60GHz quadrature LO signals are generated by the 20GHz PLL and the QILFT. In order to further improve the I/Q matching performance of the 60 GHz LO signals, a Polyphase Filter (PPF) is inserted after the PLL to generate 20 GHz quadrature LO signals as the input of the QILFT. Fig. 4 shows the schematic of the proposed QILFT. Top-injected coupled resonator [5] is adopted in the QILFT to improve the injection locking range. The coupled resonator should be carefully designed to optimize the injection efficiency and the phase noise. M_1 and M_2 generates the third-order harmonic signal, which is injected into the coupled resonator. The dc current of the harmonic generator and the oscillator can be optimized individually. It utilizes in-phase coupling technique to reduce the I/Q mismatch and improve the phase noise performance [6]. Two identical injection-locked oscillators are coupled through the coupling network, which is formed by the four diode-connected transistors. Since the coupling network is simple and symmetrical, the mismatch caused by the layout parasitic can be reduced to achieve the superior I/Q matching performance. Besides, the DiCAD transmission lines are inserted, which could be used to calibrate the I/Q mismatch in the receiver and transmitter.

2019 IEEE Radio Frequency Integrated Circuits Symposium

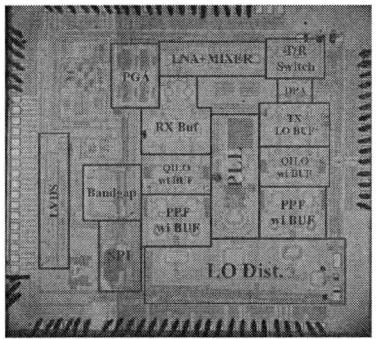

Fig. 5. Chip microphotograph of the 60 GHz transceiver.

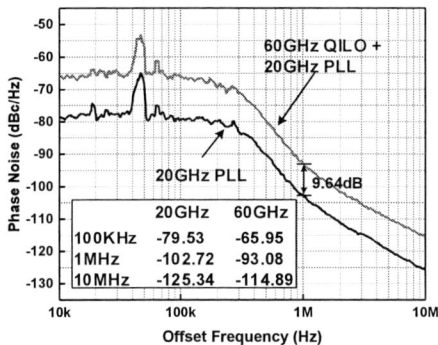

Fig. 6. Measured phase noise at 20 GHz and 60 GHz.

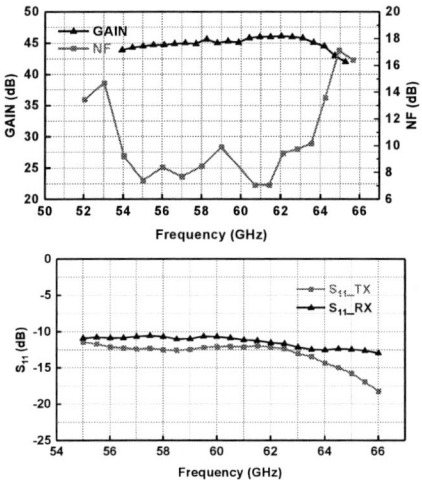

Fig. 7. Measured RF performance of the RX.

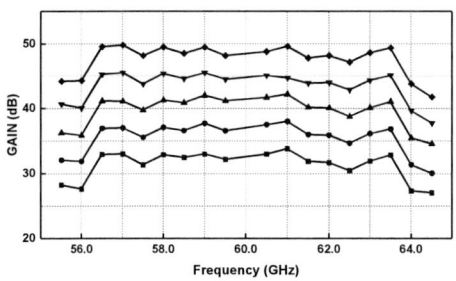

Fig. 8. Measured cascaded gain of the RX (from the T/R switch input to the PGA output).

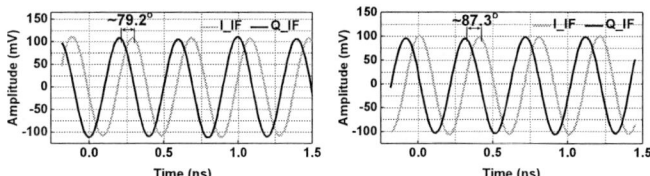

Fig. 9. Measured receiver I/Q output waveforms before and after I/Q calibration (f = 2.5 GHz).

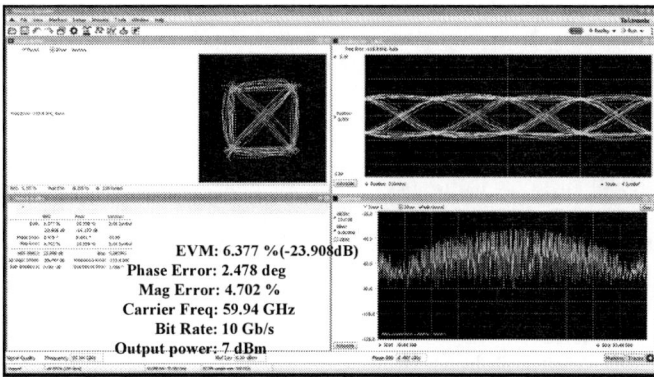

*the insertion loss of cable (7dB) and attenuator (10dB) are included

Fig. 10. Measured output spectrum, constellation and EVM for 10 Gb/s QPSK signals (the output power is 7dBm).

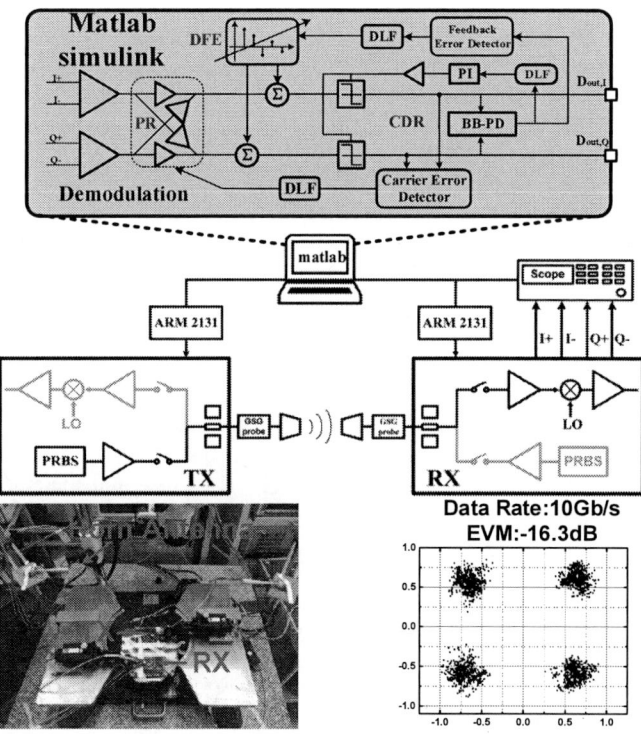

Fig. 11. TRX modulation/demodulation measurement.

III. MEASUREMENT RESULTS

The proposed transceiver has been implemented in 65nm CMOS. Fig. 5 shows its microphotograph, occupying 2.5mm x 2.6mm die area, including bonding pads and ESD IOs. As shown in Fig. 6, the measurement tuning range of the PLL covers from 17.388 GHz to 22.404 GHz. The phase noise of the 20 GHz carrier measured at the PLL output and the 60 GHz carrier at the TX output are -102.72 dBc/Hz and -93.08 dBc/Hz at 1 MHz offset, respectively. The phase noise degradation of the frequency tripler is about 9.64dB, which is close to the 9.5 dB theoretical value.

The receiver performance is measured at the input of the T/R switch on the probing station. As shown in Fig. 7, the return loss (S11) is less than -10dB over 55-66 GHz. The cascaded minimum NF is 7.1 dB at 61 GHz. Besides, the bandwidth of the receiver front-end is also tested by sweeping the input test tone and the LO frequency simultaneously and observing the PGA output. The measured receiver front-end -3-dB bandwidth (BW_{3dB}) is about 10GHz with 1.5 dB in-band ripple. Limited by the tuning range of the PLL and QILFT, the actual receiver front-end BW_{3dB} is wider than 10 GHz. Fig. 8 shows the cascaded gain of the receiver, which is tested by sweeping the input test tone frequency while fixing the LO frequency and observing the PGA output. It achieves 47 dB maximum gain with >7.0 GHz double sideband BW_{3dB} and the PGA provides about 15 dB dynamic range, where the link bandwidth is mainly limited by the PGA. Fig. 9 shows the measurement results of the I/Q imbalance calibration by adjusting the DiCAD control code in the QILFT. A 62.5 GHz test tone is injected into the receiver input and the LO frequency is fixed to 60 GHz, the 2.5 GHz I/Q PGA outputs are observed with the oscilloscope. The phase error is improved from 10.8° to 2.7° after the calibration.

The performance of the transmitter with on-chip QPSK modulation can be measured by the oscilloscope. The baseband signals are generated by the on-chip PRBS7 generator, which can be set to 2.5Gb/s, 5Gb/s and 10Gb/s data rate. As shown in Fig. 10, the error vector magnitude (EVM) of the transmitter is -23.908 dB for 10 Gb/s 60 GHz QPSK signals at 7 dBm output power, and the EVM is improved to -28.057 dB and -31.722 dB when the data rate is decreased to 5Gb/s and 2.5Gb/s, respectively.

Fig. 11 shows the Over-the-Air (OTA) modulation-demodulation system measurements. Two TRX boards are connected to the horn antennas with the cables on the probe stations, respectively. The left TRX board is used as a transmitter to radiate the 60 GHz QPSK signals and the right TRX board is used as a receiver. The IF outputs of the receiver are sampled with the oscilloscope and demodulated by the QPSK demodulator realized in MATLAB Simulink tool. The behavioural-level demodulator consists of a carrier recovery loop, a clock data recovery loop, a decision feedback equalization loop and an adaptive threshold voltage loop, which will be integrated in the future TRX chip. The EVM of the OTA modulation-demodulation system measurements is -16.3 dB with 10 Gb/s data rate for 60-GHz QPSK signals.

Table 1 summarizes the performance of the presented 60 GHz transceiver and makes a comparison with the state-of-the-art. Concluded from the comparison, our transceiver achieves better performance in RX BW_{3dB}, EVM and power.

Table 1. Performance Comparison of the State-Of-The Art

Index	This work	[2]	[7]
Tech	65nm	65nm	28nm
Integration	T/R switch PLL Mod	T/R switch PLL Mod	No T/R switch No PLL Mod/demod
NF(dB)	7.1	6.1	~7.7
RX BW_{3dB}(GHz)	7 (D-side)	5 (D-side)	3 (D-side)
Modulation	QPSK	QPSK	QPSK/16QAM
EVM(dB) TX	-23.9@10Gb/s	-21.9@5Gb/s	N/A
EVM(dB) OTA	-16.3@10Gb/s	N/A	-12.5@16Gb/s -18@27.8Gb/s
Pout(dBm)	7	6.4	N/A
PN@1MHz (dBc/Hz)	-93(60GHz)	-97.2(40GHz)	N/A
Power (mW)	TX+LO:124.3 RX+LO:146.13	TX+LO: 135 RX+LO: 176	TX+LO:210 RX+LO:110

IV. CONCLUSION

A fully integrated 60 GHz 10 Gb/s QPSK transceiver with digital transmitter and T/R switch in 65nm CMOS is presented. Link bandwidth of the TRX is expanded to ~10GHz by using the magnetically coupled resonator based matching network. The QPSK modulation is directly realized in the I/Q digital power amplifier. A 20GHz integer-N PLL and a QILFT are integrated on-chip to generate 60 GHz quadrature LO signals. The QILFT utilizes in-phase coupling technique to improve the LO I/Q matching and phase noise performance. The receiver achieves 7.1 dB noise figure and 25-47 dB dynamic gain range. The LO phase noise measured at the transmitter output is -93 dBc/Hz at 1-MHz offset from 60 GHz carrier. The EVM of the transmitter is -23.9 dB for 10 Gb/s QPSK signals at 7 dBm output power. The EVM of the Over-the-Air (OTA) modulation-demodulation system is -16.3 dB for 10 Gb/s QPSK signals.

REFERENCES

[1] R. C. Daniels, *et al.*, "60 GHz Wireless: Up Close and Personal," *IEEE Microwave Magazine*, vol. 11, no. 7, pp. 44-50, Dec. 2010.

[2] L. Kuang *et al.*, "A Fully Integrated 60-GHz 5-Gb/s QPSK Transceiver With T/R Switch in 65-nm CMOS," *IEEE Transactions on Microwave Theory and Techniques*, vol. 62, no. 12, pp. 3131-3145, Dec. 2014.

[3] X. Liu, *et al.*, "A 13 pJ/bit 900 MHz QPSK/16-QAM Band Shaped Transmitter Based on Injection Locking and Digital PA for Biomedical Applications," *IEEE Journal of Solid-State Circuits*, vol. 49, no. 11, pp. 2408-2421, Nov. 2014.

[4] M. Vigilante and P. Reynaert, "On the Design of Wideband Transformer-Based Fourth Order Matching Networks for ${E}$ -Band Receivers in 28-nm CMOS," in *IEEE Journal of Solid-State Circuits*, vol. 52, no. 8, pp. 2071-2082, Aug. 2017.

[5] H. Jia *et al.*, "A 77 GHz Frequency Doubling Two-Path Phased-Array FMCW Transceiver for Automotive Radar," *IEEE Journal of Solid-State Circuits*, vol. 51, no. 10, pp. 2299-2311, Oct. 2016.

[6] X. Yi, *et al.*, "A 57.9-to-68.3 GHz 24.6 mW Frequency Synthesizer With In-Phase Injection-Coupled QVCO in 65 nm CMOS Technology," *IEEE Journal of Solid-State Circuits*, vol. 49, no. 2, pp. 347-359, Feb. 2014.

[7] K. Dasgupta *et al.*, "A 60-GHz Transceiver and Baseband With Polarization MIMO in 28-nm CMOS," *IEEE Journal of Solid-State Circuits*, vol. 53, no. 12, pp. 3613-3627, Dec. 2018.

A 60 GHz Polarization-Duplex TX/RX Front-End with Dual-Pol Antenna-IC Co-Integration in SiGe BiCMOS

Yao Liu, Arun Natarajan

School of Electrical Engineering and Computer Science, Oregon State University, Corvallis, OR, USA

liuyao@oregonstate.edu, nataraja@eecs.oregonstate.edu

Abstract — In this work, a 60 GHz simultaneous transmit and receive (STAR) TRX front-end with co-integrated antennas to achieve efficient polarization-duplex mm-wave front-end is presented. The proposed antenna approach provides broadside radiation through the substrate and is compatible with low-res silicon substrates. On-chip slot structures are driven by PAs with antenna power combining for increased output power and efficiency. The orthogonal-polarization feeds provide >40 dB simulated isolation between TX and RX around 60 GHz. Subsequent TX self-interference cancellation (SIC) at the LNA output is achieved with part of the TX signal coupled to a cancellation path, includes a reflection-type attenuator (RTA) and reflection-type phase shifter (RTPS) that provides >20 dB gain variation and full 360° variable phase shift. Overall, total average SIC >40 dB is achieved for 1.07 GHz RF bandwidth at 60 GHz in the presence of a reflector.

I. INTRODUCTION

Dual-polarization (dual-pol) transmitter/receiver (TX/RX) capability in mm-wave transceivers (TRX) promises increased data rates for given antenna area as well as the ability to perform dual-pol imaging in radar transceivers. Integrated dual-pol TRX have been demonstrated in SiGe [1], [2] and CMOS [3]–[5] at 28 GHz, 60 GHz and 94 GHz. Dual-pol TX/RX driving dual-pol antennas that provide orthogonality have been implemented using off-chip antennas [1]. The ability to isolate signals occupying different polarizations has also been used to implement polarization-duplex links with TX and RX signals simultaneously operating on orthogonal polarization [4], [5]. Such an approach is attractive for low-latency links and relays and has been demonstrated using off-chip antennas [4] and on-chip antennas [5]. An on-chip antenna implementation is desirable to eliminate the challenging mm-wave antenna IC interface and the approach in [5] leverages high substrate resistance (unavailable on several process technologies) to achieve high antenna efficiency. The end-fire radiation pattern of the antenna approach also presents a challenge from a link packaging perspective for arrays.

A wafer-scale dual-pol antenna-cointegration approach has been demonstrated in [6], [7] where on-chip feed line is aperture coupled through on-chip slot to a patch antenna through the substrate. In this approach, detailed in Section II, the feed co-integration eliminates mm-wave IO to/from the IC simplifying packaging for mm-wave arrays while achieving overall efficiency comparable to efficiency including package/routing in off-chip antenna arrays. In the following, a 60 GHz integrated polarization-duplex front-end

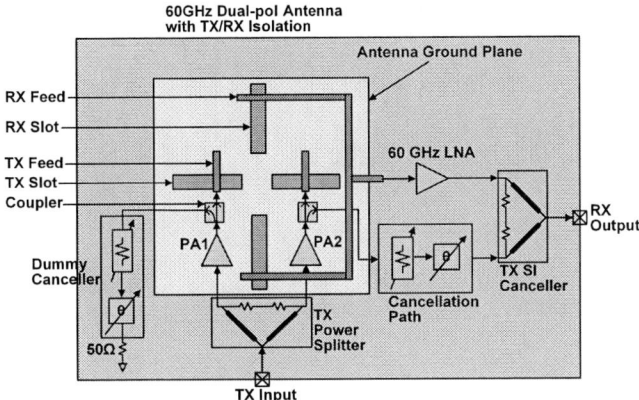

Fig. 1. Proposed 60 GHz STAR TRX Architecture: Dual-polarization antenna with high TX/RX isolation and passive canceler are used to provide two stage TX self-interference cancellation.

is demonstrated (Fig. 1) with an antenna-cointegrated approach that is (a) compatible with low-resistivity substrates and (b) provides broadside radiation for both polarizations. The dual-pol antenna feed isolation is combined with TX self-interference cancellation through a variable gain/phase shift path. The approach achieves 27 dB to 38 dB isolation between antenna feeds alone across 56 GHz to 68 GHz. The passive cancellation path provides additional SIC, with total average SIC >40 dB for ~1 GHz RF bandwidth at 60 GHz.

II. ANTENNA DESIGN

Fig. 2(a) shows the proposed antenna structure with cross-pol feed and dual-pol slots in the ground plane of the IC. The antenna feeds are implemented in top metal layer while the ground plane (with slots) is implemented in lower-metal layers M1 through M3. The dimensions of the antenna and the ground plane are shown in Fig. 2(a). The antenna structure is based on an approach where the 10 Ω-cm silicon substrate is thinned to 75 μm. A 175 μm LCP layer with the patch antenna is bonded to the substrate using an non-conductive adhesive layer. While the LCP substrate (thickness = 175 μm, dielectric constant = 3.1, loss tangent = 0.003) is relatively low-loss at mm-wave [8], the lossy silicon substrate (resistivity of ~ 10 Ω-cm) leads to coupling losses. Thinning the silicon to 75 μm can reduce this efficiency degradation. Simulated antenna efficiency including the substrate and LCP loss is 49% for the RX and 64% for the TX (Fig. 2(b)), with the difference due to the longer length of the RX feed network. Integrating the antenna feed with the circuits provides design

978-1-7281-1702-7/19 $31.00 © 2019 IEEE

2019 IEEE Radio Frequency Integrated Circuits Symposium

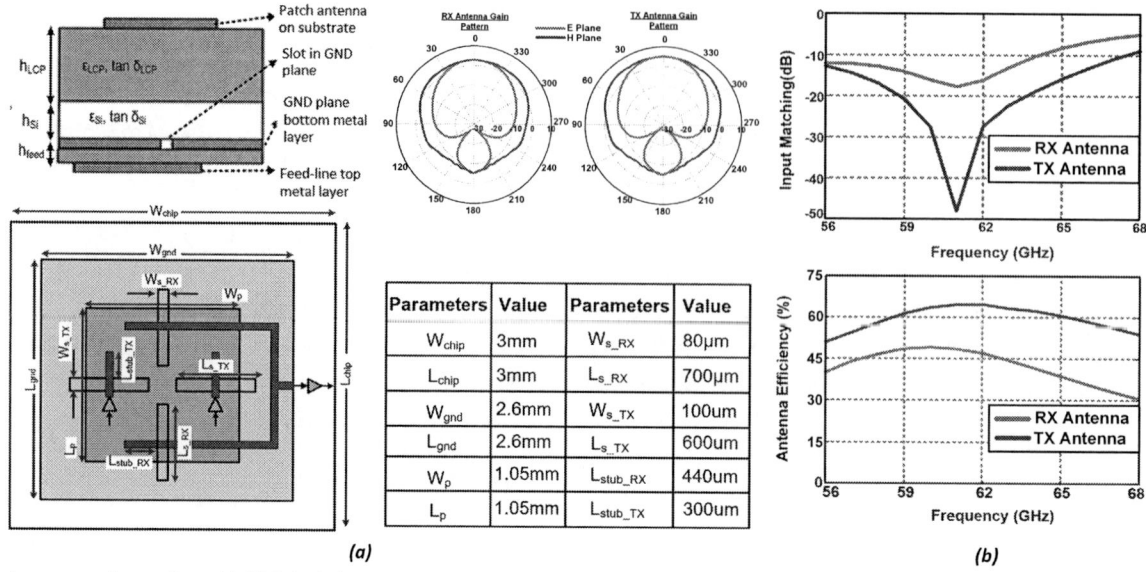

Fig. 2. (a) Antenna co-integration with high isolation dual-pol on-chip feeds and slot that are aperture-coupled to antenna on LCP through the backside of the die. (b) Simulated antenna input matching and efficiency including losses in the on-chip feed networks.

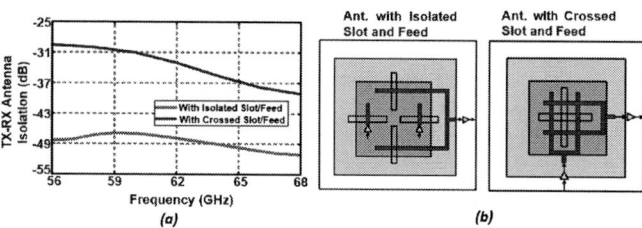

Fig. 3. (a) Simulated TX-RX antenna isolation for antenna with isolated slot/feed and crossed slot/feed. (b) Top view of two different antenna design.

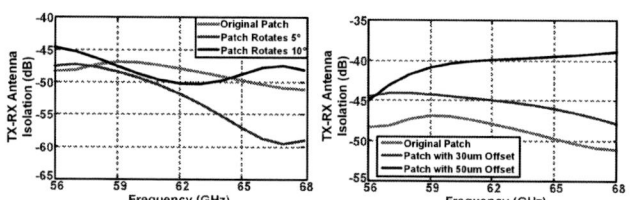

Fig. 4. Simulated TX-RX antenna isolation for nominal design and with offsets/rotation introduced between patch antenna and on-chip slot.

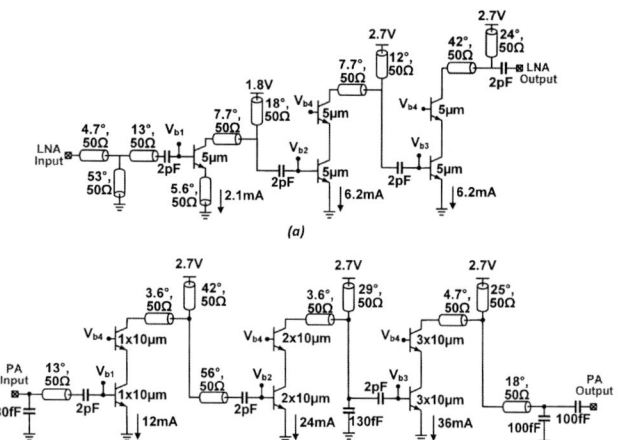

Fig. 5. (a) 60 GHz 3-stage LNA with 14 dB gain. (b) 60 GHz 3 stage class AB PA providing ~8 dBm output power.

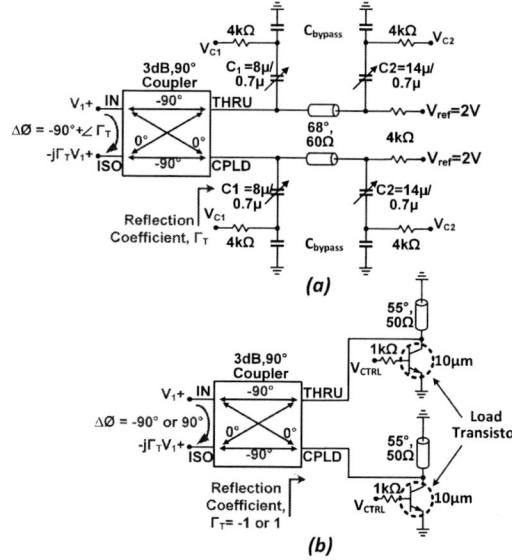

Fig. 6. (a) 60 GHz RTPS: Coupler with reflective load provides >180° continuous phase shift with low gain variation, (b) 60 GHz RTA: Coupler terminated by tunable resistor for 0/180° discrete phase shift.

freedom for impedance matching to the LNA and PA. For example, the TX antenna feed layers are matched to 29-j*28 Ω across 56 GHz to 67 GHz (Fig. 2(b)). Integrated approach also enables design optimization with precise metallization. Fig. 3 compares the isolation achieved using crossed slot approach in [7] with an approach where TX are integrated close to the slot feeds, demonstrating higher isolation with the proposed approach. The simulated TX-RX isolation shows >40 dB isolation between feeds across the 60 GHz (Fig. 4).

Sensitivity to packaging errors is considered in Fig. 4 where isolation is simulated when that the center of the patch on the LCP is offset with the center of the slots on the IC, and with the patch rotated with respect to the slot. While there is

978-1-7281-1702-7/19 $31.00 © 2019 IEEE 128

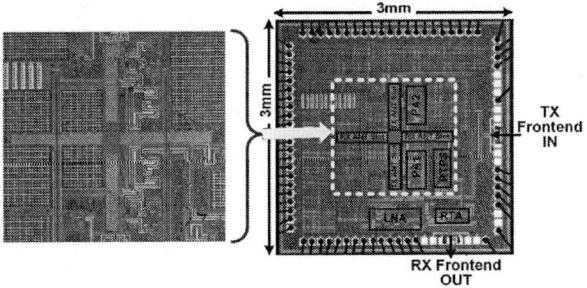

Fig. 7. 60 GHz STAR TRX die photo showing ground plane extensively reused by active circuits implemented in TowerJazz SiGe process technology.

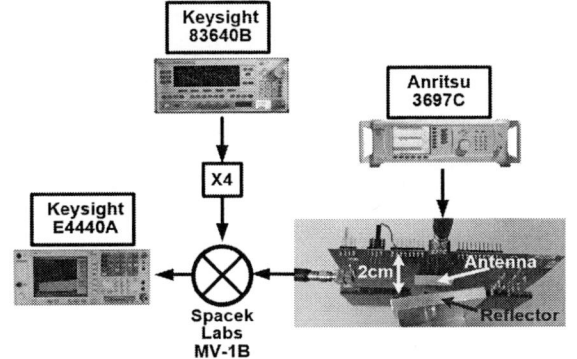

Fig. 8. Wireless measurement setup with reflector following antenna co-integration as described in Fig. 2.

some degradation, 1-mil offsets and 5° offsets still provide acceptable isolation. A critical challenge is area-efficiency of the antenna structure due to the ground plane. In the aperture-coupled structure, the feed layer is separated from the antenna layer by the ground plane. This allows circuits to be designed using t-lines that are implemented with respect to the ground plane leading to a compact implementation.

III. 60 GHz TRX Frontend with SIC

The 60GHz STAR TRX architecture based on the antenna co-integration in Fig. 2(a) is shown in Fig. 1. TX SIC following initial isolation provided by the dual-pol approach is feasible with part of the TX signal provided to the RX [4], [5]. The canceller includes variable gain and phase shift blocks to generate the required cancellation signal with changing leakage path from TX to RX with antenna reflections. The TX input drives two PA through a power splitter, which in turn drives two on-chip feed/slot structures. Similarly, feed/slot structures are combined and fed to the on-chip LNA. Fig. 5(a) shows the 3-stage LNA with 14 dB gain. The LNA achieves 7 dB NF based on stand-alone measurements. The first stage(common emitter) consumes 2.1 mA from 1.8 V and 2nd and 3rd stage(cascode) consume 12.4 mA from 2.7 V. Fig. 5(b) shows the TX design. A 3-stage class AB PA is used to generate ∼8 dBm output power at 60 GHz with a cascode structure in each stage. The PA consumes 72 mA from 2.7 V and provides 20 dB gain.

The TX signal is coupled to the variable gain/phase cancellation stage using a ∼-10 dB custom capacitive coupler in the top two metal layers. The coupler drives a reflection-type attenuator (RTA) whose phase shift, $\Delta\Phi$, depends upon the termination impedance, Z_T.

$$\Delta\Phi = -90° + \angle\Gamma \text{ where } \Gamma = \frac{Z_T - Z_0}{Z_T + Z_0}. \quad (1)$$

The termination impedance is varied using load transistor in Fig. 6(b). The RTA is used to provide 0 or 180° discrete phase shift based on $Z_T > Z_0$ or $Z_T < Z_0$. A stand alone attenuator with simple load transistor has also been used with simulated attenuation from -3 dB to -17 dB at 60 GHz. The output of the attenuator drives a passive reflection-type phase shifter (RTPS) (Fig. 6 (a)). The passive RTPS uses a two-capacitor load structure designed using the approach in [9] to enable low-loss variation while achieving the targeted 180° phase shift range. In simulation, the RTPS provides > 180° variable shift with 7 dB ± 1 dB insertion loss.

IV. Measured Performance

The IC is implemented in the TowerJazz 0.18 μm SiGe process with 2.8 μm thick metal layers. As discussed in Sec. II, the ground plane reuse with t-lines leads to the compact layout shown in Fig. 7. The IC is packaged with a patch antenna integrated on an LCP substrate.

The TX performance is measured wirelessly and the EIRP (de-embedding 60GHz path loss) is shown in Fig. 9. The small-signal gain (from 56 GHz to 68 GHz), the measured EIRP (including both PAs) across input power and the saturated EIRP measurement across frequency is shown in Fig. 9(a,b,c) respectively. The LNA gain based on probe measurements is shown in Fig. 9(d). A similar LNA has been measured to have 7 dB NF while consuming 12.4 mA from 2.7 V and 2.1 mA from 1.8 V. The performance of the second cancellation path in the frontend through the variable attenuator and phase shifter is summarized in Fig. 9(e)(f). As shown in Fig. 9(e), the variable attenuator provides >20 dB of attenuation range while providing full 360° phase shift between the RTPS and the discrete 0/180° stage in the attenuator (Fig. 9(f)). Given the degrees of freedom in the phase shifter and the the attenuator, Fig. 10(a) plots canceller insertion loss and phase shift for a range of > 20 dB and full 360° demonstrating that the attenuator path can be configured to match leakage signal amplitude and phase.

SIC measurements were carried out wirelessly while including a blocker to demonstrate the impact of reflections using the setup shown in Fig. 8. Measurements were carried out at 3 dBm EIRP based on Fig. 9(b). Measured isolation between the feeds (antenna alone) is shown in Fig. 10(b) demonstrating 34 dB SIC at 60 GHz. The passive canceller provides additional SIC, leading to total average SIC of >40 dB for 1.07 GHz BW at 60 GHz. The IC is compared to state-of-the-art in Fig. 11, demonstrating >40 dB SIC with antenna co-integration and broadside radiation.

Acknowledgment

We thank the Center for Design of Analog-Digital Integrated Circuits (CDADIC) for project support. We also

2019 IEEE Radio Frequency Integrated Circuits Symposium

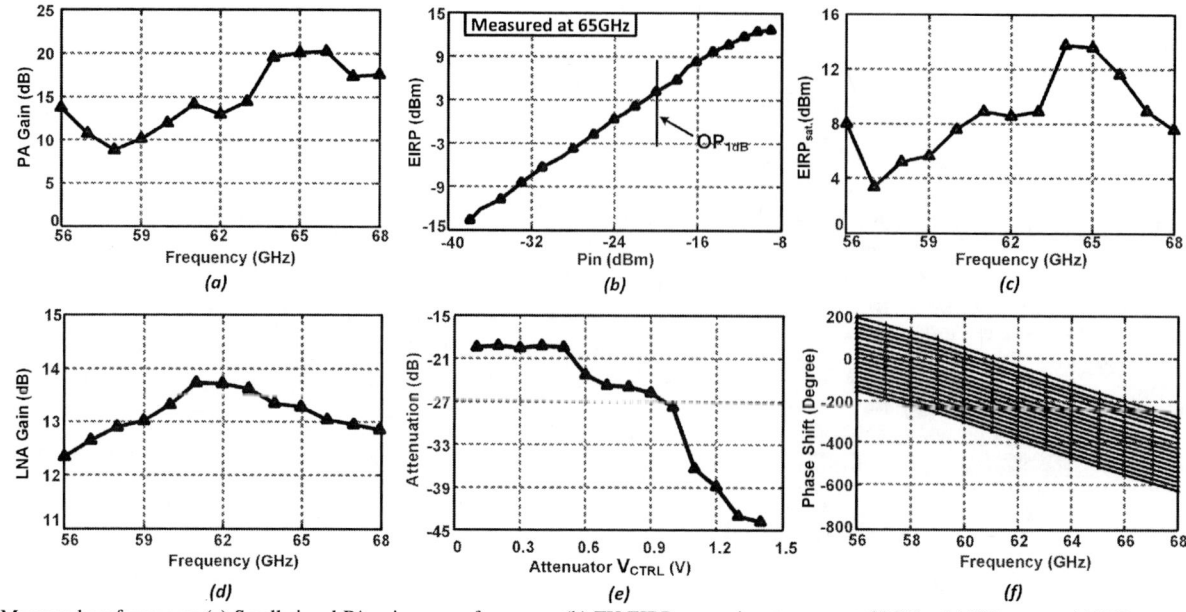

Fig. 9. Measured performance: (a) Small signal PA gain across frequency. (b) TX EIRP versus input power at 65 GHz. (c) TX saturated EIRP across frequency. (d) LNA gain across frequency. (e) Attenuator loss versus control voltage. (f) Canceller phase shift demonstrating 360° phase shift across frequency.

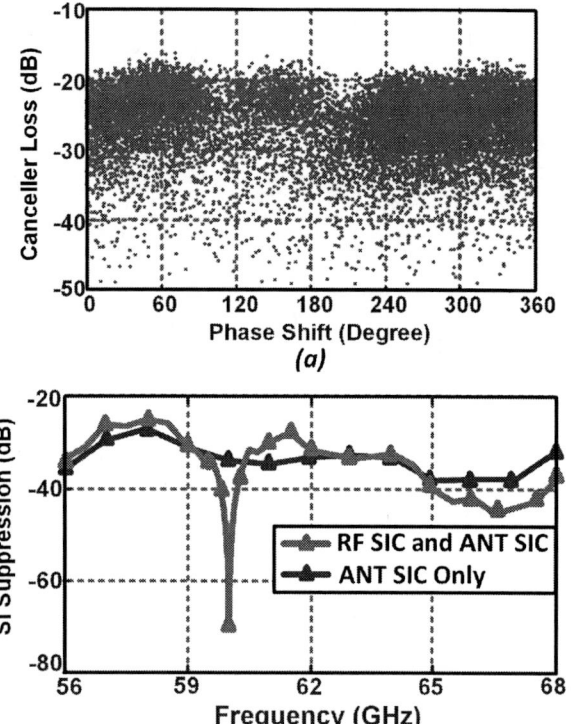

Fig. 10. (a) Measured canceller loss versus phase shift demonstrating >20 dB attenuation and 360° phase shift range. (b) SIC measurements demonstrating 34 dB antenna SIC and ~70 dB system SIC.

thank Dr. Arjun Karroy and TowerJazz for assistance with IC fabrication.

	This Work	[5]	[4]
Architecture	One Ant. Footprint	One Ant. Footprint	Dual TX/RX Ant. on PCB
Technology	0.18μm SiGe BiCMOS with 10Ω-cm sub. res.	45nm CMOS SOI with high res. sub.	45nm CMOS SOI
Ant. Type	Aperture-coupled Patch Ant.	Slot Loop Ant. on Chip	Slot Loop Ant. on PCB
Chip Area	3mm x 3mm	2.7mm x 2.7mm	1.3mm x 3.4mm (TRX Chip) 2.4mm x 3.4mm*(PCB Ant.)
Frequency Range	56-68GHz	60-75GHz	57-66GHz
Ant. SIC	27-38dB	>35dB	N/A
Avg. Ant. + RF SIC for ~1GHz BW	40dB	75dB*	75.5dB*
Canceller DC Power	0	0	44mW
Ant. SIC Loss	0	0	1.1dB(TX)/0.52dB(RX)
Ant. Radiation Pattern	Broadside	Endfire	Broadside
Radiation Efficiency	64%(TX)/49%(RX)	91%	N/A

* Graphically estimated

Fig. 11. Comparision to state-of-the-art mm-wave pol-duplex TRX

[2] B. Sadhu *et al.*, "A 28-GHz 32-Element Phased-Array Transceiver IC ... Resolution for 5G Communication," *IEEE ISSCC*, Feb. 2017.

[3] K. Dasgupta *et al.*, "A 60-GHz Transceiver and Baseband With Polarization MIMO in 28-nm CMOS," *IEEE JSSC*, Dec. 2018.

[4] T. Dinc *et al.*, "A 60 GHz CMOS Full-Duplex TRX and Link with Polarization-Based ... and RF Cancellation," *IEEE JSSC*, May. 2016.

[5] T. Chi *et al.*, "A 64 GHz Full-Duplex Transceiver Front-End with ... in One Antenna Footprint," *IEEE ISSCC*, Feb. 2018.

[6] Y. Liu *et al.*, "Millimeter-Wave IC-Antenna Cointegration for Integrated Transmitters and Receivers," *IEEE AWPL*, March 2016.

[7] ——, "60 GHz Concurrent Dual-Polarization RX Front-End in SiGe with Antenna-IC Co-Integration," *IEEE BCTM*, Oct. 2017.

[8] D. C. Thompson *et al.*, "Characterization of Liquid Crystal Polymer (LCP) .... on LCP From 30 to 110 GHz," *IEEE Trans. MTT*, Apr 2004.

[9] R. Garg *et al.*, "A 28-GHz Low-Power Phased-Array Receiver Front-End with 360° RTPS Phase-Shift Range," *IEEE Trans. MTT*, Nov. 2017.

REFERENCES

[1] A. Natarajan *et al.*, "W-Band Dual-Polarization Phased-Array Transceiver Front-End in SiGe BiCMOS," *IEEE Trans. MTT*, June. 2015.

A 180-GHz Super-Regenerative Oscillator with up to 58 dB Gain for Efficient Phase Recovery

Hatem Ghaleb[#], Christian Carlowitz[$], David Fritsche[#], Corrado Carta[#], and Frank Ellinger[#]

[#]Chair for Circuit Design and Network Theory, Technische Universität Dresden, Germany

[$]Institute of Microwaves and Photonics, University of Erlangen-Nuremberg, Germany

Abstract—This paper reports on the design of a 180-GHz super-regenerative oscillator in a 130nm SiGe BiCMOS technology. The oscillator has a tuning range of 6.5%, an output power of 0.5 dBm, and occupies an area of 0.72 mm². When operated with a periodic quench signal, the circuit requires a minimum input power of -58 dBm for a phase coherent output, and can be switched at a rate up to 10 GHz. The circuit has a dc power consumption of 8.8 mW, and a maximum regenerative gain of 58 dB. QPSK and 8-PSK modulation up to 3 Gbit/s have been demonstrated at an energy efficiency of 2.9 pJ/bit. To the best knowledge of the authors, this work is the fastest reported phase-sampling super-regenerative oscillator to date.

Keywords— 200 GHz band, super-regenerative receiver.

I. INTRODUCTION

The rising demand for high data rates in mobile wireless communication raises interest in the millimeter-wave frequency band (30-300 GHz) due to the availability of large and unallocated contiguous bandwidth. This enables high-speed communication applications at low or moderate spectral efficiency. However, it comes at the cost of high power consumption, as the transistor gain decreases and multiple cascaded amplifier stages are needed.

Thus, a transceiver architecture based on the use of a super-regenerative oscillator (SRO) with phase and amplitude sampling capability has been proposed [1] and demonstrated at 5.8 GHz [2]. The concept relies on the use of positive feedback in the oscillator to generate a very large regenerative gain [3]. The information signal is fed to the oscillator core and the phase is sampled at the turn-on instant, when the oscillator is most sensitive. The oscillation then grows exponentially to its steady-state level with the phase information preserved. To receive the following symbols, the oscillator has to be periodically quenched, thus each oscillation pulse represents one symbol. Additionally, by quenching the oscillator before a steady state is reached, the amplitude of the input signal can also be regenerated, leading to the use of the SRO with amplitude modulation, phase modulation, or both.

The advantages of this super-regenerative approach include low complexity and high power and area efficiency, since one block can replace multiple gain stages in the receiver. Through pulsed operation by repeated quenching, the power consumption is further reduced. On the other hand, the high sensitivity of the super-regenerative circuit leads to susceptibility to interferers in adjacent channels.

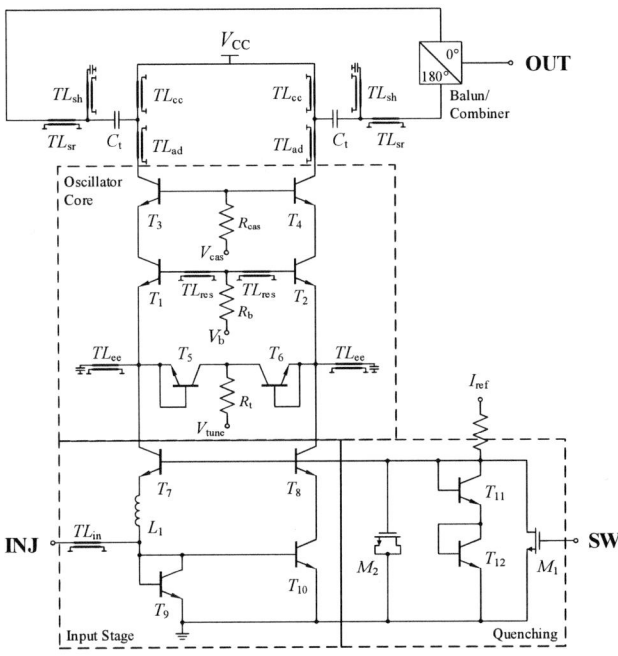

Fig. 1 Schematic of the 180-GHz SRO

This limitation is, however, relieved by the high bandwidth in mm-wave bands, as well as the high path loss, which allows for frequency reuse after a short distance. In this paper, the design and measurements of a 180-GHz SRO IC are presented.

II. CIRCUIT DESIGN AND LAYOUT

Fig. 1 shows a schematic diagram of the SRO circuit, including all input and output signals. The oscillator core is realized with the transistors T_{1-4}, with the diode-connected transistor pair $T_{5,6}$ used as varactors for oscillation frequency tuning through the tuning voltage V_{tune}.

The core is based on the differential common-collector Colpitts topology, which has demonstrated superior performance at mm-wave frequencies [4]. A differential common-base output buffer consisting of transistors $T_{3,4}$ boosts the output power delivered to the load, and ensures sufficient reverse isolation to prevent the locking of the oscillator to reflected pulses, in case of output impedance mismatch. The input transconductance stage consisting of transistors T_{7-10} acts as an active balun, injecting differential input currents into the oscillator.

978-1-7281-1702-7/19 $31.00 © 2019 IEEE

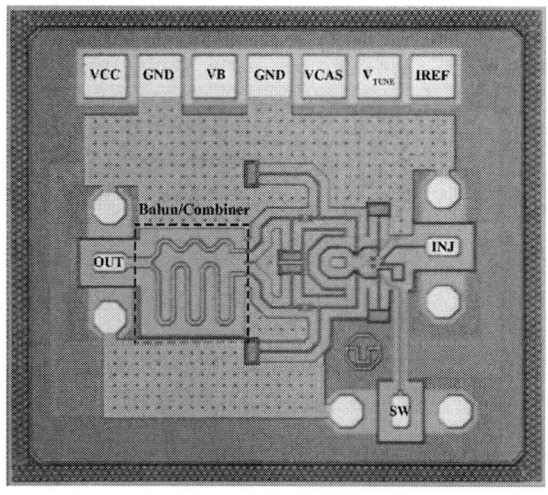

Fig. 2 Micrograph of the SRO chip with pad labels (area = 900 × 800 μm²)

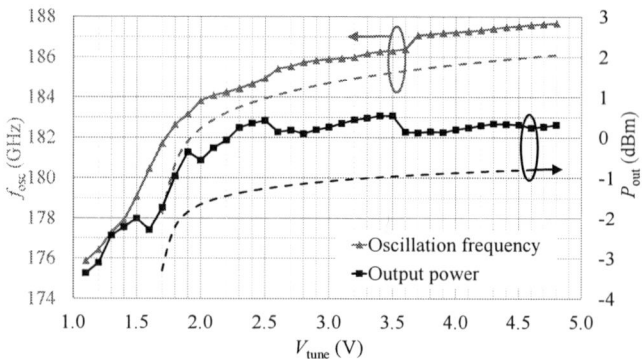

Fig. 3 f_{osc} and P_{out} vs V_{tune} from measurements (solid) and simulations (dashed)

Additionally, it helps to provide a broadband match due to the low resistive input impedance, which is transformed to 50 Ω by the short transmission line segment TL_{in} as well as the pad capacitance. A small inductor L_1 adjusts the phase balance between the two balun branches.

The diode-connected transistors $T_{11,12}$ form a current mirror with $T_{7,9,10}$, which allows the current biasing of the oscillator through I_{ref}. T_8 maintains the voltage common-mode balance between the two differential branches. By adding the NMOS transistor M_1, a signal SW can be applied to periodically quench the oscillator as required, by sinking I_{ref} and pulling down the bias voltage node.

By adding capacitance to that node, the quenching slope of the oscillator at the turn-on instant can be slowed. This has been found to improve the SRO sensitivity in terms of the minimum input power required for low-BER phase recovery $P_{inj,min}$ [5]. However, a large capacitance would limit the quench rate, thus reducing the achievable symbol rate. Therefore, an optimum value for the capacitance has been investigated and implemented with the NMOS capacitor M_2.

The supply voltage V_{CC} is provided through the feed lines TL_{cc}. The transmission lines TL_{ad}, TL_{sh} and TL_{sr}, as well as the capacitors C_t form an impedance transformation network at the output. The optimum load impedance was investigated by load-pull simulations, which used fast start-up rather than the highest output power as the optimization goal. A high inductive part was found to be desirable, whereas a high resistive part led to a higher steady-state swing, but also a higher start-up time constant. Therefore, an optimum value was chosen for the load resistance as a tradeoff between the start-up time and the steady-state output power.

All transmission lines were implemented as grounded coplanar waveguides (GCPW), and all bipolar transistors were sized at the emitter area $A_E = 2 \times 0.9 \times 0.07$ μm². The differential outputs are combined in a passive microstrip balun, which consists of one λ/4 and one 3λ/4 transmission line segments, both with a characteristic impedance of 71 Ω. The balun also performs the function of impedance transformation of single-ended 50 Ω to differential 100 Ω.

The dc routing is done using multi-layer high-capacitance zero-Ohm lines [6]. The circuit is fabricated in a 130nm SiGe BiCMOS HBT technology with f_T/f_{max} up to 300/450 GHz. The total chip area including the pads is 0.72 mm². A labeled micrograph of the SRO IC is shown in Fig. 2.

III. MEASUREMENT RESULTS

The SRO circuit was measured on-wafer with the dc and the SW pads wire-bonded on a custom PCB. The dc bias was set at V_{CC} = 4.4 V, V_{cas} = 4.1 V and V_b = 3.2 V. The circuit draws 6 mA during free-running operation. A Virginia Diodes PM4 power meter was used for signal power measurements and the calibration of measurement setup losses. For the spectrum measurements, a Radiometer Physics HM140-220 harmonic mixer was used in combination with a Rohde&Schwarz FSW67 spectrum analyzer. The CW input signal was generated using a Rohde&Schwarz ZVA-Z220 network analyzer converter.

Fig. 3 shows the measured oscillation frequency f_{osc} and the output power P_{out} as functions of V_{tune}, with both INJ and SW inputs terminated to 50 Ω. The frequency tuning range extends between 175.9–187.7 GHz, which amounts to 6.5%, or almost 12 GHz. The output power at 180 GHz is -2 dBm, whereas the maximum output power is 0.5 dBm.

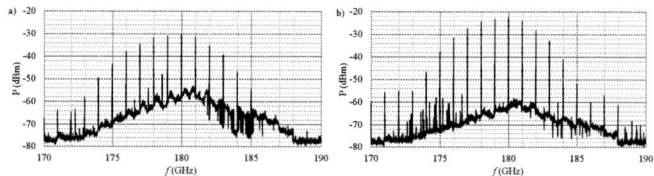

Fig. 4 Measured SRO spectrum for f_{inj} = 180 GHz, f_{sw} = 1 GHz, and (a) P_{inj} = -58 dBm, (b) P_{inj} = -48 dBm

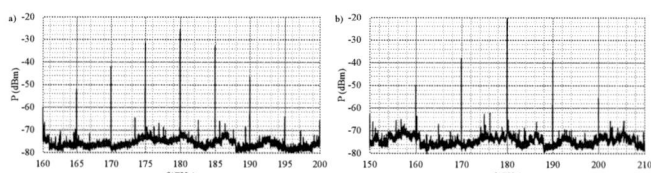

Fig. 5 Measured SRO spectrum for f_{inj} = 180 GHz, P_{inj} = -26 dBm, and (a) f_{sw} = 5 GHz, (b) f_{sw} = 10 GHz

978-1-7281-1702-7/19 $31.00 © 2019 IEEE

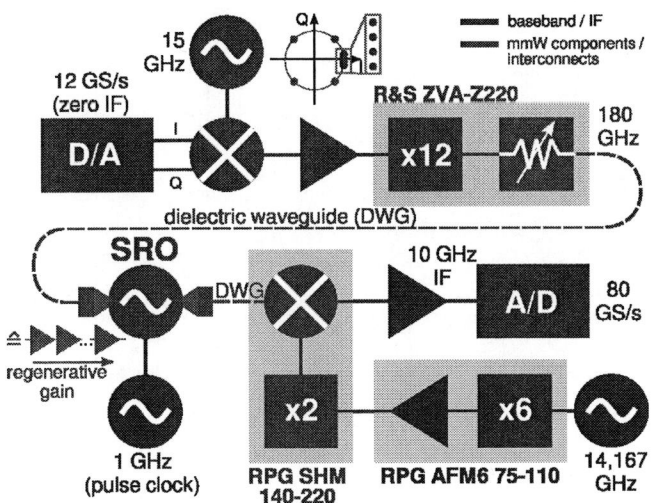

Fig. 6 Measurement setup for demonstrating SRO-based regeneration of n-PSK modulated signals

To test the regenerative sampling capability of RF and mm-wave SROs, for which direct time-domain measurements are not practical, prior work used a CW input signal of a known frequency f_{inj} and examined the spectrum as in [7]. In case of phase coherence, the spectrum consists of a sequence of Dirac delta spectral peaks at $f_{inj} - n f_{sw}$, where n is an integer, superimposed by a sinc-shaped envelope centered around f_{osc}.

Fig. 4 shows the SRO spectrum for an input CW signal at 180 GHz and a quench rate of 1 GHz. The duty cycle is set to approximately 30%, with a dc power consumption of 8.8 mW. In Fig. 4 (a), phase coherence is observed for P_{inj} = -58 dBm with clear spectral lines. In Fig. 4 (b), P_{inj} is increased by 10 dB, which leads to less sampling errors and improved phase coherence, which is observed as a slight reduction of the noise floor level. The amplitude also rises due to faster SRO start-up. In Fig. 5, quench rates of (a) 5 GHz and (b) 10 GHz are shown, with input power P_{inj} = -26 dBm.

IV. MODULATED SIGNAL REGENERATION

In order to demonstrate SRO-based regeneration of a modulated signal with high data rate (>1 Gbit/s), the setup depicted in Fig. 6 has been employed.

An n-PSK modulated signal is generated in complex baseband using a digital to analog converter and applied to a 15 GHz CW carrier using a quadrature mixer. Subsequently, the constant envelope signal is up-converted to 180 GHz with a network analyzer converter, which contains a frequency multiplier with a factor of 12. Thus, the baseband signal (blue) needs to be compressed in phase to be correctly mapped after multiplication (red; see top I/Q diagram in Fig. 6 for an example with QPSK). A subsequent attenuator is used to configure different input magnitudes for the SRO. The SRO itself is contacted with wafer probes and connected to the setup by dielectric waveguide cables (1 m, insertion loss ~10 dB each). It samples the input communication signal at the center of each symbol and amplifies it repetitively through positive feedback until the desired gain has been reached and the oscillation is turned off.

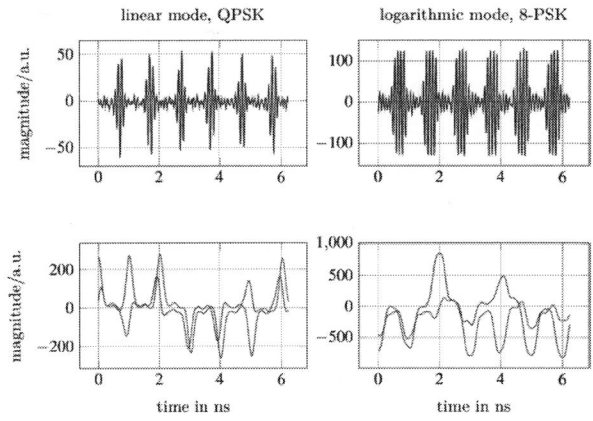

Fig. 7 Measured IF (top) and complex baseband (bottom) signals in linear and logarithmic mode, showing the regenerated communication signal

If this occurs before a steady-state amplitude is reached, the SRO is said to be operating in linear mode, due to the linear dependence between input and output amplitudes. Otherwise, it is said to be in logarithmic mode.

The SRO output signal with one pulse per symbol is down-converted to an intermediate frequency (IF) of 10 GHz using a Radiometer Physics SHM 140-220 subharmonic mixer, driven by an RF signal generator with x6 frequency multiplier.

For further processing, the intermediate frequency signal is amplified and analog-to-digital converted. Afterwards, a numerical computation software is used to remove the IF by mixing to complex baseband. Filtering is done (complex conjugate convolution) with an approximately matched SRO pulse shape filter, as well as detection of sample position and phase offset through correlation with a 1k symbol preamble. Subsequently, 10k symbols are sampled and analyzed regarding BER and EVM. Communication signals at 180 GHz with 1 Gbaud and QPSK (2 Gbit/s) and 8-PSK (3 Gbit/s) have been regenerated with a regenerative gain of up to ~20 dB. This corresponds to a highly efficient 2.9 pJ/bit for 8-PSK.

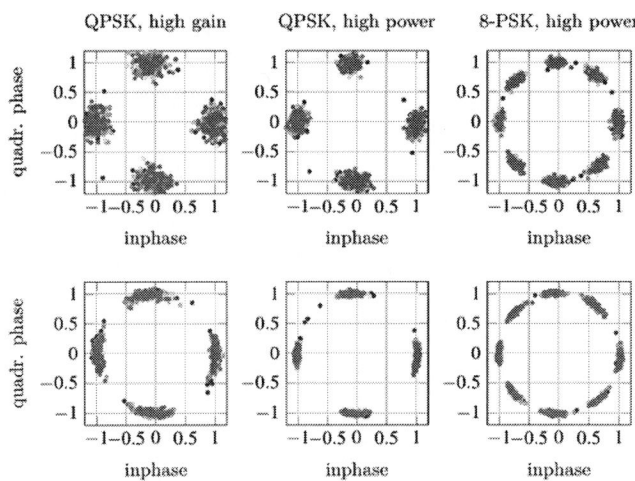

Fig. 8 Constellation diagram from measurements without SRO (top) and with SRO (bottom), for (A) QPSK with high gain (20 dB), (B) QPSK with medium gain (15 dB) and moderate compression, (C) 8-PSK with medium gain (15 dB); the colormap indicates the order of reception.

Table 1. Comparison of RF and mm-wave phase-sampling SRO ICs

Ref.	Technology	f_{max} (GHz)	f_{osc} (GHz)	Tuning range (%)	$P_{inj,min}$ (dBm)	P_{out} (dBm)	P_{DC} (mW)	f_{sw} (MHz)	f_{sw}/f_{osc} (%)	Area (mm²)
[7]	250nm SiGe	180	34.5	8	-61.5	5	122	75	0.2	0.83
[8]	180nm SOI CMOS	N.A.	2.2	4.5	-70	N.A.	1.7	0.8	0.04	1.6
[9]	250nm SiGe:C	90	6.8	N.A.	-65	-4.7	132	100	1.5	N.A.
[10]	130nm SiGe	450	25.3	5.5	-110	7.8	38	4.5	0.02	0.63
[11]	130nm SiGe	450	60	7.5	-31	7.3	107	3100	5.2	0.75
[12]	130nm SiGe	450	160	N.A.	-27	-8.4	6.6	5000	3.1	0.64
This work	130nm SiGe	450	180	6.5	-58	0.5	8.8	10000	5.6	0.72

Fig. 7 shows the measured IF signal of the SRO output before and after down-conversion to complex baseband. Both linear amplification as well as high output power with compression can be achieved. Fig. 8 depicts the constellation diagrams before and after sampling the symbols with the SRO. They show very good phase linearity, even for high-order phase modulation. The dynamic range is sufficient for up to 20 dB gain with an EVM that is better than -20 dB.

Since the gain can be easily controlled through pulse width over a wide range, gain can be traded off for EVM. This constitutes a notable advantage regarding complexity, system size and power consumption, especially at very high frequencies with very wideband signals since additional baseband VGAs can be omitted.

Table 1 shows a comparison of published SRO ICs in the RF and mm-wave frequency range with demonstrated regenerative phase sampling. Since most previous publications report CW measurement results, only these are included in the table. This work significantly advances the state of the art in several aspects. It has the highest reported oscillation frequency, quench rate, as well as fractional quench rate f_{sw}/f_{osc}. The maximum regenerative gain $P_{out} - P_{inj,min}$ of 58 dB is the best above 100 GHz, at a competitively low dc consumption of 8.8 mW in pulsed operation.

V. CONCLUSION

For energy-efficient high-speed wireless communication, the concept of super-regenerative amplification is proposed as an alternative to less efficient homodyne concepts at mm-wave frequencies. A 180-GHz super-regenerative oscillator is designed and fabricated in a 130nm SiGe BiCMOS HBT technology. The circuit demonstrates a maximum gain of 58 dB, and can be switched at a speed up to 10 GHz. With a power consumption of 8.8 mW in pulsed operation, it is far more efficient than conventional mm-wave amplifiers. The circuit was used to recover QPSK and 8-PSK modulated signals at a data rate up to 3 Gbit/s. This work presents the fastest phase-sampling SRO in terms of oscillation frequency and pulse rate, and opens new possibilities for the use of super-regenerative receivers in high-data-rate communication.

ACKNOWLEDGMENT

This work was supported by the German Research Foundation (DFG) within the frame of the research projects SPARS/MSPARS, DAAB/DAAB-TX, and 3D-LommID, as well as the programs HAEC (subproject A01) and cfAED (Resilience path).

The authors would like to thank Radiometer Physics GmbH for their assistance with the measurements.

REFERENCES

[1] C. Carlowitz and M. Vossiek, "Concept for a novel low-complexity QAM transceiver architecture suitable for operation close to transition frequency," in *2015 IEEE MTT-S International Microwave Symposium*, 2015, pp. 1–4.

[2] C. Carlowitz and M. Vossiek, "Demonstration of an efficient high speed communication link based on regenerative sampling," in *2017 IEEE MTT-S International Microwave Symposium (IMS)*, 2017, pp. 71–74.

[3] J. R. Whitehead, *Super-Regenerative Receivers*. Cambridge: Cambridge University Press, 1950.

[4] D. Fritsche, S. Li, N. Joram, C. Carta, and F. Ellinger, "Design and Characterization of a 190-GHz Voltage-Controlled Oscillator," in *Proceedings of the 46th European Microwave Conference*, 2016, pp. 493–496.

[5] J. L. Bohorquez, A. P. Chandrakasan, and J. L. Dawson, "Frequency-Domain Analysis of Super-Regenerative Amplifiers," *IEEE Trans. Microw. Theory Tech.*, vol. 57, no. 12, pp. 2882–2894, Dec. 2009.

[6] G. Tretter, D. Fritsche, J. D. Leufker, C. Carta, and F. Ellinger, "Zero-Ohm transmission lines for millimetre-wave circuits in 28 nm digital CMOS," *Electron. Lett.*, vol. 51, no. 11, pp. 845–847, May 2015.

[7] A. Strobel, C. Carlowitz, R. Wolf, F. Ellinger, and M. Vossiek, "A Millimeter-Wave Low-Power Active Backscatter Tag for FMCW Radar Systems," *IEEE Trans. Microw. Theory Tech.*, vol. 61, no. 5, pp. 1964–1972, May 2013.

[8] D.-G. Lee and P. P. Mercier, "A 1.65 mW PLL-free PSK receiver employing super-regenerative phase sampling," in *2015 IEEE Biomedical Circuits and Systems Conference (BioCAS)*, 2015, pp. 1–4.

[9] A. Esswein, R. Weigel, T. Ussmueller, C. Carlowitz, and M. Vossiek, "An Improved Switched Injection-Locked Oscillator for Ranging and Communication Systems," in *Proceedings of the 43rd European Microwave Conference*, 2013, pp. 592–595.

[10] M. V Thayyil, S. Li, N. Joram, and F. Ellinger, "A K-Band SiGe Superregenerative Amplifier for FMCW Radar Active Reflector Applications," *IEEE Microw. Wirel. Components Lett.*, vol. 28, no. 7, pp. 603–605, Jul. 2018.

[11] A. Ferchichi, H. Ghaleb, C. Carta, and F. Ellinger, "Analysis and Design of 60-GHz Switched Injection-Locked Oscillator with up to 38 dB Regenerative Gain and 3.1 GHz Switching Rate," in *2018 IEEE International Midwest Symposium on Circuits and Systems (MWSCAS)*, 2018, no. 1, pp. 340–343.

[12] H. Ghaleb, P. V. Testa, S. Schumann, C. Carta, and F. Ellinger, "A 160-GHz Switched Injection-Locked Oscillator for Phase and Amplitude Regenerative Sampling," *IEEE Microw. Wirel. Components Lett.*, vol. 27, no. 9, pp. 821–823, Sep. 2017.

A Broadband Direct Conversion Transmitter/Receiver at D-band Using CMOS 22nm FDSOI

Ali A. Farid, Arda Simsek, Ahmed S. H. Ahmed, Mark J. W. Rodwell

ECE Department, University of California Santa Barbara, CA 93106

afarid@ece.ucsb.edu, ardasimsek@ece.ucsb.edu, a_s_ahmed@ece.ucsb.edu, rodwell@ece.ucsb.edu

Abstract—This paper presents a broadband transmitter and receiver at D-band (from 123 to 146GHz) using 22nm FDSOI technology. The direct conversion receiver is implemented with a wideband fully differential LNA at the front end, using a cross coupled pair with capacitive neutralization, followed by a linear double balanced passive mixer and broadband pseudo-differential transimpedance amplifier. The direct conversion transmitter starts with an active double balanced Gilbert cell, followed by a driver amplifier. A 9:1 frequency multiplier circuit realized by two successive tripler stages provides the on-chip 135GHz Local Oscillator (LO) signal for both the Tx and Rx chains. The receiver conversion gain is 27dB with a 20GHz 3-dB bandwidth, and the P1-dB is -30dBm. The transmitter saturated output power is 2.8 dBm. Tx and Rx chains consume 196mW and 198mW respectively from a supply voltage of 0.8V.

Keywords— Broadband Transceiver, mm-wave integrated circuits, Direct conversion, capacitive neutralization, 22nm FDSOI, transimpedance amplifier, mm-wave transceivers, D-band transceivers.

I. INTRODUCTION

The rapidly emerging wave of wireless data services demands high data rate wireless links. The large available spectrum at mm-wave frequencies enables the implementation of high speed and broadband transceivers. The advance in silicon-based radio frequency integrated circuits (RFIC) enable the implementation of low cost, small form factor and low power transceivers at mm-Wave frequencies using low cost CMOS technology. Several designs have been reported at mm-wave frequency band [1][2].

In this paper, a broadband single channel transmitter and receiver at D-band are presented. These designed to serve within 135GHz MIMO transceiver arrays, hence the baseband (I, Q) transmitter input and receiver output signals will be linear superpositions of data streams which must be subsequently separated by a baseband beamformer.

A broadband LNA/PA is designed using cross coupled pairs with capacitive neutralization, and the inter-stage matching networks are stagger-tuned to realize a broadband design. A wideband pseudo-differential transimpedance amplifier (TIA), with a 20GHz 3-dB-bandwidth, is used as a baseband amplifier in the receiver chain. On-wafer characterization for the transmit and receive channels is presented in terms of the conversion gains, 3-dB bandwidth, 1-dB compression points and transmitter saturated output power.

II. TECHNOLOGY AND TRANSISTOR FOOTPRINT

The single-channel transmitter and receiver are designed using Global Foundries 22nm-FDSOI technology. The reported power gain cut-off frequency (f_{max}) and current gain cut-off frequency (f_t) for this technology are 230GHz and 240GHz, respectively, both referenced to the top metal layer [3]. The stack used provides 10 metal layers. The footprint of the core device used in this design is based on a super-low threshold voltage (V_t) NMOS with 32 fingers. The gate finger pitch is increased 2:1 above minimum to reduce C_{ds} and C_{gs} and to allow the placement of sufficient vias to satisfy electro-migration limits, when operating at 0.3mA/μm at 110^0 C. Both the drain and gate are routed up to the top metal layer. The source is directly connected to ground through the lower 4 metal layers to reduce the source inductance.

III. RECEIVER ARCHITECTURE AND BUILDING BLOCK

A direct conversion receiver (Fig. 1) consists of a 4-stage broadband LNA and a double balanced passive mixer, followed by a pseudo differential wideband transimpedance amplifier, for both in-phase (I) and quadrature phases (Q). The mixer is driven by an on-chip LO multiplier (x9), where the LO input signal is driven from external source with -3dBm input power at ~15GHz. The (I, Q) LO signals are generated by adding a (λ/4) delay line in the LO signal path, introducing a 90^0 phase shift.

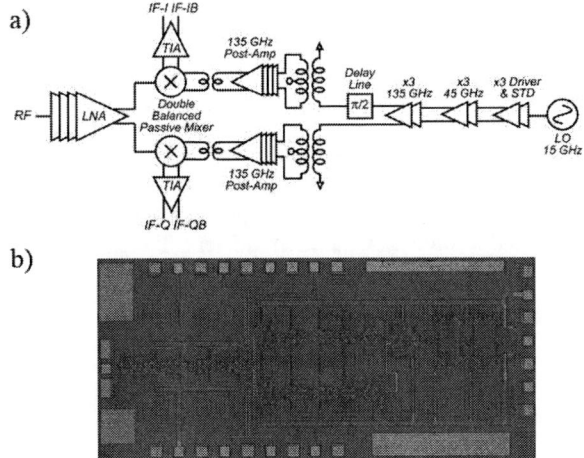

Fig. 1. D-band single-channel direct conversion receiver (a) circuit block diagram and (b) chip micrograph. The die area is 1.9mm x 0.76mm including pads.

A. D-band LNA/PA

A 4-stage fully differential common source LNA (Fig. 2) is designed using a cross coupled pair with capacitive neutralization to boost the maximum available gain. The neutralization uses alternate polarity metal-oxide-metal

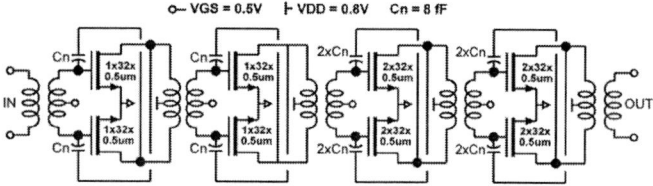

Fig. 2. Circuit diagram of a 4-stage broadband LNA/PA using staggered tuning.

(APMOM) capacitor. A center tapped transformer converts the single-ended input to a differential signal. Transformer center-taps provide DC bias feeds. For broad bandwidth, tuning of the inter-stage matching networks is staggered in frequency. The transformers use the top two wiring layers and were simulated using Keysight Momentum.

We designed this amplifier to be used as an LNA in the Rx chain and as a PA in the Tx chain, so the first stage device sizing was chosen to fulfil minimum noise figure. However, we increased the last two stages device sizing to increase the saturated output power. This LNA/PA draws 55 mA from a 0.8V supply, with a simulated gain of 16-dB and 40GHz 3-dB bandwidth. The simulated noise figure (NF) is 8.5dB.

B. Down Conversion Mixer

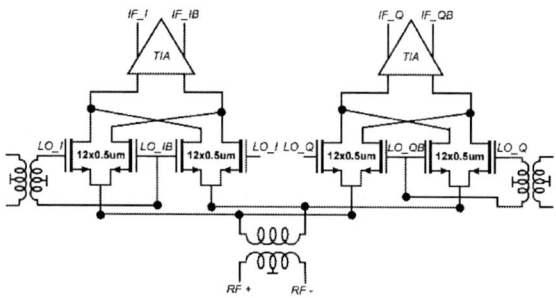

Fig. 3. Circuit diagram of the (I/Q) down conversion mixer, followed by TIA.

A pair of double balanced passive mixers (Fig. 3) down-convert the D-band signal to (I, Q) baseband. The differential output of the LNA (RF^+ and RF^-) drives both mixer inputs through a transformer, the quadrature (I, Q) LO signals are converted to differential form by transformers before driving the FET mixer gates, and the baseband (I, Q) mixer outputs are DC-coupled to transimpedance amplifiers. To ensure sufficient LO drive power, the outputs of the LO multiplier are passed through a 4-stage post-amplifier before driving the mixer LO ports (Fig. 1a).

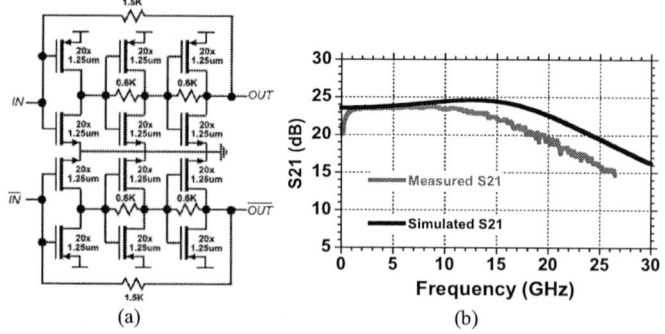

Fig. 4. Broadband TIA (a) circuit diagram of pseudo-differential TIA (b) simulation vs. measurement for single ended TIA test structure.

C. Broadband Transimpedance Amplifier

A pseudo-differential transimpedance amplifier provides the receivers' baseband gain. A three-stage voltage amplifier is first formed by an input g_m stage cascaded with two voltage-gain stages formed from g_m cells with local resistive feedback; adding global shunt resistive feedback forms a transimpedance amplifier (Fig. 4a). A test structure for a single ended TIA was measured. Fig. 4b shows the agreement in (S21) gain between simulation and measurement. There is some discrepancy in the 3-dB bandwidth between simulation (22GHz) and measurement (17GHz), which may be due to errors in parasitic extraction. The measured S-parameters shows a notch at DC, this is because the test structure is a single ended, and the supply capacitance in the test structure resonates with the DC supply probe inductance. In the Rx chain, the design is less sensitive to supply inductance, as the design is pseudo-differential.

D. Frequency Multiplier Circuit and 135 GHz LO Generation

Both transmitter and receiver employ a 9:1 frequency multiplier to generate a 135GHz LO signal; using an external reference at 15GHz with -3dBm input power. The multiplier design consists of an inverter-based single ended to differential (STD) converter, followed by two cascaded 3:1 frequency multipliers (Fig. 5). A fully differential structure reduces even harmonic generation and reduces supply coupling. The x3 frequency multipliers use a cross coupled pairs with capacitive neutralization, these driven into saturation to generate the third harmonic. The output of the first x3 multiplier is tuned at 45GHz (3rd harmonic of the input signal at 15GHz), while the second x3 multiplier is tuned at 135GHz. The topology and element values within the second 3:1 frequency multiplier are

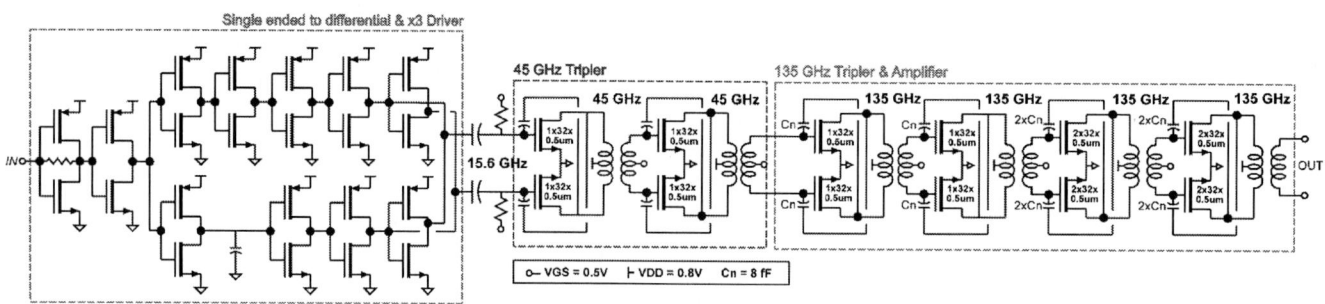

Fig. 5. Circuit schematic of the 135GHz 9:1 LO frequency multiplier.

978-1-7281-1702-7/19 $31.00 © 2019 IEEE

similar to those of the LNA/PA stages. The supply voltage of the entire chain is 0.8V. The simulated saturated output power is 3dBm and the simulated 3-dB bandwidth is 27GHz.

IV. TRANSMITTER ARCHITECTURE AND BUILDING BLOCK

In the direct conversion transmitter (Fig. 6), a pair of double-balanced Gilbert-cell mixers upconverts the (I, Q) baseband signals to D-band. The (I, Q) signals are then summed and drive a broadband power amplifier. The LO multiplier is the same as that in the receiver.

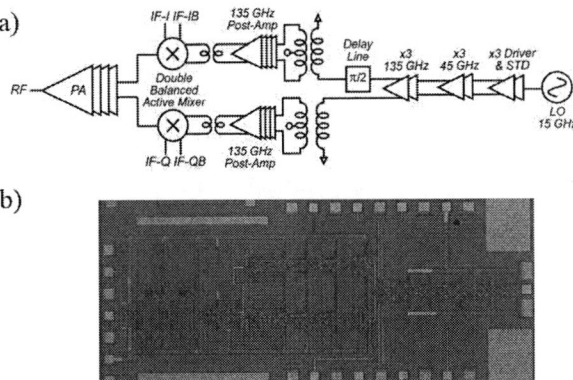

Fig. 6. D-band single channel direct conversion transmitter (a) circuit block diagram (b) chip micrograph, including pads. The die area is 1.9mm x 0.76mm.

A. IQ modulator

A pair of Gilbert cells serve as the IQ modulator (Fig. 7). The baseband inputs are DC coupled, while the LO and RF output ports are transformer-coupled. Tail bias (BIAS) of 0.35V, was chosen to maximize the modulator gain and output power (-10dBm in simulation).

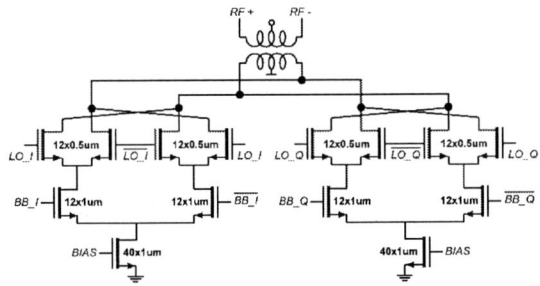

Fig. 7. IQ modulator using Gilbert cells. The (I, Q) LO ports are transformer-coupled.

V. MEASUREMENT RESULTS

Transmitter and receiver channels are fully characterized using on wafer probing. The receiver conversion gain is measured using Virginia diodes AMC 333 as the input signal source, followed by GGB (90-140 GHz) probe, to excite receiver input port. The two differential outputs are terminated by 50Ω impedances during measurement.

Fig. 8a shows the measured conversion gain with LO signal fixed at 134GHz and 135GHz, while the input signal is swept from 122GHz to 155GHz. This measures the receiver modulation bandwidth. The measured gain is 27dB, after de-

embedding probe loses and correcting for single ended to differential conversion. The 3-dB bandwidth is 20GHz. There is a good agreement between the measured and simulated gain. However, the simulated 3-dB BW is 1.5:1 larger than the measured, which might be explained as inaccuracy in apmom capacitor modelling in EM simulations. Fig. 8b shows the frequency dependent conversion gain, with a fixed baseband frequency, where the RF and LO signals are swept to keep the baseband frequency fixed at either 1GHz or 100MHz. This measures the receiver RF tuning range. The 3-dB bandwidth here is limited to 10GHz; the smaller bandwidth than in the prior measurement reflects the tuning range of the LO source.

Receiver 1-dB compression point is measured (Fig. 9) at different LO frequencies. The measured input P_{1dB} is -30dBm, which is slightly smaller than the simulated -26dBm. The receiver compression point is limited by the TIA drive capability, as this stage drives 50Ω.

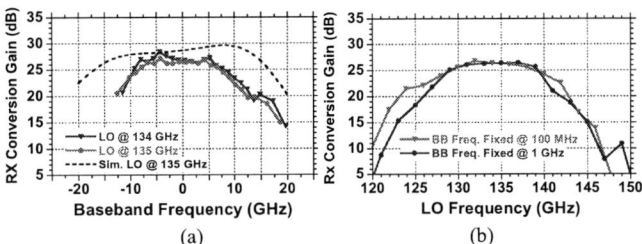

Fig. 8. a) Receiver conversion gain vs. baseband frequency with a fixed LO frequency (b) Receiver conversion gain vs. LO frequency with fixed baseband frequency.

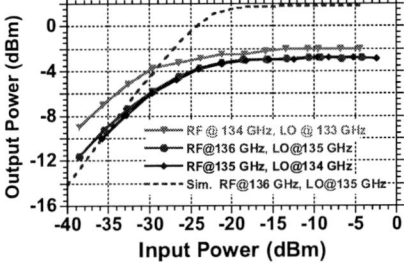

Fig. 9. Receiver normalized output power vs. input power.

Transmitter saturated output power is measured using Erickson PM4 power meter, where external signal generators drive the I and Q mixer inputs. To determine the transmitter saturated output power as a function of frequency, the transmitter was first driven by -3 dBm signals at 1, 2, or 5GHz at the (I, Q) ports, and the LO was swept from 125GHz to 145GHz (Fig. 10a). The saturated output power is 2.8dBm with a 3-dB bandwidth of 8 GHz. This determines the trasmitter frequency tuning range. Fig. 10b shows the normalized modulation sideband power with the baseband input frequency swept and the LO frequency held fixed. This measures the transmitter modulation response. The output spectrum was measured using an OML M05HWD harmonic mixer and a Rohde & Schwarz spectrum analyzer. There is a good agreement between the simulated and measured 3-dB bandwidth. Fig. 11a shows the gain compression characteristics as a function of carrier frequency. This particular measurement

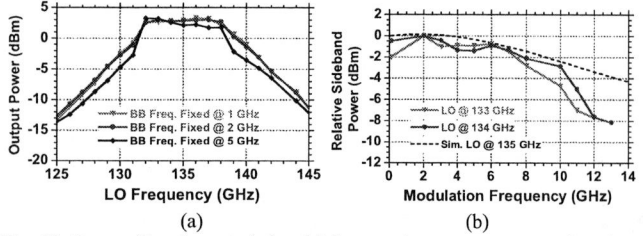

(a)   (b)

Fig. 10. Transmitter characteristics. (a) Saturated output power as a function of carrier frequency, with a -3 dBm baseband input signal, this showing an 8 GHz RF tuning range. (b) Modulation sideband power as a function of modulation frequency, this showing a ~8GHz (SSB) modulation bandwidth.

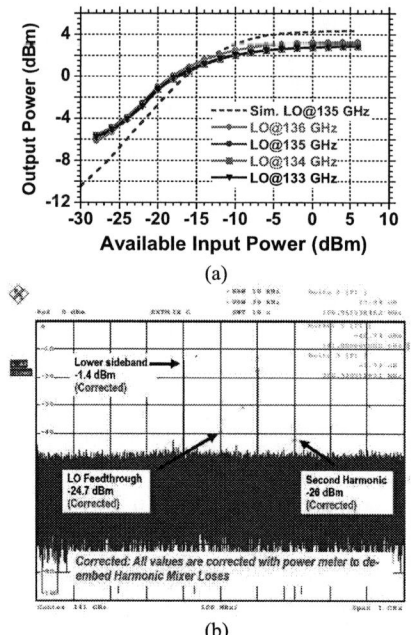

(a)

(b)

Fig. 11 (a) Transmitter output power as a function of input power, showing a typical 18dB gain (b) Transmitter output spectrum with center frequency 141GHz and 100MHz input signal.

shows approximately -7dBm LO leakage, because of incorrectly set DC levels at the transmitter baseband input ports. Fig. 11b shows the transmitter output spectrum with a 141GHz LO, a -6dBm 100 MHz input to the baseband I port, and only DC bias input to the baseband Q port, this producing I-phase but not Q-phase output modulation. With correct input DC levels (Fig. 11b), LO suppression is 23dB. The second harmonic is supressed by 24dB relative to the fundamental output signal.

VI. CONCLUSION

A broadband single-channel transmitter and receiver at D-band using CMOS 22nm FDSOI are demonstrated. Conversion gain of the entire receive channel is 27dB with a 3-dB bandwidth of 20GHz. The transmitter shows conversion gain of 18dB with a saturated output power of 2.8dBm. The transmitter and receiver consumes 196mW, and 198mW respectively from a 0.8V supply, both dominated by the 137mW LO multiplier DC power consumption. The transmitter and receiver both have bandwidth sufficient for 10 GBaud transmission.

A comparison to the state of the art transceivers at D-band is shown in Table 1. To the best of author's knowledge, this is the first sub-mm-Wave transmit/receive chain using 22nm FDSOI with the lowest supply voltage (0.8V) and highest bandwidth at the D-band.

Table 1. Comparison between state-of-the-art designs for near-140GHz transceivers.

	[4]	[5]	[6]	[7]	**This Work**
Technology	28nm CMOS	45nm CMOS	40nm SOI-CMOS	40nm CMOS	22nm SOI-CMOS
Frequency (GHz)	102-128	140	155	118	135
Conversion Gain (dB)	36-38	18 Rx - Tx	23 Rx - Tx	13 Tx	27 Rx 18 Tx
3dB Bandwidth (GHz)	18 Rx	12 Rx 8 Tx$^{\$\$}$	9 Rx - Tx	14 Tx	20 Rx 8 Tx$^{\$\$}$
NF (dB)	8.4-10.4	5.5*	20*	-	8.5*
Pdc (mW)	51	125 Rx 120 Tx	345 Tx/Rx	271	198 Rx 196 Tx
Tx Psat (dBm)	NA	-2	-10	4.5	2.8
Integration	Rx	Tx/Rx	Tx/Rx	Tx	Tx/Rx

*simulated, $^{\$\$}$single sideband

ACKNOWLEDGMENT

This work was supported in part by the Semiconductor Research Corporation (SRC) under the JUMP program (2018-JU-2778) and by DARPA (HR0011-18-3-0004). The author would like to thank Global Foundries for the 22 nm FDSOI CMOS chip fabrication. Authors also would like to thank Professor Gabriel Rebeiz, UCSD, for using his lab facilities.

REFERENCES

[1] S. Shahramian, M. J. Holyoak, and Y. Baeyens, "A 16-Element W-Band phased array transceiver chipset with flip-chip PCB integrated antennas for multi-gigabit data links," in Proc. Radio Freq. Integr. Circuits Symp., 2015, pp. 27-30.

[2] S. V. Thyagarajan, S. Kang and A. M. Niknejad, "A 240GHz wideband QPSK receiver in 65nm CMOS," 2014 IEEE Radio Frequency Integrated Circuits Symposium, Tampa, FL, 2014, pp. 357-360.

[3] M. Sadegh Dadash, S. Bonen, U. Alakusu, D. Harame and S. P. Voinigescu, "DC-170 GHz Characterization of 22nm FDSOI Technology for Radar Sensor Applications," 2018 13th European Microwave Integrated Circuits Conference (EuMIC), Madrid, 2018, pp. 158-161.

[4] T. Heller, E. Cohen and E. Socher, "A 102–129-GHz 39-dB Gain 8.4-dB Noise Figure I/Q Receiver Frontend in 28-nm CMOS," in IEEE Transactions on Microwave Theory and Techniques, vol. 64, no. 5, pp. 1535-1543, May 2016.

[5] A. Simsek, S.Kim and M.J.W.Rodwell, "A 140 GHz MIMO Transceiver in 45nm SOI CMOS," 2018 IEEE BiCMOS and Compound Semiconductor Integrated Circuits and Technology Symposium (BCICTS), San Diego, CA, USA, 2018, pp. 231-234.

[6] Y. Yang, S. Zihir, H. Lin, O. Inac, W. Shin and G. M. Rebeiz, "A 155 GHz 20 Gbit/s QPSK transceiver in 45nm CMOS," 2014 IEEE Radio Frequency Integrated Circuits Symposium, Tampa, FL, 2014, pp. 365-368.

[7] C. J. Lee et al., "A 120 GHz I/Q Transmitter Front-end in a 40 nm CMOS for Wireless Chip to Chip Communication," 2018 IEEE Radio Frequency Integrated Circuits Symposium (RFIC), Philadelphia, PA, 2018, pp.192-195

978-1-7281-1702-7/19 $31.00 © 2019 IEEE

Enhanced Passive Mixer-first Receiver Driving an Impedance with 40dB/decade Roll-off, Achieving +12dBm Blocker-P1dB, +33dBm IIP3 and sub-2dB NF Degradation for a 0dBm Blocker

Sashank Krishnamurthy, Ali M. Niknejad

University of California, Berkeley

sashank@berkeley.edu

Abstract— A "second order" passive mixer-first receiver is proposed to improve channel selectivity, linearity and noise figure in the presence of out-of-band blockers, by presenting an impedance which rolls off at 40dB/decade as the load to an N-path filter. The synthesis of this impedance is described in a step-by-step manner starting from the required impedance transfer function to its actual circuit realization. An integrated circuit prototype was fabricated in 28nm bulk CMOS as proof of concept. The receiver, capable of broadband operation from 0.2-2GHz, achieves an out-of-band IIP3 of +33dBm and a blocker P1dB of +12dBm. Additionally, it achieves a NF of 4.4dB with less than 2dB degradation in NF for a 0dBm blocker.

Keywords — N-path filter, Blocker resilience, 40dB/decade roll-off, high linearity, low NF desensitization

I. INTRODUCTION

With the advent of 5G and the continuous proliferation of sub-6GHz band mobile standards, the demand for broadband linear front-ends continues to grow. In this scenario of increasing number of bands, devices need to withstand blockers of large power without significant reduction in gain and noise performance. Mobile phones today achieve this using multiple SAW/FBAR filters for different bands, each of which is not tunable, thereby increasing area footprint and cost. The use of N-path filters, which regained popularity in [1], is a path towards the SAW-less radio receiver. It achieves tunable band-pass filtering by translating a simple first order low-pass filter to RF. There have been efforts to enhance such receivers in terms of noise performance [2], selectivity [3–5], and both selectivity and linearity [6–8]. While gyrators, cascade of N-path filters, N-path feedback paths, discrete-time techniques were used to enhance performance, the impedance translated to the RF input had a 20dB/decade roll-off in most of these receivers. Recent attempts were made to enhance selectivity and linearity by synthesizing an impedance with 40dB/decade roll-off at the output of the N-path filter [9, 10]. In this work, we provide a novel idea for a higher order N-path filter which sees an impedance with complex-pole 40dB/decade roll-off. We illustrate impedance-synthesis, from writing out what transfer function is needed for the impedance to finding out a circuit realization of the same, in a step-by-step manner. A prototype integrated circuit validates this concept.

II. IMPEDANCE SYNTHESIS

Consider the N-path filter shown in Fig. 1. In a regular first-order N-path filter, $Z_{BB}(f)$ is a shunt RC impedance,

and $v_x/v_{RF}(s')$ has a first-order roll-off (where $s' = s - j\omega_{LO}$). In order to realize steeper complex pole roll-off, the second order transfer function $v_x/v_{RF}(s')$ shown in Fig, 1, must have $Q > 1/2$. In this work, a Butterworth response ($Q = 1/\sqrt{2}$) is realized. Fig. 2 describes the step-by-step synthesis of the impedance $\gamma Z_{BB}(s) = R_S/(1 + 2s/(\omega_0 Q) + 2s^2/\omega_0^2)$ required to realize the 40dB/decade roll-off at the RF input of the receiver (where $\gamma = 2/\pi^2$ for an N-path filter driven by 4 non-overlapping LO phases).

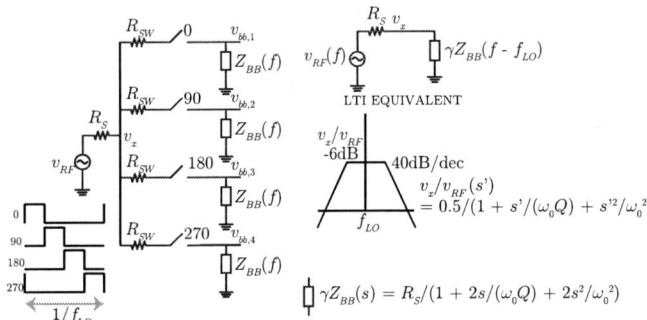

Fig. 1. 2nd Order N-path filter.

On writing the equivalent admittance $Y_{BB}(s)/\gamma$ (Fig. 2), we observe it is a parallel combination of the admittance $1/R_S$ and another admittance $Y_1(s) = (2s/(\omega_0 Q) + 2s^2/\omega_0^2)/R_S$, which can be written in the form $sC_1(1 + s/\omega_1)$, where C_1 is some capacitance. If we consider the equivalent impedance $Z_1(s) = 1/Y_1(s)$, up to $\sim \omega_1$, this rolls off like a capacitor at 20dB/decade, and beyond $\sim \omega_1$, it has a 40dB/decade roll-off. For $\omega_1 = 1/R_1 C_1$, where R_1 is some resistance, and C_1 is the same capacitance as before, using partial fraction decomposition, we observe that $Z_1(s)$ is a series combination of the capacitance C_1 and another impedance $Z_2(s)$. This $Z_2(s)$ can be realized using a parallel combination of a negative resistance and a negative capacitance $-R_1$ and $-C_1$. $Z_{BB}(s)$, thus constructed, when used as the baseband load of the N-path filter, gives the desired higher order filtering at RF. As shown in Fig. 2, this was realized by putting a resistor $R_1(A - 1)$ and $C_1/(A - 1)$ in positive feedback across an amplifier of gain A. The effects of re-radiation losses, modeled as a shunt re-radiation resistance R_{sh}, are well studied and therefore not mentioned here, and is included in the matching.

978-1-7281-1702-7/19 $31.00 © 2019 IEEE

$$\gamma Z_{BB}(s) = R_S/(1 + 2s/(\omega_0 Q) + 2s^2/\omega_0^2)$$
$$Y_{BB}(s)/\gamma = 1/R_S + (2s/(\omega_0 Q) + 2s^2/\omega_0^2)/R_S$$
$$= 1/R_S + sC_1(1+s/\omega_1)$$

$$Y_1(s) = sC_1(1+s/\omega_1)$$
$$\omega_1 = 1/R_1 C_1$$

$$Y_1(s) = sC_1(1+sR_1C_1)$$
$$Z_1(s) = 1/(sC_1(1+sR_1C_1))$$
$$= 1/sC_1 - R_1/(1+sR_1C_1)$$

$$Z_2(s) = -R_1/(1+sR_1C_1)$$

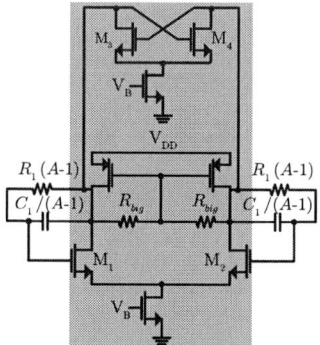

Fig. 2. Step-by-step Z-synthesis.

III. CIRCUIT DESIGN

A. Amplifier for Negative RC Synthesis ($Z_2(s)$)

The amplifier of gain A mentioned in the previous section is implemented by the circuit shown in Fig. 3. This is a simple NMOS differential pair loaded by a differential cross-coupled g_m. The gain $A \approx g_{m1,2}/g_{m3,4}$. For a fixed input transconductance $g_{m1,2}$, lower cross-coupled transconductance $g_{m3,4}$, gives higher A and therefore smaller capacitance in feedback. However, as we will see shortly, zeros in this implementation limit the frequencies up to which 40dB/decade roll-off is seen. Higher cross-coupled $g_{m3,4}$ push these zeros out to higher frequencies. In our implementation, a gain $A = 2$ was chosen. It is also interesting to note the noise implications of adding these components to synthesize the desired impedance. Looking at Fig. 2, we note that the negative R is capacitively coupled to the input of the TIA. Therefore, at low frequencies, noise from these devices do not couple to the output, and the in-band noise figure remains more or less unaffected.

B. Stability

A natural question to ask regarding the implementation of the negative R and C would be why do we have a cross-coupled g_m as the differential pair load in Fig. 3, instead of simply cross-coupling the capacitor $C_1/(A-1)$ and resistor $R_1(A-1)$ across a simple resistively loaded differential pair. The schematics in Fig. 4 answer this question. Fig. 4 abstracts the 2 cases using an amplifier of gain A and 1) negative output resistance $-R_{out}$ (denoting the amplifier in our circuit), and 2) positive output resistance R_{out} (denoting cross-coupling $C_1/(A-1)$ and $R_1(A-1)$ across a simple diff pair). In the presence of parasitic capacitance C_{par} at the output of the amplifier, input impedance in the 2 cases are computed (using the approximation $R_1(A-1) \gg R_{out}$). Note that, we are trying to implement the impedance $Z_2(s) = -R_1/(1+sC_1R_1)$ of Fig. 2. In our implementation, $Z_{in}(s)$ has a LHP pole in the presence of C_{par}, whereas in the other implementation $Z_{in}(s)$ has a RHP pole. If we continue to build up to the desired $Z_{BB}(s)$ from $Z_2(s)$ as described in Fig. 2, it can easily be seen that our implementation can be made stable, but the alternative and more obvious solution to synthesize $Z_2(s)$ can never be made stable.

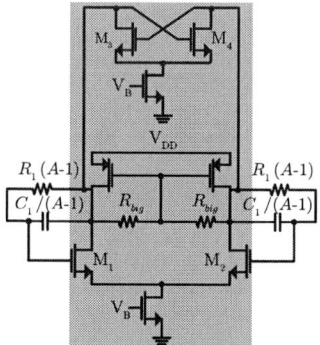

Fig. 3. Amplifier to realize negative RC.

C. Zeros - Limitation and Solution

As seen from the equation for Z_{in} in Fig. 4, even though we desired to realize $Z_2(s) = -R_1/(1 + sC_1R_1)$, there is an additional LHP pole and a RHP zero. By design, the magnitude of both the pole and the zero are much greater than $1/R_1C_1$. It is obvious that the zero is at a lower frequency than the pole. For frequencies much lower than $|1/(R_{out}(C_{par} + C_1/(A-1)))|$ (the magnitude of the zero frequency), $Z_{in}(s) \approx Z_2(s)$. The zero in $Z_{in}(s)$ eventually manifests as complex zeros in the transfer function. Therefore, it is desirable to have this zero as far out as possible from the 3-dB bandwidth of our circuit, so that we may see the 40dB/decade roll-off for a broader range of frequencies. This may be done by reducing the R_{out} as much as possible. In our implementation in Fig. 3, this implies we need to increase $g_{m3,4} \approx 1/R_{out}$ as much as possible. This explains the choice of the small gain $A = 2$.

The consequent trade-off of power in $g_{m3,4}$ and the frequency up to which 40dB/decade filtering is obtained, is relaxed by an idea partly inspired by the use of zero-nulling resistors for compensation in op-amps. If we have a resistance R_z in series with the capacitor $C_1/(A-1)$, $Z_{in}(s)$ may now be computed as shown in Fig. 5. The location of the pole is unchanged but there is an additional zero now. However, by appropriate choice of R_z, the two zeros may be pushed to a higher frequency than before. However, care must be taken with this implementation as it can be shown that too high a value of R_z will lead to instability.

2019 IEEE Radio Frequency Integrated Circuits Symposium

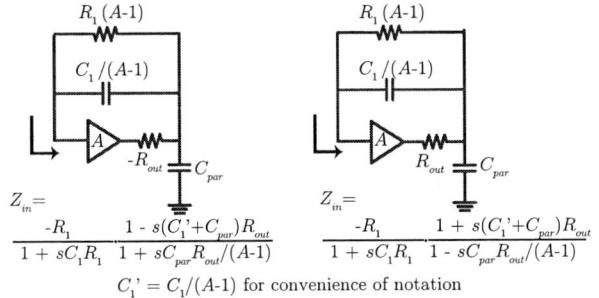

$$Z_{in}= \frac{-R_1}{1+sC_1R_1} \cdot \frac{1-s(C_1'+C_{par})R_{out}}{1+sC_{par}R_{out}/(A-1)}$$

$$Z_{in}= \frac{-R_1}{1+sC_1R_1} \cdot \frac{1+s(C_1'+C_{par})R_{out}}{1-sC_{par}R_{out}/(A-1)}$$

$C_1' = C_1/(A-1)$ for convenience of notation

Fig. 4. Comparison between cross-coupling g_m load of diff pair and cross-coupling resistor and capacitor.

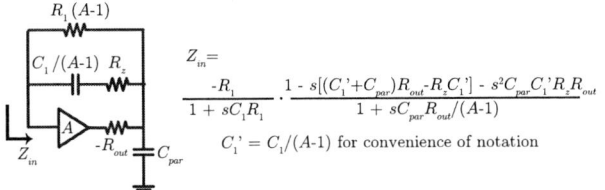

$$Z_{in}= \frac{-R_1}{1+sC_1R_1} \cdot \frac{1-s[(C_1'+C_{par})R_{out}-R_zC_1']-s^2C_{par}C_1'R_zR_{out}}{1+sC_{par}R_{out}/(A-1)}$$

$C_1' = C_1/(A-1)$ for convenience of notation

Fig. 5. Series resistance to push the zeros to higher frequency.

IV. MEASUREMENT RESULTS

A test chip was fabricated in a 28nm bulk CMOS process and wire-bonded directly onto PCB. The active area of the chip is $730\mu m \times 650\mu m$ (Fig. 6).

The measured conversion gain and input matching as a function of frequency is shown in Fig. 7 ($f_{LO} = 500$MHz). A gain of around 12.5dB and a bandwidth of 18MHz is achieved. It is observed that the gain is flat from $f_{LO} = 0.2 - 2$GHz (< 1.5dB variation). As seen from the zoomed in plot of the gain, the attenuation at 30MHz is > 20dB, illustrating that the higher order roll-off is achieved. The flat region indicates the complex zeros discussed in the previous section. This test chip does not implement programmability for R/Cs, and so, to compensate for I/Q mismatch, bias tuning was performed. This limitation and LO phase imbalance led to slightly different results for linearity and other measurements from the I/Q channels. For completeness, results from both channels are shown.

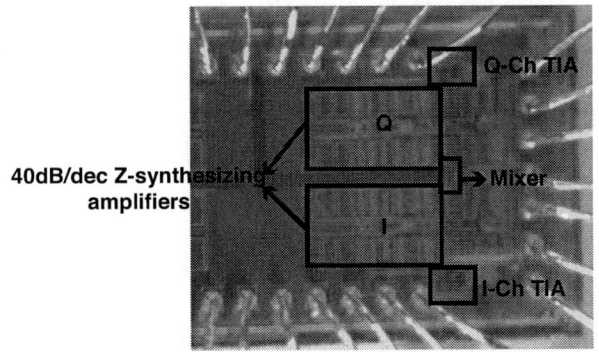

Fig. 6. Die micrograph.

Fig. 8 shows the plots of IIP3 and blocker 1-dB compression points versus offset. IIP3 was measured using

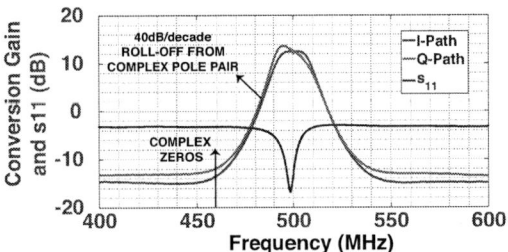

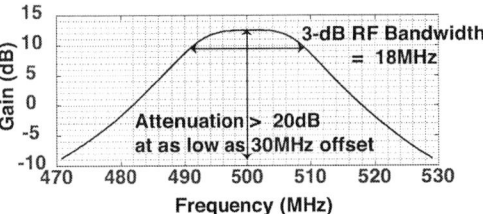

Fig. 7. Plots of measured s_{11}, gain and zoomed-in gain ($f_{LO} = 500$MHz).

a 2-tone test where tones (power = -13dBm) were placed at $f_{LO} + f_{OS} + 500$kHz and $f_{LO} + 2f_{OS} + 500$kHz, such that their IM3 product fell in-band at $f_{LO} + 500$kHz. For the blocker P1dB, the desired signal was placed at 500.5MHz, and P1dB was found as a function of the blocker offset frequency. A blocker P1dB of +12dBm at $f_{OS} = 60$MHz and IIP3 of +33.3dBm at $f_{OS} = 80$MHz confirm the benefits of the N-path filter driving a higher order impedance.

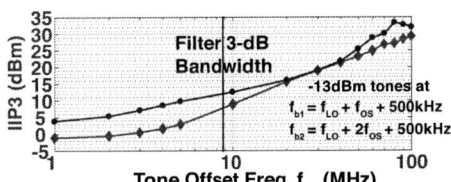

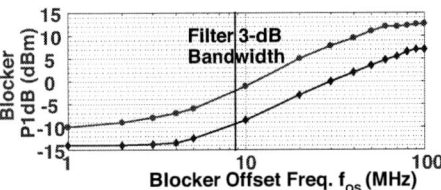

Fig. 8. Plots of IIP3 and Blocker P-1dB v/s offset frequency as measured from both I and Q channels ($f_{LO} = 500$MHz).

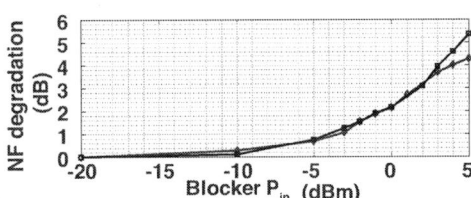

Fig. 9. Plot of 0dBm Blocker NF degradation v/s blocker P_{in} as measured from both I and Q channels ($f_{LO} = 787.5$MHz, $f_{blk} = 881$MHz).

978-1-7281-1702-7/19 $31.00 © 2019 IEEE

The noise of the on-board buffer IC driven by the chip, spectrum analyzer and the PCB losses are de-embedded, while measuring noise figure. A noise figure of around $4.3\text{dB} - 7.6\text{dB}$ was measured across the range of f_{LO} at which the chip was characterized. This is $1 - 1.5\text{dB}$ higher than our simulations, and the discrepancy could be coming from 4-phase LO overlap due to imbalances in differential input LO or due to higher parasitics at the RF input. $f_{LO} = 787.5\text{MHz}$ was chosen for the blocker noise figure measurement because that was the highest frequency such that high-Q bandpass filters were available for both the LO input at $2f_{LO}$ and the blocker at $f_{LO} + f_{OS}$, which was placed at 881MHz. The noise figure degrades only by 2dB for a 0dBm blocker at 93.5MHz offset.

V. CONCLUSION

To summarize, this work presents a novel idea for a higher order N-path filter by providing a detailed synthesis of an impedance which rolls off at 40dB/decade. It achieves highly competitive performance with respect to the state-of-the-art (Table 1), in blocker resilience, particularly, blocker P1dB, OOB IIP3 and 0dBm blocker noise figure desensitization.

ACKNOWLEDGMENT

The authors wish to thank TSMC University Shuttle Program for chip fabrication, Integrand for EMX simulator. This work was supported partially by Intel and partially by DARPA under Contract No. FA8650-18-1-7903 as part of the MIDAS program. The authors are grateful to Konstantin

Trotskovsky, Lorenzo Iotti and Nima Baniasadi of BWRC for valuable inputs during design and measurement.

REFERENCES

[1] C. Andrews *et al.*, "A passive mixer-first receiver with digitally controlled and widely tunable RF interface," *IEEE Journal of Solid-State Circuits (JSSC)*, vol. 45, no. 12, pp. 2696–2708, 2010.

[2] D. Murphy *et al.*, "A blocker-tolerant, noise-cancelling receiver suitable for wideband wireless applications," *IEEE Journal of Solid-State Circuits (JSSC)*, vol. 47, no. 12, pp. 2943–2963, 2012.

[3] M. Darvishi *et al.*, "Design of active N-Path filters," *IEEE Journal of Solid-State Circuits (JSSC)*, vol. 48, no. 12, pp. 2962–2976, 2013.

[4] R. Chen *et al.*, "Reconfigurable receiver with radio-frequency current mode complex signal processing supporting carrier aggregation," *IEEE Journal of Solid-State Circuits (JSSC)*, vol. 50, no. 12, pp. 3032–3046, 2015.

[5] I. Madadi *et al.*, "A TDD/FDD SAW-less superheterodyne receiver with blocker-resilient band-pass filter and multi-stage HR in 28nm CMOS," in *2015 Symposium on VLSI Circuits (VLSI Circuits)*. IEEE, 2015, pp. C308–C309.

[6] Y. Lien *et al.*, "A high-linearity CMOS receiver achieving +44dBm IIP3 and +13dBm B1dB for SAW-less LTE radio," in *Solid-State Circuits Conference (ISSCC), 2017 IEEE International*, 2017, pp. 412–413.

[7] C.-k. Luo *et al.*, "0.4–6 GHz, 17-dBm B1dB, 36-dBm IIP3 channel-selecting, low-noise amplifier for SAW-less 3G/4G FDD receivers," in *2015 IEEE Radio Frequency Integrated Circuits Symposium (RFIC)*, 2015, pp. 299–302.

[8] Y. Xu *et al.*, "A switched-capacitor RF front end with embedded programmable high order filtering and a +15dBm OB-B1dB," in *2015 IEEE Radio Frequency Integrated Circuits Symposium (RFIC)*, 2015, pp. 291–294.

[9] Y.-C. Lien *et al.*, "Enhanced-selectivity high-linearity low-noise mixer-first receiver with complex pole pair due to capacitive positive feedback," *IEEE Journal of Solid-State Circuits (JSSC)*, vol. 53, no. 5, pp. 1348–1360, 2018.

[10] E. C. Szoka *et al.*, "Circuit techniques for enhanced channel selectivity in passive mixer-first receivers," in *2018 IEEE Radio Frequency Integrated Circuits Symposium (RFIC)*, June 2018, pp. 292–295.

Table 1. Comparison with State-of-Art

	JSSC10 [1]	JSSC12 [2]	JSSC13 [3]	JSSC15 [4]	ISSCC17 [6]	RFIC15 [7]	JSSC18 [9]	RFIC18 [10]	**This work**
Architecture	Mixer-First	Frequency Translation Noise Canceling	Active N-path Filter	LNTA + Passive Mixer + Baseband notches	Bot. Plate Mixing in N-path filter	N-path BPF/BRF Feedback Filter	N-path Filter with positive cap. feedback	N-path filter with shunting notch	**N-path filter driving Z w/ 40dB/dec. roll-off**
Technology	65nm	40nm	65nm	65nm	28nm	32nm SOI	45nm SOI	130nm SiGe BiCMOS	**28nm**
f_{RF} (GHz)	0.1-2.4	0.08-2.7	0.1-1.2	0.5-3	0.1-2	0.4-6	0.2-8	2-11	**0.2-2**
RF Input	Single Ended	Single Ended	Differential	Differential	Differential	Differential	Differential	Single Ended	**Single Ended**
Gain(dB)	40-70	72	25	50	16	11	21	10-24	**13**
Bandwidth	20MHz	4MHz	8MHz	2-60MHz	13MHz	15MHz	20MHz	80-260MHz	**18MHz**
OOB-IIP3 (f_{OS}/BW)	25dBm (5)	13.5dBm (20)	26dBm (6.3)	-4.8dBm (4)	44dBm (6.3)	36dBm (3.3)	39dBm (4)	20dBm (1.75)	**33.3dBm (4.4)**
B1dB (f_{OS}/BW)	5dBm (5)	-2dBm (20)	7dBm (6.3)	-4dBm (4)	13dBm (6.3)	17dBm (6.7)	12dBm (4)	1.8dBm (1.75)	**12dBm (3.3)**
NF (dB)	4 ± 1	1.9	2.8	3.8-4.7	6.3	3.6-4.9	2.3-5.4	11 ± 1	**4.3-7.6**
0dBm Blk. NF desense (dB)	NA	2.2 (f_{OS}/BW = 20)	NA	NA	0.6 (f_{OS}/BW = 6.1)	NA	2.2 (f_{OS}/BW=4)	NA	**2 (f_{OS}/BW=5)**
Power (mW)	37-70	35.1-78	18-57.4	Rx:76-168 LO:54-194	38-96	81-209	50 + 30/GHz	Rx:656 LO:1466 - 1494	**Active Z:100 TIA:43 LO:3.6-36**
Supply (V)	1.2/2.5	1.3	1.2	1.2/2.5	1.2/1	2	1.2	4.5/2.5	**1.2**
Area (mm^2)	2.5	1.2	0.27	7.8	0.49	0.28	0.8	3.4/5	**0.48**

2019 IEEE Radio Frequency Integrated Circuits Symposium

A Code-Domain RF Signal Processing Front-end for Simultaneous Transmit and Receive with 49.5 dB Self-Interference Rejection, 12.1 dBm Receive Compression, and 34.3 dBm Transmit Compression

Hussam AlShammary, Cameron W. Hill, Ahmed Hamza and James F. Buckwalter

University of California, Santa Barbara

Abstract— This paper demonstrates a code-domain transceiver that incorporates a transmit (TX) modulator and receive (RX) RF signal correlator. We propose a transmission gate switch with 12.1-dBm/23.1-dBm P1dB/IIP3 for both RF correlation and filtering and rejects TX self-interference by 49.5 dB. An integrate-and-dump N-path filter improves rejection by 8 dB. The RX power consumption is 18 mW at 1 GHz. The TX modulator operates to RF power levels up to 34.3 dBm and consumes less than 40 mW for 300 Megachip-per-second (Mc/s). Over the air testing demonstrates synchronization with Barker codes.

Keywords — N-Path Filter, Transmit Rejection, Code-Reject Filter, Full-Duplex Communication, SOI CMOS.

I. INTRODUCTION

Simultaneous transmit and receive (STAR) within the same signal band offers the potential to estimate the channel response and support full-duplex (FD) communication. Several proprietary defense systems use spread-spectrum communication but cannot support FD operation since the co-located transmitter (TX) typically has peak power levels exceeding 33 dBm and produces significant receive (RX) band interference. Even with 23 dB isolation between the TX and RX, self-interference (SI) power levels exceed 10 dBm in the RX path, demanding high in-band linearity (>20 dBm) and high-rejection (>40 dB) at the LNA.

Prior work on STAR has been demonstrated [1], [2], [3], [4], [5], [6], [7]. In [8], a combination of electrical balanced duplexer with a 3-dB TX power penalty, and analog cancellation before and after the LNA offers 70 dB of rejection, but is limited to low TX SI power levels. Furthermore, analog cancellation does not reject other multiple access interference. An alternative approach based on code-domain receiver has been demonstrated at LNA/mixer with 38.5 dB of rejection [1], however, without the capability of handling strong TX SI (in-band P1dB = -11.8 dBm). Code-domain approaches [1], [5] also assume TX coding prior to the power amplifier (PA) which increases the required PA bandwidth to several hundred MHz which limits the overall system efficiency.

To solve these issues a fully-integrated code-domain transceiver capable of signal processing at the antenna is proposed. This interface is able to overlay a code-domain STAR system onto narrowband components leading to savings in ADC/DAC resolution and baseband power while maintaining efficiency by modulating the TX after the PA.

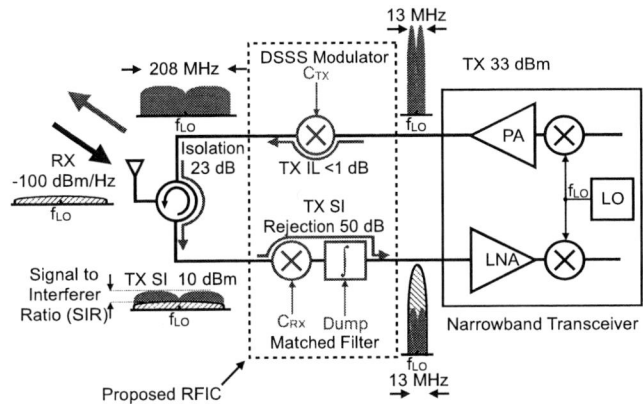

Fig. 1. Proposed DSSS modulator and matched filter with RF code correlator and integrate and dump.

Additionally, filtering before the LNA mitigates the linearity requirements of the active circuits.

The proposed approach is shown in Fig. 1. In the TX path the direct-sequence spread-spectrum (DSSS) modulator is a fast high-power BPSK modulator and modulates the original TX signal by a code sequence C_{TX} and spreads the signal bandwidth by a factor of L. The TX signal is then isolated from the RX by more than 20 dB using a circulator or duplexer technique. In the RX path, the RF signal is correlated with code C_{RX}. For matching codes, the RX signal is increased by the processing gain and concentrated within the signal bandwidth. For orthogonal codes, the RF signal is spread out of band (OOB). Consequently, the remaining OOB spectral components are rejected by the receive correlator as shown in Fig. 1.

This paper demonstrates the first fully integrated code modulator and demodulator operating in the RF domain. TX modulation after the PA is accomplished using a reflection phase shifter and high power stacked switches which can achieve modulation speeds of up to 300 Mc/s while handling an input power of 34.3 dBm. The receive correlator is implemented using an N-path filter, chosen for its high level of integration and tunability. Compared to previous N-path implementations, which have insufficient linearity to handle strong interferers present in a STAR system [9], [10], this work uses a hybrid transmission gate switch with optimum biasing to achieve a record setting in-band P1dB of 12.1 dBm. Additionally, the N-path filter has reset circuitry to implement

978-1-7281-1702-7/19 $31.00 © 2019 IEEE   143

an integrate and dump (IAD) feature which can be used to reduce inter-symbol interference (ISI) and lower the system noise. The integration of the TX and RX components of the system onto one chip is advantageous for reducing the lag time for synchronization of the TX and RX codes and achieving the highest SI cancellation.

The following section reviews the circuit techniques for the high-power DSSS modulator, the linear code correlator, and IAD filter. The measurements demonstrate the rejection of the receiver as well as the insertion loss and power handling of both the TX and RX circuitry. Finally EVM measurements indicate that the proposed solution can handle TX power of 30 dBm with nearly 50 dB of SI rejection while the RX still achieves an EVM under 20% for a QPSK signal.

II. CODE-DOMAIN TRANSCEIVER CIRCUIT

Fig. 2 shows a block level diagram of the proposed transceiver chip with peripheral components. The TX DSSS modulator includes two on-chip 35-dBm RF switches connected to the isolated and coupled port of an off chip 90° hybrid coupler. When turned off, the switches create an in-phase reflection (open). When turned on, the switches produce a 180° reflection (short). To handle the high voltage on TX side, the RF switch is based on a 12-stack CMOS SOI FET with 1.8mm NMOS and auxiliary PMOS/NMOS switches shown in Fig. 3 to distribute the voltage swing on the switch devices and prevent breakdown while improving the switching speed [11].

In the RX path, the incoming RF signal is modulated with a chopper to correlate the RF signal against a desired code sequence. The signal is then passed through a shunt 8-path filter that incorporates an IAD to reset the N-path filter after every code correlation completes. The processed signal is available at the RF output port or for the baseband amplifiers to sample the voltage across the N-path capacitors and directly receive the desired signal or aid in the synchronization process. The chip also includes circuitry to generate non-overlapping clocks to drive the switches of the N-path. The code and dump signal generation produces signals to drive the RX chopper, dump switches and DSSS modulator. On-chip logic generates Walsh (for a better rejection) and Barker codes (for synchronization phase) as well as a bypass for an off chip code.

Schematics of the RX chopper and IAD N-path filter are shown in Fig. 3. The chopper resembles a FET ring mixer but the gates are driven with relatively low speed code signals. The IAD feature is built into a standard N-path filter as a switch which equalizes the differential voltage on the N-path when the dump signal is active to reset accumulated charge to avoid ISI in the correlation. As shown in Fig. 3, the IAD N-path filter has a hybrid capacitor bank with high-density capacitors and shunt capacitors to improve OOB rejection and common-mode noise generated on chip (e.g. LO leakage, dump leakage, etc.).

To improve the linearity of these receive circuits, we propose a transmission gate (TG) rather than a single n-FET

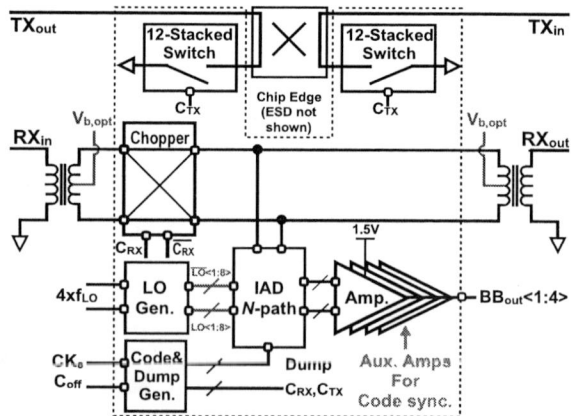

Fig. 2. Block diagram of the code domain transceiver for STAR applications.

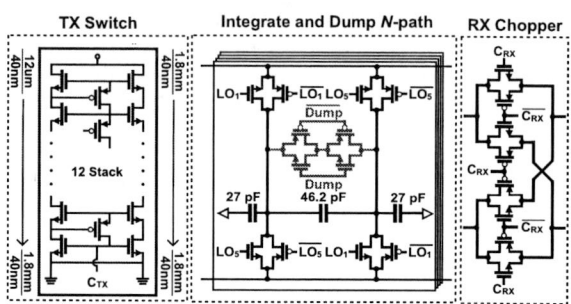

Fig. 3. Circuit schematics for critical RF front-end blocks including the 34-dBm TX switch, the 10-dBm RX chopper, and a single path of the integrate and dump N-path filter.

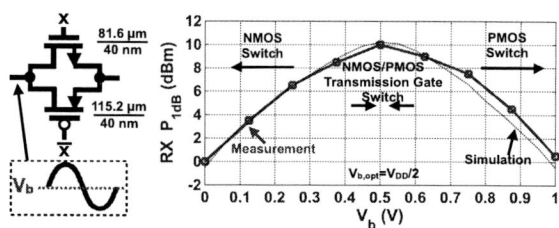

Fig. 4. Implementation of the RF transmission gate switch with comparison of the simulated and measured 1-dB power compression.

for the RF switches. Recent work also proposed using TG switches and measured an in-band 1-dB power compression (P1dB) of -17 dBm ($\Delta f/BW = 0$) [12]. Our measurements and simulations of the TG RF switch based on a 45-nm CMOS SOI design in Fig. 4 demonstrate that the TG extends the in-band P1dB to 10.1 dBm with an optimum bias ($V_b = V_{b,opt}$) that maintains the proper state of the switches (i.e. on/off) under high RF swing. Additionally, the on-resistance of the switch remains constant over a larger signal swing which improves the switch linearity. All TGs are of the same size as shown in Fig. 4.

III. MEASUREMENTS

The proposed system was fabricated with GlobalFoundries 45-nm RF SOI process. The die is shown in Fig. 5 and occupies an area of 2.59 mm^2 including the pads and ESD circuitry while the active area is 1.12 mm^2. The RX power consumption was less than 18 mW including digital circuitry at $f_{LO} = 4f_{RF} = 4$ GHz and code rate of 200 Mcps from 1-V supply and four baseband amplifiers from 1.5-V supply.

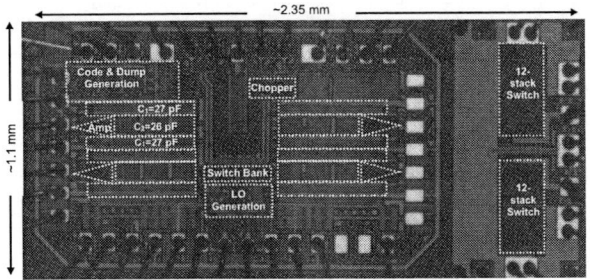

Fig. 5. Chip micro photograph of the code-domain transceiver.

Fig. 6 compares the RX performance in the presence of an in-band blocker at the same frequency as the desired RX signal at $f_{RF} = 0.5$ GHz. The spectrum is plotted for matching and orthogonal codes and illustrates a rejection of 47 dB for a QPSK signal with a spreading BW while the signal data rate is 3.125 Mbps (code lengths of M = 64). The measured rejection is plotted on Fig. 6 (bottom) for varying data rates but fixed spreading bandwidth of 100 MHz. The rejection is 49.5 dB for a 0.5 GHz CW signal and reduces to 42 dB at 1 GHz with the on-chip DSSS modulator operating at 30 dBm.

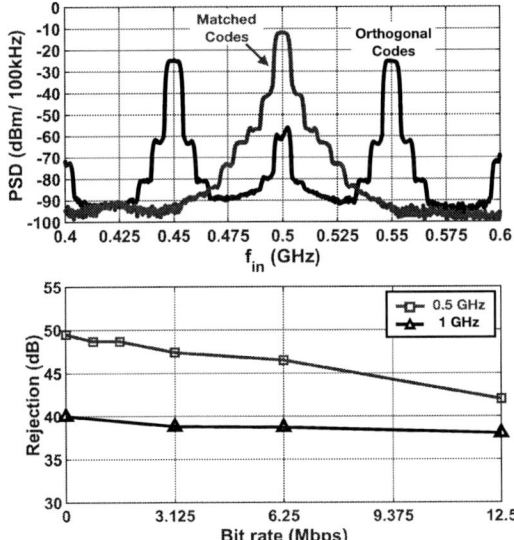

Fig. 6. Measured spectrum for matched and orthogonal codes (top) and orthogonal rejection as a function of data rate.

The TX modulator 1-dB bandwidth extends from 400 MHz to 1 GHz and insertion loss (IL) is less than 1.6dB up to a 1-dB power compression of 34 dBm at 600 MHz as shown in Fig. 7 (top). The switching time is under 3ns which enables 300 MHz spread bandwidth. The TX IIP3 is measured to be 50 dBm using a two-tone measurement setup.

The IL of the RX correlator is 0.8-3.55 dB across a tuning range from 0.4 to 1.1 GHz with P1dB of 10.1 dBm and 12.6 dBm depending on the balun ratio as shown in Fig. 7 (bottom).

The in-band IIP3 is better than 23.1 dBm across the tuning rane. The NF for RX is measured at RX_{out} between 2.6 and 5.6 dB depending on f_{LO}. The gain of the auxiliary amplifiers is measured to be 30 dB. The measured RF BW of the RX correlator is 13 MHz.

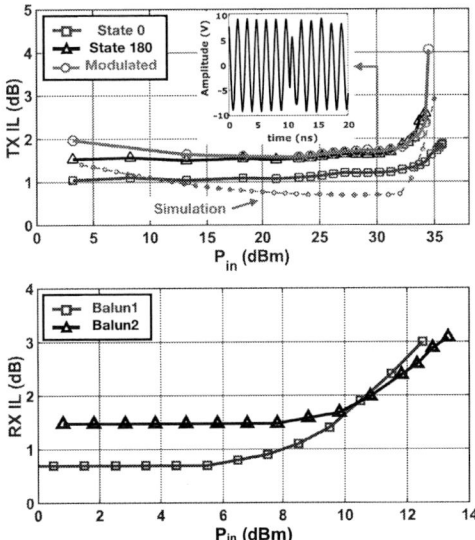

Fig. 7. Measured in-band IL compression point of the transceiver RFIC: TX (top) and RX (bottom).

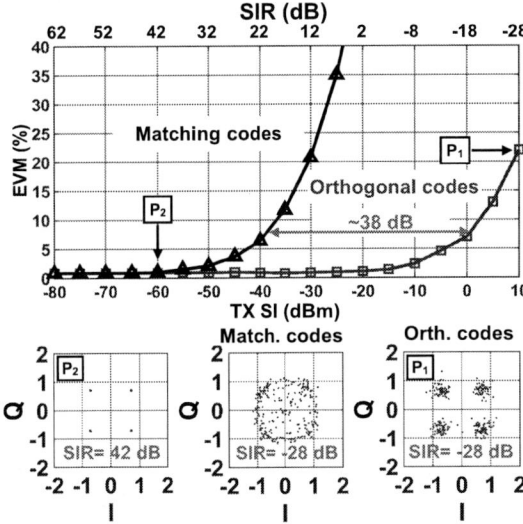

Fig. 8. EVM as a function of transmit power and normalized to signal-to-interference ratio (SIR) and QPSK constellations at selected points.

The EVM is plotted as a function of TX SI power as well as signal-to-interference ratio (SIR) in Fig. 8 for a RX signal power of -18 dBm when the TX and RX are both at 0.5 GHz. For SIR greater than 30 dB, the noise floor limits the EVM (i.e. SNR 30 dB) to around 1%. In the absence of any TX rejection, the EVM degrades to 10% as the SIR reduces to 18 dB. When RF correlation is applied, the EVM curve is shifted by 38 dB such that the same EVM is tolerated for an SIR of -20 dB. The QPSK (bit rate of 12.5 Mbps for RX and 12.4 Mbps for TX) constellation is shown for TX SI of +10 dBm.

Table 1. Comparison Against State-of-the-Art

Spec.	[1] RFIC '17	[2] ISSCC '17	[3] RFIC '18	[4] MWCL '19	[5] CICC '18	[6] IMS '18	[7] ESSCIRC '18	This work RFIC '19
CMOS Process	65 nm	65 nm	180 nm	45 nm	45 nm	45 nm	65 nm	GF 45 nm SOI
Frequency (GHz)	0.30-1.40	0.61-0.97	0.86-1.08	0.30-0.675	1.10-2.50	0.90-1.10	0.40-1.00	0.40-1.10
BW (MHz)	1	20	6.8-37.4	10	1	300	NA	13
TX-RX isolation (dB)	0	20	25-40	0	31	0	NA	0
Pre-LNA SI Rejection (dB)	0	0	0	23.2	20	21.9	33	49.5
Post-LNA SI Rejection (dB)	38.5	0	0	0	0	0	0	0
BB SI Rejection (dB)	0	0	0	0	0	0	0	0
TX SI (dBm)	NA	-26	-5 to +10	NA	NA	NA	NA	+10
IB RX P1dB (dBm)	-11.8	NA	21	2	NA	NA	NA	+12.1
IB RX IIP3 (dBm)	>17	-26	36.9	6.3	NA	21.6	NA	>+23.1
RX Power Consumption (mW)	37	36	NA	28.2	50	9.37	24	18
TX Power Consumption (mW)	NA	59	170	NA	NA	NA	24	<40
TX Power Handling (dBm)	NA	8	+30.7	NA	15	NA	NA	+34.3
TX IL (dB)	NA	1.8-3.2	2.1	NA	NA	NA	NA	1.6
Active Area (mm^2)	0.31	0.94	16.5	0.9	1.4	1.77	3.1	1.12

Finally, we demonstrate the STAR operation over a short-range link with two prototype RFIC circuits mounted on PCB. We investigate code synchronization for the receiver which was not fully addressed in earlier work [1]. Fig. 9 demonstrates an over-the-air (OTA) measurement of the proposed STAR system when the transmit and receive signal are at 900 MHz. Transmitter TX$_A$ transmits a CW tone at 900MHz modulated with a 100 Mcps Barker code of length 11. An off-the-shelf circulator isolates the TX and RX. After propagating over a 1-m channel, the signal is received at receiver RX$_B$ and correlated against the 11 possible timing lags for the Barker code through the observation receiver. The delay mismatch between the two transceivers (T$_1$) can be estimated by interpolation between the highest two peaks in the autocorrelation function. The variation in the receive power indicates a processing gain of the RX signal of at least 12 dB. The asymmetry across the peak in Fig. 9 is attributed to the non-ideal RC shape of the filter compared to an ideal integrator. Nevertheless, this filter shape is accurate to identify the highest peak.

Table 1 shows a comparison with state of the art work on full-duplex. The proposed code-domain transceiver provide higher rejection at the antenna while also handling higher power levels and higher levels of integration. Furthermore, this work demonstrates the highest N-path linearity in terms of 1-dB compression for in-band blockers.

IV. CONCLUSION

We present a code-domain transceiver consisting of a TX DSSS modulator and RX RF matched filter with a correlator and integrate-and-dump N-path filter. The RFIC demonstrates the highest spread spectrum modulator power handling and receive path P1dB and IIP3 for in-band blockers.

ACKNOWLEDGMENT

The work presented here was supported under the DARPA SPAR Program managed by Dr. Troy Olsson and Dr. Timothy Hancock through cooperative agreement HR0011-17-2-0003. The content of this paper does not necessarily reflect the pos-

-ition or the policy of the Government and no official endorsement should be inferred. We thank the support of Ned Cahoon and Arvind Sharma of GlobalFoundries for providing access to GF 45RFSOI chip fabrication.

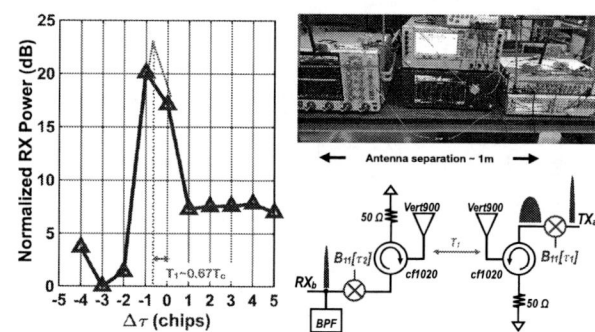

Fig. 9. Over-the-air measurement for code synchronization and its set-up.

REFERENCES

[1] A. Agrawal et al., "A 0.3 GHz to 1.4 GHz N-path mixer-based code-domain RX with TX ..." in RFIC, June 2017, pp. 272–275.

[2] N. Reiskarimian et al., "Highly-linear integrated magnetic-free circulator-receiver for full-duplex ..." in ISSCC, Feb 2017, pp. 316–317.

[3] A. Nagulu et al., "Fully-integrated non-magnetic 180nm SOI circulator with >1 W P1dB ..." RFIC, vol. 53, no. 6, pp. 1607–1617, June 2018.

[4] A. Hamza et al., "A series N-path code selective filter for transmitter rejection in ..." IEEE MWCL, vol. 29, no. 1, pp. 38–40, Jan 2019.

[5] Z. Chen et al., "A full-duplex transceiver front-end RFIC with code-domain spread spectrum ..." in CICC, April 2018, pp. 1–4.

[6] H. AlShammary et al., "A λ/4-Inverted N-path filter in 45-nm CMOS SOI for transmit rejection..." in IMS, June 2018, pp. 1370–1373.

[7] N. Mousavi et al., "A 0.4-1.0GHz, 47MHop/s frequency hopped TXR front-end with 20dB in-band ..." in ESSCIRC, Sep. 2018, pp. 66–69.

[8] K. Chu et al., "A broadband and deep-TX self-interference cancellation technique for full-duplex..." in ISSCC, Feb 2018, pp. 170–172.

[9] Y. Lien et al., "24.3 a high-linearity CMOS receiver achieving +44dBm IIP3 and +13dBm B1dB for ..." in ISSCC, Feb 2017, pp. 412–413.

[10] A. Ghaffari et al., "Tunable high-Q N-path band-pass filters: Modeling and verification," JSSC, vol. 46, no. 5, pp. 998–1010, May 2011.

[11] C. Hill et al., "A 30.9 dBm, 300 MHz 45-nm SOI CMOS power modulator for spread-spectrum ..." in IMS, June 2018, pp. 423–426.

[12] Y. Xu et al., "A chopping switched-capacitor RF receiver with integrated ..." JSSC, vol. 53, no. 6, pp. 1607–1617, June 2018.

2019 IEEE Radio Frequency Integrated Circuits Symposium

A CMOS 0.5-2.5GHz Full-Duplex MIMO Receiver with Self-Adaptive and Power-Scalable RF/Analog Wideband Interference Cancellation

Yuhe Cao, Jin Zhou
University of Illinois at Urbana-Champaign, IL, 61801, USA

Abstract—**A 65nm CMOS 0.5-2.5GHz full-duplex (FD) MIMO receiver (RX) with self-adaptive ≥24dB RF/analog interference cancellation across 20MHz BW is presented. An LMS adaptive circuitry is co-designed with and partially embedded in a wideband RF/analog interference canceller and a gain-boosted mixer-first RX. With the adaptive circuitry fully integrated and consuming 14mW, the cancellers adapt themselves to an unknown channel in 1µs. The FD MIMO RX is also power-scalable -- when used as a digital beamformer, the NF-power scalability of the FD RX enables a nearly constant canceller DC power per element, despite a quadratic increase of cancellers.**

I. INTRODUCTION

Existing integrated full-duplex (FD) radios with RF self-interference cancellation (SIC) (e.g. [1-3]) are characterized in static environments. Maintaining RF SIC on chip in the presence of a dynamic wireless channel remains challenging. A blind source algorithm and exhaustive search are used in [1] and [2], respectively, to configure the RF cancellers, but result in long tuning time. During the adaptation process, desired signals cannot be received and hence how quickly an RF canceller can tune itself is an important metric.

Microsecond-scale adaptive RF SIC has been reported for FD wireless using off-the-shelf-component-based cancellers [4-7]. In [4], 24µs overall adaptation overhead is shown based on self-interference (SI) channel estimation using digital signal processing (DSP). However, the algorithm in [4] relies on accurate knowledge of canceller *S*-parameters; it is hence questionable whether it can maintain microsecond-scale adaptation when applied to mass production using CMOS which is subject to process, voltage, temperature variations, since canceller calibration is likely required in the field.

Self-adaptive RF SIC using least-mean-square (LMS) algorithm doesn't require canceller calibration and *has shown fast adaptation in FD radios [5-7]*. However, LMS algorithm uses the information at *each* tap, and the works in [5-7] use many off-the-shelf down-converters to access the baseband (BB) data at each RF canceller tap (see Fig. 1). Each LMS down-converter in [5-7] consumes 1.5W DC power which is too high for battery-powered applications. In addition, these LMS down-converters add significant complexity to FD MIMO systems [4]. A CMOS LMS adaptive RF SIC is reported in [8] with 25µs tuning time, but has a narrow SIC bandwidth (BW) due to the lack of on-chip delay generation and operates at a fixed frequency.

To the best of our knowledge, the work in this paper is the first to integrate LMS adaptive RF SIC demonstrated for FD wireless in [5-7] on a CMOS chip. *In contrast to [5-7], in this work, the LMS adaptive circuitry is co-designed with and partially embedded in a wideband RF/analog SIC and a gain-*

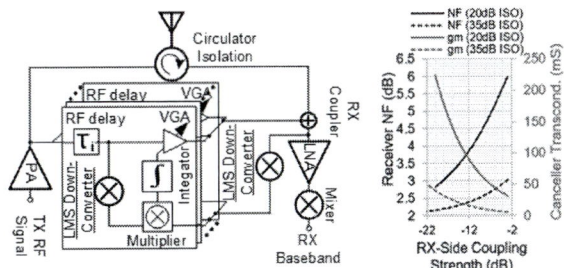

Fig. 1. A wideband RF SIC using LMS-based self-adaptive tuning (left) and SIC power-NF scalability with high and moderate antenna isolation (right).

boosted mixer-first receiver (RX). This *new RFIC-algorithm co-design method* significantly reduces the adaptation overhead by minimizing the number of canceller taps and by fusing multiple functions into one circuit block.

In measurement, the proposed self-adaptive SIC provides ≥24dB RF/analog interference cancellation across 20MHz BW and is widely tunable (0.5-2.5GHz). With the adaptive learning circuitry fully integrated and consuming 14mW, the interference cancellers adapt themselves to an unknown wireless channel in a time period that is as short as 1µs. Furthermore, using a circulator and a custom-made impedance tuner, we demonstrate joint SI suppression across the antenna interface and our self-adaptive RF SIC, achieving an overall SIC of nearly 50dB on average across 20MHz BW with *antenna VSWR up to 3:1*.

Enabled by the proposed self-adaptive SIC, a *real-time CMOS FD 2×2 MIMO RX* is implemented. Furthermore, the FD MIMO RX is *power-scalable*. Each canceller can be reconfigured with lower DC power and stronger passive coupling to RX as the number of elements grows. While stronger coupling result in higher element-level noise (Fig. 1), array-level noise can be reduced by averaging noise across elements [9]. Assuming a MIMO digital beamforming operation [10], we demonstrate the feasibility of having a linear increase in canceller DC power when scaling to a large array, despite a quadratic increase in the number of cancellers.

II. LMS-BASED ADAPTIVE RF SIC

SI or echo cancellation using LMS algorithm is a well-studied topic: transmitted signals with different delays multiply with the residual SI or echo and then the products are integrated to obtain the weights for all the delay taps (see Fig. 1). Down-conversion of the transmitted RF signals to BB not only allows an efficient LMS operation [5-7] but also mitigates the stability issue in an RF LMS feedback loop [8] as a given time delay generated by layout routing corresponds to a much smaller phase shift at BB than that at RF.

978-1-7281-1702-7/19 $31.00 © 2019 IEEE   147

2019 IEEE Radio Frequency Integrated Circuits Symposium

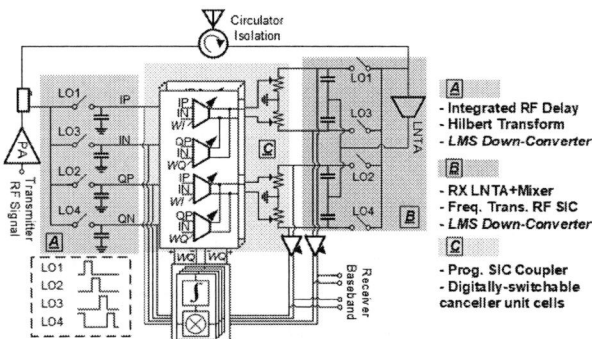

Fig. 2. Proposed self-adaptive and power-scalable wideband RF SIC.

LMS BB multiplication/integration can be realized either by using analog circuitry or DSP. A pure analog implementation results in faster responses but is at expense of consuming large silicon area. DSP in scaled CMOS processes is relatively inexpensive, and digital multiplication and integration can be very compact. However, additional analog-to-digital converters (ADCs) are needed which add more DC power and complexity.

III. PROPOSED ARCHITECTURE

A. Low-Power, Low-Complexity Self-Adaptive RF SIC

To reduce the power and complexity associated with self-adaptive RF SIC, minimizing the number of canceller taps is essential as LMS algorithm needs to access *each* canceller tap [5-7]. A wideband SIC based on RF finite-impulse-response (FIR) filter usually requires ≥5 taps [1,4] and hence needs many down-converters, multipliers, and integrators. In this work, a Hilbert-transform-based wideband RF SIC is utilized for fewer canceller taps given a fixed cancellation BW thanks to the introduction of complex (I/Q) canceller weights [2,5,6].

Nanosecond-scale RF delay for wideband SIC has been demonstrated on chip using N-path filters lately [3]. In this work, the down-conversion in delay-generating N-path filters is repurposed to translate RF signals to analog BB for LMS self-adaptive operations. In addition, the I/Q down-converter in the N-path filter is essentially a widely-tunable Hilbert transformer. *Interestingly, a simple switched-capacitor N-path circuit, as in Fig. 2, simultaneously acts as a delay generator, a Hilbert transformer, and an LMS TX-side down-converter.*

In contrast to [2,3] where an up-conversion mixer is used in the canceller for SIC at RF, the BB cancellation signals in this work are directly injected at the RX BB but the cancellation is transparent to RF by using a gain-boosted mixer-first RX (Fig. 2). Furthermore, the mixer-first RX is repurposed as the LMS down-converter at the RX side. *In essence, one passive mixer wrapping around an inverter-based LNTA (Fig. 2) replaces two mixers, one for the canceller up-conversion and one for RX down-conversion, in [2,3] while concurrently serves as the LMS RX-side down-converter.*

B. Power-Scalable MIMO FD RX

The number of cancellers in a MIMO FD radio could increase quadratically with the number of elements [4]. In

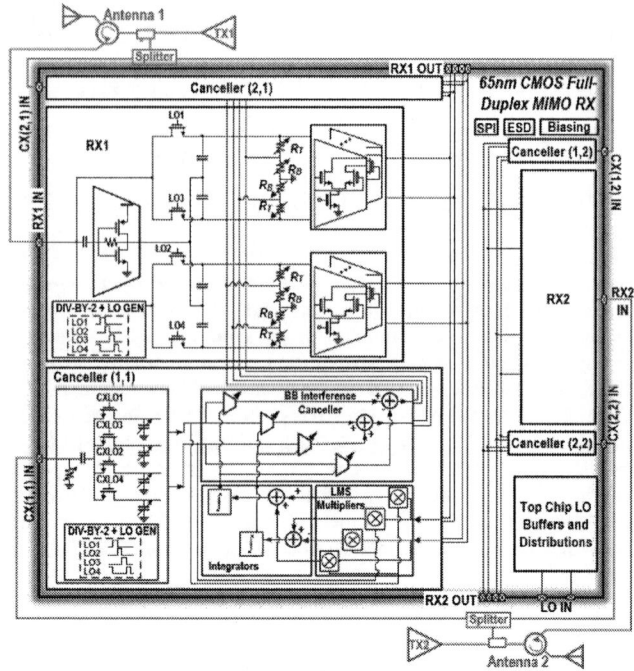

Fig. 3. Schematic of the 65nm CMOS prototype 2×2 FD MIMO RX.

addition to SI, a MIMO FD radio creates cross-interference (XI) due to the coupling among antennas. To cancel SI and XI, 4 cancellers are required in a 2×2 MIMO radio, and hence it is pivotal to have cancellers that are power scalable.

In a SISO FD radio, a weak canceller coupling is usually applied at the RX side to reduce the NF degradation [1-3]. However, this results in a high DC power with a moderate antenna interface isolation (ISO), say 20-25dB, as large power has to be coupled from the transmitter side or to be spent on generating strong cancellation signals from the cancellers (manifested by the canceller's transconductance g_m in Fig. 1). In this work, the coupling strength is made programmable via adjusting potentiometers at RX BB (Fig. 2), and a stronger passive coupling can be used to save DC power. *While the single-element NF degrades with a stronger coupling, the array-level NF can be lowered by averaging the independent noise among different array elements in a MIMO RX [9,10].*

IV. IMPLEMENTATION

Fig.3 shows the schematic of the proposed 2×2 FD MIMO RX implemented in 65nm CMOS. There are 4 cancellers to suppress SI and XI. Each canceller consists of an N-path filter, BB variable gain transconductance amplifiers (VGAs), and a VGA weights self-adaptive circuitry. The BB VGAs are split into 8 digitally-controlled unit cells for MIMO power-NF scalability. The self-adaptive circuitry includes analog multipliers implemented as choppers, analog integrators, and VGA buffers for LMS loop gain control. The canceller outputs are injected at the BB of a 4-path gain-boosted mixer-first RX and the cancellation is frequency-translated to RF. Finally, each RX BB resistor is split into two digitally controlled resistor banks (R_T and R_B in Fig. 3) with the cancellation point in between for MIMO power-NF scalability.

2019 IEEE Radio Frequency Integrated Circuits Symposium

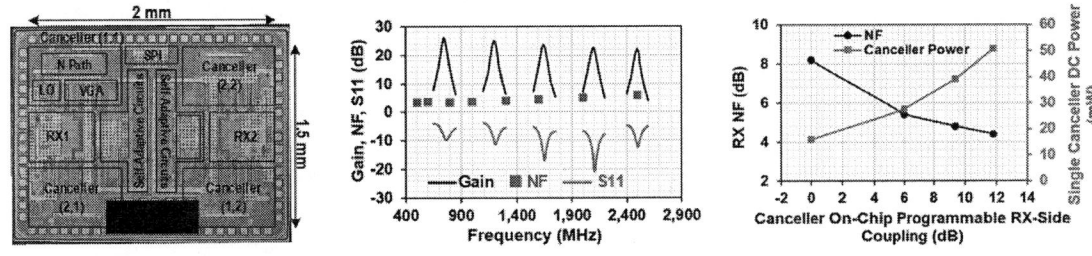

Fig. 4. FD MIMO RX die photo (left); measured RX gain, NF, and input matching from 0.5-2.5GHz (center); measured RF SIC power-NF scalability (right).

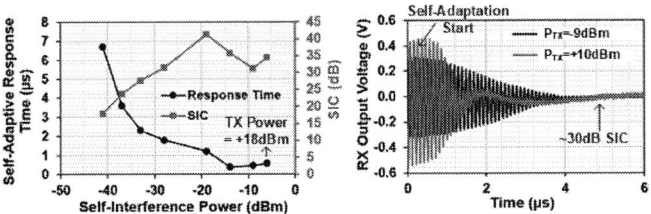

Fig. 5. LMS adaptive RF/analog SIC and its response time measurement.

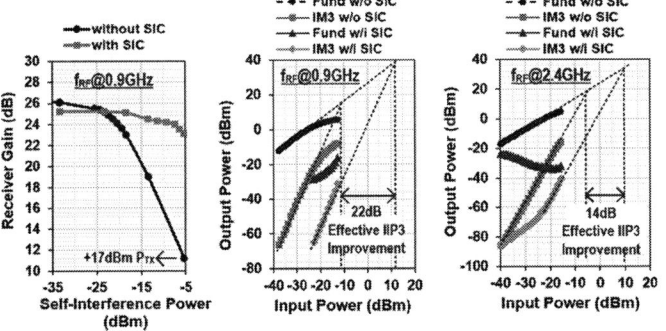

Fig. 6. SIC linearity measurement at 0.9GHz and 2.4GHz.

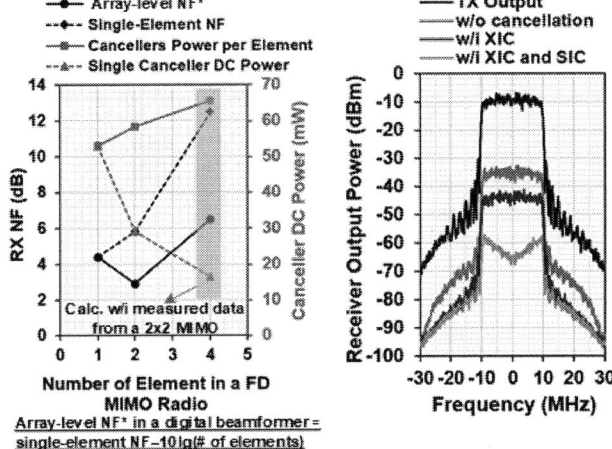

Fig. 8. FD MIMO RX power scalability, SIC and XIC measurement.

V. MEASUREMENTS AND WIRELESS DEMONSTRATION

The prototype 65nm CMOS FD MIMO RX operates over 0.5 to 2.5GHz in measurement (Fig. 4). Each RX element has measured gain of 26dB, NF of 3.1dB, and IIP3 of –10.6dBm with cancellation circuitry disabled. Power-NF scalability is measured using a single canceller. The RX NF degrades to 4.4dB with the weakest RX-side coupling and all eight canceller VGA unit cells on consuming 52mW; NF of 8.2dB is measured with the strongest RX-side coupling and two canceller VGA unit cells on consuming 16mW.

The canceller self-adaptive operation is first measured with a single canceller enabled. Using a single tone TX signal

at +18dBm with antenna interface ISO of 24dB, the canceller adapts itself to 35dB SIC in 0.6µs. The self-adaptive response time increases with weaker TX power levels, and is measured at 6.7µs with TX power of –17dBm (Fig. 5). LMS multiplier, integrator, and buffers consume 14mW in total. *All subsequent cancellation measurements use self-adaptation without any manual weights tuning.*

FD radio linearity tests are depicted in Fig. 6. SIC of up to –5dBm of SI power results in small gain compression (~2dB) of a desired signal, as opposed to nearly 15dB of compression in the absence of SIC. In a two-tone test at 0.9GHz, the SIC improves the effective in-band RX IIP3 from –11dBm to +11dBm. *This 22dB improvement in IIP3 is similar to those in other works using SIC directly at RF, indicating that the BB SIC in this work is transparent to RF and enhances the RF input-side linearity.* Two-tone test with and without SIC at 2.4GHz is also performed and has a 14dB IIP3 improvement.

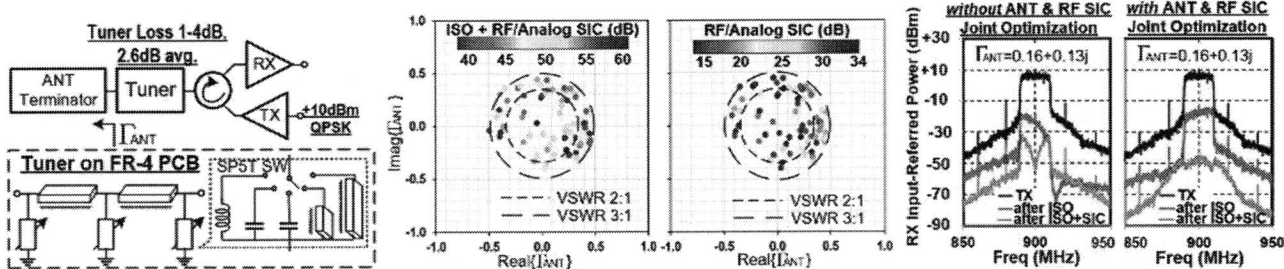

Fig. 7. Setup for joint SI suppression (left); measured SI suppression (center two); SI suppression example with and without joint optimization (right two).

978-1-7281-1702-7/19 $31.00 © 2019 IEEE   149

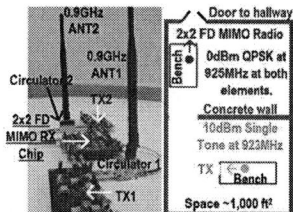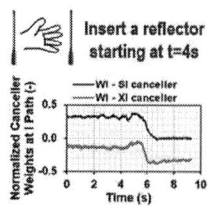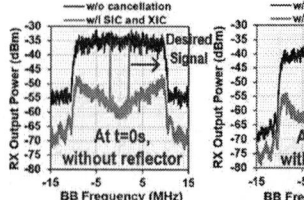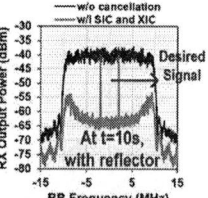

Fig. 9. Demonstration of real-time FD MIMO self-adaptive interference cancellation using low-cost off-the-shelf antennas in a dynamic wireless environment: setup (left two); real-time XIC and SIC weights on cancellers I-paths (center); interference suppression before and after having a reflector (right two).

Table 1. Measurement summary and comparison with state-of-the-art CMOS RF SIC work.

	# of input/ output	Freq. range (GHz)	NF w/o SIC (dB)	Antenna interface ISO(dB)/delay	RF/analog SIC(dB)/ BW	NF degra-dation (dB)	Eff. RX IB IIP3 (dBm)	Max. SI power (dBm)	SIC DC power (mW)	SIC NF-power scalable	RF SIC Adapt. Algorithm	RF SIC Adapt. time (µs)
[1]	1/1	1.7–2.2	2.5	30-35/ 0.3 ns	30(RF)+20(BB) / 40 MHz	1.55	+17	−10	11.5	No	Blind source	N.R.
[2]	1/1	0.9	9.6	50 / 1.4 ns	23 / 80 MHz	1.4	+3.9^	N.R.	13	No	Exhaustive search	N.R.
[8]	1/1	0.8	1.4+	56 / 132 ns	14 / 1 MHz	1.3	N.A.	−30	30.5	No	LMS	25
This work	1/1	0.5–2.5	3.1	22 / 3 ns	29 / 20 MHz	1.2	+11	−5	16-52$	Yes	LMS	1
	2/2			24 / 6 ns	24 / 20 MHz	1.3			32-104$			

^: simulated result.   +: LNA only. $: at 1GHz and not including 14mW LMS power.   N.R.: not reported.   N.A.: not applicable.

Joint SI suppression across the antenna and RF domains using a 900MHz off-the-shelf circulator, a custom-made PCB impedance tuner, and our self-adaptive RF SIC is shown in Fig. 7. *The impedance tuner is configured to ensure the best overall SI suppression (with joint optimization in Fig. 7) rather than the highest circulator ISO (without joint optimization in Fig. 7).* 45dB worst-case overall SI suppression, 51dB average, across 20MHz is measured at 30 different antenna impedance values with VSWR up to 2:1. With larger VSWR up to 3:1, 39dB worst-case suppression, 46dB average, is obtained. The number of antenna impedance points is limited by the number of tuner control bits. *Subsequent analog/digital BB SIC is often performed without joint optimization with RF SIC thanks to the much greater flexibility available at BB than that at RF [1,4,5,6].* Nevertheless, the canceller group delay here (which is pre-defined and held constant like in [1-6]) is made digitally programmable and can be used together with LMS loop gain control to perform joint SI suppression with BB SIC if needed.

A test of simultaneous cancellation of SI and XI is depicted in Fig. 8. The SI channel exhibits a group delay of 2.5ns and 30dB ISO from a circulator, while the XI channel is mimicked by 24dB fixed attention with 5.5ns delay. The XIC is enabled first and provides 6dB suppression, and the SIC further suppresses the total interference by another 18dB. When used as a digital beamformer, the NF-power scalability enables a nearly constant canceller DC power of 52-65mW per element, despite a quadratic increase of cancellers (Fig. 8).

Real-time interference cancellation using low-cost off-the-shelf antennas is demonstrated (Fig. 9). *With the presence of both SI and XI, the FD MIMO RX cancellers adapt themselves in real time and provide a robust ≥20dB suppression in a dynamic wireless environment.* When compared with the state of the art (Table I), our work is the first FD radio with integrated LMS adaptive wideband RF/analog SIC with microsecond-scale response time and is also widely-tunable. In addition, it is the first integrated FD MIMO beamforming RX whose DC power scales almost linearly with the number of antennas thanks to the canceller power-NF scalability.

VI. CONCLUSION

While FD LMS adaptive RF SIC has been demonstrated using off-the-shelf components, it is fully integrated with a 2×2 FD MIMO RX in this work for the first time. A new RFIC-algorithm co-design method is presented which significantly reduces the adaptation overhead by minimizing the number of canceller taps and by fusing multiple functions into one circuit block. A power-NF scalability makes our design the first FD MIMO beamforming RX whose DC power scales almost linearly with the number of antennas.

REFERENCES

[1] T. Zhang, et al., "Wideband Dual-Injection Path Self-Interference Cancellation Architecture for Full-Duplex Transceivers," IEEE JSSC, Jun. 2018.

[2] A. El Sayed et al., "A Hilbert Transform Equalizer Enabling 80 MHz RF Self-Interference Cancellation for Full-Duplex Receivers," in IEEE Transactions on Circuits and Systems I: Regular Papers. (Early Access)

[3] J. Zhou, et al., "Integrated Reconfigurable Wideband Cancellation of Transmitter Self-Interference in the RF Domain for FDD and Full-Duplex Wireless," IEEE JSSC, Dec. 2015.

[4] D. Bharadia, et al., "Full Duplex MIMO Radios", USENIX Conf. on Networked Sys. Design and Implementation, 2014.

[5] T. Huusari, et al., "Wideband Self-Adaptive RF Cancellation Circuit for Full-Duplex Radio: Operating Principle and Measurements," 2015 IEEE 81st VTC Spring, Glasgow, 2015.

[6] D. Korpi, et al., "Full-duplex mobile device: pushing the limits," IEEE Communications Magazine, Sep. 2016.

[7] S. Kim, et al., "A 2.59-GHz RF self-interference cancellation circuit with wide dynamic range for in-band full-duplex radio," IEEE IMS, 2016.

[8] V. Aparin, et al., "An Integrated LMS Adaptive Filter of TX Leakage for CDMA Receiver Front Ends", IEEE JSSC, May 2006.

[9] K. Trotskovsky et al., "A 0.25–1.7-GHz, 3.9–13.7-mW Power-Scalable, −10-dBm Harmonic Blocker-Tolerant Mixer-First RF-to-Digital Receiver for Massive MIMO Applications," in IEEE SSCL, Feb. 2018.

[10] L. Zhang, et al., "Scalable Spatial Notch Suppression in Spatio-Spectral-Filtering MIMO Receiver Arrays for Digital Beamforming", IEEE JSSC, Dec. 2016.

978-1-7281-1702-7/19 $31.00 © 2019 IEEE

A 0.5-to-3.5 GHz Self-Interference-Canceling Receiver for In-Band Full-Duplex Wireless

Ali Ershadi and Kamran Entesari

Analog and Mixed Signal Center, Texas A&M University, College Station, TX, 77843, USA

Abstract — This paper proposes an in-band full-duplex self-interference canceling receiver that achieves transmitter leakage cancellation by scaling the in-phase and quadrature components of TX replica, and injecting the approximated leakage to the RX interface. More than 35 dB cancellation is measured for modulated TX samples. The receiver structure is 8-phase passive-mixer-first with high linearity and on-chip sharp rejection of out-of-band blockers. The receiver has an IB-IIP3 of 6 dBm at 1 MHz offset from 2 GHz carrier, and OB-IIP3 of 27 dBm. The NF at 5 MHz baseband frequency is 3.3 dB at TDD mode. At FD mode, i.e., RX and TX are operating simultaneously at the same frequency band, only 2-2.5 dB noise degradation is observed. The NF reaches 5.3 dB in FD mode. As a proof of concept prototype is fabricated and measured in 65 nm CMOS. The system is functional from 500 MHz to 3.5 GHz.

Keywords — Full duplex (FD), harmonic-rejection mixer (HRM), in-band full-duplex (IBFD), interference cancellation, mixer-first, receiver, self interference cancellation (SIC), wideband.

I. INTRODUCTION

RF spectrum scarcity and the ever-increasing demand for higher bandwidth has drawn tremendous attention to change the today's wireless communication paradigm. Full-duplex (FD) communication offers two-time higher exploitation of spectrum. Specifically, wideband FD paves the way for frequency allocation of a spectral efficient cognitive radio. The challenging task of the receiver in this scenario is to extract the desired signal in the presence of substantial unwanted TX leakage right on top of the desired band. In addition to have a mechanism for leakage cancellation, very linear receivers are suited for such radios so as to minimize the inter-modulation (IM) of the RX and TX-leakage in order to preserve the otherwise corrupted signal [1, 2]. Self-interference-cancellation (SIC) path from TX replica to RX interface needs to adjust the required magnitude scaling and phase-shift of TX replica in a highly linear way. Distorted adjustment creates *residue* after subtraction from TX-leakage. Both the residue and the undesired RX and TX-leaked IM prove to be very difficult, if not impossible, to cancel in subsequent stages (Fig. 1). Accordingly, they can raise the noise floor to an unacceptable high level, and mask the desired signal. Passive cancellation techniques [3] can maintain the cancellation path linearity requirements, hence avoid the residue problem, while keeping the added noise to TX replica to a low level, as compared to active counterparts [4].

Different domains, namely propagation, RF, analog-baseband, and digital may be used for cancellation of leaked TX. In the RX chain, the earlier leakage cancellation occurs, not only the nonlinearity related issues are mitigated,

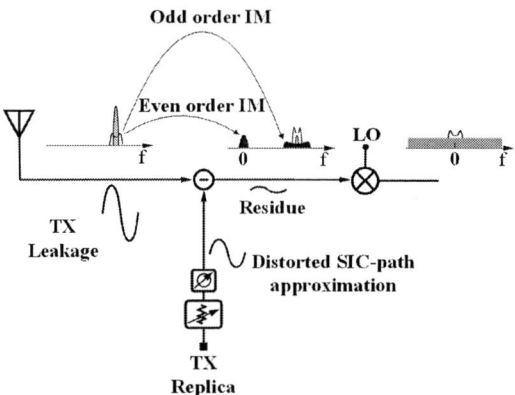

Fig. 1. Nonlinearity mechanisms leading to noise floor degradation in a SIC direct-conversion receiver.

but also a more realistic duplicate of TX-leakage can be approximated. Effects such as TX PA nonidealities, and PA noise skirt can be captured at RF stage more accurately as compared to all back-end cancellation techniques. In this work, high RF down-converted cancellation is aimed. Due to linearity requirements of an FD-RX, mixer-first architecture is chosen, and voltage gain before baseband down-conversion is avoided. Unlike previous implementation trades between RX-linearity and noise performance [5, 6], NF is not sacrificed at the cost of linearity, and an average 35 dB SIC in a single domain is achieved.

II. THE RECEIVER ARCHITECTURE

The SIC block diagram used in this work is shown in Fig. 2a. By means of an RLC network, differential quadrature components are generated from the TX replica. These components are related to each other as

$$\frac{V_I}{V_Q} = \frac{V_{I_p} - V_{I_n}}{V_{Q_p} - V_{Q_n}} = \frac{s^2 + \frac{R}{L}s - \frac{1}{LC}}{-s^2 + \frac{R}{L}s + \frac{1}{LC}}. \quad (1)$$

The amplitude and phase response is plotted in Fig. 2b. Amplitude response is flat irrespective of frequency, however, there is around 20° phase deviation from intended orthogonal components. This deviation is dealt with at SIC synthesis stage. Following the generation of differential I-Q signals from the low-Q RLC network, I-Q pairs' amplitude are scaled by the attenuator to generate the desired components. In order to minimize the added noise, capacitive attenuators are used for each branch which are controlled with two sets of separate 7- bits, respectively for I and Q paths. The differential

978-1-7281-1702-7/19 $31.00 © 2019 IEEE

2019 IEEE Radio Frequency Integrated Circuits Symposium

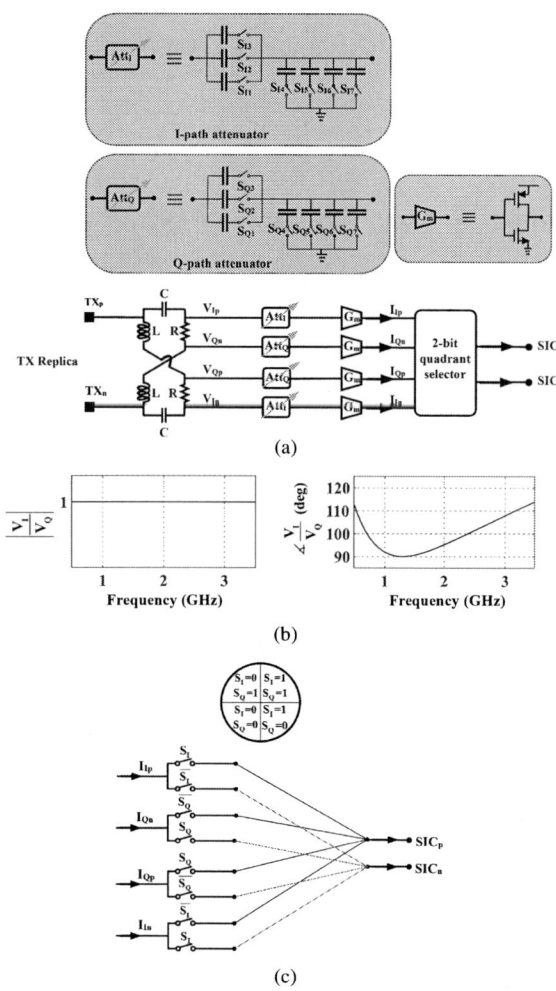

Fig. 2. a) The SIC block: adjustment of TX replica in Cartesian plane for leakage cancellation. b) Amplitude and phase response of $\frac{V_I}{V_Q}$. c) The quadrant selector diagram.

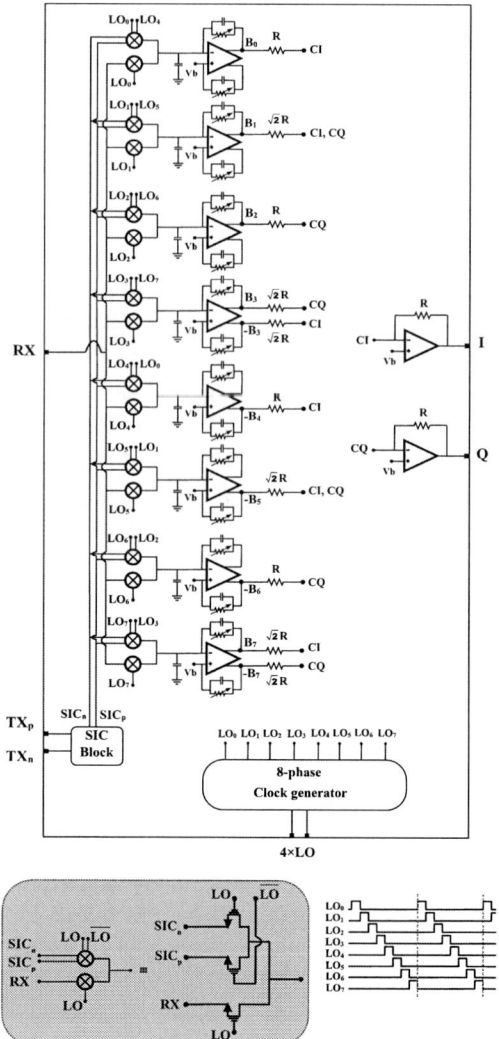

Fig. 3. Circuit diagram of the implemented SIC receiver.

quadrature voltages are then converted into current using transcoductance (Gm) cells. Class-AB Gm-cells are used to this end which can be designed with linear V-to-I conversion, provided the width of devices are chosen appropriately. The pair of I-Q currents are summed by the quadrant selector to generate the approximated leakage. This signal is injected into the current-driven mixers at RX-SIC interface to cancel the leakage. Two bits at the end of the SIC path (quadrant selector) are deciding which I-component (I_{I_p} or I_{I_n}) needs to be added to which Q-component (I_{Q_p} or I_{Q_n}), thereafter, the 4-quadrants are covered. The block diagram of the quadrant selector, used in the SIC block, is shown in Fig. 2c. The differential SIC currents are down-converted by the differential mixers at RX-SIC interface.

The entire SIC receiver circuit diagram is depicted in Fig. 3. The receiver port is connected to 8 passive-mixers, implemented by single-ended NMOS switches. Each RX mixer uses a 12.5% duty cycle LO. The generated SIC

currents with opposite polarity are down-converted by the differential mixers, which use LOs with opposite polarity for the differential branches. Since the LOs have an opposite fundamental Fourier series coefficient polarity, the SIC currents are subtracted; differential SIC currents are converted to single-ended. The cancellation accordingly takes place from the RX after down conversion.

Input matching at the RX is provided by the combination of mixer's switch on-resistance and the transimpedance amplifier (TIA) feedback resistor. Such matching constraint does not hold for the SIC-mixers and the size of switches are optimized for boosting signal-to-noise-and-distortion-ratio. The TIA noise characteristic has a direct impact on the RX NF. Of crucial importance is the TIA flicker noise. Large PMOS input devices are used for TIA so as to alleviate the noise penalty. TX-leakage cancellation occurs in current domain at the input of the TIA. The down-converted base-band signals (Fig. 4) are combined with appropriate weights to produce baseband quadrature components from the eight 45° apart

978-1-7281-1702-7/19 $31.00 © 2019 IEEE   152

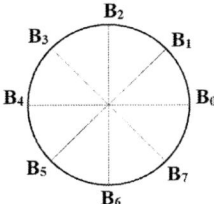

Fig. 4. Down-converted baseband signals at output of the TIAs.

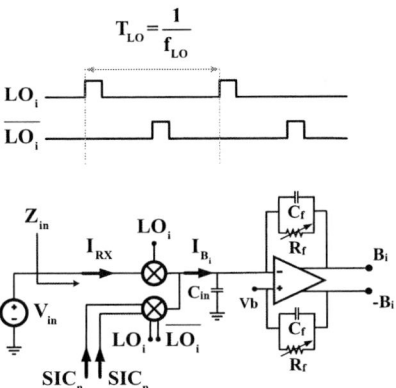

Fig. 5. RX-SIC interface model at close-in offset from f_{LO}.

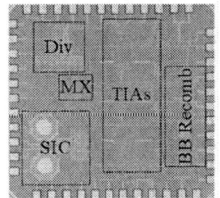

Fig. 6. Die micrograph.

baseband signals at the TIA outputs as follows

$$I = B_0 + \frac{\sqrt{2}}{2}B_1 - \frac{\sqrt{2}}{2}B_3 - B_4 - \frac{\sqrt{2}}{2}B_5 + \frac{\sqrt{2}}{2}B_7 \quad (2)$$

$$Q = \frac{\sqrt{2}}{2}B_1 + B_2 + \frac{\sqrt{2}}{2}B_3 - \frac{\sqrt{2}}{2}B_5 - B_6 - \frac{\sqrt{2}}{2}B_7. \quad (3)$$

After down-conversion all blocks ahead use feedback, hence linearity at the back-end is maintained at a high level. Open-loop recombination of down-converted components for baseband I-Q generation, such as transconductance current-summation technique used in [2]-[4], suffer linearity compared to closed-loop recombination used here (Fig. 3). Four times the LO clock is taken off-chip and is divided on-chip to produce the 8-nonoverlapping LO phases. The divider circuit is designed carefully to minimize the effect of LO phase noise, and skew on RX NF. The SIC block uses 16 control bits to adjust the TX replica for SIC. The optimum setting for switches (Fig. 2) is obtained from successive approximation algorithm that alternates between I and Q states at each step to find the minimum noise floor at the RX output, which translates to the closest TX leakage approximation.

Assuming the received signal from the antenna around the frequency of LO is composed of the desired signal (s) and TX leakage (l). In very close vicinity of f_{LO}, to be considered in-band channel, $Z_{in}(f_{LO}) = Z_0$, thus the current conducted by the ith mixer, as depicted in Fig. 5, is equal to

$$I_{RX} = \frac{V_{in}}{Z_0} = s + l. \quad (4)$$

This current is multiplied by LO_i, and is down-converted to the baseband. For the case of SIC currents, since LO_i and $\overline{LO_i}$ are $\frac{T_{LO}}{2}$ apart, their fundamental Fourier coefficients are the negative of one another. Therefore, I_{B_i} is calculated as

$$I_{B_i} = \sum_{\pm} \mathcal{F}_{i\pm 1}\{S + [L - (SIC_p - SIC_n)]\}_{(f \mp f_{LO})}, \quad (5)$$

$$\mathcal{F}_{i\pm 1} = \frac{sin(\frac{\pi}{8})}{\pi}e^{\mp j\frac{\pi(2i+1)}{8}}, \quad i = 0, 1, 2, ..., 7 \quad (6)$$

where uppercase signals represent the Fourier transform, and $\mathcal{F}_{i\pm 1}$ is the fundamental Fourier coefficient of LO_i. As shown in (5), in case of precise SIC approximation, i.e., $L = SIC_p - SIC_n$, only the desired down-converted signal will flow through R_f. C_{in} and C_f act as open circuit here. The baseband voltage at the output of the TIAs is given by

$$B_i = \sum_{\pm} \mathcal{F}_{i\pm 1}R_f S_{(f \mp f_{LO})}. \quad (7)$$

Which are the desired down-converted signals from $\pm f_{LO}$ to the baseband. These components are summed with the required weightings, obtained in (2) and (3), to generate the down-converted baseband quadrature components. This simplified analysis is only valid for close-in offset from the LO where the leakage cancellation occurs. Many aspects such as the filtering effect of capacitors on in- and out-of-band blockers, harmonics folding, and the limited gain-bandwidth of the TIA are neglected, and are beyond the scope of this paper. These are to be treated in future extension.

III. MEASUREMENTS RESULTS

A test chip was fabricated in 65 nm CMOS. The entire SIC-RX occupies an active area of 1.5 mm². The die micrograph is shown in Fig. 6. The measured performance of SIC-RX is shown in Fig. 7. NF measurement is illustrated in Fig. 7a, in TDD mode NF reaches 3.3 dB which is comparable with LNA-first receivers with superior linearity. Assuming 10 dBm TX output power and 20 dB attenuation in propagation domain (antenna-isolation in 2-antenna or circulator's isolation in shared antenna systems), -10 dBm leakage is imposed on RX (FD mode). By performing the leakage cancellation there is only 2-to-2.5 dB NF degradation compared to TDD mode, and NF reaches 5.3 dB. Over the whole frequency of operation, S_{11} is better than -15 dB (Fig. 7b). As the input matching is function of the value of RX-mixer on-resistance, TIA open-loop gain, and TIA feed-back resistor which all are process-dependent, 4-bits for controlling the value of TIA

Table 1. Measurement summary and comparison with recently published full-duplex receivers.

Specification	[2]	[3]	[6]	This work
RF frequency [GHz]	0.61-0.975	1.7-2.2	0.15-3.5	0.5-3.5
RX gain [dB]	28	20-38	24	30-50
OB-IIP3 [dBm]	+15.4	N/A	+22	+27
IB-IIP3 [dBm]	-18.4	-5	+8/ +16*	+6
NF in TDD mode [dB]	6.3	4	6.3	3.3
NF in FD mode [dB]	8	5/ 5.5**	10-12.3	5.3
Cancellation domain	Circulator + RF	RF + Analog BB	RF down-converted	RF down-converted
Number of cancellation domain(s)	2	2	1	1
SI suppression [dB]	40	55	27	35
RX power consumption [mW]	72	22	23-56***	15-38***
Canceler power consumption [mW]	36	11.5	N/A	4-12
Supply voltage(s) [V]	1.2/2.4	1.2/1.8/2.4	1.2	1.2/2.4
Active area [mm^2]	0.94	3.5	2	1.5
Technology	65 nm CMOS	40 nm CMOS	65 nm CMOS	65 nm CMOS

* Negative conductance off/ on
** Estimated (from Fig. 22 of [3]), w/ and w/o LO sideband suppression
*** Includes LO generation power consumption

feed-back resistor (R_f in Fig. 5) are used to accommodate these variations and achieve acceptable matching at the RX input. Two-tone test results are illustrated in Fig. 7c, IB-IIP3 of +6 dBm is measured at 1 MHz offset frequency, and OB-IIP3 is +27 dBm. Un-calibrated IIP2 of better than +70 dBm was measured. All the two-tone tests are carried at maximum gain. The SIC-BW is higher than 10 MHz, and accommodates the intended modulation BW. Fig. 7d depicts the conversion gain of the RX. The RX-BW is designed for 10 MHz, same as the BW of the emulated TX signals. As shown in Fig. 8a, more than 35 dB cancellation was measured for 100 samples of TX replica emulated by 10 MHz-BW 64-QAM signals, the power of the TX replica is normalized to 0 dB to illustrate the cancellation. The packaged chip and ports' connection are shown in Fig. 8b.

IV. CONCLUSION

Table 1 compares this work with other recently published full-duplex self-interference-canceling receivers. NF is lower than any other reported mixer-first, and many of LNA-first receivers. Due to the interference cancellation mechanism, and high linearity of the receiver, there is only a small noise-distortion degradation in the presence of large TX-leakage, which shows a high dynamic range is achievable in FD wireless receivers.

REFERENCES

[1] D. Yang, et al., "A Wideband Highly Integrated and Widely Tunable Transceiver for In-Band Full-Duplex Communication," *IEEE J. Solid-State Circuits*, vol. 50, no. 5, pp. 1189-1202, May 2015.

[2] N. Reiskarimian, et al., "Highly-Linear Magnetic-Free Circulator-Receiver for Full-Duplex Wireless", *ISSCC Dig. Tech. Papers*, pp. 316-317, Feb. 2017.

[3] T. Zhang, et al., "Wideband dual-injection path self-interference cancellation architecture for full-duplex transceivers," *IEEE J. Solid-State Circuits*, vol. 53, no. 6, pp. 1563-1576, Jun. 2018.

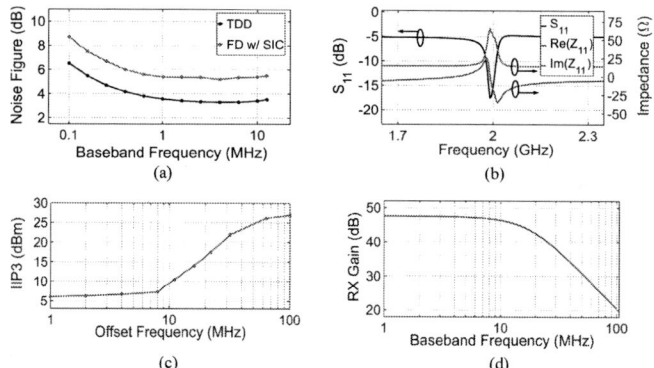

Fig. 7. Measured SIC-RX performance at 2 GHz RF frequency. a) NF in TDD and FD mode. b) RX input matching. c) IIP3 vs offset frequency in the two-tone test. d) RX conversion gain.

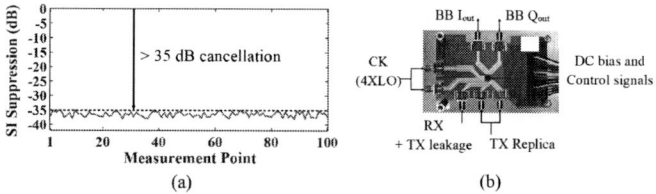

Fig. 8. a) Self interference (SI) suppression of 100 samples of TX replica. b) Packaged chip, and ports' connection in SIC measurement set up.

[4] J. Zhou, et al., "Reconfigurable Receiver with >20MHz Bandwidth Self-Interference Cancellation Suitable for FDD, Co-Existence and Full-Duplex Applications," *ISSCC Dig. Tech. Papers*, pp. 342-343, Feb. 2015.

[5] C. Andrews and A. Molnar, "Implications of passive mixer transparency for impedance matching and noise figure in passive mixer-first receivers," *IEEE Trans. Circuits Syst. I*, vol. 57, no. 12, pp. 3092-3103, Dec. 2010.

[6] D. J. van den Broek, et al., "An in-band full-duplex radio receiver with a passive vector modulator downmixer for self-interference cancellation," *IEEE J. Solid-State Circuits*, vol. 50, no. 12, pp. 3003-3014, Dec. 2015.

A Baseband-Matching-Resistor Noise-Canceling Receiver Architecture to Increase In-Band Linearity Achieving 175MHz TIA Bandwidth with a 3-Stage Inverter-Only OpAmp

Anoop Narayan Bhat[#1], Ronan van der Zee[#2], Salvatore Finocchiaro[*], Francesco Dantoni[*], Bram Nauta[#3]

[#]University of Twente, Enschede, The Netherlands

[*]Texas Instruments

[1]a.n.bhat@utwente.nl, [2]ronan.vanderzee@utwente.nl, [3]b.nauta@utwente.nl

Abstract—In this paper we propose a baseband noise-canceling receiver architecture to increase in-band linearity. Key feature of the architecture is that all active circuits are in baseband, including the LNTA. The receiver targets high IF bandwidths, enabled by a TIA composed of an OpAmp using only inverters. The receiver is fabricated in 22nm FDSOI CMOS. Measured results show an in-band IIP3 of > 9dBm for an IF bandwidth of 175MHz with sub-5dB NF across 1-6GHz LO.

Keywords—Base-station, in-band linearity, noise-canceling, wide-band IF, Inverters-only OpAmp, TIA, LNTA, IIP3

I. INTRODUCTION

High in-band linearity has become important in many sub-10GHz CMOS receiver applications: 1) High in-band linearity is necessary in applications where the band of interest may contain many signals, such as in cognitive radio, base station applications [1] and intra-band carrier aggregation scenarios [2]. 2) Most self-interference cancellation techniques used for in-band full-duplex receivers involve significant cancellation in the digital domain [3]. Therefore RF and analog receiver front-ends need to be sufficiently linear for the success of such techniques. 3) MIMO applications involving beam-forming either in digital or combination of analog and digital domains need high in-band linearity [4].

Also, most of the above applications are increasingly targeting higher IF-bandwidths in order to fulfill higher data-rate requirements. For example, recent works on base-station receiver designs [5], [6] have targeted high IF-bandwidth to support all 3GPP bands.

Fig. 1 shows various mixer-first receiver architectures in the context of in-band linearity. The receiver in Fig. 1(a) [7] does not achieve high in-band linearity due to significant signal swing at the input of the TIA, and similar is the case for [8]. Even though [1] shown in Fig. 1(b) can achieve high in-band linearity, it is more noisy due to the dedicated 50Ω matching resistor. In the noise-canceling topology of Fig. 1(c) [9] , linearity is limited by the LNTA, which operates at RF frequencies. Fig. 1(d) [10] shows another way to increase in-band linearity, but due to its lack of input matching it is not practical in many applications.

We propose a noise-canceling receiver architecture where all active circuits work at baseband frequencies. It features

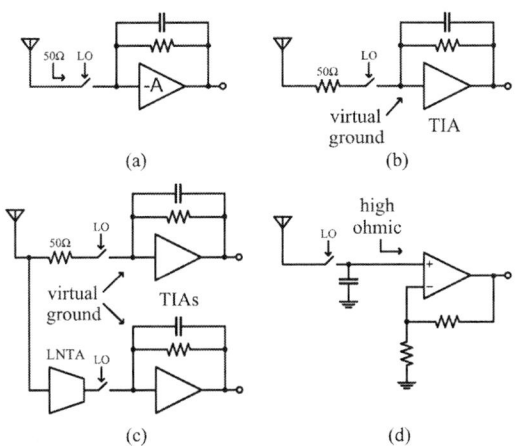

Fig. 1. Representative mixer-first receiver architectures (a) [7], (b) [1], (c) [9] , and (d) [10] for comparing their in-band linearity along with noise, matching, and OoB linearity performances

high bandwidth, high in-band linearity, OoB filtering and good input matching. The coming sections discuss the architecture, the circuits and measurements on a prototype chip.

II. ARCHITECTURE

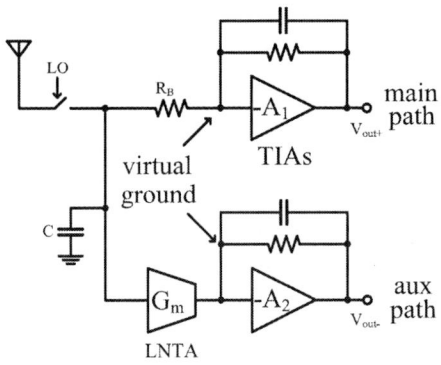

Fig. 2. Proposed baseband-matching-resistor noise-canceling receiver architecture

The proposed receiver is shown in Fig. 2. It is a noise-canceling architecture with the feature that all active circuits work at baseband frequencies. Input matching is provided by R_B, whose impedance is frequency translated to the input by the passive mixer. An auxiliary path containing an LNTA with transconductance G_m cancels the noise of this resistor. Note that the N-path filter formed by the source impedance and the capacitor C rejects OoB interferers.

The proposed receiver architecture achieves higher in-band linearity mainly because of the virtual ground at the input of the TIAs and the LNTA operating in baseband. The virtual ground at the input of the TIAs not only reduces the swing at the input of the TIAs, but also allows the loop-gain to be > 1 unlike in the case of the architecture shown in Fig. 1 (a). Higher loop-gain further reduces the distortion produced in the TIAs.

Additionally, operating the LNTA in baseband enables the use of feedback to achieve the desired linearity. Feedback not only helps to improve the linearity of the LNTA, but also makes the linearity robust to PVT changes. Most RF LNTAs do not have this luxury, such that they dominate the overall non-linearity, with linearization techniques suffering from variation across PVT as explained in [11].

Since this noise-canceling architecture has the matching resistor and all noise-canceling circuits in baseband, we refer to this architecture as BaseBand Noise-Canceling (BBNC) in this paper.

III. CIRCUIT DESIGN

The receiver targets multi-band and multi-channel applications with a bandwidth of more than 100MHz, requiring high in-band linearity and low noise over a wide range of LO frequencies.

A. Mixer and LNTA

The circuit in Fig. 2 is realized 4 times sharing one differential antenna input, providing quasi differential I and Q paths, with the passive mixer switches operating at 25% duty cycle. The passive mixer switch sizes are chosen such that they have small on-resistance to minimize noise which cannot be canceled. The N-path filter capacitors C are 2pF each. Fig. 3(a) shows the LNTA. The value of R_{LNTA} must be chosen low enough to limit their noise contribution, while $(g_{Mp} + g_{Mn}) \cdot R_{LNTA}$ must be sufficiently high to achieve the required in-band IIP3. This requires large transistor sizes. In [9], where the LNTA operates at RF, larger input transistors would degrade input matching. However, in the proposed architecture, large transistor sizes only (slightly) affect the bandwidth of the N-path filter formed by the source resistance, mixer switches and effective capacitor C shown in Fig. 2. The bias voltages are set by the replica bias circuit in Fig. 3(b), which keeps the dc output voltage of the LNTA near mid-supply. This is also the input dc voltage of the TIA to which the LNTA output is connected.

B. TIA

Designing for a wide IF-bandwidth in mixer-first receivers boils down to the design of a wide-band TIA. The TIA OpAmp is realized with only inverters as gain stages. This

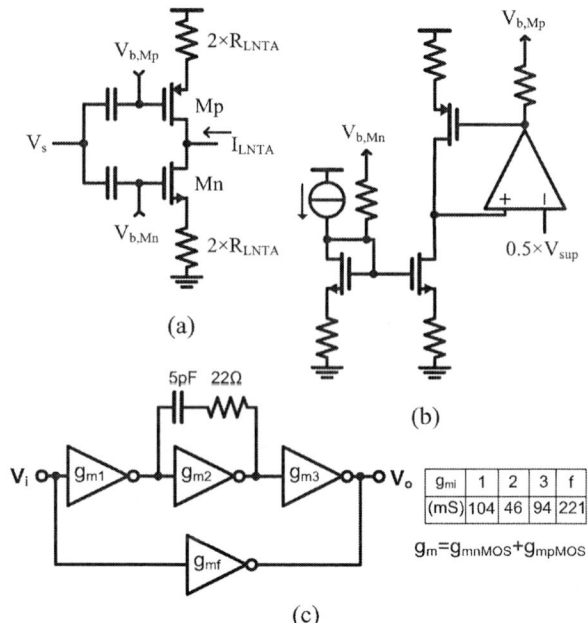

Fig. 3. (a) LNTA schematic, (b) its biasing circuit, and (c) 3-stage inverter-only OpAmp architecture

avoids unnecessary internal nodes such that the bandwidth can be high. It also offers other advantages like current re-use and rail-to-rail output swing [12],[13].

Fig. 3(c) shows the 3-stage inverter-only OpAmp designed for a UGB of 6GHz. It consists of three cascaded inverters in the main path and a feed-forward inverter path. The main path gives sufficient gain till 175MHz to achieve the required linearity. A combination of Miller compensation with right half plane zero removal and a high frequency feed-forward inverter path stabilizes the OpAmp. Note that the circuit parameters mentioned in Fig. 3(c) are for the OpAmp $-A_2$ of Fig. 2. For OpAmp $-A_1$, the corresponding values are scaled to obtain the required loop-gain.

The high loop gain of the OpAmps leads to a good virtual ground at the input, which reduces the input voltage swing, improving linearity. Also distortion caused by output voltage swing V_{out} is reduced by the high loop gain.

Note that common mode control is not necessary, since the OpAmps work independently and there is no coupling between any of the I/Q/+/-/main/aux paths.

IV. EXPERIMENTAL RESULTS

The receiver was realized on chip in a 22nm FDSOI CMOS process. The active chip area is $0.48mm^2$ and it works at a supply voltage of 0.83V. Placement of the various receiver blocks in the chip is shown in Fig. 4. Mixer switches are placed near to the bond-pad so that RF routing is minimal. The four capacitors are placed near the mixer switches to provide short return paths for the high frequency currents. The clk block consists of a $\div 2$ circuit and a 4-phase 25% duty-cycle generation circuit.

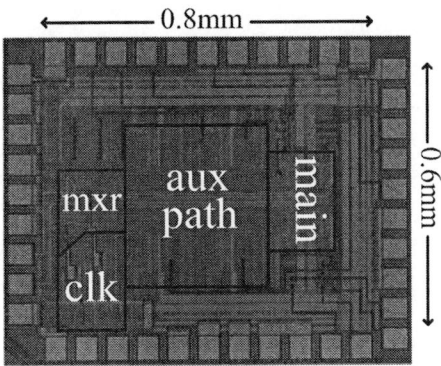

Fig. 4. Chip photo showing various receiver blocks

A. Test Setup

Fig. 5 shows the test setup used to measure the receiver. The measurements were performed with a single-ended source, followed by a passive balun driving the receiver. The differential output voltage is measured by an active differential probe. On-chip common-source amplifiers and all-pass voltage attenuator circuits are used at the output to measure NF and IIP3 respectively such that noise and distortion of the active differential probe do not dominate the respective measurements. The corresponding gain and attenuation were de-embedded. Although Fig. 5 shows circuits to measure NF of the I-path and IIP3 of the Q-path, the receiver has provisions to measure both NF and IIP3 of both the I and Q paths.

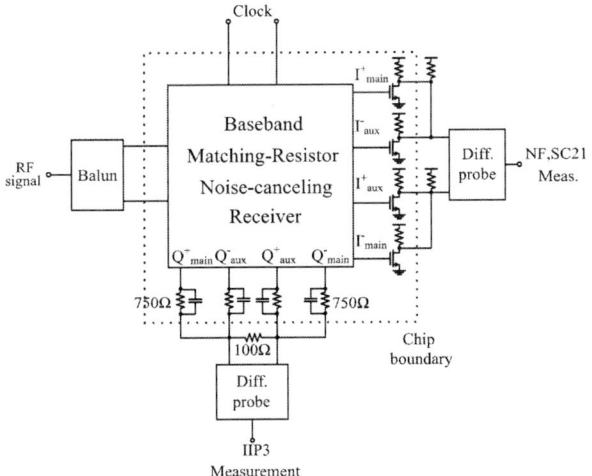

Fig. 5. Test setup to measure the receiver

B. Measurement Results

The bandwidth of the receiver is measured to be 175MHz. Fig. 6(a) shows the IIP3 measured using two-tone tests for both in-band and OoB. In case of in-band IIP3 measurement, two tones f_1 and f_2 are at $\Delta f - 2MHz$ and $\Delta f + 2MHz$ respectively. For the OoB IIP3 measurement, two tones f_1 and f_2 are at Δf and $2\Delta f - 50MHz$ such that the IM3 products

are always at 50MHz. In-band IIP3 is >9dBm for all Δf within the TIA bandwidth.

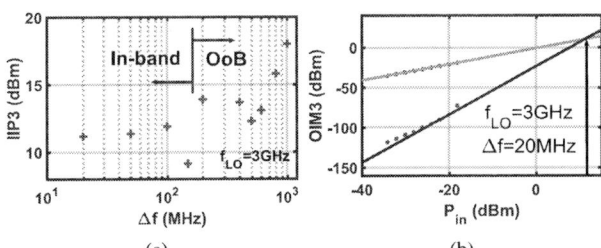

(a)   (b)

Fig. 6. Measured (a) IIP3 across interferer offset frequencies and (b) OIM3 at 20MHz offset

Fig. 6 (b) shows the measured IM3 curve for Δf=20MHz. This measurement shows that the IIP3 is valid till an input power of -20dBm. Fig. 7 (a) shows the measured S_{11} at 3GHz LO frequency. The N-path filtering action can be observed in the measured S_{11}. Fig. 7 (b) shows the measured noise figure and conversion gain (SC_{21}) across IF frequencies at 3GHz LO frequency. The < 3dB NF at 80MHz confirms the noise-canceling properties of the circuit. This matches with the simulation which shows 2.75dB, compared to 7.4dB without the auxiliary (noise-canceling) path at same offset (80MHz) and LO (3GHz) frequencies.

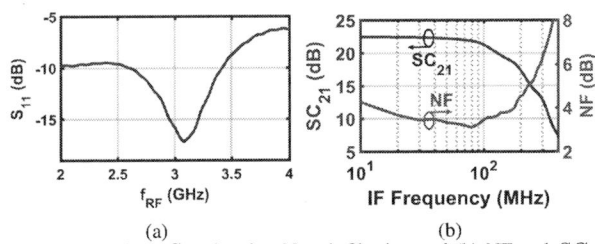

(a)   (b)

Fig. 7. Measured (a) S_{11} showing N-path filtering and (b) NF and SC_{21} at 3GHz LO

Fig. 8 shows the flexible nature of the receiver for multi-band and multi-channel operation. It shows the measured NF, SC_{21} and IIP3 at 50MHz offset and S_{11} across LO frequencies from 1-6GHz. Note that the large input capacitance of the LNTA does not degrade the S_{11} of the receiver, as explained in section III. NF and IIP3 stay around 3dB and 10dBm respectively for the measured LO sweep of 1-6GHz. Though the receiver is functional till 8GHz in extracted simulations, we measured it till 6GHz due to the frequency-limitation of the 2×LO source (12.75GHz) feeding the *clk* block.

C. Comparison

Table I lists in-band IIP3 and TIA bandwidth (along with other performance parameters) of state of the art receivers and compares it to our receiver performance. The proposed receiver has the highest IIP3 except for [6], which achieves a band-edge IIP3 of 12dBm, but this is after de-embedding the off-chip LNA, which is reflected in the higher noise figure

978-1-7281-1702-7/19 $31.00 © 2019 IEEE

Table 1. Result summary and comparison with prior art

	This Work	**JSSC16[2]**	**JSSC12[9]**	**JSSC18[8]**	**ISSCC16[5]**	**ISSCC18[6]**
Architecture	BBNC	Mixer-first	FTNC	Mixer-first	ZIF RX	ZIF RX
Technology	22FDX	28nm	40nm	45nm SOI	45nm	65nm
In-band IIP3 (dBm)	9	7 [*]	-3 [‡]	-10 [‡]	-4 [‡]	12 [††]
IF bandwidth (MHz)	175	50	2	10	50	100
f_{RF} **(GHz)**	1-6	0.4-3.5	0.08-2.7	0.2-8	0.4-4	0.4-6
NF (dB)	2.5-5	2.4-2.6	1.9	2.3-5.4	2	12
Power (mW)	172	38-75	35.1-78	$50+30$ [°] $\times f_{LO}$	200	6600 [**]
Area (mm^2)	0.48	0.23	1.2	0.8	49 [**]	68.7 [**]

[*] Measured for 20MHz bandwidth, baseband supply = 1.5V
[†] De-embedding off-chip LNA
[°] mW/GHz
[‡] taken from band-edge IIP3
[**] Full SOC

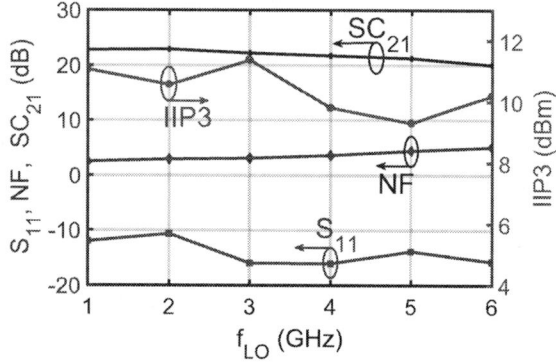

Fig. 8. Measured S_{11}, NF, SC_{21}, and IIP3 across LO frequencies

of 12dB. [2] reports an in-band IIP3 of 7dBm, however this is measured for a TIA bandwidth of 20MHz and uses a dual analog supply with a higher 1.5V supply for the baseband circuits. Our bandwidth also compares favourably to other work. We mainly target base-station applications and our power numbers are lower compared to [5] and [6] which target the same application.

V. CONCLUSION

The proposed receiver can achieve high in-band linearity over a wide RF frequency range of 1-6GHz. This is mainly because all active circuits operate in baseband and can be designed using feedback, both the LNTA and the TIA. Due to the noise-canceling properties, input matching is good with low NF. An inverter-only multi-stage OpAmp enables the wide IF bandwidth of 175MHz.

ACKNOWLEDGEMENTS

We would like to thank Texas Instruments for funding this project and Global Foundries for silicon donation. We thank H. de Vries for help during measurement and G. Wienk for CAD assistance.

REFERENCES

[1] D. Mahrof, E. Klumperink, Z. Ru, M. Alink, and B. Nauta, "Cancellation of opamp virtual ground imperfections by a negative conductance applied to improve rf receiver linearity," *IEEE J. Solid-State Circuits*, vol. 49, May 2014.

[2] C. Wu, Y. Wang, B. Nikolić, and C. Hull, "An interference-resilient wideband mixer-first receiver with LO leakage suppression and i/q correlated orthogonal calibration," *IEEE J. Solid-State Circuits*, vol. 64, pp. 1088–1101, Apr. 2016.

[3] D. Broek, E. Klumperink, and B. Nauta, "An in-band full-duplex radio receiver with a passive vector modulator downmixer for self-interference cancellation," *IEEE J. Solid-State Circuits*, vol. 50, pp. 3003–3014, Dec. 2015.

[4] H. Krishnaswamy and L. Zhang, "Analog and RF interference mitigation for integrated MIMO receiver arrays," *Proc. IEEE*, vol. 104, no. 3, pp. 561–575, Mar. 2016.

[5] N. Klemmer *et al.*, "A 45nm CMOS RF-to-bits LTE/WCDMA FDD/TDD 2×2 MIMO base-station transceiver SoC with 200MHz RF bandwidth," in *IEEE ISSCC Dig. Tech. Papers*, Feb. 2016, pp. 164–165.

[6] D. McLaurin *et al.*, "A highly reconfigurable 65nm CMOS RF-to-bits transceiver for full-band multicarrier TDD/FDD 2G/3G/4G/5G macro basestations," in *IEEE ISSCC Dig. Tech. Papers*, Feb. 2018, pp. 162–164.

[7] C. Andrews and A. Molnar, "A passive mixer-first receiver with digitally controlled and widely tunable RF interface," *IEEE J. Solid-State Circuits*, vol. 45, pp. 2696–2708, Dec. 2010.

[8] Y. Lien, E. Klumperink, B. Ternbroek, J. Strange, and B. Nauta, "Enhanced-selectivity high-linearity low-noise mixer-first receiver with complex pole pair due to capacitive positive feedback," *IEEE J. Solid-State Circuits*, vol. 53, pp. 1348–1360, May 2018.

[9] D. Murphy, H. Darabi, A. Abidi, A. Hafez, A. Mirzaei, M. Mikhemar, and M. Chang, "A blocker-tolerant, noise-cancelling receiver suitable for wideband wireless applications," *IEEE J. Solid-State Circuits*, vol. 47, pp. 2943–2963, Dec. 2012.

[10] M. Soer, E. Klumperink, Z. Ru, F. Vliet, and B. Nauta, "A 0.2-to-2GHZ 65nm CMOS receiver without LNA achieving >11dBm iip3 and <6.5dB NF," in *IEEE ISSC Dig. Tech. Papers*, Feb. 2009, pp. 222–223,223a.

[11] H. K. Subramaniyan, E. Klumperink, V. Srinivasan, A. Kiaei, and B. Nauta, "RF transconductor linearization robust to process, voltage and temperature variations," *IEEE J. Solid-State Circuits*, vol. 50, pp. 2591–2602, Nov. 2015.

[12] B. Nauta, "A CMOS transconductance-C filter technique for very high frequencies," *IEEE J. Solid-State Circuits*, vol. 27, pp. 142–153, Feb. 1992.

[13] R. Veldhoven, R. Rutten, and L. Breems, "An inverter-based hybrid $\Sigma\Delta$ modulator," in *IEEE ISSCC Dig. Tech. Papers*, Feb. 2008, pp. 491–493.

2019 IEEE Radio Frequency Integrated Circuits Symposium

A 350mV Complementary 4-5 GHz VCO based on a 4-Port Transformer Resonator with 195.8dBc/Hz Peak FOM in 22nm FDSOI

Omar El-Aassar[#1], Gabriel M. Rebeiz[#2]

[#]Department of Electrical and Computer Engineering, University of California, San Diego

[1]oelaassa@eng.ucsd.edu, [2]rebeiz@ece.ucsd.edu

Abstract— This paper presents an ultra-low voltage and power complementary VCO topology based on a 4-port transformer resonator. The design benefits from the high current efficiency of a CMOS topology and the low phase noise and supply voltage of the NMOS/PMOS-only structure without sacrificing reliability. A 4-port transformer resonator is used to provide two differential-mode (DM) and two common-mode (CM) harmonic impedances for lower phase noise, voltage supply, and sensitivity to CM tuning. The CMOS VCO is implemented in 22nm FDSOI with a core area of 0.19 mm^2. The VCO dissipates < 0.45mW from 350mV supply while achieving a peak figure-of-merit (FOM) of 192-195.8 dBc/Hz across the 20% continuous tuning range of 4.06-to-4.96 GHz. To the authors knowledge, the 4-port resonator-based CMOS VCO has the highest reported FOM for oscillators with sub-0.5mW power consumption and the lowest supply voltage (350mV) for complementary designs.

Keywords — Voltage-controlled oscillator (VCO), figure-of-merit (FOM), low phase noise, transformer, ultra-low voltage/power, FDSOI CMOS.

I. INTRODUCTION

The advancements in ultra low voltage and power (ULV/P) circuit techniques enable near battery-free energy harvesting and IoE applications. In these systems, the phase noise (PN) requirements are generally relaxed, and efficient ULV/P operation sets the target for LC oscillators. The figure-of-merit (FOM) is, thus, the parameter to be optimized rather than the PN. The peak FOM can be expressed as [1]:

$$\text{FOM}_{\text{MAX}} \propto 20\log(Q) + 10\log(\eta) + 10\log(1 + \alpha/Av), \quad (1)$$

where Q is the effective tank quality factor and is primarily dependent on the technology, η is the DC-to-RF efficiency ($P_{\text{RF}}/P_{\text{DC}}$) which can be optimized by waveform engineering for minimum I/V overlap in switching FETs, α is the effective transconductor current noise shaped by the impulse sensitivity-fuction (Γ), and Av is the voltage gain across the transconductor which can suppress device noise contribution for high Av values. High Av values are usually achieved using a step-up transformer between the drain and gate of the transconductor in high-performance low-PN NMOS topologies [2]. The gate voltage swing, however, greatly surpasses $2V_{\text{DD}}$ posing reliability concerns and forcing the usage of thick-oxide core devices which limit the loop gain, η, and the tuning range. Reducing V_{DD} to reliable levels results in a reduced loop gain and penalizes the PN and FOM [3].

In this work, we propose a reliable and ULV/P complementary VCO suitable for energy harvesting IoE applications when V_{DD} is < 400mV and P_{DC} is < 0.5mW

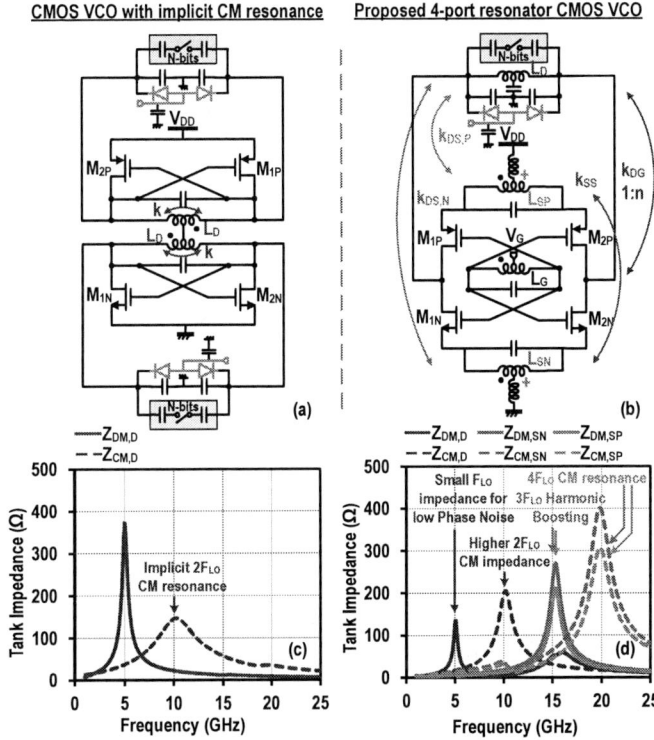

Fig. 1. A CMOS VCO with implicit CM resonance and its composite tank impedance (a & c); Proposed CMOS VCO based on a 4-port resonator and its composite tank impedance (b & d).

while maintaining an excellent FOM of 192-195.8dBc/Hz across the achieved 20% continuous tuning range.

II. COMPLEMENTARY VCO DESIGN

We first consider the CMOS VCO with implicit common-mode (CM) resonance where an NMOS and a PMOS VCOs with identical tanks are tied from their center-tap to form a CMOS structure [1] (Fig. 1(a)). A complementary VCO topology is favored over NMOS or PMOS-only in ULP applications since it results in $\sim 4\times$ less P_{DC} for the same tank (and 6-dB higher PN) under the same FOM. The CMOS topology maintains the CM resonance around the second harmonic ($2F_{\text{LO}}$) of individual tanks resulting in a smaller effective impulse-sensitivity-function (Γ_{eff}) and PN, while improving the current efficiency at the expense of the achieved voltage swing. The structure achieves an FOM up-to-196dBc/Hz with 0.5mW P_{DC}, however a V_{DD} of 0.7V

978-1-7281-1702-7/19 $31.00 © 2019 IEEE

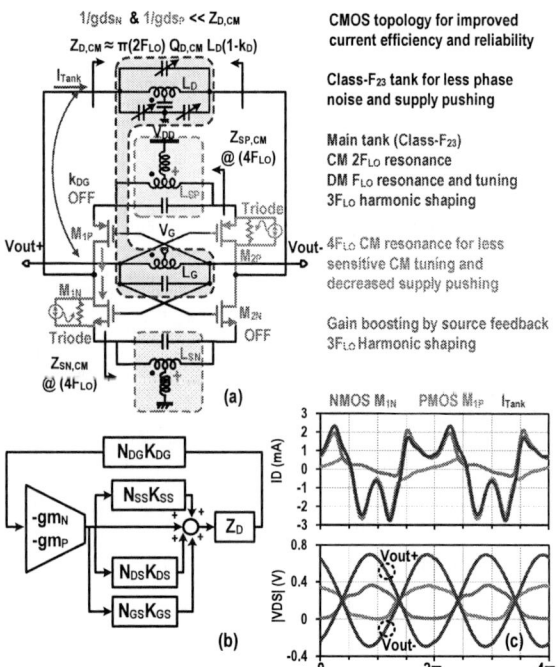

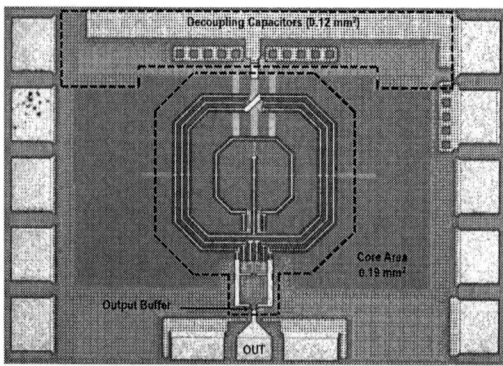

Fig. 3. Die micrograph of the fabricated CMOS VCO chip in 22nm FDSOI.

Fig. 2. (a) Proposed CMOS VCO switching mechanism. (b) Block diagram for the DM of operation showing the different paths for gain boosting and harmonic shaping. (c) Simulated time domain waveforms at 0.35V supply: drain tank current, drain-to-source current and voltage for the switching NMOS and PMOS, and the output gate voltage.

is needed to accommodate the complementary operation and reliable start-up, which is not feasible in harvesting systems.

By modifying the two similar tanks to form a single step-up class-F_{23} transformer [2], between the drains and gates of the CMOS transconductors ($gm_{N/P}$), the resonator becomes a 2-port network and FOM_{MAX} can be improved by up-to 2.2dB [1] (Fig. 1(b)). The drain swing is half that of an NMOS only topology so thin-oxide devices can be used for better start-up and η. The source feedback paths between $L_D/L_{SN/P}$ and L_{SN}/L_{SP} are used for gain boosting and $3F_{LO}$ harmonic shaping for higher η and lower PN. The $L_{SN/P}$ inductors allow voltage swings lower than ground and above V_{DD} for higher gm, and thus, the swing on the main tank is maximized.

The class-F_{23} tank generates a higher CM impedance at $2F_{LO}$ for less PN compared to the implicit CM resonance, and generally requires dedicated CM tuning in addition to DM (differential mode) tuning by controlling the ratio of differential to single-ended capacitors at the L_D side. In this design, however, the source feedback transformer is adjusted to provide a CM resonance around $4F_{LO}$ for both NMOS and PMOS. The CM resonance at $4F_{LO}$ is generally broad since it is limited by the finite Q of the capacitors and does not alter the CM $2F_{LO}$ resonance at the L_D side. The $4F_{LO}$ resonance slightly improves the PN, supply pushing, and renders the VCO less sensitive to CM tuning across the operating frequency range.

Several feedback mechanisms (Fig. 2(b)) improve the DM loop gain and allow the VCO to operate from a V_{DD}=0.35V (half of the CMOS design with implicit CM resonance

in [1] with nearly same PN and FOM). Under large signal swing, the CM resonance at the L_D side forces the noise of the triode NMOS/PMOS to circulate inside their channel resistance instead of reaching the main tank (Fig. 2(a)). The push-pull structure ensures voltage reliability, and improves the tank current efficiency (where current is either pushed-in or pulled-from the main tank each half cycle). Also, the gate swing is ∼ 1V peak-to-peak to provide rail-to-rail drive for the on-chip buffer to reduce its power consumption (Fig. 2(c)).

III. IMPLEMENTATION

The CMOS VCO is fabricated in GlobalFoundries 22nm FDSOI with a core area of 0.19 mm² (Fig. 3). Thin-oxide low-VT devices are used in the VCO core to provide large loop gain, reliable start-up and better PN performance in the thermal region. The 4-winding transformer, shown in Fig. 4(a), is implemented on the top 3 metal layers (a 2.9μm aluminum and two 3μm copper layers). The coupled windings L_G and L_D are implemented in a planar fashion while tying the two upper metals together. In addition to lowering the series resistance, this is particularly useful in reducing the mandatory metal fill imposed by the technology in the upper metal layers which can reduce Q. The source windings L_{SN} and L_{SP} form a single turn broadside transformer with k≈0.75 at 5 GHz. The center taps, for the supply and ground nodes, are realized from the same side for minimum current return path and easy decoupling. A patterned ground plane is added on the lowest metals to act as a shield from the low resistivity substrate ($\rho \approx 7\Omega$-cm). EM-structures are modeled using Integrands EMX simulation software. The EM-simulated Q of L_G, L_D, L_{SN}, and L_{SP}, are 11.1, 11.6, 15, and 12.7, respectively, at 5 GHz.

The VCO is combined with a 3-bit differential switched capacitors bank and thick-oxide N-varactors placed in parallel with the drain tank for discrete and continuous tuning. The drain side (L_D) CM resonance minimizes AM-PM noise conversion from the varactor, while the lower swing (compared to L_G side) allows the usage of thin-oxide NMOS switches in the capacitors bank for their lower $R_{ON}C_{OFF}$. The switched capacitors are controlled by two digital supply levels to minimize R_{ON}, in the ON-state, and avoid accidental switch turn-on, in the OFF-state, while ensuring reliable voltage swing across the the thin switches (Fig. 4(b)).

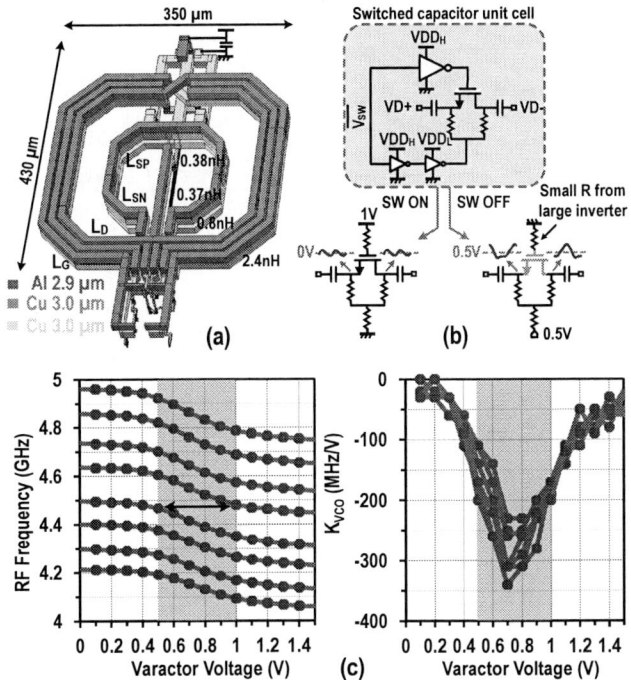

Fig. 4. (a) 4-port transformer based resonator layout. (b) Switched capacitor unit cell and its biasing scheme in ON and OFF states. (c) Measured tuning curves and VCO gain (K_{VCO}).

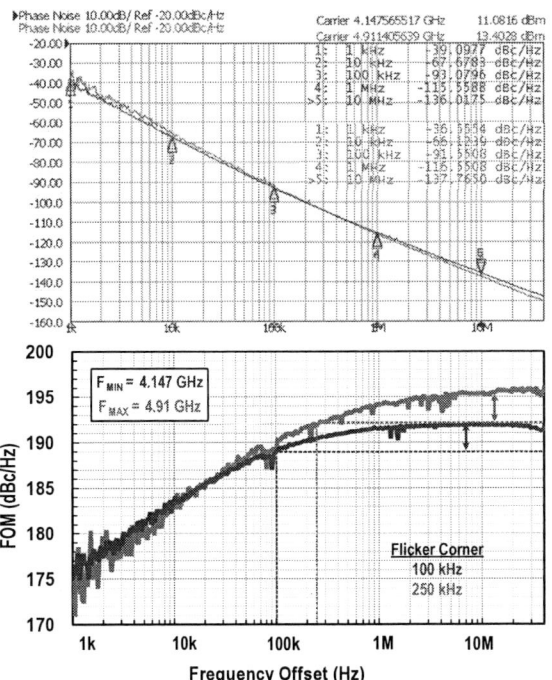

Fig. 5. Measured phase noise and FOM versus offset frequency for the high band (red) and the low band (blue) at 0.35V supply.

IV. MEASUREMENT RESULTS

A signal source analyzer (Keysight E5052B) is used for frequency, PN, and power measurements. The analyzer output DC power port is used as the VCO supply to accurately measure the supply current and the supply voltage frequency pushing. The pseudo-differential output buffer has one output terminated on-chip and the other one is RF probed through a GSG pad. The rest of the biasing and controls are provided from a dedicated LDOs board. The CMOS VCO achieves a continuous tuning range of 20% from 4.06-to-4.96 GHz with > 40% bands overlap and a VCO gain (K_{VCO}) from 120-to-340 MHz/V (Fig. 4(c)).

The PN and FOM are plotted in Fig. 5 versus the offset frequency for the highest (F_{MAX}=4.91 GHz), and lowest (F_{MIN}=4.15 GHz) bands when using a 350mV supply. At F_{MAX}, the VCO achieves a PN of -137.7, -116.6, and -91.6 dBc/Hz at 10MHz, 1MHz, and 100kHz frequency offsets, respectively, while dissipating < 0.42 mW. The peak FOM is 195.8 dBc/Hz in the thermal region and better than 195, 194, and 190.5 dBc/Hz at 10MHz, 1MHz, and 100kHz offsets, respectively. The flicker corner frequency, where the FOM is 3-dB lower than the peak value, is 250 kHz. The simulated flicker corner for low-VT thin-SOI FETs is between 5 and 10MHz, showing, thus, the effect of the CM resonance on close-in VCO PN and FOM improvement in deeply-scaled technology nodes. At F_{MIN}, with a 0.44 mW P_{DC}, the flicker-corner is 100kHz, while the peak FOM is reduced due to the decreased tank impedance (loop gain) and the loading of the capacitors bank switches, but remains better than 192 dBc/Hz with a PN of -136 dBc/Hz at 10 MHz offset.

The PN and FOM (P_{DC}) are measured across the tuning range while keeping the varactor bias at 0V for the smallest K_{VCO} (Fig. 6(a & c)). The results are consistent with F_{MAX} and F_{MIN} measurements, where the PN and FOM are most improved near the high band, when the loop gain is high, while the flicker corner is most reduced near the lower band. The flicker corner is always better than 250kHz without employing a dedicated CM tuning capacitors bank, which needs to be single ended and thus more lossy. This is attributed to the extra $4F_{LO}$ CM resonance at the source nodes which is inherently low Q. In addition, the gate bias, separated from the supply node, provides an extra degree of freedom to minimize both FETs flicker up-conversion by reducing the DC component of Γ_{eff}. This is done by modifying the noise-modulation-function of both FETs to ensure symmetry and is simply obtained by changing the V_G bias.

The varactor effect on the PN and FOM is separately studied at F_{MAX} and F_{MIN} while fixing V_G in Fig. 6(b & d). As expected, the FOM is most sensitive at large K_{VCO} values but remains better than 190 dBc/Hz at 1MHz offset across the frequency range. Adding extra bits in the capacitors bank can yield less varactor impact.

A drawback in the typical CMOS cross-coupled VCO is the high frequency pushing (supply sensitivity). The 4-port transformer in this design decouples the voltage-dependent gate-to-source capacitance (C_{gs}) of the NMOS from the supply variation and results in less pushing. Also, the supply inductor, L_{SP}, with a CM resonance at 4LO provides complementary frequency pushing reduction. The measured pushing is < 60 MHz/V for the high band when sweeping V_{DD} from 300-to-400mV (Fig. 7). For the low band, the minimum V_{DD}

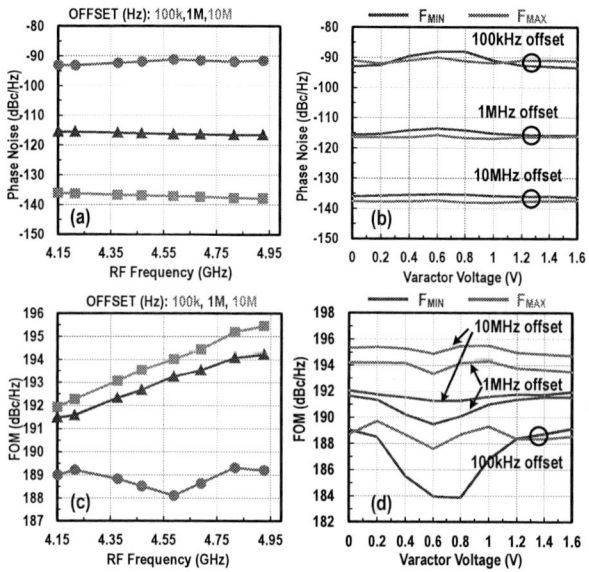

Fig. 6. Measured phase noise and FOM versus oscillation frequency (a & c); and versus the varactor tuning voltage for the upper and lower bands (b & d) at 100k, 1M, and 10MHz offsets for a 0.35V supply.

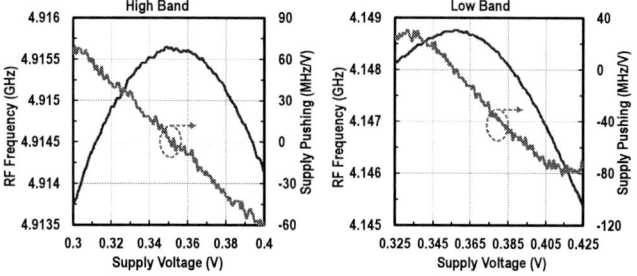

Fig. 7. (a) Measured frequency pushing for the highest and lowest bands.

is increased to 325mV to cope with gain reduction and the measured pushing is < 80 MHz/V when sweeping to 425mV.

Table 1 compares the 4-port resonator-based CMOS VCO with state-of-the art low-power oscillators. The VCO achieves the lowest PN and highest FOM for sub-0.5mW oscillators. The VCO has also the lowest V_{DD} (350mV) for published complementary designs. The proposed design delivers nearly the same PN and FOM performance as the CMOS VCO with implicit CM resonance in [1] with half the supply voltage and 45% more tuning range while abiding by the metal filling and reliable voltage limits of the technology rules. The inverse Class-F design [4] shows slightly higher FOM but a 3× higher P_{DC} and with two separate capacitor banks to simultaneously tune DM and CM resonances, with one capacitors bank placed between the high swing gate nodes. The folded design in [5] operates from only 100mV supply but has a slightly lower FOM for 0.5mW P_{DC} and a larger chip area.

V. CONCLUSION

This work presented a complementary VCO based on a 4-port transformer resonator. The proposed design combines the advantages of high current efficiency of the CMOS

cross-coupled topology and the low supply voltage and PN of the NMOS design while not sacrificing the reliability. The VCO employs CM resonance and DM harmonic shaping for low flicker and improved switching efficiency. A broad $4F_{LO}$ resonance helps maintaining the close-in PN performance across the tuning range with no dedicated tuning for the CM. The prototype, 22nm FDSOI VCO achieves a 195.8 dBc/Hz peak FOM, 20% tuning range from 4.06-to-4.96 GHz, with < 0.45 mW P_{DC} while operating from a 350mV supply.

ACKNOWLEDGMENT

The authors would like to thank Analog Devices for their financial support, Integrand for EMX electromagnetic simulator and GlobalFoundries for chip fabrication.

REFERENCES

[1] D. Murphy, H. Darabi and H. Wu, "Implicit Common-Mode Resonance in LC Oscillators," in *IEEE Journal of Solid-State Circuits*, vol. 52, no. 3, pp. 812-821, March 2017.

[2] M. Shahmohammadi, M. Babaie and R. B. Staszewski, "A 1/f Noise Upconversion Reduction Technique for Voltage-Biased RF CMOS Oscillators," in *IEEE Journal of Solid-State Circuits*, vol. 51, no. 11, pp. 2610-2624, Nov. 2016.

[3] C. Li et al., "19.6 A 0.2V trifilar-coil DCO with DC-DC converter in 16nm FinFET CMOS with 188dB FOM, 1.3kHz resolution, and frequency pushing of 38MHz/V for energy harvesting applications," in *2017 IEEE International Solid-State Circuits Conference (ISSCC)*, San Francisco, CA, 2017, pp. 332-333.

[4] C. Lim, J. Yin, P. Mak, H. Ramiah and R. P. Martins, "An inverse-class-F CMOS VCO with intrinsic-high-Q 1st- and 2nd-harmonic resonances for 1/f2-to-1/f3phase-noise suppression achieving 196.2dBc/Hz FOM," in *2018 IEEE International Solid - State Circuits Conference - (ISSCC)*, San Francisco, CA, 2018, pp. 374-376.

[5] O. El-Aassar and G. M. Rebeiz, "A 0.1-to-0.2V Transformer-Based Switched-Mode Folded DCO in 22nm FDSOI With Active Step-Down Impedance Achieving 197dBc/Hz Peak FoM and 40MHz/V Frequency Pushing," *2019 IEEE International Solid- State Circuits Conference - (ISSCC)*, San Francisco, CA, USA, 2019, pp. 416-418.

Table 1. Comparison with State-of-the-Art Low-Power Oscillators

Design			This Work	[1] JSSC'17	[4] ISSCC'18	[5] ISSCC'19	
Topology			CMOS 4-port resonator	CMOS CM resonance	CMOS Inverse Class-F	CMOS Folded	
Technology process			22nm FDSOI	28nm	65nm	22nm FDSOI	
Frequency (GHz)			4.06-4.96	4.7-5.4	3.49-4.51	4.15-4.97	
Tuning Range (%)			20%	13.8%	25.5%	18%	
VDD (V)			0.35	0.7	0.6	0.2	0.1
Power (mW) @ f_{min} / f_{max}			0.44/0.415	0.5	1.2/1.14	2.28/1.77	0.54/0.46
Phase noise (dBc/Hz)	@ 100 kHz	f_{min}	-93.1/	-91.5#/	-102.4/	-100/	-91.3/
		f_{max}	-91.6	-92#	-98.5	-96.1	-92.3
	@ 10 MHz	f_{min}	-136/	-139#/	-145.6/	-142.2/	-133/
		f_{max}	-137.8	-137#	-143.7	-144.3	-136.5
FOM (dBc/Hz)	@ 100 kHz	f_{min}	189/	188*/	192.5/	188.9/	186.4/
		f_{max}	189.2	187.5*	191	187.6	189.6
	@ 10 MHz	f_{min}	192/	195.6*/	195.6/	191/	188/
		f_{max}	195.5	194*	196.2	195.8	193.7
1/f³ Corner (kHz)			100-250	160-400*	100-300	70-700	50-180
Frequency pushing (MHz/V)			< 80	N.R.	4.5-15	< 40	
Core Area (mm²)			0.19	0.18	0.14	0.272	

N.R. = Not Reported   * Estimated from plots
^{1}FOM = -PN + $20\log_{10}(f_c/\Delta f)$ -10log(P_{DC}/1mW)   # Calculated based on the provided P_{DC}

X-band NMOS and CMOS Cross-Coupled DCO's with a "Folded" Common-Mode Resonator Exhibiting 188.5 dBc/Hz FoM with 29.5% Tuning Range in 16-nm CMOS FinFet

R. Levinger, D. Ben-Haim, I. Gertman, S. Bershansky, R. Levi, J. Kadry and G. Horovitz

Communications Devices Group, Intel, Petach-Tikva, Israel

run.levinger@intel.com

Abstract— **This paper presents two X-band state of the art digitally controlled oscillators (DCO's), one utilizes CMOS and the other NMOS as cross-coupled pairs. Both designs include a "folded" common-mode resonator in order to enhance performance while minimizing the required area. The implemented DCO's cover 9 to 12.1 GHz and 9.3 to 12.4 GHz for the NMOS and CMOS designs respectively, achieving a frequency tuning range (FTR) of 29.5%. Measured phase noise referred to 6.1 GHz and at 1 MHz offset is -118.5 and -114 dBc/Hz for NMOS and CMOS respectively. The NMOS design consumes 3.6 mW while the CMOS design consumes 1.1 mW both from a 0.8 V supply, obtained figure of merit (FoM) for both designs is higher than 188.5 dBc/Hz. Both designs are very compact with an area less than 0.054 mm2.**

Keywords— **DCO, low-power, phase-noise, CMOS, FoM, common-mode resonance, quality factor**

I. INTRODUCTION

Modern communication standards the likes of 5G and 802.11ax are extremely demanding in terms of performance, both in the radio and the modem domains. In order to facilitate handheld devices to operate these complex schemes, power consumption and size reduction considerations are also critical. The required high integration suggests a very aggressive technology node, while this allows us to tap advanced signal processing that benefits us greatly, it also adds challenges in RF design and especially in oscillator design due to increased flicker noise, metallization parasitics and tank quality factor (Q) degradation. Utilizing a common-mode resonance have been demonstrated in previous works [1], [2] to be very effective and to get the best balance between performance and power, that is,

the best FoM. In [2] the authors demonstrate that it is possible to take advantage of an implicit common-mode resonance originating from the main LC tank thus reducing the area needed for this topology. As the operating frequency and tuning range increases this method becomes less effective and forces us to a limited domain of inductor choices.

In this paper we propose a novel approach to implement the common-mode resonance by "folding" the additional inductor needed within the inductor of the main resonator. We implement this on both NMOS and CMOS based DCO's topologies implemented in a commercially available 16-nm finfet technology, and demonstrate state of the art performance surpassing 188.5 dBc/Hz FoM at 12.2GHz while covering 30% FTR.

II. CIRCUIT DESCRIPTION

A. Oscillators Topology

The DCO's schematic is shown in Fig. 1, both NMOS (Fig. 1 (a)) and CMOS (Fig. 1 (b)) are based on a voltage biased oscillator with common-mode resonance. It has been shown in [2] and [3] that using this approach can maximize oscillator efficiency and bring the FoM closest to its fundamental limit shown in Eq. 1:

$$FoM_{Max} = 176.8 + 20log_{10}Q \qquad (1)$$

It has been demonstrated by [1]-[4] that as long as we have a sufficiently high impedance between the drain and source nodes of the active devices at $2f_o$ than we are able to avoid Q degradation usually experienced in voltage biased oscillators.

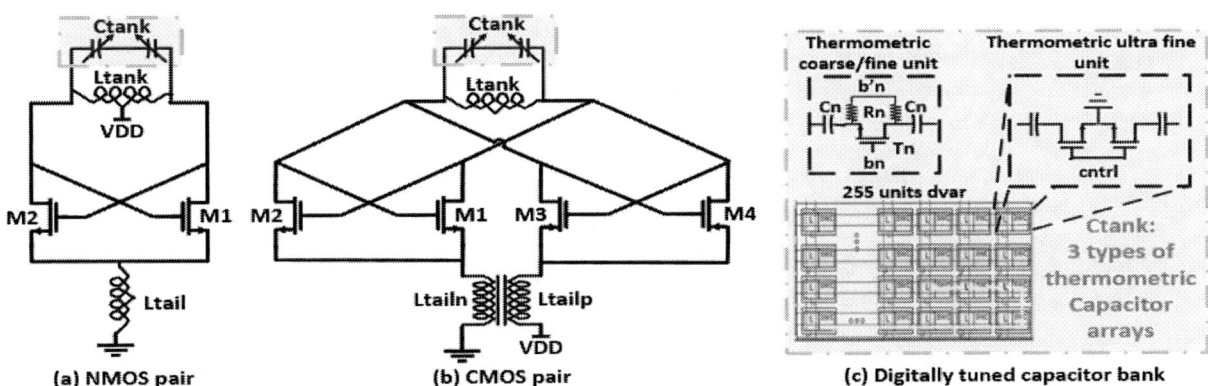

Fig. 1. (a) NMOS pair with common mode resonator (b) CMOS pair with common mode resonator (c) Digitally tuned capacitor

978-1-7281-1702-7/19 $31.00 © 2019 IEEE

This allows us to extend the voltage swing at the tank without adding noise and hence improve the oscillator phase noise. Another major benefit of using a common-mode resonance is the reduction of the even harmonics in the wave form, thus reducing flicker noise up-conversion as rigorously demonstrated in [5].

The NMOS design was targeted for lower phase noise and consumes more power, whereas the CMOS design was targeted for a moderate phase noise performance but with lower power consumption, both were designed to retain the same FoM and support the same frequency range and as such has similar values for the tank. Both designs use high loop gain since it has been demonstrated by [2] that this leads to a more robust design and reduces sensitivity to the common mode capacitance. Since the NMOS and CMOS designs common-modes behave differently it is required to use a different approach in the common-mode resonators implementation. and not as a paragraph, single author block centered to the page must be used instead of multiple author blocks). You will be asked to fix these format problems if any such format deviations are detected.

B. Folded Common-Mode Resonators

As explained in [2] using the implicit common-mode resonance, present in the LC tank due to the non-ideal coupling, can reduce the area of the oscillator and achieve (under some constraints) similar performance compared with a design having an explicit common-mode resonator. However, as the oscillation frequency gets higher and the FTR gets larger, the obtainable impedance at $2f\_o$ when using this method becomes limited. In order to maximize the common-mode impedance, considering the fact that it is required to be at 24 GHz, it is beneficial to use an explicit common-mode resonator. In order to reduce the area we have placed the additional inductor needed within the main inductor in what we refer to as "folded" common mode resonator.

Fig. 2 shows a 3D illustration of the inductor structure used in the NMOS design, separating the different modes graphically to demonstrate the "folding" concept. Fig. 2 (a) emphasizes the main inductor, it has two loops each with three current paths to maximize Q while still obtaining the required inductance. The K of the main inductor was tuned in order to have the implicit common-mode resonance coincide with the explicit one further boosting the common-mode impedance. Fig. 2 (b) emphasizes the common-mode inductor, it was designed as a quadrupole in order to minimize cross-talk and Q degradation of the main inductor [6]. An additional closed loop shield is present in order to better isolate the oscillator from potential aggressors. Fig. 3 shows a 3D illustration of the inductor structure used in the CMOS design in a similar fashion as in the NMOS design. Fig. 3 (a) shows the main inductor, it also has two loops but it does not have a center tap, reducing its implicit common-mode inductance significantly. Since the CMOS design requires a high impedance at $2f\_o$ for both the NMOS and PMOS devices in order to be effective [4], we need to implement two inductors. In order to save area and increase Q we have opted to use a quadrupole transformer, much like the NMOS design, it is "folded" within the main inductor as demonstrated in Fig. 3 (b). The CMOS design also has a shield present, but in this design it was chosen to be in the same plane as the inductor itself.

C. Digitally Controlled Capacitor

The capacitor bank is divided into three capacitor sub-arrays, these arrays are thermometric to lower mismatch. Unit cells can be seen in Fig. 1 (c). The coarse tuning array has 4 capacitors while the fine tuning array has 8 capacitors, conceptually we obtain 45 sub-bands that allow us to cover the wanted tuning range. The coarse/fine arrays degrade the Q significantly at X-band, and therefore the switches used are differential and use only one device in order to reduce switch resistance at the expense of additional parasitic capacitance.

NMOS design inductor

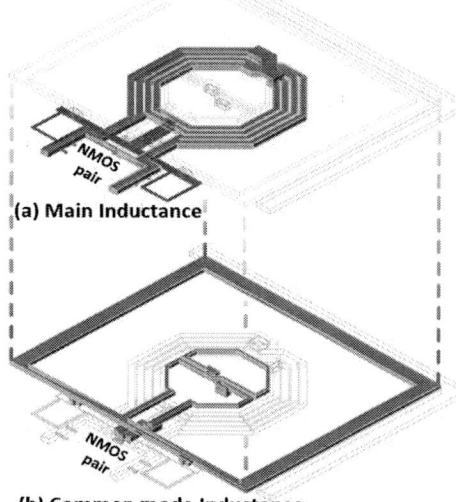

Fig. 2. A 3D illustration of the NMOS design inductor and shield: (a) main resonance inductance (b) common mode resonance explicit inductance

CMOS design inductor

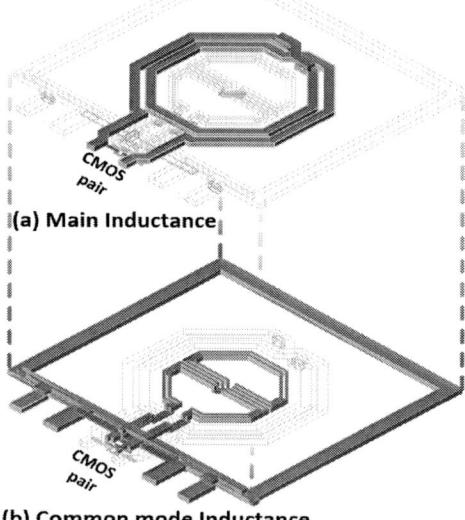

Fig. 3. A 3D illustration of the CMOS design inductor and shield: (a) main resonance inductance (b) common mode resonance explicit inductance

The ultra-fine array is essentially a digital varactor, to be controlled by the loop of the all-digital phase locked loop (ADPLL). As seen in Fig. 1 (c), it has 256 thermometric capacitors, the switch used is single ended to reduce the parasitic capacitance and remove the resistors reducing area. Additional capacitors within the digital varactor are placed so it is possible to further modulate the capacitance and increase resolution (quantization noise shaping).

III. Measurements

Die photographs of the fabricated DCO's are shown in Fig. 4. The DCO's were integrated within ADPLL's and were packaged in a flip-chip ball grid array (FC-BGA). The DCO's were measured at half the oscillation frequency using an on chip frequency divider and an output buffer chain. The NMOS design occupies an active area of 0.054 mm2 while the CMOS design occupies an area of 0.046 mm2. Both frequency and phase noise were measured using a Rohde & Schwarz FSWP-B1 signal analyzer while current consumption was measured using a Keysight 34401A digital multi-meter. Measured current consumption of the NMOS DCO is 4 mA and measured current consumption of the CMOS DCO is 1.4 mA, both from a 0.8V supply.

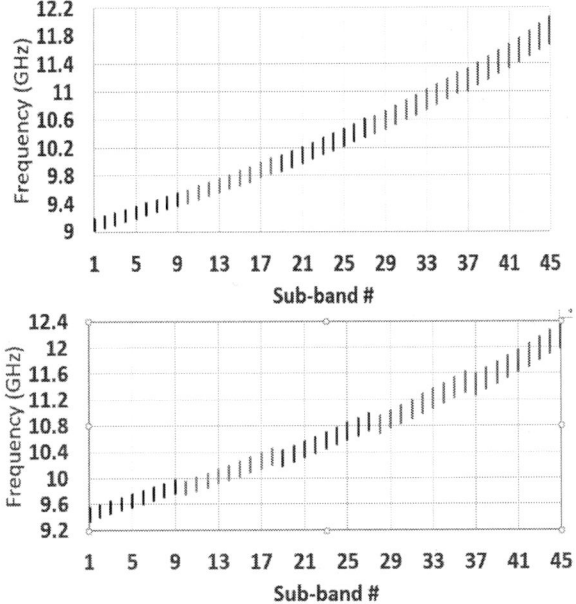

Fig. 4. Die Photos: (a) NMOS design (b) CMOS design

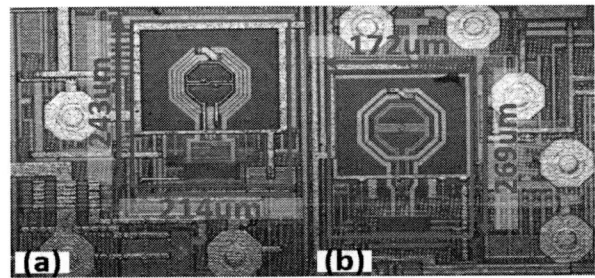

Fig. 5. Measured DCO sub-bands: (a) NMOS design (b) CMOS design

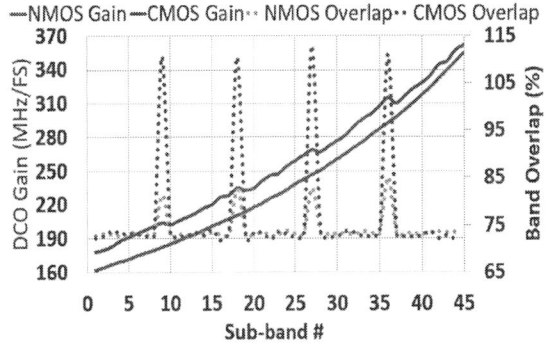

Fig. 6. Sub-band overlap and DCO gain for both designs

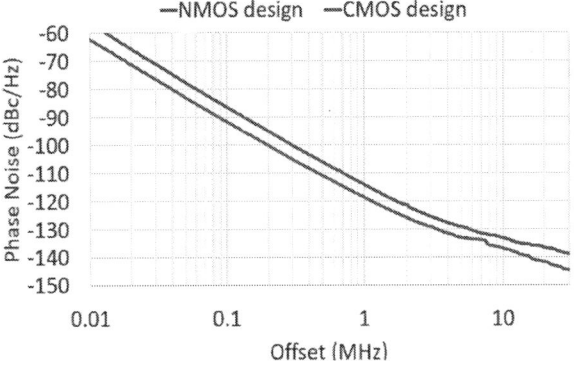

Fig. 7. Phase noise for both designs at 12.1GHz measured at half the oscillation frequency

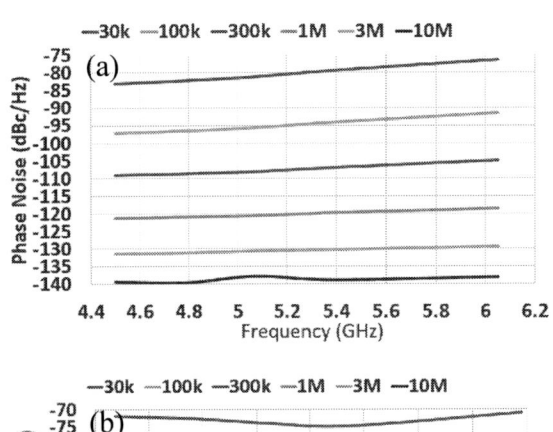

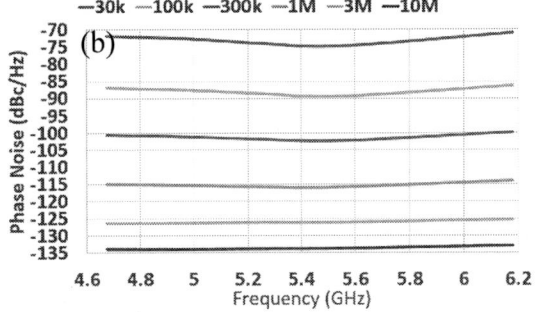

Fig. 8. Phase noise as a function of frequency measured at half the oscillation frequency: (a) NMOS design (b) CMOS design

Fig.5 shows the DCO's frequency coverage and sub-band partitioning. Each color represents a different coarse tuning capacitor. The NMOS design has a 29.5% FTR, covering 9 to 12.1 GHz, the CMOS design has 28.5% FTR, covering 9.3 to 12.4 GHz. Fig. 6 shows the band overlap and gain per sub-bands for both designs. The average overlap is 74.09/76.35 % for the NMOS and CMOS designs respectively and a minimum overlap value of 71 % is obtained, a sufficiently high value to provide a robust temperature behaviour and extend temperature drift immunity for the ADPLL. The DCO's gain vary between 161 to 361 MHz for a full scale of the digital varactor. Fig. 7 exhibit measured phase noise of both oscillators at 6.1 GHz. Measured flicker corner is in the vicinity of 600 KHz, and measured phase noise is -118.5 and -114 dBc/Hz referred to 6.1 GHz for the NMOS and CMOS designs respectively.

Fig. 8 (a) shows phase noise at various offsets as a function of frequency for the NMOS design and Fig. 8 (b) shows it for the CMOS design. It is possible to see that the phase noise at lower offsets degrades more when the oscillation frequency gets higher but this effect is much more moderate within 1MHz to 10MHz, this is probably due to additional common-mode capacitance added by the relatively large switched capacitors. Fig. 9 (a) and (b) shows the FoM of the NMOS and CMOS designs as a function of frequency and for different offsets. At 1MHz both designs exceed 188.3 dBc/Hz and the peak is obtained at the middle frequency. Table I shows a comparison to state of the art CMOS oscillators operating at or near X-band.

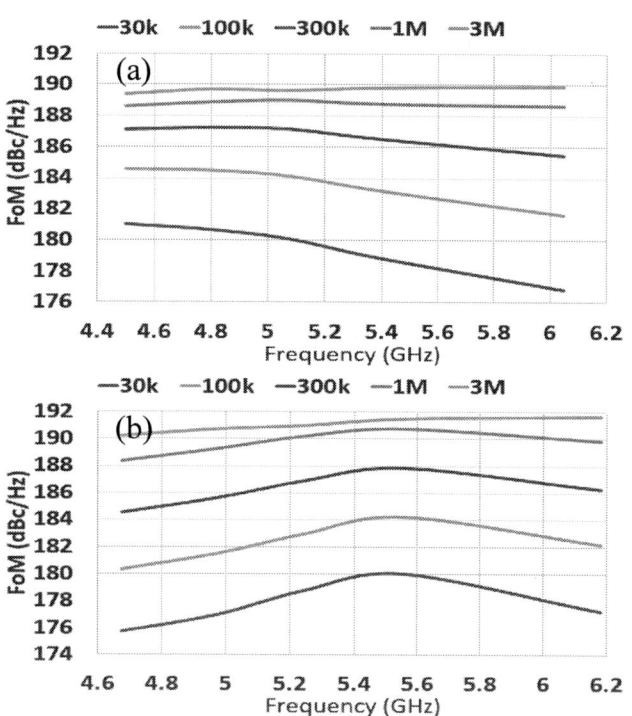

Fig. 9. Figure of merit as a function of frequency measured at half the oscillation frequency: (a) NMOS design (b) CMOS design

It is possible to see that this design obtains state-of the art results in terms of FoM, lower than -188.5 dBc/Hz for the NMOS design and -189.5 for the CMOS design, with a very small area due to the "folding" scheme.

IV. CONCLUSION

This paper demonstrates a novel method of implementing a "folded" common-mode resonator that can be useful at X-band and does not increase design area. Two flavors, NMOS and CMOS based, were implemented at a 16nm Finfet commercially available technology. Both designs exhibit state of the art performance even when compared to lower frequency designs, with a FoM of less than 188.5 and FTR of above 28.5 %, the designs are very compact taking as little as 0.054mm2 .

Table 1. Comparison to state of the art CMOS oscillators

	This work		[2]	[4]	[7]
	nmos	cmos			
Tech.	16nm finfet		28nm	55nm	16nm finfet
Supply (V)	0.8		0.9	1.5	0.4
Freq. (GHz)	9–12.1	9.3–12.4	2.8-3.7	7.4-8.4	3.2 - 4
PN* (dBc/Hz)	-112.5	-108	-117.3	-115.7	-110.3
Power (mW)	3.6	1	6.6	6.3	3.8
Area (mm²)	0.054	0.046	0.15	0.19	0.11
FoM*	188.8	189.7	190.8	189.4	186.2
FoMt*	198.8	198.8	199.5	191.6	193
FoMa*	202.1	203	199	196.9	195.8

* Ref to 12.1GHz @ 1MHz offset

$$FoM = -PN + 20log(f_o/f_{offset}) - 10logP_{DCmW}$$
$$FoM_t = FoM + 20log(FTR(\%)/10)$$
$$FoM_a = FoM + 10log(\frac{area}{1mm^2})$$

REFERENCES

[1] E. Hegazi H. Sjoland and A. A. Abidi, "A filtering technique to lower LC oscillator phase noise", IEEE J.Solid-State Circuits, vol. 36, no. 12, pp. 1921-1930, December 2001.

[2] D. Murphy, H. Darabi and H. Wu, "Implicit Common-Mode Resonance in LC Oscillators", IEEE J. Solid-State Circuits, vol. 52, no. 3, pp. 812-821, March 2017.

[3] M. Garampazzi, S. D. Toso, A, Liscidini, D. Manstretta. P. Mendez, L. Romano and R. Castello, "An Intuitive Analysis of Phase Noise Fundamental Limits Suitable for Benchmarking LC Oscillators", IEEE J. Solid-State Circuits, vol. 49, no. 3, pp. 635-645, March 2014.

[4] M. Garampazzi, P. M. Mendes, N. Codega, D. Manstretta and R. Castello, "Analysis and Design of a 195.6 dBc/Hz Peak FoM P-N Class-B Oscillator With Transformer-Based Tail Filtering", IEEE J. Solid-State Circuits, vol. 50, no. 7, pp. 1657-1668, July 2015.

[5] F. Pepe and P. Andreani, "A General Theory of Phase Noise in Transcondactance-Based Harmonic Oscillators", IEEE Tran. on Circuits and Systems-I: Regular Papers, vol. 64, no. 2, February 2017.

[6] A. Poon, A. Chang, H. Samavati and S. S. Wong, "Reduction of Inductive Crosstalk using Quadrupole Inductors", IEEE J. Solid-State Circuits, vol. 44, no. 6, pp. 1756-1764, June 2009.

[7] C. Li, M. Yuan, C. Chang, Y. Lin, C. Liao, K. Hsieh, M. chen and R. B. Staszewski "A 0.2V Trifilar-Coil DCO with DC-DC Converter in 16nm FinFet CMOS with 188dB FoM, 1.3kHz resolution, and Frequency Pushing of 38MHz/V for Energy harvesting Applications", IEEE Int. Solid-State Conf., February 2017.

A 18.2-29.3 GHz Colpitts VCOs bank with -119.5 dBc/Hz Phase Noise at 1 MHz Offset for 5G Communications

F. Quadrelli[#], F. Panazzolo[#], M. Tiebout[#], F. Padovan[#], M. Bassi[#], A. Bevilacqua[$]

[#]Infineon Technologies AG, Villach, Austria

[$]University of Padova, Italy

{fabio.quadrelli, matteo.bassi}@infineon.com, andrea.bevilacqua@dei.unipd.it

Abstract — This paper describes a bank of four SiGe BiCMOS oscillators tailored to cover the 18.2-29.3 GHz frequency range, needed to tackle the needs of 5G communications. The Colpitts oscillator topology is leveraged to achieve a lower absolute phase noise compared to Class-C oscillators, at the expense of deteriorated figure of merit. Benefiting from the proper technology choice, a careful tank design was carried out to maximize Q, minimize K_{VCO} and phase noise variations. The four oscillators feature state-of-the-art phase noise, ranging from -119.5 dBc/Hz to -116.5 dBc/Hz at 1MHz offset for 18.2 GHz and 29.3 GHz carrier frequency, respectively. For each oscillator the phase noise variation is only <2 dB over -20°C to 85°C temperature range and <6 dB over the whole 47% tuning range.

Keywords — Colpitts, 5G, BiCMOS, voltage controlled oscillators, phase noise

I. INTRODUCTION

The development of next-generation 5G communication systems is moving rapidly to market, and becoming a reality. A key block is the local oscillator, because of the wide tuning range required by the application, and stringent phase noise due to the need of supporting efficient modulation formats, such as 256 QAM or Orthogonal Frequency Division Multiplexing (OFDM).

In this scenario, SiGe BiCMOS technology is a very attractive option. In fact, maximum oscillation frequency (f_{max}) of 400 GHz, cut-off frequency (f_t) of 250 GHz and a dedicated back-end-of-the-line (BEOL) with several thick metal layers make it an outstanding candidate to achieve excellent mm-wave performance [1]. Despite high-performance oscillators are being realized in GaAs or other III-IV technologies, the interest in developing Si-based oscillators is increasing. They offer lower fabrication costs and the Si-based part scales more easily. On top of that, the marriage between high-performance HBTs and CMOS digital devices allows for enhanced mm-wave performance together with digital integration capabilities [2]. Over pure CMOS, BiCMOS offers higher transconductor efficiency, lower 1/f noise and higher breakdown voltage for the same f_{max}. Therefore, they can employ higher supply voltage and achieve larger amplitude of oscillation and consequently lower phase noise levels. On the other hand, bipolar transistors do not offer good quality switches and operation in saturation must be avoided. Also, when pn-junctions are employed as varactors to tune the oscillation frequency, careful design must be carried out to ensure they never turn on, otherwise the tank quality factor gets easily degraded.

In 5G systems, one of the main band of operation ranges from 24.25 to 30.5 GHz. To allow for a flexible IF range of 3-6 GHz, LO needs to provide a very wide tuning range, from ∼18 GHz to ∼28 GHz. Over this tuning range, the phase noise must be as constant as possible, making the design of such an oscillator very challenging.

In this work, we strictly focus on achieving the best absolute phase noise performance possible over the whole 18-28 GHz tuning range with minimum variations. For this reason, we decided to employ the Colpitts oscillator topology, which allows for a better phase noise level with slightly increased power consumption compared to the Class-C topology. Instead of typical bipolar implementations where the frequency tuning element is made of a pn-junction, in the proposed design the MOS varactors present in the BiCMOS technology are leveraged to achieve a more constant phase noise performance over the tuning range, with comparable quality factor. The four oscillators operate from 18.2 to 21.8 GHz, 20.9 to 24.7 GHz, 23.4 to 27.8 GHz and 25.4 to 29.3 GHz, respectively. Supplied at 3.3 V with constant 122 mW power consumption, they show a phase noise as low as -119.5 dBc/Hz at 1 MHz offset at 18.2 GHz, increasing to -115.5 dBc/Hz only at the highest operation frequency. For each oscillator, the phase noise variation over the corresponding tuning range is <4 dB, while it is <6 dB over the entire aggregate 47% tuning range covered by the VCOs bank. Degradation over the -20°C to 85°C temperature range is <2 dB, making this oscillators bank suitable for integration in the next-generation 5G products.

II. VOLTAGE CONTROLLED OSCILLATOR DESIGN

The main goal of this work is to achieve the minimum possible phase noise over the targeted 18-28 GHz range while guaranteeing the performance over the entire tuning voltage and over temperature with no major variations. As it is well known, the phase noise sideband of an LC oscillator $\mathcal{L}(\Delta\omega)$ at offset $\Delta\omega$ is described with good approximation by Leeson's equation:

$$\mathcal{L}(\Delta\omega) = 10\log_{10}\frac{kT}{A_0^2}\frac{L}{Q}\frac{\omega_0^3}{\Delta\omega}(1+F) \qquad (1)$$

where ω_0 is the oscillation frequency, Q is the overall tank quality factor, k the Boltzmann's constant, T absolute temperature, A_0 the amplitude of oscillation, L the inductance, Q the overall tank quality factor, and F the oscillator noise factor, which depends on the selected topology. The

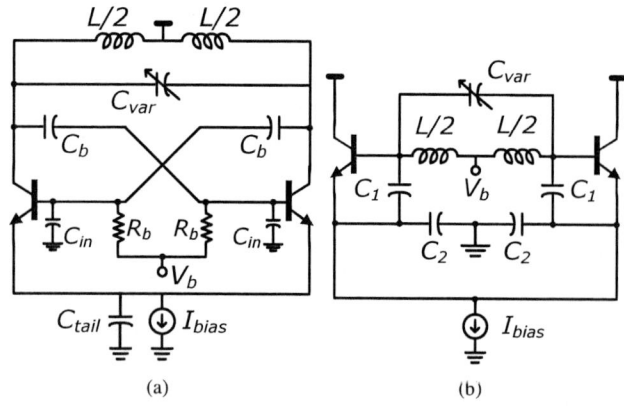

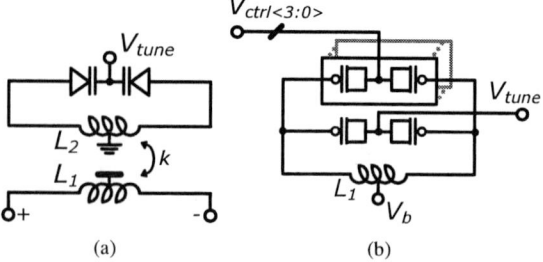

Fig. 3. (a) transformer coupled pn-junction varactor, (b) direct coupled MOS varactor with digital coarse tuning.

(a)

(b)

Fig. 1. Schematic of: (a) differential Class-C LC oscillator, (b) differential Colpitts oscillator.

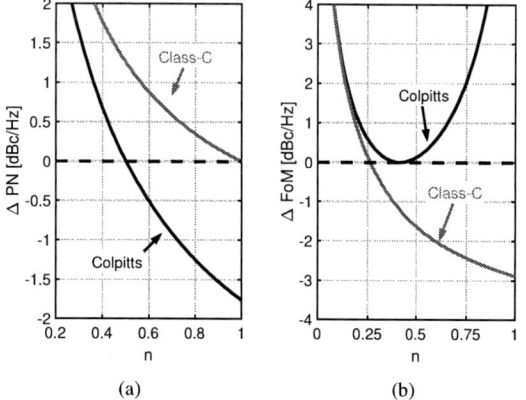

(a)

(b)

Fig. 2. Phase noise (a) and FoM (b) for Class-C and Colpitts topologies as a function of n normalized to minimum theoretical Class-C phase noise and minimum Colpitts FoM, respectively.

following next two sub-sections are dedicated to the choice of the best oscillator topology, to minimize F, and best tank configuration, to minimize L/Q [3].

A. Class-C vs Colpitts topologies

The two most popular oscillators topologies at high-frequencies ($>10\,$GHz) for bipolar VCOs are the Class-C and the Colpitts (schematics in Fig. 1(a) and Fig. 1(b), respectively). In Class-C oscillators, the presence of a significant capacitance C_{tail} at the common emitter node of the active pair results in a better conversion of the bias current into the first harmonic of the tank current, leading to a lower phase noise for the same current consumption compared to Class-B [4]. However, the Colpitts oscillator, which has a simple oscillator core, is the most favoured topology for low phase noise. To give more insight, the phase noise of the two topologies can be expressed as:

$$\mathcal{L}_{CC}(\Delta\omega) = 10\log_{10}\frac{kTR_p}{A_{max}^2}\left(\frac{\omega_0}{Q\Delta\omega}\right)^2\left(1+\frac{1}{2n}\right)$$

$$\mathcal{L}_{Colpitts}(\Delta\omega) = 10\log_{10}\frac{kTR_p}{A_{max}^2}\left(\frac{\omega_0}{Q\Delta\omega}\right)^2\left(1+\frac{1-n}{2n}\right)$$
(2)

where R_P is the equivalent tank resistance at resonance, A_{max} is the maximum oscillation amplitude allowed by technology and $n = C_b/(C_b+C_{in})$ and $n = C_1/(C_1+C_2)$ the feedback factor for the Class-C and Colpitts topologies, respectively.

The phase noise, normalized to the Class-C phase noise for $n=1$, is plotted in Fig. 2(a) along with the FoM, normalized to the Colpitts minimum FoM, in Fig. 2(b). For the same amplitude of oscillation, and if the same LC tank is employed, the two topologies differ in the phase noise factor F, only. Clearly, phase noise is always lower in Colpitts oscillator. On the other hand, FoM is equal or better in Class-C oscillator. It is also worth noticing that the Colpitts oscillator has a minimum value in the FoM as a function of the parameter n. Since the focus of this work is to achieve the lowest possible phase noise, the Colpitts topology was selected. Additionally, this topology, in its common-collecor variant, allows to tap the output signal directly from the collectors without loading the tank, whereas the Class-C VCO always requires an output buffer, thus increasing power consumption of the bank.

B. Tank Design

There are several ways to combine the varactor with an inductive element and make the LC tank. Since the technology used in this design is BiCMOS, two different approaches can be pursued: a pn-junction varactor, shown in Fig. 3(a), or a MOS varactor, illustrated in Fig. 3(b).

If a pn-junction varactor is to be used, care must be taken that the voltage across the junction is not positive, otherwise it would forward bias the diode, thus degrading the tank quality factor and the phase noise. To circumvent this issue, magnetic coupling can be used to couple the pn-junction varactors to the tank [4]. The advantage is that the transformer turn ratio can be chosen such that the swing across the varactor is reduced as compared to the output swing of the VCO, thus limiting the possibility that the junction turns on even for large amplitudes of oscillation. Moreover, in this way the anode can be biased at $0\,$V and complete tuning is achieved for $V_{tune} < V_{CC}$.

On the other hand, two main drawbacks must be taken into account. First, as reported by previous works [4], [1], still the pn-junction varactor gets forward biased for low V_{tune} voltages, leading to a relevant phase noise increase up to $\sim 12\,$dB that reduces the effective tuning range of the oscillator. Second, to reduce the frequency sensitivity, K_{VCO}, the pn-junction cannot be easily fragmented to sub-varactors

978-1-7281-1702-7/19 $31.00 © 2019 IEEE

2019 IEEE Radio Frequency Integrated Circuits Symposium

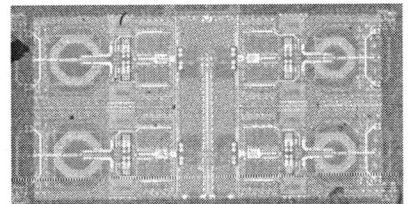

Fig. 4. Chip micrograph.

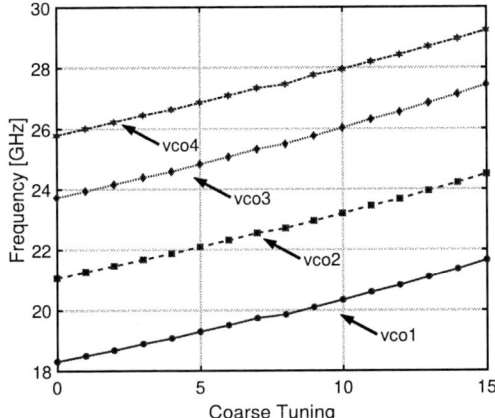

Fig. 5. Measured tuning range for $V_{CC} = 3.3V$ and fixed V_{tune} for the four VCOs as a function of the coarse tuning code.

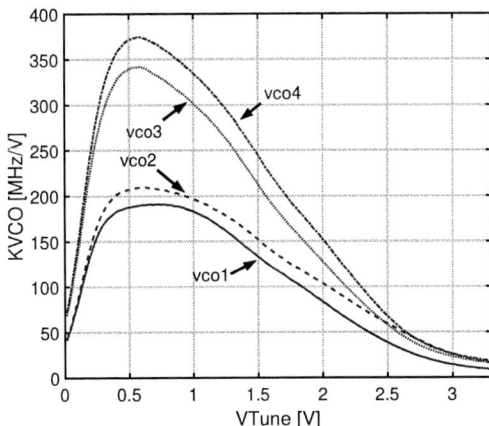

Fig. 6. Highest measured K_{VCO} for each of the four VCOs as a function of V_{tune}.

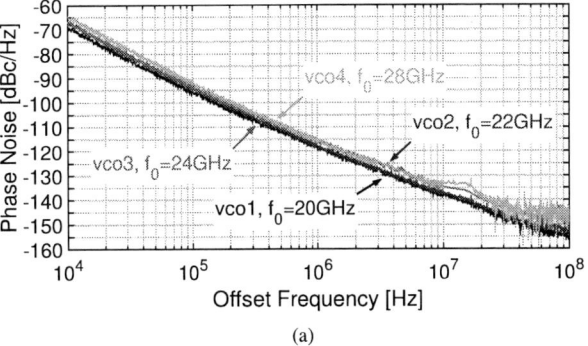

(a)

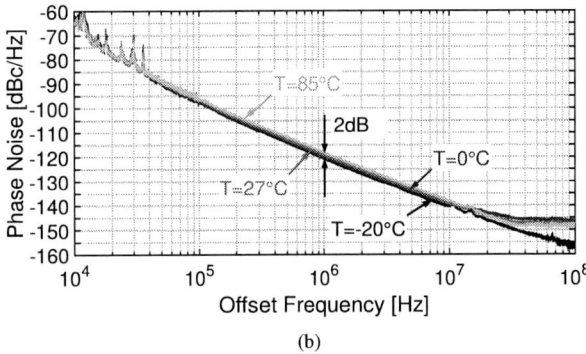

(b)

Fig. 7. (a) measured phase noise for the four VCOs (b) measured phase noise for VCO1 at -20°C, 0°C, 27°C and 85°C.

in a small area to fit into a high-Q tank because of the minimum spacing.

In light of the foregoing considerations, and thanks to the use of a BiCMOS technology, this design employs MOS varactors without magnetic coupling, as depicted in Fig. 3(b). Measurements show that the quality factor of the MOS varactors is comparable with the one of the pn-junction. The tank, shown in Fig. 3(b) was built around the Colpitts oscillator topology, where n=0.7 was selected to achieve a phase noise as low as possible. Moreover, tail resistive bias was preferred to decrease the upconversion of the tail current noise to phase noise.

III. Measurements

The proposed bank of four Colpitts oscillators has been implemented in 130 nm BiCMOS technology. The chip micrograph is show in Fig. 4. Each VCO occupies a tiny area of 0.087 mm², while the total active area, including the measurement buffers, is 0.423 mm², only.

Fig. 5 shows the measured oscillation frequency with fixed V_{tune} when the digital capacitor bank code is swept from 0 to 15. The four VCOs have a tuning range of 18, 22, 18 and 14%, respectively, for an aggregate tuning range of 47%. The corresponding total frequency range is from 18.2 GHz to 29.3 GHz.

The K_{VCO} is a key parameter when the design of the whole PLL is taken into account. In this design, thanks to the use of the combination of digital bank for coarse tuning and a

small varactor, the K_{VCO} can be kept as low as 350 MHz/V, as shown in Fig. 6, that reports the measured K_{VCO}.

The measured phase noise sidebands for the four oscillators are shown in Fig. 7(a). The measured phase noise level is -119.5, -118.5, -117.5 and -116.5 dBc/Hz at 1 MHz offset and -69, -68, -67 and -65 dBc/Hz at 10 kHz offset. The measured phase noise of the oscillators is also reported over the -20°C to 85°C range: the maximum variation is <2 dB.

It is important that the oscillator phase noise shows limited variation over the overall frequency range. Fig. 8 shows the

978-1-7281-1702-7/19 $31.00 © 2019 IEEE

2019 IEEE Radio Frequency Integrated Circuits Symposium

Table 1. Performance Summary and Comparison with the state-of-the-art

	Tech.	f_0 [GHz]	TR [%]	V_{CC} [V]	Eq. Best $\mathcal{L}$(1 MHz) @20 GHz [dBc/Hz]	Eq. Worst $\mathcal{L}$(1 MHz) @20 GHz [dBc/Hz]	Eq. T=85°C $\mathcal{L}$(1 MHz) @20 GHz [dBc/Hz]	P_{dc} [mW]	FoM(1 MHz) [dBc/Hz]	Area [mm²]
[5]	SiGe HBT	22.1	16	3.3	-117	-111†	NA	33	-188	0.048
[4]	SiGe HBT	20	13	3.3	-115	-111†	NA	23	-187	0.048
[6]	SiGe BiCMOS	38.4	9.7	3.3	-113	NA	NA	106	-180	0.18
[3]	SiGe BiCMOS	20.5	19	2.5	-119	-117	NA	52.8	-187	0.045
[1]	SiGe BiCMOS	21	19.8	4	-119	-107	NA	70	-188	0.069
[7]	SiGe BiCMOS	27	16	3	-114	NA	NA	111	-183	0.144
[8]	SiGe BiCMOS	31	19	4.5	-114	NA	NA	188	-181.4	0.223
[9]	SiGe BiCMOS	10.5	22	2.5	-116.5	-112	NA	75	-184	0.488
This work VCO1	SiGe BiCMOS	20	18	3.3	-119.5	-115.5	-117.5	122	-185	0.087
This work VCO2	SiGe BiCMOS	22	22	3.3	-119.5	-115.5	-117.5	122	-185	0.087
This work VCO3	SiGe BiCMOS	24	18	3.3	-119.5	-115.5	-117.5	122	-185	0.087
This work VCO4	SiGe BiCMOS	28	15	3.3	-119.5	-115.5	-117.5	122	-185	0.087
This work VCO1-4	SiGe BiCMOS	23.75	47	3.3	-119.5	-115.5	-117.5	122	-185	0.423

†Measured at 10 MHz frequency offset and referred to 1 MHz frequency offset.

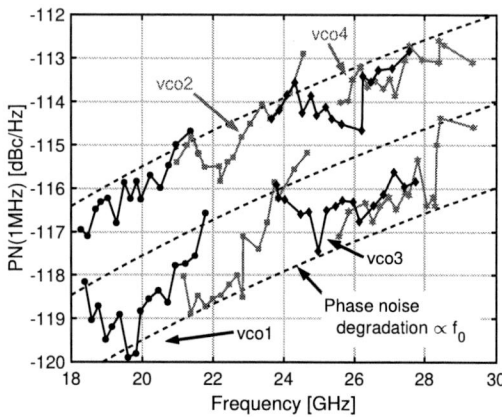

Fig. 8. Worst and best measured phase noise for each of the four VCOs by varying V_{tune}, as a function of the coarse tuning code oscillation frequency.

worst and best measured phase noise for each of the four VCOs by varying V_{tune}, as a function of the coarse tuning code oscillation frequency. The maximum variation is less than 4 dB on each oscillator and 6 dB only on the overall very large 47% tuning range. Worth noticing, 4 dB variation are naturally given from the frequency variation.

Table 1 summarizes the measured performance and compares it to the state-of-the-art for bipolar oscillators at $f_0 > 20$ GHz. The best and worst equivalent phase noise over the entire tuning range, normalized to a 20 GHz carrier is reported for fair comparison. The presented oscillator shows the lowest worst-case phase noise, with the only exception of [3], which however is not reporting any temperature measurement and is employing a 55 nm node.

IV. CONCLUSION

This work presents a bank of four VCOs for 5G applications targeting the lowest possible phase noise with minimum variations over 18-29 GHz. The proposed oscillators leverage a careful tank design with MOS-based varactor

elements to minimize typical degradation of phase noise over tuning range of the *pn*-junction varactors. The measured oscillators feature state-of-the-art phase noise, ranging from -119.5 dBc/Hz to -116.5 dBc/Hz at 1 MHz offset, respectively. The variation over the whole 47% tuning range is less than 6 dB and 2 dB over temperature.

REFERENCES

[1] F. Boscolo, F. Padovan, F. Quadrelli, M. Tiebout, A. Neviani, and A. Bevilacqua, "A 21GHz 20.5%-tuning range Colpitts VCO with -119 dBc/Hz phase noise at 1MHz offset," in *ESSCIRC 2017 - 43rd IEEE European Solid State Circuits Conference*, pp. 91–94, Sep. 2017.

[2] E. Mammei, F. Loi, F. Radice, A. Dati, M. Bruccoleri, M. Bassi, and A. Mazzanti, "8.3 A power-scalable 7-tap FIR equalizer with tunable active delay line for 10-to-25Gb/s multi-mode fiber EDC in 28nm LP-CMOS," in *2014 IEEE International Solid-State Circuits Conference Digest of Technical Papers (ISSCC)*, pp. 142–143, Feb 2014.

[3] N. Lacaita, M. Bassi, A. Mazzanti, and F. Svelto, "A K-Band low-noise bipolar Class-C VCO For 5G backhaul systems in 55nm BiCMOS technology," in *2017 IEEE Bipolar/BiCMOS Circuits and Technology Meeting (BCTM)*, pp. 150–153, Oct 2017.

[4] F. Padovan, M. Tiebout, K. L. R. Mertens, A. Bevilacqua, and A. Neviani, "Design of Low-Noise *K*-Band SiGe Bipolar VCOs: Theory and Implementation," *IEEE Transactions on Circuits and Systems I: Regular Papers*, vol. 62, pp. 607–615, Feb 2015.

[5] F. Padovan, M. Tiebout, K. Mertens, A. Bevilacqua, and A. Neviani, "A SiGe bipolar VCO for backhaul E-band communication systems," in *2012 Proceedings of the ESSCIRC (ESSCIRC)*, pp. 402–405, Sept 2012.

[6] Y. Chen, Y. Pei, and D. M. W. Leenaerts, "A dual-band LO generation system using a 40GHz VCO with a phase noise of -106.8dBc/Hz at 1-MHz," in *2013 IEEE Radio Frequency Integrated Circuits Symposium (RFIC)*, pp. 203–206, June 2013.

[7] T. Nakamura, T. Masuda, K. Washio, and H. Kondoh, "A Low-Phase-Noise Low-Power 27-GHz SiGe-VCO using Merged-Transformer Matching Circuit Technique," in *2007 IEEE Radio Frequency Integrated Circuits (RFIC) Symposium*, pp. 413–416, June 2007.

[8] C. Lee, T. Yao, A. Mangan, K. Yau, M. A. Copeland, and S. P. Voinigescu, "SiGe BiCMOS 65-GHz BPSK transmitter and 30 to 122 GHz LC-varactor VCOs with up to 21Integrated Circuit Symposium, 2004., pp. 179–182, Oct 2004.

[9] E. C. Wagner and G. M. Rebeiz, "A 9.4–11.7 GHz VCO in 0.12 μm SiGe BiCMOS with -123 dBc/Hz Phase Noise at 1 MHz Offset for 5G Systems," in *2018 IEEE Radio Frequency Integrated Circuits Symposium (RFIC)*, pp. 16–19, June 2018.

978-1-7281-1702-7/19 $31.00 © 2019 IEEE

A 9.6 mW Low-Noise Millimeter-Wave Sub-Sampling PLL with a Divider-less Sub-Sampling Lock Detector in 65 nm CMOS

Hao Wang[#1], Omeed Momeni[#2]

#University of California, Davis, USA

[1]ucdwang@ucdavis.edu, [2]omomeni@ucdavis.edu

Abstract— A 40.5 GHz sub-sampling phase-locked loop (SSPLL) with only 9.6 mW power consumption is presented. The proposed Sub-Sampling Lock Detector (SSLD) samples the output signal with on-chip generated 900 MHz reference, and can automatically detect and rectify the unlock or locked-to-wrong-harmonic states. This is done without using traditional power-consuming divider-based frequency-locked loop (FLL). The proposed SSPLL hence achieves low power, low in-band phase noise and robust operation simultaneously.

Keywords — millimeter-wave, mm-wave, sub-sampling PLL, frequency synthesizer, low-power, low-noise, lock detector, sensor.

I. INTRODUCTION

In the recent decade, on-chip Sub-Sampling PLL (SSPLL) based frequency synthesizer has become popular for its lower in-band phase noise than the frequency divider based counterparts [1]. However, one of the drawbacks of SSPLL is the possible locking to wrong harmonics of input reference. A frequency-locked-loop (FLL) based on divider-chain must be adopted to produce a correct output frequency [1]. With frequency synthesizer entering mm-wave realm, dividers, which are typically based on injection-locking structures, suffer from limited locking range and high power consumption [2]. To save power, prior-arts on SSPLL claim to turn off FLL after sub-sampling loop is locked to the correct reference harmonic [1],[3]. However, this procedure is manual, and lacks a reliable mechanism to automatically detect unlock/locked-to-wrong-harmonic states and re-enable the FLL.

In this paper, we propose a low-power SSPLL structure, achieving frequency-locking at 40.5 GHz through a novel Sub-Sampling Lock Detector (SSLD) without dividers. As a result, the SSPLL can automatically detect and rectify wrong output frequency while consuming little power. The proposed synthesizer is ideal for low-power and low-noise applications such as Internet-of-Things (IoT) systems, 5G devices, sensor networks, and specialties such as atomic clock excitation.

II. SYSTEM ARCHITECTURE

Fig. 1 shows the proposed SSPLL structure. A Type-II integer-N x9 Ring-VCO based PLL as High-frequency Reference Generation (HRG) block generates a 900 MHz signal from the input 100 MHz reference. This high-frequency reference will be used to lock the SSPLL to correct frequency. It is also used as a reference for the Sub-Sampling Lock Detector (SSLD) to detect unlock/locked-to-wrong-harmonic states.

The SSPLL takes either 100 MHz or 900 MHz as input reference, and therefore has two different configurations in charge pump and loop filter (LPF) to accomodate the two references. The 900 MHz reference is only used when the SSPLL loses lock or locks to wrong harmonics of 100 MHz. With such a high-frequency reference, the loop can only lock to one possible output frequency within the limited LC-VCO tuning range under a certain varactor bank setup.

The SSLD continuously monitors the loop, and if the SSPLL loses lock or locks to a wrong frequency, the SSLD connects the 900 MHz reference to the SSPLL for frequency aquisition. As soon as the correct frequency is acquired, SSLD will hand over the reference to 100 MHz. Therefore, the SSPLL can benefit from the low phase noise performance of low-frequency crystal references. If for any reason the loop goes out of lock, the same procedure will repeat until the loop acquires the correct frequency.

Designed for a mercury-ion atomic clock, the proposed SSPLL output center frequency is 40.5 GHz, which is a harmonic of both 900 MHz and 100 MHz. The 900 MHz can be modified for any other targeted SSPLL output frequencies and applications.

III. SUB-SAMPLING LOCK DETECTOR

Traditional PLL lock detectors use digital-intensive circuits, and need frequency dividers. To accomodate the divider-less sub-sampling structure, we propose an analog lock detector interface which directly samples the SSPLL output (V_{out}) with the 900 MHz reference, as shown in Fig. 2. When SSPLL locks to the correct frequency, which is an integer multiple of 900 MHz, the sampled phase is ideally constant. However, if the SSPLL loses lock or locks to a wrong frequency, the sampled phase varies with time. Our proposed SSLD detects these variations and determines the lock status of the SSPLL accordingly. A sampler converts the sampled phase to discrete voltage V_{sam}, and later V_{sam} is integrated to a full-scale signal V_{deta} by a transconductor amplifier. The chain is followed by a Schmitt trigger to rectify the analog output to digital signal, V_{det}. If the SSPLL is locked to a correct frequency, V_{det} remains at either 0 or 1, reflecting the constant sampled phase. Fig. 2 shows the calculated f_{det} as a function of f_{out}. The blue points represent the correct frequencies (multiples of 900 MHz) with $f_{det} = 0$. The tuning range of the VCO at a specific capacitor bank configuration should be kept below 900 MHz to avoid wrong harmonic locking. If the SSPLL exhibits a wrong frequency, V_{det} toggles

978-1-7281-1702-7/19 $31.00 © 2019 IEEE

2019 IEEE Radio Frequency Integrated Circuits Symposium

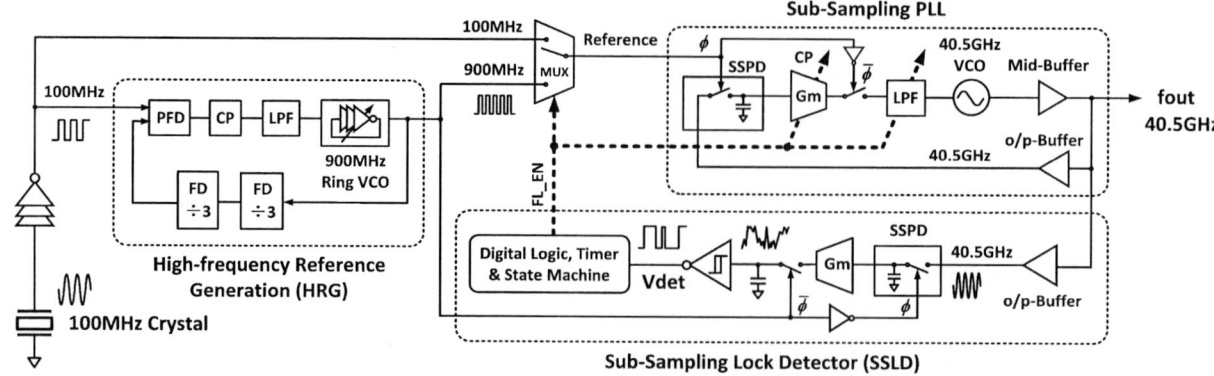

Fig. 1. Proposed Sub-Sampling PLL with Sub-Sampling Lock Detector (SSLD) and High frequency Reference Generation (HRG) loop.

between 0 and 1 at the frequency of f_{det} shown in Fig. 2. The period of f_{out} is only 25 ps, and therefore inevitable jitters from the SSPLL's output and the 900 MHz reference will cause variations in the sampled phase. That means even if f_{out} is the correct frequency, V_{det} may still toggle stochastically. This issue can be solved by the logic design explained next.

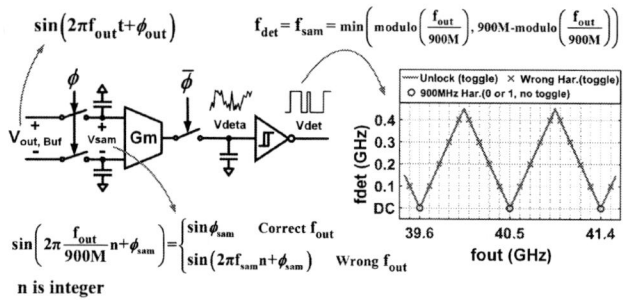

Fig. 2. Analog interface of the proposed Sub-Sampling Lock Detector.

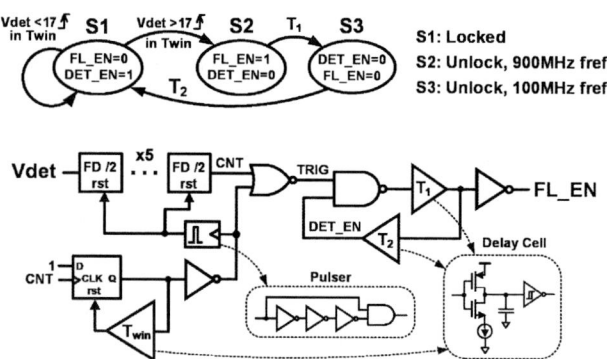

Fig. 3. State machine of the proposed Sub-Sampling Lock Detector.

Fig. 3 shows the digital logic and the state machine of the SSLD. We first evaluate the V_{det} frequency by counting its toggling within a time window of T_{win}. The jitter-induced toggling is much less frequent than those in unlock or wrong-harmonic states. If toggling exceeds a certain frequency, it can be shown that the SSPLL has a wrong f_{out}. A state machine is then designed as follow. When a wrong f_{out} is detected, the control signal FL_EN is set to 1 to switch

the SSPLL reference to 900 MHz for frequency aquisition. Meanwhile, the lock detection is disabled by setting DET_EN to 0. After a sufficient aquisition time T_1, frequency aquisition is achieved. FL_EN will then be set to 0 to switch the SSPLL reference back to 100 MHz. After a sufficient locking time T_2, the SSPLL will lock to 100 MHz with correct f_{out}. Detection will then resume through setting DET_EN to 1. This detecting and rectifying procedure is automatic. The digital part of the SSLD is almost off in the locked state, reducing the power consumption even further.

IV. PLL CIRCUITS IMPLEMENTATION

A. High-frequency Reference Generation (HRG) Loop

As shown in Fig. 1, HRG is a x9 Type-II Charge Pump PLL. Since it operates at a relatively low frequency of 900 MHz, delay cell based Ring-VCO is adopted, rather than area-consuming LC-VCO. Pseudo-differential load-controlled delay cell is used for its lower thermal noise than current-biasing delay cells. As mentioned previously, HRG 900 MHz output needs to have sufficiently low jitter to function well for SSLD. With 25 ps SSPLL f_{out} period, we choose to limit the 900 MHz RMS jitter within 1 ps. HRG loop gain and bandwidth are optimized for low jitter, good stability and low power consumption.

B. SSPLL

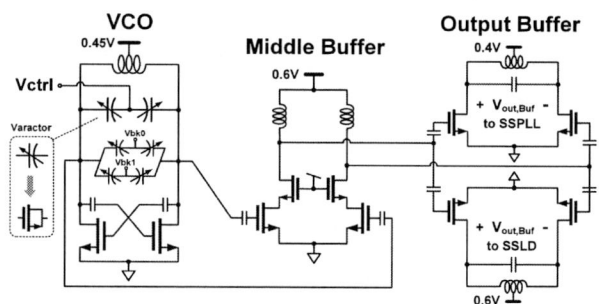

Fig. 4. VCO and buffers of the proposed Sub-Sampling PLL.

Fig. 4 shows the VCO and buffers of the SSPLL. The 40.5 GHz LC-VCO is of cross-coupled structure with 2 varactor

978-1-7281-1702-7/19 $31.00 © 2019 IEEE   172

banks for frequency calibration. Fig. 5 shows the measured VCO tuning range under various varactor bank setups. For correct operation, the charge pump output transistors need to work in saturation region. The V_{ctrl} thus has a corresponding effective region as shown in Fig. 5. Under each bank setup, the effective VCO tuning range within the effective V_{ctrl} region is around 900 MHz. A middle buffer with cascode amplifying stage is inserted between VCO and output buffers for better isolation. The two separate output buffers are connected to SSPLL and SSLD respectively. The VCO and buffers are biased in class AB operation for lower power consumption.

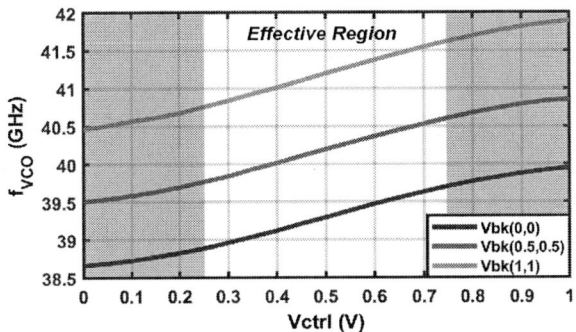

Fig. 5. Measured LC-VCO tuning range.

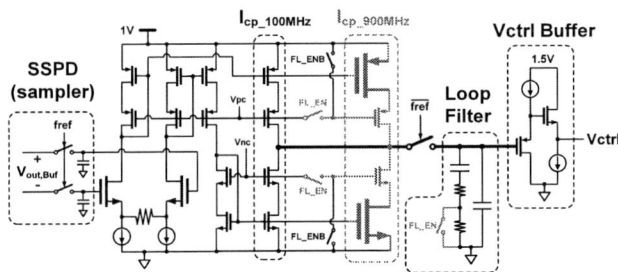

Fig. 6. SSPD, charge pump and loop filter of the proposed Sub-Sampling PLL.

Fig. 6 shows the SSPD, charge pump and loop filter of the SSPLL. Two sets of charge pump and loop filter are designed to accomodate the two references. When taking 900 MHz reference, the loop gain and bandwidth are both high to enhance aquisition range, ensure fast locking and achieve good stability. When the SSPLL shifts reference from 900 MHz to 100 MHz, the SSPLL will have to re-lock to 100 MHz to rectify the initial phase difference between output and 100 MHz. To make sure f_{out} won't sway to the adjacent 100 MHz harmonics during this procedure, loop gain and bandwidth are both designed relatively small for 100 MHz reference compared to those for 900 MHz. Switching between the two loop gains is through turning off the cascode transistors of charge pump output branches. A switch bypasses a segment of resistor in the loop filter to change the loop bandwidth. Time sequence logic is designed for smooth switching without disturbing biasing. A V_{ctrl} buffer is added to avoid VCO back-charging on the loop filter.

V. EXPERIMENTAL RESULT

The SSPLL is fabricated in a 65 nm CMOS process. Die photo is shown in Fig. 7, with core area of 0.6 mm². Total power consumption of the chip is 9.58 mW, out of which 4.99 mW, 2.76 mW and 1.83 mW are consumed by SSPLL, HRG and SSLD respectively. A 100 MHz crystal oscillator is used as input reference. The 40.5 GHz output signal is measured through a GSG probe. Measured HRG 900 MHz RMS jitter is 1.0 ps (integrated from 10 KHz to 100 MHz). With 3 MHz loop bandwidth, the SSPLL 40.5 GHz output spectrum and phase noise are shown in Fig. 8 and Fig. 9 respectively. Phase noise at 1 MHz (in-band) and 10 MHz offsets are -96.6 dBc/Hz and -106.9 dBc/Hz respectively. The integrated RMS jitter (10 KHz to 100 MHz) is 228 fs. Reference spur is below -42 dBc. Notice that the spurs at 107 KHz and 138 KHz are from the power supply. In comparison, we can see the output with 100 MHz reference has 5 dB lower in-band phase noise than that with 900 MHz. This is expected as the cascaded PLL structure has added noise from the HRG charge pump and ring-VCO. Table 1 shows the measured performance summary of the proposed SSPLL in comparison with the state-of-the-art. To the best of our knowledge, the proposed SSPLL has the lowest power consumption and the best FOM among 40–60 GHz frequency synthesizers with crystal references.

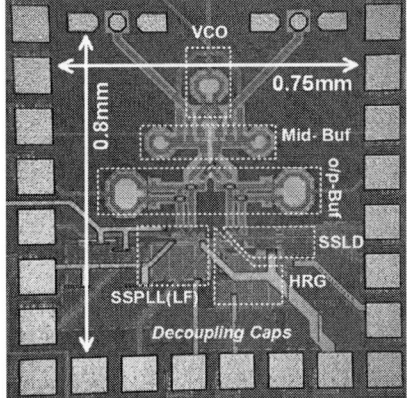

Fig. 7. Die photo. The core area is 0.6 mm².

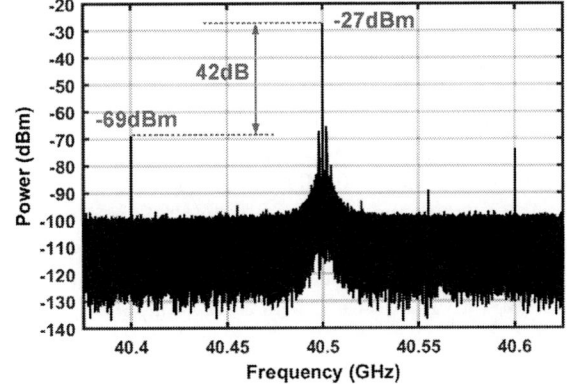

Fig. 8. Measured frequency spectrum of the SSPLL 40.5 GHz output

978-1-7281-1702-7/19 $31.00 © 2019 IEEE

Table 1. Proposed SSPLL Performance Summary in Comparison with the State-of-the-art.

Ref.	Topology	Tech. (nm)	f_{ref} (MHz)	f_{out} (GHz)	Norm. PN (dBc/Hz) @1MHz	@10MHz	σ_{rms} (fs)	Ref. Spur (dBc)	P_{DC} (mW)	Area (mm^2)	FOM$_1$[a]	FOM$_2$[b]
[2] JSSC '16	SSPLL	65	36/40	60	-95.4	-125.4	290	-73	32	0.7	-172.5	-236
[3] JSSC '15	SSPLL	40	40	60	-95.4	-104.4	200	<-40	42	0.16	-171.3	-237.7
[4] RFIC '14	PFD/CP	65	195	50/100	-95.8	-111.8	-	-40	14	0.39	-176.5	-
[5] RFIC '17	PFD/CP	250[c]	50	40	-100.6	-124.9	103[d]	-73	323	0.45	-167.7	-234.7
[6] JSSC '14	PLL +ILFM	65	100	42	-108.3	-108.5	-	-31	148	2.09	-178.8	-
This Work	**SSPLL**	**65**	**100**	**40.5**	**-96.6**	**-106.9**	**228**	**<-42**	**9.6**	**0.6**	**-179.3**	**-243.0**

[a] FOM$_1$ = $L\{\Delta f\}$ + 20 log ($\Delta f/f_{out}$) + 10 log (P_{DC}/1mW), Δf=1MHz

[b] FOM$_2$ = 20 log (jitter/1s) + 10 log (P_{DC}/1mW)

[c] 0.25 um BiCMOS

[d] 100 Hz to 1 MHz

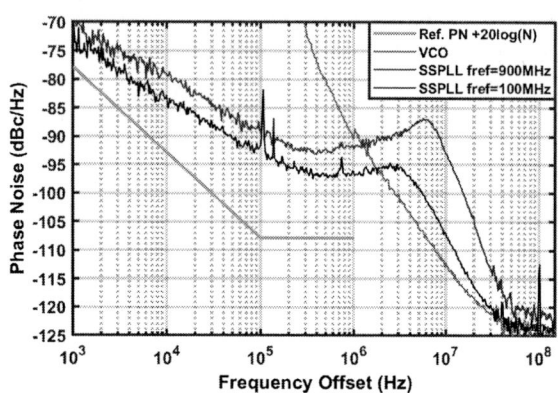

Fig. 9. Measured phase noise of the SSPLL 40.5 GHz output, compared with reference limit and VCO phase noise.

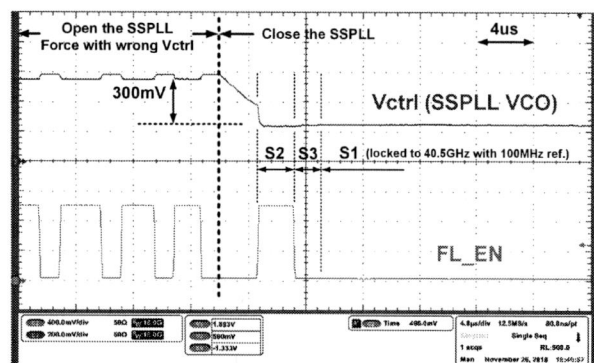

Fig. 10. Measured automatic re-lock procedure of the proposed SSPLL.

The SSPLL automatic re-lock procedure is measured and shown in Fig. 10. We first open the SSPLL's loop and force the VCO control voltage V_{ctrl} to be 300 mV higher than the locked value, to mimic the unlocked SSPLL. The gap of 300 mV is chosen because it contains several V_{ctrl} values corresponding to wrong 100 MHz harmonics. Next, we close the loop and examine the re-lock procedure by observing V_{ctrl} and reference switching signal FL_EN with oscilloscope. The SSLD can detect the unlock state and go through the procedure to re-lock the SSPLL to 40.5 GHz. SSLD digital logic states are labelled referring to Fig. 3. The procedure of locking, to 900 MHz in S2 and to 100 MHz in S3, can be observed through V_{ctrl}. FL_EN settles at 0, meaning the SSPLL is locked to the 100 MHz reference. V_{ctrl} also settles at the correct value corresponding the correct 100 MHz harmonic.

VI. CONCLUSION

An SSPLL with a Sub-Sampling Lock Detector is presented. By avoiding high-frequency dividers, the loop implements frequency acquisition with an on-chip generated high-frequency reference, so that total power consumption and robustness are both improved. Thanks to the novel SSLD, this SSPLL can automatically detect unlock and locked-to-wrong-reference-harmonic states, and make the loop re-lock to the correct output frequency. As a result, the SSPLL consumes only 9.6 mW and achieves 228 fs (10 KHz to 100 MHz) RMS jitter.

ACKNOWLEDGMENT

The authors would like to thank DARPA and NASA Jet Propulsion Laboratory (JPL) for supporting this research. The views, opinions, and findings expressed are those of the authors and should not be interpreted as representing the official views or policies of the Department of Defense or the U.S. Government.

REFERENCES

[1] X. Gao, E. Klumperink, and B. Nauta, "Sub-sampling pll techniques," in *2015 IEEE Custom Integrated Circuits Conference (CICC)*, Sep. 2015, pp. 1–8.

[2] T. Siriburanon, S. Kondo, M. Katsuragi, H. Liu, K. Kimura, W. Deng, K. Okada, and A. Matsuzawa, "A low-power low-noise mm-wave subsampling pll using dual-step-mixing ilfd and tail-coupling quadrature injection-locked oscillator for ieee 802.11 ad," *IEEE Journal of Solid-State Circuits*, vol. 51, no. 5, pp. 1246–1260, 2016.

[3] V. Szortyka, Q. Shi, K. Raczkowski, B. Parvais, M. Kuijk, and P. Wambacq, "A 42 mw 200 fs-jitter 60 ghz sub-sampling pll in 40 nm cmos," *IEEE Journal of Solid-State Circuits*, vol. 50, no. 9, pp. 2025–2036, 2015.

[4] Y. Chao, H. C. Luong, and Z. Hong, "A 0.6/1.2-v 14.1-mw 96.8 ghz-to-108.5 ghz transformer-based pll with embedded phase shifter in 65-nm cmos," in *Radio Frequency Integrated Circuits Symposium, 2014 IEEE*. IEEE, 2014, pp. 93–96.

[5] Y. Chen, L. Praamsma, N. Ivanisevic, and D. M. Leenaerts, "A 40ghz pll with- 92.5 dbc/hz in-band phase noise and 104fs-rms-jitter," in *Radio Frequency Integrated Circuits Symposium (RFIC), 2017 IEEE*. IEEE, 2017, pp. 31–32.

[6] A. Li, S. Zheng, J. Yin, X. Luo, and H. C. Luong, "A 21-48 ghz subharmonic injection-locked fractional-n frequency synthesizer for multiband point-to-point backhaul communications," *IEEE Journal of Solid-State Circuits*, vol. 49, no. 8, pp. 1785–1799, 2014.

A -40-dBc Integrated-Phase-Noise 45-GHz Sub-Sampling PLL with 3.9-dBm Output and 2.1% DC-to-RF Efficiency

Sangyeop Lee[#1], Kyoya Takano[#], Shinsuke Hara[$], Ruibing Dong[#], Shuhei Amakawa[#],
Takeshi Yoshida[#], and Minoru Fujishima[#]

[#]Hiroshima University, Japan
[$]National Institute of Information and Communications Technology, Japan
[1]sangyeop@ieee.org

Abstract—**This paper presents a millimeter-wave (mmW) sub-sampling PLL in 40 nm CMOS. Sub-sampling PLL reduces the in-band phase noise due to the charge pump lower than the ordinary $N^2\times$ when frequency is multiplied by N. Two sub-sampling phase detectors (SSPD) and charge pumps (SSCP) are employed to cancel mixing products due to sub-sampling around the VCO output tone and to enhance loop gain. The out-of-band phase noise, dictated by the VCO phase noise, is reduced by employing a VCO consisting of transmission-line resonators, large MOSFET switches, and inverse-class-F output matching. The proposed PLL, operating at 45 GHz, achieves -40-dBc integrated phase noise (0.1 kHz–40 MHz), 3.9-dBm output power, and 2.1% DC-to-RF efficiency.**

Keywords—**Sub-sampling, phase-locked loop, low phase noise, energy efficient, CMOS.**

I. INTRODUCTION

The phase noise (PN) of a signal source usually increases in proportion to the output frequency squared (PN $\propto f_{\text{out}}^2$). If frequency is multiplied by N, noise floor goes up by $20\log_{10} N$ decibels (dB) [1]. LO phase noise of a transmitter and a receiver both aggravate system signal-to-noise ratio (SNR) in mmW/sub-THz wideband radio systems. Therefore, mmW/sub-THz signal source must have low in-band (or near-carrier) and out-of-band phase noise.

In this paper, we present a mmW frequency multiplier based on a sub-sampling phase-locked loop (SSPLL), which is an effective means of reducing in-band noise caused by a charge pump [2]. To reduce out-of-band noise, we employ a high-power buffer-less VCO [3] and nonlinear harmonic-resonance control techniques [4], [5].

II. PROPOSED SSPLL-BASED FREQUENCY MULTIPLIER

Fig. 1(a) shows a conventional sub-sampling PLL (SSPLL) [2]. Frequency locking can be achieved using by a conventional charge-pump PLL (CPPLL). The CPPLL employs a phase/frequency detector (PFD) with phase-insensitive dead zone (DZ), a CP, a loop filter (LF), a divider chain (division ratio: N), and a VCO. After frequency locking, sub-sampling operation for phase locking occurs using a sub-sampling phase detector (SSPD) with a pulser and a sub-sampling charge pump (SSCP). Usually, in-band phase noise of the conventional CPPLL is dominated by multiplied ($N^2\times$) noise from large noise current of the CP, provided the phase noise due to reference signal is relatively

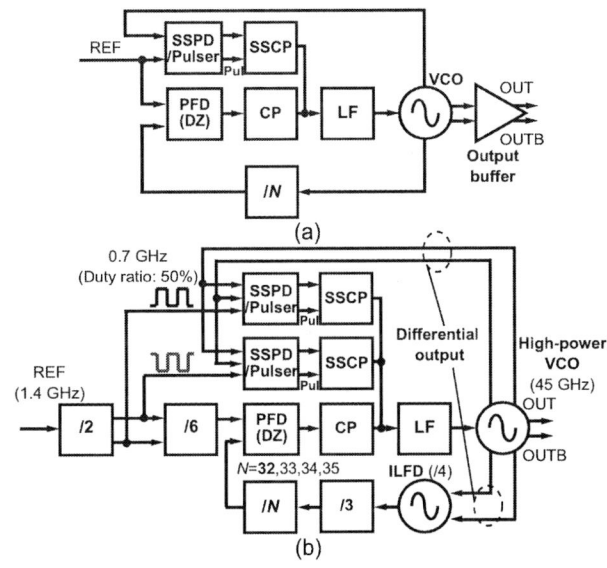

Fig. 1. (a) Conventional sub-sampling PLL [2]. (b) Proposed sub-sampling PLL employing a low-noise VCO and a pseudo frequency-doubled SSPD/CP.

low. The SSPLL can achieve low in-band phase noise because the SSPD and SSCP noise is not multiplied by N^2, thanks to sub-sampling operation [2]. Finally, a VCO output buffer is commonly employed to drive a certain load impedance, Z_{L}.

Fig. 1(b) depicts the proposed mmW sub-sampling PLL. A buffer-less high-power VCO oscillating at about 45 GHz is used. Then, a 1.4-GHz reference signal (REF) is supplied from an external source to generate a high-frequency signal with low phase noise characteristics. As a prescaler, a ring-VCO-based injection-locked frequency divider (ILFD) whose division ratio is 4 is employed. The reference signal is divided by a frequency divider (division ratio: 2) to generate a clock that has duty cycle of 50%, since the duty cycle of the clock can affect the spurious characteristics of the PLL output. Two SSPDs with pulsers and two SSCPs are adopted to balance amplitude and phase of high-frequency VCO differential outputs [6] and to enhance loop gain of the SSPLL, which has the same effect as doubling the sampling clock frequency.

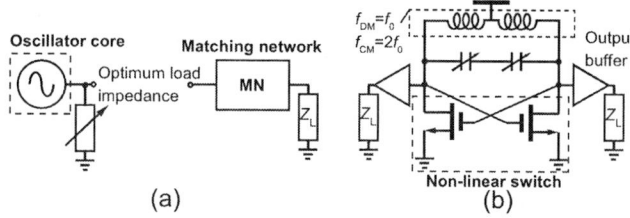

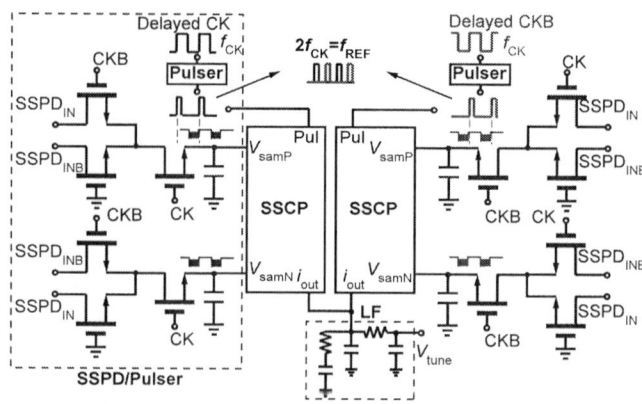

Fig. 2. (a) High-power, energy-efficient oscillator without employing an output buffer [3], (b) an oscillator based on class-D/inverse class-F topology for flicker-noise-upconversion reduction [5].

Fig. 4. Proposed pseudo frequency-doubled SSPD/CP (FD-SSPDCP).

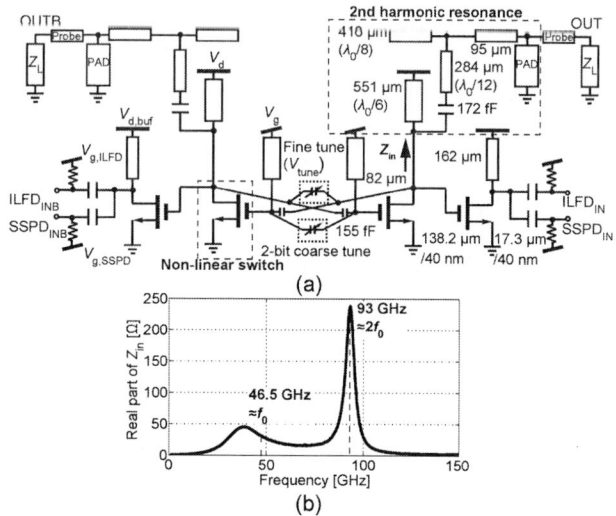

Fig. 3. (a) Proposed VCO, (b) real part of input impedance of the second-harmonic-resonance output matching network ($Z_L = 50\,\Omega$).

oscillator tank is employed to reduce the corner frequency $\Delta f_{1/f^3}$. In Fig. 2(b), the oscillator uses an inductor that resonates at the second-harmonic ($f_{CM} = 2f_0$, $f_{DM} = f_0$) and large nonlinear switches (NMOSFETs) to control the harmonic tones of the oscillator [5].

Fig. 3(a) shows the proposed VCO. It employs second-harmonic-resonance (i.e., inverse class F) output matching network [9], [10] based on on-chip transmission lines. Also, it uses large NMOSFETs (138.2 μm/40 nm) to increase the output power in the presence of harmonic tones. After determining parameters of buffer amplifiers that drive the ILFD and SSPD/pulsers, the length of TLs for gate-bias feeding into the core NMOSFETs, the size of cross-coupled capacitors (155 fF), and the size of varactors for fine and 2-bit coarse tuning are optimized to balance the power fed into the VCO to maintain oscillation and the output power. Fig. 3(b) shows the simulated input-impedance of the output matching network, which shows resonance near $2f_0$.

Based on post-layout simulation results, the fine tuning range is from 45.35 GHz ($V_{tune} = 0.9$ V) to 45.73 GHz ($V_{tune} = 0$ V). And, output frequency range can be shifted at the maximum by about 2.3 GHz with 2-bit coarse tuning (i.e., 43.03 GHz–43.38 GHz).

A. High-power VCO with reduced upconverted flicker noise

Leeson's phase-noise model (1) is commonly used to estimate phase noise of an oscillator.

$$S(\Delta f) = \frac{2FkT}{P_S}\left[1 + \left(\frac{f_0}{2Q_L\Delta f}\right)^2\right]\left(1 + \frac{\Delta f_{1/f^3}}{|\Delta f|}\right), \quad (1)$$

where F is an empirical parameter, k is Boltzmann's constant, T is the absolute temperature, P_S is the average power dissipated in the resistive part of the tank, f_0 is the oscillation frequency, Δf is the offset frequency from the carrier frequency, $\Delta f_{1/f^3}$ is the corner frequency between the $1/f^3$ and $1/f^2$ regions, and Q_L is the loaded quality factor of the oscillator tank [7], [8]. It shows that the phase noise $S(\Delta f)$ is inversely proportional to the signal power level P_S and can be reduced with decreasing the corner frequency $\Delta f_{1/f^3}$ and increasing the quality factor of the oscillator. These three factors are considered when designing a VCO.

Recently, high-frequency buffer-less oscillators having high output power, low phase noise, but good DC-to-RF efficiency, were reported. Fig. 2(a) shows the design concept [3], in which the choice of an optimum load impedance for the oscillator core and output matching are important. In another approach, harmonic resonance in the

B. Loop-gain enhancement in sub-sampling PLL

In Fig. 1(a), two SSPD/pulsers and two SSCPs are employed, considering amplitude and phase balance of high-frequency VCO outputs [6]. Fig. 4 shows the details of the proposed pseudo frequency-doubled SSPD/CP (FD-SSPDCP) including two SSPDs, two pulsers, and two SSCPs. In the SSPD, ground-biased NMOSFETs are employed to cancel LO leakage when the switch is in hold operation [11]. Frequency-divided 50%-duty-cycle clocks (CK & CKB) are used for sub-sampling operation of the switches.

However, usually, phase noise characteristics due to SSCP in an SSPLL is inversely proportional to the sampling-clock frequency and the pulse width: $S_{in\text{-}band,CP,SS} \propto 1/(f_{ref}\tau_{pul})$, where f_{ref} is the frequency of the reference signal, and τ_{pul} is the width of pulses generated from the pulser [2]. As a result, the proposed FD-SSPDCP has another important role

978-1-7281-1702-7/19 $31.00 © 2019 IEEE

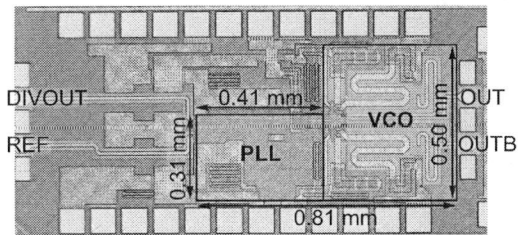

Fig. 5. A chip micrograph of the proposed PLL.

of increasing the open loop gain H_{open} ($\propto f_{\mathrm{ref}}$) using pseudo frequency-doubled pulses that are generated from both rising and falling edges of the sampling clock. Moreover, the loop gain can be increased using both SSCPs and the CP for the CPPLL simultaneously [12].

III. MEASUREMENT RESULTS

Fig. 5 shows a chip micrograph of the proposed PLL, which was fabricated using a 40 nm CMOS process. Fig. 6 shows measurement results, in which either V_g or V_d of the VCO were swept under the free-running condition while PLL components were turned off ($V_{\mathrm{tune}} = 0\,\mathrm{V}$, $V_{\mathrm{d,buf}} = 0\,\mathrm{V}$, $V_{\mathrm{g,SSPD}} = 0.51\,\mathrm{V}$). The VCO shows the best 10-MHz-offset phase noise of $-137.6\,\mathrm{dBc/Hz}$ when $V_g = 0.7\,\mathrm{V}$ and $V_d = 0.65\,\mathrm{V}$ at the oscillation frequency of 45.2 GHz. The FoM$_{\mathrm{VCO}}$ is $-193.0\,\mathrm{dBc/Hz}$ from $\mathrm{FoM}_{\mathrm{VCO}} = \mathrm{PN} - 20\log_{10}\left(f_0/\Delta f\right) + 10\log_{10}\left(P_{\mathrm{DC,VCO}}/1\,\mathrm{mW}\right)$, where $P_{\mathrm{DC,VCO}}$ is the DC power consumption of the VCO. Fig. 7 shows measured phase noise of the VCO when $V_g = 0.7\,\mathrm{V}$ and $V_d = 0.65\,\mathrm{V}$. It shows that the measured corner frequency, $\Delta f_{1/f^3}$, is approximately 600 kHz.

Fig. 8(a) shows the output power spectrum of the free-running VCO when $V_g = 0.75\,\mathrm{V}$ and $V_d = 0.65\,\mathrm{V}$. It shows that the single-ended output power of the VCO was 2.2 dBm after compensating for cable and probe losses and difference between the measured result using a spectrum analyzer and that using a power meter. Fig. 8(b) shows the output power spectrum of the locked VCO when both the SSPLL and the CPPLL are in operation. A 1.41-GHz-reference signal was supplied (39.7-fs RMS jitter (0.1 kHz–40 MHz)). In this condition, the PLL components excluding the VCO core operated under the supply voltage of 0.9 V and consumed 50.7 mW including CPPLL components. The single-ended output power was 0.9 dBm and the reference spur due to the CPPLL was $-50.4\,\mathrm{dBc}$. The reference spurs due to the SSPLL were lower than $-55\,\mathrm{dBc}$.

Fig. 9 shows measured phase noise in three different modes: (i) free-running mode with operating the ILFD, (ii) conventional CPPLL mode, and (iii) proposed SSPLL mode (with operating the CPPLL simultaneously). The proposed SSPLL presents a 10-kHz-offset phase noise of $-98.7\,\mathrm{dBc/Hz}$, and a 40-MHz-offset phase noise of $-138.8\,\mathrm{dBc/Hz}$.

Table 1 gives a performance summary and comparison with other PLLs operating above 25 GHz.

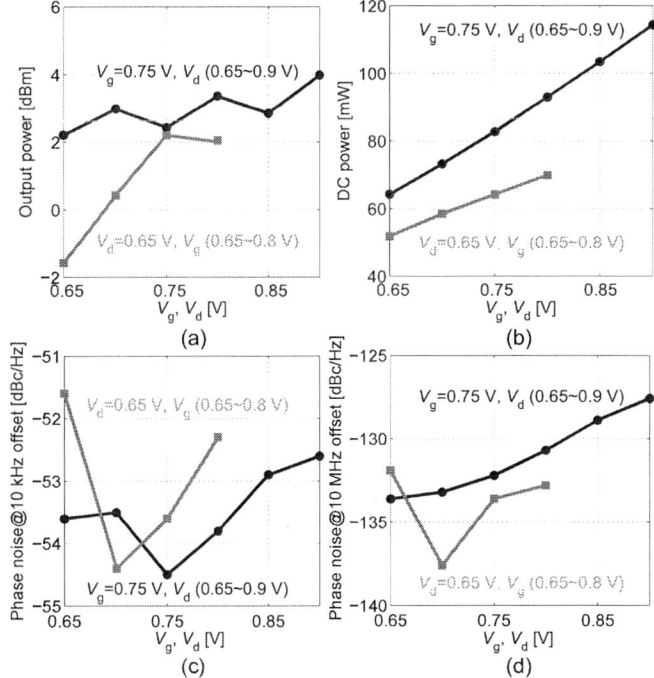

Fig. 6. Measured results of the free-running VCO ($V_{\mathrm{tune}} = 0\,\mathrm{V}$, $V_{\mathrm{d,buf}} = 0\,\mathrm{V}$, $V_{\mathrm{g,SSPD}} = 0.51\,\mathrm{V}$, and PLL components were turned off): (a) Output power (one single-ended output), (b) DC power consumption, (c) 10-kHz-offset phase noise, and (d) 10-MHz-offset phase noise.

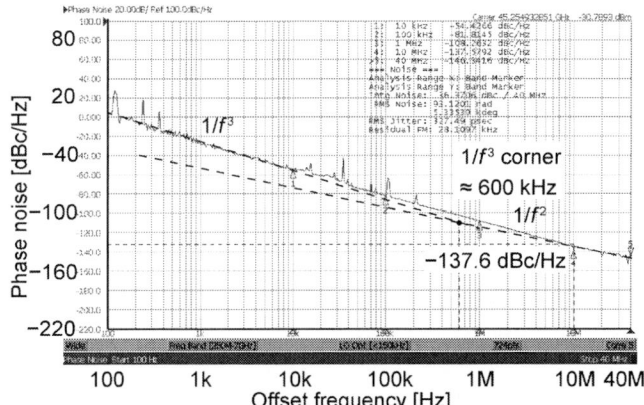

Fig. 7. Measured phase noise of the free-running VCO ($V_g = 0.7\,\mathrm{V}$, $V_d = 0.65\,\mathrm{V}$, $V_{\mathrm{tune}} = 0\,\mathrm{V}$, $V_{\mathrm{d,buf}} = 0\,\mathrm{V}$, $V_{\mathrm{g,SSPD}} = 0.51\,\mathrm{V}$, and PLL components were turned off).

IV. CONCLUSION

We demonstrated a 45-GHz sub-sampling-PLL-based frequency multiplier ($N = 32$) that can be applied to mmW and sub-THz systems. It provides good performance of -40-dBc integrated phase noise (0.1 kHz–40 MHz), 3.9-dBm output power, and 2.1% DC-to-RF efficiency. This was achieved by (i) signal power enhancement and (ii) flicker noise reduction, which are based on large switches and harmonic resonance enabled by inverse class F output matching networks. Moreover, the proposed FD-SSPDCP was used to reduce LO leakage and spurs due to sub-sampling. Pseudo

Table 1. Performance summary and comparison of PLLs above 25 GHz.

Ref.	Tech.	f_out [GHz]	f_ref [GHz]	N	PN$_\text{in-band}$ [dBc/Hz]	PN$_\text{out-band}$ [dBc/Hz]	IPN/J$_\text{RMS}$[†] [dBc]/[fs]	P_out [mW]	P_DC [mW]	P_out/P_DC [%]	FoM[‡] [dB]	Area [mm²]
This work	40 nm CMOS	45.0	1.41	32	−98.7 @10 kHz	−138.8 @40 MHz	−40.5/47.4 (0.1 k–40 M)	2.5*	114.6**	2.1	−245.9**	0.40 (core)
[13]	40 nm CMOS	62.6	0.04	1566	−89.7 @200 kHz	−120△ @50 MHz	−23.9/230 (1 k–100 M)	-	42	-	−236.5	0.16 (core)
[14]	65 nm CMOS	60.5	0.040	1512	−78.5 @100 kHz	−122 @10 MHz	−28.8/290 (10 k–40 M)	0.1	32	0.31	−235.7	1.43▽
[15]	65 nm CMOS	29.3	2.25	13	−92.8 @10 kHz	−128.8 @40 MHz	−39.1/85.6 (1 k–100 M)	-	24.3	-	−247.5	0.47▽
[16]	250 nm BiCMOS	40.0	0.050	800	−92.5 @100 kHz	−128△ @40 MHz	−34.7/104 (0.1 k–1 M)	-	323	-	−236.2	0.45 (core)

† Integrated phase noise (IPN), RMS jitter (J$_\text{RMS}$).   ‡FoM = $20 \log_{10}$ (J$_\text{RMS}$/1 s) + $10 \log_{10}$ (P_DC/1 mW).   *Assumes that differential outputs have same output power.
**Includes DC power for RF outputs.   △Estimated from measurement results.   ▽Chip area excluding pads.

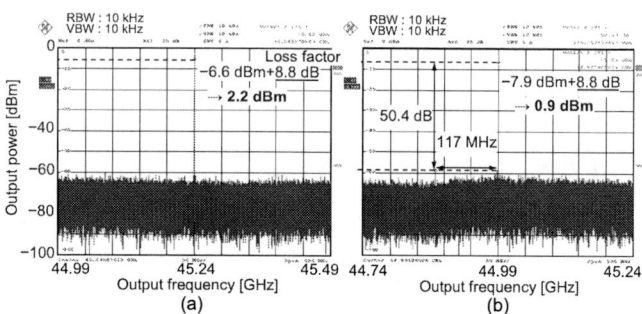

Fig. 8. Measured spectrum of the VCO output under (a) the free-running condition ($V_g = 0.75$ V, $V_d = 0.65$ V, $V_\text{tune} = 0$ V, $V_\text{d,buf} = 0$ V, $V_\text{g,SSPD} = 0.51$ V, and PLL components were turned off), and (b) the sub-sampling-locked condition.

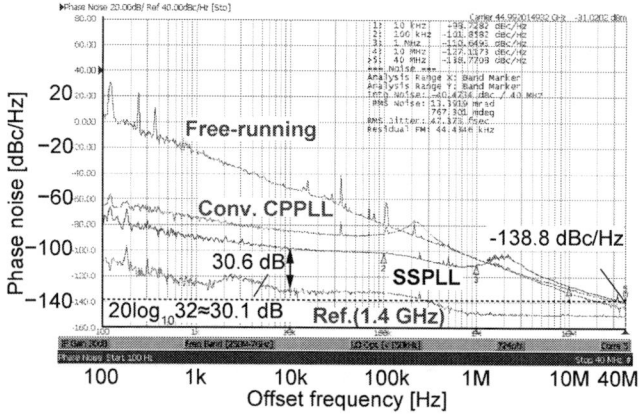

Fig. 9. Measured phase noise of the free-running VCO ($V_g = 0.75$ V, $V_d = 0.65$ V, $V_\text{d,buf} = 0.9$ V, $V_\text{g,SSPD} = 0.51$ V, and the ILFD was turned on), conventional CPPLL, proposed SSPLL, and the reference signal.

frequency-doubled pulses for the FD-SSPDCP, generated from both rising and falling edges of the frequency-divided reference signal, also help to compensate loop gain.

Acknowledgment

This work was supported in part by the Ministry of Internal Affairs and Communications of Japan, JSPS KAKENHI Grant Number JP18H03781, and VLSI Design and Education Center (VDEC), the University of Tokyo in collaboration with Cadence Design Systems and Mentor Graphics, Inc.

References

[1] J. Chen et al., "Does LO noise floor limit performance in multi-gigabit millimeter-wave communication?," *IEEE Microwave Wireless Compon. Lett.*, vol. 27, no. 8, pp. 769–771, Aug. 2017.

[2] X. Gao et al., "A low noise sub-sampling PLL in which divider noise is eliminated and PD/CP noise is not multiplied by N^2," *IEEE J. Solid-State Circuits*, vol. 44, no. 12, pp. 3253–3263, Dec. 2009.

[3] H. Khatibi et al., "An efficient high-power fundamental oscillator above $f_\text{max}/2$: A systematic design," *IEEE Trans. Microw. Theory Tech.*, vol. 65, no. 11, pp. 4176–4189, Nov. 2017.

[4] L. Fanori and P. Andreani, "Class-D CMOS oscillators," *IEEE J. Solid-State Circuits*, vol. 48, no. 12, pp. 3105–3119, Dec. 2013.

[5] M. Shahmohammadi et al., "A $1/f$ noise upconversion reduction technique for voltage-biased RF CMOS oscillators," *IEEE J. Solid-State Circuits*, vol. 51, no. 11, pp. 2610–2623, Nov. 2016.

[6] X. Gao et al., "Spur reduction techniques for phase-locked loops exploiting a sub-sampling phase detector," *IEEE J. Solid-State Circuits*, vol. 45, no. 9, pp. 1809–1821, Sep. 2010.

[7] D. B. Leeson, "A simple model of feedback oscillator noise spectrum," *Proc. IEEE*, vol. 54, no. 2, pp. 329–330, Feb. 1966.

[8] A. Hajimiri and T. H. Lee, "A general theory of phase noise in electrical oscillators," *IEEE J. Solid-State Circuits*, vol. 33, no. 2, pp. 179–194, Feb. 1998.

[9] A. Grebennikov and N. O. Sokal, *Switchmode RF Power Amplifiers*, Newnes, 2007.

[10] T. Heima, et al., "A new practical harmonics tune for high efficiency power amplifier," *IEEE Eur. Microw. Conf.*, pp. 271–274, Oct. 1999.

[11] S. Ikeda et al., "A 0.52-V 5.7-GHz low noise sub-sampling PLL with dynamic threshold MOSFET," *IEEE A-SSCC*, pp. 365–368, Nov. 2014.

[12] C. Hsu et al., "A sub-sampling-assisted phase-frequency detector for low-noise PLLs with robust operation under supply interference," *IEEE Trans. Circuits Syst. I*, vol. 62, no. 1, pp. 90–99, Jan. 2015.

[13] V. Szortyka et al., "A 42mW 230fs-jitter sub-sampling 60GHz PLL in 40nm CMOS," *ISSCC*, pp. 366–367, Feb. 2014.

[14] T. Siriburanon et al., "A low-power low-noise mm-wave subsampling PLL using dual-step-mixing ILFD and tail-coupling quadrature injection-locked oscillator for IEEE 802.11ad," *IEEE J. Solid-State Circuits*, vol. 51, no. 5, pp. 1246–1260, May 2016.

[15] S. Yoo et al., "A PVT-robust -39dBc 1kHz-to-100MHz integrated-phase-noise 29GHz injection-locked frequency multiplier with a 600μW frequency-tracking loop using the averages of phase deviations for mm-band 5G transceivers," *ISSCC*, pp. 324–325, Feb. 2017.

[16] Y. Chen et al., "A 40GHz PLL with -92.5dBc/Hz in-band phase noise and 104fs-RMS-jitter," *IEEE RFIC Symp.*, pp. 31–32, Jun. 2017.

978-1-7281-1702-7/19 $31.00 © 2019 IEEE

A High Efficiency 39GHz CMOS Cascode Power Amplifier for 5G Applications

Hyun-chul Park[1], Byungjoon Park, Yunsung Cho, Jaehong Park, Jihoon Kim, Jeong Ho Lee, Juho Son,
Kyu Hwan An, Sung-Gi Yang

Samsung Electronics Co., Ltd. Suwon, S. Korea
[1]hcpark.park@samsung.com

Abstract— **We present a 39GHz CMOS cascode power amplifier (PA) with a two-step (*L-C* and *C-L*) second harmonic termination. We analyze the distortion mechanism in a cross-coupled capacitor neutralization technique and suggest the termination scheme to enhance both linearity and efficiency of PAs. This scheme can suppress the second harmonic feedback components generated in the middle of the cascode cell and extend the range of linear output power (P_{out}). Our two-stage PA shows power gain of >24.5dB, saturation output power (P_{sat}) of >16.2dBm and power-added-efficiency (PAE) of >32% over the full 39GHz band from 37 to 40GHz. Under the 5G new radio modulation, the PA achieves linear P_{out} of 8.2dBm/10.3dBm and PAEs of 11.9%/16.5% at EVMs of -30.0dB/-25.3dB, respectively.**

Keywords— **Power amplifier, CMOS, cascode, mm-wave, 5G, 39GHz, 28nm, second harmonic termination, linearity, feedback.**

I. INTRODUCTION

Both 28 and 39GHz bands are allocated for the millimeter-wave (mm-wave) fifth generation (5G) 3GPP new radio (NR) standard [1]. A considerable path loss and poor transistor performance in mm-wave frequency make RF front-end circuit designs challenging, in particular in a CMOS power amplifier (PA) design. Although a phased array for a beam-forming can boost an equivalent isotropic radiated power and reduce a required output power (P_{out}) per transmitter element, the PA design is still difficult with the 5G NR OFDM signals which have a high peak-to-average power ratio (PAPR) of ~10dB. Both linear P_{out} and efficiency of PAs can be significantly degraded under the large back-off conditions to meet the stringent error-vector-magnitude (EVM) requirement, e.g., EVM <8% for the 64-QAM OFDM [1]-[5].

Due to limited usages of digital pre-distortion and load-modulation schemes in the mm-wave 5G system, the linear P_{out} and efficiency need to be maximized by reducing the required back-off from a saturation power (P_{sat}) [2],[5]. Recently several state-of-the-art CMOS PAs have been reported for the mm-wave 5G applications with the improved AM-AM and AM-PM (or inter-modulation) distortions, e.g., inductive source degeneration [2], AM-PM compensated Class-AB [3], continuous-mode hybrid Class-F and Class-F[-1] [4], and second harmonic termination at the output matching network [5]. However, given the high PAPRs of 7-10dB, these PAs are still required to operate under the large power back-off conditions (close to their PAPRs) from P_{sat} to achieve an EVM of better than -25dB (5-6%) [1]-[3],[5].

In this work, we present a PA design approach to lessen the required power back-off level. We propose a two-step (*L-C* and *C-L*) second harmonic termination scheme for CMOS cascode PAs with the cross-coupled capacitor neutralization

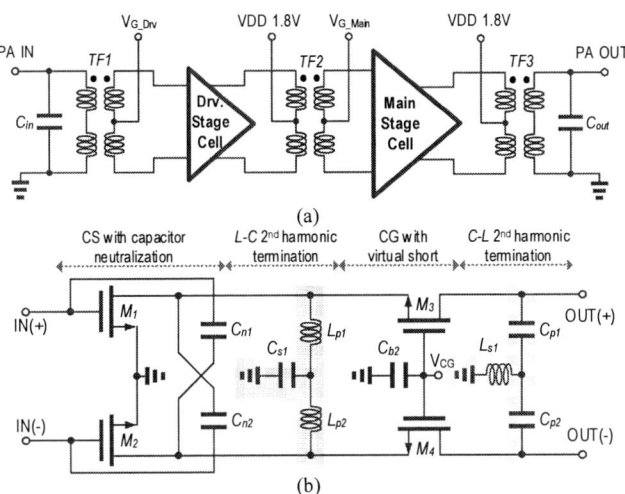

(a)

(b)

Fig. 1. Schematic of the two-stage cascode PA. (a) Two-stage PA including input/inter-stage/output transformers, and (b) the proposed differential cascode cell with the two-step (*L-C* and *C-L*) second harmonic termination.

which has been popularly utilized in the mm-wave frequency band PAs to improve both gain and stability [2]-[5]. We have analysed why the neutralization capacitors affect the third-order intermodulation distortion (IMD3) performance and how to improve it. The measured results from our 39GHz PA show only 6.5dB power back-off requirement to meet EVM of -25dB for the signal of the fully occupied 1.4GHz modulation bandwidth and 64-QAM OFDM with 10dB PAPR.

II. PA DESIGN AND ANALYSIS

A. PA Configuration and Proposed Cascode Cell

Fig. 1(a) shows a schematic of the two-stage differential cascode PA, which consists of three transformer-based matching networks at input (TF_1), inter-stage (TF_2), and output (TF_3). In Fig. 1(b) each differential cascode cell includes differential common source (CS) stage with the capacitor neutralization, differential common gate (CG) stage with a virtual short at the center node (V_{CG}), and the proposed two-step (*L-C* and *C-L*) second harmonic termination. The *L-C* second harmonic termination circuit (*orange box*) is added at the middle of CS and CG with shunt inductors ($L_{p1,2}$=180pH in this design) and a capacitor (C_{s1}=25.5fF in this design) and can suppress the second harmonic components at that node. Additionally, for the fundamental components, the shunt inductors ($L_{p1,2}$) with a virtual ground at the center node can prevent parasitic leakage currents to the substrate and can improve both gain and efficiency of the PA [6]. The *C-L*

978-1-7281-1702-7/19 $31.00 © 2019 IEEE

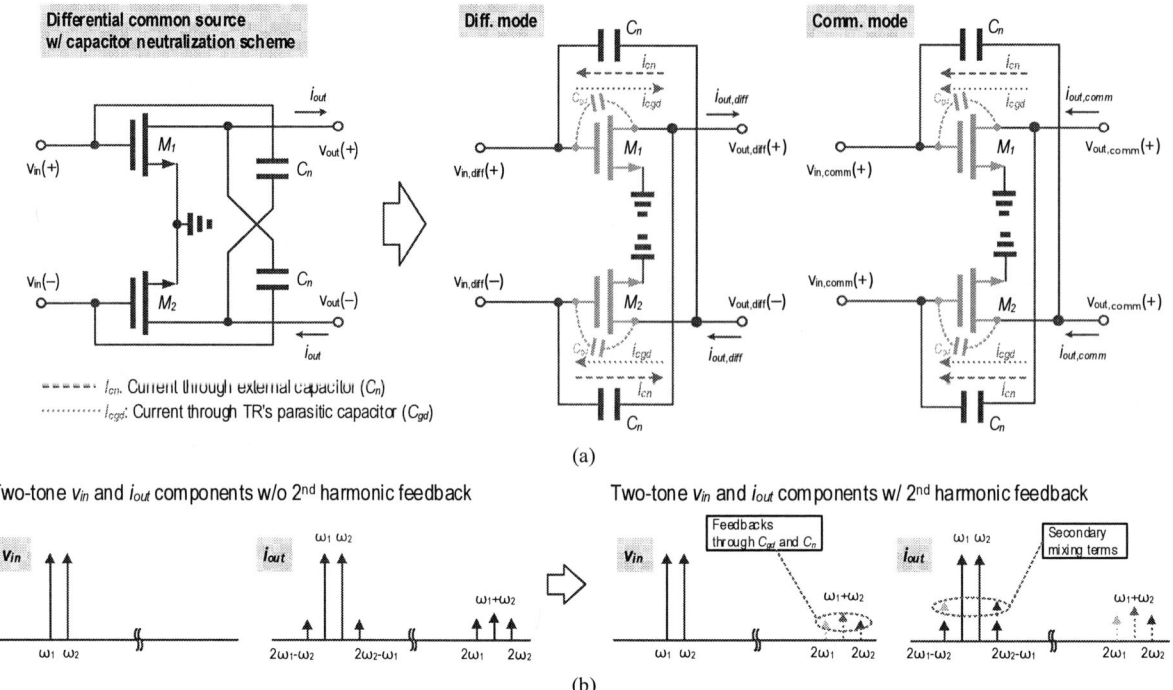

Fig. 2. Analysis of additional IMD3 mechanism in the differential CS with the cross-coupled capacitor neutralization. (a) Differential CS and two separate circuits for differential/common-modes, and (b) additional IMD3 mechanism with the second harmonic feedbacks through the neutralization capacitors.

second harmonic termination (*yellow box*) is added at the output of CG with shunt capacitors ($C_{p1,2}$) and an inductor (L_{s1}). Although the shunt capacitors reduce the bandwidth of the PA, both linearity and efficiency of the PA can be enhanced by the output second harmonic termination circuit [5],[7]. Bypass capacitors (>1pF) are added at the center taps of the transformers to minimize signal imbalances.

B. Additional IMD3 by Neutralization Capacitors

Fig. 2 shows an analysis of additional IMD3 mechanism in the differential CS with the cross-coupled capacitor neutralization. This differential CS cell can be analysed with two separate circuits of differential-mode (e.g., fundamental frequency) and common-mode (e.g., two-times fundamental frequency). Based on the two circuit modes in Fig. 2(a), additional IMD3 mechanism is analysed in Fig. 2(b) with the third-order Taylor series (memoryless) especially for the second harmonic common-mode feedback components through the neutralization capacitors (C_n). This capacitor neutralization technique works only in the differential-mode with an appropriate capacitance (C_n=47.5fF in this design). In the differential-mode, two fundamental current components of i_{cgd} and i_{cn} become the almost same at the input (and output) nodes, $v_{in,diff}$ (and $v_{out,diff}$), and then, the gate-drain capacitances (C_{gd}) in transistors ($M_{1,2}$) are properly cancelled out with the neutralization capacitors (C_n). On the other hand, in the common-mode, the neutralization capacitors (C_n) with the gate-drain capacitances (C_{gd}) make the second harmonic feedback currents from output nodes to input nodes almost doubled. As results, two-tone v_{in} at the input nodes becomes $v_{in}+\alpha \cdot g_{m2} \cdot v_{in}^2 \cdot Z_{out,2nd}$, where α (~0.5) is a voltage dividing ratio from output drain to input gate through both C_n and C_{gd}, g_{m2}

($0.77mA/V^2$) is the second-order trans-conductance of the transistors ($M_{1,2}$) and $Z_{out,2nd}$ (~4Ohm) is the output impedance of the second harmonic frequency at the CS drain nodes. The additional $2\omega_1$ and $2\omega_2$ frequency components at the input nodes are re-mixed with ω_1 and ω_2 frequency components through the second-order non-linearity ($g_{m2} \cdot v_{in}^2$) of the transistors, which generate secondary IMD3 components at the CS outputs as shown in Fig. 2(b). Hence, we introduce the second harmonic termination circuit at the middle of CS and CG to minimize the second harmonic feedback components by reducing the $Z_{out,2nd}$ (<1.5Ohm) at that node. In our analysis and design, the total transistor width of 384μm (differential) with an initial bias condition of 36.5μA/μm has been used in the main stage PA.

III. SIMULATION AND MEASUREMENT

A. PA Simulation and Comparison

To verify our approach and analysis, three different cascode PA configurations (main stage only) are compared in Fig. 3. Without any second harmonic terminations (removing both C_{s1} and L_{s1} in Fig. 1(b)), degraded AM-AM of 0.7dB and AM-PM of 4° were observed (*orange-lines* in Fig. 3(a) and (b)), which results in IMD3 of >-30dBc (*orange-lines* in Fig. 3(c)). In this condition, lower bias conditions can improve AM-AM, but cause higher AM-PM distortions. With the second harmonic termination only at the output nodes of the CG stage (removing C_{s1} only), improved AM-AM of 0.3dB and AM-PM of 1.8° were obtained by terminating the output second harmonic distortion components generated by the entire cascode cell (*blue-lines* in Fig. 3(a) and (b)), which results in IMD3 of -36dBc (*blue-lines* in Fig. 3(c)). Using the

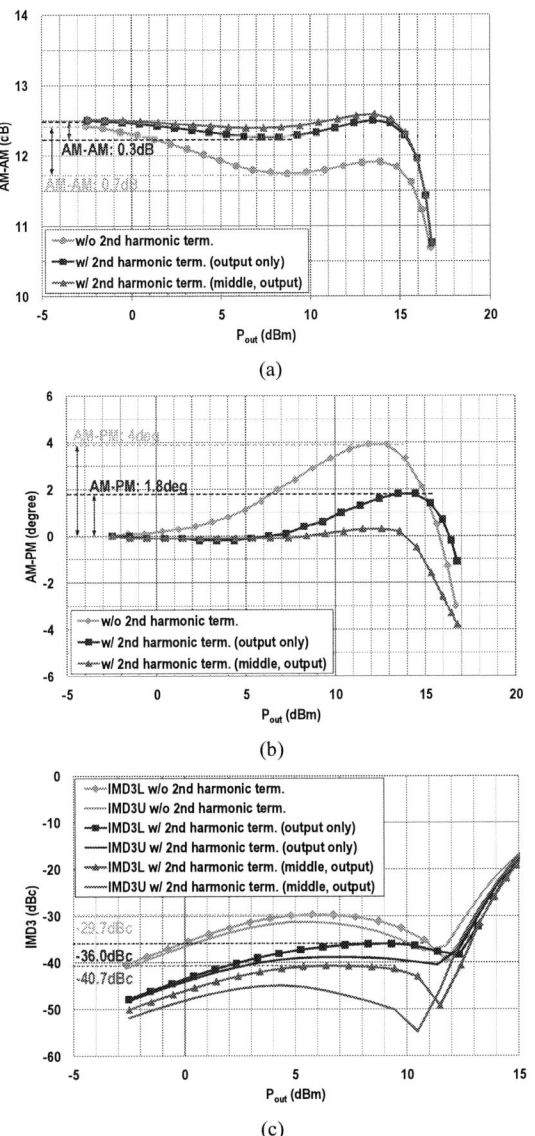

Fig. 3. Simulation comparisons for the three different cascode PA configurations with (and without) second harmonic terminations at the middle and output nodes. (a) AM-AM, (b) AM-PM, and (c) IMD3 vs. P_{out}.

proposed two-step second harmonic termination, further enhanced AM-AM of 0.15dB and AM-PM of 0.3° were achieved (*red-lines* in Fig. 3(a) and (b)), which results in IMD3 of <-40dBc (*red-lines* in Fig. 3(c)). The improved linearity leads to superior EVM performance.

B. Measurement Results

Fig. 4 shows the measured PA performance with single-tone and two-tone signals with a 700MHz tone-spacing. All measurements have been done using on-wafer tests. In Fig. 4(a), one-tone large signal test results show power gain of >24.5dB, P_{sat} of >16.2dBm, 1-dB power compression point (P_{1dB}) of >14.9dBm, and peak power-added-efficiency (PAE) of >32% over the full 39GHz band from 37 to 40GHz. In Fig. 4(b), two-tone large signal test results show a very good IMD3 performance of better than -36dBc up to P_{out} of >9dBm for the

target frequency ranges. In particular, the PA achieves a superior IMD3 performance of better than -45dBc up to P_{out} of >9.0dBm owing to the proposed two-step (*L-C* and *C-L*) second harmonic termination scheme.

Fig. 5 shows the measured PA band performance with the 5G NR modulation signal, which has a modulation bandwidth of 1.4GHz (14-CC×100MHz) with sub-carrier spacing of 60kHz and PAPR of 10dB. At the center frequency of 38.5GHz, the PA exhibits P_{out}/PAE of 8.2dBm/11.9% and 10.3dBm/16.5% at EVMs of -30dB and -25dB, respectively. At the P_{out} of 10.3dBm, the PA shows ACLRs of -32.9/-30.0dBc and DC power consumption of 63.1mW. It is verified that our proposed PA satisfies EVM requirement of -25dB with the power back-off level of as small as 6.5dB from the P_{sat} of 16.8dBm. As results, our PA can have significantly improved linear P_{out} and PAE at the mm-wave bands.

For comparisons, the performance of this work and the recently reported 5G CMOS PAs at the mm-wave bands were summarized in Table 1. Our PA has been characterized and verified with the fully occupied 1.4GHz modulation bandwidth at the 39GHz bands for the first time. Although the PA in this work shows a slightly lower peak PAE of 32.9%, it performs the highest PAE of 16.9% at the EVM of -25.3dB and P_{out} of 10.3dBm for the 10dB PAPR, 64-QAM OFDM signal. In addition, the two-stage PA shows the highest power gain of 25.8dB and FOM of 89.5 among the reported PAs. The PA IC was implemented using a 28 nm bulk CMOS process and Fig. 6 shows a die micrograph. The IC has a size of 670×470µm² including RF/DC pads and has a very compact size of 450×140µm² for an active area.

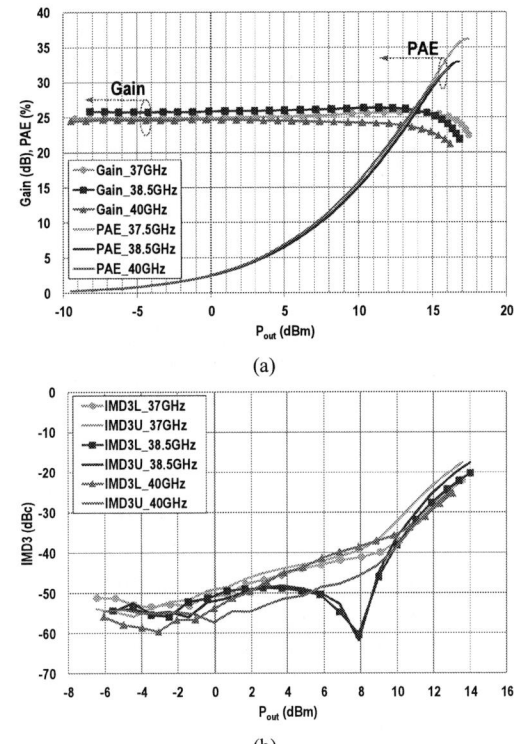

Fig. 4. Large-signal PA measurement results. (a) One-tone gain and PAE results, and (b) two-tone IMD3 results with a 700MHz tone spacing vs. P_{out}.

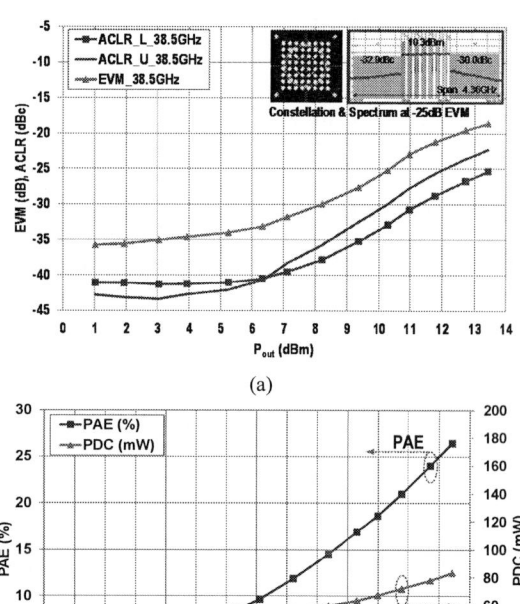

(a)

(b)

Fig. 5. Measured band performance under the 5G NR modulation. (a) EVM and ACLR including the 64QAM OFDM constellation and spectrum at -25dB EVM, and (b) PAE and P_{DC} vs. P_{out}.

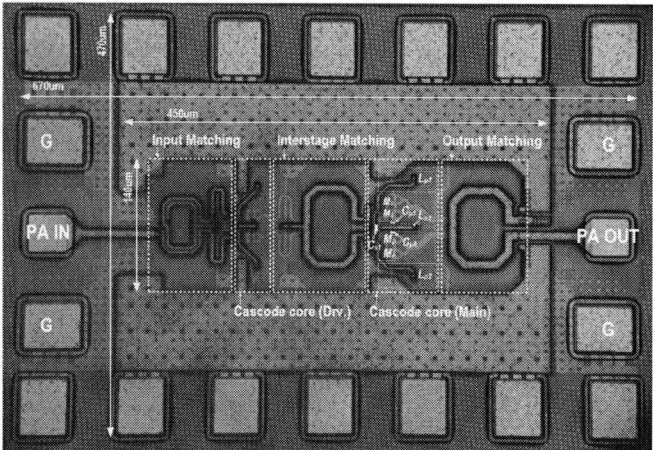

Fig. 6. Die photo of the two-stage cascode PA in 28nm bulk CMOS.

IV. CONCLUSION

We have reported a 39GHz band CMOS cascode PA IC for the mm-wave 5G applications. Using a newly proposed two-step (*L-C* and *C-L*) second harmonic termination circuit, the two-stage PA IC achieved P_{sat} of >16.2dBm with peak PAE of >32% (one-tone CW), IMD3 of better than -45dBc up to P_{out} of 9dBm (two-tone with a 700MHz tone-spacing), and linear P_{out} of 10.3dBm with PAE of 16.5% at EVM of -25.3dB (the 5G NR modulation).

Table 1. The comparison table of the recent state-of-the-art mm-wave CMOS PAs for the 5G applications.

	This Work	[2]	[3]	[4]	[5]
Technology	28nm CMOS	28nm CMOS	28nm CMOS	45nm SOI	28nm CMOS
Frequency (GHz)	38.5	30.0	34.0	39.0	28.0
PA Design Topology	Diff. cascode	Diff. CS	Diff. CS	Diff. cascode	Diff. stacked
Supply (V)	1.8	1.15	0.9	2.0	2.2
# of Stages	2	2	2	1	1
Gain (dB)	25.8	16.3	20.8	10.5	13.6
P_{sat} (dBm)	16.8	15.3	16.6	18.5	19.8
P_{1dB} (dBm)	14.9	14.3	13.4	16.3	18.6
PAE (%) at P_{sat}	32.9	36.6	24.2	41.2	43.3
PAE (%) at P_{1dB}	28.8	10.9	12.6	35.0*	26.0
PAPR (dB), Mod	10.0, 64QAM OFDM	9.6, 64QAM OFDM	8.3, 64QAM	-	7.5, 64QAM LTE
Mod. BW (GHz)	1.40	0.25	0.337, 1.35	-	0.02
EVM^{+} (dB)	-25.3	-25.0	-25.0	-	-25.0*
P_{out} (dBm) at EVM^{+}	10.3	5.3	10.1, 5.9	-	15.0*
PAE (%) at EVM^{+}	16.9	9.6	5.8, 2.3	-	27.5*
ACLR (dBc) at EVM^{+}	-32.9/ -30.0	-28.3 / -26.4	-32.1, -36.9	-	-30.0 / -30.0*
FOM**	89.5	76.8	81.9	82.5	78.7
Active Area (mm²)	0.07	0.16	0.16	0.14	0.07*

*Graphically estimated from the reported figures.
**FOM=P_{sat}[dBm]+Gain[dB]+20log(f_c[GHz])+10log(PAE_{max}[%]).

ACKNOWLEDGMENT

The authors would like to thank Y. Yoon, D. Kang, J. Lee and Prof. Y. Yang for their reviews and suggestions. The authors also thank D. H. Shin for EMX supports and GUC/TSMC for 28nm CMOS process.

REFERENCES

[1] 3GPP 5G NR, "Base Station (BS) radio transmission and reception," *3GPP TS 38.104 version 15.2.0 Release 15*, July 2018.

[2] S. Shakib et al., "A 28GHz Efficient Linear Power Amplifier for 5G Phased Arrays in 28nm Bulk CMOS," *ISSCC*, pp. 352-353, Feb. 2016.

[3] M. Vigilante et al., "A 29-to-57GHz AM-PM Compensated Class-AB Power Amplifier for 5G Phased Arrays in 0.9V 28nm Bulk CMOS," *IEEE RFIC*, pp. 116-119, June. 2017.

[4] T. Li et al., "A Continuous-Mode 23.5-41GHz Hybrid Class-F/F-1 Power Amplifier with 46% Peak PAE for 5G Massive MIMO Applications," *IEEE RFIC*, pp. 220-223, June. 2018.

[5] B. Park et al., "Highly Linear mm-wave CMOS Power Amplifier," *IEEE TMTT*, vol. 64, no. 12, pp. 4535-4544, Dec. 2016.

[6] H. Dabag et al. "Analysis and Design of Stacked-FET Millimeter-Wave Power Amplifiers," *IEEE TMTT*, vol. 64, No. 4, pp. 1543-1556, Apr. 2013.

[7] J. Ko et al., "A High-Efficiency Multiband Class-F Power Amplifier in 0.153µm Bulk CMOS for WCDMA/LTE Applications," *IEEE ISSCC*, pp. 40-41, Feb. 2017.

A Compact E-Band PA with 22.37% PAE 14.29 dBm Output Power and 26 dB Power Gain with Efficiency Enhancement at Power Back-off

Liang Chen, Lei Zhang, Li Zhang, and Yan Wang

Institute of Microelectronics, Tsinghua University, China

zhang.lei@tsinghua.edu.cn

Abstract — This paper presents a compact power amplifier (PA) with efficiency enhancement technique at power back-off region in 65-nm CMOS technology for E-band applications. Cross-coupled transistor pair, generating negative impedance is introduced to enhance power back-off efficiency and small signal power gain. Neutralization and transformer-based matching networks are employed to improve gain and stability. The measured P_{sat}, OP_{1dB} and peak PAE are 14.29 dBm, 12.03 dBm and 22.37%, respectively. A 26 dB power gain with 3 dB bandwidth of 5 GHz is achieved. The measured peak gain varies from 26 dB to 31 dB and the P_{1dB} power added efficiency increases from 6% to 13.3% by adjusting the control voltages of cross-coupled transistor pair. Compact layout of the PA yields a core area of 0.025 mm^2, and the total DC power consumption is 120 mW, enabling compact and efficient integration into phased array transceivers.

Keywords — E-band, CMOS, power amplifier (PA), efficiency enhancement, power back-off, variable gain, compact layout, phased array transceivers.

I. INTRODUCTION

Thanks to wider bandwidth and less atmospheric absorption, E-band (71-76 GHz, 81-86 GHz) systems have great potential to improve capacity and have been allocated for point to point communication by Federal Communications Commission (FCC), offering the opportunity for high speed long range wireless communications combined with phased-array systems [2]. One of the essential building blocks to develop efficient phased-array transceivers is the power amplifier, used as either a transmitter block before the antenna or a driving circuit for frequency multiplier chains. With the trend of increasing data throughput, spectrum-efficient modulation schemes with a high peak-to-average power ratio (PARR) are typically employed by modern wireless communication systems. Consequently, power amplifier needs to operate in the power back-off (PBO) region most of the time in such systems [4]. However, PA efficiency degrades dramatically as the output power scales back since the output matching network cannot provide a proper match across a wide power range.

Designing efficient mm-Wave power amplifier in advanced CMOS process poses many challenges such as high passive losses, low supply voltage and large layout parasitics. To alleviate those issues, several state-of-the-art PA topologies such as Doherty, outphasing and digital PA are proposed in [4]-[7]. In [4], an E-band transformer-based Doherty power amplifier for high PBO efficiency is proposed, achieving a peak and P_{1dB} efficiency of 12% and 11.1%, respectively. The first implementation of outphasing PA at 60 GHz is presented in [5]. Fabricated in 40-nm bulk CMOS, the outphasing TX

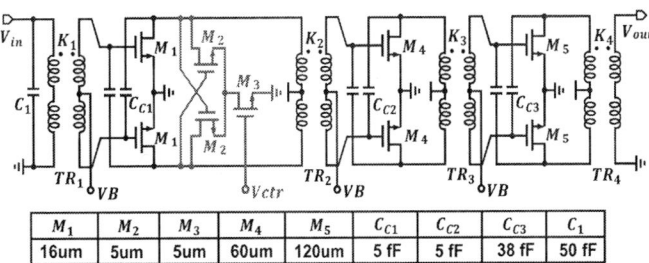

Fig. 1. Schematic of the proposed PA.

M_1	M_2	M_3	M_4	M_5	C_{C1}	C_{C2}	C_{C3}	C_1
16um	5um	5um	60um	120um	5 fF	5 fF	38 fF	50 fF

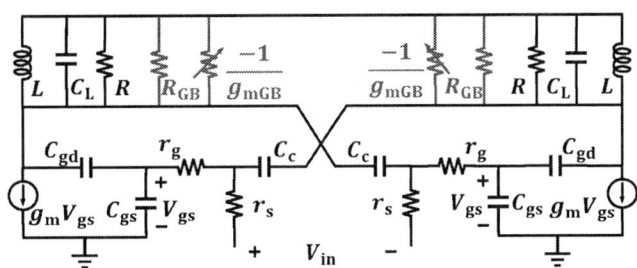

Fig. 2. Simplified small-signal model of the gain-boosting unit.

delivers 12.5 dBm output power and 15% average efficiency. A 60 GHz Cartesian architecture digital PA is proposed in [7] and an average efficiency of 16.5% is achieved when the I/Q PAs transmits 6 Gb/s 16 QAM signals at 6 dB power back-off.

However, complex input preconditioning and extra sophisticated power combiner at the PA output are required in those techniques and too much chip area is occupied, which is not beneficial to efficient integration into phased array transceivers. In this paper, efficiency enhancement technique at PBO region based on cross-coupled transistor pair (CCTP) is proposed. Without complex preconditioning and power combiner, a compact high efficiency power amplifier is achieved.

II. CIRCUIT ANALYSIS AND DESCRIPTIONS

Fig. 1 shows the proposed power amplifier schematic. Three stage common source pseudo-differential NMOS pairs with neutralization capacitors are adopted to achieve better common-mode rejection ratio (CMRR) compared to single-end structure and improve stability issues caused by transistor parasitic capacitance C_{gd}. Cross-coupled transistor pair M_2, M_3, generating negative impedance is employed in the first stage of the proposed PA to boost effective quality factor of the load impedance tank. Variable efficiency and

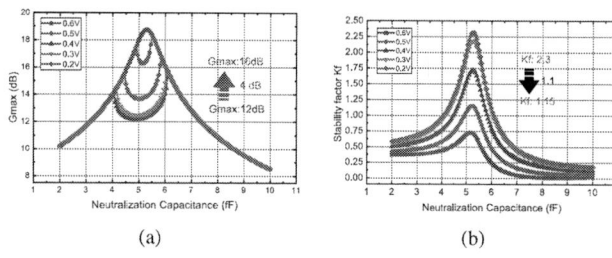

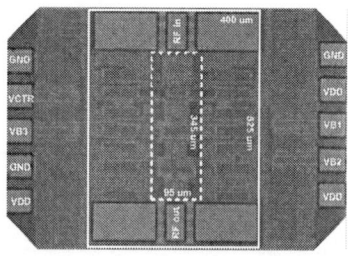

Fig. 3. Simulated (a) $G_{\max}$ and (b) stability factor K_f under different control voltage V_{ctr}.

Fig. 5. Die photograph of the proposed power amplifier. The core area is 345 µm × 95 µm.

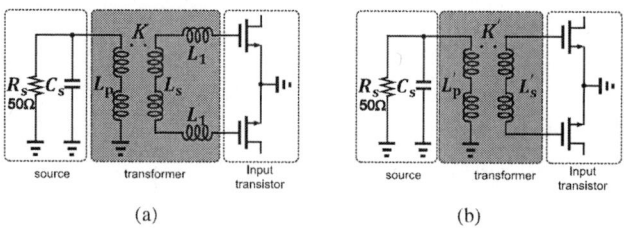

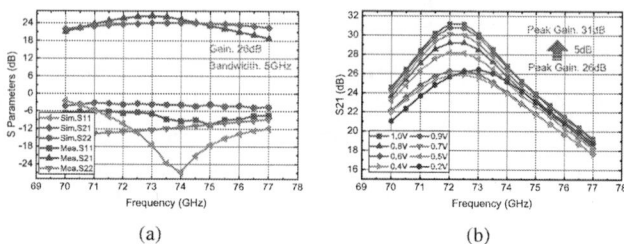

Fig. 4. (a) Simplified input matching network and (b) its equivalent schematic.

Fig. 6. (a) Measured and simulated S-parameters (b) Measured gain enhancement under different control voltage V_{ctr}.

gain enhancement can be achieved by adjusting the gate control voltage V_{ctr} of transistor M_3. Transistor width of the output stage M_5 is determined by the desired output power and the input and driver stages are sized to improve efficiency while not affecting linearity. Custom made parallel plate MOM capacitors are utilized for better accuracy and higher quality factor. All matching networks are realized by stack transformers with additional parallel capacitors.

A. Efficiency Enhancement Technique

Low transistor power gain and large insertion loss of passive devices such as on chip inductors pose great challenges on the design of E-band power amplifiers, resulting in low efficiency especially at power back-off region, where efficiency is mainly determined by the small signal power gain. Consequently, cross-coupled transistor pair is employed as the gain-boosting unit to offer variable efficiency and gain enhancement. Small signal power gain and output power at PBO region is boosted by turning on the gain-boosting unit, hence the efficiency is enhanced. The equivalent small signal model of the PA first stage with the cross-coupled transistor pair is shown in Fig. 2, where $-1/g_{\mathrm{mGB}}$ and R_{GB} are the trans-conductance and output resistance of the CCTP, respectively. By turning on the gain-boosting unit and assuming that its output capacitance has little effect on the total output capacitance C_L of the driver stage due to its small transistor size, the load impedance Z_L and quality factor Q_L of the parallel RLC tank can be expressed as

$$Z_L = \frac{RR_{\mathrm{GB}}}{R + R_{\mathrm{GB}} - g_{\mathrm{mGB}}RR_{\mathrm{GB}}} \quad (1)$$

$$Q_L = \frac{R_{\mathrm{GB}}}{R + R_{\mathrm{GB}} - g_{\mathrm{mGB}}RR_{\mathrm{GB}}} Q \quad (2)$$

where Q represents the quality factor of the parallel RLC tank when the gain-boosting unit is turned off and equals to $R/\omega_0 L$. From (1) and (2), the load impedance Z_L and

quality factor Q_L of the parallel RLC tank is enhanced by the gain-boosting unit. In order to analysis the impact of gain-boosting unit on the small signal power gain and power back-off efficiency, neutralization capacitor C_C, corresponding to the over and under neutralization is calculated.

$$C_C = \begin{cases} C_{\mathrm{gd}} + r_s \left(C_{\mathrm{gs}} + 2C_{\mathrm{gd}} \right)/Z_L & C_C > C_{\mathrm{gd}} \\ C_{\mathrm{gd}} - r_s \left(C_{\mathrm{gs}} + 2C_{\mathrm{gd}} \right)/Z_L & C_C < C_{\mathrm{gd}} \end{cases} \quad (3)$$

when the parasistics capacitance C_{gd} is exactly neutralized, $C_C = C_{\mathrm{gd}}$. The simulated G_{MAX} of the driver stage varies with C_C under different gain-boosting control voltage V_{ctr} is shown in Fig. 3 (a) and a 4 dB gain enhancement is achieved when the V_{ctr} voltage changes from 0.2 V to 0.5 V. Based on (1) and (3), the neutralization capacitance C_C gets close to each other with Z_L becoming large and equals to C_{gd} when Z_L is large enough, which agrees well with the simulation results in Fig. 3 (a). Based on Y- parameters of the small-signal equivalent circuit model shown in Fig. 2, stability factor K_f of the driver stage can be determined as

$$K_f = \frac{2 + \omega^2 \left(C_{\mathrm{gd}} - C_C \right)^2 r_g Z_L}{\omega |C_{\mathrm{gd}} - C_C| r_g Z_L \sqrt{\omega^2 \left(C_{\mathrm{gd}} - C_C \right)^2 + g_m^2}}$$

$$= \frac{2/Z_L + \omega^2 \left(C_{\mathrm{gd}} - C_C \right)^2 r_g}{\omega |C_{\mathrm{gd}} - C_C| r_g \sqrt{\omega^2 \left(C_{\mathrm{gd}} - C_C \right)^2 + g_m^2}} \quad (4)$$

Based on (4), stability factor K_f is inversely proportional to the load impedance Z_L and K_f decreases when the gain-boosting unit turns on. Fig. 3 (b) shows the simulated stability factor K_f varies with neutralization capacitance C_C under different gain-boosting control voltage V_{ctr} and K_f decreases from 2.3 to 1.15 when the V_{ctr} voltage changes from 0.2 V to 0.5 V. In order to ensure the unconditional stable

978-1-7281-1702-7/19 $31.00 © 2019 IEEE

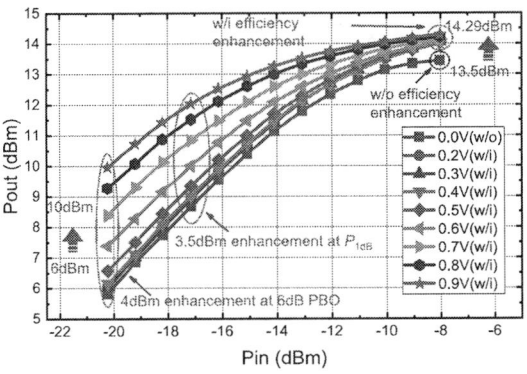

Fig. 7. Measured PA output power under different control voltages of gain-boosting unit.

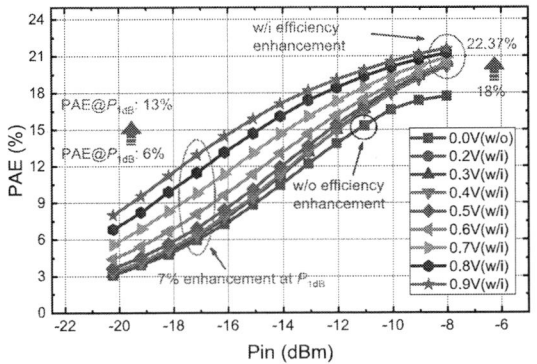

Fig. 8. Measured PAE under different control voltages of gain-boosting unit.

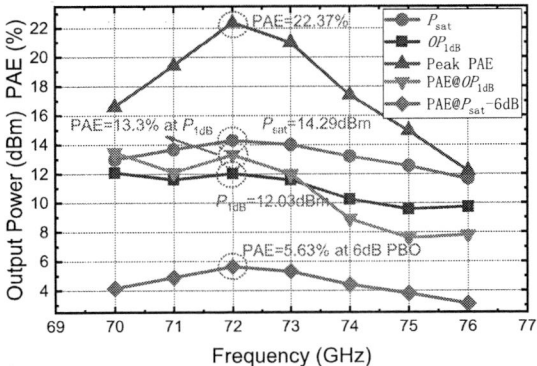

Fig. 9. Measured PA large signal performance over frequency.

state of the driver stage, transistor dimensions and control voltage V_{ctr} of the gain-boosting unit must be determined. From (4), when the stability factor K_f is larger than one, the load impedance Z_L of the driver stage can be expressed

$$Z_L < \frac{2}{r_g \omega C_{net} \left[\sqrt{\omega^2 C_{net}^2 + g_m^2} - \omega C_{net} \right]} \quad (5)$$

where $C_{net} = |C_{gd} - C_{gs}|$, representing the net capacitance between transistor drain and gate terminal after neutralization. Combining (1) (4) and neglecting R_{GB} for simplicity, the trans-conductance of the gain-boosting unit can be calculated as

$$g_{mGB} < \frac{2 - R_L r_g \omega C_{net} \left[\sqrt{\omega^2 C_{net}^2 + g_m^2} - \omega C_{net} \right]}{2 R_L} \quad (6)$$

Based on the trans-conductance, transistor dimensions and control voltage of the gain-boosting unit can be determined under all process corners to ensure the driver stage unconditional stable.

B. Matching Networks

Transformer based matching networks are adopted in this paper because they can provide a high-order LC network with almost the same chip area as a single inductor and efficient single-ended to differential conversion. The output matching network is designed to present an optimal load impedance for maximizing output power and efficiency of the

output stage. Load-pull simulation reveals that the optimal load impedance is $10+j20$ for the output stage at 73 GHz and a stacked transformer with a diameter of 16 μm is designed based on the design procedure proposed in [10]. For the input and inter-stage matching networks, conjugate matching is employed to maximize power gain and improve efficiency. However, due to the large quality factor (Q_{load}) of the input impedance, caused by the small device size of input transistor, the series parasitic resistor R_s of the transformer cannot be neglected for the reason that the equivalent parallel resistor of R_s is comparable with that of the input impedance. For transformer based matching networks, the effective trans-resistance can be expressed as

$$Z_{21} = \frac{1}{2} \frac{R_L L (1 \pm k_m)}{R_L R_s C + L (1 \pm k_m)} \quad (7)$$

where R_L is the equivalent parallel resistor of the input impedance. In order to reduce the effect of parasitic resistor R_s on the input matching network, a series inductor L_1 is inserted at the gate terminal of the input transistor to lower the input impedance quality factor Q_{load}. Fig. 4 presents the simplified input matching network and its equivalent schematic. Based on [10], L_p', L_s' and k' can be expressed as

$$L_p' = L_p, L_s' = L_s + L_1, k' = k \sqrt{\frac{L_s}{L_s + L_1}} \quad (8)$$

III. MEASUREMENT RESULTS

The proposed amplifier is fabricated in 65-nm CMOS process and the die photo is shown in Fig. 5. Dimensions of the whole chip excluding DC supply pad are 0.4 mm × 0.525 mm, which is only required to cover GSG pad-to-pad distance and the core integration area of the proposed PA is 0.032 mm². The compact chip area allows efficient integration of this PA into low-cost phased-array transceivers. Fig. 6 (a) shows the measured and simulated S-parameters. The PA achieves a peak gain of 26 dB with a 3 dB bandwidth of 5 GHz. S_{11} <-10 dB and S_{22} <-4 dB in the frequency band of interest. By adjusting the control voltage V_{ctr} of gain-boosting unit, the measured power gain S_{21} increases from 26 dB to 31 dB, which is shown in Fig. 6 (b).

Due to the limitation of testing equipment, the maximum input power is -8 dBm, and the measured output power under

Table 1. PA performance summary and comparison with the state-of-the-art works.

	[1]	[2]	[3]	[4]	[6]	[8]	[9]	**This Work**
Technology	28nm CMOS	14nm CMOS	65nm CMOS	40nm CMOS	45nm CMOS	65nm CMOS	45nm CMOS	**65nm CMOS**
VDD [V]	0.9	1.0	1.0	0.9	1.1	2	1.1	**1.2**
Frequency [GHz]	79	73	60	77	60	77	60	**73**
BW [GHz]	21.5	7.4	8.5	-	17	-	17	**5**
Gain [dB]	17	16.7	16	9.5	6	20.9	6	**26$\sim$31**
P_{1dB} [dBm]	8.25	2	5	15.2	11	13	11	**12.03**
P_{sat} [dBm]	12.3	7.4	11.5	16.2	13.8	15.8	13.8	**14.29**
PAE_{MAX} [%]	13.8	8.9	12	15.2	7	15.2	7	**22.37**
PAE_{1dB} [%]	6.5*	-	11.1	6	3.5	6	3.5	**13.3**
PAE @ P_{sat}-6dB [%]	3.5*	4.5	4	5.7	2.3*	4.8	2.3*	**5.63**
Paths Combined	1	1	1	**2**	1	**2**	1	**1**
DC Power [mW]	-	-	50	-	-	246	-	**120**
Core Area [mm^2]	0.13**	0.1	0.0533	0.1	0.0559	0.21	0.0559	**0.025**
FOM***	78.65	70.86	74.87	74.19	63.8	76.25	63.8	**91.05**

* Graphically estimated from reported figures
** Core area including pads
*** FOM=P_{sat}[dBm]+Gain[dB]+20log(freq[GHz])+10log(PAE_{MAX}[%])

different V_{ctr} voltages is shown in Fig. 7. At 72 GHz, the measured P_{sat} and OP_{1dB} are 14.29 dBm and 12.03 dBm, respectively. Fig. 8 presents the measured PAE under different V_{ctr}. The peak and OP_{1dB} power added efficiency are 22.37% and 13.3%, respectively. By turning on the gain-boosting unit and V_{ctr} changes from 0.2 V to 1.0 V, the measured output power and PAE enhancement at OP_{1dB} are 3.5 dBm and 7%, respectively. A 4 dBm enhancement of the output power is measured at 6 dB power back-off region, which is shown in Fig. 7. The measured PAE is still as high as 5.63% at 6 dB power back-off region.

The large signal frequency response of the proposed PA is presented in Fig. 9. The PA maintains a decent performance over the 70-76 GHz bandwidth, where P_{sat} and OP_{1dB} are above 12 dBm and 10 dBm, respectively. The overall performance of the designed compact PA is summarized and compared with other state-of-the-art PAs in Table 1. Thanks to the proposed efficiency enhancement technique and careful design of the transformer based matching networks, the measured output power, PAE and power gain are maintained high, resulting in highest FOM compared to previously published CMOS PAs.

IV. CONCLUSION

In this paper, a compact E-band power amplifier with efficiency enhancement technique at power back-off region in 65-nm CMOS is presented. In order to enable efficient integration into phased-array transceivers, cross-coupled transistor pair based efficiency enhancement technique without complex preconditioning and power combiner is introduced and a measured 7% PAE enhancement and 5 dB gain improvement at PBO region are achieved. The proposed E-band PA achieves a measured P_{sat}, OP_{1dB} and peak PAE of 14.29 dBm, 12.03 dBm and 22.37%, respectively.

ACKNOWLEDGMENT

This work is supported by the National High Technology Research and Development Program of China 2018YFB0105002, 2017YFB0102601, 2016YFB0101001, and Tsinghua University Initiative Scientific Research Program.

REFERENCES

[1] A. Medra, V. Giannini, D. Guermandi and P. Wambacq, "A 79 GHz variable gain low-noise amplifier and power amplifier in 28-nm CMOS operating up to 125°C," in *Proc. IEEE ESSCIRC*, 2014, pp. 183-186.

[2] S. Callender, et al., "A 73 GHz PA for 5 G phased arrays in 14-nm FinFET CMOS," in *Proc. IEEE RFIC*, 2017, pp. 402-405.

[3] W. L. Chan and J. R. Long, "A 58-65 GHz Neutralized CMOS Power Amplifier With PAE Above 10% at 1-V Supply," *IEEE J. Solid-State Circuits*, vol. 45, no. 3, pp. 554-564, Mar. 2010.

[4] E. Kaymaksut, D. Zhao and P. Reynaert, "E-band transformer-based Doherty power amplifier in 40-nm CMOS," in *Proc. IEEE RFIC*, 2014, pp. 167-170.

[5] D. Zhao, S. Kulkarni and P. Reynaert, "A 60 GHz Outphasing Transmitter in 40-nm CMOS," *IEEE J. Solid-State Circuits*, vol. 47, no. 12, pp. 3145-3183, Dec. 2012.

[6] K. Khalaf, V. Vidojkovic, J. R. Long and P. Wambacq, "A 6x-oversampling 10 GS/s 60 GHz polar transmitter with 15.3% average PA efficiency in 40-nm CMOS," in *Proc. IEEE ESSCIRC*, 2015, pp. 348-351.

[7] J. Chen et al., "A Digitally Modulated mm-Wave Cartesian Beamforming Transmitter with Quadrature Spatial Combining," in *ISSCC Dig. Tech. Papers*, 2013, pp. 232-233.

[8] J. Oh, B. Ku and S. Hong, "A 77 GHz CMOS Power Amplifier With a Parallel Power Combiner Based on Transmission-Line Transformer," *IEEE Trans. Microw. Theory Techn.*, vol. 61, no. 7, pp. 2662-2669, Jul. 2013.

[9] K. Raczkowski et al., "50-to-67 GHz ESD-Protected Power Amplifiers in Digital 45-nm LP CMOS," in *ISSCC Dig. Tech. Papers*, 2009, pp. 382-383.

[10] H. Jia, C. C. Prawoto, B. Chi, Z. Wang and C. P. Yue, "A 32.9% PAE, 15.3 dBm, 21.6-41.6 GHz power amplifier in 65-nm CMOS using coupled resonators," in *Proc. IEEE A-SSCC*, 2016, pp. 345-348.

An E-Band Compact Power Amplifier for Future Array-Based Backhaul Networks in 22nm FD-SOI

Umut Çelik[#1], Patrick Reynaert[#2]

[#]MICAS, KU Leuven, Belgium

[1]umut.celik@esat.kuleuven.be, [2]patrick.reynaert@esat.kuleuven.be

Abstract—**This paper presents a compact high output power, linear power amplifier (PA) for array based small to medium range mm-Wave base stations. Transformer based matching and cascode architecture is used to achieve record level power density in 22nm FD-SOI. The PA achieves 17.8dB gain, 17.8dBm saturated output power (P_{SAT}) with 17.3% power-added efficiency (PAE). Core area is as low as 0.052 x 0.38mm², making this PA an ideal candidate for large array systems. AMPM is less than 2.1 degrees at 1dB compression point (P1dB) at 76GHz and less than 3 degrees from 60GHz to 85GHz. To the authors' knowledge, this PA has the highest power density achieved in E-band CMOS PAs.**

Keywords— **power amplifier, e-band, CMOS, PMOS, AM-PM, millimeter wave circuits, 5G mobile communication, cascode.**

I. INTRODUCTION

The exponential increase of data traffic requires new approaches for 5G both at mobile to base station (BS) as well as BS to BS communication. Mm-Wave frequencies 57-66GHz, 71-76GHz and 81-86GHz are proposed to be used for 5G backhaul communication because they are license-free or light-licensed, they enable high frequency reuse, and they are available almost worldwide. Although mm-Wave provides huge bandwidths that are necessary for high speed data communication, atmospheric losses at those frequencies limits the distances between BS to BS links. Especially in urban areas mm-Wave BS are needed with 100m-300m distance for high speed communication. The cost of these 5G backhaul modules becomes paramount if there needs to be a BS within such short distances [1].

Directive array-based antenna designs can provide around 30dBi in a small space for mm-Wave frequencies. Such high-gain antennas reduce the output power requirements of PAs to 15-18dBm [1]. These output power levels are in the range of Silicon based technologies. The 22nm FD-SOI CMOS technology presents unique opportunities for the industry as a cheaper alternative to Finfets, but allowing higher integration density than SiGe. Nevertheless, there are several key challenges to implement a mm-wave PA in 22nm FD-SOI, such as a low breakdown voltage and large back-end parasitics compared to older CMOS nodes.

Recently, a PMOS 3-stacked PA was proposed as an alternative to NMOS PAs, since PMOS transistors exhibit a higher voltage handling capability with only slightly worse f_{max} [2]. Although the PA in [2] achieved 19.6dBm output power, the small signal gain was only 12dB. This means to operate at P_{SAT}, the PA already needs an input power close to 10dBm which limits the practicality of this PA in a transmitter module.

In this work, A PMOS cascode PA with an NMOS cascode driver stage is proposed. This way, the output stage benefits from the large voltage handling capability of PMOS transistors to achieve high output power, while the driver benefits from the higher f_{max} of NMOS transistors to achieve high gain. Driver stage is biased in Class AB to achieve good efficiency and linearity while PMOS power stage is biased to Class A to achieve high output power as well as high gain. This PA achieves 17.8dB gain, 17.8dBm P_{SAT} with 17.3% PAE at 76GHz. The core area is only 0.02 mm² resulting in a power density of 3049mW/mm². To the authors' knowledge, this is the highest power density achieved in E-band CMOS PAs. The compact size of the PA makes it a perfect candidate for array-based transmitter modules.

II. DESIGN OF TRANSISTORS & PASSIVES IN 22NM FD-SOI

22nm FD-SOI uses gate-first High-K Metal gate. Most of the active devices are built on thin-SOI such as core NMOS and PMOS whereas passive devices are built on conventional bulk CMOS substrate. The resistivity of the substrate is not high and no trap rich layers are used as in RF-SOI. Therefore, substrate losses are similar to bulk CMOS. In this design, a 10-metal RF stack is used that consists of thick AL layer, M10 and M9 are ultra-thick copper layers and M8 is thick copper layer. Dual patterning is used to scale thinnest M1/M2 pitch [3]. Although digital performance and power consumption is much better compared to old nodes, nominal voltage of 22nm FD-SOI is 0.8V per transistor. This makes designing high output power PAs challenging. Moreover, due to the use of double patterning, via placement is more restricted compared to the older CMOS technologies. The limited number of vias have a profound impact on the parasitic source and drain resistances. Additional source resistance causes degeneration resulting in a reduction of gain. Additional drain resistance mainly reduces the efficiency of the PA.

When designing very large transistors for PAs, the optimum finger width is an important parameter that requires careful optimization. The number of vias, the length of the transistor, and gate resistance contribution from vertical and horizontal connections have significant impact on the optimal layout. In Fig. 1, the proposed transistor layout for 22nm FD-SOI technology is shown. M3 and M4 are used for source distribution to decrease the resistance at source which can be seen as green in Fig 1. M1 and M2 are used stacked together seen as blue to decrease the gate resistance. Double contact from both sides of the transistor is used to further reduce the

978-1-7281-1702-7/19 $31.00 © 2019 IEEE

gate resistance. Source contacts are put on two sides of the fingers while drain contacts are put in the centre.

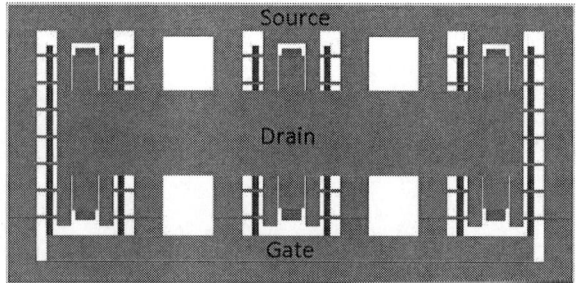

Fig. 1. Layout of the 1μmx30x3 transistor. Red is M8, green is M4 and blue is M2.

Several layouts with different finger width were designed and parasitically extracted. Fig. 2 shows the simulated f_{max} values for a 90μm PMOS and NMOS transistor with the same current density. From this comparison, a 1μm finger width was selected, resulting in an f_{max} of 236GHz for the NMOS while PMOS transistors achieving 181GHz.

In the final design, the PMOS transistors are sized as 1μmx45x3 for the power stage while the NMOS transistors are sized as 1μmx30x3 for the driver stage. RF models are used up to M1 and parasitic extraction is used for the rest of the metal stack in the design.

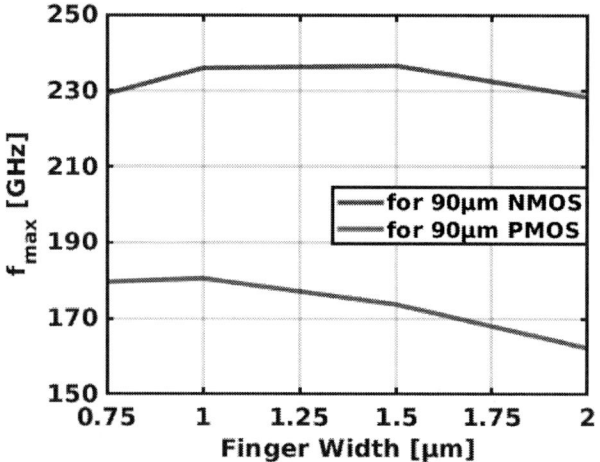

Fig. 2. Simulated parasitically extracted transistor f_{max} values up to the M8.

Special care has been given to the PA output matching since it has the most significant impact on the PAE and output power. A stacked transformer with M8 and M10 inductors in parallel is used at the transistor-side whereas a stacked M9 and AL inductor is used at the transformer output to decrease the losses and increase the coupling factor. AL layer is used to supply voltage.

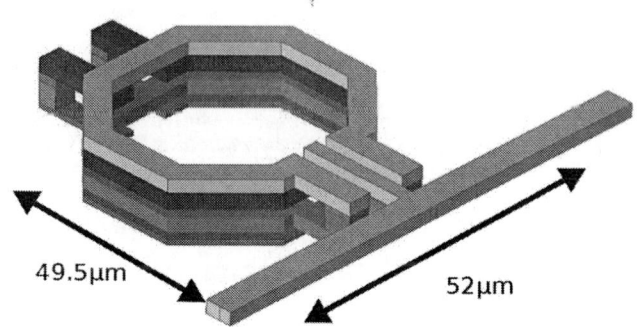

Fig. 3. Stacked output transformer for the power stage

Fig. 3. shows the optimized output transformer. It achieves 0.51dB loss at 70GHz with conjugate matching. Since the output of the PA is matched for output power, total loss becomes 0.85dB. Coupling factors of the inter-stage matching and input matching kept relatively smaller to achieve higher bandwidth.

III. CIRCUIT IMPLEMENTATION

The schematic of the proposed PA can be seen in Fig. 4. The NMOS differential cascode driver stage uses 1.6V supply voltage while 2V is used for the PMOS differential cascode power stage. Neutralization MOM (Metal-On-Metal) capacitors are used for both stages in order to improve differential stability and gain. A MOM capacitor is also used at the output of the power stage for matching. 1kΩ resistors are used to bias the gate of the common source transistors. Full Electro-magnetic simulation is used to fully characterize the PA. Special care has been given to gate nodes of the common gate transistors in order to ensure stability. M5 up to M9 are used for the gate connection. Matching is not used between common source and common gate transistors in order to save area.

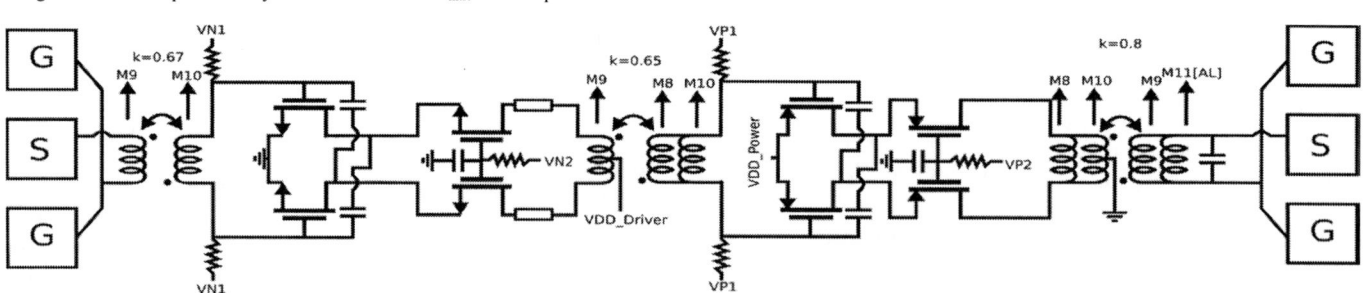

Fig. 4. Schematic of the 2-stage PA. NMOS cascode driver stage and PMOS cascode power stage with stacked transformers

978-1-7281-1702-7/19 $31.00 © 2019 IEEE

2019 IEEE Radio Frequency Integrated Circuits Symposium

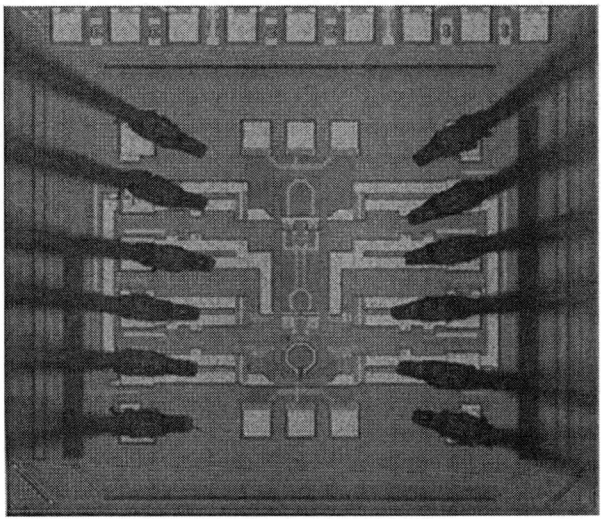

Fig. 5. Die photo of the chip

The die photo can be seen in Fig. 5. The core area is 0.052x0.38 mm² and highlighted in the die photo.

IV. MEASUREMENT RESULTS

Simulated and measured S-parameter results are shown in Fig. 6. The PA achieves maximum of 19.1dB gain at 71GHz.

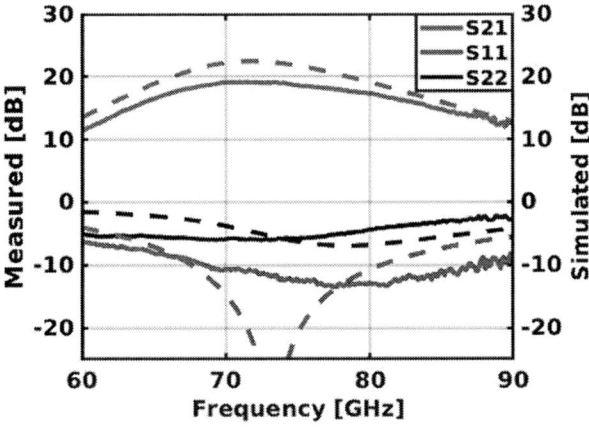

Fig. 6. S-parameter results. Dotted lines represent simulation results, full lines represent measured results.

3dB small signal bandwidth is from 64.6GHz to 82.5GHz. There is 3.1dB difference between measured and simulated gain. This discrepancy is likely due to the use of RF models that are not modelled up to the E-band. S_{11} is below -9dB from 67.4GHz to 89.3GHz.

The PA achieves 17.8dB gain, 13.3dBm Output P1dB and 17.8dBm P_{SAT} at 76GHz with 17.3% peak PAE. AMPM distortion is less than 2.1 degrees at P1dB. Fig. 7. shows the large signal measurement results. Both S-parameter and large signal measurements are done with the same bias and supply conditions.

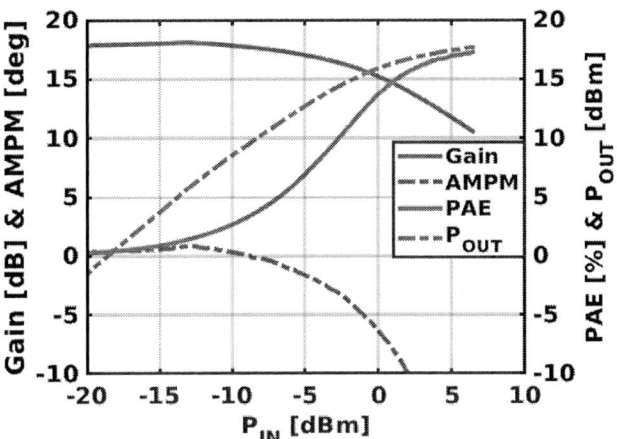

Fig. 7. Large signal measurements at 76GHz.

Fig 8. shows DC power consumption versus input power. For small signal input levels, total PA consumes 260mW power. At P_{SAT}, the power consumption rises to 310mW. Majority of the power consumption is from Class A power stage. Driver power consumption is less than 30mW at small signal and 62mW at P_{SAT}.

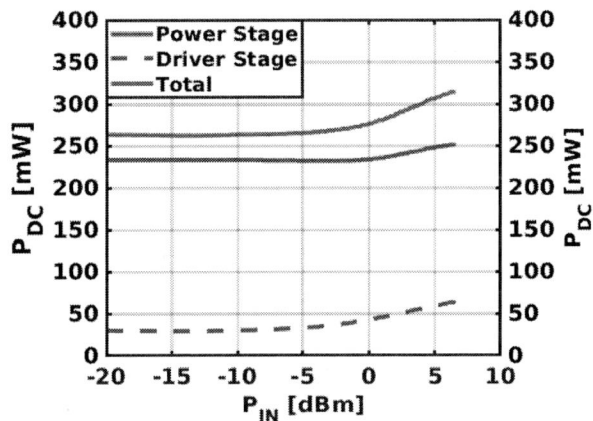

Fig. 8. DC power consumption versus input power.

Fig. 9. shows the PA performance across frequency. The PA achieves 1dB P_{SAT} bandwidth from 63GHz to 81GHz. AMPM at P1dB is less than 3 degrees from 60GHz to 85GHz. PAE is higher than 10% up to 85GHz.

Table 1 shows a comparison with the published state-of-the-art. To the authors' knowledge, this PA occupies smallest core area while delivering 17.8dBm P_{SAT} resulting in a power density of 3049mW/mm². This is the highest power density reported in CMOS based E-band PAs. AMPM performance of the PA shows that it is a very good candidate to achieve high modulation schemes such as 64QAM with high average output power.

978-1-7281-1702-7/19 $31.00 © 2019 IEEE   189

2019 IEEE Radio Frequency Integrated Circuits Symposium

Table 1. Comparison of the Silicon based state-of-the-art E-band PAs

	This Work	[4]	[5]	[2]	[6]	[7]	[8]
Technology	22nm FD-SOI	22nm Finfet	28nm Bulk	32nm SOI	40nm Bulk	55nm SiGe	120nm SiGe
Frequency [GHz]	76	75	79	78	73	80	70
Supply [V]	1.6/2	1	0.9	4.5	1.8	1.8/2.3	2
Supply per Transistor [V]	0.8/1	1	0.9	1.5	0.9	0.9/2.3	2
Gain [dB]	17.8	16.6	17	12	25.3	21	25
OP1dB [dBm]	13.3	5.7	8.25	N/A	18.9	18	12
P_{SAT} [dBm]	17.8	12.8	12.3	19.6	22.6	19	16.1
PAE_{max} [%]	17.3	26.3	13.8	18	19.3	23	16.5
AMPM at P1dB [deg]	2.08	1.5#	N/A	N/A	N/A	N/A	N/A
Core Area [mm²]	0.02	0.054	0.022*	0.05	0.25	0.024*	0.138*
Power Density [mW/mm²]	3049	353	772	1824	728	3315	296
FOM/Area [dB/mm²]	4332	1502	3575	1620	392	3820	656

FOM=P_{SAT} [dBm]+Gain[dB]+20log(freq[GHz])+10log(PAE_{max}[%]) *Estimated from the die photo. #Estimated from a graph.

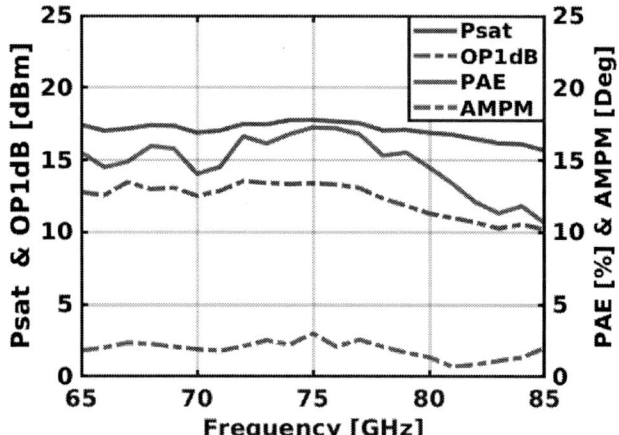

Fig. 9. Large signal PA performance from 65GHz to 85GHz.

V. CONCLUSION

A PMOS differential cascode PA with NMOS differential cascode driver stage utilizing neutralization capacitors was implemented in 22nm FD-SOI technology. The PA achieves 17.8dBm P_{SAT} in 0.02mm² core area resulting in record level 3049mW/mm² power density at 76GHz. AMPM distortion at P1dB is less than 2.1 degrees at the same frequency. It is a perfect candidate for E-band array-based backhaul communication due to its compact size and excellent linearity.

ACKNOWLEDGMENT

This work was supported by Analog Devices Inc, Limerick, Ireland. Authors want to thank Mike Keaveney from Analog Devices.

REFERENCES

[1] C. Dehos, et al., "Millimeter-wave access and backhauling: the solution to the exponential data traffic increase in 5G mobile communications systems?," in *IEEE Communications Magazine*, vol. 52, no. 9, pp. 88-95, September 2014.

[2] J. A. Jayamon, J. F. Buckwalter and P. M. Asbeck, "A PMOS mm-wave power amplifier at 77 GHz with 90 mW output power and 24% efficiency," *2016 IEEE Radio Frequency Integrated Circuits Symposium (RFIC)*, San Francisco, CA, 2016, pp. 262-265.

[3] R. Carter et al., "22nm FDSOI technology for emerging mobile, Internet-of-Things, and RF applications," 2016 IEEE International Electron Devices Meeting (IEDM), San Francisco, CA, 2016, pp. 2.2.1-2.2.4.

[4] S. Callender, S. Pellerano and C. Hull, "A Compact 75GHz PA with 26.3% PAE and 24GHz Bandwidth in 22nm FinFET CMOS," 2018 IEEE Radio Frequency Integrated Circuits Symposium (RFIC), Philadelphia, PA, 2018, pp. 224-227.

[5] A. Medra, et al., "A 79GHz variable gain low-noise amplifier and power amplifier in 28nm CMOS operating up to 125°C," *ESSCIRC 2014 - 40th European Solid State Circuits Conference (ESSCIRC)*, Venice Lido, 2014, pp. 183-186

[6] D. Zhao and P. Reynaert, "A 40-nm CMOS E-Band 4-Way Power Amplifier With Neutralized Bootstrapped Cascode Amplifier and Optimum Passive Circuits," in IEEE Transactions on Microwave Theory and Techniques, vol. 63, no. 12, pp. 4083-4089, Dec. 2015

[7] J. Zhao, et al., "2.6 A SiGe BiCMOS E-band power amplifier with 22% PAE at 18dBm OP1dBand 8.5% at 6dB back-off leveraging current clamping in a common-base stage," *2017 IEEE International Solid-State Circuits Conference (ISSCC)*, San Francisco, CA, 2017, pp. 42-43.

[8] E. C. Wagner and G. M. Rebeiz, "An 8-Way Combined E-Band Power Amplifier with 24 dBm Psat and 12% PAE in 0.12 μm SiGe," 2018 IEEE/MTT-S International Microwave Symposium - IMS, Philadelphia, PA, USA, 2018, pp. 1342-1344.

2019 IEEE Radio Frequency Integrated Circuits Symposium

An E-band Fully-Integrated True Power Detector in 28nm CMOS

Valdrin Qunaj, Patrick Reynaert

KU LEUVEN-MICAS, Kasteelpark Arenberg 10, 3000 Leuven, Belgium

{valdrin.qunaj, patrick.reynaert}@esat.kuleuven.be

Abstract—This paper presents the design of a power amplifier with a low-power fully-integrated E-band true power detector in a 28nm CMOS technology. The power detector is able to measure the true output power with a dynamic range of 27.2dB at a frequency of 75GHz for a linearity error of ±0.5dB. The detector has a low power consumption of 66μW and occupies an active area of $54 \times 130 \mu m^2$. Furthermore, the integrated detector is able to measure antenna load variations. The power amplifier design uses capacitive neutralization for gain and stability enhancement and achieves a peak gain of 23.1dB. The measured Psat, OP1dB and peak PAE at 75GHz are 11.6dBm, 9.7 dBm and 22.8% respectively.

Keywords—Power detector, Power amplifiers, CMOS integrated circuits, E-band, Millimeter wave integrated circuits, Built-in self-test.

I. INTRODUCTION

The large available bandwidth at millimeter wave (mm-wave) frequencies has made highly integrated E-band phased array systems very attractive for numerous applications such as data communications, radar and imaging. Integrated power detection in phased array systems is of critical importance in order to measure and adjust the transmitted power, detect antenna failures, measure antenna load variations, and automated built-in self-test (BIST) for each power amplifier (PA) becomes possible.

Since the demand for higher data rates requires highly linear and efficient PA's, monitoring and adjusting the transmitted mm-wave power accurately in phased array systems is crucial to optimize the overall system performance and power consumption. Another advantage of having an integrated power detector in each transmit element is that antenna failure and load variations can be detected. Subsequently, the beam can be adjusted to compensate for the defective element [1]. For these purposes, being able to measure the true output power over a wide dynamic range is key in phased array systems.

This paper presents a fully-integrated low-power mm-wave CMOS power detector that is able to measure the true output power of the PA operating at E-band frequencies. Previously published mm-wave power detectors measure voltage to determine the mm-wave output power [2]–[5]. However, by only taking the output voltage into account, the detector only gives the true output power for a fixed real load. In this design both voltage and current are measured, and used to determine the true mm-wave output power, as shown in Fig. 1.

In Section II-A of this paper the PA architecture is briefly discussed. Section II-B details the working principle of the power detector. Finally, detailed measurements of the PA

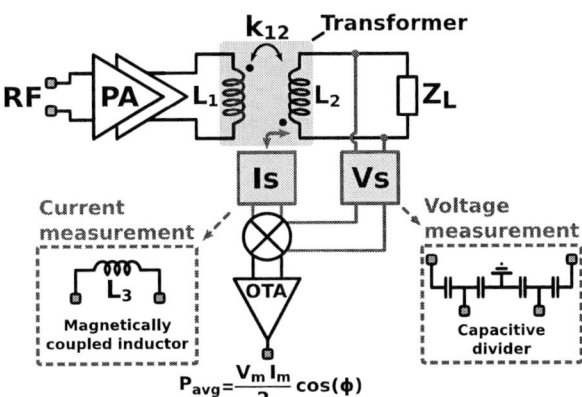

Fig. 1. System diagram of the power detector, the RF voltage and current are measured using two sensors, they are subsequently multiplied using an on chip mixer generating the average output power.

and power detector, even under antenna load variations, are presented in Section III.

II. CIRCUIT IMPLEMENTATION

A. Power Amplifier

The schematic of the PA is shown in Fig. 2. It is realized using a common source topology, the last stage has a W/L=3x33.6μm/28nm while the driver is W/L=2x25.2μm/28nm. First and last stage are biased at V_g=0.45V and V_g=0.6V respectively to improve linearity [6]. Using capacitive neutralization the power gain, stability and the reverse isolation is improved.

B. Power Detector

The architecture of the power detector is shown in Fig. 1, by measuring both the RF voltage and current, the true output power can be derived without any prior knowledge of the load impedance. The RF current is measured using a sensing coil L_3 that magnetically couples with the secondary winding L_2 of the output matching transformer. Due to this coupling, a sensing current I_{sense} is generated that is proportional to the output current. The RF voltage is measured with a simple capacitive divider. Multiplying these two signals using an

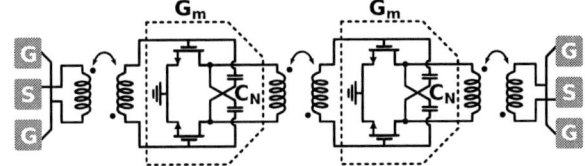

Fig. 2. Schematic of the two stage power amplifier with capacitive neutralization.

2019 IEEE Radio Frequency Integrated Circuits Symposium

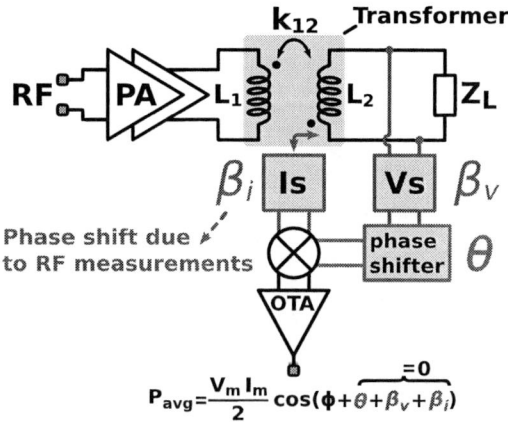

Fig. 3. Phase shift due to the RF measurements: Phase shift introduced by the RF measurements increases error of the detector, adding a phase shift θ such that $\theta + \beta_v + \beta i = 0$ reduces the error of the detector.

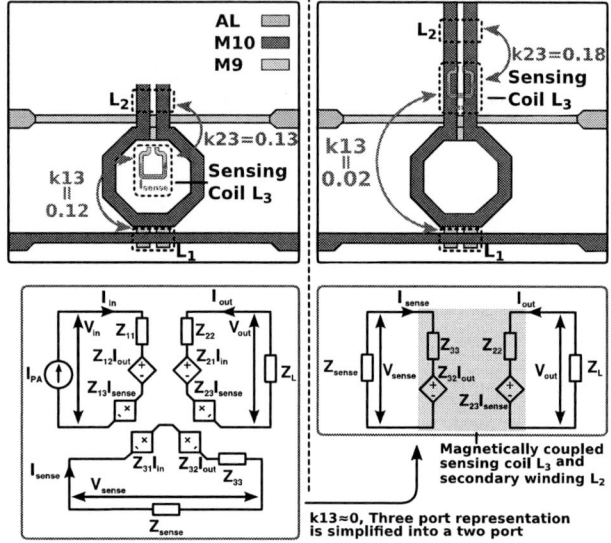

Fig. 4. Coupling factor of the sensing coil, moving the sensing coil outside the transformer reduces the unwanted coupling.

on chip mixer yields the instantaneous RF power. An OTA amplifies and filters the signal resulting in the average output power and is given by

$$P_{out}(t) = v(t) \times i(t) \quad (1)$$

$$= V_m I_m \left(\frac{\cos(\phi)}{2} + \frac{\cos(2\omega t + \phi)}{2} \right) \quad (2)$$

$$\Downarrow \text{ After the OTA}$$

$$P_{avg} = \frac{V_m I_m}{2} \cos(\phi) \quad (3)$$

where ϕ represents the relative phase difference between the voltage and current signal.

This principle was demonstrated in [7] at a frequency of 5GHz. However, when operating at mm-wave frequencies, there are certain problems that emerge when using this architecture.

Firstly, the RF voltage and current measurements introduce a phase shift β_v and β_i respectively. This phase shift is more

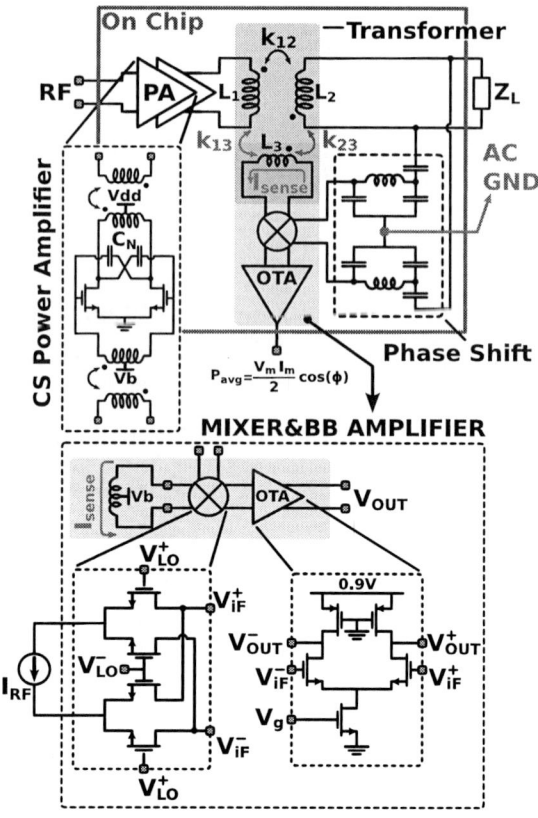

Fig. 5. The full schematic of the power detector, passive mixer, and OTA.

pronounced at mm-wave frequencies and causes inaccuracies in the detection of the output power, this is illustrated in Fig. 3. In order to reduce the error of the detector at mm-wave frequencies, a phase shift θ in the RF voltage path is introduced. This cancels out the phase shift $\beta_v + \beta_i$ such that $\theta + \beta_v + \beta_i = 0$, consequently reducing the error of the power detector such that P_{avg} equals (3), as shown in Fig. 3.

Secondly, there is an unwanted coupling k_{13} from the primary winding L_1 towards the sensing coil L_3, as aforementioned in [7]. By moving the sensing coil outside the output matching transformer and placing it underneath the transmission line, the coupling k_{13} is substantially reduced, as seen in Fig. 4. As a results of this, the sensing coil L_3 and the secondary winding of the output matching transformer L_2 form a set of coupled inductors and can consequently be described using a two port Z-parameter representation. The relationship between the output current, I_{out}, and the sensed current, I_{sense}, is given by

$$I_{sense} = -\frac{Z_{32}}{Z_{33} + Z_{sense}} I_{out} \quad (4)$$

This shows that I_{sense} is proportional to the output current and the relationship depends on the coupled inductors parameters(Z_{32} and Z_{33}) and the impedance seen at the output of the sensing coil.

The complete schematic of the power detector is shown in Fig. 5. A passive mixer topology is used because of it's low noise and power consumption. The OTA amplifies the DC

2019 IEEE Radio Frequency Integrated Circuits Symposium

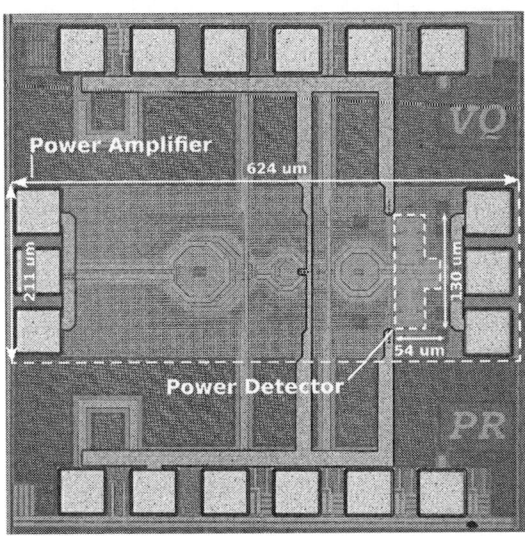

Fig. 6. Die microphotograph of the PA with integrated power detector.

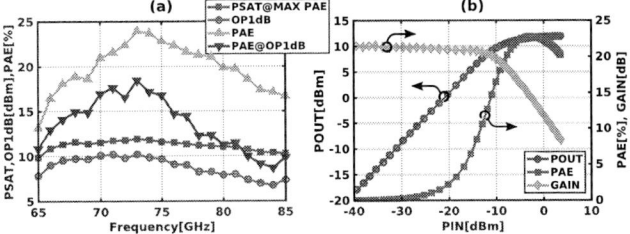

Fig. 7. Measured small-signal S-Parameters and stability factor versus frequency.

Fig. 8. Measured CW large-signal performance: (a) large-signal performance versus frequency; (b) large-signal performance versus input power at 75GHz.

signal and filters out the second harmonic, it has a bandwidth of 120MHz.

III. MEASUREMENTS

A. Power Amplifier

The PA design with integrated power detector is implemented in a 28nm CMOS technology using a digital metal stack, the core area is $0.624 \times 0.211 \text{mm}^2$ as shown in Fig. 6. The PA is measured using direct probing on the GSG pads. The S-parameter measurements were done using a VNA with extenders operating from 60-90GHz. The measured small-signal S-parameters versus frequency are shown in Fig. 7, the 3dB bandwidth is 9GHz (82GHz-73GHz). The peak small-signal gain is 23.1dB at a frequency of 77GHz. The measured stability factors μ and μ' are also shown in Fig. 7, indicating that the PA is unconditionally stable.

The continuous wave (CW) large-signal performance of the PA is shown in Fig. 8. At a frequency of 75GHz the PA achieves a saturated output power P_{sat} of 11.6dBm, 22.8% peak PAE, 9.7dBm output power at 1dB compression, and 16.77% PAE at 1dB compression.

B. Power Detector

In order to characterize the power detector, the input power of the PA was swept using a CW signal from 60GHz-90GHz.

At the same time the output power of the PA was measured using a power meter while simultaneously measuring the DC output voltage V_{det} of the integrated power detector. This was done for multiple frequency points. The output power versus the detected voltage V_{det} for multiple frequency points is shown in Fig. 9. The best linear fit for each frequency point was calculated from the measured data using linear regression. The integrated non-linearity (INL) versus the output power is calculated with reference to the best linear fit for each frequency point and is shown in Fig. 9. The dynamic range of the power detector is defined for a linearity error of ± 0.5dB. At a frequency of 75GHz, the dynamic range of the detector is 27.2dB.

In Fig. 10, the dynamic range versus frequency is shown for a linearity error of ± 0.5dB. The detector shows a quasi-flat response across a frequency range of 69GHz-83GHz with less then ± 1dB variation in dynamic range. In addition, Fig. 10 shows the detected voltage V_{det} if the output power is kept constant with frequency. The power detector is clearly able to work over a very wide frequency range, covering the entire bandwidth of the PA, while maintaining the same performance.

In order to emulate the effect of antenna variations, the load of the PA is changed using an attenuator terminated by a short. By adjusting the attenuation, both real and imaginary

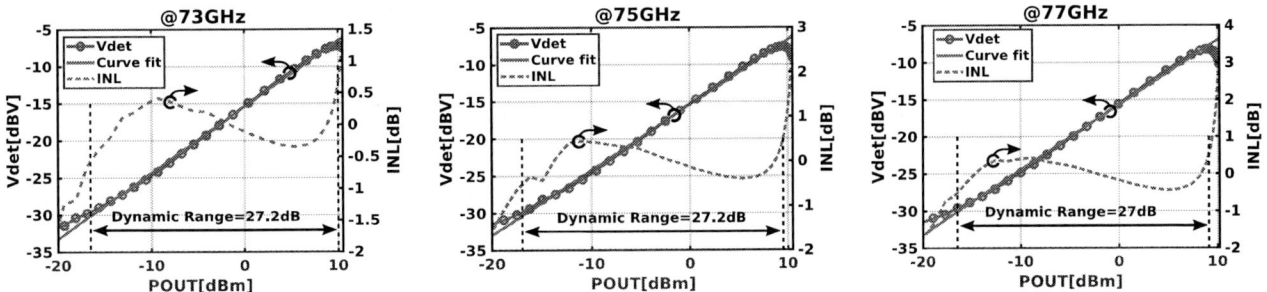

Fig. 9. Output power of the PA versus the detected voltage of the power detector, the dynamic range is also ploted for an INL of $\pm$ 0.5dB.

978-1-7281-1702-7/19 $31.00 © 2019 IEEE   193

Table 1. Comparison of state-of-art mm-wave power detectors.

	This work	[4]	[3]	[2]	[5]
Technology	**28nm CMOS**	$0.13\mu m$ SiGe	$0.12\mu m$ SiGe	65nm CMOS	$0.18\mu m$ BiCMOS
Frequency	**75 GHz**	60 GHz	71 GHz	60 GHz	20 GHz
PA integrated	**Yes**	Yes	Yes	No	No
Coupling topology	**Capacitive & Inductive**	Capacitive	Capacitive	Capacitive	Capacitive
Detector principle	**Voltage & Current**	Voltage	Voltage	Voltage	Voltage
Dynamic Range[dB]	**27.2**	8.5	10-12	25	20
INL[dB]	**±0.5**	±0.5	±0.5	/	±0.5
Power Consumption[µW]	**66**	/	/	60#	180
Active Area[μm^2]	**7000**	11030	12600	6400	385000

#Power consumption of buffer not included

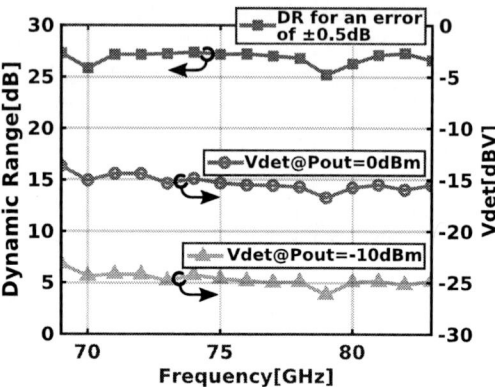

Fig. 10. Dynamic range of the power detector versus frequency for a linearity error of ±0.5dB and the detected voltage for a constant output power.

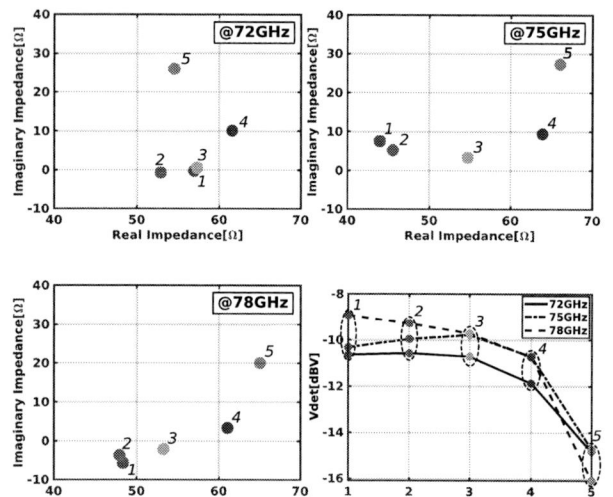

Fig. 11. Measured detected voltage under antenna load impedance variations.

part of the load impedance of the PA changes. For each impedance point, the detected voltage V_{det} is measured. This is done for multiple frequency points, as shown in Fig. 11. As the load deviates further from the nominal load, the detected voltage goes down substantially. The same behaviour can be observed for multiple frequency points. Clearly the power detector is able to measure these variations in complex load impedance.

IV. CONCLUSION

The design of a PA with a low-power fully-integrated E-band true power detector has been presented in a 28nm CMOS technology. The fully integrated detector uses a magnetically coupled sensing coil to measure the output current and a capacitive divider to measure the output voltage. These two signals are multiplied with a passive mixer giving the true output power. The measured dynamic range for a linearity error of ±0.5dB is 27.2dB at 75GHz. The power consumption of the detector is only 66µW and has an active area of 54x130µm².

The presented power detector achieves the best linearity compared to previous work and is the first fully-integrated true power detector in CMOS at mm-wave frequencies. The presented work could enable BIST, the detection of antenna failures and calibration for beamforming and array systems.

REFERENCES

[1] S. Shahramian, M. J. Holyoak, and Y. Baeyens, "A 16-element w-band phased-array transceiver chipset with flip-chip pcb integrated antennas for multi-gigabit wireless data links," *IEEE Transactions on Microwave Theory and Techniques*, vol. 66, no. 7, pp. 3389–3402, July 2018.

[2] J. Gorisse, A. Cathelin, A. Kaiser, and E. Kerherve, "A 60ghz 65nm cmos rms power detector for antenna impedance mismatch detection," in *2009 Proceedings of ESSCIRC*, Sep. 2009, pp. 172–175.

[3] R. B. Yishay, O. Katz, R. Carmon, B. Sheinman, R. Levinger, N. Mazor, and D. Elad, "High power sige e-band transmitter for broadband communication," in *2013 European Microwave Integrated Circuit Conference*, Oct 2013, pp. 73–76.

[4] U. R. Pfeiffer and D. Goren, "A 20 dbm fully-integrated 60 ghz sige power amplifier with automatic level control," *IEEE Journal of Solid-State Circuits*, vol. 42, no. 7, pp. 1455–1463, July 2007.

[5] T. Zhang, W. R. Eisenstadt, R. M. Fox, and Q. Yin, "Bipolar microwave rms power detectors," *IEEE Journal of Solid-State Circuits*, vol. 41, no. 9, pp. 2188–2192, Sep. 2006.

[6] Y. Zhang and P. Reynaert, "A high-efficiency linear power amplifier for 28ghz mobile communications in 40nm cmos," in *2017 IEEE Radio Frequency Integrated Circuits Symposium (RFIC)*, June 2017, pp. 33–36.

[7] B. Franois and P. Reynaert, "A fully integrated transformer-coupled power detector with 5 ghz rf pa for wlan 802.11ac in 40 nm cmos," *IEEE Journal of Solid-State Circuits*, vol. 50, no. 5, pp. 1237–1250, May 2015.

2019 IEEE Radio Frequency Integrated Circuits Symposium

A Coupler-Based Differential Doherty Power Amplifier with Built-In Baluns for High Mm-Wave Linear-Yet-Efficient Gbit/s Amplifications

Huy Thong Nguyen[1] and Hua Wang[2]
Georgia Institute of Technology
[1]huythong@gatech.edu, [2]hua.wang@ece.gatech.edu

Abstract—**We propose a Doherty power amplifier architecture with impedance inverting balun and impedance scaling balun to support Doherty active load modulation and Power Back-Off (PBO) efficiency enhancement at high mm-Wave frequencies. Unlike transformer baluns often used at RF frequencies, mm-Wave coupler-based baluns are employed to provide well balanced differential-to-single-ended conversion and absorb device output parasitic capacitance. Moreover, this paper reports that coupler-based balun with 45° electrical length exhibits impedance inverting behavior, which is utilized to construct Doherty active load modulation network in the reported PA design. Adaptive biasing circuit further enhances the Main/Auxiliary PA cooperation. The measured Doherty PA exhibits 20.1dBm P_{sat}, 19.3dBm P_{1dB}, and 26% peak PAE. At 7dB PBO, the measured PAE is 16.6%, demonstrating 1.45× efficiency enhancement compared to an ideal class-B PA with the same peak PAE at P_{1dB}.**

Keywords—**60GHz, adaptive bias, coupler, coupler-based balun, CMOS, dynamic bias, Doherty, mm-Wave, power amplifier**

I. INTRODUCTION

Mm-Wave Power Amplifier (PA) with high Power Back-Off (PBO) efficiency is of critical importance for efficient transmission of 5G modulation signals with large peak-to-average-power-ratio (PAPR). Among various techniques for PBO efficiency improvement, Doherty PA is a particularly promising candidate that can support large modulation bandwidth and demand low baseband digital signal processing overhead. Despite the success at RF or low mm-Wave frequencies, few Doherty PAs are demonstrated at or above the 60-80GHz band. The Doherty PAs in [1][2] exhibit degraded peak PAE and marginal PBO efficiency enhancement, mostly due to the lossy Doherty output networks and imperfect Main and Auxiliary (Aux) PA interactions. While the antenna-based Doherty radiator shows substantial PBO efficiency boost [3][4], various applications still necessitate mm-Wave PAs that can deliver power to a fixed 50Ω load.

Additionally, baluns are indispensable for differential PAs that allow power combining and broadband capacitive neutralization to substantially improve the device gain/stability; this achieves superior device-level performance over single-ended PAs, which is essential for high mm-Wave designs. State-of-the-art mm-wave PAs extensively utilize transformer-based baluns to convert differential to single-ended signal [2][5]. However, designing an efficient transformer-based balun at the upper mm-wave frequency range is a challenge, mainly due to its limited self-resonance frequency, small size and strong capacitive coupling. These factors often result in poor loss and unbalanced impedance at PA output network at high mm-wave frequencies.

In this work, we employ coupler-based baluns for differential-to-single-ended conversion, impedance matching, and active load modulation at the PA output. We study the property of balun under active load modulation and propose a Doherty architecture in section II. In section III, we report a close-form solution for an impedance inverting balun that is critical for the proposed active load modulation PA. Section IV shows a proof-of-concept design and measurement results.

II. IMPEDANCE INVERTING/SCALING BALUNS

A. Balun impedance under active load modulation

Balun is a 3-port network extensively employed at PA output to convert differential to single-ended signal. Output load of a class-AB PA is fixed during the entire PA operation; thus, balun simply performs the impedance matching from a fixed Z_L to an optimum load Z_{pa} and resonates out the device parasitic capacitor C_{par} (Fig. 1).

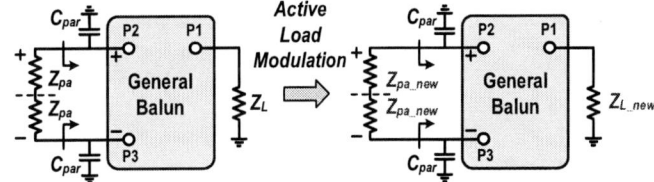

Fig. 1. Balun impedance under active load modulation.

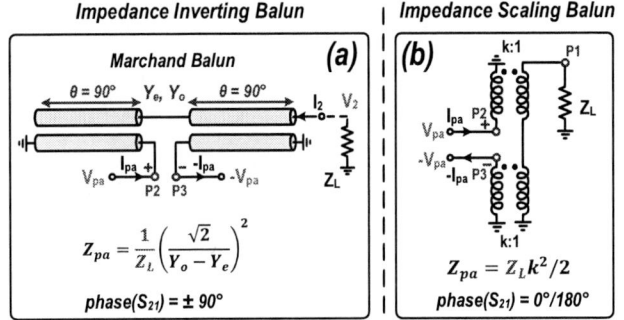

Fig. 2. (a) Marchand balun as an impedance inverting balun (IIB) and (b) idealistic transformer-based balun as an impedance scaling balun (ISB)

Meanwhile, the impedance Z_L in active load modulation PAs changes versus PBO level, subsequently the impedance Z_{pa} seen by PA must also vary (Fig. 1). The relation between these impedances depends on balun types. For example, the load seen

978-1-7281-1702-7/19 $31.00 © 2019 IEEE   195

by differential ports in Marchand balun is inversely proportional to Z_L [Fig. 2(a)], and Marchand balun acts as an impedance inverting balun (IIB). Nonetheless, in transformer-based baluns with no inter-winding capacitor, the differential impedance Z_{pa} is proportional to Z_L [Fig. 2(b)], and this type of balun acts as an impedance scaling balun (ISB).

Mathematically, it can be proved that the relationship between the output load (Z_L) and the source impedance seen by PA (Z_{pa}) is entirely dependent on the phase delay- phase(S_{21}) - from the differential ports to the single-ended port. Particularly, a balun is an IIB if phase(S_{21}) = ±90° or ISB if phase(S_{21}) = 0°/180° (Fig. 2). The observation of IIB and ISB can be useful for PA problems involving load variation, and in this work, we focus on a high mm-Wave Doherty PA design.

B. Proposed Doherty PA architecture

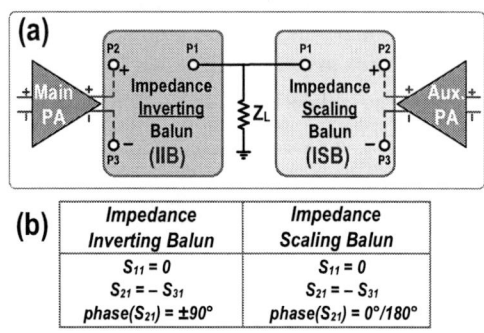

Fig. 3. (a) Proposed Doherty PA architecture with IIB and ISB. (b) Summary of IIB and ISB equations.

To support the active load modulation of differential PA, we propose a Doherty network with embedded baluns [Fig. 3(a)] with an IIB connected to Main PA and an ISB connected to Aux PA. The phase delay phase(S_{21}) of the output balun must be dedicatedly designed to achieve the desirable inverting/scaling behaviours [Fig. 3(b)].

The operation of the proposed network is as follows. At low P_{out}, the Aux PA is off, the ISB scales the high impedance of the inactive Aux PA, and the Main PA load is high. At high P_{out}, the Aux PA is on, and the Main PA path sees an increasing load at the parallel summing node, while the IIB reduces Main PA load to complete the Doherty load modulation.

III. DESIGN OF PROPOSED COUPLER-BASED DOHERTY PA

While transformer-based balun is intensively employed at RF frequencies, its limited self-resonant frequency, small size, and strong capacitive coupling cause poor loss and unbalanced impedance at high mm-wave. To perform the differential-to-single-ended conversion at PA output at this frequency range, we explore the coupler-based balun in this work [Fig. 4(a)]. It consists of 2 coupled-lines and 2 lumped capacitors, and the fundamental of this balun can be found in [6][7].

A. Impedance inverting balun

The Main PA necessitates an IIB with ±90° phase delay to perform active load modulation. We report the specific 45° electrical length of the coupled line to formulate an IIB [Fig.4(b)]. Particularly, we derive the closed-form Y-matrix of

this network as shown in Fig. 4. The inductive admittance of the coupled-lines equals to $\frac{Y_o+Y_e}{2}$ and resonates out the lumped capacitors. The phase delay from differential to single-ended port is exactly 90°, and the driving impedance of the differential ports is inversely proportional to the load Z_{Load}, resulting in a desirable IIB balun, as

$$Z_1 = \frac{1}{Z_{Load}}\left(\frac{2}{Y_o - Y_e}\right)^2$$

Moreover, this low-loss IIB is particularly conducive to high mm-wave PAs, as it naturally absorbs the PA output capacitance inside the lumped capacitor and enables AC grounds for feeding PA DC supply with no choke. The impedance inverting behaviour is specifically critical for active load modulation PAs. It is worth mentioning that Marchand and the proposed balun are both IIBs. However, the proposed balun can absorb the device parasitic cap and is only half of the size of Marchand balun (45° compared to 90° electrical length).

3D EM simulation of the proposed IIB demonstrates 0.2dB amplitude mismatch and 1° phase imbalance at 60GHz [Fig. 4(c)]. Moreover, simulated phase(S_{21}) at 60GHz equals to 90° and follows the IIB equations defined in section II.

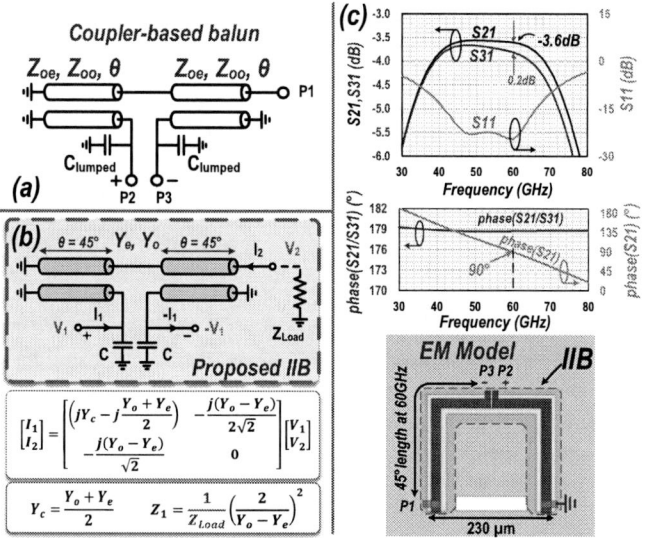

Fig. 4. (a) Schematic of the coupler-based balun (b) Proposed IIB with 45° electrical length, derivation of [Y] matrix of the network, closed-form formula for the capacitance values, impedance transformation ratio and impedance inverting characteristic (c) EM model of the IIB and 3D EM simulated phase and amplitude response of the proposed IIB.

B. Impedance scaling balun

The Aux PA demands an ISB for the proposed Doherty PA architecture. While designing a balun satisfying all 3 ISB equations can be a challenge, it is worth mentioning that a matched balun with phase(S_{21}) ≤ ±10° can relatively maintain its impedance scaling behavior over the Doherty active load modulation. In this work, we design a coupler-based balun with a short 20° electrical length to mimic this ISB.

C. EM model and active load modulation

Overall, the coupler-based Doherty network occupies only a single transformer footprint of $230 \times 230 \mu m^2$ to combine

978-1-7281-1702-7/19 $31.00 © 2019 IEEE

differential Main and Aux signals to a single-ended 50Ω port. Figure 5(a) depicts the full EM model of the Doherty PA output network. Among various possible coupled-line technologies, we implement the on-chip coupler-based baluns from the broadside coupled-line structure with two metals vertically stacking on each other. The odd-mode characteristic impedance Z_{oo} of this type of coupled-line is strongly correlated to the width W of the metal trace, and the even-mode characteristic impedance Z_{oe} is highly dependent on the ground gap G from the signal to the distributed ground. The width W and gap G are vastly orthogonal, subsequently this eases the EM tunning of desirable couplers exhibiting various even-/odd-mode Z_{oe}, Z_{oo} characteristic impedances. 3D EM simulation verifies desirable Doherty active load modulation with ~80% passive efficiency over all PBO level as shown in Fig.5(b).

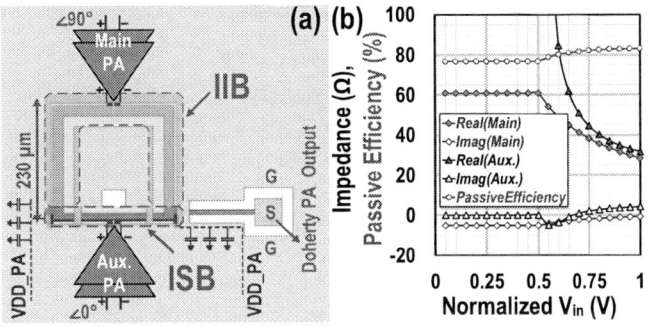

Fig. 5. (a) Full EM model of the Doherty PA output network (b) simulated active load modulation and passive efficiency of the proposed network.

D. PA schematic

Figure 6 depicts the schematic of the identical Main/Aux. PAs. Each consists of a common source driver stage and a cascode PA stage. Neutralization capacitors are employed to enhance reverse isolation, improve the stability, and boost the gain. Intra T-Line is designed between the common source and cascode stage to improve the intra-stage current gain. Transformers are employed at the input and inter-stage matching for compact layout.

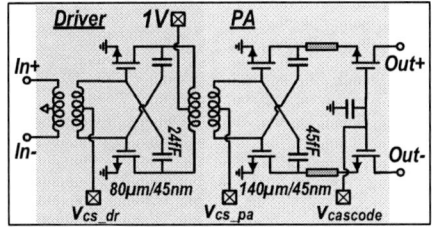

Fig. 6. Schematic of the Main/Aux. PAs.

E. Adaptive biasing circuit

Doherty PA architecture requires Main and Aux PAs to cooperate and follow the desirable Doherty current trajectory. The Aux. current is almost zero below 6dB PBO and rapidly increases to the Main current level at 0dB PBO. Similar to [2-4], we apply adaptive biasing circuit (Fig. 7) for both driver and PA stage such that the Aux. branch is completely turned off below 6dB PBO and fully turned on at 0dB PBO.

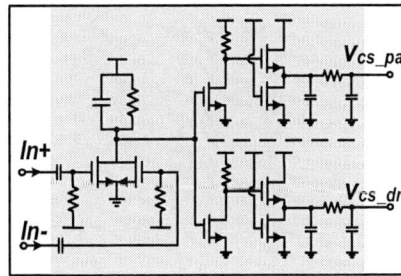

Fig. 7. Schematic of adaptive biasing circuit for the Aux. branch.

F. Top-level schematic

Fig. 8 illustrates the top-level schematic of the proposed Doherty PA. The single-ended input is split into differential 0°/90° signals via an input balun and 90° couplers. The differential 0°/90° signals are then fed to the Main and Aux path differential PAs. At outputs, the coupler-based Doherty network performs Doherty operations and delivers the combined power to a single-ended load.

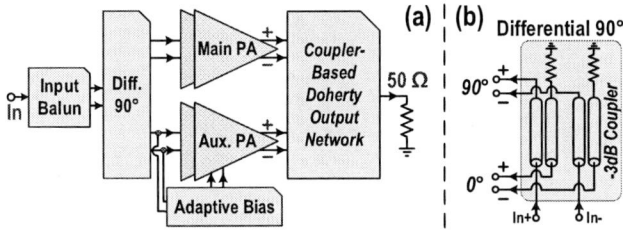

Fig. 8. (a) Top-level schematic of the proposed Doherty PA (b) Input feeding network.

IV. MEASUREMENT RESULTS

We fabricate a proof-of-concept 60GHz differential Doherty PA in 45nm Global Foundries CMOS SOI process. The chip (Fig. 9) occupies an active area of $1000 \times 760 \mu m^2$.

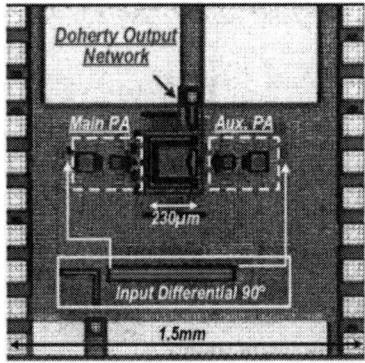

Fig. 9. Die micrograph of the proposed 60GHz Doherty PA.

Small-signal measurement [Fig.10(a)] shows broadband input matching with $|S_{11}|<10dB$ from 55-67GHz, peak 12.9dB gain, and 3dB S_{21} bandwidth from 57-63GHz. Measured large-signal performance at 60GHz [Fig.10(b)] demonstrates 20.1dBm P_{sat}, 19.3dBm P_{1dB}, and 26% peak PAE. The measured PAE at 7dB PBO is 16.6%, achieving 1.45× efficiency boost compared to a reference class-B PA with the same peak PAE at P_{1dB}.

978-1-7281-1702-7/19 $31.00 © 2019 IEEE   197

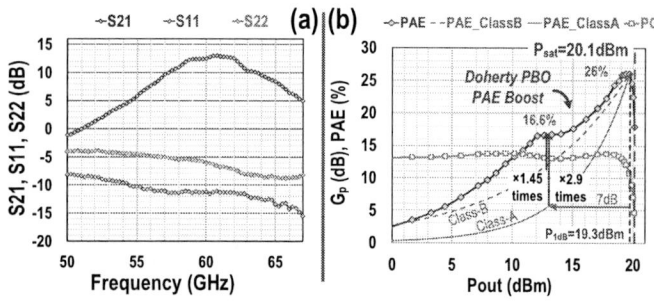

Fig. 10. (a) Measured small-signal S-parameter of the Doherty PA (b) Measured large-signal performance of the Doherty PA at 60GHz carrier frequency.

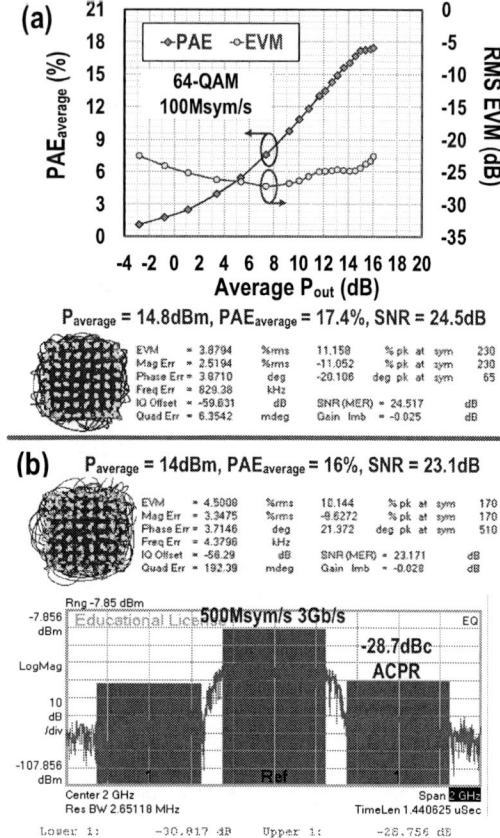

Fig. 11. Measured dynamic performance of the Doherty PA at 60GHz carrier frequency (a) 100Msym/s (b) 500Msym/s

Figure 11 depicts the modulation performance. The PA is first tested using 100Msym/s 64-QAM modulated signal for their averaged PAE and EVM versus PBO. Without any digital pre-distortion, the Doherty PA demonstrates 14.8dBm average P_{out} and 17.4% average PAE at -24.5dB RMS EVM. Additionally, it supports 14dBm/13.8dBm with 16%/15.7% average PA at -23.1dB/-22.3dB RMS EVM for 500Msym/s and 1Gsym/s 64-QAM modulated signal, respectively. Compared to reported 60GHz PAs with PBO efficiency enhancement [1][2] [8-10], the proposed coupler-based Doherty PA achieve the highest PAE at 6-7dB PBO and the highest average PAE for 64-QAM modulations (Table 1).

Table 1. Comparison to 60GHz PAs/Radiators with PBO PAE enhancement

| | This work | 60GHz PAs with PBO PAE boost | | | | 60GHz Radiators | |
		[2] Kaymaksut, TMTT'15	[1] Greene, JSSC'17	[9] Chappidi, JSSC'17	[10] Zhao, JSSC'12	[3] Nguyen, JSSC'18	[4] Nguyen, ISSCC'19
Technology	45nm CMOS SOI	40nm CMOS	130nm SiGe BiCMOS	130nm SiGe BiCMOS	40nm CMOS	45nm CMOS SOI	45nm CMOS SOI
Topology	Coupler-Based Doherty	Doherty	Doherty	Asymmetric Combiner	Outphasing	On-Antenna Doherty	3-Way Doherty
Frequency (GHz)	60	72	62	55	60	65	57
Supply (V)	2	1.5	2.8	4	1	1.9	2
Power Gain	12.9	20	19	23^	--	12	12.9
P_{SAT} (dBm)	20.1	21	17.5	23.6	15.6	19.4	21.2
P_{1dB} (dBm)	19.3	19.2	17.1	19.9	15.6^	19.2	20.7
Peak PAE	26%	13.6%	23.7%	27.7%	25%	28.3%	21.8%
PAE @ P_{1dB}	25.9%	12.4%	23.7%	15.7%	--	27.5%	20.6%
PAE @ PBO	16.6% @7dB PBO	7% @6dB PBO	13% @6dB PBO	7%^ @6dB PBO	9%^ @6dB PBO	20.1% @6dB PBO	18% @9dB PBO
PBO PAE Enhancement Ratio†	1.45	1.13	1.10	0.89^	0.72^	1.45	2.34
Modulation	64-QAM	64-QAM		64-QAM	16-QAM	64-QAM	64-QAM
Data Rate	3Gb/s	0.6Gb/s		3Gb/s	0.5Gb/s	6Gb/s	3Gb/s
RMS EVM	-23.1dB	-25.6dB		-21dB	-22dB	-23dB	-23.5dB
Average Pout	13.8dBm	15.9dBm		14.8dBm	12dBm	14.2dBm	13.5dBm
Average PAE	15.7%	7.2%		--	15%	20.2%	17.3%
Active Area(m²)	0.76	0.19	0.6	1.02	0.33	3.23*	3.61*

^ Estimated from reported figures
*Including pads and on-chip antenna
†Compared to a reference class-B PA with the same PAE at P_{1dB}

V. CONCLUSION

We present a Doherty PA topology with IIB and ISB to support active load modulation of differential PAs. Examples of IIB and ISB are illustrated, and an equation on the phase delay from the differential ports to the single-ended is imposed to determine the inverting/scaling behavior of baluns. Coupler-based baluns are employed at the PA output network to perform impedance matching, differential to single-ended conversion, invert/scale functionality, and ultimately active load modulation. The special case of coupler-based balun with 45° electrical length is analysed and proved as a canonical form of IIB. With a proof-of-concept 60GHz PA design, measurement result demonstrates a substantial PAE boost at PBO and linear-yet-efficient amplification of 64-QAM modulated signals.

ACKNOWLEDGMENT

This is in part supported by DARPA and Global Foundries.

REFERENCES

[1] K. Greene, et al., "A 60-GHz Dual-Vector Doherty Beamformer," *JSSC*, May 2017.

[2] E. Kaymaksut et al., "Transformer-Based Doherty Power Amplifiers for mm-Wave Applications in 40-nm CMOS," *TMTT*, Apr. 2015.

[3] H.T. Nguyen, et al., "A Linear High-Efficiency Millimeter-Wave CMOS Doherty Radiator Leveraging Multi-Feed On-Antenna Active Load Modulation," *JSSC*, Dec. 2018.

[4] H.T. Nguyen, et al., "A mm-Wave 3-Way Linear Doherty Radiator with Multi Antenna Coupling and On-Antenna Current-Scaling Series Combiner for Deep Power Back-Off Efficiency Enhancement", *ISSCC*, Feb. 2019.

[5] F. Wang, et al., "A Highly Linear Super-Resolution Mixed-Signal Doherty Power Amplifier for High-Efficiency Multi-Gbit/s Mm-Wave 5G Communication," *ISSCC*, Feb. 2019.

[6] K. S. Ang, et al., "Analysis and design of miniaturized lumped-distributed impedance-transforming baluns," *TMTT*, Mar. 2003.

[7] H.T. Nguyen, et al., "A 60GHz CMOS Power Amplifier with Cascaded Asymmetric Distributed-Active-Transformer Achieving Watt-Level Peak Output Power with 20.8% PAE and Supporting 2Gsym/s 64-QAM Modulation", *ISSCC*, Feb. 2019.

[8] H. Wang, et al., "Power Amplifiers Performance Survey 2000-present"

[9] C.R. Chappidi et al., "Frequency Reconfigurable Mm-Wave Power Amplifier with Active Impedance Synthesis in an Asymmetrical Non-Isolated Combiner: Analysis and Design," *JSSC*, Aug. 2017.

[10] D. Zhao, et al., "A 60-GHz Outphasing Transmitter in 40- nm CMOS," *JSSC*, Dec. 2012.

VSWR Robust Linearizer to improve Switch IMD by >20dB

Thomas Meier[#1], Atif Mehmood[#], Jonas Kaps[#]

[#]RF Innovation, Germany

[1]thomas.meier@rfinnovation.de

Abstract— **A novel circuit design approach is able to reduce intermodulation products of FET based RF switches by more than 20dB. Isolation remains unaffected, insertion loss adder is 0.1dB only and robust VSWR performance is shown. The necessary additional die area is very small keeping the cost adder at a minimum.**

Keywords— switches, predistortion, intermodulation distortion, FET integrated circuits, carrier aggregation, SOI

I. Introduction

Modern telecommunication user equipment requires more and more switches. The increasing number of bands, the introduction of massive MIMO and carrier aggregation made RF front ends of mobile devices extremely complex. In addition, 5G requires drastically increased bandwidth and linearity. All of these requirements challenge the performance of every component inside the RF front end, linearity of switches being among the most challenging and important parameters.

High effort has been spent across the whole industry to upgrade existing technologies like SOI to provide the required performance. For example, the introduction of trap rich material in SOI substrates improved linearity of larger, metal covered areas like coplanar waveguides [1], [2], but does not address the switch performance itself. Other approaches based on circuit techniques [5] are able to cancel out nonlinearities but have limitations in power range. Table 1 compares different state-of-the-art approaches for linearization.

This paper describes a novel approach that is able to reduce intermodulation products of switches by more than 20dB. Robust performance is achievable over load VSWR changes.

Table 1. State of the art switch linearization techniques.

Ref.	Method	Improvement	Comment
5)	Switch gate bias adjust	3rd Harmonic	Limited power range 6-12dB
1),2)	Trap rich layer	IMD, Harmonic	Substrate effects only
7)	Nonlinear MOS-Cap to Ground	IMD, Harmonic	VSWR sensitive
8)	Anti-parallel PIN Diodes	IMD, Harmonic	High Power (>100W) Tx
9)	Digital Interference	IMD	System approach

II. Intermodulation in Switches

A. Sources of Nonlinearities

A commonly used basic form of a switch is a single-pole-double-throw (SPDT) configuration. In its simplest implementation, a SPDT consists of just two FETs connecting an input to either output one or two like in figure 1. Turning on the upper FET (typically applying positive gate bias voltage) will connect the input to output 1. Putting the FET in "OFF" condition (typically zero or negative gate bias voltage) will disconnect the output form the input. The same applies to the lower branch 2.

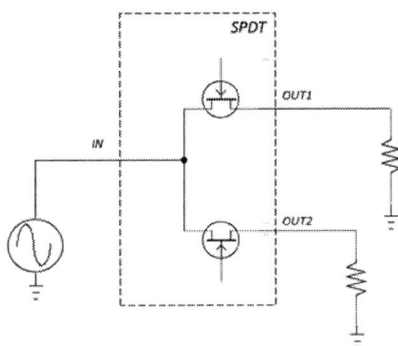

Fig. 1. Basic circuit of a single pole double throw switch.

The IV curve of a FET in "ON" and "OFF" condition is shown in figure 2. For a switch branch in "ON" condition, the FET is used in the low ohmic part of its IV curve. The main source of distortion is the slightly nonlinear part of the IV curve in the ohmic region, which predominantly creates distortion of third order.

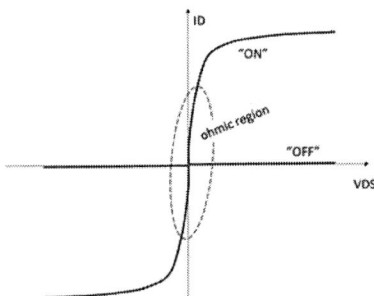

Fig. 2. IV curve of switch FET in "ON" and "OFF" condition.

In "OFF" condition, there is essentially no contribution from the IV characteristic to nonlinearities – the main source for distortion are the nonlinear parasitic gate- and source-drain capacitances. These "OFF"-branch nonlinearities are typically 15…20dB below the ones of the "ON" arm, making the "ON" branch the predominant source for intermodulation distortion. Designers can reduce these distortions by the following straight forward approaches:

1) Increase the size of the FET

2) Use several FETs in series.

The drawback of approach 1) is a reduction in isolation. The drawback of approach 2) is an increase of insertion loss. Both approaches have in common the disadvantage of an increase in necessary die size.

B. State of the Art Switches and Performance

Typical switch designs used in today's mobile communication RF front ends are based on SOI technology. Each switch arm usually consists of a series configuration of several FETs with a gate width of a few millimeters each. The number of FETs is dictated by the maximum transmit power and thus the voltage swing across the string of transistors of the "OFF" arm and the breakdown voltage of each transistor. The gate width determines the "ON" resistance and thus the insertion loss, but also the parasitic "OFF" capacitance which determines the achievable isolation.

Examples for this category of SPDT switches are Infineon´s BGS12PN10 with an IIP3 of 75 dBm [3] or PSemi´s PE42426 with an IIP3 of 83dBm [4].

III. DESIGN APPROACH

The novel approach described in this paper aims to achieve the following goals:

- Significant improvement of intermodulation performance
- Lowest cost, i.e. smallest die size
- Little to no impact on other RF parameters
- No necessity of additional DC supply
- Robustness to VSWR, process, voltage, temperature variation.

The basic circuit of this approach is shown in figure 3. It can be arranged as either a pre- or post-distortion element next to the actual switch. It consists of a reactive series element, which can be either an inductor or capacitor and an extremely small FET in parallel configuration to this inductor or capacitor.

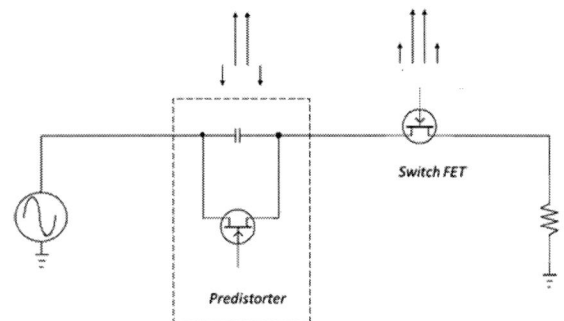

Fig. 3. Fundamentals of intermodulation reduction circuit.

The sizes and values of the components chosen are derived as follows:

- The impedance of the reactive element is very small compared to the characteristic impedance of 50 ohms, typically less than .5 Ohm for the frequency of interest. This ensures the losses introduced are negligible.
- The FET size of the predistorter is extremely small compared to the FET size of the switch thus creating a

high distortion level even with the small voltage drop across the reactance. When properly sized, the effective intermodulation signal created by this predistortion circuit is of the same amplitude as the distortion of the actual switch.

Since current and voltage in any reactive element have a 90° phase difference, the resulting phase of the distortion created by the predistorter is exactly 180° out of phase compared to the distortion of the actual switch. If both, the amplitude and phase conditions are met, a full cancellation of the intermodulation is achieved.

One of the main advantages of this configuration is the robustness to load VSWR changes. Both, the switch FET as well as the distorter FET are in series configuration, thus changing their distortion level under VSWR changes the same way keeping the cancellation intact. But it comes with the disadvantage of a bandwidth limitation: the distortion level of a FET is essentially frequency independent for both, the predistorter as well as the switch itself. But the impedance of the reactance and thus the RF signal amplitude across the distortion FET follows a 6dB per octave law. This results in a 6dB per octave characteristic of the whole predistorter. The bandwidth, where the magnitude of the predistorter and the switch are similar enough to cancel out, is approximately 25%.

This limitation can be overcome by using more than one reactive element as shown in figure 4.

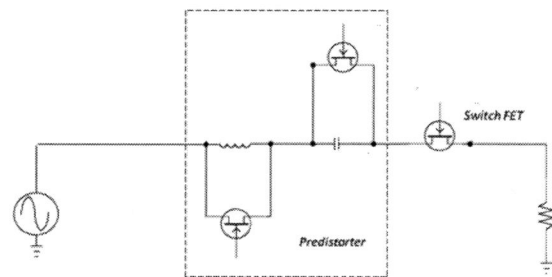

Fig. 4. Predistortion circuit with increased bandwidth

Each of the subsections with either an inductor or capacitor as the reactive element has a 25% bandwidth. With choosing the appropriate component values, the two frequency ranges can be added, resulting in about 50% useful bandwidth.

Multi-band antenna switches can also use multiple, band specific predistortion circuits for different switch arms, eliminating essentially all bandwidth limitations.

IV. DEMONSTRATOR

The principle described in the previous section can be applied on any FET based switch circuit. With that in mind, several technologies like bulk CMOS, GaAs pHEMT and SOI were used in simulations, all of them showing very similar results in terms of improvements achievable using the predistortion circuit. With SOI being today´s working horse for switches, a commercially available 180nm SOI foundry process was chosen for the implementation of the demonstrator hardware.

The main test vehicle is a single-pole-double-throw (SPDT) switch configuration. The baseline switch design uses 12 FETs in series for each arm of the SPDT with a gate width of 3.5mm for each FET. The predistorter uses FETs with a gate width of 3% of the gate width of the actual switch FETs. C- and L-values are in the 0.1…0.5nH and 30…60pF range, respectively, depending on the desired frequency range and the original distortion level of the baseline switch. No separate ESD protection is foreseen since the circuit will share the existing protection of the baseline switch. A photograph of the demonstrator hardware is shown in figure 5.

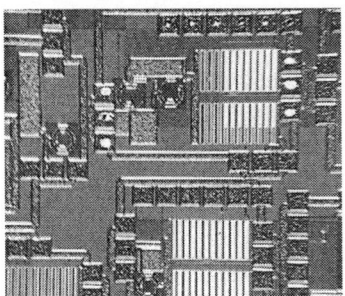

Fig. 5. Photograph of test die

The frequency range targeted for the demonstrator was 1.2-2.0 GHz. The baseline SPDT switch was designed to achieve an IIP3 of 75dBm targeting an IIP3 of >85dBm for the switch with predistortion.

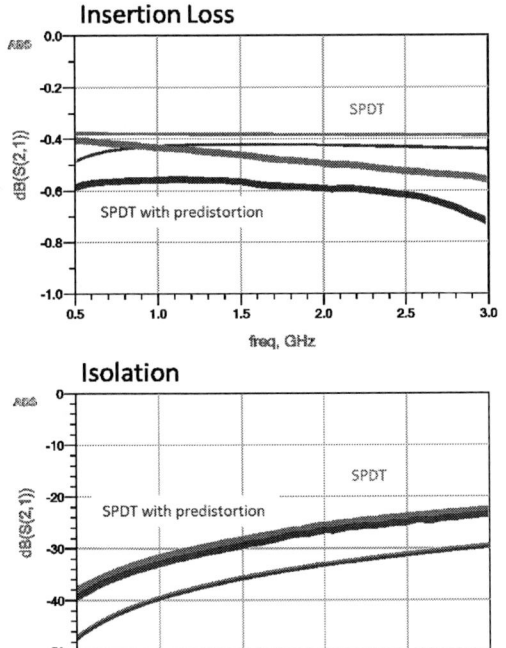

Fig. 6. Small signal results: bold line = test result, thin line = simulation

Figure 6 shows the small signal results of insertion loss and isolation. The graphs show a comparison of the baseline SPDT (red) and the SPDT with predistortion (blue). Bold lines represent the measurement results and thin lines the corresponding simulation results. The insertion loss adder for

the predistortion is tested with about 0.1dB, slightly higher than simulations predict. Isolation is unaffected by the predistorter, but simulations are about 7dB too optimistic. Most probable root cause for this difference is a not perfect modelling of substrate coupling in simulation.

Large signal results are depicted in figure 7. The upper graph shows a sweep of input power at a frequency of 1.4GHz with a carrier frequency offset of 20MHz. The lower graph shows the performance over frequency. Overall, up to 18dB of IM3 improvements are demonstrated on hardware. Simulations show a potential of >20dB of IM3 improvement over a bandwidth of 50%.

The comparison of large signal measurement results versus simulation shows several dB difference in absolute values, main reasons for this deviation will be discussed in paragraph V. The relative improvement created by the predistorter is still nicely reflected in both, measurement and simulation.

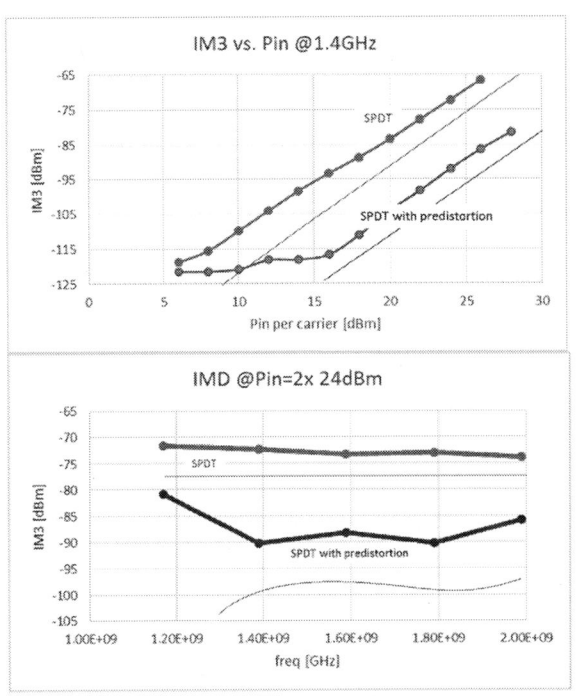

Fig. 7. Large signal results: bold line = test result, thin line = simulation

All measurements were carried out using on wafer probing for both, large signal as well as small signal measurements. This approach allows the implementation of various test circuits on the very same test die.

A lot of effort went into the test bench itself to achieve enough dynamic range to be able to measure intermodulation distances of up to 130dB. The most important aspects are:

- Excellent isolation between the two signal sources
- Notch filters (>60dB) to suppress the carrier tones at the spectrum analyser input.
- Low PIM (passive intermodulation) RF probes like T. McKay et.al. already described in [6].

Overall, an IP3=90dBm was achieved for the measurement setup making sure, IP3 levels of 85dBm for the DUT can be measured with enough margin.

Load VSWR changes can be a significant challenge for IM3 performance. As described earlier, this novel concept is very robust to load changes. Simulation results in figure 8 show that the IM3 improvement stays intact over load VSWR.

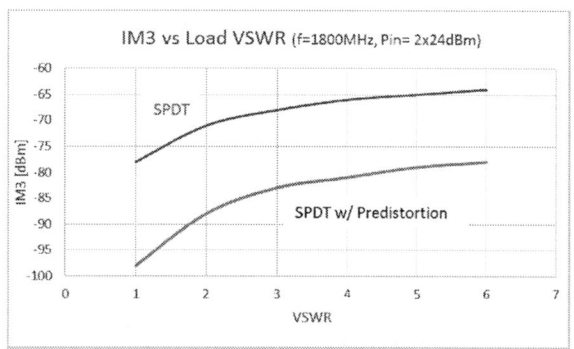

Fig. 8. Simulated IM3 vs. load VSWR

V. Current and Future Work

Based on these encouraging results redesign activities were commenced. In depth analysis of the existing hardware and extensive simulations identified already the main challenges to be addressed in the redesign phase:

- Extensive EM simulation turned out to be key for accurate simulation results
- Scaling of the large signal model over a device size factor of ten or higher appears to be a challenge. Precise results will probably require test structure arrays to cover the required device size range.
- In order to achieve tight manufacturing tolerance and robust temperature behavior especially the types and combinations of the various available resistor and FET types need to be carefully chosen.

The ongoing work also indicates that
- The same concept can be used for reduction of the 3rd harmonic.
- IM3 reduction of >20dB can be achieved over >50% bandwidth per updated simulation shown in figure 9.
- The insertion loss adder for the predistorter can be reduced to <0.05dB.
- The insertion loss for the switch itself can be reduced to approx. 0.25dB.
- Performance can be maintained over a temperature range of -25…+85degC.
- The design is robust over process tolerances.

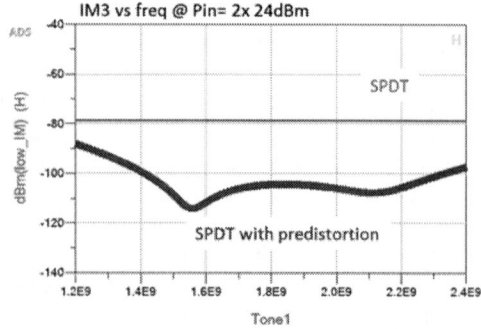

Fig. 9. Updated simulation results

VI. Conclusion

A novel predistortion circuit is shown to be able to reduce intermodulation products by approximately 20dB over a 40…50% bandwidth. Other RF parameters like isolation and insertion loss show little to no changes. The necessary die area is very small keeping the cost adder at a minimum. Robust performance over VSWR is shown. The principle can be applied to any process like bulk CMOS, SOI or pHEMT, as long as the switch function is FET based.

A demonstrator implemented on an 180nm SOI process shows results that are in line with expectation.

Specifically RF front ends of 4G or 5G user equipment may benefit from this new approach, but any other switch application requiring high linearity can get improvements on IMD performance as well.

References

[1] C. Didier, E. Desbonnets, "White Paper RF-SOI Wafer Characterization", Soitec, Mar. 2016.
[2] P. A. Rabbeni, A. Joseph, T. Letavic, A. Bandyopadhyay, "RF SOI Revolutionizing RF System Design", *Microwave Journal*, Oct. 2015.
[3] Infineon, data sheet BGS12PN10, July 2016.
[4] PSemi, data sheet PE42426, Sept. 2016
[5] D. J. Kelly, "Canceling harmonics in semiconductor RF switches", U.S. Patent 8 081 928, Dec. 20, 2011.
[6] T. McKay, P.R. Verma, S. Zhang, J.S. Wong, J. Brunner, "Switch Branch in Trap-Rich RFSOI with 84dBm OFF State IP3", 2015 IEEE SOI-3D-Subthreshold Microelectronics Technology Unified Conference (S3S), Oct. 2015
[7] Cebi et.al., "Switch Linearization by Non-Linear Compensation of a Field Effect Transistor", U.S. Patent 9059702, June 16, 2015.
[8] C. Fluhrer, "Elektronischer Hochfrequenz-Schalter", DE Patent DE103605361, July 2008
[9] R.Rimini, "Adaptive Non-linear Interference Cancellation for Intermodulation Distortion", U.S. Patent 9231801, Jan 2016

A Blocker-Tolerant Two-Stage Harmonic-Rejection RF Front-End

Faizan Ul Haq[#], Mikko Englund[*], Yury Antonov[#], Kari Stadius[#], Marko Kosunen[#], Kim B. Östman[$],
Kimmo Koli[*], Jussi Ryynänen[#]

[#]Dept. of Electronics and Nanoengineering, Aalto University, Finland
[*]Huawei Technologies Oy Finland Co. Ltd.
[$]Nordic Semiconductor Finland
[1]faizan.ulhaq@aalto.fi

Abstract — SAW-less wideband receivers need to operate linearly in the presence of strong out-of-band blockers. In this paper, we introduce a blocker tolerant harmonic rejection RF front-end which is able to suppress blockers present at the local oscillator harmonics. The suppression is achieved by applying harmonic rejection in two stages, such that the first harmonic rejection already occurs at the output of LNA. The proposed front-end achieves this harmonic rejection with simpler 6-phase LO clocking and reduced number of base-band signal paths compared to 8-phase HR architectures. Further, the proposed design does not require any precise gain coefficients and implementing the harmonic rejection in two stages makes it more mismatch tolerant. In addition, near-band blocker linearity is improved by implementing a third order base-band feedback response which acts in conjunction with N-path filtering. Implemented in a 28nm FDSOI process, the front-end demonstrate 18-37dB harmonic rejection from the first stage and around 46-53dB of harmonic rejection from the second stage with a state-of-the-art blocker compression point of 2.5dBm for a third harmonic blocker and a near-band blocker compression point of -6.5dBm.

Keywords — Harmonic rejection, blocker tolerance.

I. INTRODUCTION

The emergence of multiple radio access standards have arisen the demand for wideband receivers which should be able to operate on multiple frequency bands. Such a receiver will need to perform in presence of strong out-of-band (OB) blockers which potentially saturate the receiver. Traditionally, these OB blockers have been attenuated through off-chip surface acoustic wave (SAW) filters which are not suitable for wideband receiver scenario due to their non-tunability. What is needed is an on-chip tunable filtering alternative for blocker suppression.

One such widely implemented on-chip filtering technique is N-path filtering, which offers moderate Q factor with a wide tuning range [1]. However, one of the shortcomings of original N-path filtering concept is its inability to suppress blockers at harmonics of the local-oscillator (LO) frequency. Various works [2], [3], [4], [5] have proposed harmonic rejecting N-path filtering. However, in order to achieve higher harmonic rejection (HR), these techniques implement a higher number of paths together with the requirement of precise gain coefficients. Higher number of paths, inevitably lead to a complicated design with increased area and power consumption, while precise gain coefficients suffer from implementation inaccuracies. In addition, HR is usually implemented later in the receiver chain when gain has already been applied [3], [6]. Blocker at LO harmonics will thus already saturate the first stages before the HR is even takes place.

Some recent works [4], [5] have proposed HR at the output of first gain stage to improve receiver linearity for blockers at LO harmonics. Though these techniques achieve promising linearity improvements for blockers at LO harmonics, they still implement a higher number of N-paths with the requirement of implementing precise gain coefficients.

In this paper, we propose a two-stage harmonic rejection RF front-end which achieves HR at the output of the first gain stage. This helps to improve front-end linearity for blockers at LO harmonics. The proposed architecture is simple as it utilizes a six-phase LO clock, compared with other implemented works which use eight phases or more [4], [5]. In addition, the proposed architecture does not require any precisely tuned gain coefficients while at the same time implementing the HR in two stages makes it more mismatch tolerant. We also propose a third-order base-band (BB) response together with an N-path filtering, which helps to greatly suppress near-band blockers.

The organization of this paper is as follows: Section II details the proposed two stage HR front-end while Section III elaborate the circuit design and measured results. Section IV ends in conclusion.

II. PROPOSED ARCHITECTURE

The proposed two-stage harmonic-rejection front-end is presented in Fig. 1. The architecture implements three base-band downconversion paths each of which is 60° degrees phase shifted from the adjacent path. This decreases the number of BB paths by one as compared to 8-phase N-path downconversion [4], [5], [2]. From the three 60° degree phase shifted BB outputs, conventional I and Q outputs can easily be derived in the digital domain after an ADC.

The first stage of HR rejection is implemented by two parallel low-noise amplifiers (LNA) together with six differential passive mixers clocked by six non-overlapping LO phases. The need for two LNA's and six mixers arise to create HR at the LNA output without using overlapping LO clocks which would otherwise decrease the LNA gain. The LO clocking for the passive mixers is presented in Fig. 2 which can be altered to either the 6-phase HR LO clock (blue),

2019 IEEE Radio Frequency Integrated Circuits Symposium

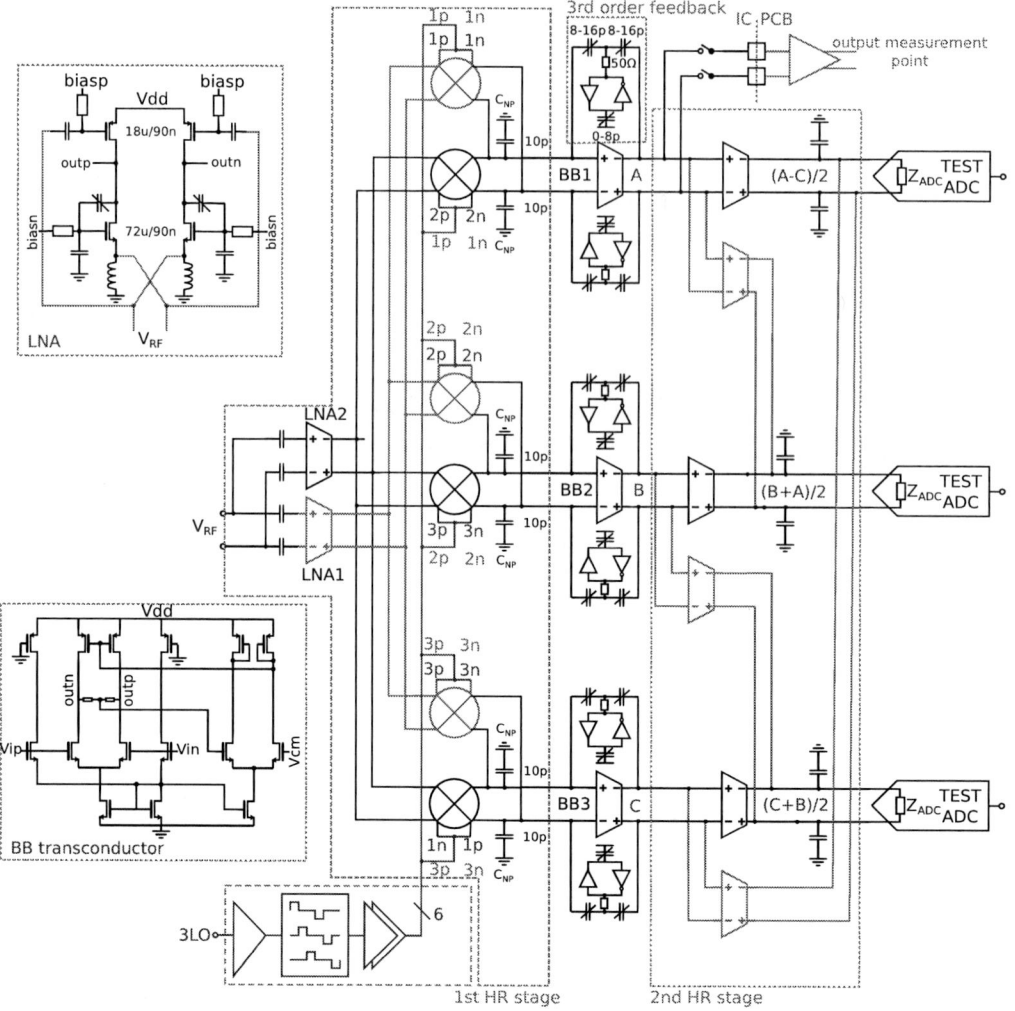

Fig. 1. Block diagram of proposed two stage harmonic rejection front-end.

or conventional 6-phase clock for lower noise contribution front-end (grey). Conventional LO clock is identical to having one equivalent LNA and three mixers while in the HR clocking arrangement, the functionality can be explained as follows: During LO phases 1p and 2p, LNA1 and LNA2 get alternatively connected to BB1. In phase 3p and 3n neither LNA1 nor LNA2 are connected while in phases 1n and 2n, LNA1 and LNA2 get connected to BB1 with opposite polarity creating an effective LO at the input of the BB1 as shown in Fig. 2. Fourier analysis of HR LO waveform reveals ideally no third harmonic and therefore, creates third harmonic rejection with lower number of clock phases. Further, as the HR occurs exactly where the gain is applied, i.e. at the LNA output, front-end linearity in presence of blockers at the 3rd LO harmonic also improves. However, complete elimination of third harmonic is not possible due to device mismatches and LO non-idealities. Therefore, to improve the HR further second HR stage is required.

The second stage makes the effective downconverted signal more closely resemble a sine wave. To do this, we implement cross-coupled BB stages between three BB paths as

shown in Fig. 1. This produces signals (A-C)/2, (B+A)/2 and (C+B)/2, where A, B and C are the signal outputs from the first BB stages of three parallel BB paths. The generated signals (A-C)/2, (B+A)/2 and (C+B)/2 resemble a sine wave more closely and simple Fourier analysis shows increased harmonic rejection.

In order to improve the near-band blocker attenuation we propose a third-order BB filtering stage as shown in Fig. 1. The response is implemented with a third-order feedback impedance, where the equivalent inductance has been implemented with gyrator transconductors for on-chip fabrication. The created response works in conjunction with N-path filtering to attain a sharp third-order band-pass response at LNA output, consequently, improving the near-band blocker linearity.

III. Circuit design and Measurements

The proposed front-end was fabricated in a 28nm FDSOI technology for an operating frequency of 0.7-2GHz. The front-end was part of complete receiver, with $\Delta\Sigma$ ADC and digital processing. The chip photograph of complete receiver

978-1-7281-1702-7/19 $31.00 © 2019 IEEE

2019 IEEE Radio Frequency Integrated Circuits Symposium

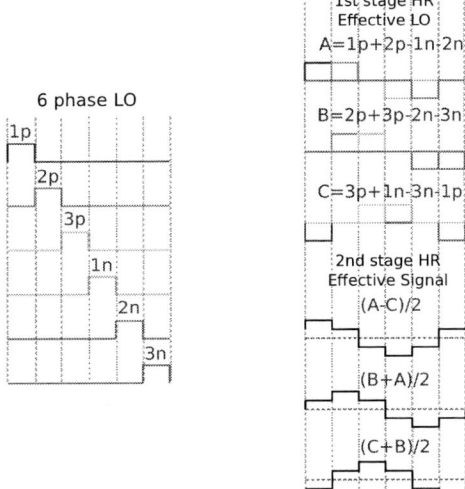

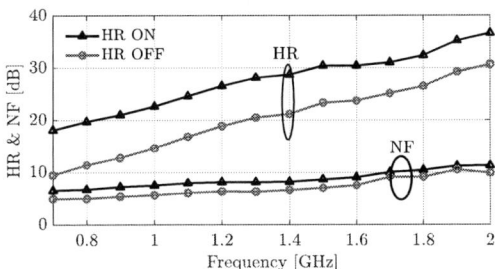

Fig. 2. Conceptual representation of proposed two stage HR with the generated effective LO at each HR stage.

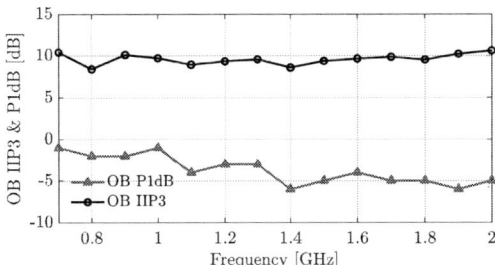

Fig. 3. NF and the first stage HR of the proposed front-end, for HR LO clocking (black) and conventional LO clocking (grey).

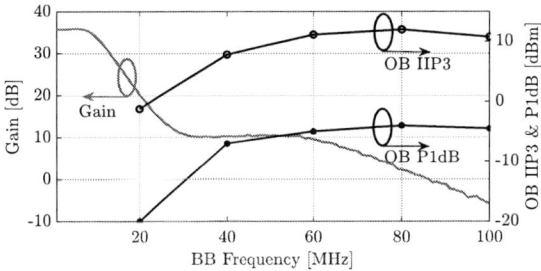

Fig. 4. BB filtering response, OB IIP3 and 1dB compression point vs. BB frequency for an f_{LO} = 1.5GHz. Third order BB filtering response gets upconverted to LNA output nodes with N-path filtering consequently helping to suppress near-band blockers.

Fig. 5. OB IIP3 and P1dB versus front-end operating frequency. The blocker is placed at f_{LO}+100MHz

IC is presented in Fig. 7, in which the front-end part together with LO generator occupies an active area of 0.65mm². In order to measure the performance of the first HR stage, a test signal from the output of the first BB amplifier was routed outside the IC. RF performance was then measured by buffering this signal with an external operational amplifier and observing response at output measurement point as shown in Fig. 1.

On the circuit level, the implemented LNA's consist of push-pull common-gate common-source amplifiers with capacitive feedback for impedance matching, while the passive mixers were implemented with large aspect ratio transistors to ensure small switch resistance. The BB integrators were implemented with a dynamically-biased differential pair where the transconductance of the first BB stage was designed to be eight times higher than transconductance of later stage for lower noise contribution. The capacitors C_{NP} provide additional blocker rejection at higher frequency offsets where the bandwidth of the first BB stage is not sufficient for a reasonable feedback response. For LO generation, an external signal at $3f_{LO}$ was provided to LO chain which was then divided by three to create six non-overlapping LO clocks with 16.7% dutycycle, and finally routed to passive mixers through inverter buffers.

Fig. 3 shows the measured NF and HR of the first stage across the operating frequency range. HR in the range of 18 to 37 dB is observed for HR LO clock. One of the consequence of the HR LO clock is noise circulation from one base-band path to other. This happens because a single LNA is connected to two base-band paths at all time instances, resulting in creation of the noise circulation path and a higher NF front-end. To validate the noise circulation, we have also implemented an LO clocking mode where each BB branch is only connected to one LNA at a time, thus eliminating the noise circulation path. With this LO arrangement around 3dB improvement in front-end NF is observed in measurements. The increase of NF at high frequencies was due to non-optimal layout design of LO chain which resulted in overlapping of LO signals at higher frequencies resulting in increased NF. A careful re-design of LO chain should reduce this increase of NF.

In Fig. 4, the third order BB filtering response, OB IIP3 and P1dB compression points are plotted versus BB frequency offset. One can observe third order filtering near the BB bandwidth of 10MHz. This third order response is upconverted to LNA output nodes with N-path filtering, consequently, helping to suppress near-band blockers. For example, the measured P1dB compression point for a near-band blocker located at 40MHz offset from f_{LO} is -6.5dBm which is just 1.5dB less than P1dB measured at a 100MHz offset. To observe the front-end linearity behavior across the operating frequency Fig. 5 shows the measured OB IIP3 and P1dB versus frequency for a blocker at 100MHz offset from LO.

978-1-7281-1702-7/19 $31.00 © 2019 IEEE   205

2019 IEEE Radio Frequency Integrated Circuits Symposium

Table 1. Performance summary and comparison

	This work	[4]	[5]	[6]	[7]
Carrier Frequency (GHz)	0.7-2	0.2-1	0.1-2.4	0.7-1.4	0.5-3
HR at first gain stage (dB)	18-37	20	52-54	NA	35
Second stage HR^2 (dB)	46-53	51-52	NA	40-67	NA
Architecture	LNA first	LNA first	Mixer first	LNA first	LNTA first
N-path phases	6-phase	8-phase	8-phase	8-phase	8-phase
Blocker input P1dB (dBm)	-1 to -5@10f$_{BW}$	-2.4@10f$_{BW}$	-6 to 2.5@27f$_{BW}$	-8.5 to 10@10f$_{BW}$	-
Blocker input P1dB@3f$_{LO}$ (dBm)	2.5@10f$_{BW}$	-2.8	-8	NR	NR
Near band blocker P1dB (dBm)	-6.5@4f$_{BW}$	NR	NR	NR	-22 to -4@4f$_{(LO}$
Noise Figure (dB)	5-11	5.4-6	1.7	1.5-8	3.8-4.7
OB IIP3 (dBm)	8-11@10f$_{BW}$	9@10f$_{BW}$	10	1-20.5@10f$_{BW}$	-20 to -4.8@4f$_{BW}$
Gain (dB)	36	36	NR	20.8-36.8	50
BB bandwidth (f$_{BW}$) (MHz)	10	2	0.2-3	10	10
Power (mW) @ Supply voltage (V)	54@1	26-32@1.2, 2.5	36.8-62.4@1	52@ 0.8, 1, 1.2	76-168@NR$^{(1}$
Process	28nm FDSOI	65nm	28nm	65nm	65nm

NR: not reported, NA: not applicable. 1) signal path 2) 1st and 2nd stage combined HR.

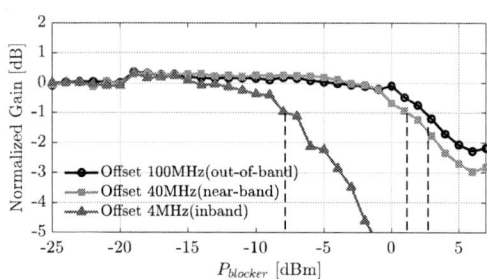

Fig. 6. Measured signal gain under the presence of strong blocker at 3f$_{LO}$. For a blocker at 3f$_{LO}$+100MHz, blocker P1dB of 2.5dBm is observed.

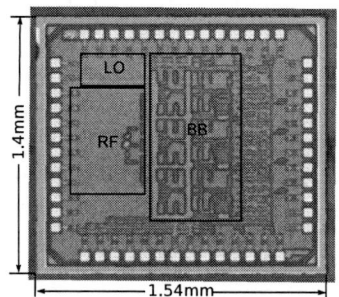

Fig. 7. Chip photograph of the proposed harmonic rejection receiver.

For observing front-end linearity in the presence of blockers at 3f$_{LO}$, linearity was measured with three offset frequencies, 100MHz, 40MHz and 4MHz. These three offsets represents downconversion of the blocker to out-of-band, near-band, and in-band reception frequencies. The measurement was performed by sweeping the blocker power and determining the P1dB point of the receiver. We observe P1dB points of -7, 1 and 2.5 respectively as shown in Fig. 6 achieving state-of-the-art results [4], [5].

IV. CONCLUSION

Blocker tolerance is one of the key specifications for SAW-less wideband receivers. In this paper, we have proposed a blocker tolerant harmonic rejection front-end which is able to efficiently suppress blockers present at LO harmonics together with suppression of near-band blockers. This is achieved by implementing HR in two stages: Output of the LNA and in the second BB stage. Further, a third order BB impedance is implemented which acts in conjunction with N-path filtering for near-band blocker attenuation. The proposed architecture uses a lower number of LO phases and BB paths compared to 8-phase HR architectures and achieves HR without implementation of precise gain coefficients. Measured results from a fabricated prototype implemented in 28nm FDSOI technology demonstrate 18-37dB HR from the first stage and around 46-53dB of HR from second stage with a near-band blocker compression point of -6.5dBm and an impressive BCP of 2.5dBm for third harmonic blocker.

ACKNOWLEDGMENT

This work has been supported by the Academy of Finland, and European Unions Horizon 2020 Research and Innovation Programme under the Marie Sklodowska-Curie Grant 704947.

REFERENCES

[1] A. Ghaffari, E. A. M. Klumperink, M. C. M. Soer, and B. Nauta, "Tunable high-Q N-path band-pass filters: Modeling and verification," *IEEE Journal of Solid-State Circuits*, vol. 46, no. 5, pp. 998–1010, May 2011.

[2] X. Zhang, B. Chi, and Z. Wang, "A 0.11.5 GHz harmonic rejection receiver front-end with phase ambiguity correction, vector gain calibration and blocker-resilient TIA," *IEEE Transactions on Circuits and Systems I: Regular Papers*, vol. 62, no. 4, pp. 1005–1014, April 2015.

[3] A. Nejdel, M. Törmänen, and H. Sjöland, "A 0.7 3.7 GHz six phase receiver front-end with third order harmonic rejection," in *2013 Proceedings of the ESSCIRC (ESSCIRC)*, Sep. 2013, pp. 279–282.

[4] Y. Xu, J. Zhu, and P. R. Kinget, "A blocker-tolerant RF front end with harmonic-rejecting N-path filter," *IEEE Journal of Solid-State Circuits*, vol. 53, no. 2, pp. 327–339, Feb 2018.

[5] D. Murphy, H. Darabi, and H. Xu, "A noise-cancelling receiver resilient to large harmonic blockers," *IEEE Journal of Solid-State Circuits*, vol. 50, no. 6, pp. 1336–1350, June 2015.

[6] J. Musayev and A. Liscidini, "Quantized analog RX front-end for SAW-less applications," in *ESSCIRC 2018 - IEEE 44th European Solid State Circuits Conference (ESSCIRC)*, Sep. 2018, pp. 306–309.

[7] R. Chen and H. Hashemi, "Reconfigurable receiver with radio-frequency current-mode complex signal processing supporting carrier aggregation," *IEEE Journal of Solid-State Circuits*, vol. 50, no. 12, pp. 3032–3046, Dec 2015.

A Low Noise Figure 28GHz LNA in 22nm FDSOI Technology

Chi Zhang[*1], Frank Zhang[*], Shafiullah Syed[*], Michael Otto[#], Abdellatif Bellaouar[*]

[*]GLOBALFOUNDRIES, USA
[#]GLOBALFOUNDRIES, Germany
[1]Chi.Zhang@globalfoundries.com

Abstract—This paper presents a 28GHz low noise amplifier (LNA) implemented in 22nm FDSOI technology. The LNA is based on inductively degenerated common source topology with cascode device. With several special layout techniques, the LNA achieved best in class noise figure (NF). At 28GHz, the LNA has a gain of 12dB, input referred third-order intercept point (IIP3) of 3.0dBm and input reference 1-dB compression point (IP1dB) of -7.6dBm. The 1-dB and 3-dB bandwidth of the LNA is 8.5GHz and 15.1GHz, respectively. The lowest achieved NF is 1.46dB at 24GHz with power dissipation (P_{DC}) of 9.8mW. Another LNA with larger device width was also implemented to achieve better NF. For this LNA, the lowest achieved NF is 1.35dB at 24GHz with P_{DC} of 13.0mW. It could also achieve 1.8dB NF at 28GHz at 5.0mW P_{DC}.

Keywords—low noise amplifier, FDSOI, mm-Wave, layout techniques, gate resistance

I. INTRODUCTION

With the trend toward 5G wireless communications, high performance and mm-Wave low noise amplifiers (LNAs) are in high demand. Traditionally, these high performance LNAs are built with non-silicon processes such as GaAs [1], GaN [2], SiGe, etc. However, these processes are normally expensive and their low integration level limits possibility of integrating RF front end (LNAs and PAs) with RF transceiver (Mixers, LO, ADC, etc.). Moreover, the massive MIMO architecture of mm-Wave 5G system requires very low power dissipation. Achieving the optimal balance among Performance, Power and Area (PPA) becomes more and more critical for circuit architecture decisions in 5G applications.

The 22nm FDSOI process offers high performance mm-Wave devices in terms of F_t, F_{max} and NF_{min} [3]. And its high integration level is perfect for RF front-end components integration. This paper presents a LNA design using 22nm FDSOI process achieving high performance, low power at the same time.

The paper is organized as follows. Section II presents circuit design methodology and layout techniques. Experimental results and comparison to LNAs designed in other processes are discussed in Section III. Conclusion of this paper is summarized in Section IV.

II. CIRCUIT DESIGN

The basic cascoded common source with inductive degeneration topology was used in the 28GHz LNA design. Schematic view of the design is shown in Fig.1. The die photo of the prototype LNA is shown in Fig.2. The device sizes are the same for the common source device (N_0) and the cascode device (N_1). Off-chip bias-T was used to provide the gate bias

for common source device (V_{G0}). Cascode device gate (V_{G1}) was biased through a pad with DC probe. All passive components shown in the schematic were designed on-chip. Both input matching inductor L_G and output matching inductor L_D were 1.5 turn inductors in thick copper metal layer. Degeneration inductor L_S was realized by transmission line with ground shield. To route the output signal to the GSG signal pad for testing, another transmission line with ground shield (L_O) was used. Cascode gate decoupling capacitance (C_{G1}) and output match capacitance (C_O) were realized by metal-oxide-metal (MOM) capacitors. The design values for all passive components are shown in Fig. 1.

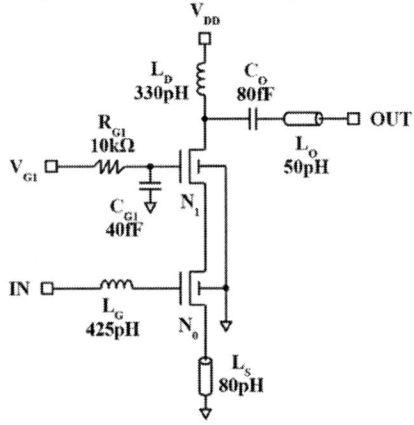

Fig. 1 Schematic view of the 28GHz LNA

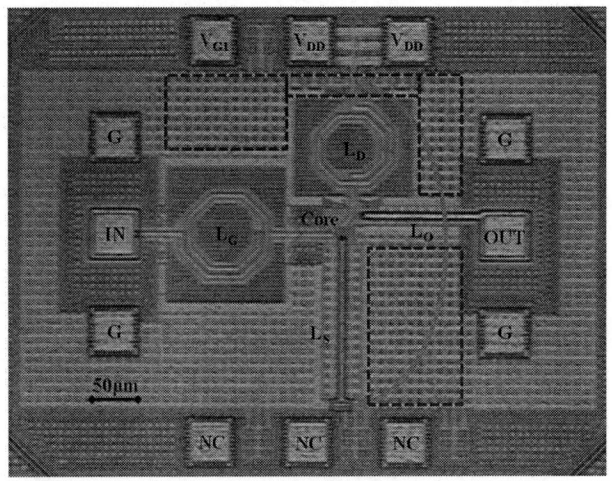

Fig. 2 Die photo of the 28GHz LNA

A. Device Sizing and Operating Point

Multiple parametric sweep simulations were conducted on N_0 device to find out the optimal bias point and device size for noise figure consideration. For a given device size, lowest NF_{min} could be achieved at around 0.4V gate bias (V_{G0}) which is 0.2V above V_t. Therefore, 0.4V was chosen as the nominal gate bias for N_0. Total device width of N_0 was determined based on trade-offs among input power matching (S11), noise matching (G_{opt}), and power dissipation (I_{DD}).

In this design, the optimal V_{G0} bias for common source gm linearity was also around 0.4V. Linearity performances including IIP3 and IP1dB were gds limited. Therefore, V_{DS} headroom was critical to achieve better IIP3 and IP1dB. It became a trade-off between linearity and power dissipation (V_{DD}). In this design, V_{DS} for both N_0 and N_1 were biased at 0.65V (V_{DD}=1.3V) to achieve good IIP3 and reasonable power dissipation. Gate bias for N_1 (V_{G1}) was then determined by V_{DS} and I_{DD}.

B. Device Layout Techniques

Series gate resistance (R_G) of the common source device is one of the major contributors of noise. Several layout techniques were used in order to reduce R_G of the device. First, the optimal finger width of the device is about 300~500nm. This is due to the vertical and horizontal gate resistance trade-off. In this design, 500nm finger width was used to achieve low R_G as well as lower number of fingers compared to 300nm finger width.

Second, this design used multiple devices with smaller number of fingers instead of single device with huge number of fingers. In most advanced processes, the lowest metal layer (M1) is used to collect multiple fingers (gates). Large number of finger results in higher gate resistance due to longer M1 routing. In this design, the number of finger was set to 10 for each device. Number of devices (multiplicity or M) is 12. Total width is 60μm.

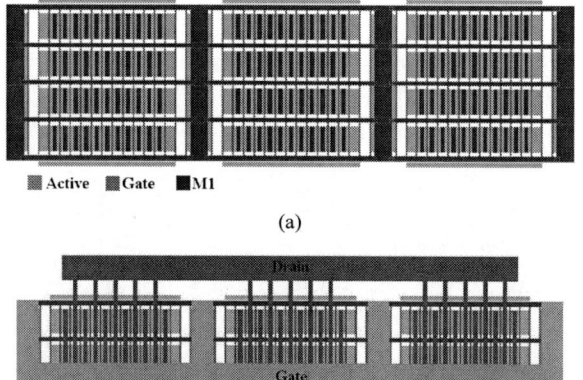

■ Active ■ Gate ■ M1

(a)

■ Active ■ Gate ■ M1 ■ Middle Level Metal ■ Upper Metal

(b)

Fig. 3 Layout views of LNA common source device: (a) up to M1 level, (b) up to upper metal level

Third, a vertical gate sharing layout style was used. The 12 devices were constructed into a 4x3 (4 rows, 3 columns) array. The 4 devices in each column would share the gate connections (M1) in between, as shown in Fig.3a. M1 gate connections were extended horizontally to both sides of the column and picked up by full metal/via stack up to 1-um thick upper metal layer which is running across the entire array and connecting to the L_G inductor directly (Fig.3b). Drains/sources of the devices in the same column were picked up by middle level metal/via stack vertically then connected horizontally using upper metal layer outside the array. R_G was significantly reduced by sharing the gate connection between rows and picking up the gate connection both vertically and horizontally using upper thick metals. Moreover, parasitic C_{gs} and C_{gd} capacitances were reduced by moving the common drain and source connections away from the gate connection thanks to the extra "height" provided by the array configuration. Because R_G of the cascode device was not a major contributor to noise, the gate connection for N_1 used only middle level metal. This would further reduce the C_{gs} and C_{gd} parasitic capacitance for the cascode device.

Forth, 2x CPP (double the standard gate-to-gate pitch) layout style was used to further reduce C_{gs} and C_{gd}. Wider gate pitch allows more space between drain/source metal/via stack to gate fingers which reduced parasitic capacitance.

C. Passive Components

This design used 10 layers metal stack with two ultra-thick copper layers. This BEOL option allows the passive components (mainly inductors) to achieve very good quality factor (Q). At 28GHz, the input matching inductor L_G is 425pH differential, with Q of 23; the output matching inductor L_D is 330pH single-ended, with Q of 18; both source degeneration inductor L_S (80pH, Q=14) and output routing inductor L_O (50pH, Q=12) were realized by ground-shielded transmission lines. MOM capacitor was used as the output matching capacitor. Only upper thin metal layers were used in the MOM capacitors to reduce terminal to substrate parasitic capacitance. It is worth noting that L_G is connected as a series component in the circuit. Both differential and single-ended L and Q model cannot accurately model the shunt capacitance seen from the two terminals of L_G. A full S-parameter model with ground reference or a π model should be used for more accurate simulation results.

D. EM and Extracted View Co-simulations

All passive components were simulated with ground plane around using both EMX and Peakview. All interconnection between inductors and devices were absorbed as part of the inductors. The LNA core was extracted as a whole block, including N_0, N_1, C_{G1} and bias resistor R_{G1}. A full EM simulation was conducted including all inductors, transmission lines, ground planes and probe pads. This resulted in a 17-port S-parameter file. Top level simulation was done using this EM file with the LNA core extracted view. The simulation would include the effects such as ground inductance, coupling between inductors, pad capacitance, etc. It is also a good way to check stability of the circuit.

978-1-7281-1702-7/19 $31.00 © 2019 IEEE

E. Ground Plane and Decoupling Capacitors

Ground plane was constructed in a tiled fashion including all upper level metals which were slotted to meet maximum density rules. Supply decoupling capacitor was also built by putting MOM capacitors underneath the ground tile. In mm-Wave design, it is critical to have a very solid common ground plane with very low inductance, especially for single-ended circuit design. In inductively degenerated common source LNA, ground inductance can be counted as part of the L_S. In 28GHz application, L_S value is quite low which makes it more vulnerable to parasitic inductance. In this design, L_G and L_O traces almost cut the ground plane into two halves, which might result in higher ground return inductance. V_{DD} decoupling capacitors were distributed across the whole chip for both top and bottom half of the layout (Fig. 2). AC ground return path from L_S ground termination to the V_{DD} supply pad was significantly reduced. Ground inductance was minimized. Therefore accurate L_S value could be achieved. This is essential for LNA input matching and stability.

Decoupling capacitor for cascode device gate (C_{G1}) was carefully designed to ensure circuit stability. For low frequency design, AC ground (large C_{G1}) is desired at this node which is not true at mm-Wave frequency. K factor (K_f) at higher frequency (tens of GHz) became a strong function of C_{G1}. In general, higher the C_{G1}, lower the K_f. Small C_{G1} is preferred for stability. However, lower C_{G1} reduces effective gm of the cascode device. LNA gain drops and NF increases. This is an important design trade-off. Additionally, any series inductance presented with C_{G1} would create a LC resonant frequency which would significantly affect the stability of the circuit in forms of S22 peaking. The interconnections from C_{G1} top plate to the N_1 gate and from C_{G1} bottom plate to ground plane were very carefully laid out to minimize series inductance.

III. Experimental Results

The prototype LNA (LNA1) was fabricated using GLOBALFOUNDRIES 22nm FDSOI process. Silicon area with and without pads is 0.21mm^2 and 0.12mm^2, respectively. The LNA was measured using PNA-X 2-port system with on wafer probing.

Measured S-parameters on multiple sites are shown in Fig. 4. Simulated S-parameters are also shown for comparison. Excellent model-to-hardware correlation (MHC) was achieved without any model optimization or back-fitting. The gain (S21) of LNA1 is measured to be 12.0dB at 28GHz. Both input and output return losses (S11 and S22) are below -10dB at 28GHz. 1-dB and 3-dB bandwidth is 8.5GHz and 15.1 GHz, respectively. The power dissipation from 1.3V V_{DD} is 9.8mW.

Measured IIP3 and IP1dB on multiple sites at different frequencies are shown in Fig.5. IIP3 was measured based on 2-tone test with 100MHz tone spacing and extrapolated at -23dBm input power for each tone. At this power level, the IM3 curve showed perfect 3dB/dB slope. At 28GHz, the LNA achieved IIP3 of 3.0dBm and IP1dB of -7.6dBm. The difference between measured IIP3 and IP1dB is about 10dB.

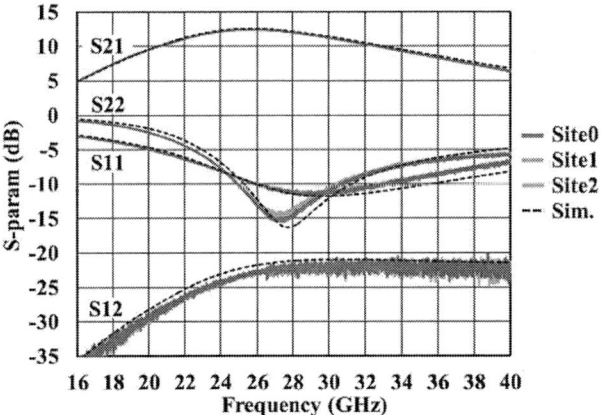

Fig. 4 Measured and simulated S-parameters of LNA1

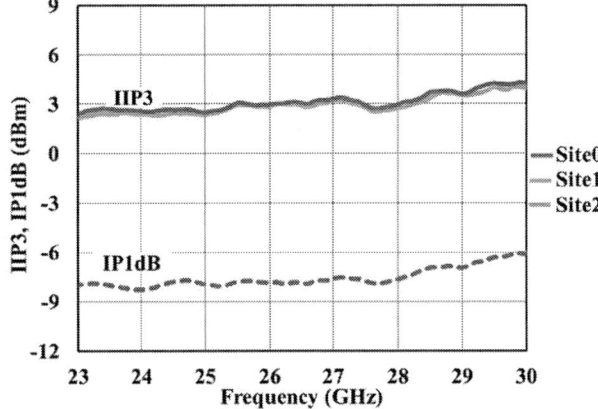

Fig. 5 Measured IIP3 and IP1dB of LNA1

In order to further improve NF of the LNA, another version was also fabricated with larger device size. The second version (LNA2) has total width of 80μm with M=16. Input series inductance L_G was modified to re-match for bigger device size. The DC bias condition was kept the same as the LNA1. Measured NF results of the two LNAs on different sites were shown in Fig. 6. The lowest achieved NF were 1.46dB and 1.35dB for LNA1 and LNA2, respectively, both at around 24GHz. The power dissipation of LNA2 from 1.3V V_{DD} is 13.0mW. It shows a clear trade-off between P_{DC} and noise performance. At 28GHz, the LNA2 achieved IIP3 of 1.4dBm and IP1dB of -7.9dBm, which were slightly lower compared to LNA1.

To investigate the NF and P_{DC} trade-off, the common source gate voltage (V_{G0}) was decreased for LNA2 to reduce the power dissipation. V_{DD} level was kept at 1.3V for consistent IIP3 and IP1dB performance. The LNA2 NF at 26GHz and 28GHz at different P_{DC} are shown in Fig. 7. NF level of 1.8dB could be achieved at P_{DC} level as low as 5mW. Further power saving can be achieved by reducing V_{DD}. However, it will degrade the IIP3 and IP1dB performance. This design is suitable for low power application where NF requirement is relaxed. Beam-forming is one of the good examples where power dissipation is more of a concern.

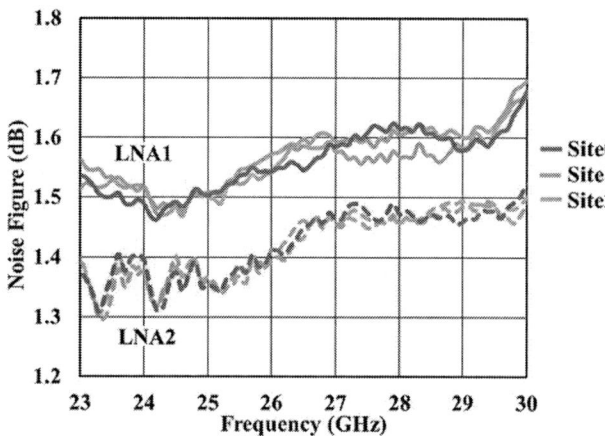

Fig. 6 Measured NF of LNA1 and LNA2

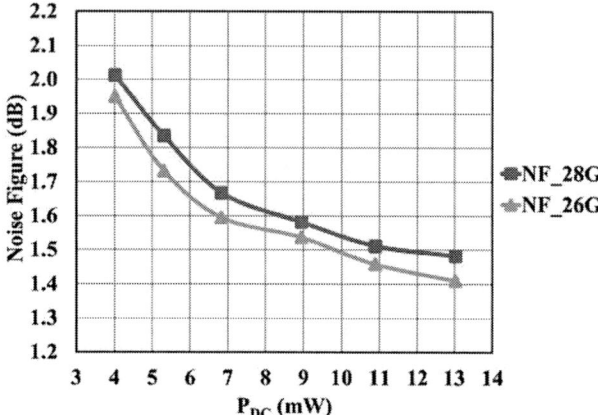

Fig. 7 Measured NF of LNA2 at different PDC

Table 1 compares the key performance parameters of this work with other Ka-band LNAs designed in different technologies. This work shows comparable noise and linearity performance with the lowest power dissipation, highest 3-dB bandwidth and very small chip area. It is also worth noting that 22nm FDSOI is using conventional substrate resistivity under the thick oxide layer (BOX) of 12 Ohm-cm. Considering the trade-off among noise, linearity and power dissipation, this work demonstrates the advantage of 22nm FDSOI process in 5G design space.

IV. CONCLUSIONS

22nm FDSOI technology offers devices with superior noise performance and also provides excellent BEOL metal stack for high Q passive components. These are much desired factors in mm-Wave LNA design for low NF. Layout optimizations are also critical to achieve low NF and to fully realize the potential of the process. Measurement on prototype LNAs showed promising results in NF, linearity and power dissipation which verified process capability. Measured results on different sites also showed excellent model-to-hardware correlation and small process variations. As a conclusion, 22nm FDSOI process is very suitable for 28GHz LNA design for 5G applications.

REFERENCES

[1] D. P. Nguyen, B. L. Pham, T. Pham and A. Pham, "A 14–31 GHz 1.25 dB NF enhancement mode GaAs pHEMT low noise amplifier," *2017 IEEE MTT-S International Microwave Symposium (IMS)*, Honolulu, HI, 2017, pp. 1961-1964.

[2] X. Tong, S. Zhang, P. Zheng, J. Xu and X. Shi, "18–31 GHz GaN MMIC LNA using a 0.1 um T-gate HEMT process," *2018 22nd International Microwave and Radar Conference (MIKON)*, Poznan, 2018, pp. 500-503.

[3] S. N. Ong *et al.*, "A 22nm FDSOI Technology Optimized for RF/mmWave Applications," *2018 IEEE Radio Frequency Integrated Circuits Symposium (RFIC)*, Philadelphia, PA, 2018, pp. 72-75.

[4] C. Li, O. El-Aassar, A. Kumar, M. Boenke and G. M. Rebeiz, "LNA Design with CMOS SOI Process-l.4dB NF K/Ka band LNA," *2018 IEEE MTT-S International Microwave Symposium (IMS)*, Philadelphia, PA, 2018, pp. 1484-1486.

[5] M. Elkholy, S. Shakib, J. Dunworth, V. Aparin and K. Entesari, "A Wideband Variable Gain LNA With High OIP3 for 5G Using 40-nm Bulk CMOS," in *IEEE Microwave and Wireless Components Letters*, vol. 28, no. 1, pp. 64-66, Jan. 2018.

Table 1. Comparison of this work to Ka-band LNAs designed in different processes

	This work	This work	[1]	[2]	[4]	[5]
Tech.	22nm FD-SOI	22nm FD-SOI	0.15μ GaAs	0.1μ GaN	45nm SOI^	40nm Bulk&
Gain (dB)	12	12.6	30	25	12.8	18.4
BW3dB (GHz)	15.1	16.7	12	13	11	9.3
Freq. (GHz)	19-34	23-40	18-30	18-31	20-31	25-34
NF (dB)	1.46	1.35	1.25	1.27	1.4	3.4
IIP3 (dBm)	3	1.4	-1.5	NA	5	-5.1
IP1dB (dBm)	-7.6	-7.9	-11.5	-7	NA	-13.4
PDC (mW)	9.8	13	212	280	15	21.5
Area (mm²)	0.21	0.21	1.87	2.3	0.3	0.26+

^ Using high-resistivity substrate
& Variable gain LNA, numbers are at lowest gain setting
+ Chip area without pads

2019 IEEE Radio Frequency Integrated Circuits Symposium

A 1.7-dB Minimum NF, 22-32 GHz Low-Noise Feedback Amplifier with Multistage Noise Matching in 22-nm SOI-CMOS

Bolun Cui[*], John R. Long[*], and David L. Harame[#]

[*]RF/MMIC Group, ECE Department, University of Waterloo, Canada
[#]GLOBALFOUNDRIES, 01109 Dresden, Germany

Abstract — A transformer-feedback low-noise amplifier (LNA) implemented in 22-nm SOI-CMOS with interstage noise matching is described. The LNA peak gain is 21.5dB at 22GHz, with a -3dB bandwidth (BW) of 19-36GHz. Minimum noise figure (NF) is 1.7dB centered at 28GHz, and remains below 2.2dB across 10GHz. Third-order input intercept (IIP₃) is -13.4dBm at peak gain when dissipating 17.3mW. Input and output return losses are >10dB across 22-32GHz (effective BW). Modulation of the FET backgate voltage increases NF by <0.5dB, while reducing power consumption to just 5.6mW.

Keywords — Low-noise amplifier (LNA), lossless RF feedback, mm-wave transformer, broadband, FD-SOI CMOS.

I. INTRODUCTION

Improvements in the bandwidth, noise figure, and power consumption of silicon integrated mm-wave receiver front-ends enables the development of advanced wireless communication systems. Low-noise amplifiers (LNAs) implemented in production bulk-CMOS technologies typically offer 3GHz bandwidth, 3dB minimum noise figure (NF), and an input intercept (IIP₃) between -15 and -20dBm in the 28GHz band [1]. The feedback LNA developed in this work benchmarks the performance of a 22-nm FD-SOI CMOS technology (GF-22FDX [2]) in the same band. We target sub-2dB NF, better than 10-dB input return loss across >10-GHz bandwidth, 20-dB peak gain, and power consumption well below 20mW, while maintaining linearity comparable to circuits implemented in bulk-CMOS.

The paper is organized as follows. Design of the input and output stages, and interstage interfacing of the proposed LNA are described in Section II. Modeling, implementation, and tuning of the interstage transformer for minimum noise figure are also detailed. Measurement results are presented in Section III, and concluding comments are drawn in Section IV.

II. LNA DESIGN

A schematic of the proposed 2-stage LNA is shown in Fig. 1. The first stage consists of a common-source amplifier with source-gate feedback via transformer T_1. Both the input and optimum source impedances for minimum noise figure seen at the gate of M_1 are modified by negative feedback to realize impedance and noise matching simultaneously. Transformer T_2 interfaces the two stages, and it is designed to minimize the noise added by M_2. It also improves the gain by stepping-up the voltage between the first and second stages. The frequency response of the cascode second-stage is extended using inductive peaking (L_{1-4}). Resistor R_F biases

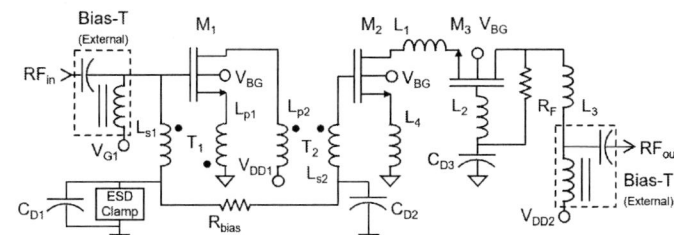

Fig. 1. Schematic of the 2-stage, wideband LNA prototype

M_3 and trims the LNA output impedance to match a 50-Ω load. The backgate voltage of the fully-depleted (FD-SOI) transistors (V_{BG} in Fig. 1) is used to adjust the LNA power consumption between low-power (LP) and low-noise (LN) modes of operation. On-chip metal-metal capacitors (C_{D1-3}) decouple key bias points, and a bidirectional ESD clamp at the cold side of T_1 protects the LNA input from electrostatic discharge. Bias-Ts facilitate characterization with a 2-port vector network analyzer (VNA).

A. Input Stage

Common-source transistor M_1 is biased at the optimum bias current for minimum noise figure (i.e., NF_{min1}). A unit-transistor size of 1μm×20nm is selected to reduce noise contributed by its extrinsic gate resistance. M_1 is comprised of 80 gate fingers in total, and the real part of its optimum source resistance for minimum noise figure ($R_{n,opt}$) is 50Ω. With V_{G1} and V_{BG} set to 0.25V and 2V, respectively, M_1 operates at 0.15μA/μm of gate width, and I_{DS1} is 12mA. V_{G1} is fed to the gate via an external bias-T for testing due to the limited number of testpads available.

Negative feedback makes the LNA input impedance dependent on transconductance gm_1 of M_1 and the transformer turns ratio n/k [3]. The first-stage bandwidth (BW) is determined by the coupling factor k_1 of T_1 [4]. Planar windings, consisting of a 1-turn primary and a 2-turn secondary coil, are implemented in second-metal copper (3-μm wide, 3.3-μm thick). The gap between turns is uniform at 2μm. The resulting transformer k_1 is 0.55 and it occupies 66×66μm² of chip area. Primary (L_{p1}) and secondary (L_{s1}) self-inductances are 90pH and 250pH, respectively, giving an electrical turns ratio $n/k = 3$. Reduced metal fill minimizes parasitic capacitance and extends the transformer's high-frequency response. The shunt parasitic capacitance of the RF input pad resonates with L_{s1} to reduce insertion loss.

978-1-7281-1702-7/19 $31.00 © 2019 IEEE

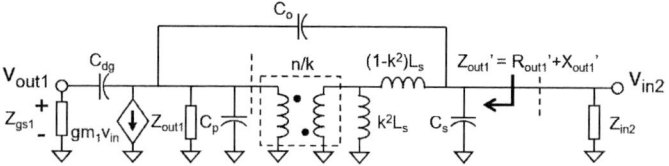

Fig. 2. Interstage coupling transformer model with leakage shifted to the secondary ($Z_{out1}=R_{out1}+jX_{out1}$)

The transformer secondary winding is also a low-impedance path to ground for static discharge (ESD) events. Large-signal simulations for the human-body model (HBM) predict that input voltages ranging from -3kV to +600V can be safely handled at the LNA input.

The first stage consumes 5mW from a 0.42-V supply (V_{DD1}). Backgate voltage V_{BG} may be adjusted to trade-off gain and NF for lower power consumption in the first stage.

B. Interstage Noise Matching

The LNA developed in this work utilizes interstage transformer coupling to set the second-stage noise factor F_2 equal to F_{min2} (i.e., noise matching). In this way, SNR degradation by the second stage is minimized. Other LNA designs have maximized power gain by impedance matching stages, e.g., [5]-[7].

Consider the lumped-element transformer model of Fig. 2. The turns ratio (n/k) of T_2 may be used to synthesize $Z_{n,opt2}$ at the second stage input to yield F_{min2} from transistor M_2. Assuming low insertion loss, the transformer also provides an opportunity for passive voltage gain between stages due to mismatch between the real part of the first stage's output impedance (R_{out1}), and the real part of the input impedance to the second stage (i.e., $R_{n,opt2}$), which is larger.

An additional inductor is needed to realize the reactance for noise matching (i.e., $X_{n,opt2}$), as FET drain reactance X_{out1} is capacitive, and the reactance required to realize F_{min2} is inductive. Leakage inductance $(1-k^2)L_s$, provides most of the inductance required to realize $X_{n,opt2}$. The impedance seen looking into the secondary output should be $Z_{out1}'=R_{n,opt2}+jX_{n,opt2}$ across the desired bandwidth, but there is a limitation imposed by the transformer's frequency response.

Fig. 3a shows the top-copper layers used to realize the interstage transformer, and the physical layout is shown in Fig. 3b. Similar to T_1, the 3-μm wide, 3.3-μm thick second-metal copper windings with 2-μm gap between turns realizes a compact layout (i.e., $71\times71\mu m^2$). The primary and secondary self-inductances are 170pH and 330pH, respectively, yielding an electrical turns ratio of 1:2.5 and a k-factor of 0.56. The inductance of the interconnect in the physical layout and the transformer leakage inductance create the reactance required to realize noise matching at the second stage input. The drain supply voltage of M_1 is fed via the transformer primary, and the parasitic interconnect inductance of the supply lines are absorbed in the transformer tuning. Shunt tuning capacitances across the primary (C_p) and secondary (C_s) are 127fF and 93fF,

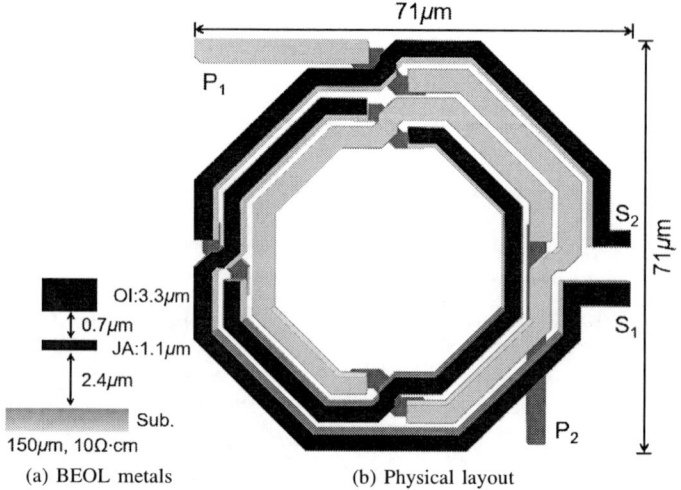

Fig. 3. (a) BEOL Stack-11 top copper layers, (b) Physical layout of the interstage transformer (primary: P_1/P_2, secondary: S_1/S_2 are both in OI with JA underpass)

respectively. They are comprised of parasitic capacitances at the drain of M_1 and gate of M_2, together with shunt parasitics from the transformer. Capacitive tuning of the transformer's self-inductance synthesizes the desired impedance seen at the second stage input ($Z_{n,opt2}$).

C. Output Stage

Gate areas for M_2 and M_3 are chosen to realize the desired in-band gain (20dB) and NF (<2dB). Increasing the width of M_2 lowers $R_{n,opt2}$ and the turns ratio of the interstage transformer, which leads to lower voltage gain from the interstage mismatch. Simulations predict that gate area optimization for the second stage yields only minor performance gains for NF and gain. Therefore, uniform device sizes are used (80μm, 1μm×20nm/finger) to simplify the layout and performance verification for the 22-nm prototype. All transistors are biased for lowest noise factor. The gates of M_2 and M_1 are set to the same voltage (i.e., 0.25V) via isolation network R_{bias} and C_{D2}.

Inductors L_{1-3} (165, 150, 270pH, respectively) peak the second-stage frequency response. Shunt feedback around M_3 forces the real part of the LNA output impedance to 50Ω when R_1 is 145Ω. The reactive portion is nulled between 24 and 32GHz by inductance L_3 and the output bondpad capacitance.

A 1.05-V supply for the output stage (V_{DD2}) maximizes headroom and linearity from the common-gate amplifier. Power consumption of this stage is 12mW. The bias-T at the output facilitates characterization with a 2-port network analyzer (VNA). A folded-cascode output using a PMOS device for M_3 is a potential alernative when interfacing the LNA to another stage on a chip. Power consumption and linearity trade-offs are also possible with a folded cascode, as the load impedance on-chip could be set larger than 50Ω.

2019 IEEE Radio Frequency Integrated Circuits Symposium

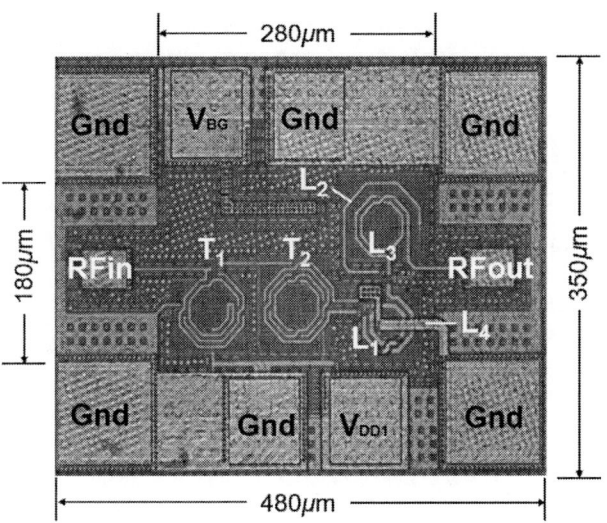

Fig. 4. Die micrograph of the LNA prototype

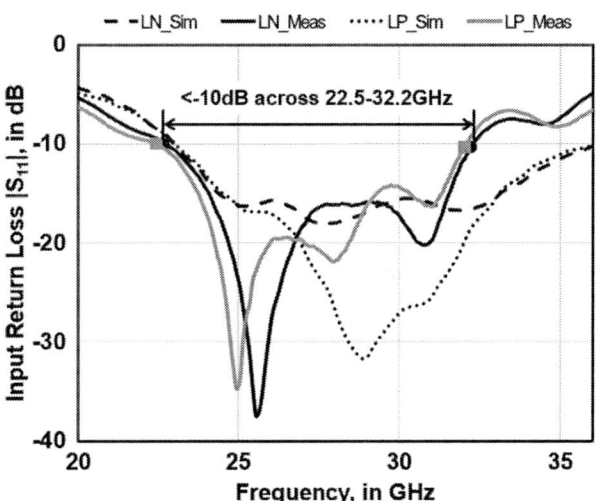

Fig. 6. Measured and simulated input return loss

D. Stability

Circuit stability must be considered in feedback amplifier design, especially at out-of-band frequencies where transistor gains may be relatively large. The secondary winding of T_1 is a low-impedance path to ground for frequencies below the passband. Therefore, the LNA does not require additional damping at the input to ensure stability (which may degrade gain and NF). The high-gain second stage is stabilized by source degeneration inductor L_4 (100pH). The μ-stability test predicts that the LNA is unconditionally stable, with $|\mu| > 1$ across the entire frequency range from simulations.

III. MEASUREMENT RESULTS

A micrograph of the LNA prototype fabricated in Global Foundries 22FDX® process is shown in Fig. 4. The total and active chip areas are 0.17mm² and 0.05mm², respectively. The S-parameters, NF, and third-order intercept point (IP₃) of multiple LNA samples were measured on-wafer using a wideband VNA.

The backend of 22FDX (BEOL, Stack-11) is comprised of 10 interconnect metals and optimized for RF/mm-wave applications with 3 thick layers (2 copper, 1 aluminum) at the top of the stack. A quality factor of 17 is achieved for inductors L_{1-4}. Control of the FD-SOI transistor backgate enables trade-offs on the fly between RF performance (low-noise mode) and power consumption (low-power mode) for the LNA.

Fig. 5 shows measured and simulated small-signal forward ($|S_{21}|$) and reverse ($|S_{12}|$) gains. In low-noise mode (LN), V_{BG} is set to 2V and the peak gain is 21.5dB at 22GHz. The measured -3dB BW is 19 to 36.2GHz (i.e., 17.2GHz of BW), giving a gain-bandwidth product (GBW) of 174GHz. The DC power consumption in LN mode is 17.3mW. When V_{BG} is reduced to 0.62V (i.e., LP mode), the power consumption falls to 5.6mW. Gain decreases by 3.6dB to a peak of 17.9dB at 22.5GHz, and -3dB BW is 19.6 to 35.6GHz (GBW of 113GHz). Very good agreement between measurement and simulation is seen for both modes.

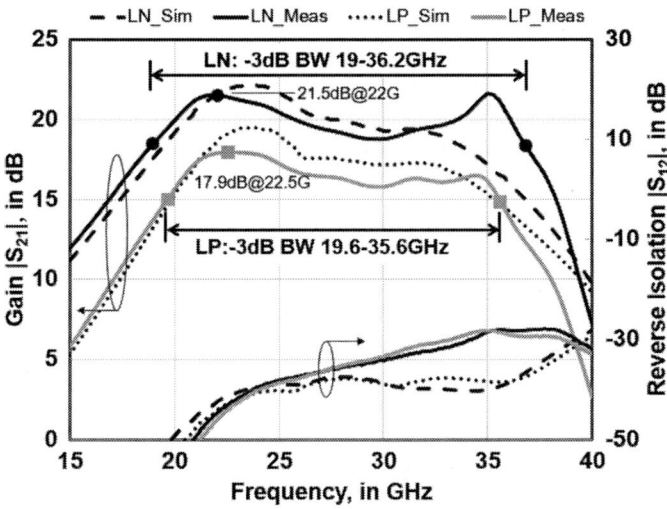

Fig. 5. Measured and simulated small-signal gain and reverse isolation

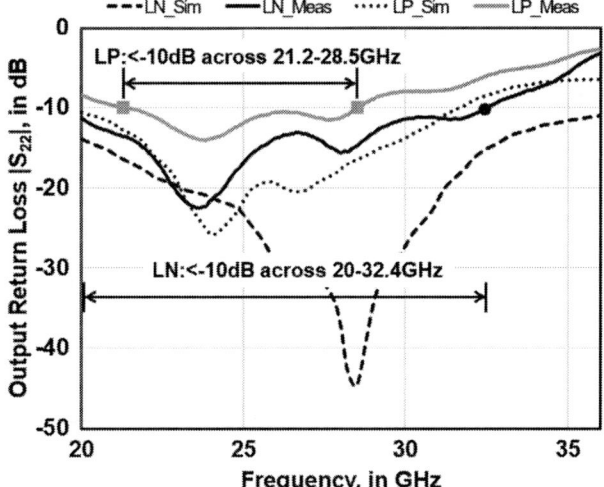

Fig. 7. Measured and simulated output return loss

978-1-7281-1702-7/19 $31.00 © 2019 IEEE

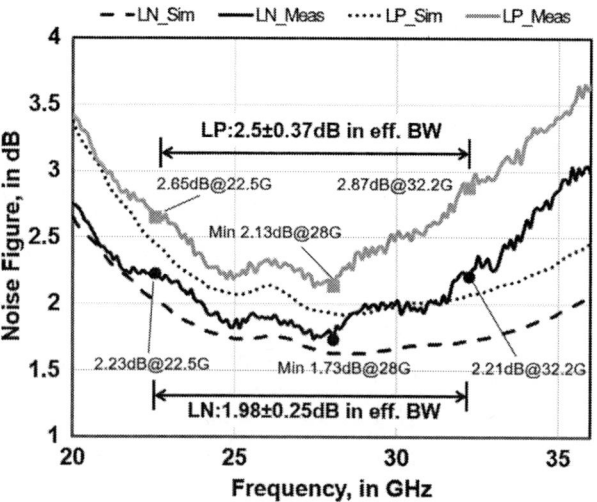

Fig. 8. Measured and simulated noise figure

Table 1. Wideband mm-Wave LNA Performance Comparison

	This work		GSMM 2018 [5]	RFIC 2018 [6]	MWCL 2017 [7]
	LN	**LP**			
BW_{eff}^{*} (GHz)	22-32 (10)	22-32 (10)	27-47.5 (20.5)	29-37 (8)	26-32.7 (7.7)
Gain (dB)	20.1±1.4	17±0.9	18.5±1.5	27±1.5	25.6±1.5
Peak Gain (dB)	21.5@ 22GHz	17.9@ 22GHz	20@ 28GHz	28.5@ 32GHz	27.1@ 27GHz
NF (dB)	1.7-2.2	2.1-2.9	4.2-5.5	3.1-4.1	3.3-4.3
IIP_3 (dBm)	-13.4@ 22GHz	-14.4@ 22GHz	-9.4@ 28GHz	-12.5@ 32GHz	-12.6@ 27GHz
P_{DC} (mW)	17.3	5.6	58	80	31.4
Act. Area (mm^2)	0.05	0.05	0.2	0.21	0.26
Process	22-SOI CMOS	22-SOI CMOS	45-SOI CMOS	0.25-μm SiGe	40-Bulk CMOS

*gain is within the -3dB BW for $|S_{21}|$ and $|S_{11}|<$-10dB

Figs. 6 and 7 show measured and simulated input ($|S_{11}|$) and output ($|S_{22}|$) return losses, respectively. Again, simulation and measurement agree well. Input and output impedances are near 50Ω (i.e., >10dB return loss) from 22.5 to 32.2GHz in LN mode and 21.2 to 28.5GHz for LP mode. Note that $|S_{22}|$ varies between power modes, which is expected due to the change in transistor drain currents.

We define an effective bandwidth (BW_{eff}) for the LNA as the frequency range where $|S_{21}|$ gain is within -3dB of its peak and $|S_{11}|$ is below -10dB. The measured BW_{eff} is 9.7GHz (22.5 to 32.2GHz) in both LN and LP modes.

The 50-Ω noise figure (NF) is plotted in Fig. 8. The measured NF at 28GHz (passband center) in LN and LP modes is 1.73dB and 2.13dB, respectively. Within BW_{eff}, the NF in low-noise mode is 1.98±0.25dB, and increases by 0.5dB in LP mode. The discrepancy between simulation and measurement beyond 28GHz is likely due to two factors. Firstly, noise matching between stages is affected by interwinding parasitic capacitance of the transformer and transistor parasitics. Also, uncertainty in the measured NF increases with increasing frequency.

Two input tones near 22GHz (i.e., at peak gain) with a frequency separation of 500kHz are applied at the LNA input to determine the IP_3. Equal power levels of -35dBm (i.e., near small-signal) were used for the fundamental tones. The measured IP_3 referred to the input (IIP_3) is -13.4dBm and -14.4dBm in LN and LP modes, respectively. IIP_3 degrades as the bias current is decreased between modes, but only by 1dB.

Table 1 compares the performance of the LNA prototype in LN and LP modes with other LNAs operating in the 28-GHz band. In LN mode, the prototype realizes substantially lower NF while consuming less DC power and chip area than the other amplifiers listed. Variation in NF is also reduced to just 0.5dB across 10GHz bandwidth. Moreover, backgate modulation enables a 3x reduction in DC power consumption, while maintaining low NF and wide BW. The narrowband

cascode amplifier in [8] reported a lower NF of 1.45±0.15dB across 24 to 28GHz. However, the single-stage design has ~7dB less gain (13.4±0.6dB), and the input and output return losses do not exceed 10dB.

IV. Conclusions

A 22-nm FD-SOI CMOS technology developed for RF/mm-wave applications enables broader bandwidth, lower noise, and trimmable power consumption in a 2-stage LNA. The prototype achieves 21.5-dB peak gain, 1.7-dB minimum NF and 10-GHz BW when consuming 17mW. When power consumption is lowered to just 5.6mW, 18-dB gain, 2.1-dB minimum NF and 10-GHz BW are maintained.

Acknowledgments

The authors gratefully acknowledge financial support from the Natural Sciences and Engineering Research Council of Canada. Technology access was facilitated by MOSIS.

References

[1] (2018) AWMF-0157 28 GHz silicon 5G Tx/Rx quad core IC. [Online]. Available: http://www.anokiwave.com/products/

[2] S. Ong, et al., "A 22nm FDSOI technology optimized for RF/mmWave applications," in 2018 IEEE-RFIC, Philadelphia, PA, Jun.10–12, 2018.

[3] M. T. Reiha and J. R. Long, "A 1.2 V reactive-feedback 3.1-10.6 GHz low-noise amplifier in 0.13 um CMOS," IEEE J. Solid-State Circuits, vol. 42, no. 5, pp. 1023–1033, May 2007.

[4] J. R. Long, "Monolithic transformer for silicon RF IC design," IEEE J. Solid-State Circuits, vol. 35, no. 9, pp. 1368–1381, Sep. 2000.

[5] V. Chauhan and B. Floyd, "A 24–44 GHz UWB LNA for 5G cellular frequency bands," in 2018 11th Global Symposium on Millimeter Waves, Boulder, CO, USA, May22–24, 2018.

[6] Z. Chen, H. Gao, D. Leenaerts, et al., "A 29–37 GHz BiCMOS low-noise amplifier with 28.5 dB peak gain and 3.1-4.1 dB NF," in 2018 IEEE-RFIC, Philadelphia, PA, USA, Jun.10–12, 2018.

[7] M. Elkholy, S. Shakib, J. Dunworth, et al., "A wideband variable gain LNA with high OIP3 for 5G using 40-nm bulk CMOS," IEEE Microw. Wireless Compon. Lett., vol. 28, no. 1, pp. 64–66, Dec. 2017.

[8] C. Li, O. El-Aassar, A. Kumar, et al., "LNA design with CMOS SOI process-1.4db NF K/Ka band LNA," in 2018 IEEE/MTT-S IMS, Philadelphia, PA, Jun.10–15, 2018.

A 112-GS/s 1-to-4 ADC front-end with more than 35-dBc SFDR and 28-dB SNDR up to 43-GHz in 130-nm SiGe BiCMOS

X.-Q. Du[1], M. Grözing[1], A. Uhl[1], S. Park[1], F. Buchali[2], K. Schuh[2], S.T. Le[2] and M. Berroth[1]

[1]Institute of Electrical and Optical Communications Engineering, University of Stuttgart, Germany

[2]Nokia Bell Labs, Stuttgart, Germany

Abstract—**A 112 GS/s 1-to-4 ADC front-end in IHP 130 nm SiGe BiCMOS based on charge sampling is presented. In experimental tests, the ADC front-end achieves more than 35 dBc SFDR and more than 28 dB SNDR up to 43 GHz. Furthermore, sampling of 100 Gbaud (=200 Gb/s) PAM-4 signals with an EVM of 11.3% for 400k received symbols is demonstrated.**

Keywords—Charge sampling, analog-to-digital converter (ADC), pulse amplitude modulation (PAM)

I. INTRODUCTION

Emerging Rx sampling circuits with three-digit gigasample rates are based on voltage-mode sampling architectures [1-3]. The sampling in these architectures in its simplest form is realized by a hold capacitor and a switch that is instantaneously triggered by either the rising or falling clock edge (Fig. 1a)). For slow-varying input signals, this sampling technique exhibits excellent robustness against clock jitter. For fast-varying input signals, however, it causes strong SNR degradations, as small time deviations in the clock transition can already result in large sampling errors. As shown in [4], the signal-to-noise ratio (SNR) varies with the input frequency f by

$$SNR_{VS} = \left(\frac{1}{2\pi f t_{j,rms}} \right)^2. \tag{1}$$

Eq. (1) highlights that the high-frequency SNR of a voltage-mode sampler only can be improved by reduction of its root mean square (rms) sampling clock jitter $t_{j,rms}$. If, for instance, a rms sampling jitter of 80 fs is assumed, a SNR of better than 35 dB theoretically can only be achieved up to 35 GHz. To cover a larger frequency band, the voltage-mode sampler will require an even lower rms sampling jitter, which is a challenge. An alternative approach to mitigate clock jitter requirements can be realized with charge sampling such as proposed in [5]. For the same rms clock jitter (= 80 fs), this sampling approach can enable SNR improvements of up to 3 dB. The maximum frequency for a SNR better than 35 dB can therewith be increased by 14 GHz from 35 GHz to 49 GHz, which is shown in the later course of this paper.

The charge-sampling concept is used to implement a 112 GS/s 1-to-4 ADC front-end in 130 nm SiGe BiCMOS. Sections II and III of this paper explain the sampling concept of the front-end circuit and how it can enable better SNR performance. Section IV introduces the circuit implementation of the ADC front-end and compares the achieved measurement results against the state of the art. Towards the end of this paper, electrical sampling of four-level pulse amplitude modulated signals (PAM-4) up to 100 Gbaud is presented.

II. CHARGE SAMPLING

A. Operation principle

Charge-sampling circuits acquire their samples in the current domain by signal integration. Fig. 1b) depicts the schematic of a simplified charge sampler and illustrates the underlying operation principle: The input voltage v_{IN} is first converted into an equivalent input current by a transconductance amplifier with linear V-I transfer characteristics. Subsequent sample acquisition is then performed in three phases:

1. Reset (R)

At the beginning of each sample acquisition, the integration capacitor C_I is discharged. By closing S_2, the charge information of the previous sample is removed and the initial condition for the following current integration set.

2. Integration (I)

After the discharge, both switch states are reversed: S_1 is closed and S_2 is opened. The current of the transconductance amplifier is directed to C_I and then is integrated.

3. Hold (H)

To induce the hold phase, S_1 is opened. As S_2 is still opened, no discharging currents can flow and v_{OUT} is held constant until the next sample is acquired.

The sequence for the acquisition of the next sample then repeats gain: The capacitor is discharged (R), integrated (I) and finally decoupled to induce the hold phase (H). The hold phase consumes half of the sample period T_S and each of the other two phases (R, I) half of the remaining time.

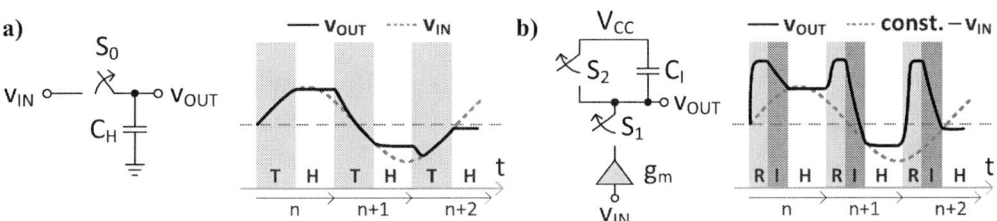

Fig.1. a) Voltage-mode sampler with Track (T) and Hold (H) operations. b) Charge sampler with Reset (R), Integration (I) and Hold (H) operations.

978-1-7281-1702-7/19 $31.00 © 2019 IEEE

B. Signal-to-Noise performance

Under the assumption that the output current of the trans-conductance amplifier in Fig. 1b) is $i_{IN} = \cos(2\pi f t)$, the AC output voltage of the charge sampler at sampling instant kT_S can be expressed by

$$v_{OUT}(kT_S) = -\frac{1}{C_I}\int_{kT_S+T_1}^{kT_S+T_2}\cos(2\pi f\tau)\,d\tau, \qquad (2)$$

where T_1 denotes the start and T_2 the end of the current integration. Derivation of eq. (2) to T_1

$$\frac{dv_{OUT}}{dT_1} = \frac{1}{C_I}\cdot\cos(2\pi fkT_S + 2\pi fT_1) \qquad (3)$$

and subsequent rms calculation give an output noise voltage of

$$v_{N1,rms} = \left(dv_{OUT,T_1}\right)_{rms} = \frac{1}{\sqrt{2}C_I}(dT_1)_{rms} \qquad (4)$$

due to jitter in the rising clock edge (dT_1). Equivalently, the rms output noise voltage $v_{N2,rms}$ due to jitter in the falling clock edge (dT_2) can be obtained by

$$v_{N2,rms} = \left(dv_{OUT,T_2}\right)_{rms} = \frac{1}{\sqrt{2}C_I}(dT_2)_{rms}. \qquad (5)$$

If both clock edges are uncorrelated and have the same rms clock jitter, i.e. $(dT_1)_{rms} = (dT_2)_{rms} = t_{j,rms}$, the output noise voltages in eqs. (4) and (5) can be combined by

$$v_{N,rms} = \sqrt{v_{N1,rms}^2 + v_{N2,rms}^2} = \frac{1}{C_I}\cdot t_{j,rms}. \qquad (6)$$

For infinitesimally small sampling steps T_S, the rms of the output signal in eq. (2) over the integration period T_I ($T_1 = 0$ and $T_2 = T_I$) is

$$v_{S,rms} = \sqrt{f\cdot\int_0^{f^{-1}}v_{OUT}^2(kT_S)dkT_S} = \frac{\sqrt{1-\cos(2\pi fT_I)}}{2\pi fC_I}. \qquad (7)$$

The SNR of the charge sampler can thus be predicted by

$$SNR_{CS} = \frac{v_{S,rms}^2}{v_{N,rms}^2} = \left(\frac{\sqrt{1-\cos(2\pi fT_I)}}{2\pi ft_{j,rms}}\right)^2. \qquad (8)$$

Eq. (8) highlights two important design observations:

1. At low input frequencies the clock jitter will restrict the maximum achievable SNR of the charge sampler by

$$SNR_{CS,max} = \lim_{f\to 0}SNR_{CS} = \left(\frac{T_I}{\sqrt{2}t_{j,rms}}\right)^2. \qquad (9)$$

   Every doubling of the integration period T_I will improve the theoretical SNR limit by 6 dB.

2. At high input frequencies, i.e. $f \geq (4T_I)^{-1}$, the SNR of the charge sampler for the same rms clock jitter is better than that of a voltage-mode sampler, as the increase of the denominator is counteracted by the increase of the numerator with increasing input frequency f (cf. eq. (1)).

Fig. 2 illustrates the derived SNR prediction of eq. (8) for a Gaussian clock jitter of $t_{j,rms} = 80$ fs and an integration period of $T_I = 8.9$ ps. Behavioral simulations of the charge sampler in Fig. 1b) are performed to support the correctness of the proposed equation. The simulated SNR values are obtained by discrete Fourier transformations (DFT) of the charge sam-

pler's time-domain output signal (DFT length: 8192). As illustrated in Fig. 2, SNR improvements of up to 3 dB are achievable at mm-wave frequencies, highlighting the attractiveness of the presented charge-sampling concept for ultra-broadband sampling applications.

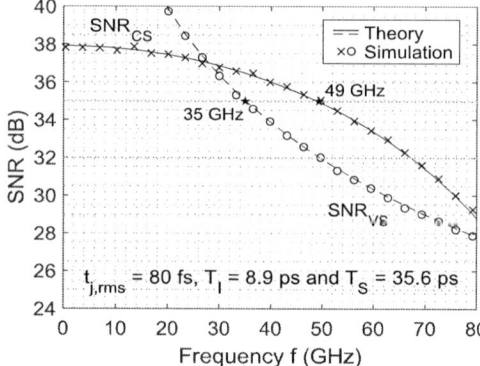

Fig. 2. Charge sampling vs. voltage sampling according to eqs.(8) and (1).

III. TIME-INTERLEAVED CHARGE SAMPLING

To increase the sample rate of the charge sampler in Fig. 1b), time-interleaving is exploited. Fig. 3a) depicts a simplified charge sampler with four time-interleaved cores and illustrates the obtained output signals for a sinusoidal input signal. Time-interleaved operation is achieved with an analog current demultiplexer. The demultiplexer consecutively deinterleaves the input current to four integration capacitors. Reset (R), integration (I) and hold (H) operations at each capacitor then deliver the output samples, as explained in Section II. For successive sampling, each channel operation is delayed by a quarter of its sample period, i.e. $1/(4f_S)$. In case that the output waveforms are digitized by four sub-ADCs, a quantized version of the input signal can be reconstructed by digital data interleaving (indicated by orange waveform in Fig. 3a)).

The analog current demultiplexer is operated by three differential clocks with 50% duty cycle each. The phase relations

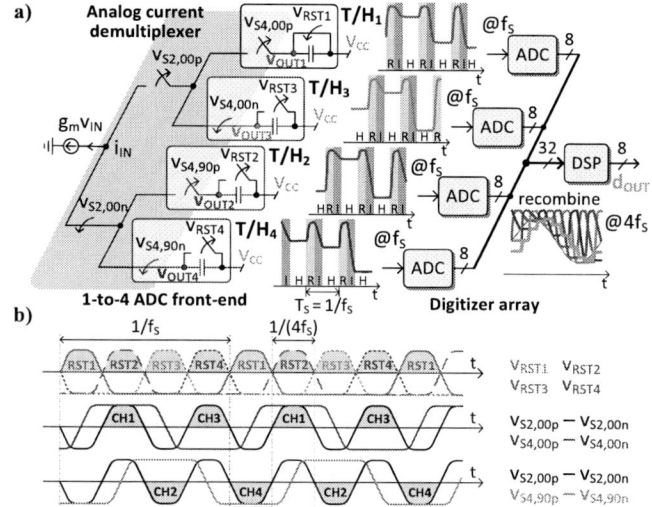

Fig. 3. a) Time-interleaved charge sampler. Full reconstruction (orange) of sinusoidal input data (magenta) after recombination of four sampler outputs. b) Clock (CLK) and reset (RST) signals for control of the charge sampler.

of these signals are depicted in Fig. 3b). A half data rate clock ($v_{S2,00}$) controls the first switching level and two quadrature quarter data rate clocks ($v_{S4,00}$ and $v_{S4,90}$) the second switching level of the demultiplexer. The half data rate clock is centered in both quarter data rate clocks and ensures that the switching in the second level is always settled before the switching in the first level begins. In case that a positive half wave is applied on the first switching level, the input current is steered to the odd-numbered outputs (v_{OUT1}, v_{OUT3}) and in the other case to the even-numbered outputs (v_{OUT2}, v_{OUT4}). The same approach is used on the second switching level to further refine the deinterleaving to a single output (e.g. v_{OUT1}).

IV. A 112 GS/s 1-TO-4 ADC FRONT-END

A. Circuit implementation

Fig. 4a) depicts the block diagram of the charge-sampling ADC front-end. For even-order harmonic suppression, the complete circuit is designed fully-differentially in current mode logic. The core circuit of the ADC front-end requires seven control signals that each are generated on-chip: The half data rate clock ($v_{S2,00}$) is provided by the regenerated input clock and the two quadrature quarter data rate clocks ($v_{S4,00}$ and $v_{S4,90}$) by following divide-by-2 frequency division (FD). The four 25% duty cycle reset signals (v_{RST1}, v_{RST2}, v_{RST3} and v_{RST4}) are obtained by logic OR operations of both quarter data rate clocks. Their phase relations as depicted in Fig. 3b) can individually be tuned by external bias control signals.

An insight into the schematic design of the core circuit is given in Fig. 4b). The linear transconductance amplifier for V-I conversion is realized by an emitter-degenerated differential pair (I.1). G_m peaking is achieved by exploiting parasitic layout capacitances that build high-frequency shorts of the emitter degeneration resistors. To improve the settling time of the reset, the DC operation point of the analog demultiplexer's input current (II) is reduced by two matched PMOS current sources (I.2). This ensures that less DC current is integrated on the capacitors. Thus, lesser charges have to be discharged. The reset switches are realized by switched emitter followers (III) [5] that are controlled by four differential reset signals.

B. Die photograph

The ADC front-end is implemented in IHP 130 nm SiGe BiCMOS and consumes a die area of 1.5 mm x 1 mm (see Fig. 4a)). The technology (SG13G2) supports 300 GHz f_T and 450 GHz f_{max} for its HBTs and seven metal layers for routing.

C. Time-domain measurements

The proposed ADC front-end is wire-bonded on a RF printed circuit board (PCB) and measured by a four-channel sub-sampling oscilloscope with 70 GHz analog bandwidth. To trigger the scope, the divide-by-256 clock output of the ADC front-end is used. The input data signal is provided by a 43.5 GHz signal generator and the clock signal by a 67 GHz signal generator. Both generators are synchronized by their 10 MHz reference to ensure a common time base. Differential input signaling is enabled by two 67 GHz broadband baluns.

The captured output signals are offline resampled. Further signal processing such as filtering is not performed. With a 128-DFT, the signal-to-noise-and-distortion ratio (SNDR) and the spurious-free dynamic range (SFDR) of these signals are determined. Fig. 5 shows the obtained measurement results for different input frequencies. At all measured frequency points, the input voltage swing is calibrated to 500 mV. To ensure that the noise of the scope sampling modules won't affect the SNDR of the ADC front-end, their noise power is excluded. The raw measured SNDR improves therewith by up to 1.8 dB. The noise power of the sampling modules is determined in separate time measurements, where only 50 Ω terminations are connected to the modules. As shown in Fig. 5, all ADC front-end outputs achieve a SFDR better than 35 dBc and a SNDR better than 28 dB up to 43 GHz.

D. Comparison to the state of the art

Table 1 compares the measurement results of the ADC front-end against state-of-the art >100 GS/s Rx samplers. Up to the moment of writing, three works [1-3] present sampling circuits with three-digit gigasample rates. Of these three works, however, only [3] shows full sampling performance beyond 100 GS/s. In [1] a 32x time-interleaved 128 GS/s ADC front-end in 22 nm FDSOI CMOS with 60 GHz track bandwidth is

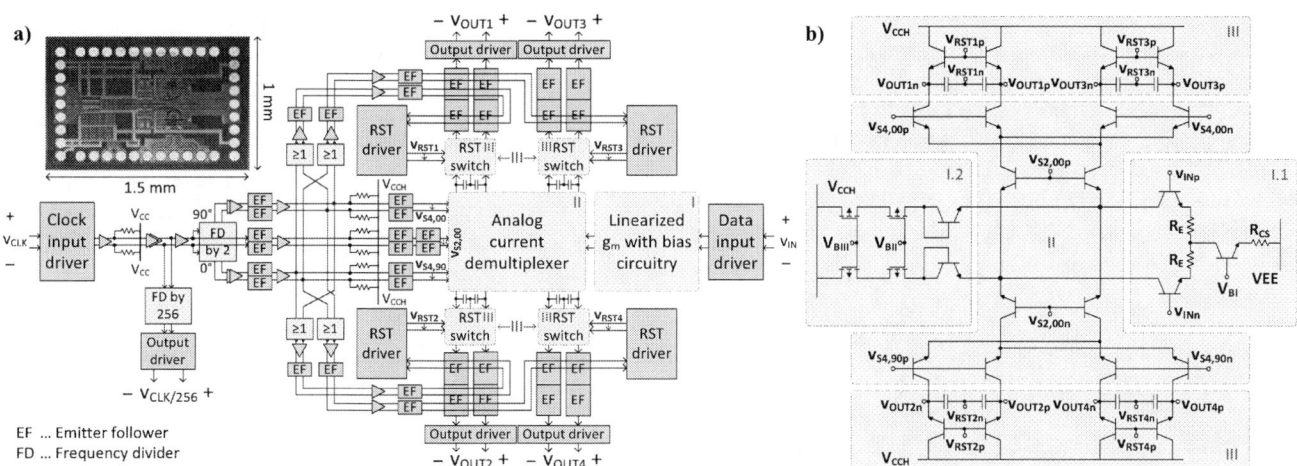

Fig.4. a) Die photograph and simplified block diagram of ADC front-end. b) Schematic of ADC front-end core with linearized transconductance amplifier (I.1.), bias circuitry (I.2), analog current demultiplexer (II) and reset switches (III).

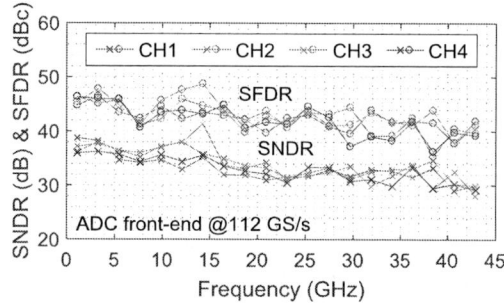

Fig.5. Measured SFDR and SNDR of ADC front-end up to 43 GHz.

presented, but sampling of an input signal is not demonstrated in this work. In [2] a 108 GS/s track-and-hold in 55 nm SiGe BiCMOS is shown. Circuit characterization is done at 90 GS/s (SFDR = 40 dBc at 15 GHz). The only work that shows full >100 GS/s Rx sampling is [3], in which a 2x time-interleaved 128 GS/s 5 bit ATC-TAC in 55 nm SiGe HBT is described (ATC: Analog-to-Thermometer Converter, TAC: Thermometer-to-Analog Converter). On-wafer measurements of this circuit show a SNDR of >26 dB up to 32 GHz. A comparison to this circuit, however, is difficult, as the SNDR of its ATC front-end circuit might be underestimated by its following 5 bit analog-to-thermometer conversion.

This work presents the first three-digit gigasample rate ADC front-end that shows sampling of input signals beyond 32 GHz. The wire-bonded ADC front-end is characterized up to 43 GHz and achieves >35 dBc SFDR and >28 dB SNDR at 112 GS/s. Note that the measured frequency range is restricted by the maximum frequency of the used sine wave generator. The record high-frequency linearity is achieved with 3.34 W at power supplies of 3.5 V & 6.5 V. In a fully-integrated time-interleaved ADC solution, the clock regeneration circuitry (0.24 W), the FD by 256 (0.55 W) with ≥67 GHz measured frequency range and the four output drivers (0.27 W) can be omitted and the total power dissipation decreases by ~1.06 W.

V. PAM-4 EXPERIMENT UP TO 100 GBAUD

To perform a first data transmission feasibility experiment, the ADC front-end is integrated into the electrical testbed of Fig. 6a). A 100 GS/s 6 bit digital-to-analog converter (DAC) [6] is used as the data source. Up to a bandwidth of 43 GHz, its effective bit resolution is better than 4.5 bits. Pre-calculated four-level signals at symbol rates up to 100 Gbaud are sent out periodically. The random PAM-4 sequence length is 16384 symbols. A raised cosine pulse shape with 0.1 roll-off is applied to limit the bandwidth below 50 GHz for symbol rates below 100 Gbaud. The electrical PAM signals are transmitted back to back to the ADC front-end, which is clocked at half of

the sampling rate of 50 GHz. It outputs four differential tributaries, which are converted to single-ended signals by a four-channel RF amplifier. The four channels are digitized by a 50 GS/s 20 GHz bandwidth real-time oscilloscope. The offline digital signal processor (DSP) consists of a filter and a multi-plexer (MUX). The system suffers from intersymbol interference (ISI) caused by the DAC and by the ADC front-end due to bandwidth limitations. The mitigation of ISI requires a 4x4 MIMO filter, which further mitigates any gain and skew differences between the four channels. The MIMO filter consists of linear FIR filters and the filter taps are optimized by minimizing the mean square error at filter output. The output tributaries are simply multiplexed to regenerate the serial sequence before assessing the performance by calculating the SNR or the error vector magnitude (EVM). After a first power optimization we measured the electrical back to back performance of a 100 Gbaud (= 200 Gb/s) PAM-4 signal of 19.2 dB SNR or 11.3% EVM for the complete multi-IC end-to-end test setup. Fig 6c) depicts the measured histogram of 400k received 100 Gbaud PAM-4 symbols. A bit error rate (BER) of 5.7e-5 is achieved without any DAC pre-emphasis.

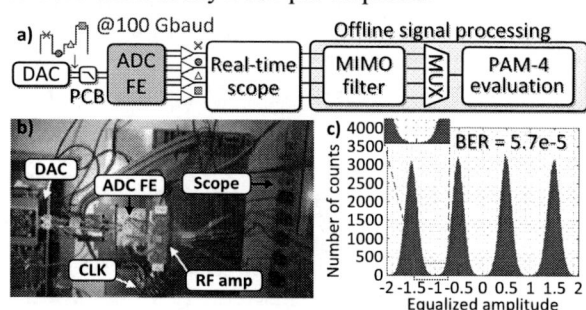

Fig. 6. a)+b) PAM-4 test setup. c) Resampled 200 Gb/s PAM-4 symbols.

VI. CONCLUSION

This paper presents the first three-digit gigasample rate ADC front-end with a record frequency band coverage of up to 50 GHz to enable 100 Gbaud PAM-4 reception. Charge-sampling is exploited to achieve a SFDR of >35 dBc and a SNDR of >28 dB up to 43 GHz at 112 GS/s.

ACKNOWLEDGMENT

The circuit design and fabrication have been supported by German DFG (grant no. BE2256/19-2) and the measurements in part by European ECSEL-JU/EU-H2020 (grant no. 737454) and German BMBF (grant no. 16ESE0210).

REFERENCES

[1] A. Zandieh et al., "128-GS/s ADC Front-End with Over 60-GHz Input Bandwidth in 22-nm Si/SiGe FDSOI CMOS," IEEE BCICTS, San Diego, CA, USA, 2018, pp. 271-274.

[2] K. Vasilakopoulos et al., "A 108GS/s track and hold amplifier with MOS-HBT switch", IEEE MTT-S IMS, San Fran., CA, 2016, pp. 1-4.

[3] A. Zandieh, P. Schvan and S. P. Voinigescu, "A 2x-Oversampling, 128-GS/s 5-bit Flash ADC for 64-Gbaud Applications," IEEE BCICTS, San Diego, CA, USA, 2018, pp. 52-55.

[4] W. Kestner, The Data Conversion Handbook. Newnes, 2005.

[5] X. Du, M. Grözing and M. Berroth, "A 25.6-GS/s 40-GHz 1-dB BW Current-Mode Track and Hold Circuit with more than 5-ENOB," IEEE BCICTS, San Diego, CA, USA, 2018, pp. 56-59.

[6] K. Schuh et al., "100 GSa/s BiCMOS DAC Supporting 400 Gb/s Dual Channel Transmission," IEEE ECOC, Germany, 2016, pp. 1-3.

Table 1: Comparison of >100 GS/s state of the art samplers.

	f_S (GS/s)	SFDR (dBc) @f_{in} (GHz)	SNDR (dB) @f_{in} (GHz)	P_{DC} (W)	Technology
This work	112, 4x28	>44@01 >43@25 >39@43	>36@01 >32@25 >28@43	3.34	130 nm SiGe BiCMOS
[3]	128, 2x64	37@01 33@21 32@32	29@01 28@21 26@32	1.8	55 nm SiGe BiCMOS

A Dual-28Gb/s Digital-Assisted Distributed Driver with CDR for Optical-DAC PAM4 Modulation in 40nm CMOS

Qiwen Liao[12], Shang Hu[3], Jian He[15], Bozhi Yin[3], Patrick Yin Chiang[34],

Jian Liu[12], Nan Qi[12] and Nanjian Wu[12]

[1]State Key Lab. of Superlattices and Microstructures, Institute of Semiconductors, CAS, China
[2]Center of Materials Science and Optoelectronics Engineering, UCAS, China
[3]Fudan University, China
[4]PhotonIC Technologies, Shanghai, China
[5]Xi'an University of Technology, China
qinan@semi.ac.cn

Abstract—**This paper presents a dual 28Gb/s modulator driver with on-chip PAM4 clock and data recovery (CDR) in 40nm CMOS. Used in the 400G Ethernet, 56Gb/s PAM4 signal is recovered by the CDR and demodulates into dual-28Gb/s NRZ data streams to drive the silicon photonic MZM DAC. Push-pull driver cells are employed to reuse the current for power saving with high-swing output. A digital-assisted distributed topology is proposed, extending the driver bandwidth and enabling low-power flexible pre-emphasis within each segment. Precise retiming is implemented by phase interpolation for the velocity match both in distributed driver segments and MZM DAC segments. Measurement results show the CDR+driver achieves 4Vpp differential voltage swing, 1.78ps RMS jitter and 1.34W power consumption (including PAM4 CDR and dual-channel driver) at 50Gb/s PAM4 input, while the standalone driver contributes to 1.5ps RMS jitter at 28Gb/s NRZ outputs.**

Keywords— **silicon photonics, MZ modulator, driver, CMOS**

I. INTRODUCTION

The ever-growing bandwidth of the datacenter interconnects is evolving from 100 to 400Gb/s. One of the promising solutions to achieve 400G (400GBase-LR8) is to build 8-channel 50Gb/s data links running in parallel with PAM4 modulation. Due to the high wavelength stability, high-linearity and wide bandwidth, the Mach-Zehnder modulator (MZM)-based link is widely employed for medium and long-range optical interconnects intra and inter the datacenters [1]. The PAM4 modulator driver is required to be linear with high-swing outputs and high-bandwidth. Standalone driver chips implemented in SiGe or III-V process are thus common, which features high power consumption and low integration level.

For data rate beyond 25Gb/s, the input channel loss introduces significant ISI and data-dependent jitter (DDJ), which could only be removed through data decision and retiming. A PAM4 CDR is needed prior to the 50Gb/s MZM driver (Fig. 1), which removes the jitter and enlarges the link margin. Moreover, the distributed amplifier (DA) topology is usually adopted to extend the bandwidth, which absorbs the circuit parasitic capacitor into a T-line [2] [3] [4]. However, besides of the intrinsic driver bandwidth, additional pre-emphasis is always needed for the MZM and E/O co-packaging, which is preferred to be flexible. In this work, a digital-assisted distributed topology is proposed, which utilizes the multi-phase clock recovered by the integrated CDR, providing in-segment

tunable pre-emphasis without power-hungry analog delay cells, as well as precisely regulating the propagating timing across all segments of the DA.

Thirdly, the PAM4 modulation is traditionally realized in the electrical domain by outputting 4-level analog signals from the driver. The limited tolerable voltage swing is further divided by 3, which leads to poor extinction ratio. An alternative way is to implement it in the optical domain, by driving an MZM DAC with multiple low-swing drivers [3]. Particularly, a two segments MZM weighted by length can be driven by a dual-NRZ driver proposed in this work for the PAM4 modulation. Precise electrical/optical velocity match across the two segments is regulated by phase interpolation and in-segment 2:1 multiplexing.

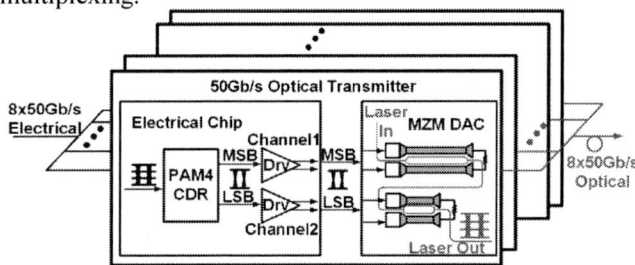

Fig. 1 8x50Gb/s optical transmitter array for 400GbE

II. SYSTEM ARCHITECTURE

Fig. 2 shows the block diagram of the proposed driver, consisting of a PAM4 CDR, a distributed driver and an interface between them. To characterize the driver performance without CDR, a PRBS pattern generator is integrated generating quadrature-phase clocks with the aid of external clock to produce dual channel 4Vpp swing outputs for MZM. In the CDR+driver mode, the CDR directly recovers clock from the 50Gb/s PAM4 signal and makes data decision by sampling the PAM4 signal with the recovered clock.

The PAM4 CDR incorporates the half-rate Bang-Bang phase detector (BB PD), charge pump (CP), loop filter (LPF), voltage controlled oscillator (VCO) and clock distribution chain. The PD receives input PAM4 signals and produces recovered data and UP/DN signals to CP, with sampling clocks from VCO and frequency divider. By tuning the CP current and resistor of LPF, 5-15MHz bandwidth is implemented.

2019 IEEE Radio Frequency Integrated Circuits Symposium

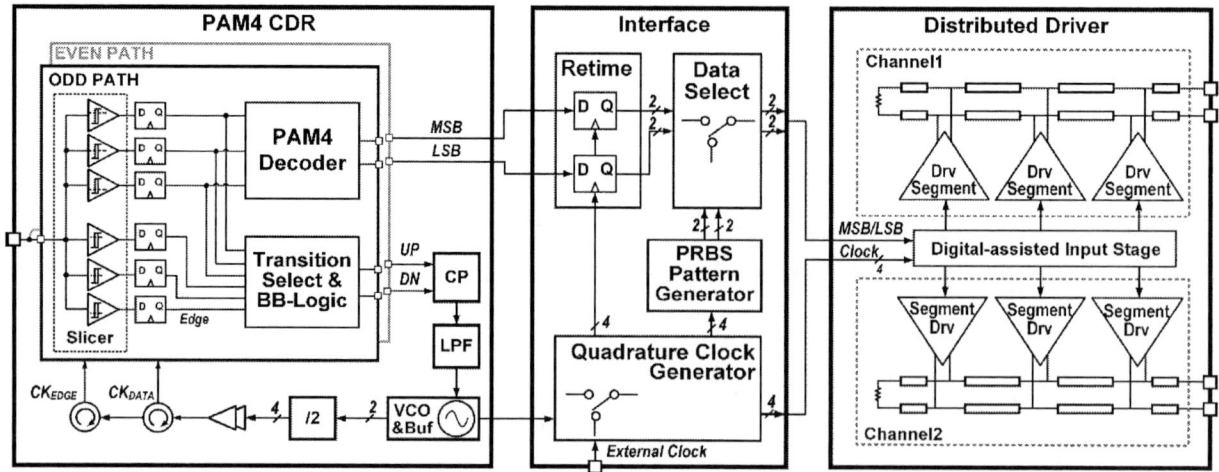

Fig. 2. Schematic of the system diagram of the proposed MZM driver

In the Interface, there are quadrature clock generator, PRBS pattern generator, data retimer and data selector. The quadrature clock generator consists of a clock selection switch and a CML high speed /2 divider. The PRBS pattern generator produces 4 bits half-rate PRBS7 data stream. The data from CDR is retimed by quadrature clock I and Q before feeding to two channels of the driver. The data fed to driver is chosen by the data selector. With the clock and data selector, the work states of the driver are configurable.

As there is optical velocity delay between two segments of MZM DAC, the retimer is precisely designed to eliminate distortion of optical PAM4 signal. The MSB data and LSB data are retimed by quadrature clock I and Q, which is digitally adjustable from 0 to 2π. The delay between retimed MSB and LSB data is optimized with good match to the optical delay.

The Distributed Driver consists of a digital-assisted input chain and two channels distributed driver, in which there are three segment drivers and the artificial transmission line.

III. DRIVER IMPLEMENTATION

There are four challenges in the design of distributed MZM driver. Firstly, a ~4Vpp large voltage swing is required, which means a large tail current in current-mode driver. The second challenge is bandwidth limit caused by large parasitic capacitance of large size transistors. Thirdly, the delay matching of input stages with output transmission line, which has great impact on output signal. The fourth challenge is accurate timing control of feed forward equalization (FFE) taps, considering the PVT variation of previous analog delay cells.

A. Distributed driver design

To achieve over 4Vpp output with 50Ω load, the push-pull amplifier in segment driver is used with current efficiency improvement than conventional CML structure [3] [5]. The proposed cascode push-pull amplifier circuit is shown in Fig. 3. The thick oxide MOS transistor M_{N3}, M_{N4}, M_{P3}, M_{P4} is used as the common gate of the cascode to protect the transistors from breakdown, considering the single-end 2Vpp swing at the output node. However, the output bandwidth is severely limited by large parasitic capacitance of these thick oxide transistors,

especially the stack of NMOS and PMOS comparing to only NMOS in CML. To extend the bandwidth, the DA structure is proposed with artificial transmission line to absorb the parasitic capacitance.

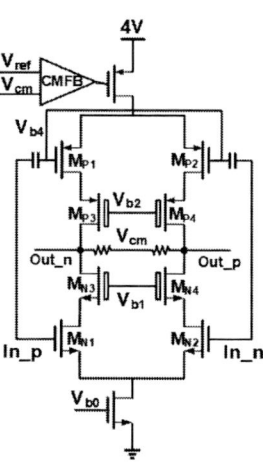

Fig. 3 The cascode push-pull amplifier in segment driver

The detailed circuits of proposed driver is shown in Fig. 4. Every segment driver shown in Fig. 2 includes a phase interpolator (PI), a 2:1 multiplexer (MUX) and a two-taps driver with main tap and post tap for FFE. The PI is used to adjust the sampling clock phase between 0 to 2π with control words. The MUX works as the serializer to produce full-rate data for driver stage. The digital-assisted input chain incorporates a quadrature clock distribution chain with three stages clock buffers (CK BUF) and two inverter-based data distribution chains for channel1 and channel2, respectively.

Based on the DA and artificial transmission line to extend bandwidth, the input delay matching is important for output signal quality. Previous works [3] use transmission line for input data distribution, which is hard to adjust the delay time and faced with attenuation, reflection and distortion. To make the input delay matching flexibly adjustable, the digitally controlled distributed amplifier was proposed [4]. In [4], the full-rate data is retimed by digitally controlled full-rate clock before feeding to predriver. By adjusting the clock phase with

978-1-7281-1702-7/19 $31.00 © 2019 IEEE   220

clock synchronization circuit, the input delay matching is digitally controlled.

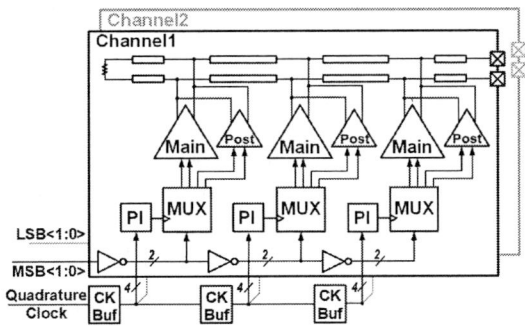

Fig. 4 Block diagrams of proposed two channels distributed driver

In this work, the digital-assisted input delay matching is utilized with half-rate architecture, including both half-rate quadrature clock and half-rate data, which is different from the full-rate architecture in [4]. One of the most significant advantages of the half-rate architecture is the large sampling margin for the retiming clock, which exhibits nearly two times wider adjusting region, shown in Fig. 5 (a). The large sampling margin is especially of great importance when the jitter of input data is large, which would compress the high SNR sampling location significantly. Additionally, the half-rate architecture is power efficient than the full-rate architecture.

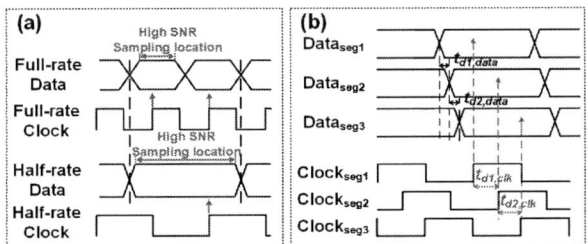

Fig. 5 (a) High SNR sampling region comparison between full-rate and half-rate architecture, (b) Timing diagram of input data and clock of three segments

The input delay calibration procedure of three segments is divided to three steps, the timing diagram of input data and clock of three segments is shown in Fig. 5 (b). Because the inverters are used to distribute the data, there are some delay of the input data between any two segments, denoted as $t_{d1,data}$ and $t_{d2,data}$. The first step is single segment calibration by tuning the polarity and strength of quadrature clock I/IB and Q/QB to get the optimized sampling range and the corresponding PI control words for every segment. At the next step, adjust the second clock phase to the center of optimized sampling range of the second segment to guarantee the global optimized sampling. The third step is tuning the clock phases of the first segment and the third segment based on the control words setting of the second clock phase to get an optimized eye diagram at the channel output. The relative clock phases of three segments is shown in Fig. 5 (b), the delays are denoted as $t_{d1,clk}$ and $t_{d2,clk}$. The accurate delay times are determined by the length of output transmission line and the delay times of the inverters and clock buffers.

B. Accurate timing control of FFE taps

Besides the transmission line for bandwidth enhancement, the feed forward equalization (FFE) is incorporated to improve the output bandwidth and compensate the output channel loss, like bonding wire, PCB trace and MZM transmission line. In previous work [3], the analog delay cells are utilized to produce one or more unit interval (UI) delay for FFE taps. However, analog delay cell is less flexible, power hungry and sensitive to PVT variation. Based on the proposed digital-assisted input chain, a 2-to-1 MUX shown in Fig. 4 is designed to produce data for main-tap and post-tap with accurate 1UI delay. Firstly, the 2 bits data is retimed by clock from PI with two D-flipflop. Then three D-flipflop are utilized to make accurate 1 UI delay between main-tap and post-tap data

IV. EXPERIMENTAL RESULTS

The proposed driver was fabricated in standard 40nm CMOS process, occupying 2.24mm² area including ESD, PADs and decoupling capacitors, the chip photograph is shown in Fig. 6. Each channel of the distributed driver occupies 0.25mm², including artificial transmission line.

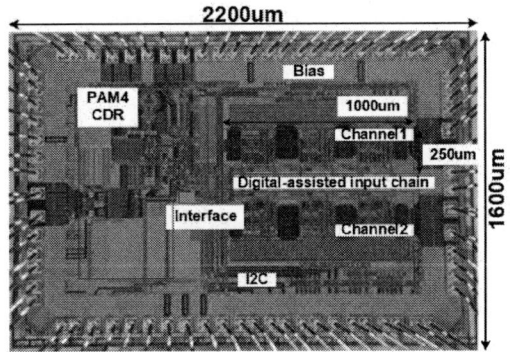

Fig. 6 MZM driver photograph

A differential full-rate clock is provided by a programmable pattern generator (Keysight M9505A) as the clock source of quadrature clock generator and on-chip PRBS generator. The differential output of the driver is firstly connected with DC block and 12dB attenuator and fed to the wide-band oscilloscope. With 25GHz and 28GHz clock in, the 25Gb/s and 28Gb/s NRZ electrical signals were produced in every channel with good input delay matching, the clear eye diagrams were shown in Fig. 7.

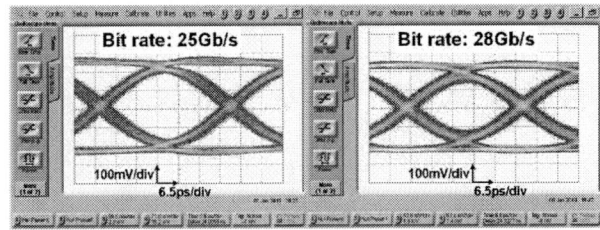

Fig. 7 Measured driver output eye diagrams at different speeds

The input delay matching is calibrated according to the calibration procedure illustrated above. To illustrate the effect of input delay mismatch, the optimized delay setting is changed with different extra delay offset. As shown in Fig. 8, comparing to the good delay matching in (a), the (b), (c), (d) are distorted with different delay offsets. Apparently, the eyes closed gradually as more segments with delay offsets.

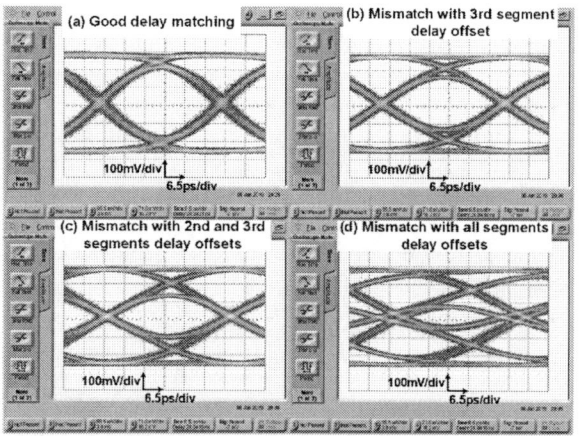

Fig. 8 Eye diagrams of 25Gb/s driver outputs with different input delay offset

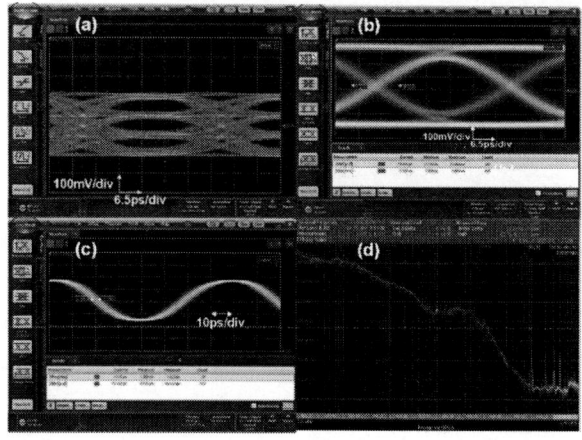

Fig. 9 (a) 25.78Gbaud PAM4 signal input, (b) driver output of recovered MSB data, (c) recovered half-rate clock, (d) phase noise of recovered clock

In the joint test of CDR+Driver, a 25.78Gbaud PAM4 signal with 300mVpp swing is fed to CDR, shown in Fig. 9 (a). The Fig. 9 (b) is eye diagram of driver output of recovered MSB data with 1.79ps jitter$_{RMS}$ and 13.3ps jitter$_{p-p}$, (c) is eye diagram of recovered half-rate clock with 1.22ps jitter$_{RMS}$ and 10.1ps jitter$_{p-p}$, (d) is phase noise curve of recovered clock with 94.8 dBc/Hz at 1MHz offset.

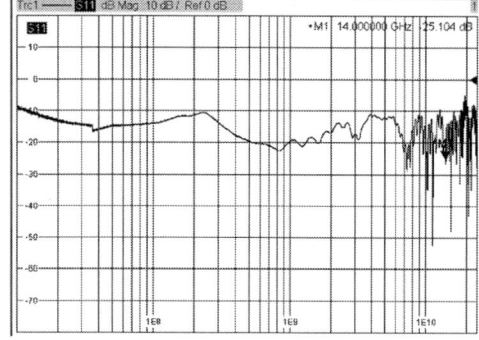

Fig. 10 Measured output return loss of proposed MZM driver (single-ended)

To eliminate the reflection of on-chip artificial transmission line, bonding wire and TW-MZM transmission line, a differential 100Ω back termination resistor is utilized. Output return loss S22 is measured with all segments open. A good impedance matching is achieved with less than -10dB S22 measured from 10MHz to over 14GHz.

Table 1 summarizes the performance metrics of this work and makes comparisons with the recent published MZM drivers.

Table 1. Performance summary and comparisons

	[2] OFC15	[4] JSSC17	[6] RFIC15	This work
Data Rate(Gb/s)	25	40	30	28
Equalization	No	No	No	**FFE**
Driver structure	DA	DA	Single	DA
Input mode	Analog	Digital	Analog	**Digital**
Output Swing(V)	6.4	6	3.3	4
Area(mm2)	No	1.8	0.072	0.25
Power(mW)	520	1920	438.5	585*/channel
Process	65nm CMOS	130nm SiGe	65nm CMOS	40nm CMOS

*Including Interface and clock distribution

V. Conclusion

A dual 28Gb/s modulator driver with on-chip PAM4 CDR is proposed in 40nm CMOS. Push-pull driver cells are employed to reuse current and generate high-swing output. The digital-assisted DA extends the bandwidth and enables low-power flexible pre-emphasis in segments. Precise retiming is implemented by phase interpolation for the velocity match both in distributed driver segments and MZM DAC segments. Based on techniques above, the driver is implemented in 40nm CMOS and tested with 4Vpp output swing, 1.5ps RMS jitter at 28Gb/s.

Acknowledgment

This work is supported in part by the CAS Pioneer 100-Talents Program, National Natural Science Foundation of China (NSFC), No. 61874115, and the Open Foundation of State Key Laboratory of Optical Communication Technologies and Networks, Wuhan Research Institute of Posts & Telecommunications (No. 2017OCTN-02).

References

[1] D. Marris-Morini, et al., "Recent Progress in High-Speed Silicon-Based Optical Modulators," in Proceedings of the IEEE, vol. 97, no. 7, pp. 1199-1215, July 2009.

[2] N. Qi, et al., "A 25Gb/s, 520mW, 6.4Vpp Silicon-Photonic Mach-Zehnder Modulator with distributed driver in CMOS," in OFC, 2015, pp. 1-3.

[3] N. Qi, et al., "A 32Gb/s NRZ, 25GBaud/s PAM4 reconfigurable, Si-Photonic MZM transmitter in CMOS." in OFC, 2016, Th1F.3.

[4] L. Vera and J. R. Long, "A 40-Gb/s SiGe-BiCMOS MZM Driver With 6-Vpp Output and On-Chip Digital Calibration," in IEEE Journal of Solid-State Circuits, vol. 52, no. 2, pp. 460-471, Feb. 2017.

[5] J. Hwang et al., "A 32 Gb/s, 201 mW, MZM/EAM Cascode Push‑Pull CML Driver in 65 nm CMOS," in IEEE Transactions on Circuits and Systems II: Express Briefs, vol. 65, no. 4, pp. 436-440, April 2018.

[6] K. Li, et al., "A 30 Gb/s CMOS driver integrated with silicon photonics MZM," 2015 IEEE Radio Frequency Integrated Circuits Symposium (RFIC), pp. 311-314.

A 77dB-SFDR Multi-Phase-Sampling 16-Element Digital Beamformer with 64 4GS/s 100MHz-BW Continuous-Time Band-Pass ΔΣ ADCs

Rundao Lu[#1], Sunmin Jang[#2], Yun Hao[#3], Michael P. Flynn[#4]

#University of Michigan, Ann Arbor, USA

[1]lurundao@umich.edu, [2]smjang@umich.edu, [3]haoyun@umich.edu, [4]mpflynn@umich.edu

Abstract— This paper tackles the fundamental limitation of distortion in large-scale digital beamforming. SNR improves by 3dB for every doubling of array size; however, distortion is correlated and so is not improved by the array gain. This work introduces the concept of multiple ADCs per element, with each sampling at a different phase, to both reduce distortion of the ADCs and RF frontend. A further advantage is that multi-phase-sampling with continuous-time band-pass delta-sigma modulators (CTBPDSMs) reduces the ADC clock-jitter sensitivity. A prototype 16-element 1GHz-IF digital beamformer employs four multi-phase-sampling sub-ADCs per element. The prototype beamformer IC integrates 64 4GS/s sub-ADCs and digital processing to generate four simultaneous beams. Bit-stream digital beamform processing efficiently handles the aggregate 0.256TS/s from the entire ADC array. The measured beamformer SNDR and SFDR are 56dB and 77dB, respectively. Multiphase sampling improves measured HD3 by 9dB.

Keywords— Digital beamforming, phased-array, receiver, linearity, delta-sigma modulator (ΔΣM).

I. INTRODUCTION

Large-scale beamforming is essential for emerging RF and mm-wave communication systems. Digital beamforming has important advantages for large arrays including accurate beam patterns, multiple simultaneous beams, and fast steering [1-2]. ADC performance is critical for large arrays but despite some progress [2], distortion from the element ADCs remains a fundamental limitation. Distortion is a limitation because although averaging in large arrays improves SNR (i.e., ideally a 3dB "array gain" for doubling of array size), averaging cannot reduce distortion. Techniques to improve ADC SNR and distortion [3] invariably lead to much higher power consumption and much larger die area. This work introduces the new approach of using multiple sub-ADCs per-element to significantly improve both SNR and distortion. We demonstrate a 1GHz IF input, 64-ADC 16-element digital beamformer with a measured SFDR of 77dB.

The Continuous-Time Bandpass Delta Sigma Modulator (CTBPDSM) ADC is very attractive for digital beamforming but is limited in SNDR and SFDR. CTBPDSMs benefit from high speed, low power, and compact area. In a digital beamformer, CTBPDSMs can directly digitize a high frequency IF, and facilitate very efficient digital bit-stream beam processing [2]. However, practical considerations such as the nonlinearity of the feedback DACs, the finite amplifier GBW, and clock jitter limit the SNDR of GHz input-frequency CTBPDSMs to ~40dB [2]. ADC techniques to mitigate these challenges consume high power and area and are prohibitive for high-bandwidth beamformer array applications [4].

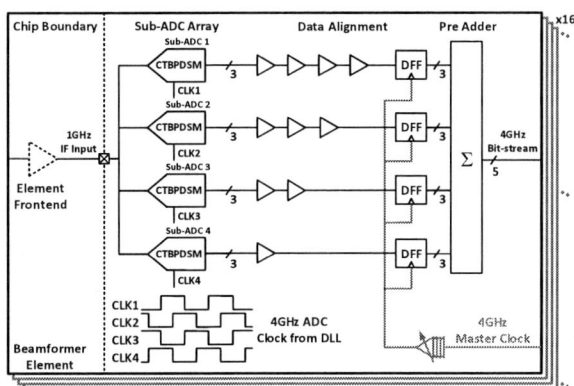

Fig. 1: Four sub-ADCs per-element sample the 1GHz IF input. The sub-ADCs bit-stream outputs are digitally aligned and added together.

Our new approach gets around the limitations of the CTBPDSM ADC by replacing the per-element ADCs in a conventional beamformer with multiple lower-performance ADCs. To circumvent the beamformer ADC performance bottleneck, we introduce two new techniques: 1) multiple parallel per-element ADCs; and 2) multi-phase parallel ADC sampling. These techniques improve ADC SNR by 7dB, improve HD3 by 9dB, and relax the clock phase noise requirement by 5.6dB.

II. MULTI-PHASE-SAMPLING CONTINUOUS-TIME BANDPASS DELTA-SIGMA MODULATOR

A. Multiple per-element ADCs and multi-phase-sampling

We replace the single, per-element ADCs in a conventional digital beamformer with multiple (i.e. four) parallel sub-ADCs per element to effectively and efficiently improve SNR and

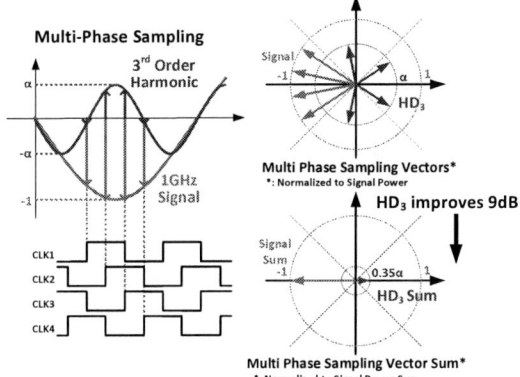

Fig. 2: ADC multi-phase-sampling attenuates odd harmonics.

2019 IEEE Radio Frequency Integrated Circuits Symposium

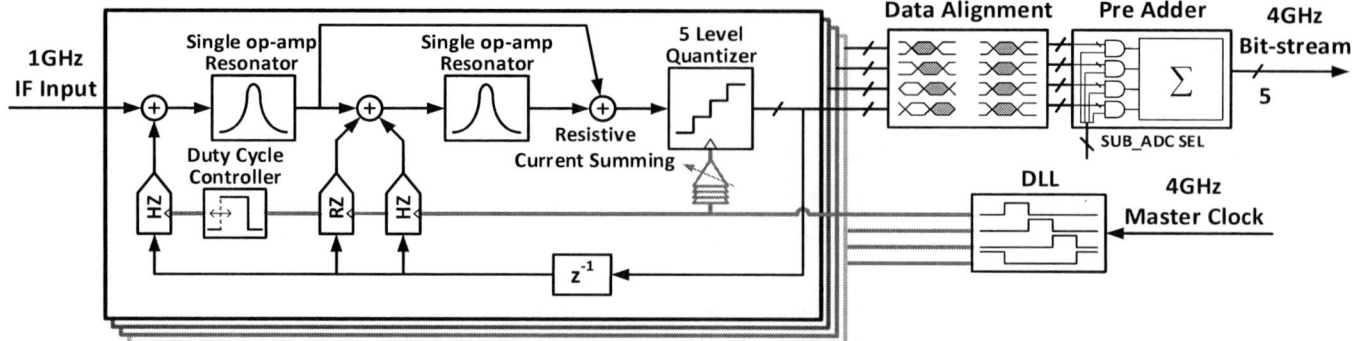

Fig. 3. Multi-phase sampling continuous-time bandpass delta-sigma modulator (CTBPDSM) sub-ADC array with duty cycle controller.

SFDR, As shown in Fig. 1, for each element, four bandpass sub-ADCs sample the IF signal from the RF frontend. Furthermore, these sub-ADCs sample at phase-shifted clock edges, instead of sampling at the same clock instant. As we will see, multi-phase sampling in the sub-ADCs and summing the result cancels distortion and suppresses jitter.

Combining the outputs of multiple ADCs that sample at the same instant improves SNR and SNDR, but this improvement is limited by ADC distortion. In multi-phase sampling, we distribute the phases of the sampling clocks of the parallel ADCs to provide two significant benefits: 1) suppression of harmonic distortion both within the ADCs and from the frontend, and; 2) averaging of in-band noise including that introduced by clock jitter.

We explain the distortion cancellation with the sampling plot and vector plots of Fig. 2. Our prototype system employs four parallel bandpass sub-ADC sampling the 1GHz IF input at 4GS/s. This sampling frequency is four times the IF carrier frequency and the sampling instants for the four sub-ADCs are spaced in 90-degree increments of the sampling clock. As depicted in Fig 2, the summation of these time-offset samples suppresses odd harmonics. The vector diagram in the figure shows that summing the time-offset samples leads to a substantial attenuation of the 3rd-order harmonic. Considering the slight attenuation of the fundamental, HD3 improves by 9.1dB. This filtering effect extends to higher-order odd harmonics (i.e. 12.6dB suppression of HD5). It filters both harmonics from the ADC and from the frontend. Suppression

of these harmonics is vital because they would otherwise alias into the band of interest. Another advantage is that multi-phase sampling attenuates uncorrelated jitter noise, and hence suppresses effects of clock jitter. Multi-phase sampling relaxes the phase noise requirement of the clock by an estimated 5.6dB.

B. Continuous-time bandpass delta-sigma modulators

CTBPDSMs are inherently suited to multi-phase sampling thanks to their small size – four sub-ADCs measure only a total of 390μm x 140μm. Another essential advantage of CTBPDSMs is that quantization noise is highly uncorrelated allowing an SNR improvement of 3dB for every doubling in the number of sub-ADCs. (In contrast, Nyquist ADCs could not be used as the quantization noise of parallel Nyquist ADCs tends to be correlated). Moreover, clock jitter-induced noise among different sub-ADCs is highly uncorrelated and therefore also benefits from a 3dB improvement every doubling of ADCs.

Beyond the benefits of multiphase sampling, there are several practical benefits to using multiple sub-ADCs per element. Although a constant ADC FoM predicts the same power when using multiple ADCs, we see the following benefits: 1) the sub-ADC power is much smaller simplifying power routing; 2) device mismatch is averaged out among the sub-ADCs; 3) the much smaller sub-ADC size means relaxed routing for time-critical feedback loops, and; 4) flexibility in trading power and performance through enabling and disabling sub-ADCs.

Fig. 3 shows the implementation of the sub-ADC array. The multi-phase sampling clocks are generated by an on-chip per-element delay-lock-loop (DLL). The ADC output bit-streams are digitally aligned. A pre-adder sums the bit-streams together, forming a combined 4G/s 5-bit stream. To facilitate testing, each sub-ADC bit-stream is gated by AND gates, so that we can observe arbitrary combinations of sub-ADCs.

The sub-ADCs are identical, except that they sample at different clock phases. The sub-ADCs employ a 4th-order architecture similar to [2]. Single op-amp resonators are adopted for power efficiency. Since the coefficient of the first-stage Return-to-Zero (RZ) DAC is small, this RZ DAC is omitted to save power and reduce noise with little stability penalty. In this work, we introduce duty-cycle control of the Half-delay return-to-Zero (HZ) feedback DAC, to optimize the ADC linearity and stability. The duty cycle of the first-stage HZ

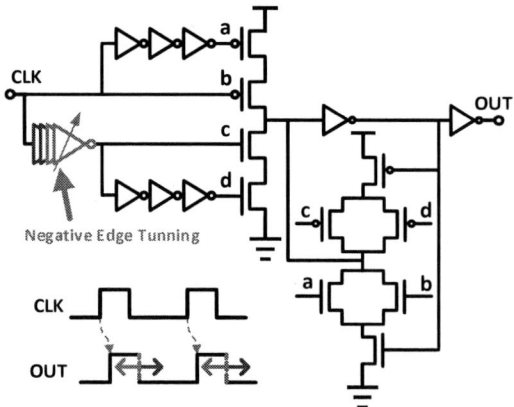

Fig. 4: Duty cycle controller first-stage HZ DAC.

978-1-7281-1702-7/19 $31.00 © 2019 IEEE   224

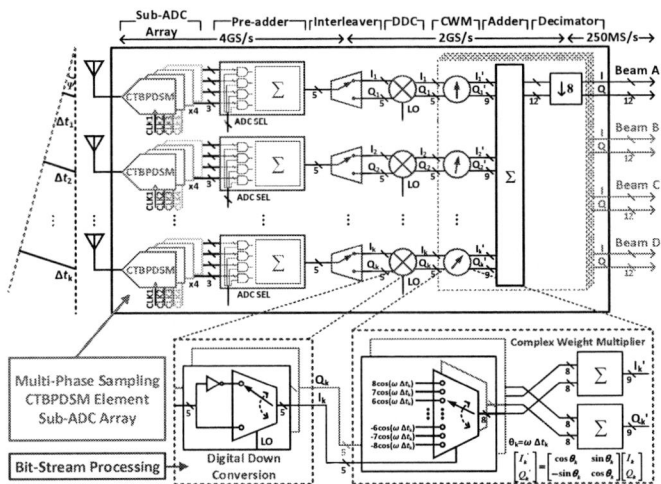

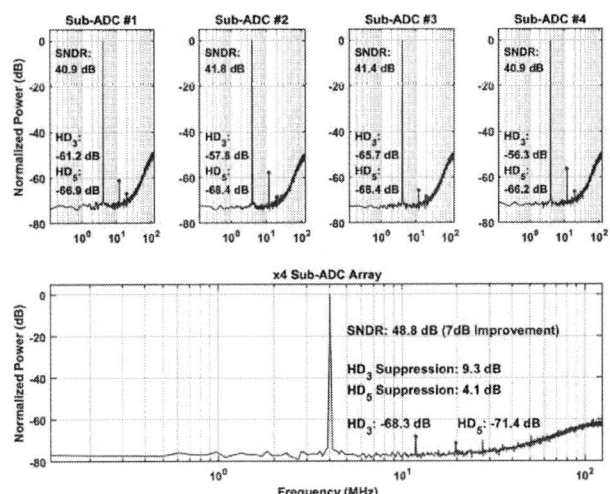

Fig. 5. System architecture of bit-stream processing digital beamformer with multi-phase sampling sub-ADC array and bit-stream processing.

Fig. 6. Measured power spectra of four sub-ADCs (top) and combined 4x sub-ADC array (bottom).

current DAC is critical to the linearity and stability of the CTBPDSM. We tune the DAC clock duty cycle with a delay chain (Fig. 4 [5]) to optimize the positions of NTF/STF zeros and poles. Behavioral simulations indicate that this duty-cycle optimization reduce harmonics related to the DAC by ~10dB.

III. DIGITAL BEAMFORMER SYSTEM ARCHITECTURE

Fig. 5 shows the system architecture of the bit-stream-processing digital beamformer. Four CTBPDSM sub-ADCs digitize each 1GHz IF input. These four ADCs share the same 1GHz IF element input, but each sub-ADC samples on a different phase (CLK1, 2, …) of the 4GHz clock. The four sampling clock phases, spaced in 90-degree increments, are generated by a Delay Lock Loop (DLL). The sub-ADC outputs are digitally synchronized to a single 4GHz digital clock domain and summed. The combined ADC output is interleaved to reduce the sample rate by a factor of two to 2GS/s. Interleaving takes advantage of the 4x relationship between the sampling rate and the IF input frequency. Because the I and Q LO mixing sequences are alternately 0, we can dispense with half of the samples fed to the I and Q down-conversion mixers. Digital Down-Conversion (DDC) down-converts the half-rate interleaved streams to baseband I and Q signals. 10-bit Complex Weight Multiplication (CWM) rotates the baseband I/Q bit-stream vectors. An adder combines 16 phase-shifted signals to generate a beam. Finally, the beam is decimated by 8 by a 4th-order CIC decimator to produce a 250 MS/s 12-bit beam output. Four sets of CWMs, summers, and decimators produce four independent instantaneous beams.

Bit-stream processing (BSP) efficiently supports the 0.256TS/s aggregate sampling rate of the 64 4GS/s ADCs. In BSP [2], the quantizer outputs of the CTBPDSMs are directly processed without filtering or decimation. BSP takes advantage of the short digital word length to implement down-conversion, and weight multiplication with simple digital MUXes, saving

power and area compared to a conventional DSP approach. The digital beamform processing for all four beams occupies 0.14mm^2 and consumes 200mW.

IV. MEASUREMENTS

The prototype 16-element four-beam beamformer is fabricated in 40nm CMOS and occupies a total area of 4.6mm^2 (Fig. 9). The active area of the 64 sub-ADCs is 0.9mm^2. Fig. 6 shows the measured power spectra for a single sub-ADC and for four combined sub-ADCs. The measurements confirm two advantages of using multi-phase sub-ADC array: 1) the overall SNDR increases by 7dB because of the thermal noise, jitter noise and quantization noise are decorrelated among sub-ADCs; and 2) the combined HD3 and HD5 are improved by 9.3dB and 4.1dB[1]. Fig. 7 shows the measured power spectrum of the entire

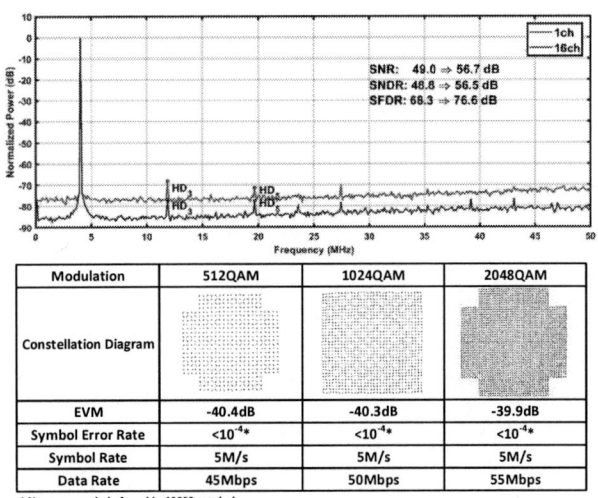

Fig. 7. Measured power spectra (top) and measured constellations (bottom) for the 16-element digital beamformer IC.

---

[1] HD3, HD5 are -59, -67.3dB if the ADCs are combined in-phase. With multi-phase sampling the measured HD3, HD5 are -68.3, -71.4dB corresponding to improvements of 9.3, 4.1dB, respectively.

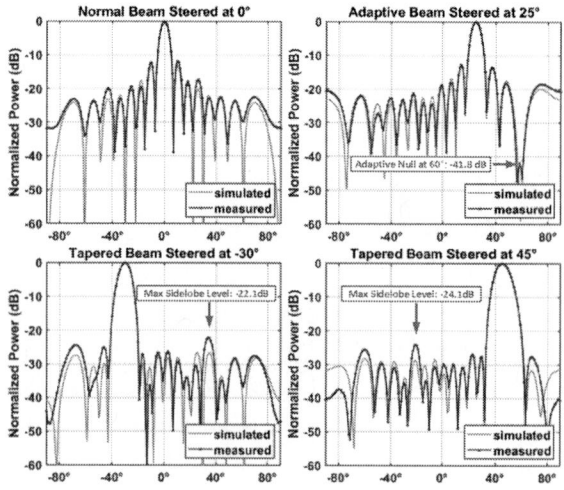

Fig. 8: Measured beampatterns overlaid on simulated beampatterns.

Table 1. Performance Summary

# of Elements	16
# of Simultaneous Beams	4
Aggregate Sample Rate	0.256TS/s
Bandwidth	100MHz
Array SNDR	56.5dB
Array SFDR	76.6dB
Coefficient Resolution	10bits
Adaptive Null	-41.7dB
Multiple Beam	Supported
Tapering Beam	Supported
Active Area	1.04mm²
ADC Power (64 sub-ADCs with DLLs + Data Alignment +Pre-adders)	1.8W
Digital Power	200mW
Total Power (64 sub-ADCs + Clock + DBF)	2W
Integration	Bandpass ADC + Digital Beamformer
Technology	40nm CMOS

16-element, 64 sub-ADC. The measured SNDR is 56dB (9.1-bit ENOB) and SFDR is 77dB when the beam is steered to 45°. This high SNDR allows the prototype beamformer to receive 2048QAM without symbol errors in 16000 test symbols. The measured EVMs are -40.4dB, -40.3dB, -39.9dB for 512QAM, 1024QAM, 2048QAM respectively. The total power consumption of the IC is 2W.

As shown in Fig. 8, the measured beampatterns are near ideal. Digital beamforming enables more advanced beampatterns such as adaptive nulls and tapered beams. As an example, the adaptive null in Fig. 8 has a measured 41.8dB rejection ratio. Also shown in Fig. 8, tapering reduces the measured nearest side-lobe power by 14dB for steering angles of -30 degrees and 45 degrees.

V. CONCLUSION

Multi-phase sampling with multiple sub-ADCs per element is introduced to overcome the ADC linearity bottleneck of the large-scale digital beamforming. The measured HD3 suppression is 9dB. The clock phase-noise requirement is relaxed by 5.6dB compared to single-element CTBPDSM. A total of 64 4GS/s sub-ADCs along with digital beamform processing are integrated onto a single prototype chip. Bitstream processing efficiently forms four beams from the aggregate 0.256TS/s data stream generated by the ADC array. The measured overall SNDR and SFDR are 56dB and 77dB, respectively. This excellent performance is verified by 2048QAM modulation testing showing a measured EVM of -39.9dB.

ACKNOWLEDGMENT

This work was supported by DARPA ACT and DARPA MIDAS.

REFERENCES

[1] S. H. Talisa et al., "Benefits of Digital Phased Array Radars," in Proc. IEEE March 2016.

[2] S. Jang et al., "A 16-Element 4-Beam 1 GHz IF 100 MHz Bandwidth Interleaved Bit Stream Digital Beamformer in 40 nm CMOS," JSSC May 2018.

[3] S. Dey, et al, "A 50 MHz BW 76.1 dB DR Two-Stage Continuous-Time Delta–Sigma Modulator with VCO Quantizer Nonlinearity Cancellation," JSSC March 2018.

[4] H. Shibata et al., "A DC-to-1 GHz Tunable RF ΔΣ ADC Achieving DR = 74 dB and BW = 150 MHz at f_0 = 450 MHz Using 550 mW," in IEEE Journal of Solid-State Circuits, vol. 47, no. 12, pp. 2888-2897, Dec. 2012.

[5] K. Agarwal and R. Montoye, "A Duty-Cycle Correction Circuit for High-Frequency Clocks," 2006 Symposium on VLSI Circuits, 2006. Digest of Technical Papers., Honolulu, HI, 2006, pp. 106-107.

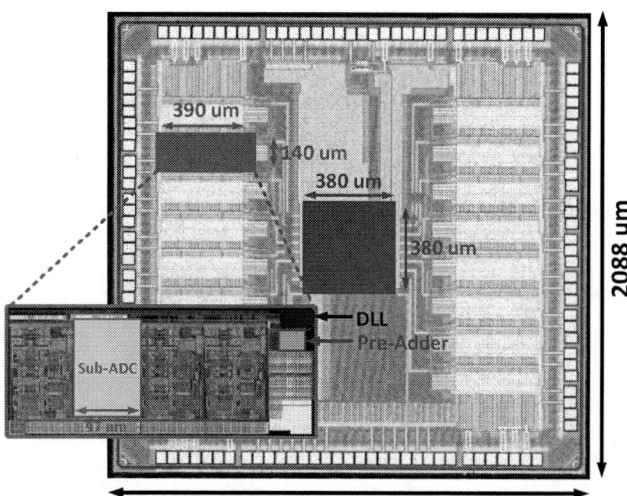

Fig. 9: Die micrograph and layout of the element sub-ADC array.

A Wideband Digitally Controllable RFIC with Gain and Wavelength Tunability and Built-in Self Test Functionalities for Optical Transceiver Modules in FTTx Applications

Sreekesh Lakshminarayanan[*1], Harman Malhotra[*], David Navara[#], Norbert Reiss[#], Klaus Hofmann[*2]

[*]Integrated Electronic Systems Lab, Technische Universität Darmstadt, Germany
[#]DEV Systemtechnik GmbH, Germany
{[1]Sreekesh.Lakshminarayanan, [2]Klaus.Hofmann}@ies.tu-darmstadt.de

Abstract— This paper presents a wideband RF mixed-signal integrated circuit for optical transceiver modules in FTTx applications. The RF transceiver has an operating frequency range of 700 MHz to 2.5 GHz and a gain which is tunable to the range of 31.5 dB in 0.5 dB steps. Built-in self testing is incorporated with a 3-bit digital-controlled ring oscillator for testing the basic functionality of the circuit. Wavelength tunability for the optical transceiver module is realized by means of a digital-controlled current source which has a $\pm$ 60 mA output current range and a resolution of 1 mA. The RF and analog parts of the circuit operate on a 3.3 V supply, whereas the digital core has a 1.2 V supply voltage. The die occupies 5.07 mm^2 in area and the IC has a maximum power consumption of 674.74 mW.

Keywords— RF Transceiver, Attenuator, Low-noise amplifier, Power amplifier, Built-in self-test, Digital-controlled oscillator, BiCMOS integrated circuits, Digital-controlled current source.

I. INTRODUCTION

With increasing demands for higher bandwidths for end-user applications, Fiber-to-the-x (FTTx) technology has found widespread deployment in recent times [1]. Optical transceiver modules play an important role in these FTTx deployments by enabling the signal conversion between the optical and electrical mediums apart from providing functionalities like signal amplification, amplitude/power control and wavelength control [2]. Currently, these transceiver modules adopt a discrete realization for most of these functionalities. The need for an area constrained implementation makes an integrated realization of the same extremely attractive. Radio frequency integrated circuits (RFIC) with digital controllability are finding wide application areas due to their high integration potential. External control loops with on-chip digital controllers and communication interfaces like SPI and I2C are commonly used along with on-chip RF circuitry for these purposes.

Testing cost of IC's remains a significant hurdle that needs to be overcome in order to reduce the overall costs involved. At present, testability is widely integrated within digital IC's and many analog IC's to minimize testing costs. However, testability is still not implemented on a wide scale in RFIC's due to the higher frequencies and the complexities involved. Commonly used approaches to implement built-in self test (BIST) in RFIC's include loopback testing and alternate test framework based approaches. Other approaches are being investigated in order to incorporate a higher degree of testability within RFIC's [3].

In this work, an RFIC for optical transceiver modules in FTTx applications, with integrated gain and wavelength tunability and a digital-controlled oscillator (DCO) based BIST functionality, is presented.

II. SYSTEM ARCHITECTURE

The overall architecture of the RFIC is shown in Fig. 1. The RF signal chain can be used in either receiver or transmitter mode of operation depending upon the specific application. The attenuator settings can be digitally controlled to adjust the gain and the dynamic range of the transceiver. The attenuation is provided by the coarse attenuator and the fine attenuator stages. The LNA provides amplification with low noise overhead and its output is coupled to the tilt attenuator, which acts as a gain equalizer. The PA provides additional signal amplification and ensures high power and good linearity at the output. An external coupler and balun can be used to couple the output power to the power detector. The power detector output is read out digitally through the ADC and the chip controller.

The digital chip controller consists of the SPI slave and interface circuits for digital read in and read out. It provides control bits for the attenuators, the RF switch, the digital-controlled oscillator (DCO) and the digital-controlled bidirectional current source (DCCS). The output from the frequency sense circuit and the 9-bit SAR ADC are also read out through the chip controller. The DCO is used to generate the test stimulus for testing the transceiver and a frequency sense (FS) circuit is added to measure the frequency of the generated test stimulus.

The DCCS, which consists of an 8-bit current steering DAC and the bidirectional VCCS presented in [4], is used as a tunable current source for the optical diode to vary the wavelength of the transmitted or received signal. A clock distribution network (CDN) is used to generate reference clocks for the ADC and the FS circuits from an external clock.

III. RF CIRCUIT IMPLEMENTATIONS

A. Attenuators

The fine and coarse attenuators are implemented using multi-stage pi-attenuators similar to the work presented in [5].

978-1-7281-1702-7/19 $31.00 © 2019 IEEE

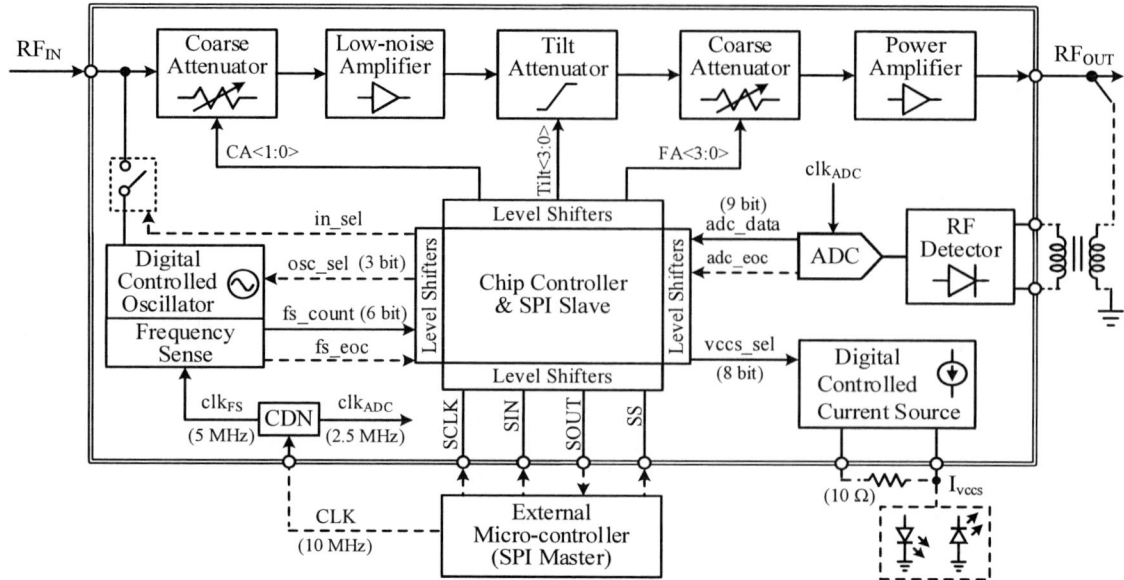

Fig. 1. Block diagram of the RF transceiver IC.

A single pi-attenuation stage is shown in Fig. 2a. The coarse attenuator consists of 2 cascaded pi-attenuators of 8 dB and 16 dB attenuation, respectively, and provides attenuation settings of 0 dB, 8 dB, 16 dB or 24 dB. The fine attenuator is realized as a cascade of 4 pi-attenuators with 0.5 dB, 1 dB, 2 dB and 4 dB attenuation stages and has an effective attenuation range from 0 dB to 7.5 dB with a step size of 0.5 dB. For frequency dependent attenuation a tilt attenuator, realized based on a bridged-T attenuation stage as shown in Fig. 2b, is utilized. Digital control is provided by means of select switches (S_{S1}-S_{S4}) which set the effective parallel resistance value R_{EFF} and thereby determine the frequency dependent attenuation value as given by the transfer function in (1) [6]. To improve the wideband matching performance, series inductors are added at the input and output of all three attenuators.

$$b/a = 1/(1 + m) \qquad (1)$$

$$where, m = Z_1 Z_2 / (Z_1 Z_4 + Z_2 Z_3 + Z_2 Z_4 + Z_3 Z_4);$$

$$Z_1 = R_{EFF}/(1 + sC_{BYP}R_{EFF});$$

$$Z_2 = Z_3 = 50; Z_4 = R_P + sL_{BW}.$$

Fig. 2. Attenuator schematics: (a) single attenuation stage; (b) tilt attenuator.

B. Amplifiers

The required RF signal amplification is provided by means of two wideband amplifiers, the low-noise amplifier (LNA) and the power amplifier (PA). The wideband LNA (illustrated in Fig. 3) is realized closer to the input side to keep the overall noise figure of the transceiver low. It is implemented using high performance SiGe HBTs in a cascode CE core stage and LC input and output matching stages [7]. A shunt feedback resistor is added to the cascode transistors to maintain good gain flatness, improve the wideband matching and to reduce the process dependencies. A resistive divider bias circuit is used to bias the cascode transistors for low noise figure and high gain performance.

The PA, illustrated in Fig. 4, consists of an active input matching stage, a cascode amplification stage with resistive and capacitive degeneration and an LCR output matching stage.. The emitter degeneration resistor R_E improves the linearity and along with the off-chip bypass capacitor C_E helps in improving the stability of the PA. The resistive shunt feedback network comprising of R_{FB11} and R_{FB21} helps in obtaining a flat gain response. The active input matching stage is designed using a combination of resistor R_{IN} and transistor Q_{IN} to obtain wide-band input matching without consuming a significant amount of chip area. The output matching condition is met using an LCR circuit consisting of L_C, CC_{OUT} and R_{OP}-R_{OS} pair. The PA is biased to simultaneously provide high gain and output compression point performance. An off-chip biasing circuit is used to avoid the area overhead needed for an on chip realization. The inductor L_C of both the LNA and PA are realized off-chip due to area considerations.

C. Power Detection

An off-chip realization of the power coupling network is considered due to the area overhead required to integrate it within the IC. The wideband logarithmic power detector demonstrated in [8] is used for power detection. The detector has a dynamic range larger than 50 dB and sensitivity lower than -60 dBm across the bandwidth of interest.

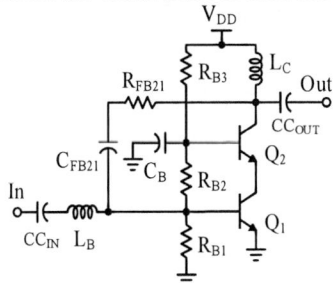

Fig. 3. Low-noise amplifier.

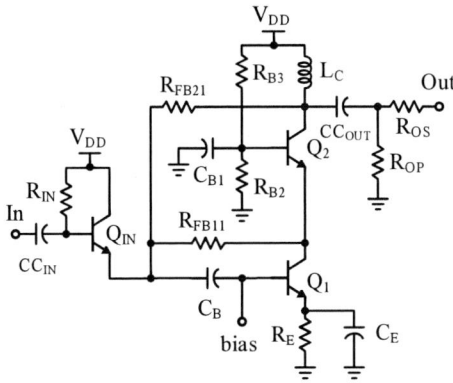

Fig. 4. Power Amplifier.

D. Switches

The switch at the output of the DCO and the switches within the attenuator stages are realized using NMOS transistors.

IV. TESTING MODE

The RF test stimulus required for the test mode of operation is generated using a 5 stage digital-controlled ring oscillator and a harmonic cancellation active buffer.

The core of the ring oscillator circuit is the CMOS inverter based delay cell as shown in Fig. 5a. The current through the core stage inverter is adjusted by setting the digital bits D_1 - D_3 and thereby the delay is varied and different frequency states which are spread across the bandwidth of interest are generated. The signals from the multi-phase ring oscillator are summed up using the buffer circuit shown in Fig. 5b, which generates the test stimulus with low harmonic content by using the principle of harmonic cancellation [9]. Along with signal summation, the buffer circuit also helps in attenuating the oscillator output to an amplitude level which is within the input dynamic range of the RF transceiver. A frequency sense circuit is added to measure the frequency of the generated signal and provide a 6 bit digital readout.

In the normal operation mode, the DCO is turned off and the switch at the output of the DCO is open. In the test mode, the switch at the output of the oscillator is closed to pass the test stimulus as input to the transceiver. The oscillator state is controlled digitally to change the frequency of the test stimulus and to enable testability across the wide bandwidth of

interest. Seven different frequencies can be generated, with one frequency each at lower end, middle region and upper end of the bandwidth of interest. The amplitude of the generated test stimulus is known. Therefore, the gain of the transceiver and the basic functionality of individual RF blocks can be verified by observing the power at the output of the transceiver for different attenuation settings.

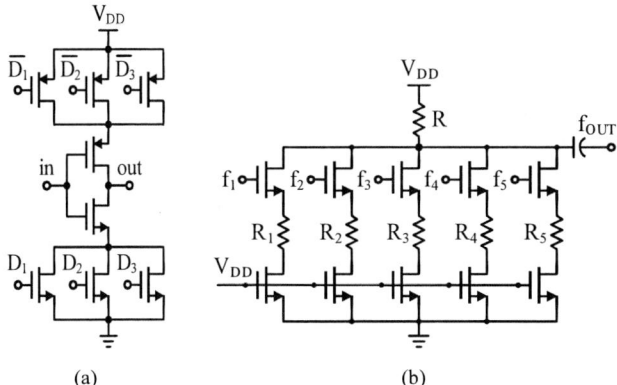

(a)   (b)

Fig. 5. DCO building blocks: (a) delay cell (b) signal summation buffer.

V. MEASUREMENT RESULTS

The IC is realized using a SiGe BiCMOS technology. EM modeling/circuit co-simulations are carried out to model the impact of signal routing, bond-pads, bond-wires, package leads and PCB routing during the design phase. The operating bandwidth of the RF transceiver is from 700 MHz to 2.5 GHz. An attenuation range of 0 to 31.5 dB with a resolution of 0.5 dB is achieved. The tilt attenuator has a tilt range of 0 to 5 dB with a resolution less than 1 dB. The LNA and PA achieve a minimum gain of 14 dB and 10.6 dB respectively, across the entire frequency range as verified by on-wafer measurements. Fig. 6 and Fig. 7 illustrate the gain of the RF transceiver path for different values of attenuation for the tilt attenuator operating in bypass mode and in maximum tilt mode, respectively. 1dB compression point measurement of the transceiver for the center frequency of 1.6 GHz is depicted in Fig. 8. The measurement results of the overall chip are summarized in Table 1. Fig. 9 shows the chip photograph. The die occupies an area of 5.07 mm^2. The overall power consumption is limited to a maximum of 674.74 mW across all modes of operation.

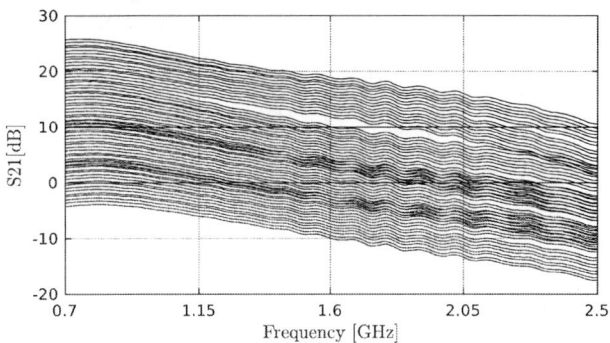

Fig. 6. RF path gain for different values of attenuation when tilt attenuator is in bypass mode.

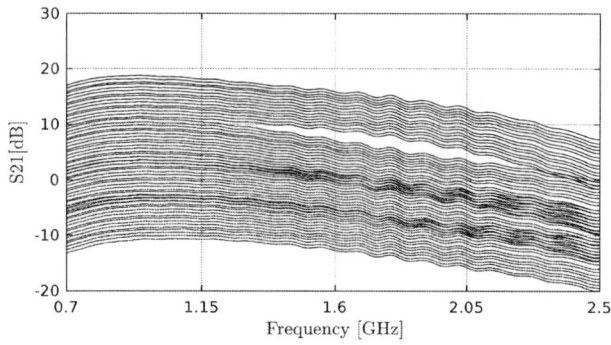

Fig. 7. RF path gain for different values of attenuation when tilt attenuator is in full tilt mode.

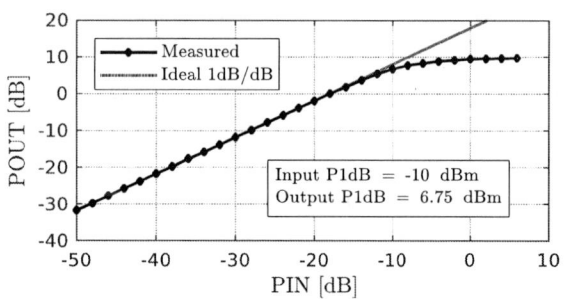

Fig. 8. Measured 1dB compression point of the RF transceiver at 1.6 GHz (attenuation set to 0.5 dB and tilt operated in bypass mode).

The overall gain and the basic functionality of the different RF blocks have been evaluated using the test mode. Example results from the test mode of operation for three different test stimuli generated by the DCO are provided in Table 2. The BIST mode can be used during the post-production phase as well as during the deployment phase to evaluate the gain performance of the RF transceiver and to monitor its functionality across its life-cycle.

Table 1. Performance summary.

Parameter	Value
Frequency range [GHz]	0.7 - 2.5
S11; S22 ; S12 [dB]	< -8 ; < -10 ; < -60
Attenuation range [dB]	31.5
Attenuation resolution [dB]	0.5
Noise figure [dB]	7.42 (at 1.6 GHz)
Output P1dB [dBm]	6.75 (at 1.6 GHz)
DCCS output current range [mA]	> +/- 60
DCCS resolution [mA]	< 1
Tilt attenuation range [dB]	5
DC Power consumption [mW]	< 674.74

Table 2. Results from the test mode of operation for 3 different test frequencies (attenuation set to 0.5 dB and tilt operated in bypass mode).

Parameter	Value		
Digital Setting	001	011	111
Frequency [GHz]	0.6	1.612	2.525
Transceiver output power [dB]	-2.4	-7.4	-15.01
Transceiver gain estimate [dB]	24.2	17.68	10.1

Fig. 9. Chip Photograph. Die area: 2.78 mm x 1.76 mm.

VI. CONCLUSION

A wideband RF transceiver IC with digital gain tuning, digital wavelength control and self testing capabilities intended for optical transceiver modules has been presented. A gain tuning range of 31.5 dB is achieved. A digital-controlled current source with +/- 60 mA output current range is integrated for wavelength tunability. The BIST mode can be used to monitor the gain performance of the RF transceiver and to study its post-production and deployment phase functionality. High integration density is achieved by utilizing digital controls, minimum number of integrated inductors and area efficient design approaches. The required board area and form factor of an optical transceiver module can be significantly reduced by means of this mixed-signal RFIC which integrates multiple functionalities into a single chip.

ACKNOWLEDGMENT

This work was supported by the German Federal Ministry for Economic Affairs and Energy (BMWi) through AiF in the context of the Zentrales Innovationsprogramm Mittelstand (ZIM) project under grant ZF4080811PR7.

REFERENCES

[1] L. Hutcheson, "FTTx: Current Status and the Future," in *IEEE Communications Magazine*, vol. 46, no. 7, pp. 90-95, July 2008.

[2] A. Brillant. *Digital and analog fiber optic communications for CATV and FTTx applications*, Washington: SPIE, 2008.

[3] J. J. Dabrowski and R. M. Ramzan, "Built-in Loopback Test for IC RF Transceivers," in *IEEE Transactions on Very Large Scale Integration (VLSI) Systems*, vol. 18, no. 6, pp. 933-946, June 2010.

[4] S. Lakshminarayanan, J. Ning and K. Hofmann, "A CMOS Bidirectional -69 mA to +63 mA Output Range Voltage Controlled Current Source for Laser Diode Current Control in FTTx Applications," *ANALOG 2018; 16th GMM/ITG-Symposium*, Munich/Neubiberg, Germany, 2018, pp. 1-4.

[5] B. Ku and S. Hong, "6-bit CMOS Digital Attenuators With Low Phase Variations for X-Band Phased-Array Systems," in *IEEE Transactions on Microwave Theory and Techniques*, vol. 58, no. 7, pp. 1651-1663, July 2010.

[6] P. G. Sulzer, "A Note on a Bridged-T Network," in *Proceedings of the IRE*, vol. 39, no. 7, pp. 819-821, July 1951.

[7] H. Chen, Y. Lin and S. Lu, "Analysis and Design of a 1.6–28-GHz Compact Wideband LNA in 90-nm CMOS Using a π-Match Input Network," in *IEEE Transactions on Microwave Theory and Techniques*, vol. 58, no. 8, pp. 2092-2104, Aug. 2010.

[8] S. Lakshminarayanan and K. Hofmann, "A wideband RF power detector with −56 dB sensitivity and 64 dB dynamic range in SiGe BiCMOS technology," *2017 IEEE International Symposium on Circuits and Systems (ISCAS)*, Baltimore, MD, 2017, pp. 1-4.

[9] R. K. Pokharel, P. Nugroho, A. Anand, K. Kanaya and K. Yoshida, "Digitally controlled CMOS quadrature ring oscillator with improved FoM for GHz range all-digital phase-locked loop applications," *2012 IEEE/MTT-S International Microwave Symposium Digest*, Montreal, QC, 2012, pp. 1-3.

A Compact Single-Ended Dual-band Receiver with Crosstalk and ISI Reductions for High-density I/O Interfaces

Jieqiong Du[#], Jia Zhou[#], X. Shawn Wang[#], Chien-Heng Wong[#],
Huan-Neng Chen[$], Chewn-Pu Jou[$], Mau-Chung Frank Chang[#]

[#]High Speed Electronics Lab, University of California, Los Angeles, USA
[$]Taiwan Semiconductor Manufacturing Co., Ltd. Taiwan

Abstract— A compact single-ended dual-band receiver is presented, consisting of one baseband using PAM-2 signaling and a coherent RF band using PAM-2/PAM-4 signaling. Exploiting the orthogonality between different bands and between adjacent PCB channels, the receiver reduces crosstalk and relaxes equalization requirements, enabling low-BER data transmission at frequencies where channel loss and crosstalk are high. The receiver adopts a passive-mixer-first architecture to improve the linearity and a flipped source-follower-based low-pass filter to improve the energy efficiency. Occupying 0.004 µm^2 in TSMC 28nm technode, the proposed single-ended receiver achieves 0.45 pJ/bit at 9 Gb/s/pin for closely spaced 5-inch FR-4 channels and 0.6 pJ/bit at 6 Gb/s/pin for lossy and strongly coupled 22-inch FR-4 channels.

Keywords — single-ended signaling, crosstalk, intersymbol interference, wireline receiver, frequency division multiplexing.

I. INTRODUCTION

The demand for the aggregated I/O bandwidth increases rapidly while the number of available I/Os increases only slightly due to the package limitations. Single-ended signaling improves the pin-efficiency and I/O bandwidth by doubling the data rate per pin, compared with differential signaling. However, as the data rate increases, crosstalk noise and intersymbol interferences (ISI) severely degrade the signal integrity. In [1] and [2], circuit solutions to reduce crosstalk and ISI are proposed at the cost of additional filters and multi-tap decision feedback equalizers, which limits the energy efficiency. Recent studies in differential links [3]–[5] show that multi-band signaling has the potential to greatly reduce equalization complexity for lossy channel. [3] also suggests that multi-band signaling reduces crosstalk because different frequency bands are orthogonal and only in-band crosstalk interferes. However, because of the reduced in-band signal power in multi-band signaling, signal-to-crosstalk ratio does not improve for overlapping quadrature-modulated RF bands at adjacent channels used in [3]–[5], which is an issue in single-ended links.

In this work, a single-ended dual-band receiver is proposed to efficiently alleviate far-end crosstalk (FEXT) from adjacent channels as well as intersymbol interferences. The receiver consists of a baseband using PAM-2 signaling and a Radio Frequency (RF) band using PAM-2/PAM-4 signaling. By applying PAM modulation, crosstalk is effectively reduced by leveraging the orthogonality properties between different frequency bands and adjacent PCB channels in the following ways: 1) RF and baseband signals are mutually orthogonal to each other in frequency domain. 2) RF band signals from adjacent PCB channels are also orthogonal to each other because of the far-end crosstalk characteristics. The proposed receiver (RX) achieves 0.45 pJ/bit at 9 Gb/s/pin with 5-inch FR-4 channels and 0.6 pJ/bit 6 Gb/s/pin with 22-inch FR-4 channels under strong crosstalk from an adjacent aggressor.

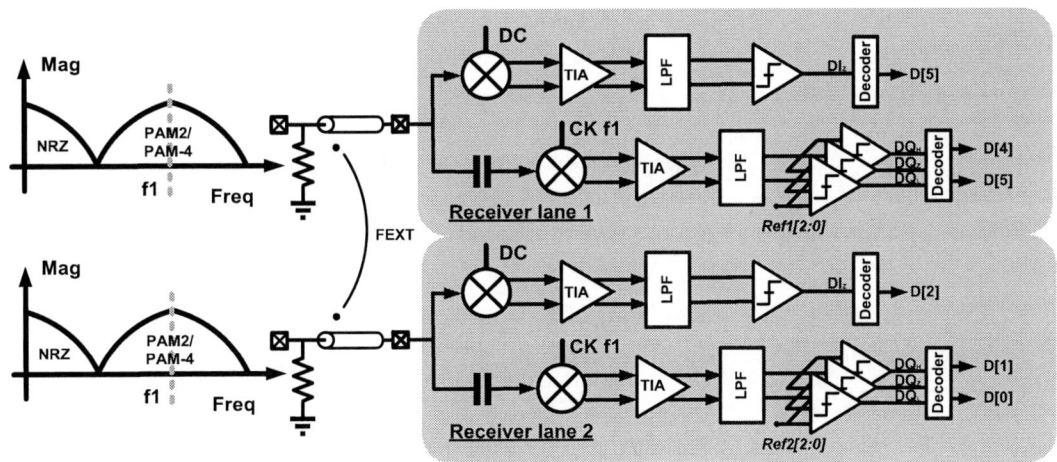

Fig. 1. Receiver Architecture.

978-1-7281-1702-7/19 $31.00 © 2019 IEEE

II. ARCHITECTURE

The proposed architecture is shown in Fig. 1. Each incoming signal utilizes two bands. The transmitters to generate such spectrum can be found in [3]–[5]. In the receiver, the dual-band signals are separated where baseband signal is recovered by filtering and the RF band signal is recovered by down-conversion with the interferences suppressed.

The dual-band architecture is designed to effectively reduce crosstalk from adjacent channels, which is usually the dominant source of crosstalk. For data transmissions in the baseband, the data bandwidth is reduced by a factor of two or three compared with conventional single band NRZ signaling. By reducing the baseband bandwidth, the signal-to-crosstalk ratio at Nyquist rate is greatly reduced because of the band-pass characteristic for FEXT [1]. Low-pass filters also provide over 30dB suppression for crosstalk coming from the RF band. For data transmissions in the RF band, signal integrity is also improved because the crosstalk and signal are designed to be orthogonal to each other and would be efficiently filtered out, even though the crosstalk energy is not reduced. For inductive-coupling dominated channels, adjacent channel FEXT transfer function is approximated as the derivative of the channel response [1]. Therefore, the crosstalk from adjacent-channel RF band will be approximately 90° different from the signal. The RF carrier phase is calibrated using a digital-controlled delay line so that the crosstalk is minimized and small deviation from 90° phase shift will not severely degrade the performance since the signal energy is parabolic around the optimal phase.

The dual-band data transmission also reduces ISI by doubling the symbol period, fitting channel characteristics and self-equalizing a low-pass channel [5]. For a low-pass wireline channel, each band has a smaller bandwidth, so that in-band loss variation is smaller. Furthermore, for the RF band, the spectrum of PAM-2/PAM-4 is double side-banded (both sidebands contain duplicated information). While the upper sideband experiences more loss at channel than the lower sideband, the in-band loss variation will be compensated after the down-conversion when both upper and lower sidebands are summed together. Therefore, the ISI is greatly reduced.

III. CIRCUIT IMPLEMENTATION

The receiver has two signal paths, each consisting of a current-mode passive mixer, a transimpedance amplifier, a 5th order low-pass filter(LPF), comparators, and decoders. The RF path has an additional capacitor serving as the ac coupling capacitor as well as a high-pass filter to reduce the baseband interferences.

The receiver adopts a passive-mixer-first architecture to achieve better efficiency, linearity and flexibility. The input impedance termination of the receiver is provided by a parallel combination of the baseband path and the RF path. Fig. 2a shows the passive mixer and the first stage of TIA. The mixer serves three purposes: 1) the mixer naturally serves as a single to differential circuit to improve PSRR; 2) the mixer translates

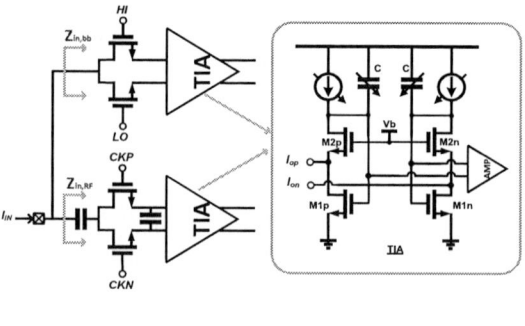

(a)

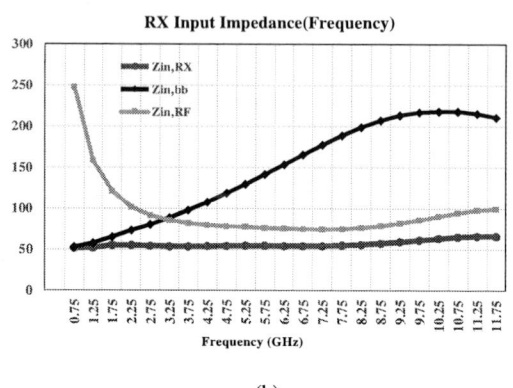

(b)

Fig. 2. a) Mixer and Transimpedance amplifier; b) simulated RX input impedance, baseband input impedance, and RF band input impedance.

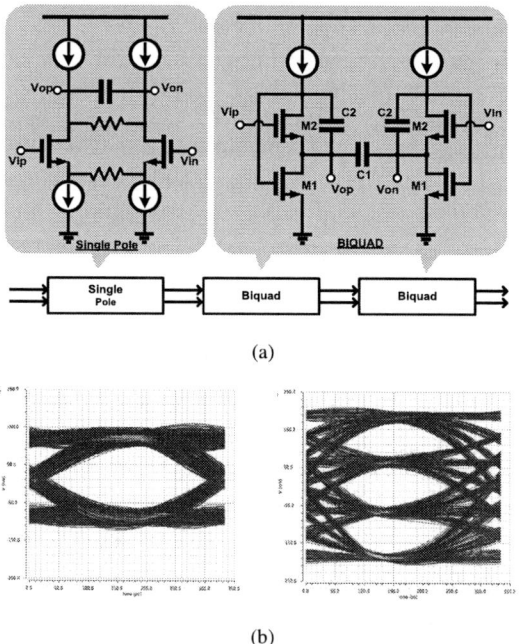

(a)

(b)

Fig. 3. a) $5^{t}h$ Order LPF; b) simulated eye diagram at LPF Output at 9 Gb/s/pin. (LEFT: baseband;Rigth: RF band)

the impedance characteristics of the transimpedance amplifier to RF band and therefore reduces the required bandwidth of the transimpedance amplifier; and 3) the mixer in the RF path down-converts the RF signal to baseband while the baseband mixer serves as a dummy to decrease data skews. The input impedance of the receiver is given by the parallel impedance of the baseband path and the RF path, and the simulated input impedance versus frequency with a carrier at 6 GHz is shown in Fig. 2b. Along the baseband path, the input impedance is determined by the input impedance of the TIA, which is band-pass shaped. Along the RF band, the input impedance equals the upconverted and scaled input impedance of the TIA in series with the switching resistance of the mixers, which is band-stop shaped [6]. With this architecture, the received baseband signal current sees a lower impedance at the receiver baseband path and therefore flows to the baseband path, while the received RF signal current flows to the RF path. The overall impedance at RX input is near 50 Ohms over frequencies of interest. This architecture also enables some flexibility in choosing the appropriate RF band frequencies.

Following the TIA, a 5th order linear-phase LPF removes out-of-band interferences from adjacent frequency band as well as the adjacent-channel crosstalk. The 5th order LPF consists of three stages as Fig. 3a shows. The first stage is a first-order amplifier and the source degeneration is used to improve the linearity. The other two stages are second-order bi-quads using flipped-source-follower based cells [7], which enables low-power and low-voltage operation. The LPF provides about 35 dB attenuation at 6 GHz. The simulated output of the LPF at 9 Gb/s/pin is shown at Fig. 3b (3 Gb/s PAM-2 at baseband and 6 Gb/s PAM-4 at 6 GHz RF Band). Continuous-time comparators then amplify low-pass filtered outputs to full scale where the thresholds are tuned by low-speed digital-to-analog converters to cancel out mismatches or support higher-order modulation like PAM-4.

IV. MEASUREMENT

During the testing of the receiver, a Tektronix AWG7122B Arbitrary Waveform Generator (AWG) serves as the transmitter and shapes the dual-band signal. The RX carrier clock is fed from an external clock source, which is synchronized with the AWG through a 10 MHz reference clock. The RX carrier phase is calibrated offline through an on-chip digital-controlled delay line. The receiver is tested with two channels – two pairs of 5-inch and 22-inch FR-4

PCB traces. The transmitted signal consists of one baseband and one RF band at 6 GHz with a 3 GHz symbol bandwidth. Uncorrelated PRBS7 patterns are used at different frequency bands and different channels.

With the 5-inch PCB traces, the receiver operates up to 9 Gb/s/pin by recovering a PAM-2 signal at baseband and a PAM-4 signal at RF band with a symbol rate of 3 GHz. The signal-to-crosstalk ratio at the Nyquist rate of the baseband is improved to 18dB, compared to transmitting NRZ signals of 9 Gb/s where the signal-to-crosstalk ratio under the Nyquist rate is 8dB. Even though the signal-to-crosstalk ratio at RF band decreased to only 4dB, the effect is negligible because of the orthogonality between different channels. Fig. 4 shows the measured insertion loss and FEXT of the 5-inch PCB traces. It also shows the demodulated PAM-4 data eye diagram measured after the continuous-time comparators from the RF band with and without crosstalk, which shows minor degradation in the presence of an aggressor.

With the 22-inch FR-4 traces, the receiver also proves to be highly tolerant to ISI. The receiver recovers a PAM-2 signal at the baseband and a PAM-2 signal at the RF band with a symbol rate of 3 GHz, adding to a total rate of 6 Gb/s. Instead of compensating 18dB loss at 3 GHz for a 6 Gb/s NRZ signaling, only 6dB equalization is required for the 3 Gb/s baseband signal with the transmitter while the RF band requires no extra equalization with a channel loss of more than 20 dB at 6 GHz. Fig. 5 shows measured insertion loss and FEXT of the 22-inch PCB traces. It also compares the crosstalk immunity of dual band signaling to that of conventional NRZ signaling through the same channels. The two mid-left figures show the equalized conventional NRZ eyediagram at 6 Gb/s with and without crosstalk. The channel is equalized using FFE in the AWG transmitter. As is seen in the bottom mid-left figure, while channel loss can be equalized by the AWG, crosstalk still closes the data eye. In comparison, the data eye for proposed dual-band signaling at 6 Gb/s (3 Gb/s at each band) is open with 0.1 UI margin for channel 1 and 0.2 UI margin for channel 2 at BER<10-12 with crosstalk.

The receiver operates up to 9 Gb/s/pin with 5-inch FR-4 traces and up to 6 Gb/s/pin with the 22-inch FR-4 while consuming 4.05mW (2.88mW for mixers, TIAs, low pass filters and comparators and 1.17mW for the digital-controlled

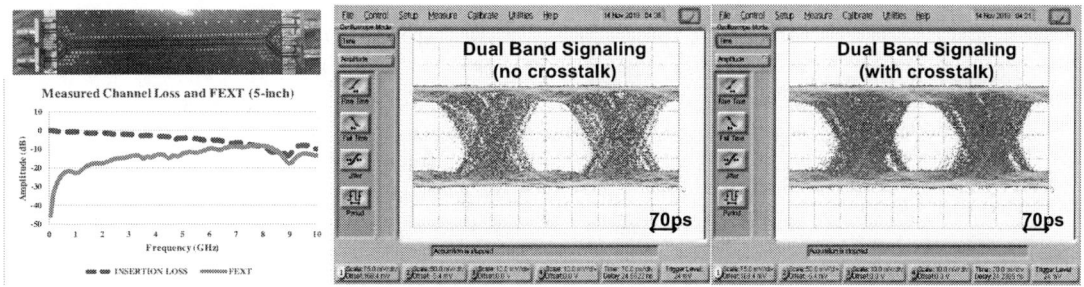

Fig. 4. LEFT: Measured channel insertion loss and FEXT; MID: Demodulated PAM-4 (LSB) data eye diagram at dual band receiver with 5-inch FR-4 without crosstalk. RIGHT: Demodulated PAM-4 (LSB) data eye diagram at dual band receiver with 5-inch FR-4 with crosstalk.

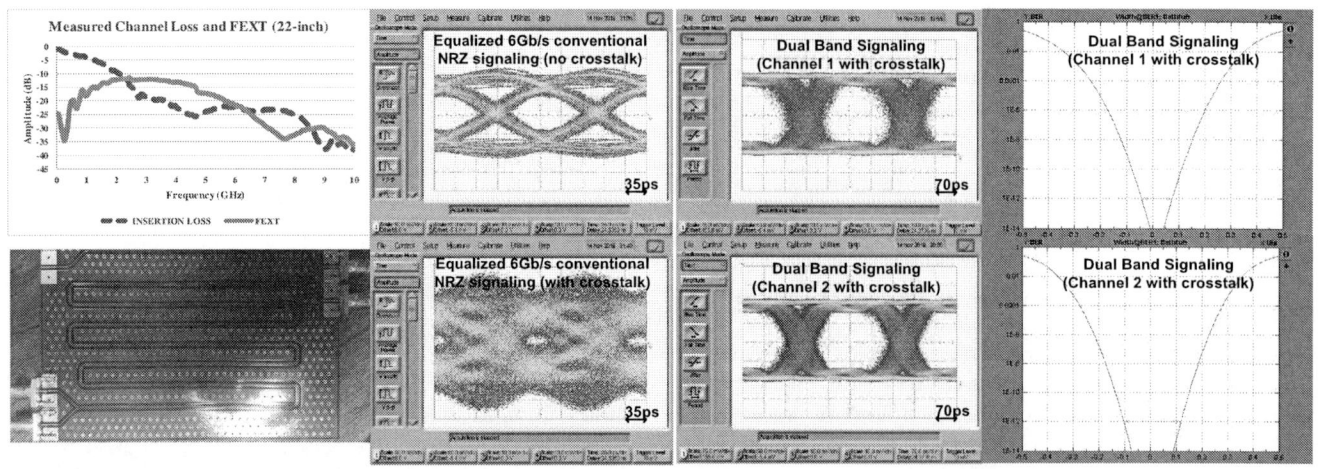

Fig. 5. LEFT: Measured Channel Loss and FEXT (22-inch FR-4); MID LEFT: Conventional NRZ signaling for comparison (from AWG to Oscilloscope); MID RIGHT: Demodulated data eye diagram from dual band receiver for both channel 1 and channel 2; RIGHT: Estimated bathtub for channel 1 and channel 2 for the dual band receiver.

delay line and clock buffers). It is fabricated in TSMC 28nm HPC technode; the active area for the proposed receiver is only 0.004 μm² per lane. Fig. 6 shows a microphotograph of the fabricated chip.

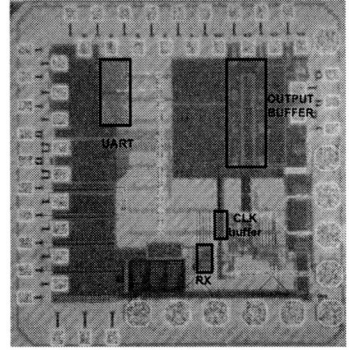

Fig. 6. Die Photo

Table 1. Performance Comparison with Prior Arts

Reference	[1]	[2]	[3]	This Work	
Technode	65nm	32nm SOI	40nm	28nm	
XTC Type	CTXC	CTXC DFXC	Multi-tone	Dual Band	
I/O Type	Single Ended	Single Ended	Diff-erential	Single Ended	
Channel	6-inch FR-4	28-inch/ 39-inch Roger	12-inch MDB	5-inch FR-4	22-inch FR-4
Data Rate (Gb/s/pin)	12	7	4.5	9	6
Area (mm)	0.036	0.012	0.0078	0.004	0.004
Energy Efficiency (pJ/bit)	1.78*	5.9	0.49*	0.45	0.6

*Calculated for the whole RX

V. CONCLUSION

In this paper, a low-power 6-9 Gb/s/pin dual-band receiver is presented. By leveraging the orthogonality between two different frequency bands and the 90-degree phase shift between adjacent channels, the receiver has effectively reduced the crosstalk from adjacent PCB channels and improved the signal integrity. The receiver also shows ISI reduction along lossy channels without requiring equalizers. Occupying 0.004 μm² in TSMC 28nm CMOS technology node and consuming 4.05 mW, the receiver shows great promises for low-power wireline data transmissions for high-density I/Os.

REFERENCES

[1] T. Oh and R. Harjani, "4×12 Gb/s 0.96 pJ/b/lane analog-IIR crosstalk cancellation and signal reutilization receiver for single-ended I/Os in 65 nm CMOS," in *2012 Symposium on VLSI Circuits (VLSIC)*, June 2012, pp. 140–141.

[2] C. Aprile, A. Cevrero, P. A. Francese, C. Menolfi, M. Braendli, M. Kossel, T. Morf, L. Kull, I. Oezkaya, Y. Leblebici, V. Cevher, and T. Toifl, "An eight-lane 7-Gb/s/pin source synchronous single-ended rx with equalization and far-end crosstalk cancellation for backplane channels," *IEEE Journal of Solid-State Circuits*, vol. 53, no. 3, pp. 861–872, March 2018.

[3] K. Gharibdoust, A. Tajalli, and Y. Leblebici, "A 4×9 Gb/s 1 pJ/b NRZ/multi-tone serial-data transceiver with crosstalk reduction architecture for multi-drop memory interfaces in 40nm cmos," in *2015 Symposium on VLSI Circuits (VLSI Circuits)*, June 2015, pp. C180–C181.

[4] Y. Du, W. Cho, P. Huang, Y. Li, C. Wong, J. Du, Y. Kim, B. Hu, L. Du, C. Liu, S. J. Lee, and M. F. Chang, "A 16-Gb/s 14.7-mW tri-band cognitive serial link transmitter with forwarded clock to enable PAM-16/256-QAM and channel response detection," *IEEE Journal of Solid-State Circuits*, vol. 52, no. 4, pp. 1111–1122, April 2017.

[5] W. Cho, Y. Li, Y. Du, C. Wong, J. Du, P. Huang, S. J. Lee, H. Chen, C. Jou, F. Hsueh, and M. F. Chang, "10.2 a 38mW 40Gb/s 4-lane tri-band PAM-4 / 16-QAM transceiver in 28nm cmos for high-speed memory interface," in *2016 IEEE International Solid-State Circuits Conference (ISSCC)*, Jan 2016, pp. 184–185.

[6] A. Mirzaei, H. Darabi, J. C. Leete, X. Chen, K. Juan, and A. Yazdi, "Analysis and optimization of current-driven passive mixers in narrowband direct-conversion receivers," *IEEE Journal of Solid-State Circuits*, vol. 44, no. 10, pp. 2678–2688, Oct 2009.

[7] M. De Matteis and A. Baschirotto, "A biquadratic cell based on the flipped-source-follower circuit," *IEEE Transactions on Circuits and Systems II: Express Briefs*, vol. 64, no. 8, pp. 867–871, Aug 2017.

A 26 dBm 39 GHz Power Amplifier with 26.6% PAE for 5G Applications in 28nm bulk CMOS

Kaushik Dasgupta, Saeid Daneshgar, Chintan Thakkar, James Jaussi, Bryan Casper

Intel Corporation, Hillsboro, USA

Email: kaushik.dasgupta@intel.com

Abstract—**Continued demand for 5G cellular connectivity in mobile handheld devices, where antenna real-estate is at a premium, necessitates high output power from individual transmitter elements. While more expensive heterogeneous and SOI CMOS process based power amplifiers (PAs) provide high P_{out} at good efficiencies, deep sub-μm bulk CMOS still remains the technology of choice for cost and integration benefits. This paper presents a 5G mm-Wave PA at 39 GHz that generates a saturated P_{out} of 26 dBm with a peak power-added efficiency (PAE) of 26.6% and a saturated power gain of 28.6 dB. The output stage utilizes compact layout & triple-well transistors to enable efficient yet reliable device stacking and a compact, 4-way, low-loss, series-parallel power combiner further enhances P_{out}. High average power measurements have been demonstrated during single-carrier as well as 5G NR OFDM modulations at competitive efficiencies. Long term reliability measurements using aging acceleration techniques demonstrate the robustness of the implemented PA. This PA achieves one of the highest ITRS figure-of-merit among reported 5G mm-Wave works in CMOS and also the highest output power among deep sub-μm (<90nm) 5G bulk CMOS PAs.**

Keywords—**5G, CMOS power amplifier, millimeter-wave.**

I. INTRODUCTION

The widespread demand for gigabit per second cellular connectivity for content-rich handheld applications has driven the development and recent standardization of the 5G New Radio (NR) specifications in the mm-Wave bands world-wide from 24 to 43 GHz [1]. As a result, there has been a surge of interest in 5G chipsets for base-station applications where the form factor allows for multi-element, large-aperture phased arrays to be realistically implemented. The peak output power (P_{max}) requirement of the individual transmitter (TX) elements for such applications is moderate (i.e. $5 - 15$ dBm [2]) and the desired link range is achieved through array gain.

However, in a hand-held form-factor with limited usable dimensions, only a small number of elements can be physically integrated [2], thus dramatically increasing the requirement on P_{max}. For an illustrative ~100 m link (for better coverage than the 30 m estimate in [2]) with moderate multipath shadowing, even an asymmetric uplink-based phased-array (e.g. $4 - 8$ element TX, $64 - 128$ element RX) transmitting a high-order constellation (e.g. 64 QAM $-$ 256 QAM) over ~ 500 MHz bandwidth would require per-element TX P_{max} of >23 dBm at 39 GHz. Generating such high P_{max} at mm-Wave frequencies in handheld devices necessitates high efficiency power amplifiers (PAs).

Recent works in SOI CMOS and III-V processes have demonstrated high P_{max} at reasonable efficiency [3], [4].

However, the high-cost and low-yield of some of these technologies make low-cost bulk CMOS a desirable process technology choice for the mobile 5G market. Additionally, integrating the TX in the same CMOS process node as the 5G modem/transceiver can present integration benefits in volume manufacturing and costs. However, bulk CMOS integration of high-power PAs brings about a set of challenges mainly due to swing limitations and long-term reliability issues affecting P_{max}.

This paper addresses these challenges through the use of (a) triple-well device stacking in bulk CMOS to reliably achieve high swings and (b) compact on-chip distributed active transformer (DAT) to achieve low-loss 4-way power combining. By using these techniques, a 3-stage, 39 GHz PA for 5G user equipment (UE) applications is implemented in 28 nm bulk CMOS, with a saturated output power (P_{sat}) of 26 dBm while maintaining a peak PAE of 26.6%. The PA delivers up to 23.5/21.5/19.5 dBm average P_{out} for SC-QPSK/16QAM/64QAM modulations respectively and is also shown to be compatible with the newly developed 5G-NR 64QAM OFDM Uplink signals at an average P_{out} of 14.7 dBm. Long-term reliability of the PA is also demonstrated through voltage and temperature stress tests over multiple samples for continuous operation at saturated output power (P_{sat}) and OFDM measurements.

II. POWER AMPLIFIER DESIGN

A. Stacking in bulk CMOS

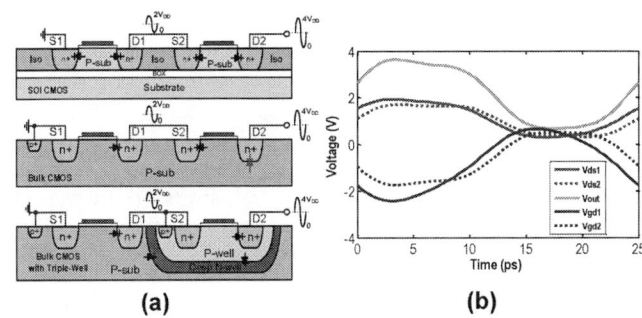

Fig. 1. (a) Cross-sections of a 2-stack cell showing critical breakdown junctions (red diodes) for SOI-CMOS, bulk CMOS, & bulk CMOS with triple-well transistors. (b) Simulated voltage swings at critical nodes.

Device stacking [5] in deep sub-μm SOI CMOS is a well-established technique to alleviate breakdown voltage limits by dividing the voltage swing across several transistors,

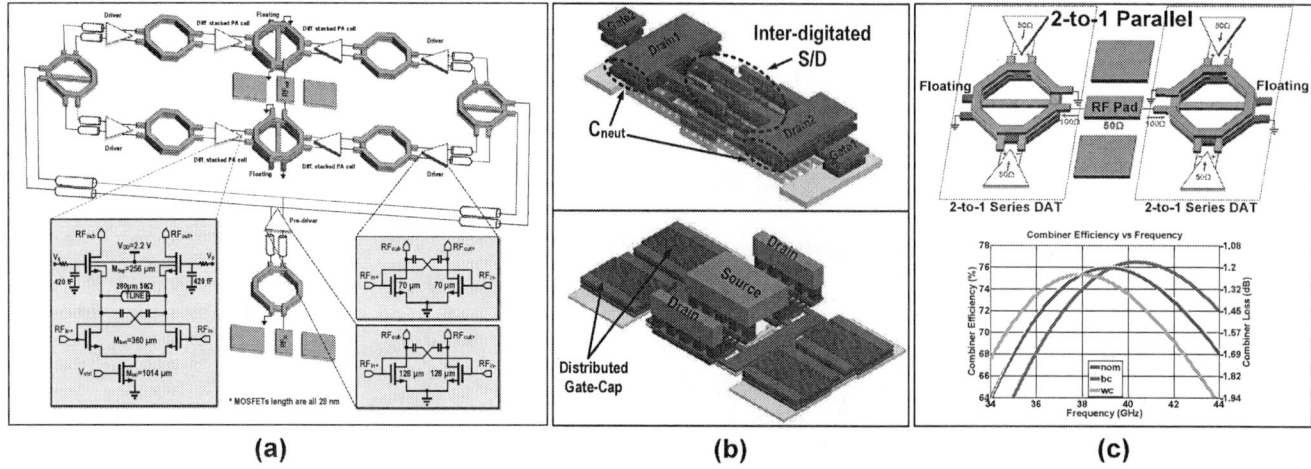

Fig. 2. (a) Block diagram of the 3-stage 39 GHz power amplifier showing schematics of the stacked output stage and common-source driver stages. (b) 3D view of the PA unit cell for the common-source device for M_{bot} and 3D view of two parallel stacked transistors with their corresponding gate capacitances as the unit cell for M_{top}. (c) 3D layout view of 4-to-1 series-parallel combiner showing output pad arrangement to minimize parasitic losses and simulated combiner efficiency over BEOL corners.

thus improving reliability. Furthermore, SOI CMOS PAs do not face explicit drain-bulk breakdown issues while sustaining higher swings at the output node, and therefore benefit dramatically from stacking (Fig. 1(a)). While stacking can be implemented in bulk CMOS, drain-bulk reverse breakdown generally restricts usable supply voltages and therefore limits swing and power enhancement (red diode in Fig. 1(a)).

To mitigate these concerns, this work utilizes triple-well transistors in a bulk CMOS process where the top stacked transistor is a triple-well device with its p-well tied to its source. Since the source on the stacked device swings in phase with its drain, shorting the source to its p-well relieves the breakdown stress on the drain-bulk diode. As an added benefit, the topology also eliminates source-bulk capacitance. Fig. 1(b) shows the simulated single-ended voltage swings across the critical drain-source and gate-drain junctions demonstrating proper stack operation.

Fig. 2(a) shows the architecture and schematics of the 3-stage PA with neutralized differential driver and pre-driver stages and a 2-stack neutralized differential output stage. Four such output stages are power combined using a 4-to-1 series-parallel DAT based low-loss combiner.

B. Power Cell Design

A compact, low-parasitic yet reliability-compliant PA transistor layout is key to maxime the efficiency of the output stage. In order to reduce source-drain parasitic capacitances, the bottom transistor differential pair is implemented using an inter-digitated layout with a shared source connection. The layout also utilizes inherent opposite-sided gate-to-drain metal overlap parasitic for gate-drain neutralization, thus improving gain and stability (Fig. 2(b)). The top transistor (M_{top} in Fig. 2(a)) is a triple-well transistor which is layed out in a distributed fashion with interleaved gate capacitors to minimize parasitic inductances (Fig. 2(b)).

Inter-stage impedance matching between the top and bottom transistors is achieved through a differential T-line. Compared to a single-ended T-line [3], a differential T-line eliminates lossy decoupling capacitors used in single-ended designs, thus improving efficiency. A single output stage with a lossless output matching network achieves a simulated P_{sat} of 21.5 dBm with drain-efficiency of 43 % from a 2.2 V supply.

C. Low-loss DAT-based power combining

Enhancing P_{sat} to higher than achievable by pure device stacking requires on-chip power combining. Amongst power combining topologies, DAT-based combiners are well established as low-loss and small form-factor combiners of differential PA stages [6]. However, since most antenna drives are single-ended, differential to-single-ended conversion in a DAT often creates severe impedance-imbalance, thus leading to significant degradation in efficiency. The imbalance is inherently created due to the asymmetry in inter-winding capacitance on the secondary coil. To mitigate this problem, the implemented DAT balances the inter-winding capacitance by utilizing an opposite floating dummy winding underneath the primary to equalize electrical coupling [7].

To further increase P_{out}, two parallel 2-to-1 series-combined DATs are arranged on two sides of the RF output pad thus forming a 4-to-1 series-parallel combiner (Fig. 2(c)). This transforms the 50 Ω load to a 50 Ω optimum impedance required by each PA. The opposite-sided arrangement also eliminates extra leads to the RF pad as in [6]. The designed 4-to-1 combiner achieves a simulated loss as low as 1.2 dB in the nominal BEOL corner (Fig. 2(c)).

In simulation, the 3-stage PA achieves a P_{sat} of 26.4 dBm with an overall drain-efficiency of 27.8%, including the driver stages. Simulations were also performed to verify the PA performance over VSWR events caused by antenna mismatches. The PA maintains a $P_{sat} > 24$ dBm over a

978-1-7281-1702-7/19 $31.00 © 2019 IEEE

3:1 VSWR (Fig. 3). The corresponding simulated drain efficiencies and peak gate-drain voltages are also shown in Fig. 3.

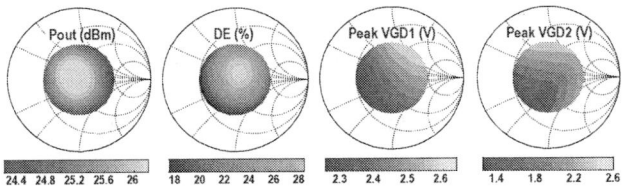

Fig. 3. Simulated performance of the PA against 3:1 VSWR mismatch.

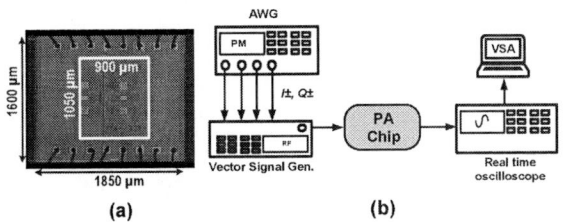

Fig. 4. (a) Die micrograph. (b) Setup for modulated measurements.

III. MEASUREMENT RESULTS

The 5G mmWave PA test-chip was fabricated in a standard 28 nm bulk CMOS process and occupies an active area of $0.95 \, \text{mm}^2$ including RF pads (Fig. 4(a)). DC supplies and biases are wire-bonded to a standard FR4 PCB while the GSG input and output signals are probed.

A. Static Performance

The measured small-signal gain is 38 dB centered at 38 GHz with an $S_{11} < -10 \, \text{dB}$ bandwidth of 37 - 42.3 GHz (Fig. 5(a)). The peaking in small-signal gain is attributed to over-neutralization in the amplifier stages. Unconditional stability was confirmed over the entire frequency range of 10 MHz to 50 GHz (Fig. 5(b)).

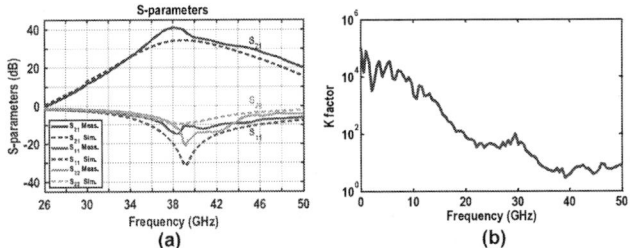

Fig. 5. (a) Measured and simulated S-paramters and (b) measured stability factor (K) showing unconditional stability over frequency.

The PA achieves a P_{sat} of 26 dBm at a peak drain and power-added-efficiency (PAE) of 26.7% and 26.6%, respectively (Fig. 6). The measured output P_{1dB} is 21.5 dBm and the 1-dB P_{sat} bandwidth is from 36.5 to 42 GHz with a >20% efficiency over the same frequency range (Fig. 6). Measurements over various supply voltages including the process nominal of 0.9 V per transistor demonstrate the flexibility of operation of the PA with a $P_{sat} > 24$ dBm even at the lowest supply voltage (Fig. 6).

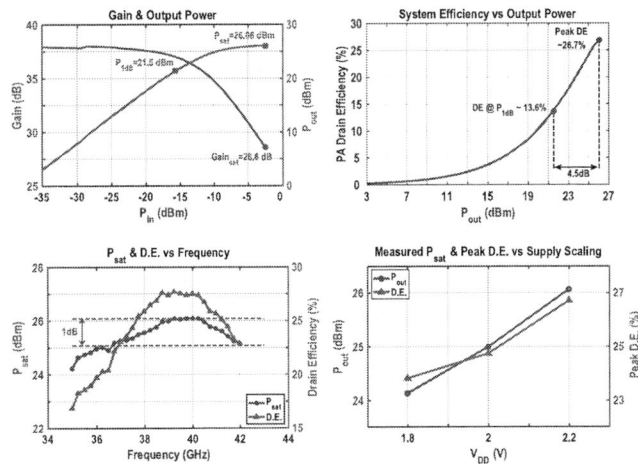

Fig. 6. Measured large signal performance.

B. Dynamic Performance

For dynamic measurements, an AWG is used to generate differential I/Q signals which are then up-converted to 39 GHz with a vector signal generator (Fig. 4(b)). The chip output is analyzed using a real-time oscilloscope and VSA software with equalization disabled. Measurements were performed to demonstrate the high average P_{out} of the PA when transmitting both single-carrier (SC) and OFDM modulated signals. A simple look-up-table based memory-less pre-distortion (DPD) was applied at high output P_{out}s for 16/64 QAM. Measured SC-QPSK/16 QAM/64 QAM modulations at average P_{out}s of 23 / 21.3 / 19.5 dBm and average efficiencies of 16.2 / 13.3 / 8.3% demonstrate the feasibility of high-power bulk CMOS PAs in mobile handheld 5G devices (Fig. 7(a)). EVM measurements at the maximum P_{out}s for the same constellations were performed across varying symbol rates (Fig. 7(b)).

Finally, the PA was measured using the recently standardized [1] FR2-5G-NR 39 GHz 50 MHz bandwidth (limited by setup) 64 QAM OFDM signal (μ=0.2). Without any DPD, an average output power of 14.7 dBm was measured with an ACPR of −33 dB (Fig. 8).

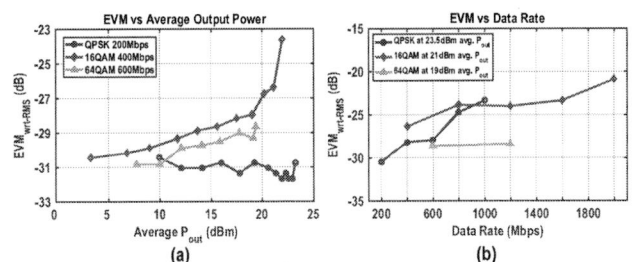

Fig. 7. (a) Measured EVM vs output power for SC-modulations. (b) Measured EVM vs data-rate for SC-modulations. RRC filtering was deployed for all SC modulation schemes with a roll-off factor of 0.35.

C. Reliability Measurements

A major concern for high P_{out} CMOS PAs is long-term reliability including transistor aging due to large voltage

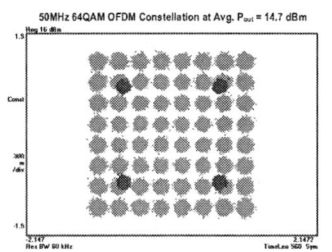
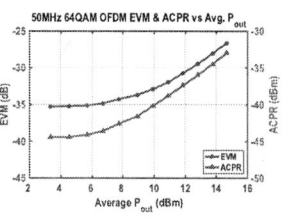

Fig. 8. Measured constellation and ACPR of the PA for a 5G NR 50 MHz 64 QAM OFDM signal.

swings. To test these concerns, the PA was operated continuously for over 22 hours at peak P_{out} without any additional thermal relief. The measured power showed no appreciable degradation (>0.1 dB) over the entire period (Fig. 9(a)). Similar measurements at the highest average P_{out} for OFDM also verified no degradation. For long term reliability evaluation, the PA was operated continuously at peak power at an elevated V_{DD} of 2.6 V and 80°C for over 66 hours. The PA lifetime can then be estimated based on the acceleration factor as shown in equation 1, where T_{J0} & T_{JS} are the nominal and stressed junction temperatures of the chip, V_0 & V_S are the nominal and stressed supply voltages and β_1 & β_2 are constants which depend on the breakdown mechanisms. A lifetime of >8 years was predicted for the PA at a P_{out}>24.8 dBm.

$$AF = e^{\beta_1 \left(\frac{1}{T_{J0}} - \frac{1}{T_{JS}} \right)} \cdot e^{\beta_2 \left(V_S - V_0 \right)} \tag{1}$$

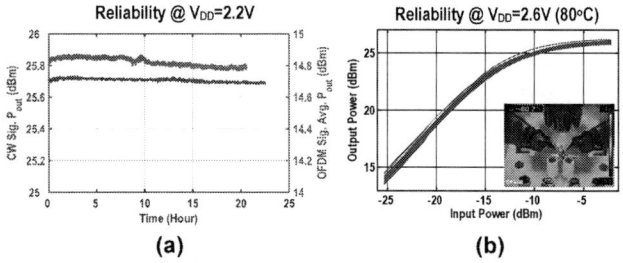

Fig. 9. (a) Measured CW and modulated performance over >20 hrs. (b) Superimposed P_{out} vs P_{in} measurements conducted every hour at V_{DD}=2.6 V and 80°C showing minimal degradation in large and small-signal performance over 66 hrs.

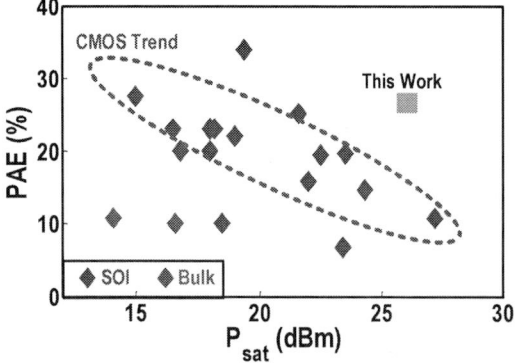

Fig. 10. State-of-art comparison against published CMOS (SOI & bulk) PAs at 35-45 GHz [11].

Table 1. Comparison with >23 dBm 5G mmWave CMOS PAs

Metric	This work	[8]	[9]	[4]	[10]
Technology	28 nm bulk	90 nm bulk	45 nm SOI	45 nm SOI	45 nm SOI
Freq. (GHz)	39	28	28	29	40
V_{DD} (V)	2.2	2.4	10.8	5.2	4.8
P_{sat} (dBm)	26	26	24.6	24.8	26.8
PAE (%)	26.6	34.1	15	26	10
OP_{1dB} (dBm)	21.5	23.2	23	21	-
1dB BW (GHz)	5.5	8	4	10	13
Gain (dB)	38	16.3	17.3	13	19.4
ITRS FoM (dB)	110	86.6	82.6	80.9	88.3
Modulations	SC-QPSK 16/64QAM OFDM	256QAM	-	-	-
Reliability Measurements	Yes	No	No	Yes	Yes

ITRS FoM=Gain(dB)+P_{sat}(dBm)+20log$_{10}$(Freq.(GHz))+10log$_{10}$(PAE(%))

IV. CONCLUSION

Compared to state of art high-power 5G mm-Wave PAs (Table 1), this work achieves the highest ITRS FoM with high unconditionally stable gain (30 dB), and can therefore be driven with minimal overhead. Compared to the state-of-art 90 nm CMOS PA [8] at a lower frequency and with unknown long-term reliability, this work achieves equal output power in a 28 nm CMOS process with >8 year predicted lifetime. This works achieves the highest PAE among published works in CMOS (bulk & SOI) with >20 dBm P_{out} at 35-45 GHz (Fig. 10). Compared to the state of art PA demonstrated with 5G NR OFDM signaling [2], this PA achieves similar modulation capability at 10× higher average P_{out}. Reliable mm-Wave high-power generation in a fine-grain bulk CMOS process, as presented in this work, would substantially simplify integration and lower cost in 5G handsets.

REFERENCES

[1] IEEE. 3GPP 5G-NR specifications. [Online]. Available: http://www.3gpp.org/DynaReport/38-series.htm

[2] S. Shakib et al., "A highly efficient and linear power amplifier for 28-GHz 5G phased array radios in 28-nm CMOS," *IEEE JSSC*, Dec 2016.

[3] H. Dabag et al., "Analysis and design of stacked-FET millimeter-wave power amplifiers," *IEEE TMTT*, April 2013.

[4] J. A. Jayamon, J. F. Buckwalter, and P. M. Asbeck, "Multigate-cell stacked FET design for millimeter-wave CMOS power amplifiers," *IEEE JSSC*, Sept 2016.

[5] S. Pornpromlikit et al., "A watt-level stacked-FET linear power amplifier in silicon-on-insulator CMOS," *IEEE TMTT*, Jan 2010.

[6] D. Zhao and P. Reynaert, "A 0.9V 20.9dBm 22.3%-PAE E-band power amplifier with broadband parallel-series power combiner in 40nm CMOS," in *IEEE ISSCC*, Feb 2014.

[7] Y. Zhao and J. R. Long, "A wideband, dual-path, millimeter-wave power amplifier with 20 dBm output power and PAE above 15% in 130 nm SiGe-BiCMOS," *IEEE JSSC*, Sept 2012.

[8] W. Huang, J. Lin, Y. Lin, and H. Wang, "A K-band power amplifier with 26-dBm output power and 34% PAE with novel inductance-based neutralization in 90-nm CMOS," in *IEEE RFIC*, June 2018.

[9] S. Helmi et al., "High-efficiency microwave and mm-Wave stacked cell CMOS SOI power amplifiers," *IEEE TMTT*, July 2016.

[10] R. Bhat, A. Chakrabarti, and H. Krishnaswamy, "Large-scale power combining and mixed-signal linearizing architectures for watt-class mmWave CMOS power amplifiers," *IEEE TMTT*, Feb 2015.

[11] H. Wang et al. Power amplifiers performance survey 2000-present. [Online]. Available: https://gems.ece.gatech.edu/PA_survey.html

A 24-43 GHz LNA with 3.1-3.7 dB Noise Figure and Embedded 3-Pole Elliptic High-Pass Response for 5G Applications in 22 nm FDSOI

Li Gao, and Gabriel M. Rebeiz
University of California San Diego, La Jolla, CA, USA
gaoliscut@gmail.com, rebeiz@ece.ucsd.edu

Abstract—This paper presents a 20-44 GHz low-noise amplifier in 22 nm FDSOI with low noise figure and low DC power consumption. The LNA is based on a 3-stage cascode amplifier which is co-designed with embedded high-pass filters so as to results in a very sharp rejection at < 16 GHz, exhibiting an elliptic filter response. This is ideal for wideband 5G amplifiers as it greatly reduces the 2nd the 3rd harmonic interference issues arising from X and Ku-band blockers (8-16 GHz). A wideband transformer-based input matching is used resulting in S_{11} < -9 dB at 20-40 GHz. The measured gain is > 17 dB at 20-43 GHz with a peak of 23 dB at 40 GHz. The LNA achieves a NF of 3.1-3.7 dB (3.4±0.3 dB) at 24-43 GHz, an in-band IIP3 of -13.2 to -19 dBm at 20-40 GHz, all at a power consumption of 20.5 mW. Operation at 12 mW is also shown, with a maximum gain and minimum NF of 18.2 dB and 3.4 dB at 24-43 GHz. To our knowledge, the LNA represents the highest FoM achieved at this frequency range to-date and includes an embedded 3-pole filter response.

Keywords— Low noise amplifier, wideband, highpass filter, CMOS SOI, 5G

I. INTRODUCTION

The 5G communication standard and systems has been becoming a reality in the past year due to increased data-rate requirements, and worldwide communication bodies (such as the FCC) has designated the 24-29 GHz and 37-43 GHz as the approved frequency bands for mm-wave communications.

Low-noise amplifiers (LNA), which are an indispensable block in receivers, have been explored for narrow-band 5G systems [1]-[2]. In order to support the 28 GHz and 39 GHz communication bands using the same system, a wideband LNA is desired [3]-[6]. However, these designs either have high power consumption or a high noise figure (NF), and are still not competitive for wideband systems.

Another issue with wideband amplifiers is their tolerance to in-band and out-of-band interferers. The in-band interferers are handled with the amplifier in-band IP1dB and IIP3, but out-of-band interferers can be very high especially at 10-16 GHz due to presence of radar systems and high-power satellite transmitters. An external passive filter is typically used to reduce such interferers, but it is best if the amplifier itself has a sharp high-pass response and a high equivalent IIP2 and IIP3 at these frequencies.

In this work, a low power and wideband LNA is proposed with 3.1-3.7 dB NF at 24-43 GHz. The RF gain bandwidth is 20-44 GHz, covering the entire 5G application bands. Section II describes the technology and transistor characterization and Section III shows the design details. Measurements are presented in Section IV, and are followed by comparison with state-of-the-art amplifiers.

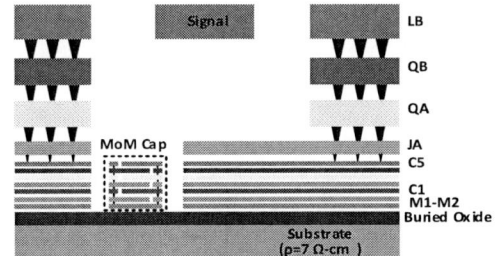

Fig. 1 Back-end stack up of the Global Foundries 22 nm FDSOI.

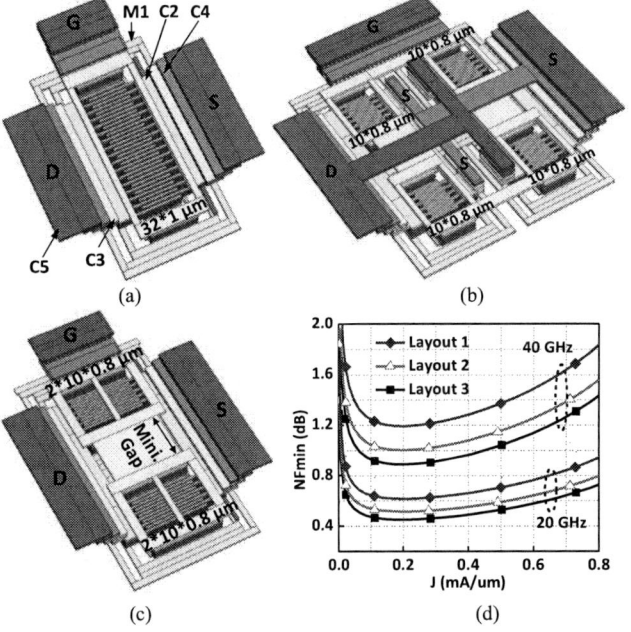

Fig. 2. Three different layouts up to C5 for a 32 μm width transistor (a) Layout 1; (b) Layout 2; (c) Layout 3 and (d) Simulated NF_{min}.

II. TECHNOLOGY

This LNA is designed in GlobalFoundries 22 nm FDSOI, which provides 10 layers of copper and one aluminum top-layer (Fig. 1). The transistors are built in a buried oxide layer that isolates the active device from the 7 Ω-cm resistivity substrate. The technology provides several types of transistors, and the super-low Vt NMOS is used in this design due to its low NF_{min} and high f_t/f_{max} (270/320 GHz).

For a transistor used in LNA design, the most important technology parameters are f_t/f_{max}, NF_{min} and R_g, with R_g being highly dependent on the gate layout. Usually a wide and multi-layer structure is used to connect the transistor fingers

978-1-7281-1702-7/19 $31.00 © 2019 IEEE

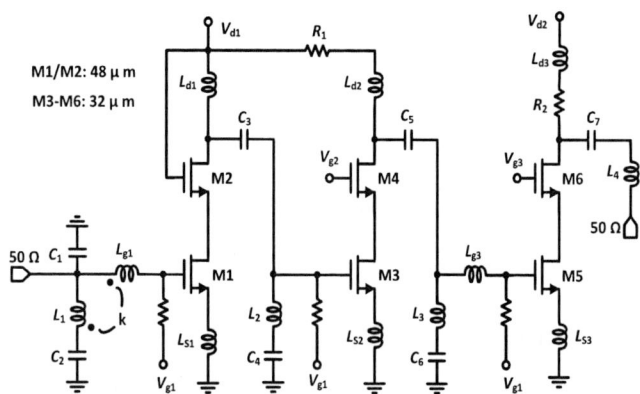

Fig. 3 Schematic of the wideband LNA with 3-pole elliptic high-pass filter.

Table 1. Component values (H or F)

L_1	L_{g1}	L_{s1}	L_{d1}	L_2	L_{s2}	L_{d2}	L_3	L_{g3}	L_{s3}	L_{d3}
400 p	400 p	100 p	220 p	400 p	20 p	300 p	400 p	400 p	20 p	200 p
L_4	C_1	C_2	C_3	C_4	C_5	C_6	C_7	R_1	R_2	
400 p	75 f	430 f	152 f	430 f	100 f	430 f	230 f	20 Ω	50 Ω	

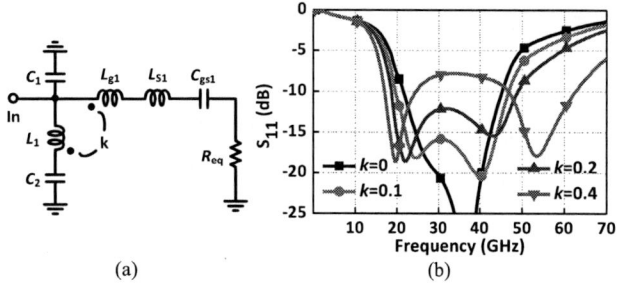

(a)   (b)

Fig. 4 Simplified input matching network and input matching versus k. (Q=16 for inductors).

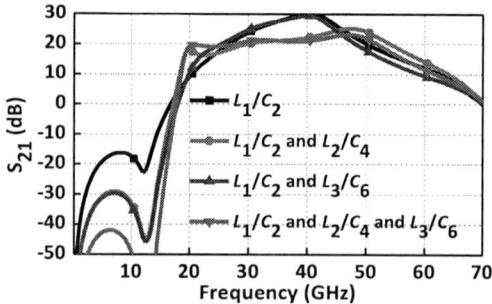

Fig. 5 Simulated S_{21} response with different filter order.

III. CIRCUIT DESIGN

The wideband LNA is shown in Fig. 3, and is based on a 3-stage cascode design for different frequency peaking values. To maintain a wideband gain response, de-Q series resistors are added in the drains of stages 2 and 3. Note that the relatively large resistor used in stage 3 (R_2=50 Ω) has minimal effect on the NF, and provides wideband output matching.

The input circuit is realized by source degeneration and a ladder-based transformer which introduces multiple resonant points for wideband impedance matching. L_{S1} increases the real part of the input impedance and eases the design of the input matching network. Fig. 4(a) presents the simplified input matching network, where $R_{eq}=g_{m1}L_{S1}/C_{gs1}$. g_{m1} is extracted to be 42 mS using g_m= Re(Y_{21}) at the bias point. C_{gs1} is extracted to be 62 fF using C_{gs}=Im($Y_{11}+Y_{12}$)/ω [2]. Thus, the calculated R_{eq} is 72 Ω and if all the reactance is resonated out, this results in an S_{11} = -14.8 dB. The coupling between L_1 and L_{g1} can be positive or zero, and is used to tune the matching network.

There are several resonances in the input matching points. The input impedance is:

$$Z_{in} = \frac{1}{j\omega C_1} \parallel (j\omega L_1 + \frac{1}{j\omega C_2}) \parallel Z_1 \quad (1)$$

$$Z_1 = j\omega(L_2 + M + L_{S1}) + \frac{1}{j\omega C_{gs1}} + R_{eq} \quad (2)$$

$$M = k\sqrt{L_1 L_2} \quad (3)$$

First, the L_1 series-resonates with C_2 and generates a transmission zero at f_1 since it is in shunt with the main signal. f_1 is given by:

$$f_1 = 1/2\pi\sqrt{L_1 C_2} \quad (4)$$

f_1 is designed to be 13 GHz to provide sharp rejection at the left side of the passband. At $f > f_1$, the series L_1 and C_2 is equivalent to an inductor L_{eq1} which parallel-resonates with C_1 to generate a pole at f_2 given by:

$$f_2 = 1/2\pi\sqrt{L_{eq1} C_1} \quad (5)$$

Also $L_{g1}+M+L_{s1}$ will resonate with C_{gs1} to generate another pole at f_3 given by:

$$f_3 = 1/2\pi\sqrt{(L_{g1} + M + L_{S1})C_{gs1}} \quad (6)$$

Therefore, the input matching network is equivalent to a two-pole network with one transmission zero, and the positions of the two poles are controlled by the transformer coupling coefficient k. A large k results in a large M and a smaller L_{eq}, which pushes f_3 to lower and moves f_2 to upper. Therefore,

together to reduce R_g. However, due to design rule constraints in deep CMOS nodes, M1 metal linewidth cannot exceed 0.1 μm, resulting in a large R_g. Therefore, if a 32 μm–wide transistor is connected as in Fig. 2(a), which employs 32 fingers each with a finger length of 1 μm, the NF$_{min}$ is relatively high due to an increased R_g. In order to reduce R_g, four smaller transistors can be connected in parallel, with each a width of 8 μm. Also, it is best to stack the M1 to C3 layers first and then use the thick and low-loss C3 layer for the routing (Fig. 2(b)). However, each transistor still has a long interconnection metal which is used to connect the back gate and the body. To further reduce R_g, multi-number of vertical gate fingers connection is provided in this technology, which eliminates part of the interconnection stack and allow the transistors to be placed closer to each other, as shown for a 2x2 cell (Fig. 2(c)).

The simulated NF$_{min}$ of the three layouts are shown in Fig. 2(d). It can be observed that layout 3 exhibits the lowest NFmin, and is 0.45/0.88 dB at 20/40 GHz, with a current density of 0.15-0.25 mA/μm. In this layout, the 32 μm transistor is realized using 4×8 μm wide transistors. A 48 μm transistor is also realized using 4×12 μm transistors in a 2x2 cell and is used in the LNA design.

978-1-7281-1702-7/19 $31.00 © 2019 IEEE

2019 IEEE Radio Frequency Integrated Circuits Symposium

Fig. 6 Microphotograph of the wideband LNA (Chip size: 0.73 mm × 0.64 mm; Core size: 0.57 mm × 0.38 mm).

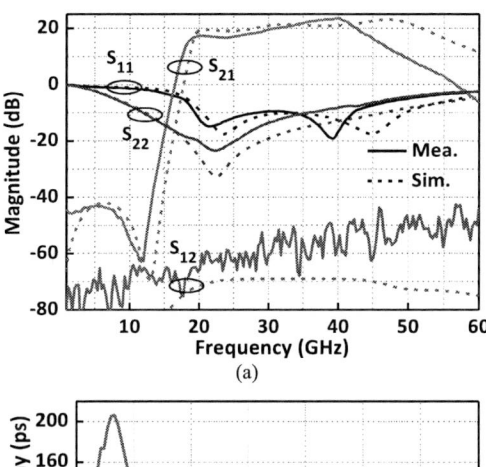

(a)

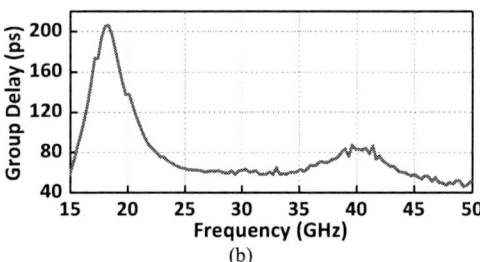

(b)

Fig. 7 (a) Measured and simulated S-parameters of the LNA; (b) Measured group delay.

the lower and higher pole frequencies are f_3 and f_2, respectively.

Fig. 4(b) presents the simulated input matching versus k. When $k=0$, one obvious pole is at 37.5 GHz and the other obscure pole is around 28 GHz (due to limited inductor Q of 16 at 30 GHz). When k is increased, the two poles are split resulting in a wide matching bandwidth but also a higher matching ripple. A $k=0.2$ is used in the wideband LNA with poles located at 22 and 44 GHz. A bandstop filter is thus embedded in the input matching network without deteriorating its performance.

However, due to the limited Q of the inductors, the rejection level is not enough if only a first-order bandstop filter is utilized. Therefore, we implemented a third-order bandstop filter in the LNA by placing three different bandstop filters at the input of the three cascode stages. This also reduces their effect on the LNA NF. Fig. 5 presents the simulated S_{21} with different bandstop filter orders. If L_1/C_2 is used, S_{21} is -15 dB at ~13 GHz and the gain shows a peaked single-pole response. When L_1/C_2 and L_2/C_4 are both used, S_{21}

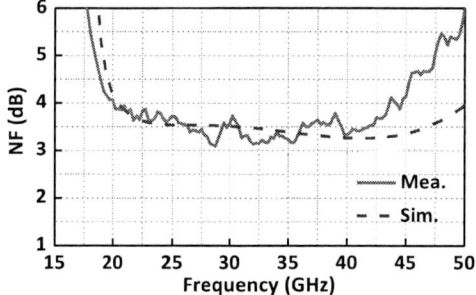

Fig. 8 Measured and simulated NF of the LNA.

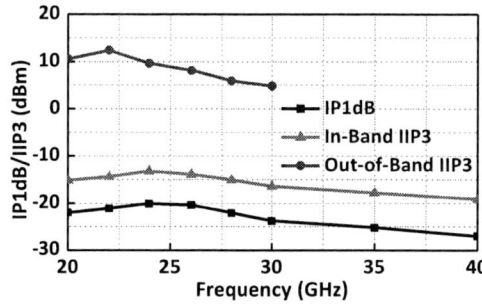

Fig. 9 Measured input P1dB, in-band IIP3 and out-of-band IIP3.

Table 2. Measured harmonic related IM3

f_1 (GHz)	f_2 (GHz)	Tones (GHz)	IM3 Tone (GHz)	P_{in} @ $f_{1,2}$ (dBm)	OP_{tones} (dBm)	OP_{IM3} (dBm)
27.5	28	Same	27 ($2f_1$-f_2)	-30	-11.4	-42.8
13.8	14	27.6 ($2f_1$) 28 ($2f_2$)	27.2 ($2*2f_1$-$2f_2$)	-10	-34.4	-109.4
9.16	9.33	27.48 ($3f_1$) 27.99 ($3f_1$)	26.97 ($2*3f_1$-$3f_2$)	-4	-39.9	-110

is -30 dB, and this dual-pole network changes the gain shape to a flat response with a sharp cutoff. When L_1/C_2, L_2/C_4 and L_3/C_6 are all used, S_{21} is < -40 dB at < 15 GHz, and the gain shows > 60 dB rejection at < 15 GHz when compared to the passband response.

IV. MEASUREMENTS

The wideband LNA is implemented in the GF 22 nm FDSOI process with a chip size of 0.73 mm × 0.64 mm (Fig. 6). All measurements including NF and linearity were performed using GSG on-chip probing and a Keysight N5247B network analyzer (PNA-X). Biasing conditions are: V_{g1}, V_{g2} and V_{g3} = 0.4, 1 and 1.4 V, respectively. V_{d1} and V_{d2} are 1 V and 1.6 V with corresponding current of 12 mA and 5.3 mA, and the DC power consumption is 20.5 mW.

The measured gain is > 17 dB at 20-44 GHz with a peak of 23 dB at 40 GHz, and the gain variation is <±1.1 dB at 24-29 GHz (S_{21av}=18 dB) and 37-42 GHz bands (S_{21av}=22 dB) (Fig. 7a). The measured results do not agree with the simulated response above 42 GHz, and is due to the mandatory dense metal-fill for all the metal layers, which results in higher parasitic capacitances and is hard to simulate accurately. The high-pass elliptic filter response is clearly observed, with a

978-1-7281-1702-7/19 $31.00 © 2019 IEEE   241

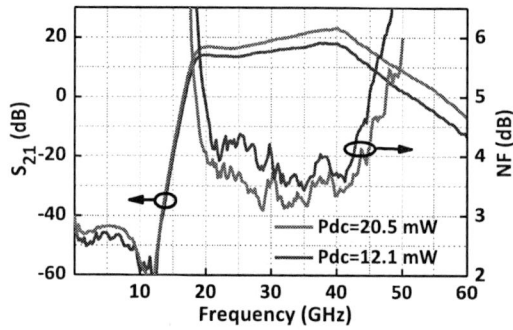

Fig. 10 Measured LNA gain and NF versus DC power consumption.

Table 3. Comparison with previous work

Ref.	Tech.	BW (GHz)	Peak Gain (dB)	NF (dB)	Pdc (mW)	FoM (dB)
TMTT 2015 [3]	0.25 μm SiGe	25-34	26.4	2.1-3.5	134	1-7.1
MWCL 2017[4]	65nm CMOS	15.8-30.3	10.2	3.3-5.7	12.4	2.9-10.4
RFIC 2018 [5]	0.25 μm SiGe	29-37	28.5	3.1-4.1	80	4.6-8.1
GSMM 2018 [6]	45 nm CMOS SOI	24-44	20	4.2-5.5	58	2.6-6.5
This work	**22 nm CMOS FDSOI**	24-43	23 18.2	3.1-3.7 3.4-4.3	20.5 12.1	19.7-21.9 17.5-20.6

gain of -20 dB at 15 GHz and < -40 dB at all frequencies below 13.2 GHz. It should be noted that the passband response is designed with a slight positive slope so as to compensate the negative slope of following mixer gain when used in a complete receiver. The measured S_{11} is < -9 dB at 20- 44 GHz. The measured S_{12} is < -50 dB over the whole band.

For wideband operation, the group delay variation is an important parameter since a large group-delay variation results in inter-symbol interference. The measured group delay is shown in Fig. 7(b). The variation is < ±13 ps when operating at an instantaneous bandwidth of 22-44 GHz.

Fig. 8 presents the measured and simulated NF and with good agreement. This is done using the PNA-X which calibrates the cable loss out and puts the reference at the chip. The measured NF is 3.1-3.7 dB at 24-43 GHz and < 4 dB at 20-45 GHz.

Fig. 9 presents the measured LNA linearity. The input P1dB is -20.4 to -27 dBm and the in-band IIP3 is -13.2 to -19 dBm at 20-40 GHz. Due to the filtering effect, the out-of-band IIP3 is ~20 dB higher than the in-band IIP3 (Fig. 9). This is measured using a tone at 15 GHz (ω_1) and another tone at 2-10 GHz (ω_2) and taking the IM3 value at 20-30 GHz ($2\omega_1$-ω_2).

Another study of the LNA out-of-band linearity is shown in Table 2. In one case, interferers at 13.8 GHz (ω_1) and 14 GHz (ω_2) are used, which generate second-harmonic components at 27.6 GHz ($2\omega_1$) and 28 GHz ($2\omega_2$) with an OP$_{tones}$ of -34.4 dBm, and which in turn mix to generate an IM3 at 27.2 GHz. This IM3 component is measured at -109.4 dBm for -10 dBm interferer power levels. Another case is

done using interferers at 9.16 GHz (ω_1) and 9.33 GHz (ω_2), which generate third-harmonic components at 27.48 GHz ($3\omega_1$) and 27.99 GHz ($3\omega_2$). An interferer power of -4 dBm is required to generate an IM3 level of -110 dBm at 26.97 GHz. This shows the effect of the 3-pole elliptic filter in the LNA.

The LNA can also be operated at lower power with competitive performance. When both V_{d1} and V_{d2} are reduced to 0.8 V, the LNA Pdc becomes 12.1 mW. In this case, the LNA exhibits a maximum gain of 18.2 dB at 40 GHz and a NF of 3.4-4.3 dB at 24-43 GHz (Fig. 10). The 3-pole high-pass elliptic response is conserved.

V. COMPARISON

Table 3 compares the 22 nm FDSOI LNA with state-of-the-art wideband LNAs. It is observed that this work demonstrates state-of-the-art gain, NF, bandwidth and power consumption, and with the FoM determined using:

$$FoM = 20\log_{10}\left(\frac{Gain[lin.] \times BW[GHz]}{Pdc[mW] \times (NF[lin.]-1)}\right) \quad (7)$$

The achieved FoM are very high compared to previous results and show the advantages of deep-CMOS SOI technology. The LNA also has a high in-band IIP3 in the 23-29 GHz range (-13.2 to -16 dBm).

VI. CONCLUSION

This paper presented a wideband 22 nm FDSOI LNA with an embedded elliptic 3-pole response. The LNA covers all mm-wave 5G bands including 23-29 GHz and 37-43 GHz with low noise figure and high linearity. A very high tolerance to out-of-band interferers is observed, removing the need of passive filters between the antenna and the LNA, and improving the system performance.

ACKNOWLEDGEMENTS

This work was supported in part by Analog Devices Inc., Boston, MA and in part by Samsung Electronics, San Jose, CA. The authors would like to thank GlobalFoundries for the 22nm FDX access and Integrand Software for supplying licenses for EMX electromagnetic modeling software.

REFERENCES

[1] K. Kibarouglu, M. Sayginer, and G. M. Rebeiz, "A low-cost scalable 32-element 28-GHz phased array transceiver for 5 G communication links based on a 2x2 beamformer flop-chip unit cell," *IEEE J. Solid State Circuits*, vol. 53, no. 5, pp. 1260–1274, May. 2018.

[2] O. Inac, M. Uzunkol, and G. M. Rebeiz, "45 nm CMOS SOI technology characterization for millimeter-wave applications," *IEEE Trans. Microw. Theory and Tech.*, vol. 62, pp. 1301–1311, June 2014.

[3] Q. Ma, et. al., "Silicon-based true-time-delay phased-array front-ends at Ka-Band", *IEEE Transactions on Microwave Theory and Techniques*, vol. 63, no. 9, pp. 2942-2952, Sept. 2015.

[4] P. Qin, and Q. Xue, "Compact wideband LNA with gain and input matching bandwidth extensions by transformer," *IEEE Microw. Compon. Lett.*, vol. 27, no. 7, pp. 657–659, Jul. 2017.

[5] Z. Chen, H. Gao, D. M. W. Leenaerts, D. Milosevic and P. G. M. Baltus, "A 29-37 GHz BiCMOS low-noise amplifier with 28.5 dB peak gain and 3.1-4.1 dB NF," in *IEEE Radio Frequency Integrated Circuits Symposium (RFIC)*, pp. 288-291, 2018.

[6] V. Chauhan, and B. Floyd, "A 24-44 GHz UWB LNA for 5G cellular frequency bands", in *Global Symposium on Millimeter Wave*, 2018.

2019 IEEE Radio Frequency Integrated Circuits Symposium

A 4-element 28 GHz Millimeter-wave MIMO Array with Single-wire Interface using Code-Domain Multiplexing in 65 nm CMOS

Manoj Johnson[#], Armagan Dascuru[$], Kai Zhan[#], Arman Galioglu[$], Naresh Adepu[$], Sanket Jain[#], Harish Krishnaswamy[$] and Arun Natarajan[#]

[#]School of EECS, Oregon State University, Corvallis, OR 97331, USA
[$]Department of Electrical Engineering, Columbia University, New York, NY 10027, USA

Abstract— Millimeter-wave MIMO systems with digitization of every element enable spatial multiplexing, virtual arrays for radar, digital beamforming (DBF) for high mobility scenarios. However, per-element digitization results in a formidable I/O challenge in large-scale tiled MIMO mmWave arrays. This work demonstrates a 28GHz 4-element MIMO RX with a single-wire interface that multiplexes the baseband signals of all elements and the LO reference through code-domain multiplexing. System considerations are presented and the approach is validated through DBF after de-multiplexing of the baseband signals from the single wire. Each element in the array achieves 16 dB conversion gain while consuming 60 mA from 1.2 V. The IC occupies 5.75 mm^2 in 65-nm CMOS.

I. INTRODUCTION

MIMO arrays represent the next paradigm for communication and radar systems, enable spatial multiplexing [1], enhanced radar resolution through virtual arrays [2] and digital beamforming for user discovery/tracking in highly mobile scenarios. Increasing maturity of mm-wave ICs has led to several demonstrations of large-scale *phased* arrays with hundreds of elements at mm-wave [3], [4]. However, MIMO arrays demand per-element digitization, resulting in I/O challenges at mm-wave for large signal bandwidths.

Integrated phased arrays have focused on (a) tiling scalable unit-cells (N tiles of M elements to create $N \cdot M$ element-arrays) and (b) single-beam or few multi-beam I/O [4]. Since practical considerations lead to separate ICs for RF and ADC/DSP, a single coaxial interface for scalable phased arrays is proposed in [5] where the combined IF (at 8.64 GHz) from a 16-element array is multiplexed with the LO reference (at 0.27 GHz) and control signals (at 2.16 GHz) on to a single wire. While this simplifies the interface for phased arrays, spatial information loss due to phase-shifting and combining prior to the interface makes multi-beam formation impossible. A code-domain path-sharing approach to re-use on-chip blocks is proposed in [6] that enables on-chip block-reuse for 2-elements at 5 GHz.

Implementing a tiled array with a single-wire approach for mm-wave MIMO requires information from each element to be preserved for subsequent digitization. Signals from multiple elements must co-exist on the single-wire interface, requiring multiplexing in the time, frequency or code-domain (Fig. 1(a)). In this paper, we present a novel 28 GHz MIMO array that uses spread-spectrum codes on a single-wire interface to enable digital beamforming while preserving the simplicity

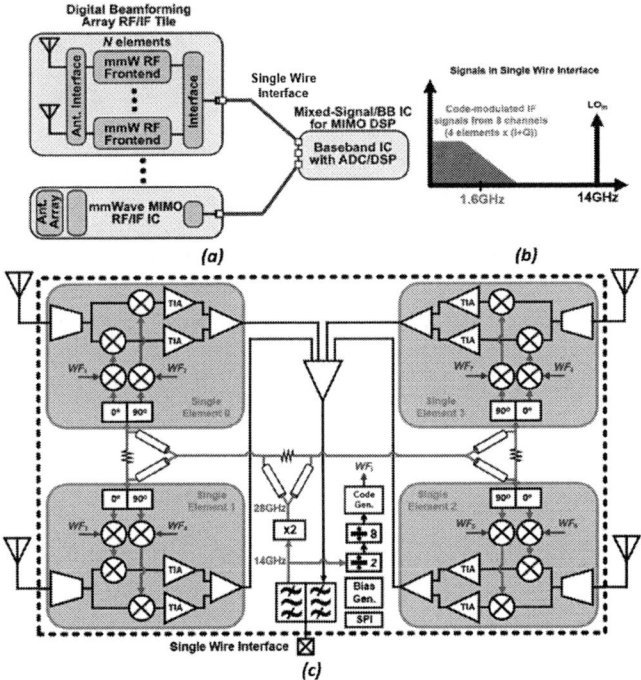

Fig. 1. (a) Proposed approach for scalable tiled MIMO Array with IF→BB interface over a single-wire (b) Frequency-domain duplexing between code-multiplexed IF and LO reference, (c) 28 GHz MIMO RX array with code-domain IF multiplexing on single-wire interface.

of the single wire connecting to array tiles (Fig. 1(a)). Spread-spectrum codes are used to multiplex each antenna's IF signals, while these IF signals are separated from the LO reference in frequency (Fig. 1(b)). Digital multi-beamforming with the 4-element array demonstrates the feasibility of the proposed approach and promises a path towards scalable mm-wave MIMO arrays.

II. CODE-DOMAIN MULTIPLEXING ON SINGLE-WIRE INTERFACE

Multiplexing IF/BB signals from multiple elements on to a single-wire requires orthogonality for subsequent de-multiplexing. In this work, the I and Q IF signals in each element are spread using orthogonal codes on the interface as shown in Fig. 2(a). The isolation and subsequent digital multi-beamforming depends upon the intrinsic orthogonality between the codes as well as relative value of the integration

978-1-7281-1702-7/19 $31.00 © 2019 IEEE   243

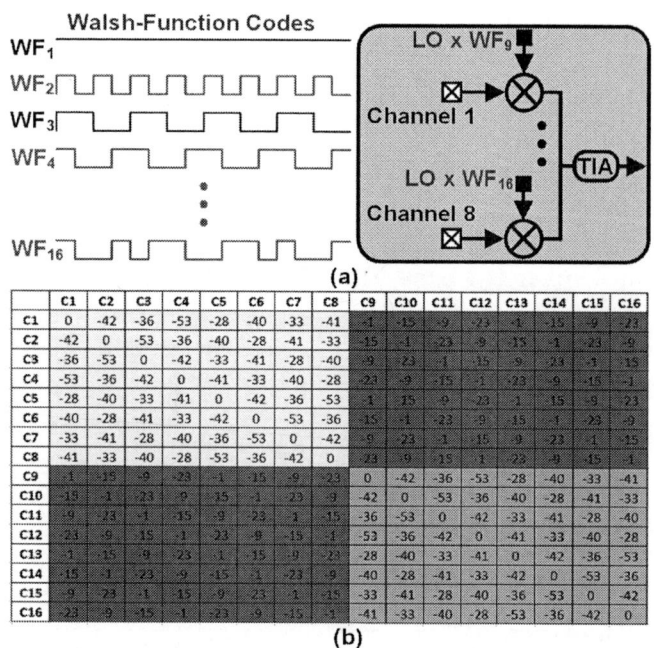

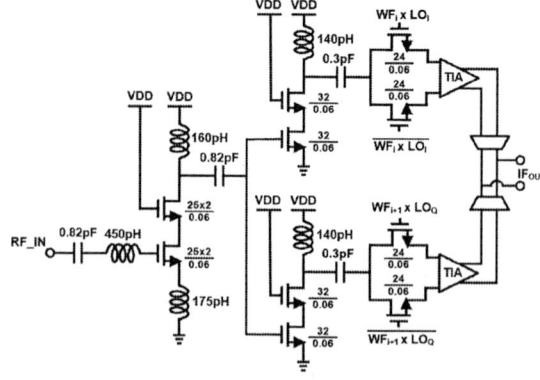

Fig. 3. 28 GHz signal path with LNA, passive mixer, TIA and summing amp that adds the I and Q signals generated by the code-modulated LO.

Fig. 2. (a) Code-domain multiplexing of IF signals to enable MIMO DSP in backend, orthogonal walsh-function codes used to ensure low cross-code leakage (b) Isolation between pairs of codes with spreading factor of 16; 8 codes selected (green) to modulate information from 8 channels.

bandwidth and chip rate. The cross-correlation between the codes, C_i, for a signal-bandwidth, BW_S is computed as,

$$R_{i,j} = \frac{\int_0^{BW_S} \mathcal{F}\left(C_i \cdot C_j\right) df}{\int_0^{BW_S} \mathcal{F}\left(C_i \cdot C_i\right) df}$$

The cumulative leakage from other code-modulated signals into the desired code, $(\sum_{j=1, j \neq i}^N R_{i,j})$, must be minimized. Walsh-Function based orthogonal codes are adopted in this work (Fig. 2(a))[7]. Simulated cross-correlation for Walsh Function based codes of length 16 and $BW_S = \frac{ChipRate}{CodeLength}$ is shown in Fig. 2(b). While code length of 16 theoretically allows for 16 channels (8 elements x 2 I,Q), this empirically leads to higher than -20 dB cumulative leakage into desired signals. However, the targeted isolation can be achieved for 8 channels (4 elements x 2 I,Q) by selecting 8 out of the 16 codes (in green) to ensure orthogonality (Fig. 2(b)). A signal bandwidth of 100 MHz translates to a spread bandwidth of 1.6 GHz for each element in the 28 GHz IC.

III. DESIGN OF 28 GHz MIMO RX

The block diagram of the 4-element 28 GHz MIMO RX array is shown in Fig. 1(c). The 28 GHz RF input is down-converted to BB with a code-modulated 28 GHz LO that results in an IF bandwidth of ~1.6 GHz. Building blocks in the signal path and LO-path are detailed in the following.

A. 28 GHz Signal Path:

The LNA in each element consists of a single-ended inductor-degenerated input stage matched to 50 Ω assuming

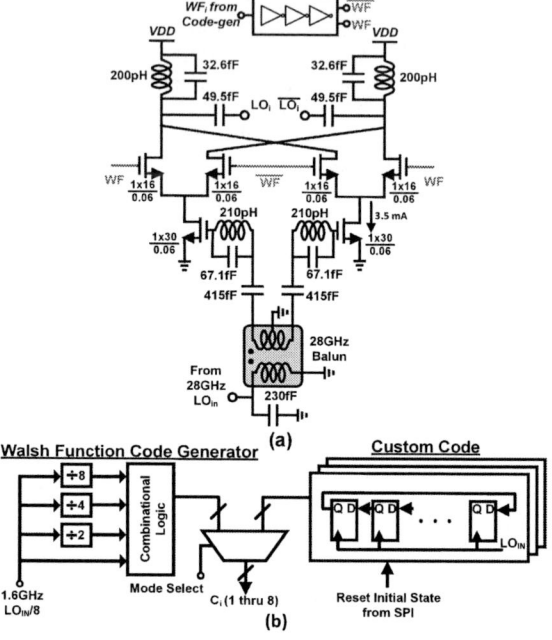

Fig. 4. (a) Current-steering LO-path modulator to create code-modulated LO, (b) code generation using divider chain/combinatorial logic or through a register array with SPI-controlled initial state.

a multi-wirebond matching network from the IC to the PCB [8]. The LNA output drives two trans-conductance stages, which in turn drive quadrature passive mixers in each element (Fig. 3). As described in Sec. III-B, the I and Q LO signals driven to each mixer are modulated with orthogonal codes (4 elements with I and Q correspond to 8 codes) (Fig. 3). Since the code-domain signals ensure orthogonality and subsequent recovery of I and Q signals, the output of the I and Q TIAs following the mixers are summed together in a source-coupled differential buffer stage. A complementary buffer stage combines the IF-signals from all elements and drives the 50 Ω impedance of the duplexer. The simulated NF of each element is ~6 dB without code-modulated LO. The 2-stage LNA, each TIA and the summing amplifier consume 12.3 mA, 7 mA and 8.4 mA from 1.2 V respectively while the

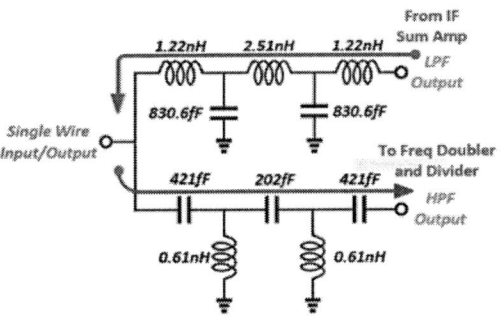

Fig. 5. Duplexer enables LO and IF signals multiplexing on single-wire interface, with high-pass path for the LO and low-pass from IF to output.

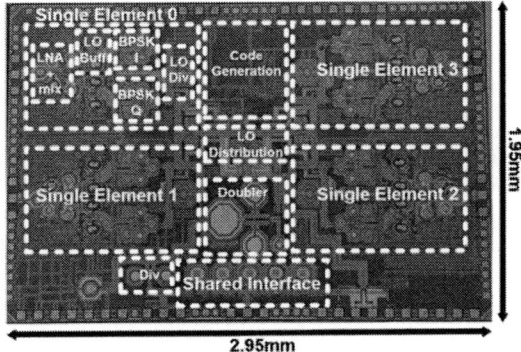

Fig. 6. Die photo of 4-element 28 GHz MIMO array.

4-element summing amplifier draws 15 mA.

B. LO Path:

The input LO signal, LO_{IN}, at 14 GHz is doubled to achieve the 28 GHz LO and divided by 8 to derive the 1.75 GHz chip rate clock required for code-generation. The frequency doubler uses an implicit common-mode second harmonic trap to increase conversion gain, providing ~3 dBm output power while consuming 22 mA. In each element, a quadrature coupler is used to generate the LO for the I and Q mixers. The LO is code-modulated through the modulator shown in (Fig. 4), where the desired code is applied to the current-steering nodes to generate a modulated LO-signal. Codes for each element can be generated on-chip using either (a) combinatorial logic based on the $LO_{IN}/8$ (1.75GHz) and an array of div/2, div/4 and div/8, or (b) shift-register loop array that are initialized to custom states using the SPI and driven by the 1.75 GHz clock (Fig. 4(b)). Each 28 GHz modulator consumes 7 mA while the code-generation draws 4.1 mA when operating at 1.6 GHz.

C. Duplexer:

The LO signals and code-modulated IF are interfaced through the duplexer in Fig. 5 which combines low-pass and high-pass filtering to/from the I/O to the LO input and IF output. Duplexer simulations indicate -2 dB insertion loss at 14 GHz and ~50 dB isolation from the IF to the LO at 14 GHz.

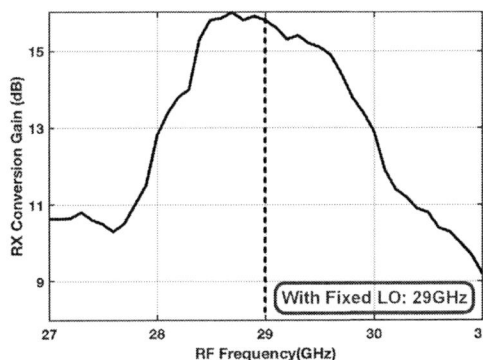

Fig. 7. Single-element RF to BB gain measured at 29 GHz LO.

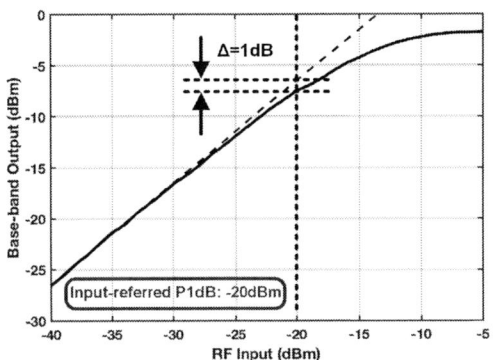

Fig. 8. Single-element linearity measurement at 29.05 GHz.

IV. MEASURED PERFORMANCE

The proposed 4-element 28 GHz array is implemented in a 65 nm CMOS process with 3.4 μm thick metal layers (Fig. 6). The IC is packaged using a chip-on-board approach with a two wirebond L-C-L scheme for the 28 GHz input [8]. A similar L-C-L scheme is used for the single-wire interface I/O that provides matching at baseband and around 14 GHz.

Each element on the IC consumes 62 mA from 1.2 V. Single-element measurements in Fig. 7 show ~16 dB

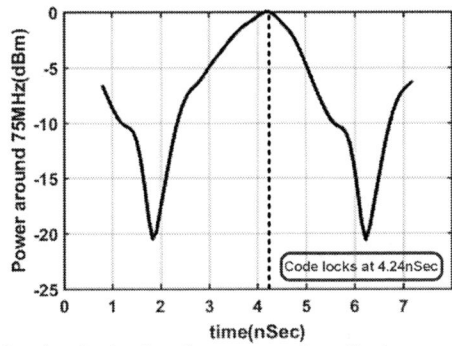

Fig. 9. Synchronization based on recovered amplitude across relative delay between IF signal and despreading.

2019 IEEE Radio Frequency Integrated Circuits Symposium

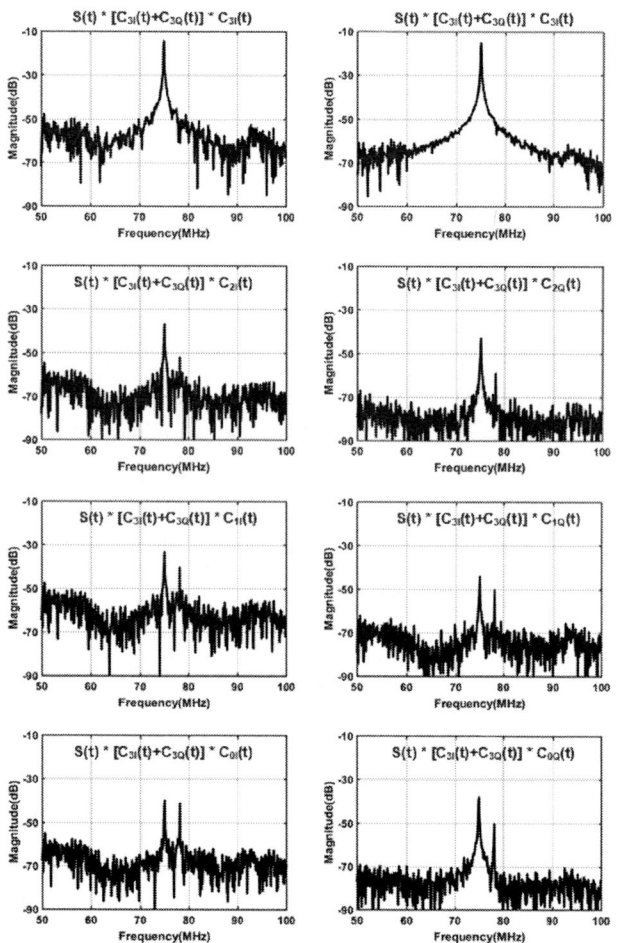

Fig. 10. Isolation measured with an input at 29.275 GHz (LO:29.2 GHz) applied to Element 3 alone. Despreading with codes corresponding to other elements leads to smaller output, while signal can be recovered by despreading with codes corresponding to Element 3I and Element 3Q.

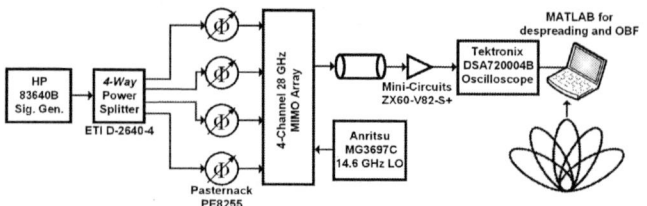

Fig. 11. Measurement setup to demonstrate digital beamforming using the 28 GHz MIMO array.

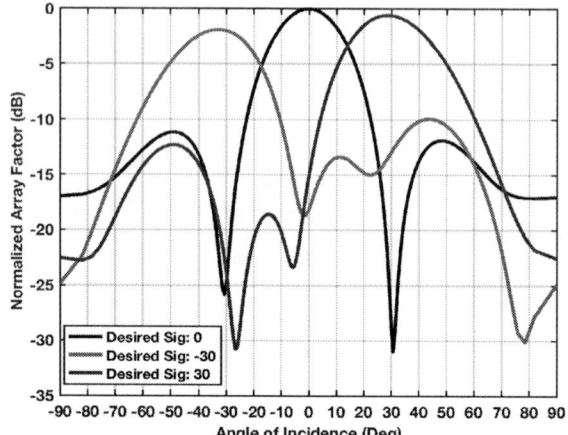

Fig. 12. Measured RX array factor using digital beamforming following despreading to recover signals from each element. Signal reception with different AoI demonstrates digital beamforming in the MIMO array following single-wire interface.

conversion gain from RX input to BB output which is close to simulations. Single-element linearity measurements demonstrate input-referred P1dB of -20 dBm (Fig. 8). NF estimated via gain method are consistent with ~10 dB single-element NF due to bias limitations.

Despreading the code-modulated signals to recover signals from each element requires synchronization with the despreading codes. Fig. 9 plots the recovered signal amplitude as the despreading code is delayed with respect to the IF signal. Coarse and fine steps are used to determine optimal delay for synchronization. Symmetric on-chip code distribution ensures synchronization with one code is sufficient for despreading all code-modulated signals.

Isolation between elements is measured using the custom-code approach in Fig. 4(b). Fig. 10 shows the results of despreading when a 29.275 GHz input (with LO:29.2 GHz) is provided to one element (corresponding to I and Q signals in element 3). The signal for the desired element can be recovered by applying the correct despreading code. The measurement setup to demonstrate digital beamforming is shown in Fig. 11. A single tone is applied to all elements

through off-chip phase shifters in order to emulate different angles of incidence. The single-wire output is captured using an oscilloscope and subsequent despreading recovers signal from each element for digital beamforming. Fig. 12 plots measured performance demonstrating the feasibility of digital multi-beamforming using the MIMO array.

ACKNOWLEDGMENT

This project was funded by DARPA MIDAS and CDADIC. We thank Dr. T. Hancock for helpful discussions and Rohde & Schwarz for equipment support.

REFERENCES

[1] E. Larsson *et al.*, "Massive MIMO for next generation wireless systems," *IEEE Commun. Mag.*, Feb. 2014.
[2] V. Giannini *et al.*, "A 192-Virtual-RX 77/79GHz GMSK Code-Domain MIMO Radar SoC," in *IEEE ISSCC*, Feb. 2019.
[3] S. Shahramian *et al.*, "A fully integrated scalable W-band phased-array module ... self-test," in *IEEE ISSCC*, Feb. 2018.
[4] B. Sadhu *et al.*, "A 28GHz 32-element phased-array TRX IC with concurrent dual polarized beams ..." in *IEEE ISSCC*, Feb. 2017.
[5] M. Boers *et al.*, "A 16TX/16RX 60GHz 802.11ad chipset with single coaxial interface and pol. diversity," in *IEEE ISSCC*, Feb 2014.
[6] F. Tzeng *et al.*, "A CMOS Code-Modulated Path-Sharing Multi-Antenna Receiver Front-End," *IEEE JSSC*, May 2009.
[7] H. F. Harmuth, "Applications of Walsh functions in communications," *IEEE Spectrum*, Nov 1969.
[8] G. Liu *et al.*, "Broadband mm-Wave LNAs (47-77 GHz and 70-140 GHz) using a T-Type Matching Topology," *IEEE JSSC*, Sept 2013.

978-1-7281-1702-7/19 $31.00 © 2019 IEEE

A 16-Element Phased-Array CMOS Transmitter with Variable Gain Controlled Linear Power Amplifier for 5G New Radio

Yunsung Cho, Woojae Lee, Hyun-chul Park, Byungjoon Park, Jeong Ho Lee, Jihoon Kim,
Jooseok Lee, Seokhyeon Kim, Jaehong Park, Sangyong Park, Kyu Hwan An, Juho Son, Sung-Gi Yang

Samsung Electronics Co., Ltd., Suwon-si, Republic of Korea
yunsung.cho@samsung.com

Abstract—This paper presents a 28-GHz 16-element phased-array transmitter for the fifth-generation (5G) new radio (NR) applications, focusing on the power amplifier (PA). The IC was fabricated using a 28-nm bulk CMOS technology with a flip-chip package, and evaluated with 5G NR signal having an 800-MHz total bandwidth. Each channel of the 16 phased-array transmitter has a 16 dBm output at 1-dB compression point and a large gain range from 24 to 63 dB. The full 16-element array achieves an average output power of 18 dBm with a record-level system error vector magnitude (EVM) of less than -33.5 dB (<2.1%) at a total power consumption of 1.63 W (102 mW of per channel) including all biases, digital control, and power-management-unit (PMU) blocks, over the frequency band from 26.5 to 29.5 GHz.

Keywords— 28GHz, 5G, beamforming, mm-wave, phased array, transceiver, transmitter.

I. INTRODUCTION

Increasing demands for high data rate systems are driving the development of the fifth generation (5G) wireless systems with data rates on the order of more than 1-10Gb/s. To accomplish the demands/needs, 5G communications will be deployed at millimeter-wave (mm-wave) bands such as 28 and 39 GHz.

Recent works have demonstrated mm-wave transceiver ICs supporting switching or combining among three 4-channel sub-array groups in 28nm bulk CMOS [1], direct conversion transceiver at 28GHz [2], phased arrays for picocells in SiGe [3], and transceiver ICs with RF interfaces for large-scale arrays in SiGe [4]-[5]. This work presents the 16-element phased-array transmitter that significantly improves the power consumption as well as the linearity in the full-array operation. It also increases transmit (TX) output power per channel element compared to the recently reported solutions.

5G systems are based on orthogonal frequency-division multiplexing (OFDM), which adds more challenges to the power amplifier (PA) design due to its high peak-to-average power ratio (PAPR) of 10-dB and wide bandwidth around 1 GHz. When the PA operates with the wide bandwidth signal, the second order nonlinearity at the envelope frequency can be particularly problematic due to the large fractional bandwidth [6]-[8]. The PA mixes up the envelope with the amplified fundamental component, which is called memory effect. In Section III-(a), a PA structure is proposed to relieve the memory effect by isolating the V_{DD} path between the final- and other-stages. A large dynamic gain range is implemented in TX and the PA is responsible for a large portion of the gain ranges. It has an advantage in noise figure (NF) for operation with a wide bandwidth signal, that will be described in Section III-(b). In Section IV, the measurement results will be shown.

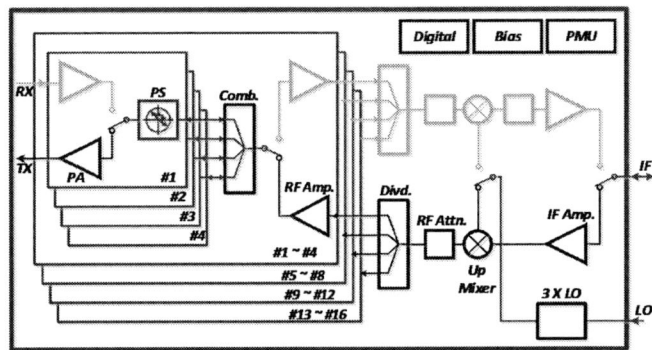

Fig. 1. A simplified block diagram of developed 5G transceiver with 16-element array

II. TRANSMITTER ARCHITECTURE

Fig. 1 shows the proposed architecture of the 16-element phased-array transceiver. In this work, we focus on the design of the TX path and its performance. The input frequency is from 10.36 to 12.04 GHz, and the transmitter converts it to an output RF frequency from 26.5 to 29.5 GHz. The TX path has a large dynamic gain range from 24 to 63 dB with a 1-dB gain step. A 1-dB step attenuator is proposed at the input matching of the PA. The PA can cover the 18 dB gain range and the rest of the gain ranges are provided by the TX intermediate frequency (IF) amplifier and RF attenuator. The TX IF amplifier has NF of less than 4.5 dB at the maximum gain-mode and covers 9 dB gain range with a 3 dB gain-step using the current control method by turning off a common-gate (CG) transistor. The RF step attenuator between the up-mixer and the combiner provides low insertion loss as well as high linearity. It has a gain range of 12 dB with a 3 dB gain-step. The 4-bit passive phase shifter is used in each channel and it has an insertion loss of less than 7.3 dB for all the phase controls. The PA which is a key block of the TX path, is composed of four amplifier stages. Both good linearity and high efficiency are achieved by the proposed PA structure with the even harmonic control method, which will be explained in Section III-(a).

III. DESIGN OF POWER AMPLIFIER

A. Structure of 4-Stage Power Amplifier

The PA in each channel is composed of four stages as shown in Fig. 2. In general, the final-stage of a PA has a deep class-AB topology to get a high efficiency but it generates large 2nd order intermodulation (IM2) components at the 2nd harmonic and envelope frequencies. The 2nd harmonic termination circuits can eliminate the mixed 3rd order intermodulation

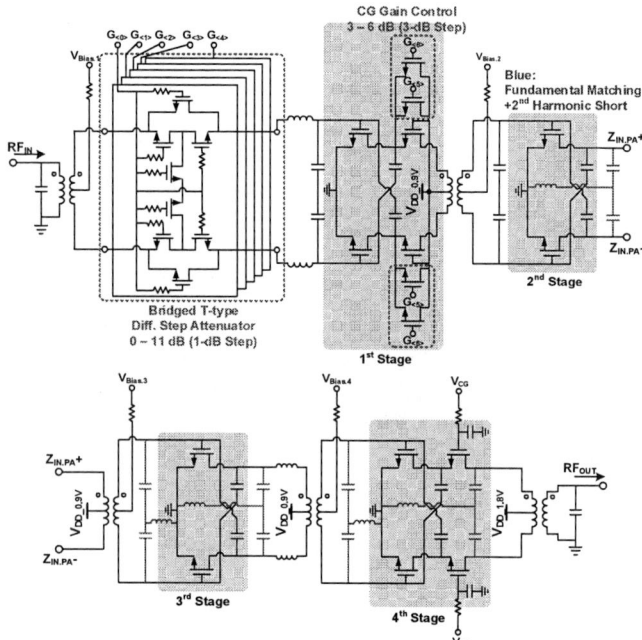

Fig. 2. A schematic of the proposed 4-stage PA in each channel.

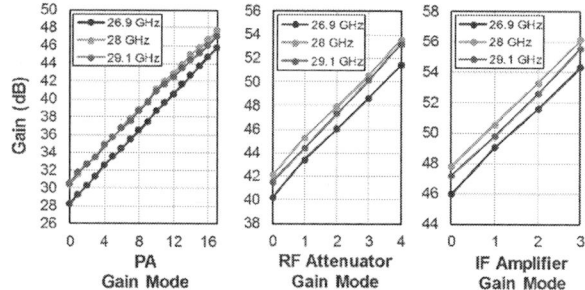

Fig. 3. Measured gain of each TX block according to gain-mode control. When the gain of one block is controlled, other block's gain-mode is fixed.

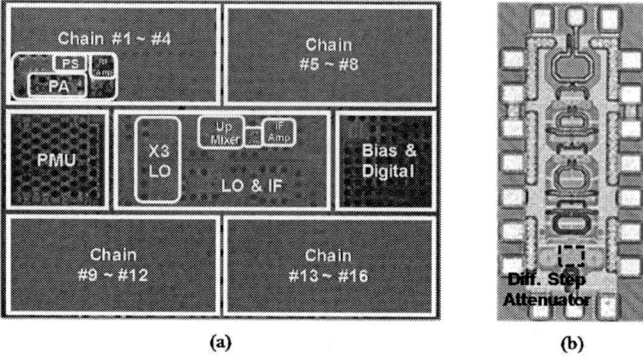

Fig. 4. (a) Die photograph of the 16-element phased array transceiver and (b) IP block of standalone PA for on-wafer probing in a 28-nm bulk CMOS process.

distortions (IMD3s) generated by mixing of the fundamental and 2nd harmonic [6]. However, this circuit does not always ensure the enhanced linearity over a wide bandwidth since the IMD3s are also generated by mixing of the fundamental and envelope nonlinear terms. When the impedance at the envelope frequency is not small enough, the asymmetric IM3 is generated by the memory effect [7]. A severe issue occurs when the envelope nonlinear components from the final-stage feedback to the 1st/2nd/3rd-stages through the V_{DD} path. It not only generates the large mixed IMD3 terms at all the stages but also amplifies the nonlinear components sequentially passing through each stage of the amplifier. Therefore, the linearity is significantly deteriorated. To lessen this problem, the impedance of the V_{DD} network at the envelope frequency should be minimized to near to short. For the 5G system, the low impedance conditions should be retained up to 1 GHz to cover the wide bandwidth of the signal. However, it is difficult to realize this condition by using on-chip passive devices. As one of the solutions for this issue, the on-chip low-drop out (LDO) for the PA is presented in [8], but it is not efficient in terms of the power consumption and size. Especially, the implementation of both the multi-array of the PAs and the LDOs on the chip are a burden. In this work, the stacked amplifier is adopted only for the final-stage and the separated drain-voltage supplies between the final- and other-stages are proposed. The V_{DD} of 0.9V is applied for the 1st/2nd/3rd-stages and the 1.8 V is supplied for the final-stage. This structure can isolate the final-stage from the other-stages at the envelope frequency. The 1st/2nd/3rd-stages have a near class-A biasing compared with the final-stage and their sizes of the transistors are small. Therefore, the IM2 components generated by the

1st/2nd/3rd-stages are relatively small and hardly affect the linearity. The V_{DD} of 1.8V domain that is only for the final-stage can be easily managed at the package side. It is terminated by the off-chip capacitors on the module-package. The dominant IM2s generated by the final-stage are supressed by the 2nd harmonic termination circuit and the separated V_{DD} path isolates the envelope components from the final stage to the 1st/2nd/3rd/-stages. This architecture has advantages in the linearity as well as the efficiency. A stacked structure enhances the gain of the final-stage compared with the common-source (CS), improving the power-added efficiency (PAE) [6]. It is also a suitable structure to obtain a large output power.

B. Large Dynamic Gain Range

The variable-gain control plays a fundamental role in optimizing the system capacity. The transmitter gain should be adjusted so that proper power is received at each terminal. In the 5G system, the signal has a wide bandwidth that expands up to ~1 GHz, so the signal-to-noise ratio (SNR) at a given signal power level is lower than that in the 4G system. For that reason, NF is also an important factor even in TX mode. The first block in a system usually has the most significant impact on the total NF because the effective NFs of the following stages are reduced by the stage gains. Consequently, the NF requirements of the subsequent stages are usually more relaxed. Therefore, when the transmitter operates as a low-gain-mode, the gain should be reduced by the final-block of the TX, preferentially.

978-1-7281-1702-7/19 $31.00 © 2019 IEEE   248

2019 IEEE Radio Frequency Integrated Circuits Symposium

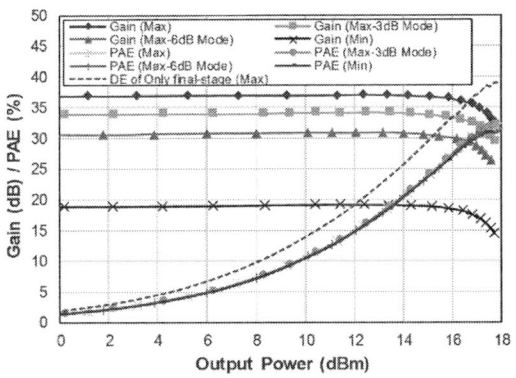

Fig. 5. Measured PAE and gain of the standalone PA for the variable gain mode with a CW signal at 28 GHz.

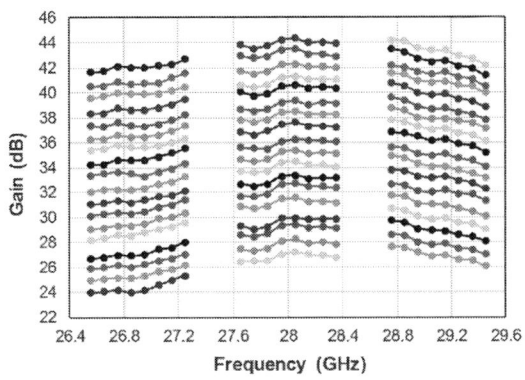

Fig. 6. Measured gain step and flatness response of the transmitter, across the frequency range from 26.5 to 29.5 GHz with the 8CC aggregated signal.

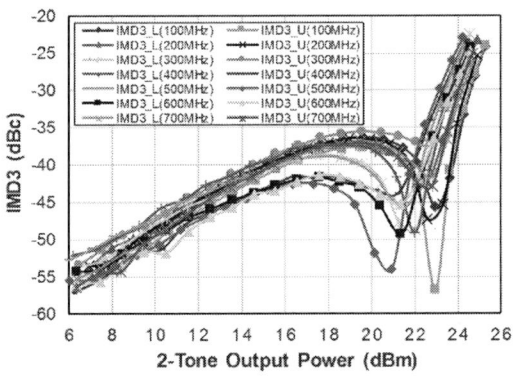

Fig. 7. Measured IMD3 of the fully operated 16-array transmitter for the 2-tone signal with a various tone-spacings at 28 GHz.

In the proposed transmitter, the PA which is a final block of the TX path has a gain range of 18 dB with a 1-dB gain-control step. A 1-dB gain step is provided by proposed differential step attenuator at the input of the first-stage, as shown in Fig. 2. It is a bridged T-type attenuator and each required resistance is realized by a parasitic R_{on} of the corresponding transistor. The R_{on} of a transistor has better tolerance than the poly resistor for the process variation. The step attenuator is merged into the input matching network and has a compact layout area. It has a sufficient linearity to drive the PA for all gain-modes. A 3-dB gain step is implemented by turning off the CG transistor at the first-stage. The combination of the attenuator and CG control

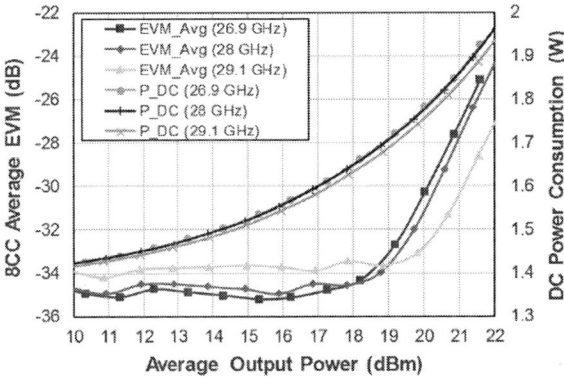

Fig. 8. Measured EVM and consuming DC power of the fully operated 16-array transmitter for the 8CC 5G NR signal at center frequency of 26.9, 28, and 29.1 GHz, respectively.

can cover the 18 dB gain range. A few gain steps in the PA can be used for channel power equalization in the multi-array system. The IF amplifier and the RF attenuator have 9 dB and 12 dB gain ranges, respectively, with a 3 dB gain-steps. Fig. 3 shows the measured TX gain for the 5G signal having an 800 MHz bandwidth, according to the gain control of each block. As a result, a large dynamic gain range from 24 to 63 dB is implemented in the TX.

IV. IMPLEMENTATION AND MEASUREMENT

The proposed transceiver was fabricated using a 28-nm bulk CMOS technology with a flip-chip package, and the total chip area is 4.7*6.4mm^2 as shown in Fig. 4(a). First, to validate the performance of the standalone PA, the intellectual property (IP) block of the PA shown in Fig. 4(b) is measured using on-wafer probing. Fig. 5 shows a PAE, a drain-efficiency (DE), and a power gain for a continuous waveform (CW) signal at 28 GHz. The 4-stage PA in the maximum gain-mode delivers a gain of 37 dB, a saturated output power of 18 dBm with PAE of 32%, and an output 1-dB compression point (OP$_{1dB}$) of 16 dBm with PAE of 29%.

The full 16-element operation of the transmitter with a supply voltage of 1.8V is measured by the conduction test using a 16-way combiner on an in-house test board. The one carrier component (CC) of the new radio (NR) signal has 64 quadrature-amplitude-modulation (QAM), 10 dB PAPR, and 100 MHz bandwidth. Eight CCs are aggregated and used for the test. The IF input frequency is two times of the local oscillation (LO) frequency and the RF frequency of the transmitter output is five times of the LO frequency in this system. Fig.6 shows the gain step and the flatness response of the transmitter for the 8CC NR signal. Only a partial gain range is plotted due to both the noise and the maximum driving power limitation of the measurement instruments. The proposed transmitter achieves a good IMD3 performance for the various tone-spacings as shown in Fig. 7. The IMD3s are maintained under -35 dBc up to a 2-tone output power of 22 dBm and a large asymmetry is not observed thanks to the proposed PA structure. Consequently, the proposed transmitter achieves an average error vector magnitude (EVM) of under -33.5 dB (<2.1%) for

978-1-7281-1702-7/19 $31.00 © 2019 IEEE

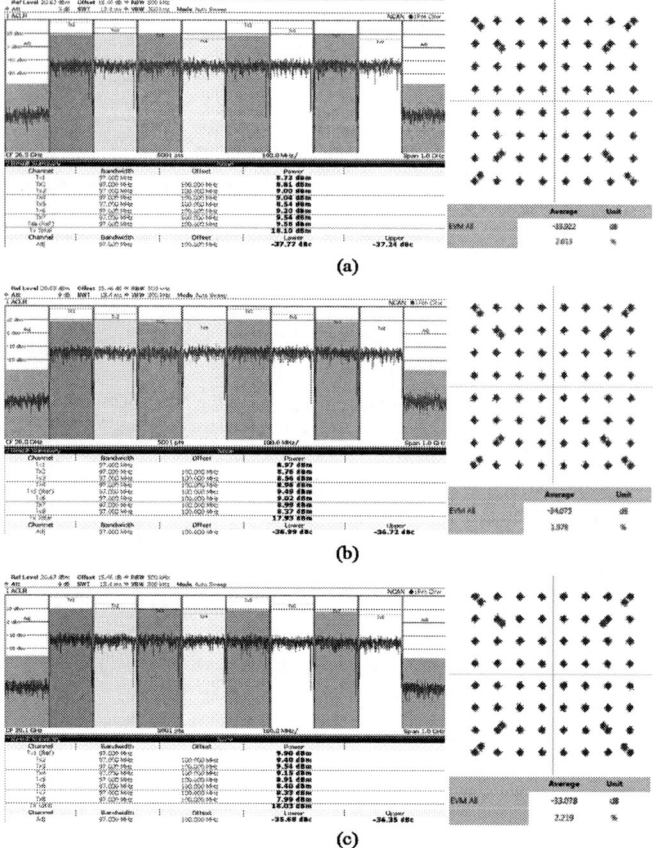

Table 1. Performance summary compared with state of the art 28 GHz phased-array transceivers in TX mode.

Parameter	This Work		ISSCC18 [1]	JSSC18 [2]	JSSC17 [3]	JSSC18 [4]
Technology	28nm RF CMOS		28nm LP RF CMOS	28nm RF CMOS	0.13um SiGe BiCMOS	0.18um SiGe BiCMOS
Front end channels per IC	16		24	8	32	4
TX Input / RX Output Interface	11.2 GHz IF		6.5 GHz IF	Analog IQ BB	3 GHz IF	RF
Measured RF Frequency	26.5 - 29.5 GHz		27.5 - 28.5 GHz	25.8 - 28 GHz	27.5 - 28.5 GHz	28 - 29 GHz
Signal & Bandwidth (BW)	5G NR OFDM 64QAM **10 dB PAPR** **800 MHz BW**	5G NR OFDM 64QAM **10 dB PAPR** **100 MHz BW**	OFDM 64QAM - 400 MHz BW	LTE 64QAM 7.5 dB PAPR 20 MHz BW	-	-
Phase Shifter Resolution	4 bit		3 bit	3 bit	5 bit	6 bit
TX Gain (dB)	24-63		34-44	48	17-47	-2 to 12
PA Psat (dBm)	>17.5		>14	10.5	16	12.5
PA OP1dB (dBm)	>16		>12	9.5	13.5	10.5
PA 64QAM PAE	7.2% (@6 dBm) 9.2% (@9.2 dBm)		7.5% (@6 dBm)	3.0%	-	-
TX Total Power (W)	1.63 (16xCH) @Pout 18 dBm		0.36 (4xCH)	0.416 (4xCH)	4.6 (32xCH)	0.8
TX OFDM 64QAM Pout/Channel (dBm)	6	6	6	3	-	2.5
TX OFDM 64QAM EVM (dB)	<-33.5	<-34.5	< -27*	-	-	-
**TX efficiency per Channel (@EVM -27 dB)	7.2% (@9.2 dBm)	8.2% (@9.9 dBm)	4.4% (@6 dBm)	-	-	-

*Graphically estimated

**TX efficiency is shown at EVM -27 dB to compare the efficiency with [1]

Fig. 9. Measured 5G 8CC NR spectra and EVM in TX mode at an average output power of 18 dBm. (a) A 1st CC EVM of -33.9 dB at 26.9 GHz, (b) a 4th CC EVM of -34 dB at 28 GHz, (c) and an 8th CC EVM of -33 dB at 29.1 GHz.

the 8CC and the total power consumption of 1.63 W (102 mW/channel) including all bias, digital control, and power-management-unit (PMU) blocks, at an average output power of 18 dBm across the 26.9 to 29.1 GHz. These data are shown in Fig. 8. The measured spectra of 5G NR 8CC and EVM constellations are depicted in Fig. 9. Table 1 summarizes the comparison of the measured results with state-of-the-art 28 GHz phased-array transceivers having at least 4 channels. This work delivers higher OP1dB and superior EVM performance in comparison with those of the state-of-the-art transceivers in TX mode. It consumes 102 mW/channel.

V. CONCLUSIONS

A 28-GHz 16-element phased-array transceiver for 5G applications is presented, focusing on the transmitter with the PA. In the design of the PA, a stacked amplifier is adopted only for the final-stage and the separated drain-voltage domain between the final- and other-stages is proposed. This architecture provides advantages for both the linearity and the efficiency. A large dynamic gain range from 24 to 63 dB is implemented in TX. The PA is responsible for a large portion of the gain ranges and it provides merits in NF for a wide bandwidth signal operation. The transceiver was fabricated using 28-nm bulk CMOS technology with flip-chip. For the 8CC of NR signal having 64QAM, 10 dB PAPR, and 800 MHz

bandwidth, it achieves superior EVM of under -33.5 dB (<2.1%) and total power consumption of 1.63 W (102 mW/channel) at an average output power of 18 dBm.

ACKNOWLEDGMENT

The authors would like to thank RF Lab. members in Samsung Electronics for technical discussions and supports.

REFERENCES

[1] J. Dunworth et al., "A 28GHz bulk-CMOS dual-polarization phased-array transceiver with 24 channels for 5G user and basestation equipment," in *IEEE ISSCC Digest Tech.*, Paper, Feb. 2018, pp. 70-72.

[2] H. Kim et al., "A 28-GHz CMOS direct conversion transceiver with packaged 2x4 antenna array for 5G cellular system," *IEEE J. Solid-State Circuits*, vol. 53, no. 5, pp. 1245–1259, May 2018.

[3] B. Sadhu et al., "A 28-GHz 32-element TRX phased-array IC with concurrent dual-polarized operation and orthogonal phase and gain control for 5G communications," *IEEE J. Solid-State Circuits*, vol. 52, no. 12, pp. 3373–3391, Dec. 2017.

[4] K. Kibaroglu et al., "A 28-GHz phased-array transceiver with series-fed dual-vector distributed beamforming," *IEEE J. Solid-State Circuits*, vol. 53, no. 2, pp. 1260–1274, May 2018.

[5] Y. Yeh et al., "A 28-GHz phased-array transceiver with series-fed dual-vector distributed beamforming," in *IEEE RFIC Symp. Dig.*, Paper, June 2017, pp. 65-68.

[6] B. Park et al, "Highly linear mm-wave CMOS power amplifier," *IEEE Trans. Microw. Theory Techn.*, vol. 64, no. 12, Dec. 2016.

[7] M. Franco et al., "Minimization of bias-induced memory effects in UHF radio frequency high power amplifiers with broadband signals," in *IEEE RFIC Symp. Dig.*, Paper, Jan. 2007, pp. 369-372.

[8] D. Jeong et al., "Linear CMOS power amplifier at Ka-band with ultra-wide video bandwidth," in *IEEE RFIC Symp. Dig.*, Paper, June 2017, pp. 220-223.

2019 IEEE Radio Frequency Integrated Circuits Symposium

A 37-40 GHz Phased Array Front-end with Dual Polarization for 5G MIMO Beamforming Applications

Ankur Guha Roy[#1], Ozgur Inac[#2], Amitoj Singh[#&], Tsvika Mukatel[*], Ohad Brandelstein[*],
Thomas W. Brown[#], Salah Abughazaleh[#], Joseph S. Hayden III[#], Byungho Park[#], Greg Bachmanek[#],
Te-Yu Jason Kao[#], Josef Hagn[%], Sidharth Dalmia[#], Doron Shoham[*], Brandon Davis[#], Iris Fisher[*],
Raanan Sover[*], Amit Freiman[*], Bin Xiao[#], Baljit Singh[#], Jonathan Jensen[#]

[#]Intel Corp, USA
[*]Intel Corp, Israel
[%]Intel Corp, Germany
[&]now with Samsung Semiconductor, USA
[1]ankur.guha.roy@intel.com, [2]ozgur.inac@ intel.com

Abstract—**This paper presents a dual polarized 37-40 GHz transmitter-receiver (TRX) phased array front-end RFIC implemented in a 28nm bulk RF-CMOS process. The TRX front-end contains 8 channels, 4-vertical (V) and 4-horizontal (H). Each transmit (TX) or receive (RX) channel consists of PA or LNA, 5-bit passive phase shifter, coarse and fine gain control amplifiers. The TX channel shows an op1dB of 10.2 dBm at the antenna bump. The RX channel shows a noise figure (NF) of 6 dB. TX and RX channels show an EVM of -32.57 dB and -29.80 dB respectively. A 4×4 antenna array was implemented on PCB using 4 TRX front-end RFICs. Beam patterns with different 5G-modulated waveforms were characterized.**

Keywords— **5G, phased array, 39 GHz, MIMO, Beam-forming, CMOS, dual-polarized, K_a band.**

I. INTRODUCTION

5th Generation (5G) wireless technology for mobile applications has been an important area of research lately. Recent announcements by FCC [1], opening up the 37-39 GHz frequency band has enabled rapid development and prototyping of 5G wireless technology for millimeter wave applications. Applications of the phased array transceivers in this frequency band will include both fixed wireless networks (e.g. customer premise equipment) and mobile wireless networks (e.g. user equipment).

This paper presents a transceiver front-end RFIC for 37-40 GHz operation. Recent works in the field of 5G research have reported dual-polarized phased array architectures in the 28 GHz band [2]–[5]. However, there are very limited prior work on the 39 GHz frequency band [6], [7]. In [6] the power amplifier was implemented on a separate chip necessitating complex board design. In [7] the transceiver was implemented for a single-channel only. The proposed work in this paper is the first fully integrated, dual-polarized, multi-chain, front-end transceiver in the 39 GHz band for 5G phased array applications. Additionally, the works reported in [2], [3] use special processes like SiGe BiCMOS. The work reported in this paper uses a bulk CMOS process, making the implementation extremely cost-effective.

II. ARCHITECTURE AND LINE-UP

The proposed front-end is part of a full system intended to serve the emerging 5G cellular applications. The block diagram of the entire system is shown in Fig. 1. The front-end RFIC is designed to work with a companion chip for up/down conversion from RF to IF. The companion chip supports two possible IFs: 4-6.8 GHz and 10.56 GHz. The IF values were chosen such that the companion chip can directly interface with the baseband/IF transceiver in [8]. In this paper only the front-end RFIC will be described.

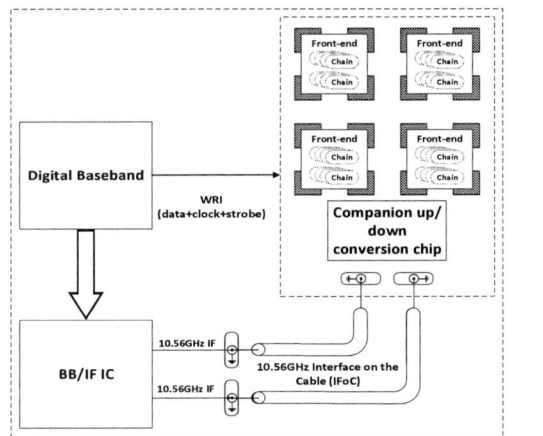

Fig. 1. Block diagram of the system in a 4×4 array configuration for CPE applications. The antenna array supports both V&H polarizations.

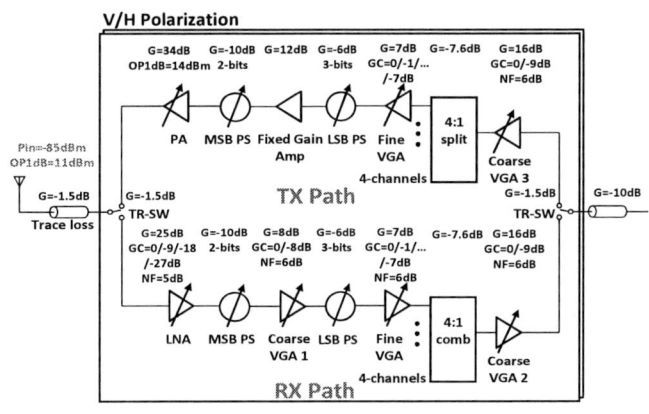

Fig. 2. Line-up of the TX and RX path.

978-1-7281-1702-7/19 $31.00 © 2019 IEEE

2019 IEEE Radio Frequency Integrated Circuits Symposium

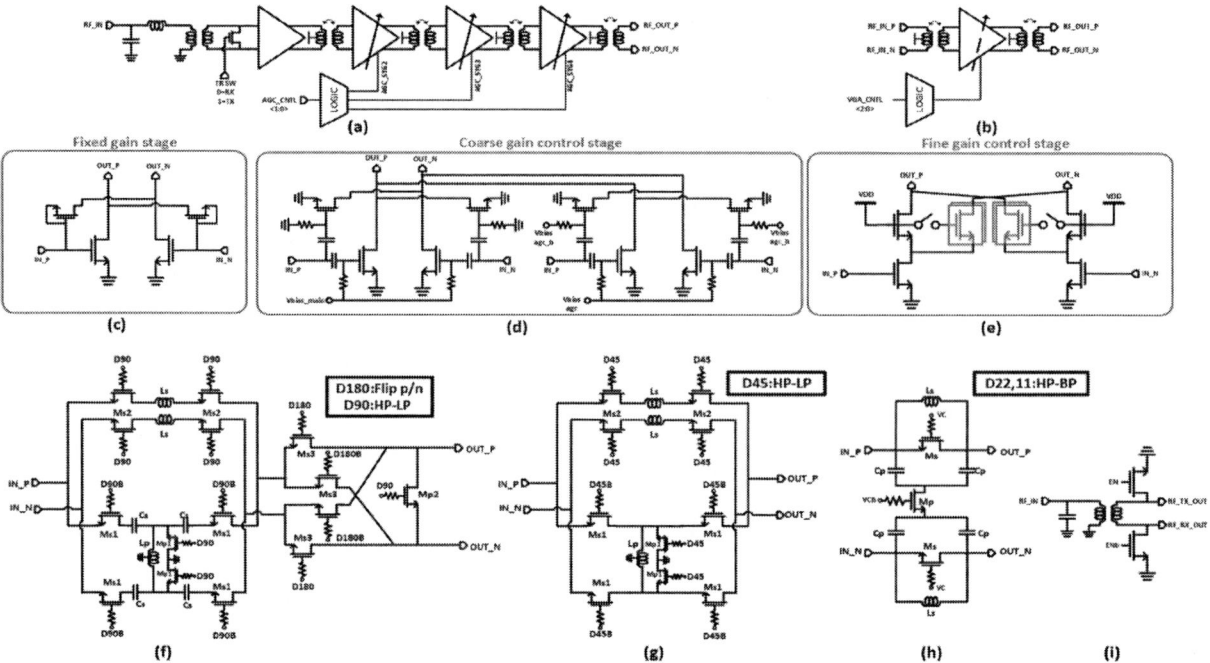

Fig. 3. (a) LNA with fixed and coarse gain control stage, (b) VGA with fine gain control stage, (c) Fixed gain stage, (d) Coarse gain stage, (e) Fine gain stage, (f) MSB (180°/90°) phase shifter, (g) LSB (45°) phase shifter, (h) LSB (22.5°/11.2°) phase shifter, (i) TX-RX shunt switch.

The RFIC is comprised of two sets of four half-duplex transceiver channels. The RFIC allows either TX or RX beamforming communication in a wide-band channel of 1200 MHz within an RF frequency band of 37-40 GHz. The four chains in each polarization are intended to be activated simultaneously to form a coherently combined beam-forming (BF) signal in a desired direction (either in reception or transmission). By using a different data stream on V and H polarization, polarization based MIMO layer can be achieved.

The line-up of the front-end RFIC is shown in Fig. 2. The RFIC is designed to support both mobile hand-set (~4-8 antennas), and fixed Customer Premises Equipment (CPE) (~16-64 antennas) applications in different hardware configurations. This necessitates an RF line-up which is highly reconfigurable and scalable. To accommodate these modes, RF line-up is optimized to have high gain and wide dynamic range while maintaining a low NF and high OP1dB throughout the system. The passive phase shifters are split into two in the line-up to have a more balanced gain/loss distribution. The coarse-gain steps (gain>7 dB) in the line-up of Fig. 2 were used for automatic gain control of the link based on the received signal strength. The fine gain stages have a ~1dB "fine" gain control resolution, and enable independent fine-tuning of each element's gain to compensate for changes in the chain and enable amplitude tapering during beamforming.

The RFIC supports pre-defined beam settings called "codebook". Required phase shift settings for steering the beam at a specific angle can be preloaded to on-chip registers. Beam-steering can be executed with a single command instead of setting multiple phase shifters, reducing the latency. The codebook uses 2KB on-chip memory. TX/RX and beam switch settling time is < 500 nsec.

The RFIC is digitally controlled via a dedicated interface and protocol called WRI (Wireless-Radio-Interface) which is a serial interface based on three wires: two data-bits (D1, and D0) and a clock (clk) signal. The WRI interface enables connecting multiple front-end devices to the same control signals. The front-end chip also supports standard SPI, which is shared over the same data and clock signals as WRI. An external hardware control selects between the two modes.

III. CIRCUIT DESIGN AND PACKAGING

A. Circuit Design

The TX and RX channel amplifiers use three different types of amplifier cores: fixed gain, coarse gain, and fine gain as shown in Figs. 3(a)-(b). The fixed gain amplifier was implemented using a neutralized common-source, pseudo-differential pair (Fig. 3(c)). The neutralization capacitor was implemented using a MOSFET to enable tracking with process and temperature variations. The coarse gain amplifier was implemented using two neutralized pseudo-differential pairs (Fig. 3(d)). The high/low gain steps were obtained by adding/subtracting the output currents of the two neutralized pairs. The fine gain amplifier was implemented with a cascode structure and switchable cascoded cells to steer the output current (Fig. 3(e)). The switched cascade cells are weighted to ensure ~1 dB gain steps.

The digital controlled analog phase shifters were implemented using high-pass/low-pass or high-pass/band-pass filter combinations (Figs. 3(f)-(h)). The 180° phase shift is obtained by simply swapping the differential signals. Fig. 3(i) shows the shunt-shunt switch used for TX/RX mode switching.

The phase shifters and the fine gain amplifier designs were re-used in both TX and RX. The power amplifier uses a scaled

978-1-7281-1702-7/19 $31.00 © 2019 IEEE

version of the fixed gain core used in the LNA to generate a higher output power level.

On-chip, lumped Wilkinson power combiners were used for combining/splitting the RF signals into different chains.

B. Packaging and Transition Modeling

The chip was packaged in a low cost, ultra-thin (< 0.88mm) 4-layer flip-chip plastic 350um pitch BGA package. The Silicon-package-PCB interconnects were modeled using HFSS (Fig. 4). The package loss was < 0.65 dB and the isolation between the V&H polarizations was <-30 dB.

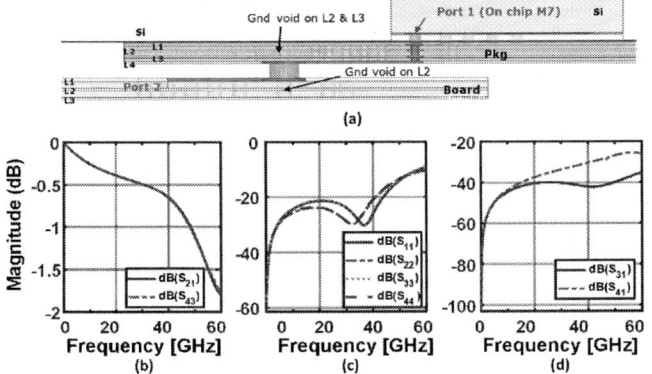

Fig. 4. Modeling of Silicon-Package-Board transition. (a) Side view of the transition structure. (b) Insertion loss, (c) input and output match characteristics, (d) cross-coupling between V&H channels.

IV. MEASUREMENTS

The TRX front-end chip was fabricated in a 28 nm bulk RF-CMOS process. The test-board is shown in Fig. 5. The die micrograph (5.18mm×3.33mm), and the 4×4 dual polarized antenna array are shown in Fig. 5(a) and (b) insets, respectively.

A. Single Channel Measurements

For the TX & RX chains, the measured average gain across

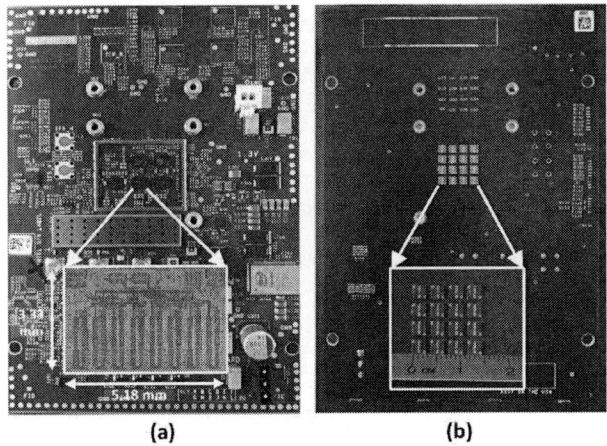

Fig. 5. Top and bottom view of the 4x4 dual polarized array test PCB. The die micrograph (5.18mmx3.33mm) and the antenna array dimensions are shown inset.

32-different phase settings was in the 45-48 dB range and 36.9-41.9 dB range respectively, using a nominal phase setting. The measured s-parameters (magnitude and relative phases) in the TX and RX modes with different phase shifter settings are shown in Figs. 6(a), (b), (e), and (f), respectively. With different phase settings, the in-band gain variation was within ±2 dB for both TX and RX path. The noise figure (NF) of the RX chain is 6 dB at highest gain setting and it increases with different gain settings as shown in Fig. 6(c). The measured results are close to the simulated numbers at 27°C. The NF increases by only ~ 8 dB with a 27 dB gain reduction in the AGC. The fine gain control steps of the RX VGA are shown in Fig. 6(g). The VGA shows a 10.2 dB range from the highest gain with steps of ~1.5 dB.

The continuous wave (CW) measurements were done by enabling one channel at a time (non-beamforming) with 5 different units. The P1dB values for the TX and RX modes are

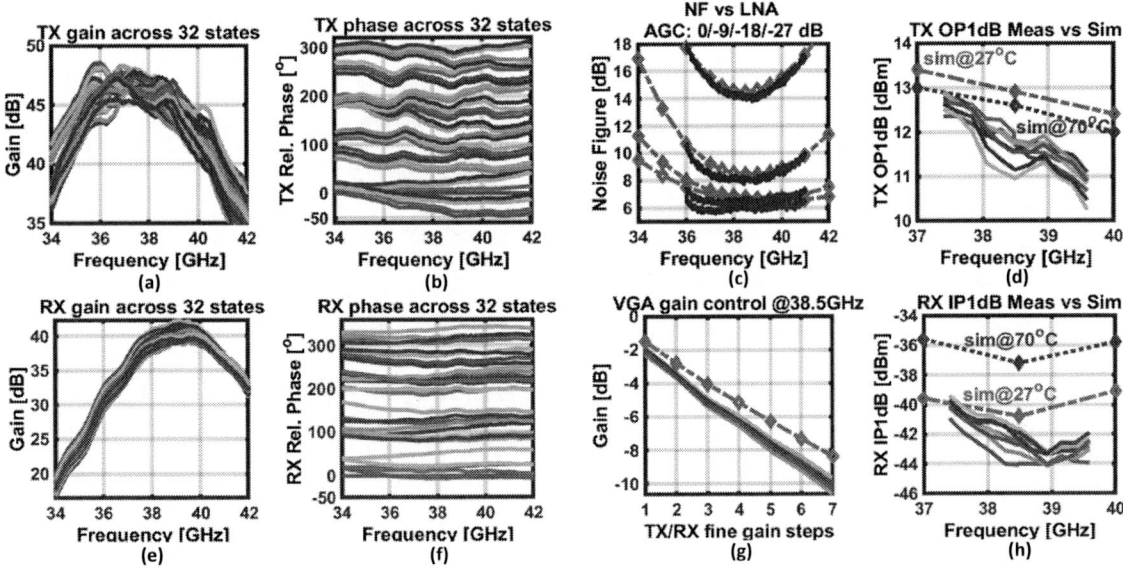

Fig. 6. Measured (a) TX gain at different phase states, (b) TX phase states, (c) RX NF with different AGC steps, (d) TX OP1dB, (e) RX gain at different phase states, (f) RX phase states, (f) TX/RX fine gain steps, and (g) RX IP1dB.

shown in Figs. 6(d) & (h) respectively. The TX and RX channels show an average op1dB>10.2 dBm and average ip1dB< -39 dBm across the entire frequency band for all the channels. The RX IP1dB can be improved by RX gain control.

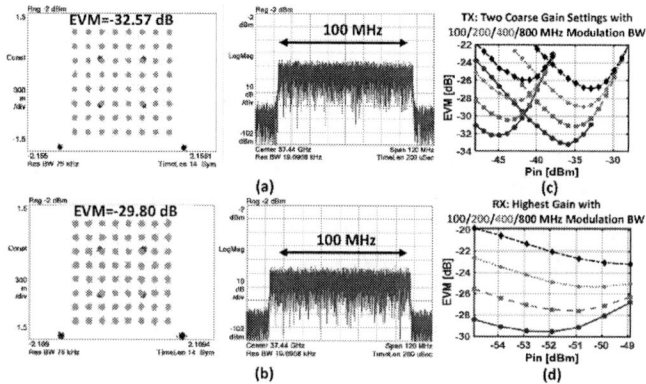

Fig. 7. Sample constellation and spectrum for (a) TX chain, (b) RX chain with a 64-QAM OFDM signal of 100 MHz BW (@4.92 Gbaud), EVM variation with input power for different modulation BWs: (c)TX chain, (d) RX chain.

Single channel EVM was measured using Pre-5G 100/200/400/800 MHz bandwidth (BW) 64-QAM OFDM modulated signals at 37.44 GHz carrier frequency. Constellation diagram and signal spectrum for the TX and RX chains are shown in Figs. 7(a)-(b), respectively. Variation of EVM in TX and RX channels with input power for different modulation BWs are shown in Figs. 7(c)-(d).

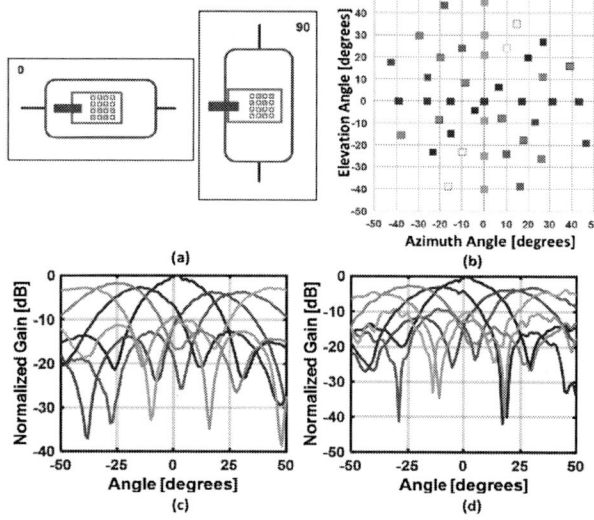

Fig. 8. (a) Two different orientations of the SV board facing the horn antenna (bird's eye view). (b) Measured main beam locations with Azimuth and Elevation angles. Each diagonal cut correspond to a particular angle of the SV board. Measured beam patterns in (c) 0° orientation, (d) 90° orientation of the SV board w.r.t. the horn antenna.

B. 4×4 Array Measurements

The 4x4 antenna array covers a 90 degree spherical sector, each beam has a 3-dB beamwidth of ~26 degrees, and beam placement accuracy (given the 5-bit phase shifter) is around 3.5-5 degrees. The 4x4 array beam patterns were measured in the TX mode using a horn-antenna placed in the far field. Due

to the symmetric layout in both RFIC and the antenna board, the amplitude and phase difference across different channels were minimal. The calculated phase settings for different beams were loaded to RFIC codebook and the beam patterns (single axis 2-D cut plane) were measured by rotating the SV board with respect to the horn antenna as shown in Fig. 8(a). The beam locations for different beams were shown in Fig. 8(b). Figs. 8(c) & (d) show the beam patterns in the 0° and 90° orientations of the SV board respectively using a 38.5 GHz input signal.

Table 1. Performance summary of the proposed TRX front-end and a comparison with the state-of-the art 5G TRX beam-forming front-ends.

Parameter	[This work]	[2]	[3]	[4]	[5]
Process/Tech. Node	28 nm CMOS	0.13 um SiGe	28 nm CMOS	0.18 um SiGe	28 nm LP CMOS
Freq. Band [GHz]	37-40	27-29	25.8-28	28-32	26.5-29.5
Channels per chip	8	32	8	4	24
Chip area [mm²]	17.2	165.4	7.3	11.7	27.8
Phase shifter resolution	5	6	3	6	3
Rx IP1dB/Channel [dBm]	-36/-44	-22.5	-	-22	-
Rx NF [dB]	6-7.6	6	6.7	4.6-5.2	4.4-4.7
Rx Pdc/Channel [mW]	78.5	206	50	130	42
Tx op1dB/Channel [mW]	10.2-12.3	13.5	9.5	10	>12
Tx Pdc/Channel [mW]	339 @p1dB	319 @psat	85 @7dB BO	200 @p1dB	122 @p1dB

Table 1 shows the comparison of the proposed TRX front-end with the state-of-the art 5G beamforming front-ends.

V. CONCLUSION

A 4×4 dual-polarized TRX beam-forming front-end RFIC for 39 GHz 5G applications using a 28 nm bulk CMOS process was reported in this work. A 16 element antenna module using the RFICs were also built to demonstrate the over the air capabilities. The proposed solution achieves comparable performance to the previously reported 5G beam-forming front-ends in the 28 GHz band. The proposed front-end is scalable and can be extended to implement larger/smaller antenna arrays.

REFERENCES

[1] FCC, "Use of Spectrum Bands above 24 GHz for Mobile Radio Services." 03-Aug-2018.

[2] B. Sadhu et al., "A 28GHz 32-element phased-array transceiver IC with concurrent dual polarized beams and 1.4 degree beam-steering resolution for 5G communication," IEEE ISSCC Dig. Tech., 2017, pp. 128–129.

[3] H. Kim et al., "A 28-GHz CMOS Direct Conversion Transceiver With Packaged 2×4 Antenna Array for 5G Cellular System," IEEE J. Solid-State Circuits, vol. 53, no. 5, pp. 1245–1259, May 2018.

[4] K. Kibaroglu et al, "An ultra-low-cost 32-element 28 GHz phased array transceiver with 41 dBm EIRP and 1.0-1.6 Gbps 16-QAM link at 300 meters," Proc. RFIC Symp, 2017, pp. 73–76.

[5] J. D. Dunworth et al., "A 28GHz Bulk-CMOS dual-polarization phased-array transceiver with 24 channels for 5G user and basestation equipment," IEEE ISSCC Dig. Tech. Papers, 2018, pp. 70–72.

[6] X. Li et al., "A 39 GHz MIMO Transceiver Based on Dynamic Multi-Beam Architecture for 5G Communication with 150 Meter Coverage," Proc. IMS, 2018, pp. 489–491.

[7] Z. Chen et al., "A 256-QAM 39 GHz Dual-Channel Transceiver Chipset with LTCC Package for 5G Communication in 65 nm CMOS," Proc. IMS, 2018, pp. 1476–1479.

[8] B. Jann et al., "A 5G Sub-6GHz Zero-IF and mm-Wave IF Transceiver with MIMO and Carrier Aggregation," IEEE ISSCC Dig. Tech. Papers, 2019.

An 802.11ba 495μW -92.6dBm-Sensitivity Blocker-Tolerant Wake-up Radio Receiver Fully Integrated with Wi-Fi Transceiver

Renzhi Liu[1], Asma Beevi K. T.[2], Richard Dorrance[1], Deepak Dasalukunte[1], Mario A. Santana Lopez[3],
Vinod Kristem[2], Shahrnaz Azizi[2], Minyoung Park[2], Brent R. Carlton[1]
[1]Intel Labs, Hillsboro, OR, USA
[2]Intel Labs, Santa Clara, CA, USA
[3]Intel Labs, Guadalajara, Mexico
[1]renzhi.liu@intel.com, [2]asma.kuriparambil.thekkumpate@intel.com

Abstract—**An 802.11ba-based wake-up radio (WUR) receiver is presented. The WUR receiver prototype is integrated within an 802.11a/b/g/n/ac Wi-Fi transceiver, occupying 0.05mm² for RF/analog frontend and 0.08mm² for digital baseband. The WUR receiver consumes 495μW standalone and consumes 667μW from Wi-Fi system supply. The receiver has a measured sensitivity of -92.6dBm and can tolerate -40.5dBm Wi-Fi adjacent channel blocker with 3dB receiver de-sensitization. The WUR receiver can operate when the Wi-Fi system is in sleep mode and can turn on Wi-Fi radio upon receiving 802.11ba-based wake-up packet over the air.**

Keywords—**802.11ba, wake-up radio, Wi-Fi**

I. INTRODUCTION

IEEE 802.11 standard based Wi-Fi is the most widely adopted connectivity solution. However, its unscheduled nature means there exists a fundamental trade-off between low-power and low-latency operations. This limitation of Wi-Fi presents significant challenges not only for achieving low-power always-connected systems, such as wireless sensor networks and Internet of Things, but also for reducing power consumption and latency for next generation Wi-Fi system for mobile/laptop platforms. To overcome this, an event-driven low power wake-up radio (WUR) can be added for monitoring the wireless channel. WUR would turn on the Wi-Fi radio only upon receiving wake-up message, thereby achieving both low power and low latency. In December 2016, a task group was formed for IEEE 802.11ba (TGba), aiming to standardize the WUR operation within IEEE 802.11 standard.

Various non-standard based WUR solutions have been proposed in 2.4GHz/5GHz bands [1-4], however, they all have major shortcomings for future Wi-Fi chip integration. Off-chip high-Q inductors are required for achieving high WUR sensitivity in [1-2] and >1mm² silicon area for RF/analog frontend is needed in [2], both obstructing WUR receiver adoption in Wi-Fi chip. In [2-3], for achieving low WUR power consumption, ultra-low voltage supply for major power consuming blocks is used. However, the power could quickly expand after system integration once supply efficiency is factored in. In [4], although no off-chip component and no low-voltage supply is needed, its sensitivity and blocker tolerance is not sufficient for the WUR replacing legacy Wi-Fi wake-up mechanism without significant loss of communication range and quality.

In this paper, we present a first-ever prototype of an 802.11ba-based WUR receiver design as well as monolithic integration of WUR and Wi-Fi transceiver to demonstrate the technology. The presented WUR receiver prototype can achieve sensitivity and blocker tolerance performance close to Wi-Fi chips [5] for comparable communication range and quality while consuming only 495μW standalone and 667μW from the switching supply in a Wi-Fi system. Moreover, the WUR only occupies 0.05mm² for RF/analog frontend and 0.08mm² for digital baseband (DBB) without requiring external components, which is crucial for the adoption of WUR in next generation Wi-Fi chips.

II. 802.11BA WAKE-UP RADIO

802.11ba is an amendment to the 802.11 standard currently in development by TGba, with a goal to standardize the WUR operation and achieve less than 1mW WUR receiver active power and low latency for Wi-Fi devices. Based on TGba Draft 1.0 [6], the 802.11ba utilizes multi-carrier-on-off-keying modulation scheme (MC-OOK). The MC-OOK uses the center 13 OFDM subcarriers among 64 total subcarriers within a 20MHz Wi-Fi channel, thus occupying 4MHz bandwidth. For a mandatory 62.5kbps PHY rate, each MC-OOK symbol is 2μs for the WUR preamble and 4μs for the WUR data field with each WUR data bit represented by 4 MC-OOK symbols with 1/4 rate Manchester-coding. A WUR packet starts with legacy 802.11 preamble to provide coexistence between the legacy 802.11 and new 802.11ba devices. The 802.11 preamble is followed by MC-OOK based WUR preamble for packet acquisition and then WUR data field. An example of 802.11ba signal is shown in Fig. 1(a).

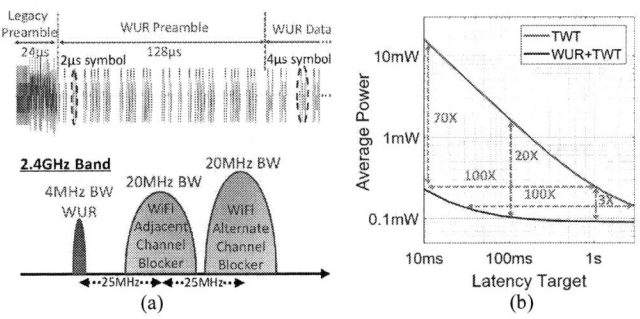

Fig. 1 (a) 802.11ba WUR signal overview in time and frequency domain; (b) TWT vs. WUR+TWT for average power and latency

802.11ba also supports duty-cycled WUR operation, allowing WUR receiver to power on and off periodically to save even more power. The WUR duty-cycle can be scheduled according to a target-wake-time (TWT) schedule. TWT is a power saving feature adopted in 802.11ah/ax, where the 802.11 station (STA) turns on Wi-Fi radio to receive data from the access point (AP) at a pre-negotiated wake-time after sleep, regardless whether there is buffered data in AP to be transferred or not. When WUR is used together with TWT, the STA only needs to turn on the WUR receiver at the schedule wake-time and only wakes up Wi-Fi radio if needed based on the received WUR packet. Fig. 1(b) shows latency target versus the average system power consumption when there is no buffered data in AP. The system power is estimated from 802.11ax power consumption model in [7] and WUR is assumed to consume 667µW active power and be active for 2ms in each TWT interval (which is also latency target). As shown in the figure, up to two orders of magnitude improvement on average power or latency can be achieved by having WUR with TWT versus TWT alone.

III. WUR IMPLEMENTATION AND WI-FI INTEGRATION

For WUR and Wi-Fi system integration, the WUR receiver needs to achieve not only low power operation, but also compact area. Fig. 2 shows a block diagram of the proposed 802.11ba-based WUR receiver. A mixer first architecture was adopted for the WUR [4]. The double balanced (DB) mixer is followed by an interleaved dynamic amplifier, which significantly improves gain and noise figure by virtue of common gate–common source (CG-CS) coupling. To achieve similar blocker tolerance of the main Wi-Fi radio, an overall 4th order baseband filtering was designed for WUR by adding another 2nd order bi-quad filter. Meanwhile, an LC-VCO is also adopted for providing sufficient phase noise performance. The LC-VCO is embedded within a frequency lock loop (FLL) which provides sufficient frequency accuracy for the MC-OOK demodulation while only using a 32.768kHz RTC reference. The RF/analog frontend is followed by a 7-bit asynchronous SAR ADC and an 802.11ba-based DBB. Integration scheme for WUR and main Wi-Fi radio is also shown in Fig. 2. The WUR RF inputs directly tap on the Wi-Fi receiver inputs. A 1.1V switching supply from system platform is used as the main power source, with micro-LDOs designed for WUR internal supplies.

A. RF/analog Frontend

Minimizing the device sizes in the passive mixer is vital for lowering overall WUR power. However, this increases the noise of zero IF receiver chain. To reduce the noise figure (NF) while keeping low power consumption for signal amplification chain, a novel mixer-embedded dynamic baseband low noise amplifier (LNA) is proposed.

The dynamic amplifier is shown in Fig.3 (a). It uses common gate–common source (CG-CS) coupling and complimentary current reuse structure to accomplish low noise and low power operation. LNAs have used CG-CS coupling for its noise cancellation properties. But in the proposed low-power baseband LNA, the input impedance of the CG devices are much higher than the source impedance, thus limiting the amount of CG-CS noise cancellation. Still, the noise reduction properties of CG-CS coupling makes it attractive for low power amplifiers.

The noise reduction can be explained by noting that due to CG-CS coupling, signal gain is twice that of the device noise gain. For the same power, this higher signal to device noise ratio at the amplifier output reduces amplifier noise contribution towards NF by a factor of 4 comparing to a non-CG-CS coupling structure. Amplifier power efficiency is further enhanced by adopting a complementary structure for current reuse.

Dynamic biasing technique is used to realize CG-CS coupled structure in baseband, as conventional AC coupling with resistor based DC biasing will either result in gigantic passive devices or significantly higher NF due to the biasing circuit. Dynamic biasing for an amplifier involves storing the required biasing voltage across a capacitor during a reset mode and later using this stored voltage to keep the amplifier in proper operating region while it amplifies the signals. Dynamic amplifier control signal is given in Fig.3 (b). Fig 3 (c) shows the reset mode of proposed amplifier during which the amplifying devices are diode connected and their gate to source voltages are stored across capacitors. Illustrated in Fig. 3 (d) is the active mode, during which the CG-CS coupled devices amplify the signal while the biasing capacitors keep them in their proper operating region. The reset mode of the dynamic amplifier causes disruption to signal amplification. Continuous signal amplification is enabled by interleaving two dynamic amplifiers, with one being active and the other being off or in reset. This only cause negligible power consumption overhead

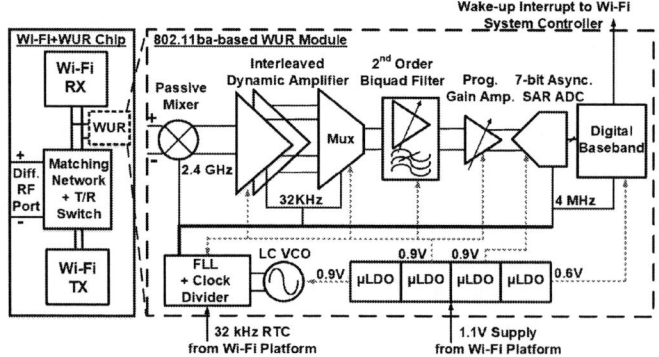

Fig. 2 WUR architecture and Wi-Fi chip integration diagram

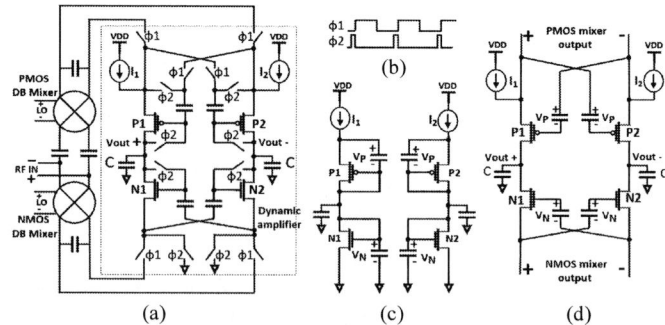

Fig.3 (a) Mixer-embedded dynamic baseband LNA design; (b) active-reset control signal; (c) reset mode; (d) active mode

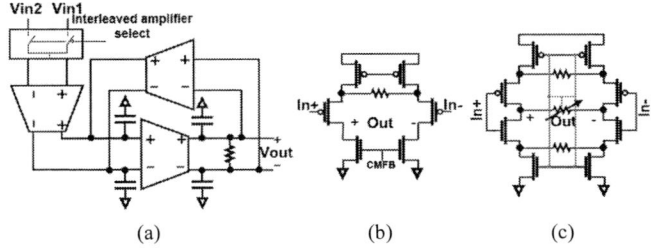

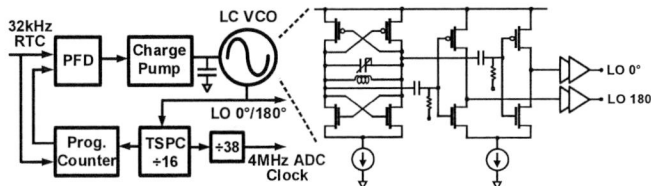

(a)　　　　　(b)　　　　　(c)

Fig.4 (a) g_mC biquad; (b) g_m cell in the biquad; (c) PGA

Fig. 5 FLL block diagram and LC-VCO design

as the inactive amplifier is turned off most of the time and only turned to reset for a small fraction of the overall time-interleaving cycle. Thanks to the circuit architecture, this mixer-embedded analog frontend has a high gain of 42dB and NF of 11dB while only consuming 37uA current. The simulated NF improvement by using CG-CS coupling is 2.7 dB, as improvement is limited by mixer noise contribution.

The complex pole filter [8] after dynamic amplifier is shown in Fig. 4 (a) (b). The biquad is followed by a programmable gain amplifier (PGA) shown in Fig. 4 (c) whose gain can be varied from 0dB to 18dB. The entire signal chain has a 60dB peak gain with 6dB tuning step and a 2MHz typical bandwidth.

B. Frequency Synthesizer

As phase coherence is not required for MC-OOK demodulation, a frequency lock loop (FLL) is sufficient for generating the WUR LO signal. The FLL architecture is shown in Fig. 5, similar to the one in [4]. The reference clock is a 32.768kHZ RTC from the Wi-Fi platform, minimizing the Wi-Fi system overhead when it is in sleep mode. The LO frequency tuning step is 0.5MHz, achieved by a programmable counter. The simulated FLL frequency variation caused by the counter's quantization error is within ±0.2MHz. As a result, the overall LO frequency error is <0.5MHz without calibration, causing <0.5dB WUR signal power loss in a 2MHz-bandwidth baseband.

Based on analysis, for achieving sufficient phase noise performance that leads to similar Wi-Fi blocker tolerance as main Wi-Fi radio, LC-VCO is a lower power option than ring-VCO for WUR receiver. The LC-VCO design is shown in Fig. 5. The design is highly area constrained, with a 10nH inductor achieving a quality factor of 6 at 2.4GHz, which limits the VCO swing. A VCO buffer is then used for amplifying the swing by about 4 times at the cost of worsened but still sufficient phase noise performance.

C. ADC and DBB

An 802.11ba-based DBB is integrated with the WUR frontend. The DBB is constantly monitoring the wireless channel for frontend automatic gain control and WUR preamble detection. Upon WUR packet acquisition, WUR data frame is demodulated and parsed.

Different from using a single bit ADC [1-3], a 7-bit asynchronous SAR-ADC is adopted for digitization. Multiple ADC output bits not only provide wider dynamic range, but also provide flexibility for DBB to correct frontend non-idealities, such as DC offset and filter out any low frequency noise. Both the ADC and the DBB is driven by an

approximately 4MHz clock, which is 1/608 of LO frequency. This results in <2% clock frequency error for various 2.4GHz bands and it is also corrected by the DBB using a frequency/timing tracking algorithm.

D. WUR and Wi-Fi Integration

As shown in Fig. 2, the WUR receiver frontend is connected in parallel to the Wi-Fi receiver, interfacing the Wi-Fi transmitter with the same T/R switch. A step-up matching network is embedded within the T/R switch, providing passive voltage gain for both WUR and Wi-Fi receiver, improving their sensitivity.

The WUR receiver requires 0.9V supply for RF/analog frontend and 0.6V supply for DBB. For WUR system integration, a 1.1V switching supply is used. Multiple micro-LDOs (μLDOs) were designed for WUR internal supplies for noise isolation. The μLDOs filter switching supply noise at WUR signal frequency (<2MHz), improving front end sensitivity, and at Wi-Fi blocker frequency (>15MHz), for LC-VCO phase noise performance. The four μLDOs combined consume 16μA quiescent current and occupy 0.01mm².

IV. MEASUREMENT RESULTS

The integrated 802.11ba-based WUR receiver and 802.11a/b/g/n/ac Wi-Fi transceiver is fabricated in 28nm CMOS technology. Fig. 6 shows photos of two sections of the WUR design (surrounding Wi-Fi blocks are not shown). The active area for the WUR frontend blocks measures 0.05mm² while the DBB measures 0.08mm². Table 1 shows a current consumption breakdown of the receiver during active WUR packet acquisition. The total current consumption is 606μA, leading to a total power consumption of 495μW with 0.9V RF/analog supply and 0.6V digital supply. While integrated within the Wi-Fi chip, the WUR receiver consumes 667μW from a 1.1V system power supply. The majority of the current consumption comes from the LC-VCO, which is limited by the adoption of a low-Q inductor for saving silicon area.

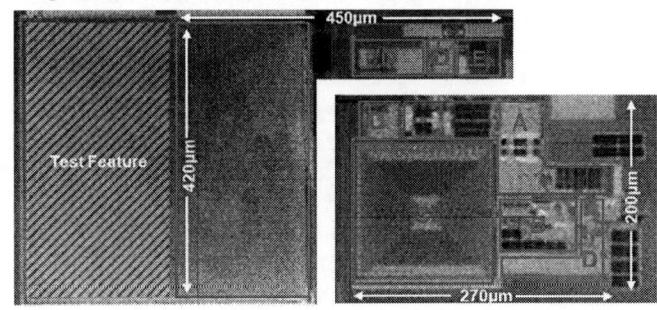

Fig. 6 Chip photo: (A) RF/analog frontend, (B) FLL, (C) VCO inductor, (D) μLDOs, (E) ADC, (F) DBB

Table 1. WUR Receiver Current Consumption Breakdown

	VCO and LO Driver	RF/analog frontend	ADC	DBB	Total
Current (μA)	325	86	25	170	606

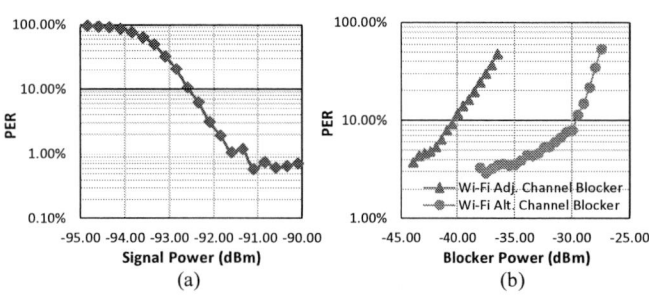

Fig. 7 (a) WUR sensitivity performance; (b) WUR receiver Wi-Fi blocker tolerance performance (-89.6dBm WUR signal input)

The WUR receiver is operated in 802.11ba mandatory mode at 2.4GHz band with 62.5kbps data rate. Fig. 7 (a) shows the receiver sensitivity performance, with WUR packet containing Wi-Fi legacy preamble, WUR preamble and a 64-bit data field. As shown in Fig. 7 (a), we can achieve -92.6dBm sensitivity for 10% PER. Fig. 7(b) shows Wi-Fi blocker tolerance of the receiver. With 3dB receiver de-sensitization, the WUR can tolerate -40.5dBm or -30dBm Wi-Fi blockers from non-overlapping adjacent channel (25MHz offset) or alternate channel (50MHz offset). This corresponds to signal to interference ratio (SIR) of -49dB and -60dB respectively.

Wi-Fi system wake-up through WUR was demonstrated over the air. The WUR can monitor the wireless channel for wake-up packet while the Wi-Fi system is in sleep mode. Upon successfully receiving an 802.11ba-based wake-up packet, the WUR receiver generates a wake-up interrupt signal that is routed to a Wi-Fi system controller and turns on the Wi-Fi main radio. When WUR is turned-off, no performance degradation is observed for the Wi-Fi radio as compared to the original Wi-Fi transceiver without WUR.

Table 2 shows a comparison for state-of-art WUR designs. This work is the first reported 802.11ba-based WUR design and also the first integration of WUR receiver with Wi-Fi chip. Compared to [2-4] which has no DBB, this work with full WUR receiver can achieve state-of-art sensitivity, blocker tolerance and power consumption without requiring external components while maintaining a highly compact area which eases Wi-Fi system integration.

V. CONCLUSION

The first-ever 802.11ba-based WUR receiver as well as its integration prototype with an 802.11a/b/g/n/ac Wi-Fi transceiver is demonstrated in this paper. WUR receiver can achieve state-of-art -92.6dBm sensitivity, >49dBr Wi-Fi adjacent channel blocker tolerance and 495μW power consumption without using external passive components while occupying 0.13mm² silicon area. This demonstration shows low-cost, high performance 802.11ba-based WUR and high performance Wi-Fi main radio can co-exist.

Table 2. State-of-art WUR receiver comparison

	[2]	[3]	[4]	This work
Process	65nm	40nm	14nm	28nm
Freq. band (GHz)	2.4	5.8	2.4	2.4
Modulation	FSK	OOK/FSK	MC-OOK	MC-OOK
Data rate (kbps)	25/50	62.5	62.5	62.5
Power supply (V)	0.6	0.5/0.95	0.95	0.6/0.9
Power consumption (μW)	466 (no DBB)	470 (no DBB)	95 (no DBB)	392 (frontend) 495 (w/ DBB)
Sensitivity (dBm)	-102*	-92.5*	-72*	-92.6**
SIR (dB)	<-52†	-24†	-20‡	-57†/-49‡
802.11ba based	No	No	No	Yes
Wi-Fi integration	No	No	No	Yes
External passive components	High Q Inductor	No	No	No
Active area (mm²)	1.5 (no DBB)	0.17 (no DBB)	0.15 (no DBB)	0.05 (frontend) 0.13 (w/ DBB)

* 0.1% BER, bit level performance only
** 10% PER, including packet acquisition and decoding performance
† Continuous wave blocker @ 20MHz offset
‡ 20MHz bandwidth Wi-Fi blocker @ 25MHz offset adjacent channel

REFERENCES

[1] C. Salazar, A. Cathelin, A. Kaiser and J. Rabaey, "A 2.4 GHz interferer-resilient wake-up receiver using a dual-IF multi-Stage N-path architecture", *IEEE J. Solid-State Circuits*, vol. 51, no. 9, pp. 2091-2105, Sep. 2016.

[2] H.-G. Seok, O.-Y. Jung, A. Dissanayake and S.-G. Lee, "A 2.4GHz, -102dBm-sensitivity, 25kb/s, 0.466mW interference resistant BFSK multi-channel sliding-IF ULP receiver", in *Proc. VLSI Circuits*, Jun. 2017, pp. C70-C71.

[3] J. Im, H.-S. Kim and D. D. Wentzloff, "A 470μW -92.5dBm OOK/FSK receiver for IEEE 802.11 WiFi LP-WUR", in *Proc. ESSCIRC*, Sep. 2018, pp. 302-305.

[4] E. Alpman, A. Khairi, M. Park, V. S. Somayazulu, J. R. Foerster, A. Ravi and S. Pellerano, "95μW 802.11g/n compliant fully-integrated wake-up receiver with -72dBm sensitivity in 14nm FinFET CMOS", in *Proc. RFIC*, Jun. 2017, pp. 172-175.

[5] S. T. Yan, L. Ye, R. Kulkarni, E. Myers, H.-C. Shih, H. Wu, S. Saberi, D. Kadia, D. Ozis, L. Zhou, E. Middleton and J. L. Tham, "An 802.11a/b/g/n/ac WLAN transceiver for 2 × 2 MIMO and simultaneous dual-band operation with +29 dBm P_{sat} integrated power amplifiers", *IEEE J. Solid-State Circuits*, vol. 52, no. 7, pp. 1798-1813, Jul. 2017.

[6] TGba, IEEE P802.11ba/D1.0, http://www.ieee802.org/11/private/Draft_Standards/11ba/index.html

[7] C. Ghosh, "Discussion on deep and shallow sleep states", *IEEE Standard* 802.11-15/1100r2, Sep. 2015.

[8] R. Kolm and H. Zimmermann, "A 3rd-order 235MHz low-pass gmC-filter in 120nm CMOS", in *Proc. ESSCIRC*, Sep. 2006, pp. 215-218.

A -80.9dBm 450MHz Wake-Up Receiver with Code-Domain Matched Filtering using a Continuous-Time Analog Correlator

Vivek Mangal, Peter R. Kinget

Columbia University, NY

Abstract — **A continuous-time, clockless analog correlator uses pulse-position-encoded analog signal processing with VCOs as integrators and pulse-controlled relaxation delays; it operates as a matched filter to despread asynchronous wake-up codes. A correlator prototype has been designed in 65nm CMOS-LP technology, consumes 37nW from 0.54V, and performs code-domain filtering with an 11-bit Barker code for a wake-up receiver. The hardware of the proposed n-bit correlator architecture scales with $O(n)$ compared to $O(n^2)$ for asynchronous switched-capacitor correlators. A -80.9dBm 40nW wake-up receiver with 9dB better sensitivity and 3dB improved selectivity to AM interference thanks to the correlator is demonstrated.**

I. INTRODUCTION AND MOTIVATION

IoT wake-up receivers operating with less than 1μW, demodulate the RF signal using a non-linear energy detector (ED) [1]. This imposes limits on their sensitivity and, their range for a given peak transmit power [2]. Improving range and the receiver sensitivity requires integrating the data pulse for a longer duration of time, and thus decreasing the data-rate and increasing the latency. ED wake-up receivers use OOK modulation and respond to a wake-up code ranging between 10 to 32bits. Standard designs [1], [2] use a *clocked, digital correlator* after the baseband comparator to detect the wake-up code, requiring synchronization with the incoming signal or $2x$ oversampling. However, in the presence of in-band, AM interference the frequency-domain selectivity is limited, and the receiver can get blocked as demonstrated in [2].

A *continuous-time (CT), clockless analog correlator* (Fig. 1) before the baseband comparator can perform matched filtering (MF); this eliminates the synchronization challenges and improves output SNR and thus sensitivity; it also provides code-domain filtering for enhanced selectivity and can suppress AM interference. The wake-up codes can be treated as direct sequence code-division multiple-access (DS-CDMA) signals, by integrating over the entire code-length using the CT-correlator, the sensitivity can be improved without lowering the data rate or increasing the latency.

Traditionally, *asynchronously sampled* DS-CDMA matched filters [3], [4] are implemented using a bank of matched-filters. In [3], a bank of 260 256bit-long recycling integrated correlators were implemented for receiving a 256-bit code. A sampled 63-bit switched-capacitor correlator was implemented in [5] for synchronization purpose using 2016 capacitors. An n-bit correlator implemented with these architectures requires hardware that scales with $O(n^2)$, which is large and power consuming.

For true *clockless* analog correlators, continuous-time analog delays are needed (Fig. 1); at very high frequencies

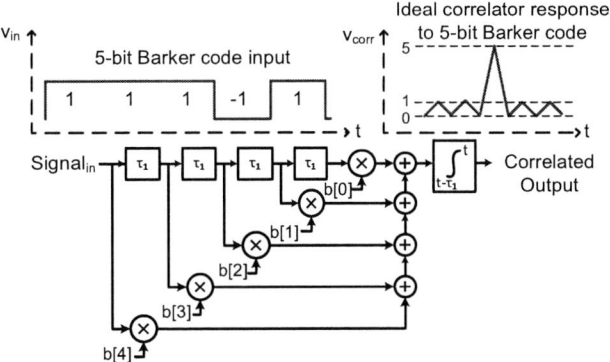

Fig. 1. Block diagram for a 5-bit CT analog correlator and its response to an incoming 5-bit Barker code.

such delays can be implemented with transmission lines, but these are infeasible at baseband. For an analog signal encoded in time domain (e.g. with PPM or PWM modulation), digital delays can be utilized to realize the CT analog delays. Clock-less CT signal processing using a CT ADC and DAC and digital delays has been used to implement analog FIR filters [6]. We utilize VCOs for the signal encoding and digital delays to realize a true clockless analog correlator. This correlator performs matched filtering for a wake-up receiver to improve its sensitivity and rejection of AM interference. For this correlator, the hardware scales with $O(n)$, since no synchronization to the incoming signal needs to be performed.

This demonstration of asynchronous matched filtering further lays the foundation for the use of clockless CDMA in wake-up systems.

II. WAKE-UP RECEIVER ARCHITECTURE

Fig. 2 shows the wake-up receiver using an analog correlator for code-domain matched filtering. The RF front end consists of a matching network followed by a gate-biased ED [2]. Its output is amplified using a one-stage current-reuse amplifier that drives the analog correlator. The analog correlator consists of VCOs that integrate and encode the signal into PPM, delay lines and DACs to implement the correlator and a 4-phase filter to suppress the VCO frequency and its harmonics. The output of the correlator is fed to a comparator that decides if the receiver should wake up.

The signal VCO and reference VCO need to be frequency locked; their outputs are compared with a PFD and fed back to the ED reference (i.e. drain node) with a charge-pump (CP) for DC feedback; with a second PFD-CP driven by delayed VCO outputs a zero is created to stabilize this second-order feedback loop.

2019 IEEE Radio Frequency Integrated Circuits Symposium

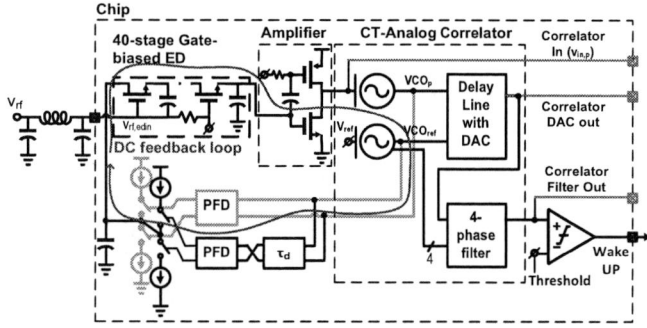

Fig. 2. Energy-detector wake-up receiver with a clockless analog correlator for matched filtering; the DC feedback loop through the gate-biased ED and part of the baseband provides automatic biasing.

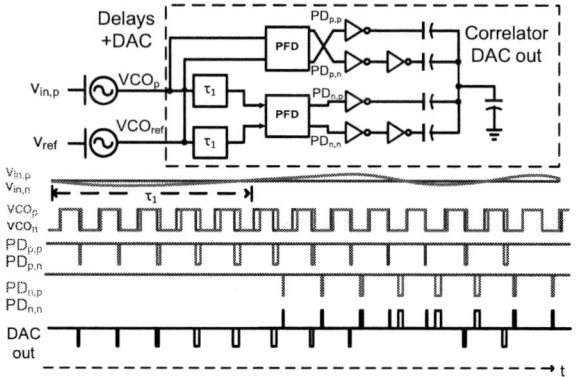

Fig. 3. Implementation and operation of a matched filter for 1-bit rectangular pulse.

III. N-BIT CT ANALOG CORRELATOR IMPLEMENTATION

The output, v_{corr}, of an ideal n-bit clockless correlator with rectangular bits with a period of τ_1 can be written as:

$$v_{corr}(t) = \int_{\tau=-n\tau_1}^{0} v_{in}(t-\tau) \, h[\tau] \, d\tau \qquad (1)$$

where $v_{in}(t)$ is the input signal, and $h[\tau]$ is a piecewise linear function representing the correlation coefficients. $h[\tau]$ is defined for n time periods corresponding to the correlation sequence, so we can write:

$$v_{corr}(t) = \sum_{k=1}^{n} \int_{\tau=-k\tau_1}^{(1-k)\tau_1} v_{in}(t-\tau)h[\tau]d\tau \qquad (2)$$

$$= \int_{\tau=-\tau_1}^{0} \left(\sum_{k=1}^{n} v_{in}(t-k\tau)h[k\tau] \right) d\tau \qquad (3)$$

Fig. 1 represents an ideal correlator for a 5-bit code as described in (3). Signal $v_{in}(t-k\tau)$ is evaluated by delaying the signal using k delay cells, bit settings $b[k]$ represent $h[k\tau]$, and the output is integrated for a time duration of τ_1.

A. Matched filter for a 1-bit rectangular shape

For a 1-bit matched filter, (3) can be rewritten as:

$$v_{corr,1}(t) = \int_{-\infty}^{0} v_{in}(t-\tau)h[\tau]d\tau - \int_{-\infty}^{-\tau_1} v_{in}(t-\tau)h[\tau]d\tau \qquad (4)$$

where $h[\tau]$ is 1 or -1 depending on the data-bit being received. Fig. 3 shows the implementation of a 1-bit matched filter.

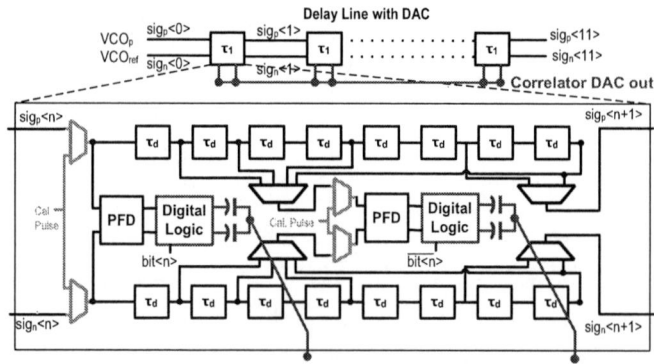

Fig. 4. Delay line and DAC implementation for an 11-bit analog correlator by cascading the delay blocks used for 1-bit matched filter in Fig. 3

The front end of the correlator consists of VCOs converting the input signal into pulse frequency modulated outputs. The pulse output position at VCO_p relative to the pulse positions at VCO_{ref} gives the integral output $\int_{-\infty}^{t} K_{vco}(v_{in,p}(\tau) - v_{ref})d\tau$. The relative pulse positions are compared with a PFD and fed to an adder implemented using capacitor-DAC to convert the signal back to voltage domain. This evaluates the first term in (4). The output pulses of the VCOs are also delayed using latch-based delay cells with delay τ_1 and the relative position is again evaluated using a PFD, thus evaluating the second term in (4). The output is subtracted using capacitor-DAC. This provides a CT windowed integrator response for a window of time τ_1 provided by the delay cell.

B. CT Analog correlator as a MF for an 11-bit code

N delay elements in Fig. 3 can be cascaded to keep track of v_{in} for the past $n\tau_1$ duration. Each delay element τ_1 implementation (Fig. 4) consists of 8 τ_d delay cells cascaded in series, PFDs to evaluate relative position of pulses and a capacitor-DAC to convert the signal back in voltage domain. The capacitor-DAC is weighted by the correlator coefficients $h[\tau]$ by swapping the input signals. 11 such delay elements are connected in series and the outputs of the DACs are connected together to evaluate the correlator output $v_{corr}(t)$.

The correlator DAC output is a weighted sum of PFD pulses and contains strong frequency components at f_{VCO} and the harmonics. A 4-phase filter uses the 4-phases of the reference VCO to sample the DAC output and averages the output over one VCO period to supress the outputs at f_{VCO} and it's harmonics.

C. Delay cell design

Each delay element τ_d consists of 3 granular τ_g delay cells as shown in Fig. 5. An input falling edge sets the SR latch to discharge C1, until it reaches threshold to turn T3 on. Once T3 is triggered, it delivers a falling edge pulse at the output and resets the latch. The delay τ_g is controlled by current mirror T1 and T2, MIM-cap C1 and the T3 threshold. Variations in transistors are controlled using current mirror trimming.

The minimum pulse width (τ_{pulse}) required for the input pulse is decided by the setup time of the latch. The maximum

978-1-7281-1702-7/19 $31.00 © 2019 IEEE   260

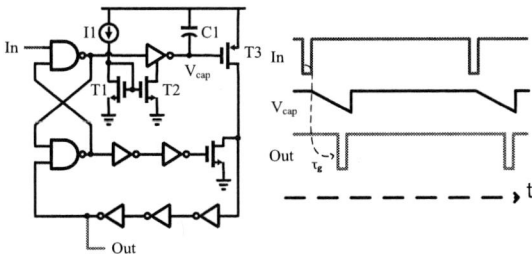

Fig. 5. Unit delay cell implementation providing a delay of τ_g

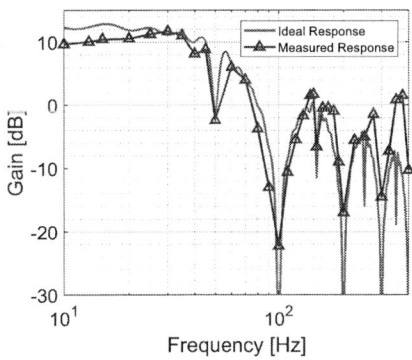

Fig. 6. Ideal vs measured frequency response of the implemented code-domain matched filter.

pulse width should be less than the delay of the unit cell. Therefore, the input-pulse instantaneous frequency must be less than $(1/(\tau_g + \tau_{pulse}))$. For $\tau_1 = 10msec$, $\tau_g = 416\mu sec$. This leads to a maximum input-pulse frequency of 2.3kHz. Due to the variations in the current mirrors controlling the delay cells and the added jitter, the operating frequency of the reference VCO is set to 1.1kHz.

IV. RECEIVER IMPLEMENTATION

A 40-stage gate-biased ED is used with an input resistance of 200kΩ. A passive, tuned L-C matching network matches the impedance of the ED to the antenna at 450.8MHz. The measured conversion gain from RF input to the ED output is 14.8mV/nW. The amplifier after the ED provides a voltage gain of 26dB. Its output is fed to the correlator with 4-phase ring oscillators operating at 1.1kHz with $K_{VCO} = 25kHz/V$. In the DC biasing loop, 2pA charge-pump currents are used with a 20pF load capacitor. A delay cell with a PFD/CP is used to introduce a zero by using $\tau_d = 1.5mSec$ and a CP current of 0.8 times the main CP current. The delay cells of the correlator are configured with $\tau_1 = 10mSec$ to receive a 100bps RZ-OOK '11100010010' Barker wake-up code. A clocked comparator operating at f_{VCO} using the reference oscillator is configured with a threshold of -20mV, corresponding to 5.8σ of the output noise for a false-alarm rate smaller than 1/Hr.

Wake-up code selection : Barker-codes are known for low auto-correlation with side-lobes no larger than $|1|$. For our measurements, we used an 11-bit Barker code representing $h[\tau]$. The spectrum of a Barker code is flat, as shown in Fig. 6, which provides ideal suppression to a single-tone interferer present at baseband. In the correlator, the signal bits for the desired code are added in voltage, and the noise is added in power, which provides a 7dB improvement in SNR for RZ-OOK symbols[1] of 0 and 1. When referred to the input of the ED receiver, the correlator can provide up to 3.5dB improvement in sensitivity compared to 1-bit threshold comparator implementation.However, since the delay cells don't provide any signal gain, the noise figure degrades with each added delay cell, which puts a limit on the sensitivity improvement.

[1]For OOK symbols of 1 and -1, the SNR improvement is 10dB. In our current prototype, the front-end has a high-pass cutoff of 30Hz, giving better performance with RZ-OOK symbols for 100bps, but the correlator architecture can also operate with OOK symbols.

V. MEASUREMENT RESULTS

The receiver die (Fig. 8d) uses an area of $0.15mm^2$ and consumes 40nW with the correlator using $0.11mm^2$ and consuming 93% of the power.

The S11 at the receiver input is measured as in Fig. 8a. The RX has a 3-dB bandwidth of 10MHz at 450.8MHz. Next, a 450.8MHz RF signal modulated with a sinusoidal AM signal with a frequency f_{bb} is fed to the input of the receiver. Frequency f_{bb} is swept from 10Hz to 400Hz; the demodulated baseband signal strength is measured at the input of the correlator and the output of the correlator DAC to evaluate the correlator frequency response. Fig. 6 shows the ideal and measured correlator frequency response, showing a good match. As expected, the measured response starts to deviate from its ideal behavior at higher frequencies when f_{bb} gets closer to f_{VCO}, but good suppression is maintained.

An RF signal modulated with the 50% RZ-OOK encoded 11-bit Barker code is applied to measure the sensitivity. Fig. 7a shows the response to a -79.3dBm input signal. The waveforms starting from the top are: baseband signal used to modulate the RF signal generator, baseband input to the correlator, correlator's filtered output and the output from the comparator. The corresponding missed detection ratio (MDR) is measured (Fig. 8b) to be < 10^{-3} at -80.9dBm RF input.

Next, signal-to-interference ratios (SIRs) are measured for a 100bps OOK modulated '1010' pattern AM interferer and a continuous-wave interferer by measuring the noise equivalent power at the output of the correlator filter (Fig. 8c). The corresponding SIRs are 1.1dB and -29.7dB at a 3MHz offset. Next, the rejection of unwanted 50% RZ-OOK encoded 11-bit codes is measured. Fig. 7b demonstrates that the correlator provides a 5dB rejection to the '10101010101' code and >10dB rejection to '11111111111' code.

VI. COMPARISON TO THE STATE OF THE ART

Compared to the state-of-the-art wake-up receivers, the receiver consumes 7dB higher power but has 9dB better latency-normalized sensitivity thanks to the gate-biasing technique in [2] and the additional conversion gain from the correlator. This leads to a 2dB better FoM compared to SoA

2019 IEEE Radio Frequency Integrated Circuits Symposium

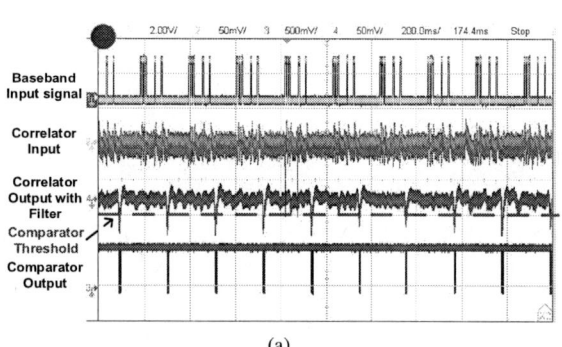
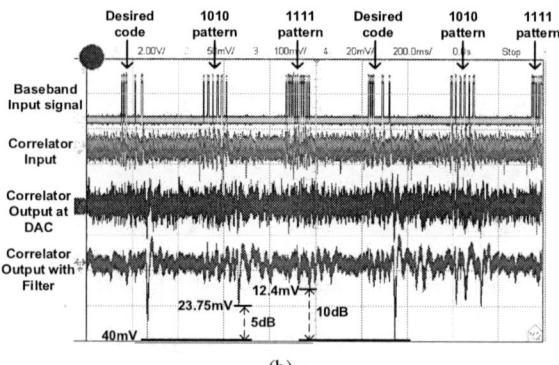

(a)   (b)

Fig. 7. (a) Receiver response to the wake-up Barker code '11100010010' sent every 220msec with the comparator threshold at -20mV; (b) receiver response to the desired wake-up Barker code, '10101010101' code and to a '11111111111' code at 100bps chip rate (comparator output not shown due to limited channels available on the oscilloscope). All codes use 50% RZ-OOK modulation.

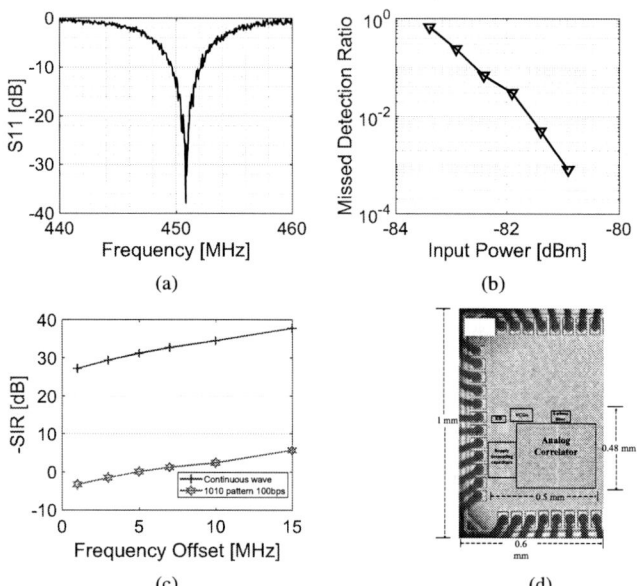

(a)   (b)

(c)   (d)

Fig. 8. Measured wake-up receiver performance (a) RF Input reflection; (b) Missed-Detection Ratio at 450.8MHz; (c) Signal to Interference Ratio for a continuous wave and a '1010' pattern AM interferer; (d) Chip micrograph.

while also providing 3dB improved rejection to in-band AM interference for an 11-bit code.

Besides Barker codes, Gold codes can also be used with the presented correlator architecture especially when going to longer length codes to further improve the performance in future work. Gold codes also have low cross-correlation and can thus be used to implement asynchronous CDMA for wake-up receivers with the presented correlator.

VII. CONCLUSIONS

A continuous-time analog correlator using pulse-position encoding has been presented that is used as a code-domain matched filter to improve the sensitivity and selectivity of a wake-up receiver. The 450MHz receiver prototype designed in 65nm CMOS-LP technology consumes 40nW from 0.54V, and demonstrates code-domain filtering for an 11-bit Barker

Table 1. Comparison with the state-of-the-art ED wake-up receivers operating at > 400MHz

Key Features		ISSCC'16	RFIC'17	CICC'17	ESSCIRC'17	ISSCC'18	This Work
		30-Stage Rectifier	Rectifier-Antenna Co-design	Self-Mixer	Active Rectifier	Passive Rectifier	CT-Analog Correlator
CMOS Tech. Node	(nm)	65	65	130	180	130	65nm
Correlator		Digital	Digital	Digital	Digital	Digital	Analog
Frequency	(MHz)	2400	2400	550	405	433	450.8
Chip Rate	(kbps)	8.2	2.5	400	0.3	0.2	0.1
Code-length	(bit)	31	32	11	16	16.5	11
Power	(nW)	236	365	222	4.5	7.4	40
Passive Gain	(dB)	NR	NR	19	18.5	NR	21
Sensitivity	(dBm)	-56.5	-61.5	-56.4	-63.8	-71	-80.9
Latency	(msec)	3.8	12.8	0.0275	53.3	82.5	110
Circuit Area	(mm²)	0.25	1.1	0.2	6	1.95	0.15
Modulation		OOK	OOK	OOK	OOK	OOK	RZ-OOK
In-band SIR(3MHz away)							
Square wave AM interferer		NR	NR	NR~	4*	NR	1.1
C-Wave Interferer		NR	-19.1	-14	-28	NR	-29.67
Normalized Sens.#	(dBm)	-68.6	-71	-79.2	-70.2	-76.4	-85.7
FoM^	(dB)	-134.9	-135.4	-145.7	-153.7	-157.7	-159.7

~Not reported for a square-wave AM jammer   #Normalized Sens.=Sensitivity + 5log(Latency/1s)
*For a PRBS AM interferer   ^FoM=Normalized Sens. + 10log(Power/1W)

code. The proposed n-bit correlator hardware scales with $O(n)$ compared to $O(n^2)$ for asynchronous switched-capacitor correlators. This demonstration of code-domain selectivity opens up the possibility of asynchronous CDMA in future wake-up receivers.

ACKNOWLEDGMENT

The authors thank partial support from Analog Devices and NSF EECS 1309721.

REFERENCES

[1] J. Moody et al., "A -76dBm 7.4nW wakeup radio with automatic offset compensation," in ISSCC, Feb. 2018, pp. 452–454.

[2] V. Mangal and P. R. Kinget, "A Wake-Up Receiver With a Multi-Stage Self-Mixer and With Enhanced Sensitivity When Using an Interferer as Local Oscillator," Journal of Solid-State Circuits, pp. 1–13, 2019.

[3] D. Senderowicz et al., "A 23mW 256-tap 8 MSample/s QPSK matched filter for DS-CDMA cellular telephony using recycling integrator correlators," in ISSCC, Feb. 2000, pp. 354–355.

[4] T. Shibano et al., "Matched filter for DS-CDMA of up to 50 MChip/s based on sampled analog signal processing," in ISSCC, Feb. 1997.

[5] E. Gantsog et al., "A 12.46μW Baseband Timing Circuitry for Synchronization and Duty-Cycling of Scalable Wireless Mesh Networks in IoT," in RFIC, June 2018, pp. 328–331.

[6] B. Schell and Y. Tsividis, "A Continuous-Time ADC/DSP/DAC System With No Clock and With Activity-Dependent Power Dissipation," Journal of Solid-State Circuits, vol. 43, pp. 2472–2481, Nov. 2008.

978-1-7281-1702-7/19 $31.00 © 2019 IEEE

A 4×4×4-mm³ Fully Integrated Sensor-to-Sensor Radio using Carrier Frequency Interlocking IF Receiver with -94 dBm Sensitivity

Li-Xuan Chuo[#1], Yejoong Kim[#], Nikolaos Chiotellis[#], Makoto Yasuda[$], Satoru Miyoshi[*],
Masaru Kawaminami[$*], Anthony Grbic[#], David Wentzloff[#], Hun-Seok Kim[#], David Blaauw[#]

[#]University of Michigan, Ann Arbor, USA

[$]Mie Fujitsu Semiconductor Limited, Yokohama, Japan

[*]Fujitsu Electronics America, Inc., Sunnyvale, USA

[1]lxchuo@umich.edu

Abstract— **Ultra-low power mm-scale IoT platforms enable newly emerging applications such as pervasive agricultural monitoring and bio-sensing. Although there is an increasing interest in sensor-to-sensor communication, as defined in Bluetooth v5.0, prior research in mm-scale wireless systems is mostly limited to asymmetric sensor-to-gateway communication. This paper introduces a 4×4×4 mm³ fully integrated radio system that integrates a transceiver chip, antenna, power management unit and baseband processor for sensor-to-sensor mesh networks. The proposed system uses a low power 32 kHz reference frequency and a carrier frequency interlocking IF receiver. It achieves -94 dBm sensitivity with 97 μW power consumption and -12.6 dBm EIRP for sensor-to-sensor mesh network communication.**

Keywords— **radio transceivers, low-power electronics, dipole antennas, wireless sensor networks, wireless mesh networks.**

I. INTRODUCTION

Newly emerging Internet of Things (IoT) applications, such as agricultural monitoring, smart cities and ubiquitous beacon services, require ultra-low power (ULP) platforms of cm- and even mm-scales. Although there is increasing interest in sensor-to-sensor mesh networks, such as the new Bluetooth v5.0 standard, prior work in mm-scale ULP wireless communication systems mostly utilize a sensor-to-gateway topology (star-network) [1]. Millimeter-scale dimensioned systems exacerbate the challenges in sensor-to-sensor transceiver design, including: 1) accurate timing and carrier frequency synchronization among ULP transceiver nodes; 2) extremely stringent link budget stemming from the poor, electrically-small, antenna efficiency; and 3) inadequate battery capacity (~10 uAh) and instantaneous battery current capacity (~100 uA) due to mm-scale battery size and constrained by VDD fluctuation.

This work proposes a mm-scale fully self-contained radio system that supports sensor-to-sensor communication, enabling mesh network formation for mm-scale devices. Conventional transceivers use a tens of MHz crystal oscillator (XO) as the frequency reference. However, their long start-up time (~1 ms) and the high power consumption of their crystal oscillator (~100 μW) and PLL (~1 mW) [2] make it incompatible with mm-size batteries. Instead, our proposed design uses only a 32.768-kHz real-time clock (RTC) with <10nW power consumption to frequency lock a power oscillator merged with a high-Q 3D loop antenna, which doubles as a DCO and operates in open loop after locking. Hence, the proposed design combines the conventional PA and PLL and eliminates the

power-hungry high frequency crystal and DCO, enabling a highly energy efficient TX design and reducing off-chip components. The locked power oscillator is then reused at the start of each RX operation to generate a calibration tone. To this tone a free-running ring oscillator (RO) is tuned such that the center of the down-converted RF signal aligns to the passband of the IF amplifier. This achieves TX-RX carrier frequency synchronization, which is critical in narrowband communication. It also eliminates the need for explicit estimation of the bandpass IF amplifier center frequency, which significantly lowers the RX power consumption and simplifies the design. Combined with the reuse of TX power oscillator as a Q-enhanced amplifier (QEA) [3, 4], we achieve good sensitivity of -94 dBm at the input of the -7.8 dBi (estimated from simulation) mm-scale antenna with 97 μW power consumption. TX consumes 3.5 mW peak power for 125μs pulse duration drawn from decaps and achieves -12.6 dBm EIRP. The complete system, including the proposed transceiver chip, high-Q antenna, 32 kHz crystal, Cortex-M0 processor and power management unit, is integrated within a 4×4×4 mm³ sensor platform and communicates at a data rate of 4 kbps.

II. CARRIER FREQUENCY INTERLOCKING IF RECEIVER

Fig. 1 shows our proposed radio architecture. Unlike conventional low-power receiver approaches such as direct conversion, low-IF, and uncertain-IF [5], the proposed architecture takes advantage of a high-Q antenna and TRX front-end sharing. A direct-conversion scheme is popular for its simplicity but suffers from flicker noise, DC offset, and LO leakage, making it hard to achieve good sensitivity. A low-IF architecture has been widely used to address these issues, but it requires a high quality RF local oscillator (LO), which is very power hungry. Recently, uncertain-IF architectures using a low-power unlocked LO were proposed where uncertain-IF is accommodated by increasing the IF amplifier bandwidth. However, widening the IF amplifier bandwidth to address IF uncertainty increases power and reduces SNR (by incorporating more noise), nullifying the benefit of narrowband communications to achieve superior sensitivity.

The proposed architecture instead uses a narrow band RX chain that combines the band-selective (Q~1000, bandwidth ~2 MHz) QEA and a bandpass IF amplifier to obtain superior SNR than typical uncertain IF architectures while maintaining low power consumption. First the digitally-controlled power

2019 IEEE Radio Frequency Integrated Circuits Symposium

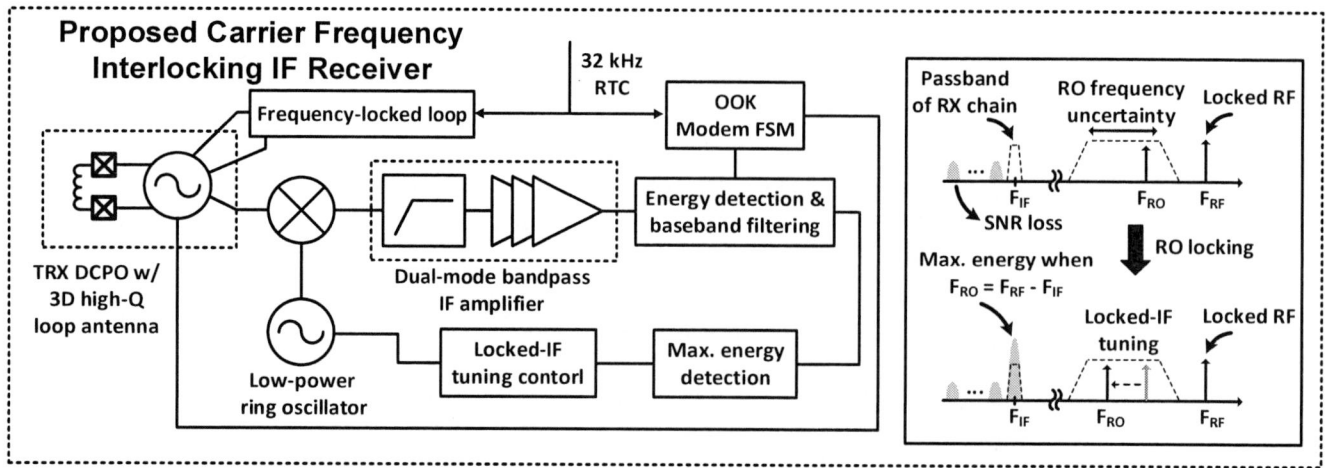

Fig. 1. The proposed carrier frequency interlocking IF radio architecture and frequency domain explanation

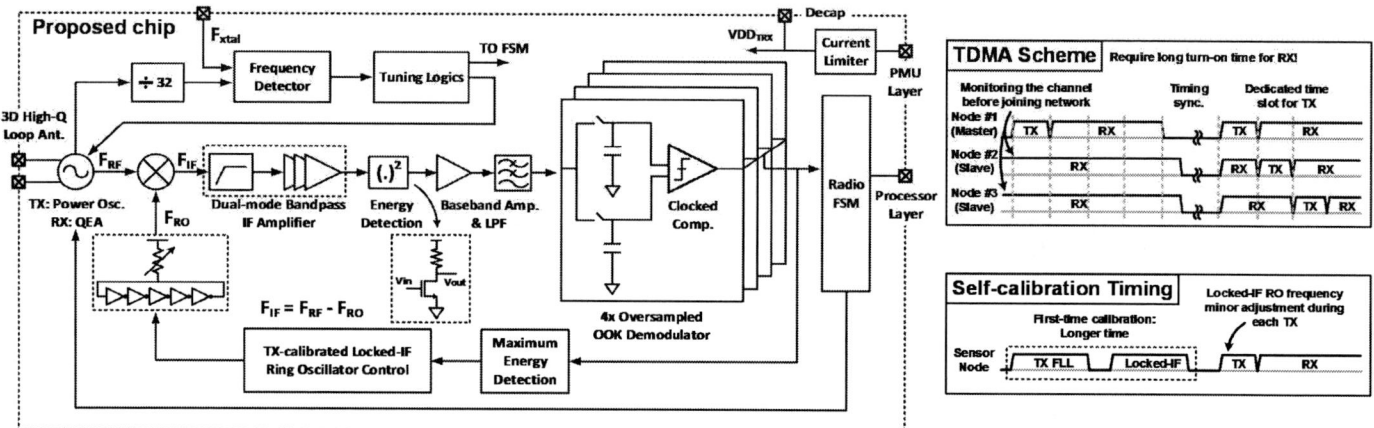

Fig. 2. Details of the transceiver circuits with calibration loop timing and the radio sensor node multiple access scheme

oscillator (DCPO) is frequency-locked using the 32 kHz crystal for TX. The DCPO also serves as the RX front-end by implementing the QEA and provides the RF reference tone for calibration to align the down-converted IF signal to the passband of the IF amplifier in the receive chain. During this calibration, the frequency-locked DCPO signal (F_{RF}) is down-converted to IF (F_{IF}) by the RO frequency (F_{RO}) and passed through dual-mode bandpass IF amplifiers. A maximum energy detection loop then sets F_{RO} such that the bandpass IF amplifier output is maximized which ensures that the RO is tuned and F_{IF} is centered in the IF-amplifier's passband for maximum sensitivity. As a result, the IF amplifier bandwidth can be reduced. The proposed architecture preserves the high sensitivity while using a low-power RO as LO (Fig. 1).

III. TRANSCEIVER CIRCUITS IMPLEMENTATION

A. Transceiver Overview and System Timing

The DCPO uses a co-designed high-Q 3D loop antenna as the resonant component, which lowers the power consumption in TX mode and also serves as a highly band-selective filter in RX mode (Fig. 2). The TX and RX front-ends share the same circuits with different bias conditions, resulting in naturally

similar carrier frequency. Though there is still frequency difference results from non-linearity of capacitors, it was found to be small compared to the RX QEA bandwidth (2 MHz). The TX DCPO frequency calibration is performed first, followed by the RX interlocking IF tuning. Depending on the initial control setting, the TX DCPO frequency-locked loop (FLL) may require a long settling time at first due to the use of 32 kHz reference. However, once the initial calibrated frequency control word is stored in the processor, the radio will only need to fine-tune and update the setting each time before communicating. The self-calibration timing is shown in Fig.2.

B. DCPO Front-End Design and Operation

During TX DCPO frequency locking, the DCPO cross-coupled transistors are biased with high current (~850 μA) to start the oscillation. The FLL is composed of a 32x divider, asynchronous counter, and frequency tuning logic (Fig.3). The counter is clocked by a divided RTC (32.768/4 = 8.192 kHz) supplied by the processor for high frequency resolution. It counts the number of divided RF signal transitions (~75 MHz) and locks the frequency accordingly. Since the carrier frequency is set by the 3D loop antenna and the (mostly) off-chip capacitance, it does not have significant VDD dependency.

978-1-7281-1702-7/19 $31.00 © 2019 IEEE 264

Thus, the FLL is mainly for compensating the antenna and off-chip capacitor variations. In RX mode, the same cross-coupled pair is biased in the Q-enhanced region for boosting the Q value of the loop antenna to have a band-selective filter response and in-band voltage gain (62.7 dB, simulation). The targeted RF front-end filter bandwidth of the receiver chain is 2 MHz (Q ~ 1000), meaning that the TX frequency synchronization should also be within 0.1% (Q ~ 1000). The measured results shown in Fig.3 indicates that the FLL calibration can synchronize the carrier frequency to within 0.1% across 5 tested chips.

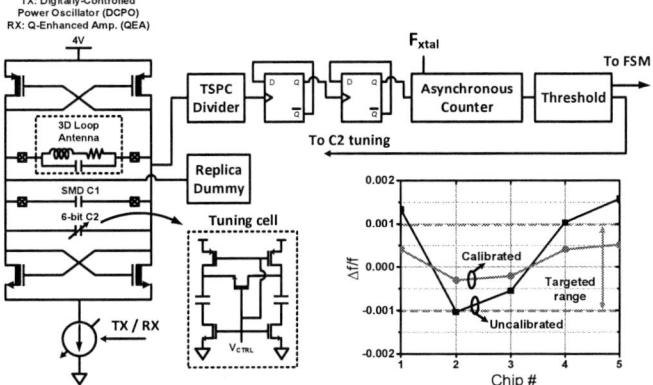

Fig. 3. Transmitter frequency-locked loop using 32-kHz RTC and the measured results, showing it can lock the frequency within 0.1% error from chip-to-chip

C. Carrier Frequency Interlocking IF Receiver Designs

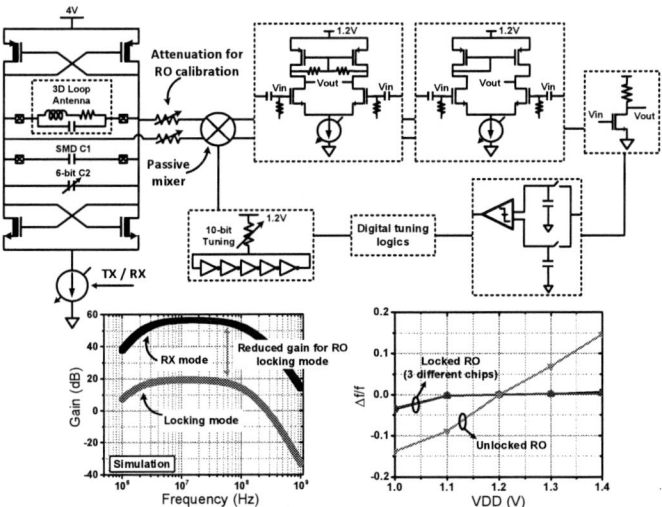

Fig. 4. Reuse the RX chain with lower gain setting for RO tuning loop and the measured RO frequency with different supply voltage

After the carrier frequency is locked, the radio starts IF calibration. The RX chain is turned on during the TX operation with the self-injected attenuated TX signal as shown in Fig.4. Because the input RF signal is strong (directly from the DCPO), the IF amplifiers were designed to have a calibration mode with 40dB less gain (Fig. 4, left, bot). This lowers power consumption during calibration and avoids saturation of the amplifier. The simulated results in Fig. 4 show that both high and low gain modes have identical bandpass frequency response but different gains. We reuse the remainder of the RX

chain (energy detector and comparator) to detect the output signal from the IF amplifiers and tune the RO frequency. The energy detector output is sampled with different RO frequency settings and compared to the prior output by reusing the data-demodulation comparator. This search continues until it finds the maximum output level at which point the amplifier passband is aligned with the RX IF signal down-converted from the DCPO TX signal. In the mm-scale sensor node design, the VDD will fluctuate due to the limited battery and decap current and capacity. As a result, it is difficult to achieve a stable low-power RO frequency with a conventional design. Fig.4 shows the measured RO across VDD variation with and without the proposed TX-RX interlocking technique. The free-running RO frequency will deviate by more than 20% (100's MHz in GHz) when VDD varies from 1.4 V to 1.0 V. With the proposed locking calibration, the RO frequency is stable within 1% across VDD, meeting the bandwidth of the IF amplifier.

When both DCPO and RO are tuned, data communication is performed using binary pulse position modulation (PPM) that compares symbol power levels at two different position, eliminating the need of an accurate reference voltage. The proposed chip supports time-division multiple access (TDMA) scheme as shown in Fig.2. Each node is given a programmable time slot (ID) for collision-free transmission.

IV. CHIP MEASUREMENT AND RADIO SYSTEM EVALUATION

The transceiver chip was fabricated in MIFS 55-nm DDC CMOS and integrated with a custom designed antenna in $4 \times 4 \times 4$ mm^3 form factor. Since this transceiver was co-designed with the antenna and was not matched to 50 ohm, its performance evaluation includes the antenna gain (-7.8 dBi). The measurement results are shown in Fig. 5.

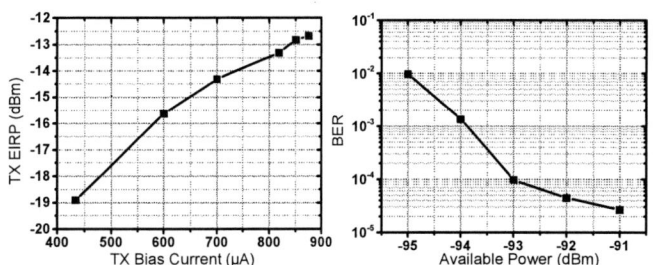

Fig. 5. Measured transceiver chip (w/ antenna) EIRP and sensitivity

In TX mode, the EIRP of the sensor node transmitting a continuous tone was measured using a horn antenna (LB-530-NF) in an anechoic chamber which ranged from -18.9 to -12.6 dBm with 1.7 to 3.5 mW TX power consumption. Due to the co-design of antenna, the radio will have emission during the calibration. However, the output power is low since the antenna has low gain and the signal does not contain the correct header for communication. The receiver has a sensitivity of -94 dBm measured at the input of a -7.8 dBi (simulation) antenna for 10^{-3} BER at 4 kbps. A measured time-domain waveform of wireless communication between two integrated nodes in Fig.6 validates TDMA based sensor-to-sensor communication. For measurement in Fig. 6, the number of TDMA slots are pre-

Table 1 Performance summary and comparison to other works

		This work	ISSCC 2017 [3]	ISSCC 2015 [5]	ISSCC 2015 [6]
Technology		55 nm	180 nm	65 nm	65nm
System Form Factor		4x4x4 mm³	3x3x3 mm³	4.6x4.6 mm² Receiver chip only	1.3x0.9 mm² Receiver chip only
# of off-chip components		5 (Antenna, xtal, decaps)	4 (Antenna, decaps)	11 (Inductors, decaps)	3 (Inductors)
Frequency		2.4 GHz	900 MHz	2.4 GHz	2.4 GHz
Sensor-to-Sensor Communication?		Yes	No	No	No
RX	Technique	Carrier Frequency Interlocking IF	Tuned-RF	Dual Uncertain-IF	RX-based FLL
	Modulation	Binary PPM	Binary PPM	OOK	OOK
	Power Consumption	97 µW	1850 µW	99 µW	227 µW
	Data Rate	4 kbps	7.8 kbps – 62.5 kbps	10 kbps	1 Mbps
	Sensitivity*	-94 dBm (@ input of -7.8 dBi antenna)	-93 dBm (@ input of -23.4 dBi antenna)	-97 dBm (@ input of AFE)	-83 dBm (@ input of AFE)
TX	Technique	RTC-FLL	Free-running	N/A (RX only)	N/A (RX only)
	Max. EIRP	-12.6 dBm	-26.9 dBm		
	Power Consumption	3.5 mW @ 4V	2 mW @ 4V		
	Data Rate	4 kbps	30 bps – 30.3 kbps		

programmed to six. Node #1 (ID=3) transmits at the third time-slot and listens at all other time-slots. The same rule applies to node #2 (ID=6) that uses the sixth time-slot for TX. The zoomed-in plot in Fig. 6 shows that the two mm-scale nodes synchronously transmit and receive using proper time slot, thus establishing successful communication links. Fig. 7 shows the fully integrated system that integrates a transceiver chip, antenna, power management unit, and baseband processor for wireless node-to-node communication measurements. Table 1 summarizes the overall performance and compares it to other low-power radio systems.

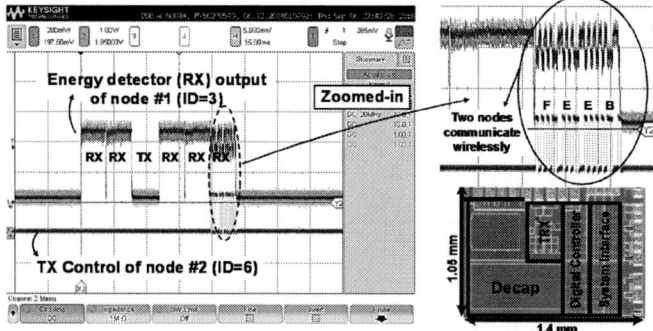

Fig. 6. Measured time domain sensor-to-sensor communication waveform using the fully integrated radio system and the transceiver chip micrograph

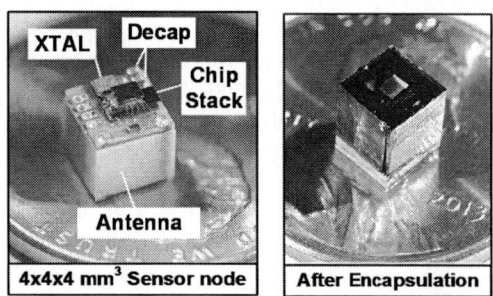

Fig. 7. The fully integrated radio system including transceiver, processor, PMU, antenna, XTAL, and decaps

V. CONCLUSION

Challenges in mm-scale node-to-node communication includes electrically small antenna (low radiation efficiency, difficult to impedance match), frequency / timing synchronization and stringent power / energy budget (limited peak current and battery capacity). In this paper, we present a 4x4x4 mm³ fully stand-alone radio system that integrates an RF transceiver chip, antenna, power management unit and baseband processor with a new carrier frequency interlocking IF receiver architecture for low-power and high sensitivity node-to-node mesh networks. The proposed mm-scale system demonstrates sensor-to-sensor communication with -12.6 dBm EIRP for TX and -94 dBm sensitivity (at the input of a -7.8 dBi antenna) with 97 µW power consumption for RX.

REFERENCES

[1] T.-C. Chang, et al., "A 30.5 mm³ fully packaged implantable device with duplex ultrasonic data and power links achieving 95 kb/s with <10⁻⁴ BER at 8.5 cm depth," in *IEEE ISSCC Tech. Dig.*, pp. 460–461, Feb. 2017.

[2] M.-S. Yuan, et al., "A 0.45V sub-mW all-digital PLL in 16nm FinFET for bluetooth low-energy (BLE) modulation and instantaneous channel hopping using 32.768kHz reference," in *IEEE ISSCC Tech. Dig.*, pp. 448–450, Feb. 2018.

[3] L.-X. Chuo, et al., "A 915MHz asymmetric radio using Q-enhanced amplifier for a fully integrated 3×3×3mm³ wireless sensor node with 20m non-line-of-sight communication," in *IEEE ISSCC Tech. Dig.*, pp. 132–133, Feb. 2017.

[4] J. Kang, et al., "A 1.2cm² 2.4GHz Self-Oscillating Rectifier-Antenna Achieving -34.5dBm Sensitivity for Wirelessly Powered Sensors," in *IEEE ISSCC Tech. Dig.*, pp. 374–375, Feb. 2016.

[5] C. Salazar, et al., "A -97dBm-Sensitivity Interferer-Resilient 2.4GHz Wake-Up Receiver Using Dual-IF Multi-N-Path Architecture in 65nm CMOS," in *IEEE ISSCC Tech. Dig.*, pp. 242–243, Feb. 2015.

[6] J.-S. Lee, et al., "A 227 pJ/b −83 dBm 2.4 GHz multi-channel OOK receiver adopting receiver-based FLL," in *IEEE ISSCC Tech. Dig.*, pp. 1-3, Feb. 2015.

2019 IEEE Radio Frequency Integrated Circuits Symposium

A 55nm SAW-Less NB-IoT CMOS Transceiver in an RF-SoC with Phase Coherent RX and Polar Modulation TX

PS. Tseng[1], W. Yang[2], MJ. Wu[2], LM. Jin[2], DP. Li[2], EC. Low[2], CH. Hsiao[1], HT. Lin[1], KH. Yang[1],
SC. Shen[1], CM. Kuo[1], CL. Heng[2], GK. Dehng[1]

[1]MediaTek Inc., Hsinchu, Taiwan
[2]MediaTek Inc., Singapore
bosen.tzeng@mediatek.com, wei.yang@mediatek.com

Abstract— A SAW-Less NB-IoT transceiver for IoT devices with >10 years battery life and >164dB MCL is presented. Tunable RX Front-end supports 26 NB-IoT bands, achieves <-140dBm Narrowband Reference Signal Received Power (NRSRP) and tolerates up to -15dBm out-of-band blocking. RX phase continuity across sub-frames reduces RX power consumption by 30%. The Polar TX achieves >+4dBm output power, <5% EVM, and <-57dB 300kHz SEM. The 55nm transceiver consumes 11.8mW/25.8mW for receiving / transmitting, and occupies 2.23mm² die area in a 5.6x5.6mm² packaged chip.

Keywords— NB-IoT, SAW-Less transceiver, polar transmitter, ADPLL.

I. INTRODUCTION

Based on LTE cellular technology, 3GPP introduces a new NB-IoT radio standard intended for long range communication with a massive number of battery operated wireless IoT devices. In contrast to LTE Cat. 4 and higher, the objectives of the NB-IoT standard are: lower cost structure than enhanced Machine-Type Communication (eMTC), more than 10 years battery life, extended coverage, and support for 50,000 user equipment per cell. The new objectives give rise to RF characteristics like Single-Input Single-Output (SISO) and half duplex system to eliminate the costly diversity antennas and duplexers, repetition and encoding for up to 16dB enhancement to SNR, and 200kHz channel bandwidth to increase capacity [1]. These introduce new challenges to NB-IoT RF design, including stringent out of band RX blocking requirement without SAW filter, increased TX-to-VCO pulling due to narrower channel bandwidth, significantly lower power consumption for >10 years battery life as well as RF SoC for low cost solution. This paper presents an RF transceiver integrated in 55nm CMOS RF-SoC to meet these new NB-IoT requirements.

The RF transceiver block diagram is depicted in Fig. 1, illustrating a self-tuned SAW-Less RX and an ADPLL-based polar modulator with linear PGA. The choice of ADPLL allows the implementation of self-injection locked (SIL) pulling mitigation to address increased TX pulling concern and helps to realize the reset LO (RSTLO) scheme necessary to ensure phase continuity when RX LO is turned off intermittently.

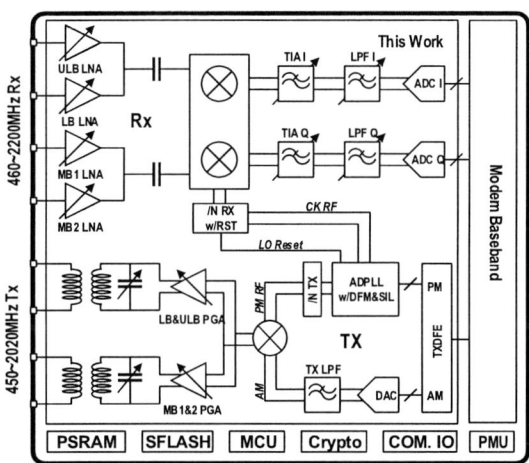

Fig. 1. RF transceiver Block Diagram. NB-IoT single chip consists of RF, Baseband, PMU, MCU, Crypto engine, and Communication IO on single die packaged with PSRAM and SFLASH.

II. SAW-LESS RX AND TX

A. Self-tuned RX Circuitry

3GPP Rel. 15 introduces 26 NB1/NB2 frequency bands, spanning from 450MHz to 2200MHz. To cover all frequency bands using only 4 RX ports, a low noise CMOS LNA (M1~M4) with self-tuned matching network is implemented. Fig. 2 illustrates the matching topology formed by an external inductor (Lext), and an integrated self-tuned array of capacitors (Ctune) and resistors (Rtune). The addition of the resistor array introduces another degree of freedom to the tuning of the matching network across all frequency bands. For each frequency band, in order to minimize process variations, Ctune and Rtune values are automatically mapped to a combination of results from on-chip filter and bias calibrations. Hence, this tunable network can simultaneously provide good input return loss and matching gain for low noise RX, and attenuates n*LO out-of-band blockers (up to -15dBm) for SAW-less operation. In Fig. 2, LNA with current mode output is connected to a single balanced passive mixer, then followed by a trans-impedance amplifier and a 2nd order bi-quad LPF for channel selectivity. And a 10b noise shaping SAR ADC is used for digitizing the received signal.

978-1-7281-1702-7/19 $31.00 © 2019 IEEE   267

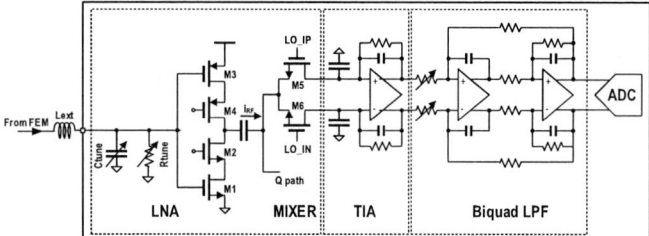

Fig. 2. Self-tuned SAW-Less RX circuitry.

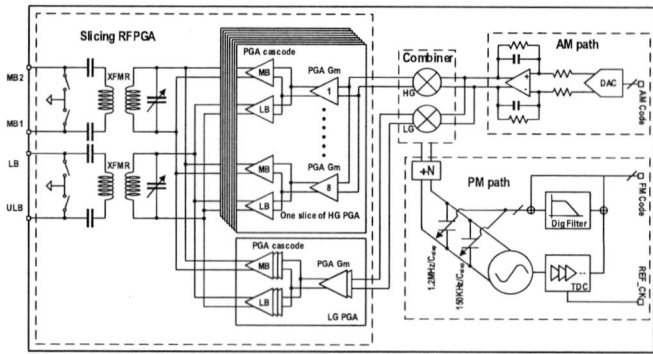

Fig. 3. Tx Polar modulator and slicing RFPGA.

B. Polar TX Circuitry

A polar modulator with linear PGA (Fig. 3) has lower TX current consumption due to the use of nonlinear modulator. It is also chosen to mitigate the increased DCO pulling risk due to the relatively narrow 200kHz channel bandwidth in NB-IoT. However, Peak-to-Average Power Ratio (PAPR) for NB-IoT is 3dB higher than EDGE system, which has the same channel bandwidth. This imposes a new challenge because the modulation bandwidth of PM and AM paths can exceed 12MHz and 2MHz respectively for 12 tones SC-FDMA QPSK signal with 15KHz Sub-Carrier Space (SCS). To address the differing wide modulation bandwidth and fine frequency step requirements, a DCO with two digitally controlled capacitor banks is used (Fig. 3). The DCO operating frequency ranges from 2792MHz to 4400MHz. The coarse and fine capacitor banks have a frequency resolution of 1.2MHz and 150kHz respectively. The PM digital codes are directly modulated to RF through these capacitor banks. Together with two point modulation for large modulation bandwidth, dynamic element matching, and sigma delta modulation of the capacitor banks, the integrated phase error of the PLL is <1° and its noise is <-60dB/30kHz RBW at 400kHz offset. The AM path of the modulator (Fig. 3) consists of a 9b current DAC and a 4th order LPF to minimize in-band group delay variation to <15ns, and out-of-band noise to <–146dBc/Hz at 10MHz offset. Both the AM and PM paths are combined together through a voltage mode passive mixer with 50% duty cycle, and then amplified by RF PGA.

In order to extend TX output power dynamic range to 63dB, the passive mixer and RF PGA are partitioned into low gain (LG) and high gain (HG) paths. Each of these paths is further divided into low band (LB) and middle band (MB) circuits to cover all 26 frequency bands using only 4 RF ports (Fig 3). The transformer of each RF PGA is connected to a capacitor array, which is calibrated for each frequency band. In this manner, the RF PGA is able to meet the linearity requirement even under 2:1 VSWR condition. Furthermore, each RF PGA can be sliced to achieve <0.5dB gain step error. This removes expensive on-chip loopback power calibration, or factory calibration for all TX power levels.

III. ADPLL SX

In NB-IoT, downlink (DL) physical channels are time-multiplexed [2], thus allowing invalid sub-frames to be declared for not receiving or expecting narrowband reference signal. It will be more power efficient to turn off RF RX and RX LO during this period. To achieve this power reduction, RX phase continuity across the invalid sub-frames must be guaranteed for cross sub-frame coherent combination to work. This poses a significant challenge for RX LO because LO divider has multiple operating states upon start-up, each out-of-phase from each other. In other words, simply turning off the RX LO intermittently causes RX phase discontinuity across invalid sub-frames. In this work, RX phase coherency is achieved in 2 steps. To resolve the phase ambiguity of RX LO, a RSTLO signal resets the divider to a known state upon exiting the invalid sub-frames (Fig. 4). The generation of RSTLO signal is illustrated in the Fig. 4 timing diagram. Except the phase accumulator of frequency control word (FCW), the entire RF RX and ADPLL are turned off during invalid sub-frames to minimize power consumption. The RSTLO signal is generated from a programmable delay that comes from the phase difference between ADPLL reference clock and FCW accumulator. In this way, at the beginning of the valid sub-frame, the RX LO phase is made coherent to the last valid sub-frame. For the case of 6 invalid sub-frames within a 20ms radio frame, RF RX power consumption is reduced by 30% through this method.

Fig. 4 also illustrates a SIL loop that automatically detects and corrects for DCO phase error caused by the pulling from the RF PGA output signal. As SIL is a blind anti-interference technique [3], there is no need for calibration to characterize the aggressor.

2019 IEEE Radio Frequency Integrated Circuits Symposium

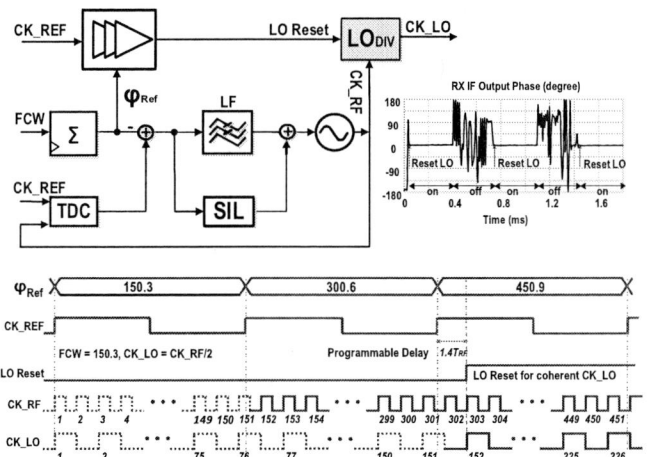

Fig. 4. ADPLL block diagram, reset LO timing diagram, and measured RX IF phase.

IV. MEASUREMENT

This NB-IoT RF transceiver with digital baseband and PMU is fabricated in 55nm CMOS process. A summary of the measured performance is tabulated in Fig. 5. The RF-SoC is able to receive a NRSRP of -140dBm, and reach 164dB maximum coupling loss for coverage enhancement. The receiver can tolerate -15dBm out-of-band blocker at 85M away from band edge. The maximum TX output power is +4dBm with low TX EVM of < 0.7% for 3.75kHz single tone UL signal, and < 3.5% for 15kHz 12-tones SC-FDMA QPSK signal. Measured out-of-band noise at 10MHz and 20MHz offsets are -139dBc/Hz and -144dBc/Hz at Tx output, that can meet <-50dBm/1MHz for NB-IoT co-existence requirement. Together with an external PA, the maximum and minimum output power at the antenna port is +23dBm and -54dBm respectively for 3.75kHz single tone UL signal. TX EVM is < 1% for all UL power levels. The RF circuitry consumes only 11.8mW for receiving and 25.8mW for transmitting. The RF-SoC consumes 20mW average power while DL signal is at -124dBm Energy per Resource Element.

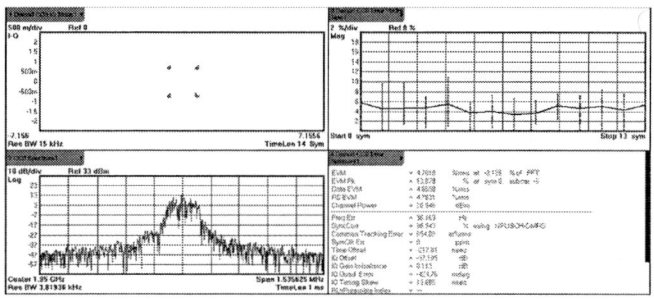

Fig. 6 Tx EVM and spectrum measured at antenna port. (20.9dBm, 15kHz 12 Tones SC-FDMA QPSK, 1950MHz carrier frequency)

The effectiveness of the SIL is demonstrated by examining the TX spectrum emission mask (SEM). At close-in out-of-band frequencies (f_{OOB}) of 100kHz and 150kHz, TX SEM improves by at least 5dB with SIL (Fig. 7)

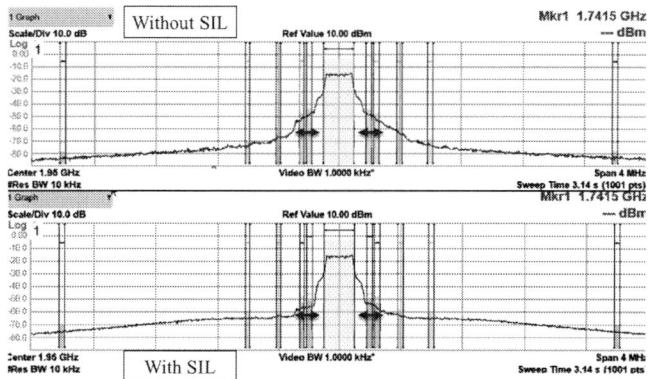

Fig. 7 TX SEM (f_{OOB} = 100/150kHz) measured at SoC output improved with SIL pulling mitigation

The die size of the transceiver is 2.23mm2 die area, and the RF-SoC is packaged in a 5.6x5.6mm^2 BGA package. The die micrograph of the RF transceiver is shown in Fig. 8.

Feature	This work	[4]	[5]	[6]	Unit
	Multi-tone NB-IoT	Single tone NB-IoT	EC-GSM-IoT	2G/3G/LTE-A	
Frequency	450~2200	750~960	-	-	MHz
VDD	0.9/1.1	1.7	-	1.2/1.8	V
Technology/Area	55n/2.23	180n/-	130n/-	40n/13.9	m/mm^2
RF SAW	No	-	Yes	Yes	
Power-RF RX	11.8	31.2~38.6	-	50	mW
Power-RF SoC	20.2[1]	-	156.6[2]	-	mW
NF	3.5~4.5	4	-	2.5 (B7)	dB
Sensitivity w/ repetition	-140	-	-121.7	-	dBm
	w/o PA	w/ PA	w/o PA	w/ PA	
Power-TX	25.8	-	59~67	-	162	mW
max TX Pout	4	-	-	24	>5 (B7)	dBm
TX EVM (ST3.75KHz)	0.73	<1	-	3.8		%
					<2	
TX EVM (MT12C)	3.5	4.7	-	-		%
Low power EVM		0.8	-	12	-	%
TX ACLR1/2	-50/-59	-	-	-	-50/-55	dBc
TX SEM@300k	-57	-	-	-	-	dBc

1. NRS EPRE: -124dBm

2. Complete RF-SoC averaged over one receive EC-MCS1 CC4 TTI

Fig. 5 Performance comparison table.

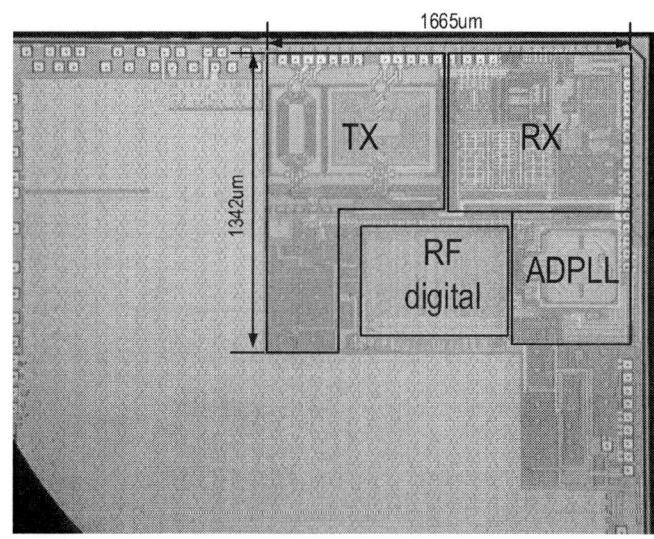

Fig. 8 Die micrograph photo of the NB-IoT RF transceiver.

978-1-7281-1702-7/19 $31.00 © 2019 IEEE   269

V. Conclusion

By eliminating the need for RF SAW filter, minimizing the overall power consumption of the RF transceiver while achieving good RF performance, this RF-SoC addresses the design challenges of low-cost, small form factor and long battery operation raised by the 3GPP NB-IoT standard.

Acknowledgment

The authors would like to thank A. Espolong, BH. Oh, RJ. Zhang, and GH. Qi for circuit and system validation, and many other colleagues who helped chip development.

References

[1] M. Chen, et al., "Narrow Band Internet of Things," IEEE Access, pp. 20557-20577, Oct. 2017.

[2] O. Liberg, et al., "Cellular Internet of Things: Technologies, Standards, and Performance," pp 235, 265, 273, 2018.

[3] C. Hsiao, et al., "Design of a Direct Conversion Transmitter to Resist Combined Effects of Power Amplifier Distortion and Local Oscillator Pulling," IEEE Trans. Microw. Theory and Tech., vol. 60, no. 6, pp. 2000–2009, June 2012.

[4] Z. Song, et al., "A Low-Power NB-IoT Transceiver With Digital-Polar Transmitter in 180-nm CMOS", IEEE Transactions on Circuits and Systems, pp. 2569-2581, Sept. 2017.

[5] B. Weber, et al., "A SAW-less RF-SoC for cellular IoT Supporting EC-GSM-IoT -121.7 dBm Sensitivity Through EGPRS2A 592 kbps Throughput," ESSCIRC, pp. 340-343, Sept. 2017.

[6] C. Chiu, et al., "A 40nm Low-Power Transceiver for LTE-A Carrier Aggregation," ISSCC, pp. 130-132, Feb. 2017.

2019 IEEE Radio Frequency Integrated Circuits Symposium

A 1.04 - 4V, Digital-Intensive Dual-Mode BLE 5.0/IEEE 802.15.4 Transceiver SoC with extended range in 28nm CMOS

Nam-Seog Kim[#1], Myoung-Gyun Kim[#], Ashutosh Verma[*], Gyungseon Seol[#], Shinwoong Kim[#],
Seokwon Lee[#], Chilun Lo[#], Jaeyeol Han[#], Ikkyun Jo[#], Chulho Kim[#], Chih-Wei Yao[*], Jongwoo Lee[#]

[#]Samsung Electronics Co., Ltd., Hwaseong-si, Korea
[*]Samsung Semiconductor, Inc., San Jose, CA, USA
[1]namseog.kim@samsung.com

Abstract— **The proposed fully-integrated digital-intensive TRX SoC allows dual-mode protocols of BLE 5.0 and IEEE 802.14.5 with extended-range wireless connectivity and simple ad-hoc mesh networks for the IoTs in smart homes. A 1.04 - 4V with dual-mode supply schemes enables the SoC to be applied for various IoT systems energy-effectively and extends battery lifetime cost-effectively. The TRX employs a low insertion loss CMOS transmit-receive switch, a class-D power amplifier with HD2 calibration, a low-IF RX architecture with ΔΣ ADC-based complex filter, and a low power ADPLL with LMS-based two-point direct frequency modulation. The TRX implemented in a 28nm CMOS process achieves the maximum output power of +10dBm while consuming 45mW at the TX, <-102dBm at IEEE802.15.4 mode while consuming 6mW at the RX.**

Keywords— **CMOS, BLE, IEEE 802.15.4, IoT, transceiver, SoC, ADPLL, transmit-receive switch.**

I. INTRODUCTION

The internet of thing (IoT) systems based on energy-efficient wireless networks can play a key role in the smart home applications. The bluetooth low energy (BLE) and the IEEE 802.15.4 are the most widely used wireless standards for the applications. The system-on-chips (SoCs) integrating dual-mode BLE and IEEE 802.15.4 protocols can support different connectivity standards, which enables the SoC to be easily integrated in the IoT smart home devices by supporting interactions with a BLE-enabled smartphone and the Zigbee mesh networks.

Most of IoT systems run with batteries of various supply voltage levels, such as 3V coin batteries, 1.5V button cell batteries, and 1.2V rechargeable batteries. Thus, the IoT systems should support a wide range of supply voltage levels with high-energy efficiency. Moreover, around 1V operation is required to extend battery lifetime. The extended wireless communication range is also desirable for a variety application including gateways, electric and gas meters, home and building automation, lighting, and security systems. The extended range can be achieved by high transmitter (TX) output power and low receiver (RX) sensitivity of the systems.

II. DUAL-MODE BLE 5.0/IEEE 802.14.5 SoC

The proposed SoC features an RF transceiver (TRX) with BLE 5.0 / 802.15.4 (ZigBee 3.0 and Thread 1.1.1) physical layer (PHY) in 2.4 GHz ISM frequency band. The PHY also provides long-range options with 500kb/s and 125kb/s data rate in BLE5.0. Moreover, the SoC integrates a digitally controlled crystal oscillator (DCXO), a Cortex-M4 microcontroller, a 216kB SRAM and a 1.25MB NOR eFlash memory, a security engine, and an extensive set of peripherals including analog signal monitors and timers.

A. RF Transceiver Architecture

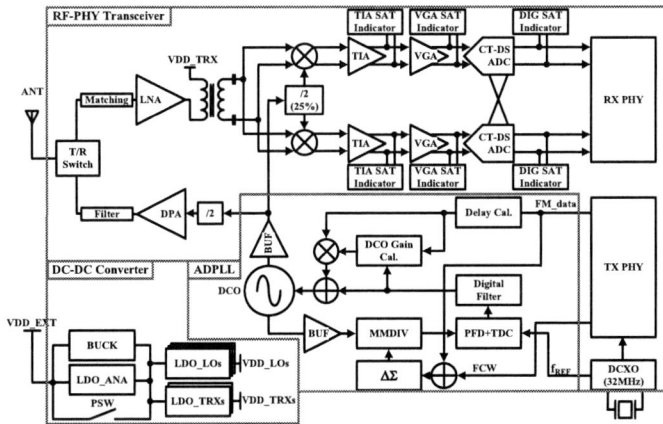

Fig. 1. The proposed transceiver architecture.

The transceiver block diagram is shown in Fig. 1. It utilizes an analog and digital low-IF receiver and direct frequency modulation (DFM) transmitter. The on-chip digitally controlled crystal oscillator (DCXO) is integrated to generate low-noise reference clocks (f_{REF}) for the transceiver and other blocks in the SoC. For wide range of the supply voltage operation, a DC-DC converter is embedded to provide two battery supply modes. In high voltage battery mode (>1.7V), the BUCK is enabled and generates 1.2V for the internal low-dropout regulators (LDOs) that isolates an LO blocks from TRX blocks. In low voltage battery mode (<1.7V), the BUCK is bypassed, and the internal LDOs are directly connected to the external supply (VDD_EXT). The internal LDOs use PMOS pass transistors for low voltage operation of the SoC, but the LDOs have slow settling time due to the high output impedance compared to NMOS pass transistor LDOs. Deep-sleep mode is achieved by switching off the internal LDOs to get sub-µW average power consumption.

B. On-Die T/R Switch

Only time division duplex (TDD) scheme is used for the BLE and the IEEE802.14.5 modes, so the RX blocks and the

978-1-7281-1702-7/19 $31.00 © 2019 IEEE   271

TX block do not operate at the same time, which allows the RX and the TX to share the same ball or package pin on the board as well as use one pad on-chip as shown in Fig. 2. The transmit-receive (T/R) switch with MOSFETs can be implemented form TX output to the pad and from the pad to RX input to isolate each other for the TDD mode. However, this causes loss along the path and consumes large area. Thus, the low-insertion loss T/R switch is adopted. The LNA and the power amplifier (PA) provide high output impedance respectively at the operation frequencies when each one does not work. In the TX mode, the RX input path should emulate high-impedance in order to minimize loss of the received RF power (Pin) and the PA output power (Pout). The TX_{ON} signal is asserted during the TX mode, and 50% of the electrostatic discharge (ESD) capacitance (C_{ESD}) at the pad and tuning cap C_{TX} ($=0.5*C_{ESD} + C_{TX}$) resonate out the matching inductor (L_G), thereby realizing a high input impedance of >300Ω. The resonant frequency is tuned to around 2.4 GHz. TX_{ON} signal is de-asserted during the RX mode, the PA load capacitor (C_{TL}) is tuned to the high output impedance of >300Ω.

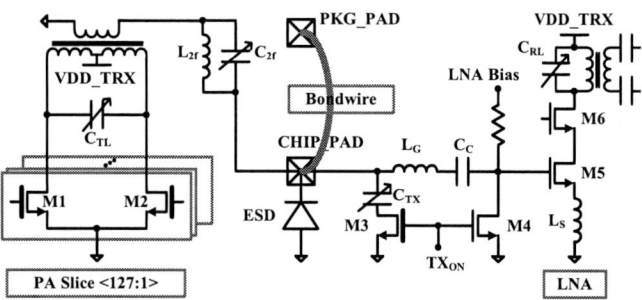

Fig. 2. Integrated PA, LNA, and T/R switch

C. Current mode Class-D PA with HD2 Calibration

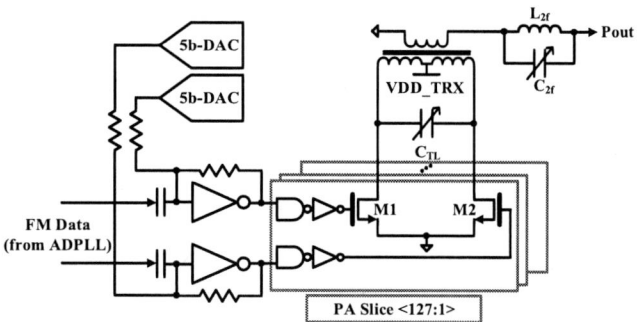

Fig. 3. Class-D power amplifier with HD2 calibration

Switching mode power amplifiers can provide high drain efficiency. Class-D amplifiers are popular switching-mode amplifiers for audio frequencies, but it shows less efficiency at the higher frequency due to parasitic reactance. The current mode class-D (CMCD) amplifier uses the square current source as opposed to the voltage mode class-D (VCMD) amplifier. Moreover, the filters are connected in parallel. Due to the filter, there is no voltage across the transistor at each switching time and zero-voltage-switching can be achieved. The parasitic drain

capacitance of the switch (C_{DS}) can be a part of the filter capacitor that is a great advantage over the VCMD amplifiers for the PAs in the gigahertz range.

Fig. 3 illustrates the CMCD PA used in the proposed transceiver. The PA consists of two thick oxide NFETs to tolerate the high Pout. The LC filter capacitor together with the C_{DS} of M1 and M2 is to tune the operating frequency of the PA. A 3:2 transformer (XFMR) is used to transform the 50ohm to match the antenna impedance. The frequency modulated (FM) data from the synthesizer are first magnified through a common input buffer stage. Following NAND and inverter gates can select the number of PA slices that is 127 units with 7-bit binary weighted controls. The sliced PA controls output power digitally. Wide Pout range can be realized by utilizing only M1 or M2 single-ended operation also. The digitally controlled PA also provides ramp-up sequence during channel hopping, which reduces parasitic coupling mechanism between the PA and digital controlled oscillator (DCO). FCC regulates that harmonic levels of the Pout are < -41dBm. For +10dBm Pout, the HD2 should be <-50dBc that is not easy for the power efficient Pas to achieve. Thus, after the PA XFMR, 2flo notch filter (L_{2f}//C_{2f}) is added to suppress the unwanted second order harmonic distortion (HD2). Moreover, the 5-bit DACs at the TX input buffers are added to calibrate the LO duty cycle to decrease the HD2 by using automatic test equipment with chip-level unmodulated TX mode.

D. Low-IF Receiver with ΔΣ ADC-based Complex filter

The proposed BLE/IEE802.15.4 receiver is an analog and digital low-IF type since it has both analog and digital complex band-pass filters to provide the required anti-aliasing filtering and image rejection as shown in Fig. 1. The LNA circuit is shown in Fig. 2. It employs an inductive degenerated common-source cascoded amplifier. The output is loaded with a 1.4:1 XFMR to achieve single-to-differential conversion and a tuned load capacitance helps resonate out the drain at around 2.4 GHz. A matching inductor L_G is used at the gate to resonate out the transistor gate capacitance and realize an input matching. A passive down-conversion double-balanced mixer is used in the RX. The Series capacitors between the LNA and the mixer are used to level shift the gate close to the common mode voltage, which helps reduce switch size and consequently lowers the LO power consumption. The trans-impedance amplifiers (TIAs) and variable gain amplifiers (VGAs) are added to reduce the required RF gain while keeping a decent NF near 6dB. The controlled gain range of RX path is >60dB for the minimum sensitivity of <-96dBm and the maximum input power of 0dBm. The automatic gain control (AGC) is done with two saturation indicators at the outputs of the TIA and VGA to be settled within 8μsec.

The proposed transceiver utilizes a complex continuous-time (CT) ΔΣ ADC followed by the VGA. The ADC is 4-bit output with 32 MHz sampling frequency. The complex band-pass filter is implemented by cross-coupled I/Q path signal using resistors to shift the IF center frequency to 1.5MHz with 2MHz of bandwidth. The ADC provides a 60 dB dynamic range while including both quantization noise and device noise, which is enough to get RX sensitivity level of <-96dBm even with in-band interference of 27dB. The ADC I/Q mismatch is managed to get image rejection ratio (IMRR) of <-32dBc. A digital complex band-pass filter is additionally required to provide sufficient image rejection.

E. Low Voltage Direct Frequency modulation with ADPLL

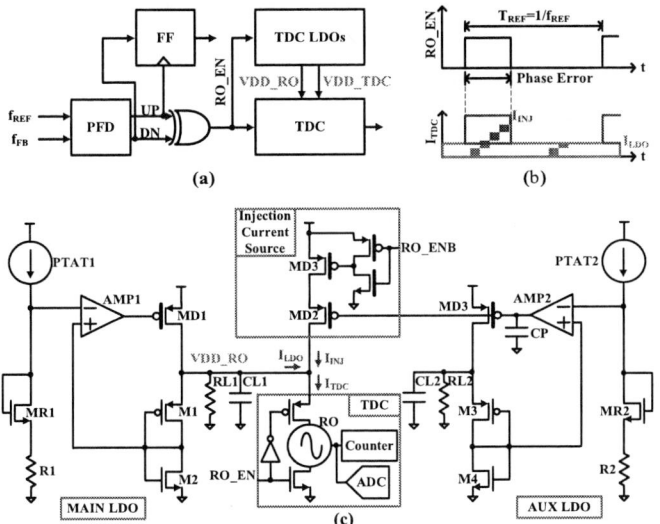

Fig. 4. Fast settling current injection LDO for low power TDC.

The proposed transceiver employs the all-digital PLL (ADPLL) that encodes frequency modulation (FM) easily. The non-linearity calibration on the time-to-digital converter (TDC) [1] lowers fractional and reference spur levels. The ring oscillator (RO) in the TDC is enabled only while RO_EN signal, the phase error of the phase frequency detector (PFD) output, is asserted as shown in Fig. 4(a). However, the PMOS-type LDO for low voltage operation degrades the TDC linearity due to its long settling time. The proposed low power ADPLL employs the current injection LDO (CILDO) to compensate the long settling time of the PMOS-type LDO as shown in Fig. 4(b). The injection time is the same with RO enable time, so the same control signal of RO_EN controls the injection current source. Moreover, the injected current amount is the same with that of the RO consumes during RO_EN assertion as shown in Fig. 4(c). The injected current amount also needs to track the RO current variation with different processes, voltages, and temperatures (PVT). The auxiliary LDO (AUX LDO) is added to generate bias condition of the injection current source, and it uses a replica bias scheme for low power consumption. The AUX LDO generates stable output voltage even with the supply voltage variation. The process variation can be compensated by using the same reference bias circuit with the main LDO. For temperature tracking, the RO current slope with temperature is different from the main bias circuit, so it requires a different slope from the bias circuit of the main LDO. The different voltage slope with temperature is achieved by using different bias resistor value (R2) from the main bias value (R1).

The FM is realized at two different points, direct modulation on the DCO and the multi-modulus divider (MMDIV) ratio control, when the ADPLL is locked to the channel frequency as shown in Fig. 1. Modulation gain at the divider input can be accurately controlled by the $\Sigma\Delta$ modulator but is band-limited by PLL bandwidth. On the other hand, the DCO modulation bears high-pass characteristic. If the modulation gain and phase of the low-pass and high-pass paths are perfectly matched, the combined modulated VCO output is independent of loop bandwidth. The phase mismatch is calibrated by adjusting FM data delay at the DCO (Delay Cal.). The DCO gain calibration requires accurate estimation of the DCO gain, K_{DCO}, which is nonlinear and highly sensitive to PVT variations. The ADPLL adopts a gain calibration circuit based on least mean square (LMS) algorithm as shown in Fig. 5. The calibration is done with correlation analysis between FM data and phase error in ADPLL. It includes 8-tap moving average filter to detect phase error magnitude exactly and supports various gain steps for adjusting calibration time and accuracy. The proposed LMS-based gain calibration runs in background, and the initial gain value comes from the one-time boot-up calibration circuit.

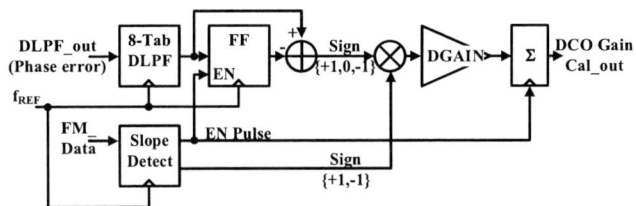

Fig. 5. LMS-based gain calibration circuit.

III. MEASUREMENT

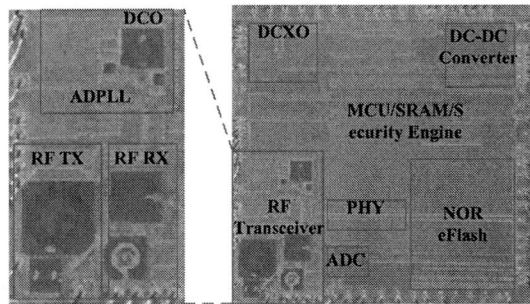

Fig. 6. Chip photograph

The SoC is fabricated in a 28nm CMOS process housed in a low-cost 48-pin QFN package. Fig. 6 shows the die photograph. The RF transceiver area is 1.21mm². The SoC also integrates a buck converter, the DCXO, PHY, MCU, security engine, eFlash memory, and a number of voltage regulators. The receiver consumes 6mW with a sensitivity of -102/-98/-96dBm in IEEE802.15.4, BLE, and BT5.0. The transmitter consumes 49mW to deliver an 11dBm output to the antenna. Fig. 7(a) shows the measured transmitter performance with the number of PA slices. The transmitter output power (Pout) is digitally adjustable with more than 36dB of range and can be programmed up to +10dBm by 7bit binary sliced Class-D PAs. The measured integrated phase noise (IPN) is < -38dBc and difference is <1.8dB with different Pouts. Fig. 7(b) shows the measured Pout and HD2/HD3 in dBm without and with the

978-1-7281-1702-7/19 $31.00 © 2019 IEEE   273

calibration shown in Fig. 3. The calibration is done at 2.44GHz, and the same setting is applied to the other channels. The Pout is not changed too much. The HD2 at 2.44GHz is reduced from -38dBm to -76dBm, and other HD2s are also reduced by > 8dB. HD3s are almost the same without and with the calibration.

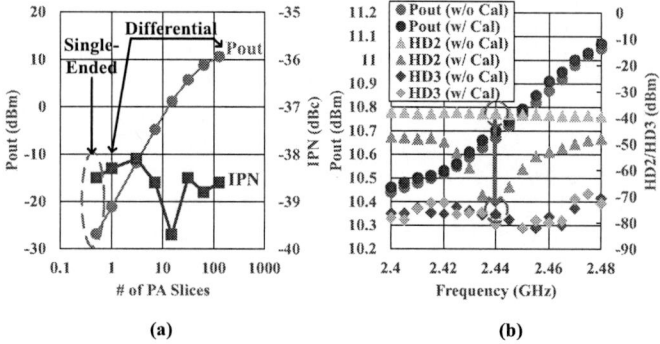

Fig. 7. (a) Pout and IPN with the number of PA slices and (b) Pout and HD2 without and with HD2 calibration.

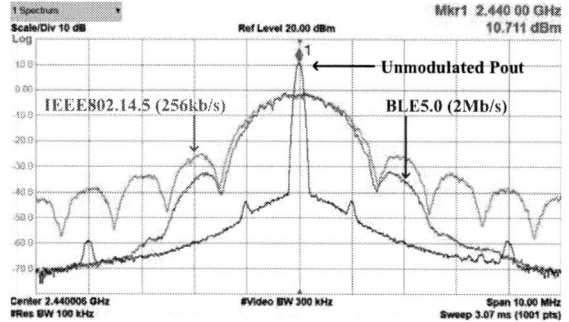

Fig. 8. Maximum Pout and BLE 5.0 / IEEE 802.15.4 modulation spectrums.

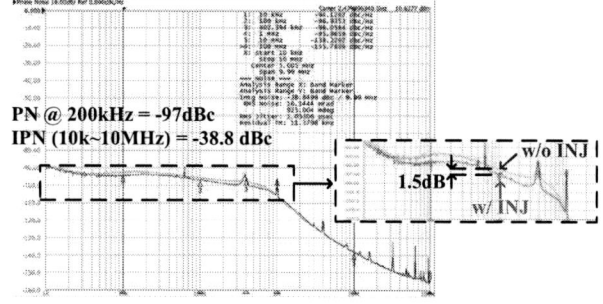

Fig. 9. TX output IPN without and with current injection LDO in TDC.

Fig. 8 shows the maximum Pout of 10.64dBm at 2.44GHz. The PA core efficiency is 30%. The GFSK modulation spectrum of the BT5.0 mode with 2Mb/s data-rate and the O-QPSK modulation spectrum of the IEEE802.14.5 mode with 250kb/s are shown in the red line and in blue line respectively at the maximum power with PRBS patterns. The GFSK modulation has 15dB margin from the BLE spectral mask and the O-QPSK modulation has 10dB margin from the IEEE802.15.4 spectral mask. Moreover, the LMS gain calibration supports stable modulation index of <1% GFSK error.

Fig. 9 shows IPNs with and without the current injection on TDC LDO. The in-band phase noise at 100kHz is improved by

2dB since the current injection compensates slow setting time of the PMOS-type LDO.

Table 1 compares the performance of the proposed SoC with prior arts having BLE and IEEE802.15.4 standards. The radio can provide a wide supply voltage range with high-energy efficiency by using both on-chip DC-DC converter and bypass mode. It achieves the highest Pout of +10dBm, the lowest HD2 of -63dBc, and comparable RX sensitivities of -96dBm at BLE5.0 mode and -102dBm at IEEE802.15.4 mode.

IV. CONCLUSION

A single-chip SoC fully compliant with dual-mode BLE 5.0 and 802.15.4 (ZigBee 3.0 and Thread 1.1.1) modes is demonstrated in a 28nm CMOS technology. The low power SoC fully integrates DCXO, MCU, SRAM, and eFlash, and a security engine. A wide range of operating supply voltage can be applied energy-efficiently by adopting DC-DC converter strategy for different voltage levels. The transceiver can provide long-range wireless connectivity with +10dBm of TX output power and -102dBm of RX sensitivity. Moreover, long-range options of the BLE5.0 with 512kb/s and 125kb/s can extend the communication rage by reducing RX sensitivity by 6dB and 12dB respectively. Thus, the proposed SoC can be applied for a wide range of home sensor networks.

Table 1. Performance summary and comparison with the BLE/IEEE802.15.4s.

		This Work	ISSCC15 [2]	ESSCIRC16 [3]	VLSI17 [4]
Standards		BT5.0 / BLE / 802.14.5	BLE / 802.14.5	BT5.0 / BLE / 802.14.5	BT / BLE / 802.14.5
Standards		SoC	TX / RX /DBB	SoC	TX / RX /DBB
Technology (nm)		28	40	40	40
Supply Voltage (V)		1.04 – 3.6	0.9 – 1.3	0.9 – 1.2	1.2
PA Type		Class-D	Class-D	Class-D	Class-A / Class-AB
Pmax (dBm)		11	1	1	4
HD2 (dBc)		-63 (w/ Cal.)	-49	-	-
RX Architecture		Low-IF	Sliding-IF	Sliding-IF	Low-IF
RX Sensitivity (dBm)	BLE	-98	-94	-92	-98
	802.15.4	-102	-97	-94	-104
ACR (2nd/3rd) (dB)		33/37	25 / 35	-	-
P$_{Active}$ (mW)	TX$_{Front}$	49 @ 10dBm	4.4 @ 1dBm	10 @ 0dBm	10 @ 0dBm
	RX$_{Front}$	6	3.3	5.3	7.8
P$_{Sleep}$ (µW)		1.5	-	1.5	-
RF Area (mm^2)		1.21	1.3	1.6	0.95
On-Chip T/R Switch		Yes	No	Yes	Yes
On-Chip Matching		Yes	Partial	Yes	No
Synthesizer Type		ADPLL	ADPLL	ADPLL	CPPLL

REFERENCES

[1] C.-W. Yao et al., "A 14-nm 0.14-psrms fractional-N digital PLL with a 0.2-ps resolution ADC-assisted coarse/fine-conversion chopping TDC and TDC nonlinearity calibration," IEEE J. Solid-State Circuits, vol. 52, no. 12, pp. 3446–3457, Dec. 2017.

[2] Y. Liu et al., "A 3.7 mW-RX 4.4 mW-TX fully integrated Bluetooth low-energy/IEEE802.15.4/proprietary SoC with an ADPLL-based fast frequency offset compensation in 40 nm CMOS," in ISSCC Dig. Tech. Papers, Feb. 2015, pp. 236–237.

[3] X. Wang et al., "A 0.9–1.2 V supplied, 2.4 GHz Bluetooth low energy 4.0/4.2 and 802.15.4 transceiver SoC optimized for battery life," in Proc. 42nd Eur. Solid-State Circuits Conf. (ESSCIRC), Sep. 2016, pp. 125–128.

[4] A. Zolfaghari et al., "A multi-mode WPAN (Bluetooth, BLE, IEEE 802.15.4) SoC for low-power and IoT applications," in Proc. Symp. VLSI Circuits, Jun. 2017, pp. 74–75.

A 24.5-43.5GHz Compact RX with Calibration-Free 32-56dB Full-Frequency Instantaneously Wideband Image Rejection Supporting Multi-Gb/s 64-QAM/256-QAM for Multi-Band 5G Massive MIMO

Min-Yu Huang[1], Taiyun Chi[2], Fei Wang[1], Sensen Li[1], Tzu-Yuan Huang[1], and Hua Wang[1]

[1]School of Electrical and Computer Engineering, Georgia Tech, Atlanta, GA 30332 USA

[2] Speedlink Technology Inc., Cupertino, CA, USA

Abstract— **This paper presents a 24.5-43.5GHz compact RX frontend achieving 32-56dB full-frequency instantaneously wideband image rejection, which can cover major mm-Wave 5G bands at 24.5/28/37/39/43 GHz. It utilizes a transformer-based IQ network able to accommodate large load impedance transformation with robust I/Q generation, which provides impedance up-scaling and passive voltage amplification to boost the LO swing. It achieves a low-loss mm-Wave I/Q LO generation with a compact size (0.14mm²) and state-of-the-art instantaneously broad bandwidth 25-50GHz without calibration or switching/tunable elements. After image rejection, the wideband desired RX signal is successfully demodulated, showing state-of-the-art 12Gb/s 64-QAM with -27.6dB EVM and 8Gb/s 256-QAM with -33.47dB EVM under wideband modulated image signals with the same modulation scheme and data rate. To the best of authors' knowledge, this paper is the first demonstration receiver frontend to support broadband modulated multi-Gb/s 64-/256-QAM image rejection with no calibration, switching/tuning elements, or external controls, enabling wideband low-latency 5G MIMOs in complex EM environments.**

Keywords— **5G communication, CMOS, image rejection, millimetre-wave, MIMO, receiver, wideband.**

I. INTRODUCTION

Wideband high-capacity wireless access nodes are essential for next-generation networks. For example, future 5G user equipment (UE) favors multi-band operation (especially at 24.5/28/37/39/43 GHz) to support multi-standard communication and international roaming. However, a large fractional bandwidth (BW) (>50%) poses challenges for frontend hardware, and image jamming often becomes a major issue in extreme spectral planning. Moreover, 5G MIMO systems are expected to concurrently handle multiple modulated signals (64-/256-QAM) at Giga-bits/s, which necessitates wideband >30dB SNR to demodulate multiple signals simultaneously and demands instantaneously wideband image rejection ratio (IRR). Additionally, for concurrent signal receiving, intermodulation distortions are significant, and thus high-linearity tunable-gain RX is highly desired to avoid decorrelations among the array elements during beamforming.

Although high-order RC-CR Poly-Phase Filters (PPFs) are popular in low RF frequency, their use in mm-Wave Image Rejection (IR) often exhibit large signal loss, highly capacitive input, limited driving capability, and employing high-order RC-CR PPFs in wideband LO demands more LO power and causes overhead penalty in LO drivers. Moreover, narrow-band but reconfigurable mm-Wave IR RXs inherently cannot support an instantaneously wide BW for concurrent multi-signal receiving

[1]. Recently, a 71-86 GHz bidirectional IR Weaver RX is reported to reduce fractional BW requirement of LO generation [2], and it rejects wideband images at Giga-bits/s for 16-/64-QAM modulations. However, it requires extensive open-loop calibrations for the capacitor tuning bank in the LO RC-CR PPF to achieve wideband IRR for signal demodulation, introducing reconfiguration latency for communication links.

To address these difficulties, this paper presents a 24.5-43.5GHz compact RX achieving calibration-free full-frequency instantaneously wideband image rejection that supports Giga-bits/s 64-QAM/256-QAM modulation. It consists of a mm-Wave wideband LNA, low-loss broadband transformer-based I/Q network for LO generation, differential I/Q double-balanced mixers, IF amplifiers, and IF 2-stage RC-CR PPF (Fig. 1). The RX frontend is co-designed with a T/R switch to support 5G MIMO transceiver. Serial-to-Parallel-Interface (SPI) controls RX conversion gain to handle various concurrent receiving scenarios in 5G MIMOs.

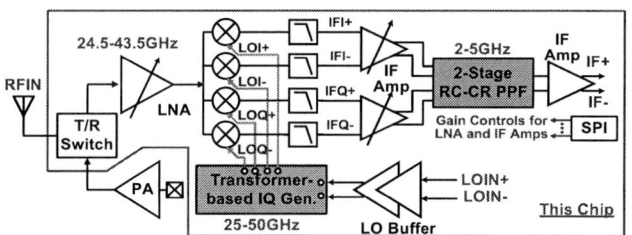

Fig. 1. Block diagram of the 24.5-43.5GHz RX with 32-56dB full-frequency instantaneous wideband image rejection.

II. RECEIVER WITH CALIBRATION-FREE INSTANTANEOUSLY WIDEBAND IMAGE REJECTION

A. Ultra-Wideband LNA with T/R Switch Co-Design

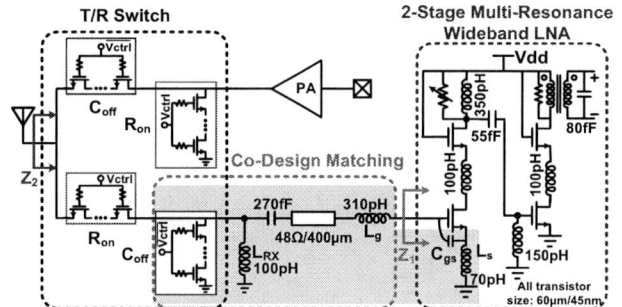

Fig. 2. Schematic of the 2-stage multi-resonance LNA with T/R switch co-design.

The LNA is designed with different resonant loads in two stages to serve as a wideband frontend (Fig.2). To mitigate the loading effect of the parasitic capacitors from the T/R switch and the off-state PA, separate shunt inductors are applied at the TX/RX inputs. The RX input shunt inductor L_{RX} is further co-designed with L_g, L_s, and C_{gs} of the first-stage LNA, which creates a high-order network for wideband input matching (Fig. 2 and Fig. 3).

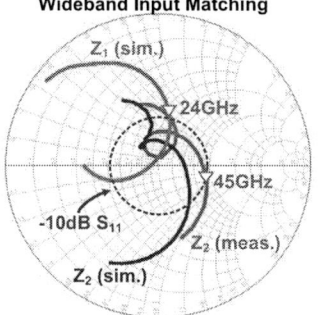

Wideband Input Matching

Fig. 3. Simulated/measured S_{11}.

B. Calibration-Free Ultra-Wideband IQ LO Generation

The mm-Wave transformer-based IQ network is designed to convert one differential LO input to two well-matched differential I/Q LO outputs over an extremely wide BW [3]. It is designed based on 2-stage one-inductor-footprint IQ transformer (Fig. 4a-4b). The measured I/Q amplitude imbalance is $<\pm0.15$dB and I/Q phase mismatch is $<\pm1.8°$ over 25-50GHz with 50Ω probing (Fig. 4c). The measured passive loss is only 2.1dB at 35GHz in additional to the 6dB inherent loss due to 1-to-4 power splitting (overall 8.1dB loss) (Fig. 4e). The transformer-based IQ network is able to accommodate large load impedance transformation with robust I/Q generation. In the actual RX system, the output loads (RL) of the transformer-based IQ network are terminated by 500Ω instead of 50Ω at the mixer switching quad inputs, which provides impedance up-scaling and passive voltage amplification to boost the LO swing (Fig. 4d). With R_L=500Ω, simulation shows 4.1dB LO swing enhancement, $<\pm0.09$dB amplitude imbalance, $<\pm1.9°$ phase mismatch over 25-50GHz without any calibration or tuning/switching element. (Fig. 4g) It achieves a low-loss mm-Wave I/Q LO generation with a compact size (0.14mm^2) and state-of-the-art instantaneously wide BW.

III. MEASUREMENT RESULTS

The 24.5-43.5GHz compact IR RX frontend is implemented in a 45nm CMOS SOI process (Fig. 5a). Measurement setup of the instantaneously wideband IRR RX chip is shown in the Fig. 5b. Multi-CW/modulated desired signals and image signals are

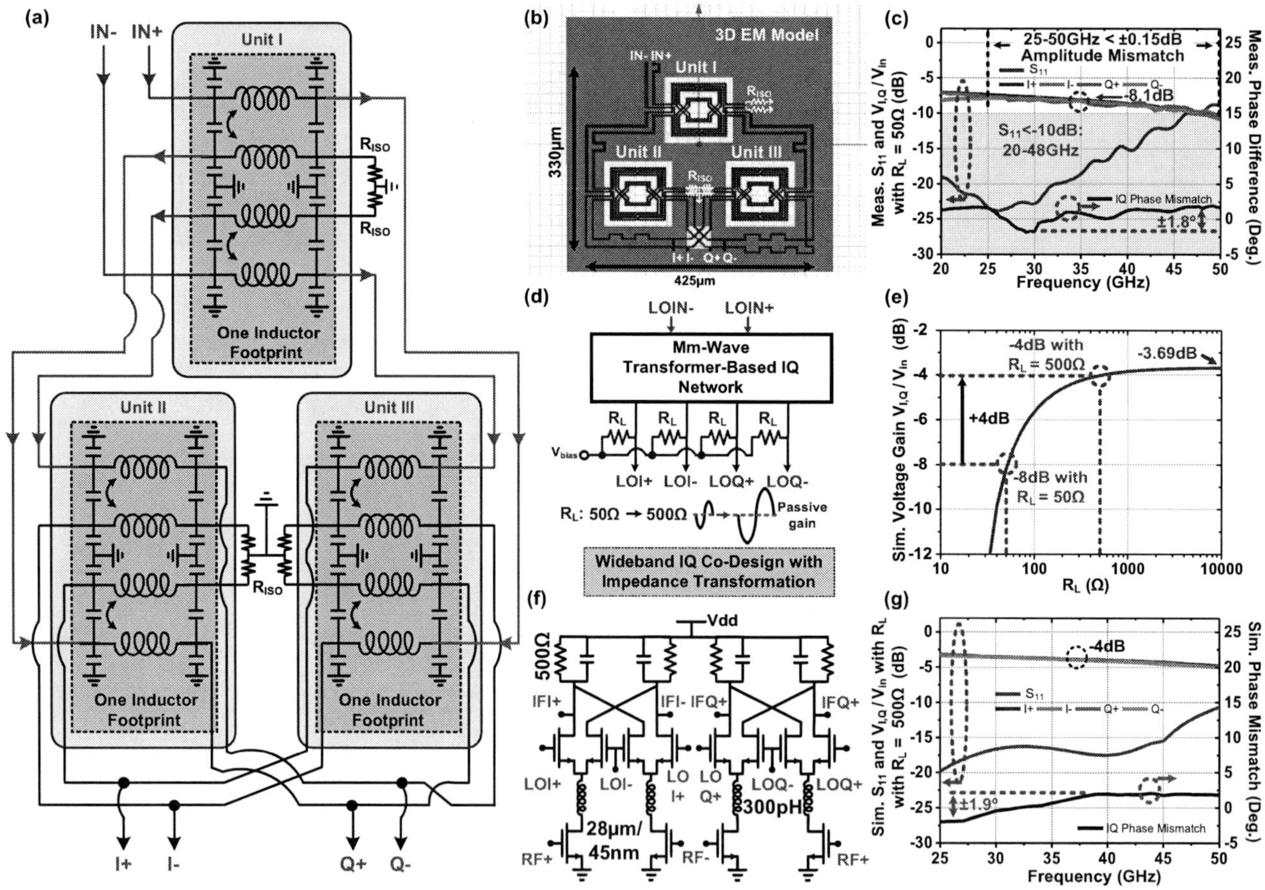

Fig. 4. (a) Schematic, (b) 3D EM model, and (c) Measurement results of the wideband low-loss transformer-based IQ network. (d) Conceptual diagram and (e) simulation of the passive gain amplification. (f) Schematic of the IQ double balanced mixer. (g) Simulation results with RL = 50Ω. RL = 500Ω is used when integrating with mixer to provide passive voltage amplification.

978-1-7281-1702-7/19 $31.00 © 2019 IEEE

2019 IEEE Radio Frequency Integrated Circuits Symposium

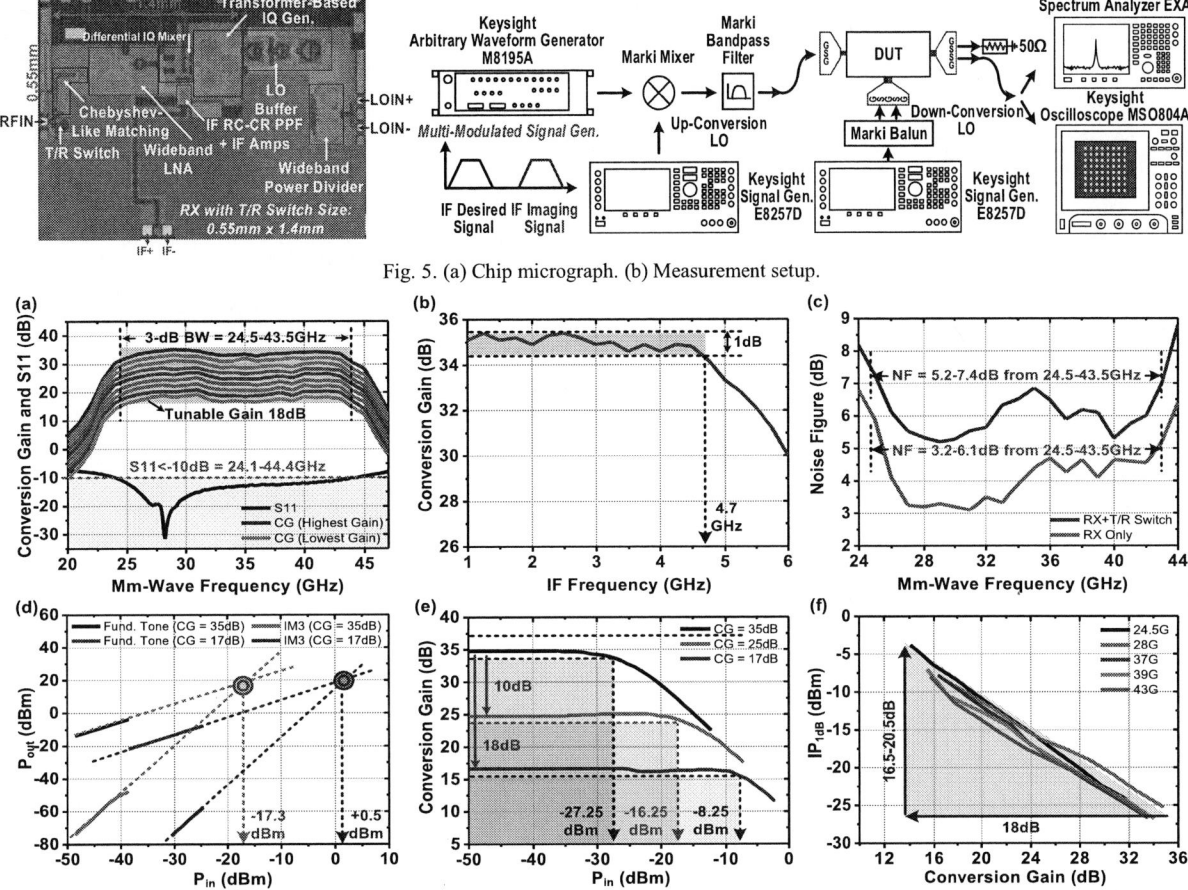

Fig. 5. (a) Chip micrograph. (b) Measurement setup.

Fig. 6. (a) Measured input matching and (b) tunable conversion gain (CG) versus mm-Wave/IF frequency. (c) Measured noise figure (NF) with/without T/R switch, (d) IIP3, and (e)-(f) IP1dB under various gain settings.

generated by a Keysight M8195A AWG, and then up-converted to mm-Wave 5G carrier as multiple input signals to the RX.

The RX is first tested for its CW performance. It achieves 3-dB conversion gain BW from 24.5 to 43.5GHz (56% fractional BW), covering the major 5G mm-Wave bands (24.5/28/37/39/43GHz) (Fig. 6a). The 35.2dB maximum conversion gain with 18dB gain control offers a wide range of RX linearity to facilitate concurrent multi-signal receiving in MIMOs. Down-converted IF signal shows a flat conversion gain <1dB variation up to 4.7GHz (Fig. 6b), enabling wideband demodulation. Figure 6c shows that the measured noise figure (NF) is 3.2-6.1dB for RX-only and 5.3-7.4dB for RX + T/R switch. RX IIP3 is +0.5/-17.3dBm (Fig. 6d) and IP1dB is -8.25/-27.25dBm with 17/35dB RX conversion gain at 28GHz (Fig. 6e). The RX linearity is then measured with 18dB gain tunability for the major 5G bands, showing 16.5-20.5dB IP1dB improvement for high-linearity MIMO (Fig. 6f).

Next, to test image rejection, a desired signal and an image tone are together sent to the RX (Fig. 7). For the desired input signal over 23-44GHz, a state-of-the-art full-frequency wideband 32-56dB IRR is achieved with fixed IF at 3.5GHz (Fig. 7). No calibration or tuning/switching element is used in these CW IRR tests.

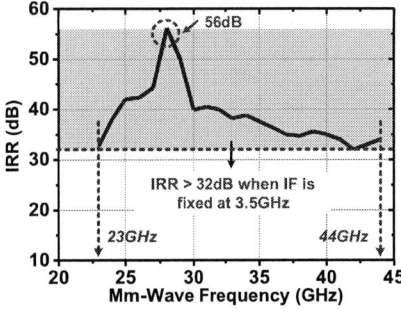

Fig. 7. Measured IRR versus mm-Wave frequency.

To evaluate image rejection during wideband demodulation, one mm-Wave modulated desired signal at $f_{LO} - f_{IF}$ and its image tone at $f_{LO} + f_{IF} + f_{spacing}$ with the same input power, modulation scheme, and data rate are together sent to the RX (Fig. 8). After down-conversion, the spectrum shows a clear wideband modulated image rejection (desired signal at 28GHz, LO at 31.75GHz, and image signal at 36.5GHz), and the desired signal is successfully demodulated for 6Gb/s 64QAM with -34.3dB EVM. Note that $f_{spacing}$ of 1GHz is only for IF spectrum illustration purpose. Next, we set $f_{spacing}$ = 0Hz, so the wideband modulated image tone and desired signal are completely overlapped after down-conversion (Fig. 8). After

978-1-7281-1702-7/19 $31.00 © 2019 IEEE   277

wideband image rejection, the desired signal shows a clear constellation and is successfully demodulated for 6/12Gb/s 64-QAM with -32.56/-27.6dB EVM and 8Gb/s 256-QAM with -33.47dB EVM, demonstrating a state-of-the-art mm-Wave instantaneously wideband GHz image rejection. Again, no calibration or tuning/switching element is used in these wideband modulation IRR tests.

Table 1. Comparison with mm-Wave RX and I/Q LO generation.

	This Work	S. Mondal ISSCC'18	N. Ebrahimi JSSC'18	F. Piri ISSCC'18	B. Welp TMTT'18
		Mm-Wave Full Receiver		Mm-Wave LO Generation	
BW (GHz)	24.5-43.5 / 25-50[a]	27-29.8	35-38.8	71-86	28-44	12.5-23[c]
Fractional BW (%)	56 / 66.7[a]	9.7	10.2	19.1	44	59.2
Conversion Gain (dB)	35.2	33	26.5	0	N/A	N/A
Gain Control Tuning (dB)	18	N/A	6	N/A	N/A
IRR (dB)	32-56	35	30	40-44	>26[c]
External Controls for IRR	No	No	Yes	No	Yes
Frequency Spacing for Signal and Image	0Hz	60MHz	2GHz	N/A	N/A
Signal Modulation Data Rate (EVM) After Image Rejection	12Gb/s 64-QAM (-27.6dB) 8Gb/s 256-QAM (-33.5dB)	Continuous Wave	9Gb/s 64-QAM (-28.3dB)	Continuous Wave	Continuous Wave
IP1dB (dBm)	-7 to -25.5	-23 to -30	N/A	N/A	N/A
NF (dB)	3.2-6.1 / 5.3-7.4[b]	5.7-8.5	>14	N/A	N/A
Power Consumption (mW)	60	52.5	150	39	242
Core Area / Element (mm²)	0.14[a] / 0.52 / 0.77[b]	1.1[c]	2.09[c]	0.2	0.087
Process	45nm CMOS SOI	65nm CMOS	90nm SiGe	55nm CMOS	130nm BiCMOS

a. Transformer-based IQ network. *b.* With T/R switch. *c.* Estimated value from the figure.

IV. CONCLUSION

Measurement results are compared to recently reported 5G mm-Wave full RXs and I/Q LO generation in Table 1. The proposed RX covers the widest mm-Wave frontend bandwidth

24.5-43.5GHz (LO generation is from 25-50GHz) for multi-band 5G communication. The RX also achieves a compact size, low noise figure 3.2-6.1dB (5.3-7.4dB with T/R switch), and superior linearity -7/-25.5dB IP1dB at lowest/highest gain. Most importantly, with low-loss wideband transformer-based mm-Wave I/Q network, this is the first demonstration of instantaneously wideband GHz image rejection with no calibration, switching/tuning elements, or external controls, enabling instantaneously wideband low-latency 5G MIMOs in complex EM environments [4]-[7].

ACKNOWLEDGEMENT

The authors thank Speedlink for project supports and GlobalFoundries for chip fabrication.

REFERENCES

[1] S. Mondal, et al., "A Reconfigurable 28/37GHz Hybrid-Beamforming MIMO RX with Inter-Band Carrier Aggregation and RF-Domain LMS Weight Adaptation," *ISSCC*, Feb 2018.

[2] N. Ebrahimi and J. F. Buckwalter, "A High-Fractional-Bandwidth, Millimeter-Wave Bidirectional Image-Selection Architecture with Narrowband LO Tuning Requirements," *JSSC*, Aug 2018.

[3] M. Huang, et al., "A 23-30GHz hybrid beam-forming MIMO receiver array with closed-loop multi-stage front-end beam-formers for full-FoV dynamic and autonomous unknown signal tracking and blocker rejection," *ISSCC*, Feb. 2019.

[4] B. Sadhu, *et al.*, "A 28-GHz 32-element TRX phased-array IC with concurrent dual-polarized operation and orthogonal phase and gain control for 5G communications," *JSSC*, Dec 2017.

[5] M. Huang, et al., "A Full-FoV Autonomous Hybrid Beamformer Array with Unknown Blockers Rejection and Signals Tracking for Low-Latency 5G Mm-Wave Links" *TMTT*, 2019.

[6] M. Huang and H. Wang., "A 27-41GHz MIMO Receiver with N-Input-N-Output Using Scalable Cascadable Autonomous Array-Based High-Order Spatial Filters for Instinctual Full-FoV Multi-Blocker/Signal Management," *ISSCC*, Feb. 2019.

[7] T. Yang, *et al.*, "A 4-channel Beamformer for 9-Gb/s MMW 5G Fixed-wireless Access over 25-km SMF with Bit-loading OFDM," *Proc. Optical Fiber Communication Conference (OFC)*, Mar. 2019.

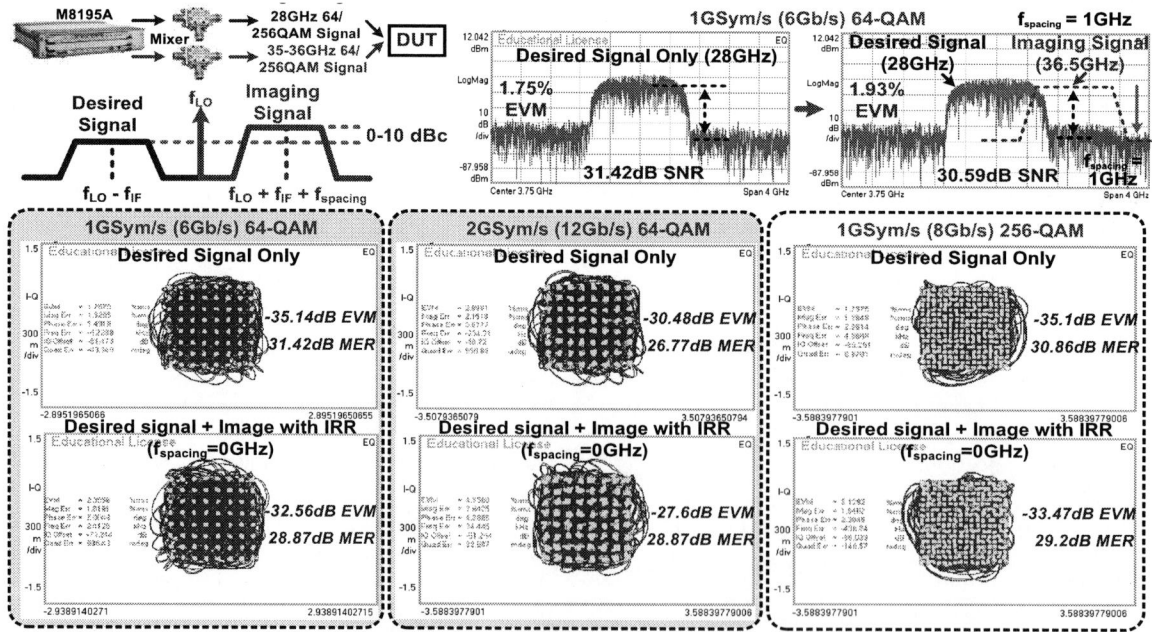

Fig. 8. Wideband modulated 64-/256-QAM image rejection is demonstrated and the desired signal is successfully demodulated.

A 39GHz 64-Element Phased-Array CMOS Transceiver with Built-in Calibration for Large-Array 5G NR

Yun Wang[1], Rui Wu[1], Jian Pang[1], Dongwon You[1], Ashbir Aviat Fadila[1], Rattanan Saengchan[1], Xi Fu[1], Daiki Matsumoto[1], Takeshi Nakamura[1], Ryo Kubozoe[1], Masaru Kawabuchi[1], Bangan Liu[1], Haosheng Zhang[1], Junjun Qiu[1], Hanli Liu[1], Wei Deng[1], Naoki Oshima[2], Keiichi Motoi[2], Shinichi Hori[2], Kazuaki Kunihiro[2], Tomoya Kaneko[2], Atsushi Shirane[1], Kenichi Okada[1]

[1]Department of Physical Electronics, Tokyo Institute of Technology, Japan
[2]NEC Corporation, Kawasaki, Japan

Abstract — This paper presents a 39GHz 64-element phased-array transceiver based on 4-element transceiver chipset with LO phase shifting architecture and built-in gain phase calibration. A phase-to-digital-convertor (PDC) and a high-resolution phase detection mechanism are proposed. The built-in calibration has an measured accuracy of 0.08-degree RMS phase error and 0.01-dB RMS gain error. The LO phase shifting based transceiver has a 0.04-dB maximum gain variation over the 360° full tuning range. The proposed pseudo-single-balanced mixer realizes -70 dBm LO-feedthrough (LOFT) cancellation and maximum 0.5° LO-to-LO isolation. The 8TX-8RX phased-array transceiver module 1-m OTA measurement supports 5G new radio (NR) 400MHz 256QAM OFDMA modulation with -30.0dB EVM. The 64-element transceiver has a $\mathrm{EIRP_{MAX}}$ of 53dBm. The 4-element chip consumes a power of 1.5W in TX mode and 0.5W in RX mode.

Keywords — 5G New Radio, 39 GHz, Phased-array, Transceiver, Calibration, LO phase shifting, LO-feedthrough (LOFT), millimeter wave, CMOS.

I. INTRODUCTION

Fifth-generation new radio (5G NR) wireless technologies will utilize millimeter-wave bands and beamforming for wider bandwidth, higher spatial efficiency and higher signal strength. To enhance the beamforming accuracy, especially for a large-scale antenna array used in basestation (BS), the antenna array excitations are required to minimize amplitude and phase mismatch. Recently, RF phase shifting and LO phase shifting phased-array transmitters and receivers are developed for 5G communication [1-6]. The RF phase shifting transceiver suffers from high gain variation over the phase tuning range, which is usually worse than 1 dB. The LO phase shifting transceiver shows high phase-tuning resolution, low gain variation and wideband characteristics. However, beamforming quality in [3] will suffer from phase drifting and gain expansion or suppression with supply and temperature changes. In this paper, a 39GHz 4-element LO phase shifting phased-array transceiver chip with built-in gain and phase calibration is presented. The on-chip calibration embedded in the proposed transceiver has an accuracy of 0.08-degree RMS phase error and 0.01-dB RMS gain error. The 39GHz phased-array transceiver supports 5G NR 400MHz 256QAM OFDMA in 1-m link over-the-air (OTA) measurement at band n260.

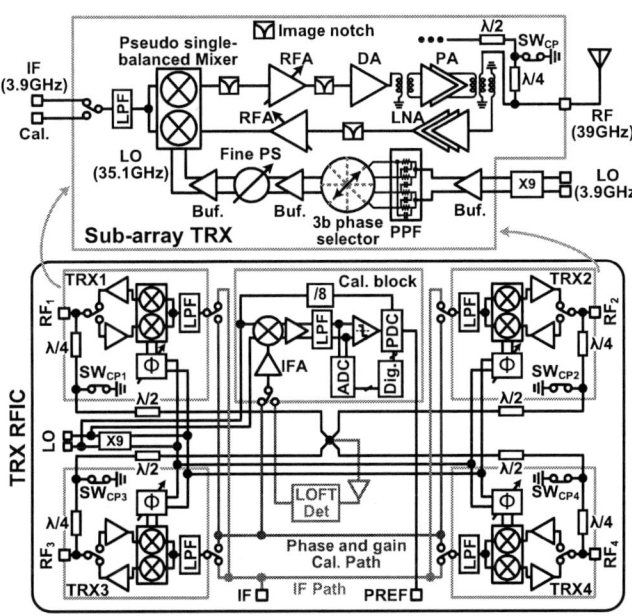

Fig. 1. Proposed 39GHz phased-array transceiver architecture.

II. CIRCUIT DESIGN

A. Transceiver Architecture

Fig. 1 shows the proposed 39GHz phased-array transceiver architecture. The transceiver chip is composed of four sub-array transceivers, an LO frequency multiplier and a calibration block. A quarter-wave length transmission line based coupling network is used to switch between transceiver mode and calibration mode. The sub-array transceiver consists of a transmitter, a receiver and a LO phase shifter chain. The transmitter includes pseudo-single-balanced passive mixer, RF-amplifier (RFA), drive amplifier (DA) and two-stage differential power amplifier (PA). The receiver has three-stage LNA, RF-amplifier and pseudo-single-balanced passive mixer. Notch filters at image frequency are added in both TX and RX for image suppression. The LO-phase-shifting based architecture is chosen since it can achieve very fine beam steering resolution and gain-invariant phase tuning. The LO phase shifter chain is realized by employing a polyphase filter (PPF), a 3-bit phase selector and a fine phase shifter. For phase and gain quantization, the output of the receiver is connected

2019 IEEE Radio Frequency Integrated Circuits Symposium

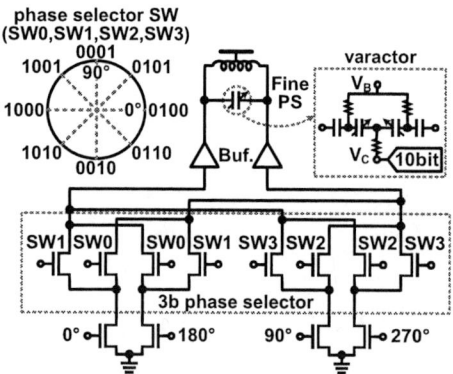

Fig. 2. Proposed phase shifter circuit schematic.

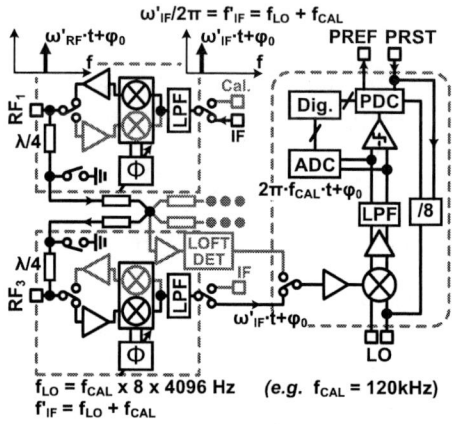

$f_{LO} = f_{CAL} \times 8 \times 4096$ Hz   *(e.g. f_{CAL} = 120kHz)*
$f'_{IF} = f_{LO} + f_{CAL}$

Fig. 3. Phase and gain calibration mechanism.

to the calibration block through a single-pole double-throw switch. The transmitting LO feed-through (LOFT) can also be detected by using a LOFT detector and the calibration block. The TX/RX switching, at the antenna side, is realized by the proposed transformer-based coupler. While at IF side the TX input and RX output are connected through the pseudo-single-balanced mixer for better linearity and phase-shifting isolation.

B. LO and Calibration Circuits

Fig. 2 shows the details of the phase shifter circuit schematic. The PPF quadrature output is connected to the 3-bit phase selector. The phase selector has eight steps with 45° phase shift of each step. Compared with the 90° quadrant phase selector [3], the proposed phase selector has a smaller phase step. Therefore, the fine phase shifter can be designed with a relaxed phase coverage, which improves the gain consistency over the varactor tuning range.

Fig. 3 shows the phase and gain calibration mechanism. A pair of TX and RX (not in the same sub-array) is used to calibrate the phase and gain characteristics. For example, RX3 is used to quantize TX1 phase gain values, and vice versa. The TX1 IF frequency f_{IF}' and LO frequency f_{LO} has a small frequency offset (*e.g.* 120kHz). The RF signal containing TX1 phase gain information will be down-converted to a low frequency f_{CAL} in the calibration block. A 10-bit

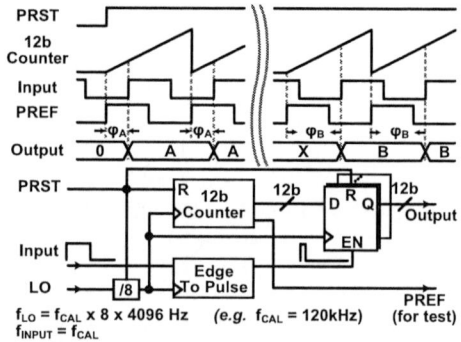

$f_{LO} = f_{CAL} \times 8 \times 4096$ Hz   *(e.g. f_{CAL} = 120kHz)*
$f_{INPUT} = f_{CAL}$

Fig. 4. Proposed PDC circuit schematic and timing diagram.

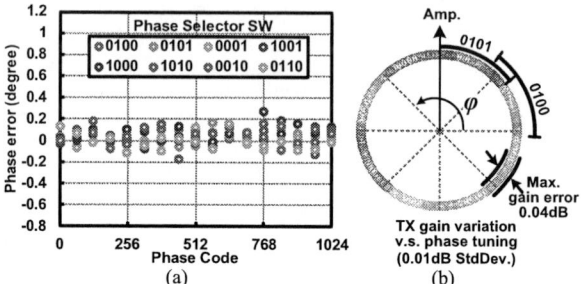

Fig. 5. Measurement results of (a) PDC quantization error, (b) TX gain variation over 360° tuning range.

analog-to-digital converter (ADC) will quantize the gain value, and a 12-bit phase-to-digital converter (PDC) will quantize the phase value.

Fig. 4 shows the PDC timing diagram, the phase input signal is quantized by a 12-bit counter, which results in a qantization resolution of 0.09°. The proposed on-chip phase quantization technique achieves a very high resolution at millimeter-wave frequency, and it is 30 times improved compared with the analog solution in [7].

Fig. 5 shows the measurement results of the PDC quantization error and TX gain variation over 360° tuning range. The measured RMS phase quantization error between PDC readout value and external oscilloscope readout value is 0.08°. The measured standard deviation of TX gain variation over the phase shifter full 360° tuning range is 0.01dB, and the maximum gain variation is 0.04dB.

C. Front-end design

The full transceiver RF front-end is shown in Fig. 1. A pseudo single-balanced mixer is proposed to cancel the LOFT and enhance the LO-to-LO isolation between each sub-array elements. Fig. 6 (a) shows the proposed mixer circuit schematic. The proposed mixer combines the signal path (LO+) and a dummy path (LO-) for the TX. The LOFT can be further reduced by adjusting the bias voltage at the LO port, IF port, and dummy IF port. In addition, the proposed mixer mitigates the issue of LO leakage to IF path, which causes LO phase shifters affecting on each other when tuning the phase. The mitigation of the LO leakage to IF path is realized by summing the LO signal in opposite phase. An R-C low-pass filter (LPF) at the mixer IF port is added

978-1-7281-1702-7/19 $31.00 © 2019 IEEE   280

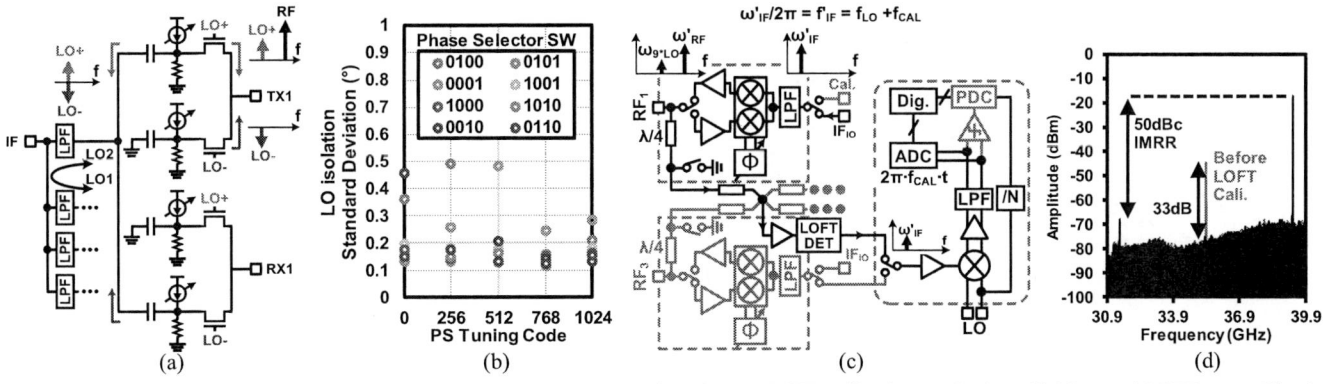

Fig. 6. (a) Pseudo single-balanced mixer circuit schematic, (b) Measured LO isolation, (c) LOFT calibration mechanism, (d) Measured LOFT auto-calibration.

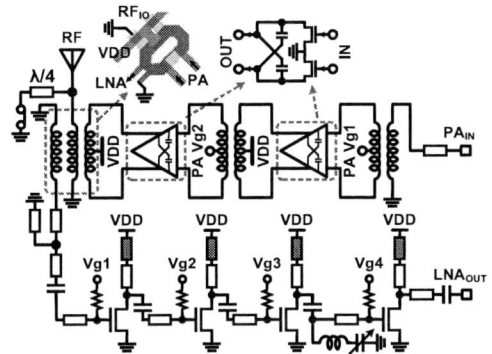

Fig. 7. PA/LNA circuit schematic.

to further enhance the LO isolation. The measured phase variation of one TX path at different phase control code shows a maximum standard deviation of 0.5°. The phase variation is measured by sweeping the phase of other TXs over 360°. Fig. 6 (c) shows the LOFT calibration mechanism. The square-law LOFT detector generates LOFT strength indication signal at f'_{IF} and quantized by ADC in calibration bock. Fig. 6 (d) shows the measured LOFT auto-calibrated result with LOFT<-70dBm.

The PA/LNA with antenna switch, shown in Fig. 7, employs the proposed stacked transformer to switch between PA mode and LNA mode. The transformer middle layer is designed to couple PA output power when PA is on, and couple to LNA input when LNA is on. The identical standalone PA/LNA circuit is fabricated for characterization. The measured LNA gain and noise figure is 33dB and 7.0dB, respectively. The measured PA achieves P_{SAT} of 15.5dBm, P_{1dB} of 13.5dBm, and peak PAE of 25.5%.

III. TRANSCEIVER MEASUREMENT RESULTS

As shown in Fig. 8, the transceiver fabricated in 65nm CMOS process occupies a chip area of 12mm². The transceiver consumes a DC power of 1.5W/chip in TX mode and 0.5W/chip in RX mode both from a 1-V supply.

Fig. 9 shows the measured RF front-end characteristics. The one-path TX is evaluated using 5G NR 400MHz bandwidth MCS10/19/27 modulated signal (OFDMA). As shown in Fig. 9 (a), the TX achieves an average output power of 3.6dBm at -24.6dB EVM while transmitting MCS19

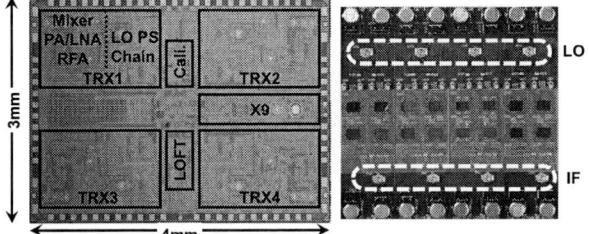

Fig. 8. Chip micrograph and 64-element module.

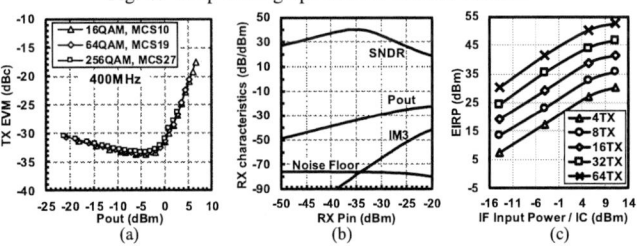

Fig. 9. (a) TX EVM, (b) RX SNDR, (c) TX EIRP.

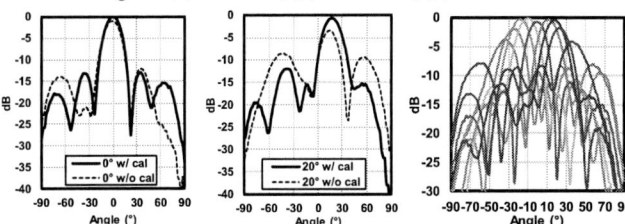

Fig. 10. Measured beam pattern.

64QAM modulation signal. Fig. 9 (b) shows the RX achieves a peak SNDR of 40dB. The RX SNDR with 400MHz signal bandwidth is calculated from the measured RX gain, IM3 and noise figure. The transceiver module with 64-element (4×16) patch antenna is implemented. The measured maximum EIRP is 53dBm for a 64-element (4×16) TX. Fig. 10 shows the radiation pattern of 1×4 TX array with and without calibration measured at 0-degree and 20-degree beam steering angle. At 20-degree angle, the main lobe strength is 3dB improved after calibration and side lobe is suppressed more than 5dB. After calibration, the radiation pattern with beam steering angle in +/- 40-degree is demonstrated.

Fig. 11 shows the measured OTA constellation and EVM performance. Both TX and RX utilize 8-element (2x4) array.

Table 1. Performance comparison of mm-wave phased array transceivers for 5G and beyond.

	This work	[1]	[2]	[3]	[4]	[5]
Frequency (GHz)	39 (n260)	28	60	28	29	28
Process	65nm CMOS	28nm CMOS LP	28/40nm CMOS	65nm CMOS	180nm SiGe	130nm SiGe
Architecture	LOPS	RFPS	RFPS	LOPS	RFPS	RFPS
PS resolution	3+10 bit / 0.05°	3 bit -	6bit / 6°	2+3+10 / 0.04°	6bit / 5.6°	1+5 bit / 5°
TX Psat/path (dBm)	15.5	14	6.5	18 (w/o SW)	12.5	16.4
Chip power dissipation (W)	1.5 / 4TX 0.5 / 4RX	0.36 / 4TX 0.17 / 4RX	8.4 / 144TX 6.6 / 144RX	1.2 / 4TX 0.6 / 4RX	0.8 / 4TX 0.5 / 4RX	4.6 / 16TX 3.3 / 16RX
Chip area (mm²)	12	28	292 (full radio)	12	12	166
calibration	phase, gain, LOFT	N/A	N/A	N/A	gain, IQ	N/A
Max gain variation (dB)	0.04	-	1.5	0.03 (RMS)	0.8	1.5
RMS phase error (°)	0.08	-	-	0.28	6	1
TX LOFT (dBm)	< -70	-	-	-	-	-
Array size	64	24	288	8	32	128
EIRP$_{MAX}$ (dBm)	53	35 (8 ele.)	51	39.8	45	57
OTA TX to RX EVM (dB)	-30.2 400MHz 64QAM	-41 (TX only) 100MHz 64QAM	-24 (TX only) 1150MS/s 16QAM	-35 800MS/s 64QAM	-27 500MHz 64QAM	N/A
5G NR evaluated	Yes	Yes	N/A	No	N/A	N/A

1m OTA Measurement	Modulation	64QAM MCS19	64QAM MCS19	64QAM MCS19
	BW	400MHz	400MHz	400MHz
	Beam direction	0°	20°	40°
	TX to RX Constellation			
	TX to RX EVM*	-30.2dB	-30.1dB	-28.6dB

1m OTA Measurement	Modulation	QPSK MCS4	16QAM MCS10	256QAM MCS27**
	BW	400MHz	400MHz	400MHz
	Beam direction	0°	0°	0°
	TX to RX Constellation			
	TX to RX EVM*	-30.7dB	-30.3dB	-30.0dB

*RMS power normalized EVM, measured with external down-conversion mixer
**MSCs are defined in Table 5.1.3.1-2 table 2 for PDSCH in 3GPP TS 38.214 V15.1.0

Fig. 11. Measured 5-m OTA constellation and EVM performance.

The TX 5G NR OFDMA IQ modulation signal is up-converted to IF by Keysight signal generator (E8267D) with vector modulation. The RX IF output signal is down-converted by an external mixer and demodulated by keysight digitizer (M9703B). The transceiver is evaluated with 400MHz QPSK, 16QAM, 64QAM and 256QAM modulation signal. At 1m OTA distance and 0-degree beam direction, the measured TX to RX RMS power normalized EVM are -30.7dB, -30.3dB, -30.2dB and -30.0dB in QPSK, 16QAM, 64QAM and 256QAM, respectively. At 20-degree and 40-degree beam direction, the measured EVM are -30.1dB and -28.6dB in 64QAM.

Table 1 shows the comparison table of millimeter-wave phased-array transceivers for 5G and beyond. This paper demonstrates a 39GHz phased-array transceiver with built-in phase, gain and LOFT calibration, which can ease the deployment of the large array. A 1-m OTA link is achieved

with 5G NR 400MHz bandwidth MCS19 64QAM modulation signal. A EIRP$_{MAX}$ of 53dBm is achieved, the LOFT is auto-calibrated to -70dBm, and image signal is 50dBc suppressed.

IV. CONCLUSION

This paper presents a 64-element phased-array transceiver for 5G NR at band n260. The proposed transceiver integrates phase gain and LOFT calibration monolithically. The on-chip phase and gain calibration error are 0.08° and 0.01dB respectively. A 1-m OTA link is supported with 5G NR 400MHz 64QAM modulation.

ACKNOWLEDGMENT

This work is partially supported by the MIC/SCOPE #175003017, STAR, and VDEC in collaboration with Cadence Design Systems, Inc., Mentor Graphics, Inc., and Keysight Technologies Japan, Ltd.

REFERENCES

[1] J. D. Dunworth et al., "A 28GHz Bulk-CMOS Dual-Polarization Phased-Array Transceiver with 24 Channels for 5G User and Basestation Equipment," in IEEE ISSCC, Feb. 2018, pp. 70–72.

[2] T. Sowlati et al., "A 60GHz 144-Element Phased-Array Transceiver with 51dBm Maximum EIRP and 60 Beam Steering for Backhaul Application," in IEEE ISSCC, Feb. 2018, pp. 66–68.

[3] J. Pang et al., "A 28GHz CMOS Phased-Array Transceiver Featuring Gain Invariance Based on LO Phase Shifting Architecture with 0.1-Degree Beam-Steering Resolution for 5G New Radio," in IEEE RFIC, Jun. 2018, pp. 56–59.

[4] K. Kibaroglu, M. Sayginer, and G. M. Rebeiz, "An Ultra Low-Cost 32-element 28 GHz Phased-array Transceiver with 41 dBm EIRP and 1.0C1.6 Gbps 16-QAM Link at 300 Meters," in IEEE RFIC, Jun. 2017, pp. 73–76.

[5] B. Sadhu et al., "A 28-GHz 32-Element TRX Phased-Array IC With Concurrent Dual-Polarized Operation and Orthogonal Phase and Gain Control for 5G Communications," in IEEE ISSCC, Feb. 2017, pp. 3373–3391.

[6] K. Scheir et al., "A 52GHz Phased-Array Receiver Front-End in 90nm Digital CMOS," in IEEE ISSCC, Feb. 2008, pp. 184–605.

[7] B. Perez et al., "360° Phase Detector Cell for Measurement Systems Based on Switched Dual Multipliers," in IEEE MWCL, May 2017, pp. 1531–1309.

2019 IEEE Radio Frequency Integrated Circuits Symposium

A 24.2-30.5GHz Quad-Channel RFIC for 5G Communications including Built-In Test Equipment

D. Dal Maistro[#], C. Rubino[#], M. Caruso[#], M. Tiebout[#], I. Maksymova[#], M. Ilic[#], P. Thurner[#],
M. Zaghi[#], K. Mertens[#], S. Vehovc[*], I. Tsvelykh[*], E. Schatzmayr[#], M. Druml[#], R. Druml[#],
M. Mueller[#], M. Anderwald[#], J. Wuertele[*], U. Rueddenklau[*]

[#]Infineon Technologies, Austria
[*]Infineon Technologies, Germany

Abstract — A wideband quad-channel beamforming RFIC for worldwide 5G infrastructure applications in 130nm SiGe BiCMOS features an Rx single channel gain and NF of 22 and 4dB respectively at a total power consumption of 1.6W. Tx performance includes a P1dB of 18dBm CW and 11.5dBm RMS output power at 3% EVM and 1.8W total power consumption. Beamforming is based on a temperature invariant lumped true time delay and a phase invariant programmable gain amplifier. Integrated built-in test equipment including LO-generation, signal injection and detection enables low-cost RF production testing and array calibration.

Keywords — phased array, 5G, true time delay, LNA, power amplifier, built-in self-test

I. INTRODUCTION

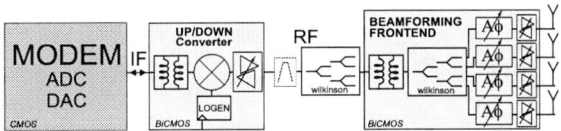

Fig. 1. System Architecture

The infrastructure deployment of mm-Wave 5G telecommunications requires cost-effective, easy and worldwide usable, long-term reliable RFICs. The system setup shown in Fig. 1 enables a tile based low-cost PCB approach, having patch antennas on one side and any other component on the opposite side of the board. An RFIC quadchannel implementation as shown in Fig. 2 has been chosen to minimize the connection losses between the 4 chip corners and the antennas, in order to enable effective cooling and thereby to support a high PA output power. The quadchannel chip is realized in 130nm SiGe BiCMOS and measures 17mm^2. Frequency up/down-conversion and RF-beamforming are split into two chips, to support worldwide emission mask requirements through external filter adaptation.

This paper presents the first RF-beamformer chip covering all 5G frequency bands between 24.2 and 30.5 GHz, featuring 18dBm Tx oP1dB and 4dB Rx NF, and integrating a complete built-in test equipment (BITE) for low-cost production test and system phased array calibration.

II. RFIC

The RF beamformer supports a direct–to-antenna interface without any external component. The frontend is designed to interface differential patch antennas, yielding optimal isolation. In order to compensate for the external filter and Wilkinson power divider losses, an amplification stage (Amp5 in Fig. 2) is inserted, increasing Tx gain up to 38dB. The common path provides gain programmability (PGA5rx/tx in Fig. 2), in order to support any beamformers to up/down-converters ratio between 1 and 16. Additionally, the RFIC features center frequency programmability, enabling the best gain flatness over the chosen frequency range (Fig. 5 and 7).

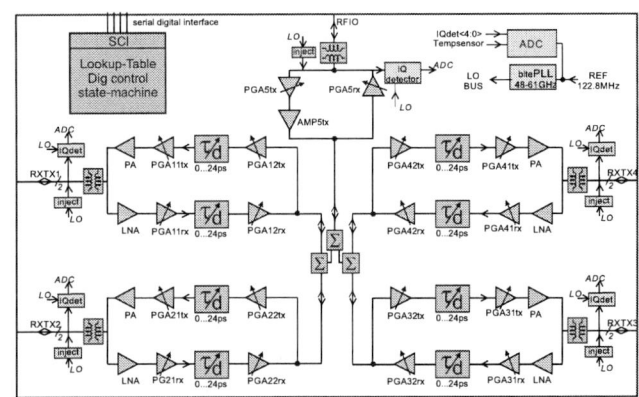

Fig. 2. RFIC Architecture

A serial control interface (SCI) is used to program the chip in order to minimize the digital bus width between the central modem and the set of beamforming chips. A fast control is crucial for scanning and following mobile users, therefore SCI operation supports 125MHz clock frequency. To enable on-the-fly beam switching, an SRAM based lookup table is used, capable of storing 512 Rx and 512 Tx amplitude and delay control bit sets. This solution bounds the communication time required to update the steering settings below 0.25μs, as just the SRAM address pointer has to be updated by using a short SCI command.

Frontend performance sets not only the overall system performance, but also its cost, since any increase in PA output power enables a reduction in the number of antennas and RFIC's, while keeping the target EIRP. LNA and PA in the frontend (Fig. 3) have been co-designed in order to avoid any lossy switch or λ/4 transmission line. The coupling capacitors and transformer have been sized in order to protect the LNA

978-1-7281-1702-7/19 $31.00 © 2019 IEEE   283

transistors and achieve the best compromise among oP1dB, NF and impedance matching. Rx/Tx switching is realized by alternatively enabling/disabling LNA or PA through the biasing inputs in Fig. 3. ESD protection at the RF pads is provided by the grounded transformer middle tap, and packaged samples withstood over 2kV in a HBM test.

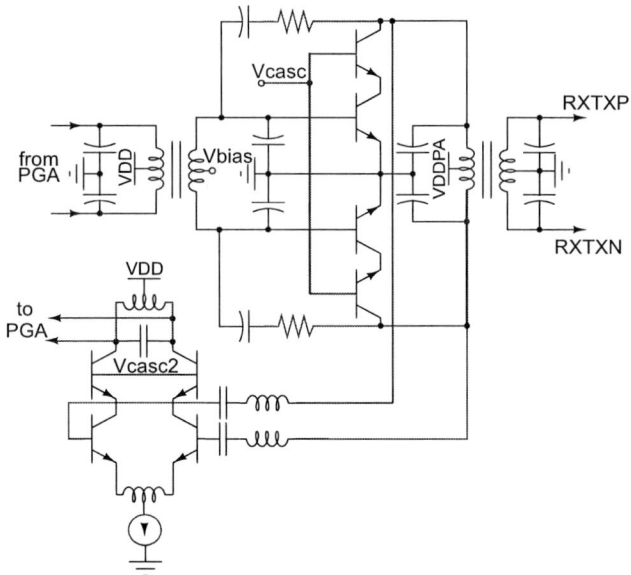

Fig. 3. Frontend Topology using LNA/PA co-design

Amplitude control is implemented using a phase invariant current-steering programmable gain amplifier (PGA) [1] offering more than 30dB gain control range in a single PGA, with a phase error below 8° up to an attenuation level of 12dB. Phase or time delay control is based on a matched lumped tunable delay line, based on transformers instead of inductors in order to minimize area and optimize worst-case insertion loss down to 10dB. The insertion loss variation versus delay setting is smaller than 5dB, and can be compensated by programming the proper attenuation settings in the lookup table. A temperature invariant implementation based on MOS-varactors (measured result in Fig. 6) has been implemented, since it avoids frequent system calibrations. Phase shifting is monotonic over the whole frequency range, and the resolution is well below 6°. The 24ps relative delay line range is chosen to support a phase shifting range over 180° degree at 24.2GHz. The remaining 180° shift is provided by each differential current steering PGA. Such delay range also supports true time delay (TTD) beamforming for a 2 by 2 subarray (azimuth=+/-60°, elevation=+/-15°), with the advantage of a frequency independent beamforming [2].

The beamforming RFIC is realized in 130nm SiGe BiCMOS process, the die size is 4.4 x 4.4mm^2 and the chip photograph is shown in Fig. 4. The package is an automotive qualified eWLB with one RDL layer, its size is 6x6mm^2 and it supports cooling from top as well as from bottom through several ground balls. The Rx single channel measurements are presented in Fig. 5 and 6. Tx measurements are shown in Fig.

7 and Fig. 8. A 64QAM modulation with 400MHz bandwidth has been used for the EVM measurement.

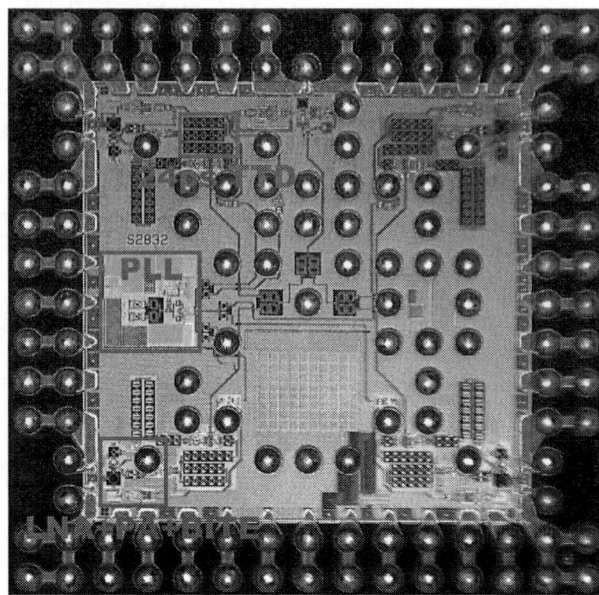

Fig. 4. Packaged Flipped Chip Photograph

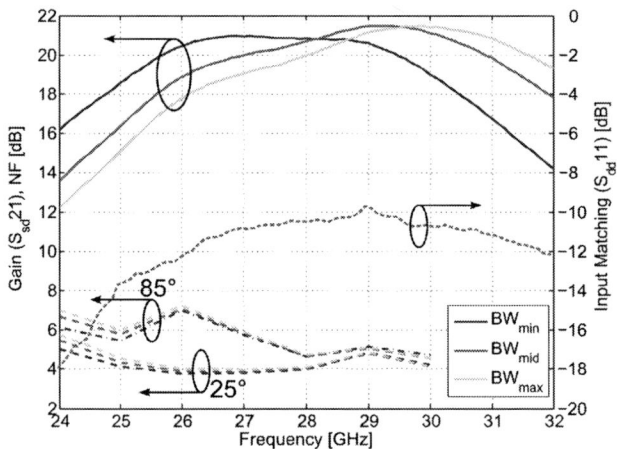

Fig. 5. Rx Gain, Impedance Matching and NF (25°C/85°C)

III. BUILT-IN TEST EQUIPMENT

Since direct RF testing above 20GHz is extremely expensive, a BITE approach adding high-frequency measurement equipment to the RFIC has been implemented. The BITE equipment (represented by the blue boxes in Fig. 2) involves test signal generation, distribution, injection, and amplitude and phase detection at each RFIC port. A 11bits ENOB SAR ADC is integrated in order to acquire the BITE outputs. BITE operation consists of injecting the test signal into one RF port, propagating it throughout the chain, and detecting amplitude and phase with an IQ downconverter located at the output RF port. The downconverter produces I and Q DC votages holding the desired information, by

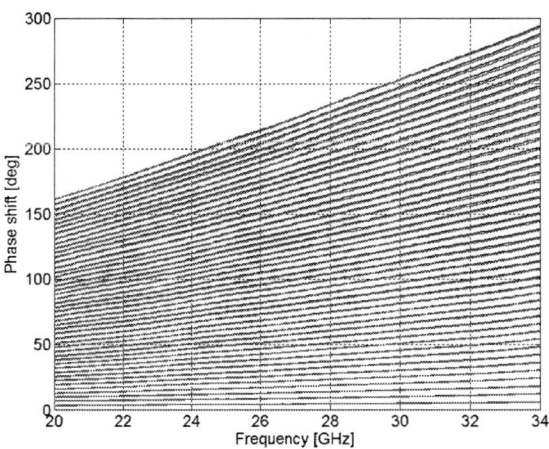

Fig. 6. Rx relative phase shift vs phase control state, 27°C/85°C

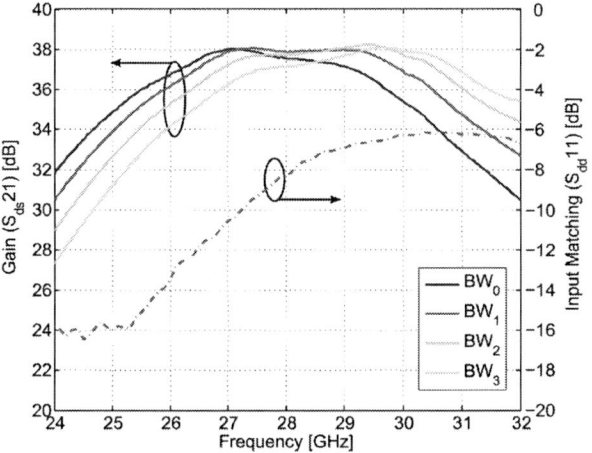

Fig. 7. Tx Gain and Impedance Matching

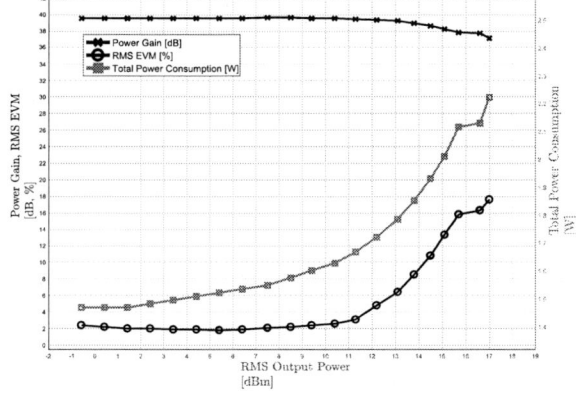

Fig. 8. Tx Gain, RMS EVM and total Pdc vs Output RMS power

mixing the test RF signal with an LO derived from the same source [3]. Such a test equipment basically implements an

on-chip vector network analyzer, digitally controlled through the serial interface. It allows, for example, to characterize the relative attenuation and phase shift introduced by the PGAs or by the delay lines when varying their state (Fig. 9-11), with no need for external mmWave test fixtures.

To avoid isolation and leakage problems between RF channels and BITE equipment, in contrast to [3], the test signal generation and distribution operate at twice the target RF frequency. The generation is carried out by an integrated analog integer-N PLL working from 48 to 62GHz, with margin for process and temperature variations. Since a single LO signal is used for both injection and detection and the delay introduced by the signal paths is small (<10ns), the phase noise of the LO is not critical at all. The complete frequency range is, therefore, robustly covered using a bank of two Colpitts VCOs designed for wide tuning range, while relaxing the phase noise target (ca. -85dBc/Hz @ 1MHz offset). A 122MHz reference clock is used to operate the PLL.

The signal injection and detection circuits are co-designed in all RF ports, in order to avoid any switch and keep the area as small as possible. Both circuits include frequency division by two: the injection circuitry employs a low power Miller divider, while in the IQ downconverters wideband and accurate quadrature LO generation is based on a CML static divider, optimized for best device matching and achieving a quadrature error below 2°.

The complete BITE design targets relative amplitude and phase measurement accuracy better than 1dB and 3° respectively, over a dynamic range of at least 20dB. The BITE accuracy is demonstrated, for a signal frequency of 27.5GHz, by the measured performance in Fig. 9-11, where the on-chip measurement has been compared to the one provided by an external VNA, when sweeping the channel phase control state: the relative quantities (i.e. normalized to state 0) are plotted. As the BITE only requires a reference clock frequency of 122MHz and SCI communication below 125MHz, it can be operated in frontend wafer testing.

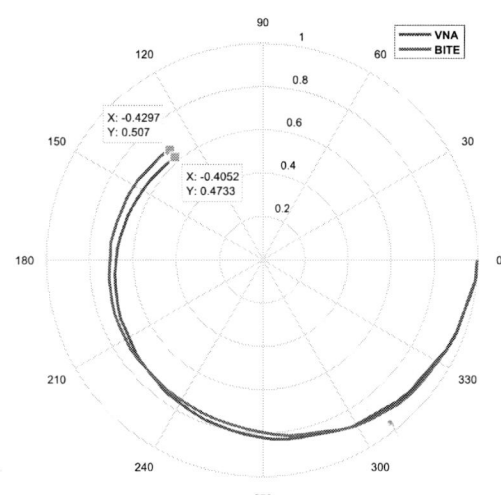

Fig. 9. BITE versus VNA: relative IQ vector

Table 1. Measurement summary and comparison with the state-of-the-art.

	This work	[4]	[5]	[6]	[7]	[8]	
Tech.	SiGe BiCMOS 130nm	SiGe BiCMOS 130nm	CMOS 65nm	SiGe BiCMOS 180nm	CMOS	CMOS 28nm	
Freq. [GHz]	24.25-30.5	27-29	28	28-32	26.5-29.5	26.5-29.5	
Channels	4	16/pol	4	4/RFIC	4	24	
Area [mm^2]	19.4	164	12	11.75	13	29	
Package	eWLB	Laminate	–	Flipped on PCB	WLCSP	Flipped on PCB	
Rx Pdc [W]	1.6	3.3/pol	0.6	0.5	1.3	0.167/4 ch	
Tx Pdc [W]	1.8 @ P1dB	4.6/pol	1.2@11dBm Pout	0.8	1.7 quiescent	0.36/4 ch	
Tx OP1db [dBm]	18	14	15.7	10	14	> 12	
NF [dB]	4	6 (frontend)	4.1	4.6	5.5	4.4-4.7	
BITE	YES	NO	NO	NO	NO	NO	

IV. SUMMARY

Allover results are summarized in Table 1 and compared to previous works. This is the first RFIC covering all the bands of interest from 24.2GHz to 30.5GHz, and provides the highest Tx output power capability and lowest Rx Noise Figure, taking also the package into account. The integrated built-in test equipment, including LO-generation,

signal injection and detection, implements relative phase and amplitude measurement capability with excellent accuracy, a key factor to enable cost effective mass production of 5G mmWave systems.

REFERENCES

[1] P. Scaramuzza et Al., "Class-AB and class-J 22 dBm SiGe HBT PAs for X-band radar systems," in *ESSCIRC 2017 - 43rd IEEE European Solid State Circuits Conference*, Sep. 2017, pp. 187–190.

[2] M. Longbrake, "True time-delay beamsteering for radar," in *2012 IEEE National Aerospace and Electronics Conference (NAECON)*, July 2012, pp. 246–249.

[3] S. Y. Kim et Al., "A 76–84 GHz 16-element phased array receiver with a chip-level built-in-self-test system," in *2012 IEEE Radio Frequency Integrated Circuits Symposium*, June 2012, pp. 127–130.

[4] B. Sadhu et Al., "A 28-GHz 32-Element TRX Phased-Array IC With Concurrent Dual-Polarized Operation and Orthogonal Phase and Gain Control for 5G Communications," *IEEE Journal of Solid-State Circuits*, vol. 52, no. 12, pp. 3373–3391, Dec 2017.

[5] J. Pang et. Al., "A 28GHz CMOS Phased-Array Transceiver Featuring Gain Invariance Based on LO Phase Shifting Architecture with 0.1-Degree Beam-Steering Resolution for 5G New Radio"," in *2018 IEEE Radio Frequency Integrated Circuits Symposium (RFIC)*, June 2018, pp. 56–59.

[6] K. Kibaroglu et. Al., "An ultra low-cost 32-element 28 GHz phased-array transceiver with 41 dBm EIRP and 1.0–1.6 Gbps 16-QAM link at 300 meters," in *2017 IEEE Radio Frequency Integrated Circuits Symposium (RFIC)*, June 2017, pp. 73–76.

[7] "Anokiwave AWMF-0157". [Online]. Available: http://www.anokiwave.com/

[8] J. D. Dunworth et Al., "A 28GHz Bulk-CMOS dual-polarization phased-array transceiver with 24 channels for 5G user and basestation equipment," in *2018 IEEE International Solid - State Circuits Conference - (ISSCC)*, Feb 2018, pp. 70–72.

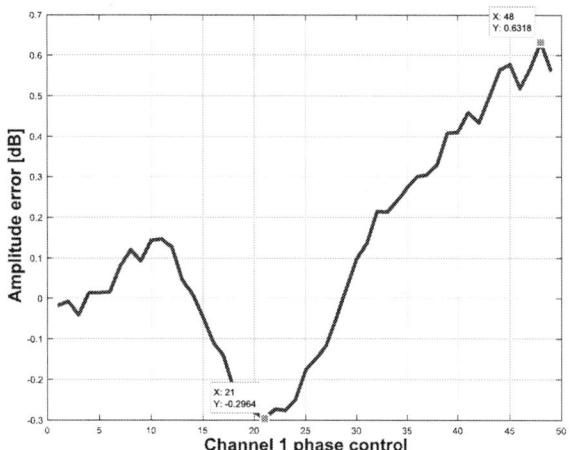

Fig. 10. BITE versus VNA: relative amplitude error

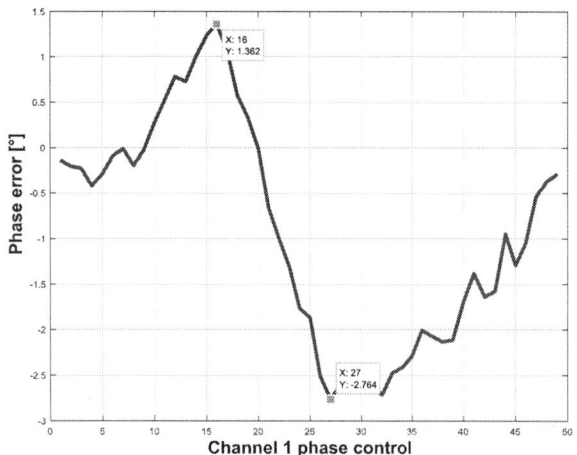

Fig. 11. BITE versus VNA: relative phase error

2019 IEEE Radio Frequency Integrated Circuits Symposium

A Highly Linear 28GHz 16-Element Phased-Array Receiver with Wide Gain Control for 5G NR Application

Youngchang Yoon[*], Kyu Hwan An, Daehyun Kang, Kihyun Kim, Sangho Lee, Jae Sik Jang,
Donggyu Minn, Bohee Suh, Jooseok Lee, Jihoon Kim, Meeran Kim, Jeong Ho Lee, Sung Tae Choi,
Juho Son, and Sung-Gi Yang
Samsung Electronics Co., Ltd., Suwon, South Korea
[*]yc.yoon@samsung.com

Abstract— This paper presents a 28GHz 16-element phased-array receiver in a 28nm bulk RF CMOS process targeting 5G NR application. The receiver implements a high-gain low-NF LNA, a linearity enhanced RF amplifier with post-distortion, and RF/IF step attenuators for a wide gain range and fine gain step control. Thereby, the presented receiver demonstrated a system NF of 3.5dB, a gain control range of 50dB (1dB step), and an EVM of -37.7/-43.1dB with 5G NR 800/100MHz signals while consuming a low power of 0.54W (equivalent to 33.5mW per channel).

Keywords— Fifth generation (5G), 28GHz, Phased-array receiver, CMOS integrated circuits, Millimeter wave integrated circuits.

I. Introduction

Recently, the evolution of millimeter-wave (mm-Wave) technologies in parallel with intensive efforts to deploy fifth-generation (5G) new radio (NR) service has introduced a few noticeable transceiver designs based on SiGe BiCMOS [1] and CMOS [2-4] for 28GHz band. In these works, not just the process technologies but also architectures, interfaces, and the number of elements in a single integrated circuit (IC) are widely varied. This paper presents a new and successful example of a 28GHz 16-element phased-array receiver (RX) implemented in a 28nm bulk RF CMOS process with superior RX performance and a low power consumption compared to the recent state-of-the-art RXs.

II. Phased Array Receiver Architecture

Fig. 1 shows the block diagram of the 28GHz 16-element phased-array transceiver. Although this transceiver is fully implemented, the main focus of this paper is the RX path, which achieves a good system noise figure (NF) and good linearity while providing wide gain range and fine resolution. The RX path consists of low noise amplifier (LNA), bi-directional passive phase shifter (PS), and 4-way lumped-element Wilkinson combiners, radio frequency (RF)/ intermediate frequency (IF) amplifiers, an RF down mixer, and RF/IF step attenuators. The path is designed to receive an RF input signal from 26.5 to 29.5GHz and convert it to an IF signal at around 11GHz with up-to 800MHz signal bandwidth. A local oscillator (LO) input is externally provided and multiplied by 3 for down conversion. The PS is a cascade of filter type phase transition providing 22.5-degree phase resolution with a 4-bit control. The receiver demonstrated excellent performances such as a good system NF, a good EVM, and a total gain range of 50dB with 1dB gain step compositely implemented by the coarse step

LNA (i.e., high and low-gain modes) and the fine step RF/IF attenuators, 3dB and 1dB step, respectively.

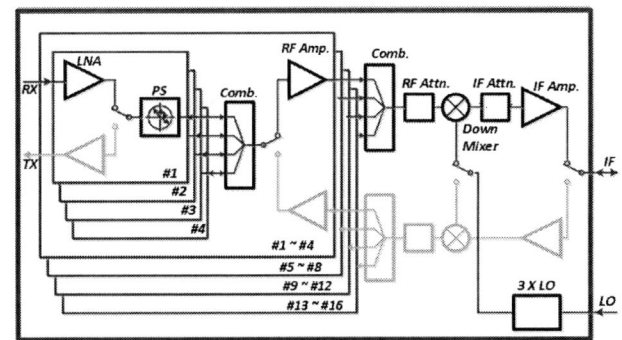

Fig. 1. The block diagram of the proposed 28GHz 16-element phased-array transceiver. The RX path is highlighted.

The superior performance of the receiver is achieved by adopting following design approaches. First, the low system NF level is achieved by a low-NF LNA and separate TX/RX ports, thereby avoiding duplexing loss which has a direct impact on the system NF. A four-stage cascaded structure is implemented for the LNA design to have a high gain and a robust full-chain NF even at high temperature. A common-source topology is used for the first and fourth stages, which makes a multi-stage amplifier to have both a good NF and linearity. The LNA supports a low-gain mode, which is implemented by turning off a common-gate transistor at second stage and using an attenuator between the third stage and fourth stage in order to increase the dynamic range of the RX system. This method does not degrade the NF at the high-gain mode because no additional devices such as transistors and resistors are utilized for implementing gain modes. Second, the newly proposed linearity improvement technique is exploited in this RX design. If an amplifier has improved linearity performance, it can meet the requirement with less power consumption. Therefore, linearity improvement techniques are essential to save overall power consumption. Section III describes the proposed linearity improvement technique in detail. Third, the RX design utilizes the newly proposed step attenuator design for gain control. The step attenuator provides low insertion loss with high linearity while occupying a compact layout area. Thanks to the step attenuators' gain control, amplifiers on the RX path have been optimized without a burden of gain control inside of

978-1-7281-1702-7/19 $31.00 © 2019 IEEE

amplifiers themselves. Section IV presents the proposed step attenuator design in detail.

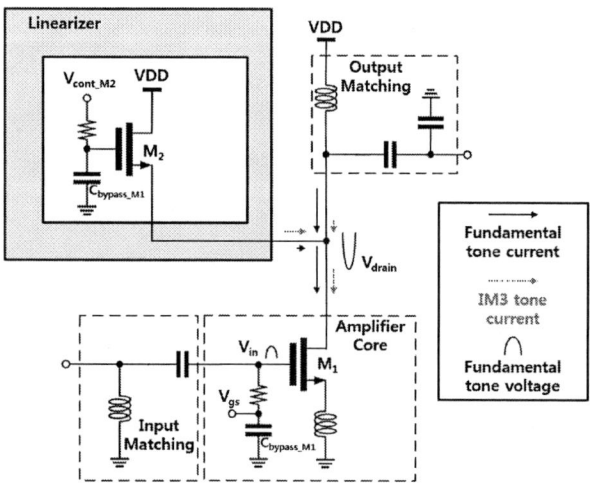

Fig. 2. Schematic of the proposed linearity improvement technique.

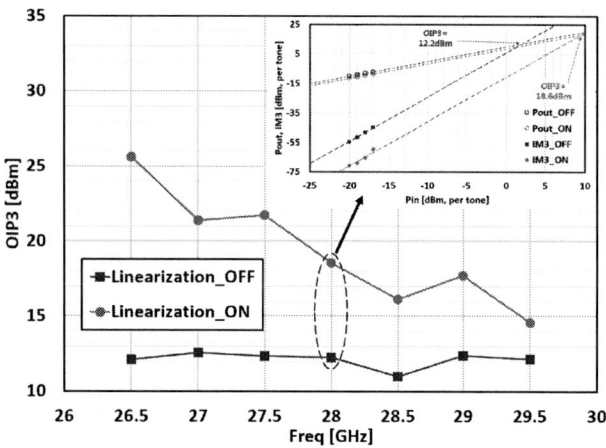

Fig. 3. Measured OIP3 over frequency at -10dBm one-tone output power level. Measured fundamental and IM3 curves at 28 GHz.

III. AMPLIFIER DESIGN WITH THE PROPOSED LINEARITY IMPROVEMENT TECHNIQUE

There have been continuous efforts to improve linearity for the RX path in 2G/3G/4G communication systems, which is still relevant to the 5G NR system. Previously reported linearity improvement techniques can be categorized into three groups listed as derivative superposition [5], modified derivative superposition [6], and post-distortion [7], [8]. Among them, a post-distortion method is more attractive for mm-Wave 5G NR application thanks to less side effects (e.g., lower input parasitic capacitance, less degraded gain and NF) compared to the other two approaches. We present a newly proposed post-distortion technique here.

The proposed post-distortion technique is presented in Fig. 2, in which the fundamental tone voltage/current and the 3rd order inter-modulation (IM3) current are indicated. The

amplifier's core transistor (M_1) generates the fundamental as well as the IM3 tone currents due to its intrinsic non-linear characteristics. Meanwhile, post-distortion transistor (M_2) is sized and biased to have equivalent IM3 tone current to that of M_1 while allowing much less fundamental tone current. By implementing this topology, most of the fundamental current flows to the output matching network direction, while most of the IM3 current flows into the M_2 transistor path. Therefore, linearity improvement can be achieved with negligible impact on the fundamental signal amplification. This property arises from the following factors. The M_2 has larger gate-source voltage swing (V_{gs}) compared to that of the M_1, and the IM3 tone of a MOS transistor increases more sharply (3rd order) with respect to the applied V_{gs} compared to that of the fundamental tone (1st order).

While this approach can be categorized as a post-distortion technique, the following features make it distinguished from the previous reported techniques in [7], [8]. First, the proposed technique does not require additional DC power consumption. Both drain and source nodes of M_2 transistor are connected to the same DC level (V_{DD}), so there is no DC power consumption due to the proposed linearity improvement circuitry. Second, the fundamental signal loss penalty (i.e., gain loss, NF degradation) due to on-resistance of M_2 can be minimized if the output-matching network is designed for both output return loss and optimum linearity matching by utilizing the on-resistance of M_2.

Fig. 3 shows the measured output third-order intercept point (OIP3) over frequency both with and without the post-distortion technique. As described in Fig. 2, the post distortion circuitry can be enabled/disabled by controlling V_{cont_M2}. When V_{cont_M2} is set to on-state, the enabled post distortion technique improves OIP3 by more than 13dB at 26.5GHz while consuming the same DC power. There is about 1.5dB gain drop when the post distortion is enabled. The inserted graph exemplarily shows the measured fundamental and IM3 tone powers at 28GHz. The inserted graph exemplarily shows the measured fundamental and IM3 tone powers at 28GHz.

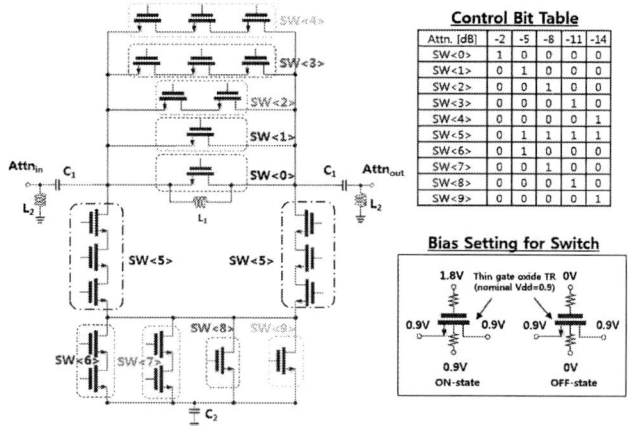

Fig. 4. Schematic of the proposed 28GHz step attenuator, control bit table, and bias setting for ON-/OFF- state switch.

978-1-7281-1702-7/19 $31.00 © 2019 IEEE   288

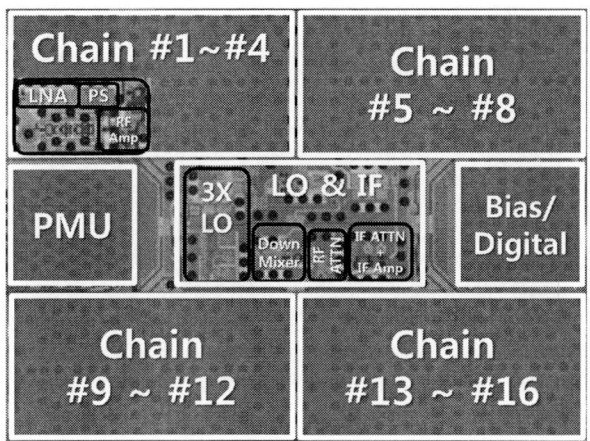

Fig. 5. Die photograph of the 28GHz 16-element phased-array transceiver. Sub blocks for receiver path are highlighted.

IV. HIGHLY LINEAR STEP ATTENUATOR

A wide gain control range with an accurate gain step is a desirable feature for the RX path. To this end, rather than conventional gain control schemes such as load switching, cascade switching, or current steering in variable gain amplifiers (VGA), an RF step attenuator in a bridged-T type is proposed in Fig. 4, which also includes the corresponding control signal table. The attenuator is optimized in consideration of low minimum insertion loss, the required attenuation range, low input/output return loss, and high linearity.

First, there is a trade-off between the minimum insertion loss (-2dB) and the attenuation range (12dB). To lower the minimum insertion loss, the width of SW<0> needs to be increased. However, the larger width of SW<0> introduces larger parasitic capacitance between the input and output nodes in its OFF-state, thereby resulting in the decreased attenuation range. The L_1 inductor is utilized to resonate out the parasitic capacitance alleviating the trade-off relationship. Second, this attenuator is placed in-between the 50-Ohm interfaces requiring the impedance matching. The L_2 inductors are used to cancel out parasitic capacitance at the input/output nodes. Last, C_1, C_2, and multiple stacked switch topology are used to improve linearity. Both ON-/OFF-state switches can degrade the attenuator's linearity. To improve the linearity of the OFF-state switch, fully reverse-biased setting method is adopted as shown in Fig. 4. C_1 and C_2 operate as a DC block to source/drain nodes of all switches. All the source/drain nodes are set to 0.9 V, and a switch's gate bias is set to 1.8V/ 0V for ON-/OFF- state, respectively. On the other hand, the linearity of the ON-state switch can be improved by utilizing a multi-stacked switch method to distribute the large voltage swing to the multiple switches.

The measured RF attenuator has the minimum insertion loss of 2dB and the input third-order intercept point (IIP3) of over 22dBm, a better or comparable insertion loss/linearity with a much compact layout size when compared to the recently published designs [5], [6]. An IF 1dB step attenuator is also

designed following the same approach and modified only for the frequency ranges.

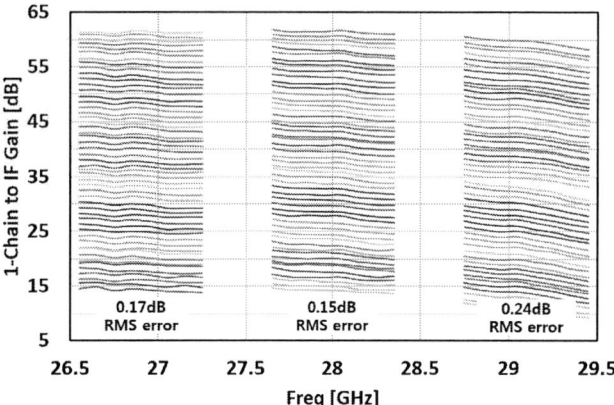

Fig. 6. Frequency response of the proposed 16-element phased-array receiver with 1-dB gain step.

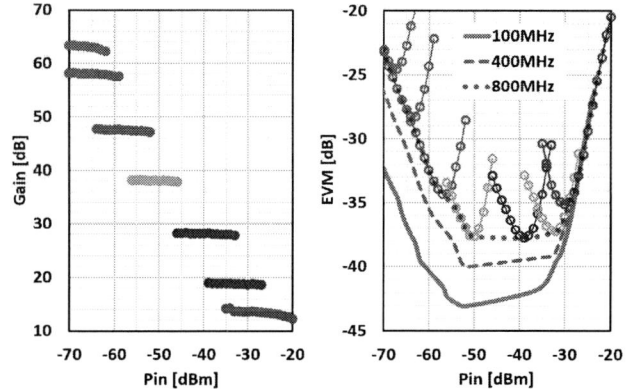

Fig. 7. Gain/EVM measured data over input power at 28 GHz for 16-element phased-array receiver. EVM results with 800/400/100MHz BW signals.

V. MEASUREMENT RESULTS OF 16-ELEMENT PHASED-ARRAY RECEIVER

Fig. 5 is the die photograph of the implemented 28GHz 16-element phased-array transceiver, and the RX path is indicated with black boxes. It is fabricated in a 28nm bulk RF CMOS process. The chip size is 4.7×6.4mm^2.

The proposed receiver is packaged and measured in a conduction mode test and all the input/output losses up to the reference points at the package level are de-embedded. The one carrier component (CC) of the NR signal has 120kHz subcarrier spacing, 64QAM, 10dB PAPR, and 100MHz BW, and the same four/eight CCs are aggregated for the 400/800MHz BW test.

Fig. 6 shows the gain step response of the 16-element RX path across the frequency range from 26.5 to 29.5GHz having three 800MHz BW signals. Each signal band's gain step is accurately controlled with the maximum RMS error of 0.24dB in its 1dB step implementation.

Table 1 . Comparison table with recent state-of-the-art receivers.

Parameter	Unit	This Work	[4] ISSCC 2018 Qualcomm	[3] JSSC 2018 LG	[2] JSSC 2018 Carnegie Mellon Univ.	[1] ISSCC 2017 IBM/Ericsson
Technology	N/A	28nm RF CMOS	28nm RF CMOS	28nm RF CMOS	65nm CMOS	0.13μm SiGe BiCMOS
Operating RF Freq.	GHz	26.5 - 29.5	26.5 - 29.5	25.8 - 28	25 - 30	27 - 29
Polarization	ea	1	2	1	1	2
Channel	ea	16	24	8	8	32
Interface	N/A	11GHz IF	6.5GHz IF	Analog IQ BB	Analog IQ BB	3GHz IF
Measured Signal	N/A	5G NR OFDM 64QAM, 10dB PAPR 800/400/100MHz BW	OFDM 64QAM, - 400MHz BW	LTE OFDM 64QAM, 7.5dB PAPR 20MHz BW	-	-
# of PS bits (PS resolution)	bit	4 (22.5deg.)	3 (45deg.)	3 (45deg.)	-	6 (5deg.)
Max RX Gain	dB	60 (16ch.)	34 (4ch.)	69 (8ch.)	34	34 (16ch.)
RX System NF	dB	3.5 - 4.3	4.4 - 4.7	6.7	7.3	6
RX EVM (Signal BW)	dB	-37.7/-40/-43.1[2] (800/400/100MHz)	-36.5[1] (400MHz)	-33.3 (20MHz)	-	-
Total RX power	W	0.539 (16ch.)	0.167 (4ch.)	0.4 (8ch.)	0.34 (8ch.)	3.3 (16ch.)
RX power per channel	mW	33.7[2]	41.8[1]	50	42.5	206.3

[1]4-channel sub-array measurement result with 400MHz BW signal (20~23 GHz external LO)

[2]16-channel measurement result with 800/400/100 MHz BW including 3X LO path EVM contribution (5.6 GHz external LO)

Gain/EVM performances over the input power range at 28GHz with different gain mode settings are presented in Fig. 7. Solid symbolic color lines represent 800MHz's EVM results depending on each gain mode, and gray lines connect the best EVM points at a given input power level with different signal bandwidths (800/400/100MHz), which can be acquired by automatic gain control. The best EVM levels for 800/400/100MHz are -37.7/-40/-43.1dB, respectively. When the signal bandwidth decreases (e.g., from 800 to 100MHz), the EVM level improves because the input signal has better SNR. This phenomenon becomes more apparent when the input power is low so that the EVM performance is dominantly decided by noise contribution (i.e., Pin < -60dBm). At these low input power ranges, 100MHz's EVM is about 9dB lower than 800MHz's, which is matched to the theoretical expectation.

Table 1 shows the summary of this work and the comparison with other state-of-the-art receivers in 28GHz band. This work shows superior performance with a 3.5~4.3dB system NF at room temperature, a 50dB gain range with 1dB step, and -37.7dB EVM with a 5G NR 800MHz signal while consuming total power consumption of 0.54W, which is corresponding to 33.7mW per channel. This work represents the best EVM performance with the lowest power consumption. When a phased-array system is scaled to larger one, a low-NF, a good linearity, a high gain and a fine step control with a low power consumption are useful features to achieve a low sensitivity level and maintain the signal quality

VI. CONCLUSION

This paper successfully demonstrates a 28GHz CMOS receiver for the 5G NR application with the state-of-the-art performance and a very low power consumption (i.e., -37.7dB EVM with a 5G NR 800MHz signal and the power consumption of 33.7mW per channel). This was achieved by adopting the low-NF LNA design, the proposed linearity

improvement techniques, and the highly linear step attenuator design.

ACKNOWLEDGMENT

The authors are grateful to all members of RF lab. in Network Business R&D team, Samsung Electronics Co., Ltd.

REFERENCES

[1] B. Sadhu, et al., "A 28GHz 32-Element Phased-Array Transceiver IC with Concurrent Dual Polarized Beams and 1.4 Degree Beam-Steering Resolution for 5G Communication," ISSCC, pp. 128-129, Feb. 2017.

[2] S. Mondal, et al, "A 25-30 GHz Fully-Connected Hybrid Beamforming Receiver for MIMO Communication," *IEEE J. Solid-State Circuits*, vol. 53, no. 5, pp. 1275-1287, May 2018.

[3] H. T. Kim, et al., "A 28GHz CMOS direct conversion transceiver with packaged 2x4 antenna arrays for 5G cellular system," *IEEE J. Solid-State Circuits*, vol. 53, no. 5, pp. 1245-1259, May 2018.

[4] J. D. Dunworth, et al., "A 28GHz Bulk-CMOS Dual-Polarization Phased-Array Transceiver with 24 Channels for 5G User and Basestation Equipment," ISSCC, pp. 70-71, Feb. 2018.

[5] T. W. Kim, B. Kim, and K. Lee, "Highly linear receiver front-end adopting MOSFET transconductance linearization by multiple gated transistors," *IEEE J. Solid-State Circuits*, vol. 39, no. 1, pp. 223–229, Jan. 2004.

[6] V. Aparin and L. E. Larson, "Modified derivative superposition method for linearizing FET low-noise amplifiers," *IEEE Trans. Microw. Theory Tech.*, vol. 53, no. 2, pp. 571–581, Feb. 2005.

[7] N. Kim, V. Aparin, K. Barnett, and C. Persico, "A cellular-band CDMA CMOS LNA linearized using active post-distortion," *IEEE J. Solid-State Circuits*, vol. 41, no. 7, pp. 1530–1534, Jul. 2006.

[8] B. Guo and X. Li, "A 1.6-9.7 GHz CMOS LNA Linearized by Post Distortion Technique," *IEEE Microw. Wireless Comp. Lett.*, vol. 23, no. 11, pp. 608–610, Nov. 2013.

[9] J. Bae, J. Lee, and C. Nguyen, "A 10–67-GHz CMOS dual-function switching attenuator with improved flatness and large attenuation range," *IEEE Trans. Microw. Theory Tech.*, vol. 61, no. 12, pp. 4118–4129, Dec. 2013.

[10] J. Bae and C. Nguyen, "A 44 GHz CMOS RFIC Dual-Function Attenuator with Band-Pass-Filter Response," *IEEE Microw. Wireless Comp. Lett.*, vol. 25, no. 4, pp. 241–243, Apr. 2015.

A Quadrature Class-G Complex-Domain Doherty Digital Power Amplifier

Shih-Chang Hung[1], Si-Wook Yoo[2], and Sang-Min Yoo[3]

Department of Electrical and Computer Engineering, Michigan State University, USA

[1] hungshih@egr.msu.edu, [2] yoosiwoo@egr.msu.edu, [3] syoo@egr.msu.edu

Abstract—An efficient digital quadrature power amplifier is presented. It shows a good system efficiency (SE) at back-off, demonstrating four efficiency peaks with the combination of a dual-supply Class-G and complex-domain Doherty (CDD) in the IQ plane. The proposed digital quadrature transmitter in 65-nm CMOS demonstrates 27.8-dBm peak output power (P_{out}) with a peak SE of 32.1%. For an 802.11ax 40-MHz (20-MHz) 1024-QAM OFDM signal with 13.1-dB (12.4-dB) peak-to-average power ratio (PAPR), it demonstrates an error vector magnitude (EVM) of -42.0 dB (-43.3 dB) at an average P_{out} of 14.7 dBm (15.4 dBm). The average SE measured with a 20-MHz single-carrier 1024-QAM signal with 6.8-dB PAPR at 21-dBm P_{out} is 18.4%.

Keywords— Class-G power amplifier (PA), digital transmitter, quadrature transmitter, digital PA (DPA), RFDAC, switched-capacitor PA (SCPA), Doherty, voltage-mode Doherty (VMD).

I. INTRODUCTION

Modern wireless communication systems require RF transceivers with very low power consumption, high linearity, and wide bandwidth to support the demand for a very high data throughput for mobile subscribers. For example, communication standards such as 802.11ax require high linearity of <−40-dB error vector magnitude (EVM) and a bandwidth of up to 160 MHz. High linearity and wide bandwidth enable a very fast communication speed, while increased energy efficiency gives an extended life for battery-powered devices. For improved power efficiency and output power (P_{out}), many power amplifier (PA) architectures such as Doherty [1], Class-G [2][3], power combining [3][4], polar [5], and digital PA (DPA) [1]–[7] have been investigated. Doherty architecture demonstrates an improved efficiency with an additional efficiency peak through load modulation, while Class-G provides multiple efficiency peaks at power back-off (PBO) with multiple supply voltages. For DPAs, an enhanced Class-G technique demonstrated an improved efficiency and linearity associated with the distributed transition of supply voltages over multiple unit cells [2].

A combination of these techniques could lead to better efficiency. For example, a polar Class-G DPA with Doherty configuration demonstrates a high average efficiency with multiple efficiency peaks at the PBO region. A recent work based on voltage-mode Doherty (VMD) [3] demonstrated an efficient transformer (XFMR) power combining by utilizing two XFMR primary windings concurrently even at the deep PBO region.

A polar transmitter has a good output power and energy efficiency, but it shows a limited bandwidth and requires a complex system implementation such as a coordinate rotation

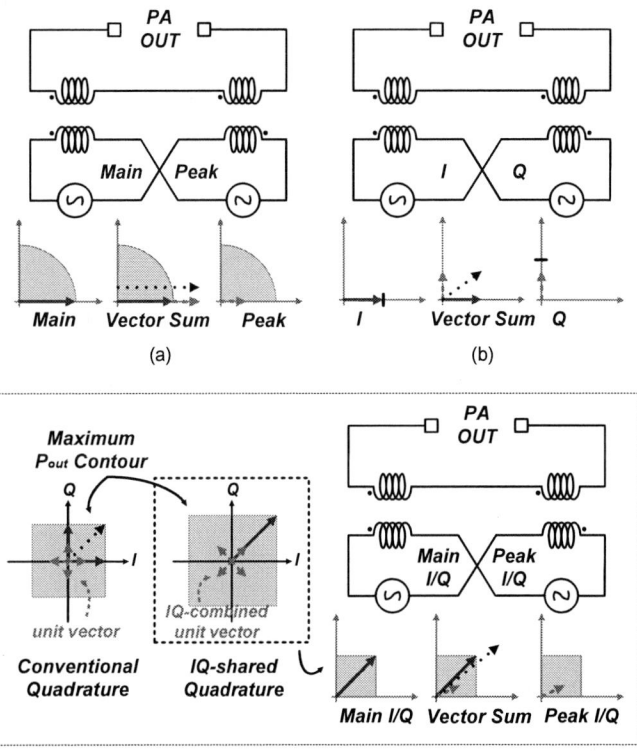

Fig. 1. PA architectures with XFMR power combining: (a) a polar PA in Doherty configuration, (b) a quadrature PA with dedicated I/Q sub-PAs, and (c) a quadrature PA in CDD configuration.

digital computer (CORDIC) and a wideband phase modulator [6]. On the other hand, a quadrature transmitter has a great advantage in its wide bandwidth and simple system architecture. Fig. 1 shows simplified examples of various power combining configurations. The solid and dashed arrows indicate the output vectors of sub-PAs, and the dotted arrow shows the vector sum at the combined PA output. A polar PA in Doherty configuration (Fig. 1 (a)) uses output vectors in the same phase in both main and peak PAs for the output impedance modulation and an extra efficiency peak at the PBO region. However, in this configuration, the phase needs to be modulated externally with a phase modulator. A quadrature PA configuration (Fig. 1 (b)) combines two orthogonal in-phase (I) and quadrature (Q) vectors using XFMR, but its P_{out} and efficiency are degraded because each sub-PA is dedicated to the orthogonal I and Q vectors.

978-1-7281-1702-7/19 $31.00 © 2019 IEEE

2019 IEEE Radio Frequency Integrated Circuits Symposium

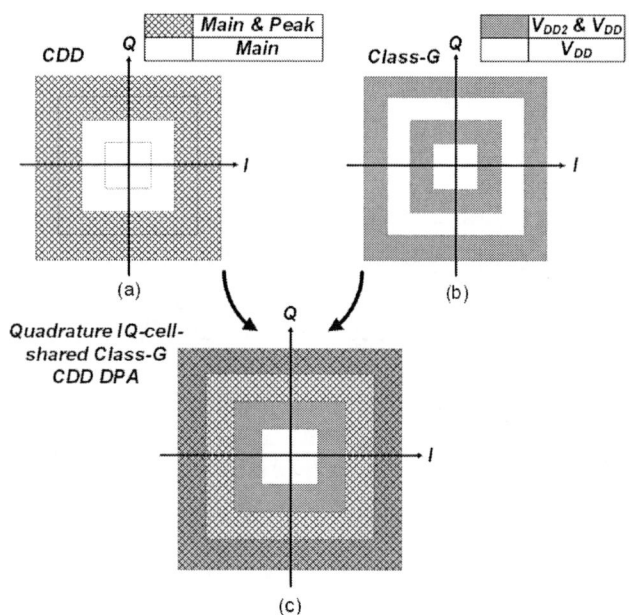

Fig. 2. Operation region vs. I/Q codes for (a) CDD, (b) Class-G, and (c) the quadrature IQ-cell-shared Class-G CDD DPA.

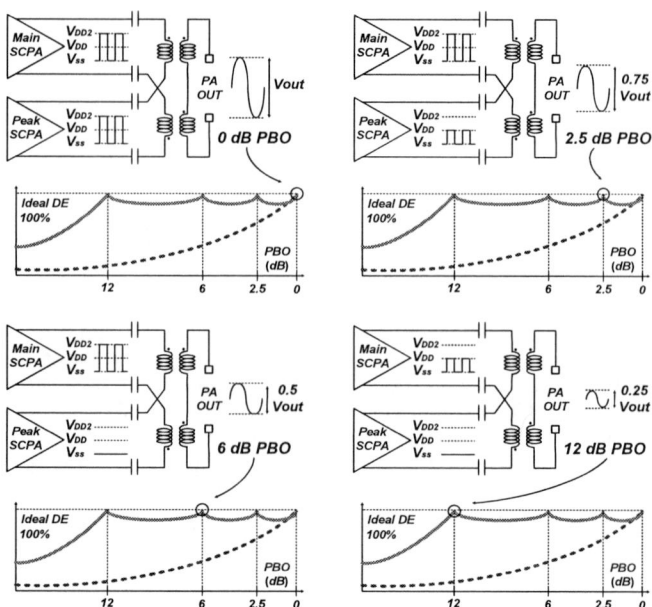

Fig. 3. Ideal DE curve for the Class-G CDD DPA for the output vector with 45°, 135°, 225°, or 315° angle.

II. DESIGN OF THE PROPOSED QUADRATURE IQ-CELL- SHARED CLASS-G COMPLEX-DOMAIN DOHERTY DPA

A. Theory of Operation

A quadrature IQ-cell-shared DPA [7] can generate an output vector in the complex domain as in the conventional quadrature configuration without a phase modulator. The maximum P_{out} and efficiency are significantly improved in comparison to the conventional quadrature configuration (Fig. 1 (b)) because the output vector is not represented by the orthogonal I and Q vectors generated by the dedicated sub-PAs as shown in Fig. 1(c). Proposed complex-domain Doherty (CDD) improves the efficiency with additional efficiency peaks at the PBO region because the peak PA operates only when the main PA generates a maximum P_{out}. The CDD provides a load modulation as in the conventional Doherty configuration for vector components with the same angle, while it expresses the complete complex domain as in the quadrature configuration for orthogonal vector components. The proposed quadrature DPA based on IQ-cell-shared Class-G switched-capacitor (SC) PA (SCPA) operates in the CDD configuration as described in Fig. 1(c). Each main and peak quadrature Class-G IQ-cell-shared SCPA delivers the maximum P_{out} at 45°/135°/225°/315° with IQ-combined unit vectors [7] as shown in Fig. 1(c). The output vectors with different angles in the main and peak Class-G PAs are coupled with XFMR in VMD configuration to achieve multiple efficiency peaks in the complex domain.

Fig. 2 illustrates the operation region in the complex domain of the proposed DPA with CDD and dual-supply Class-G architecture. In Fig. 2 (a), the area marked with a diagonal grid depicts the operation region where both main and peak PAs are turned on, and the white area shows the low-

power region where only the main PA operates. The Class-G operation is also implemented for both main and peak PAs, which adds two efficiency peaks in addition to the efficiency peak for Doherty operation. In Fig. 2 (b), the grey area indicates the region of the Class-G operation with both V_{DD} and VDD2 (=2VDD) supply voltages, and the white area represents the area with only V_{DD} supply voltage. The combined operation with CDD and dual-supply Class-G in the complex domain is illustrated in Fig. 2 (c). The detailed operation of the proposed DPA in different PBO regions is illustrated in Fig. 3. An ideal drain efficiency (DE) curve for the output vector with the 45°, 135°, 225°, or 315° angle demonstrates multiple efficiency peaks at 0-dB, 2.5-dB, 6-dB, and 12-dB PBO as shown in Fig. 3. The ideal DE shows an efficiency peak associated with the Doherty at 6-dB PBO and two additional efficiency peaks associated with the Class-G at 2.5-dB and 12-dB PBO. The efficiency curve also shows a smooth transition between different supply voltage domains and between main/peak PAs due to the continuous changes in supply voltages and impedance.

B. Overall Architecture

The proposed 12b quadrature DPA, as shown in Fig. 4, consists of 11b main and peak DPAs that are integrated with a VMD power-combining XFMR for impedance modulation in the complex domain. It operates as a standalone transmitter and generates an RF signal from digital I/Q data without a CORDIC or a phase modulator. As described in Fig. 4, a four-phase signal generator is used to generate the four IQ-combined unit vectors in the quadrature transmitter. Each main and peak PA is a quadrature IQ-cell-shared Class-G SCPA. The most significant bits of the I/Q data are assigned to select the main and peak PAs, and the next 5 and 6 bits are

978-1-7281-1702-7/19 $31.00 © 2019 IEEE

2019 IEEE Radio Frequency Integrated Circuits Symposium

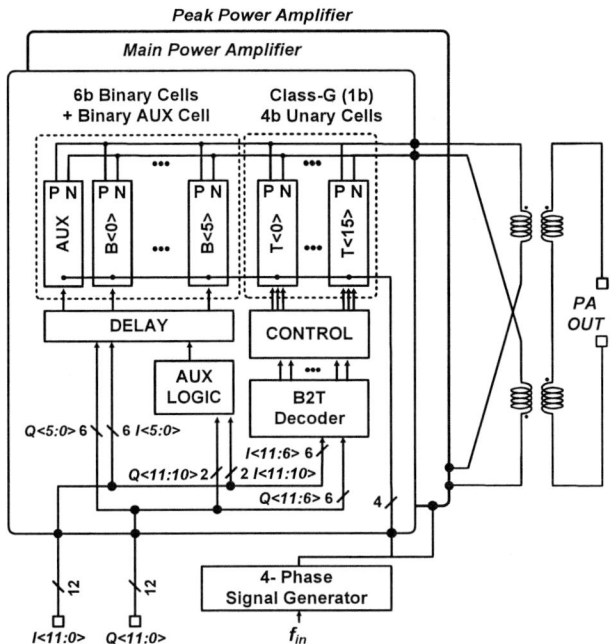

Fig. 4. Block diagram of the quadrature IQ-cell-shared Class-G CDD DPA.

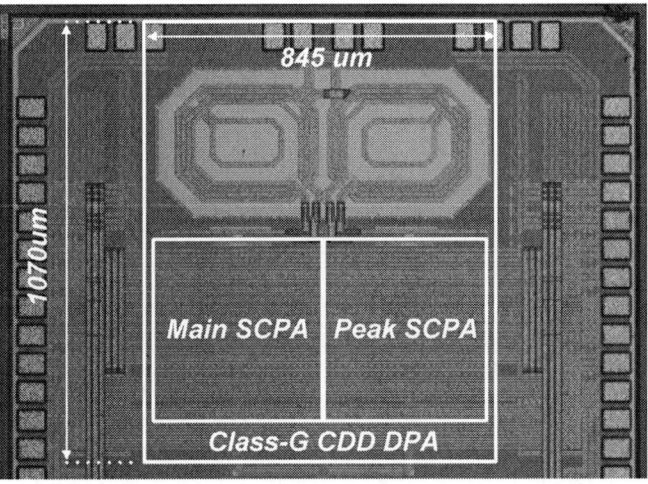

Fig. 5. Die photograph.

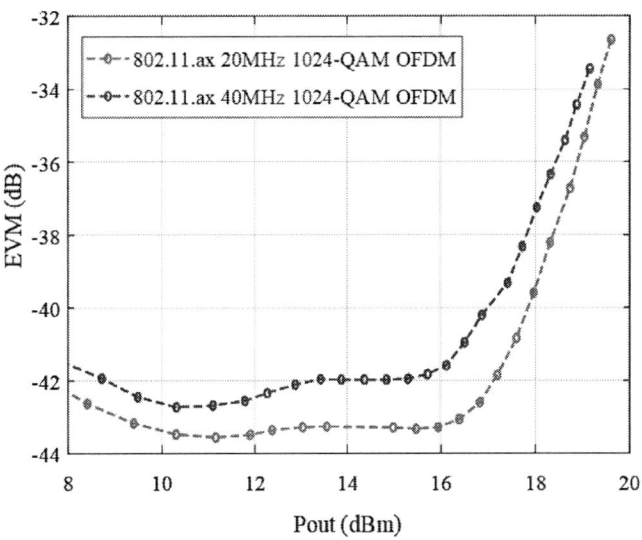

Fig. 6. EVM vs. average P_{out} for an 802.11ax 40-MHz (20-MHz) 1024-QAM OFDM signal with 13.1-dB (12.4-dB) PAPR.

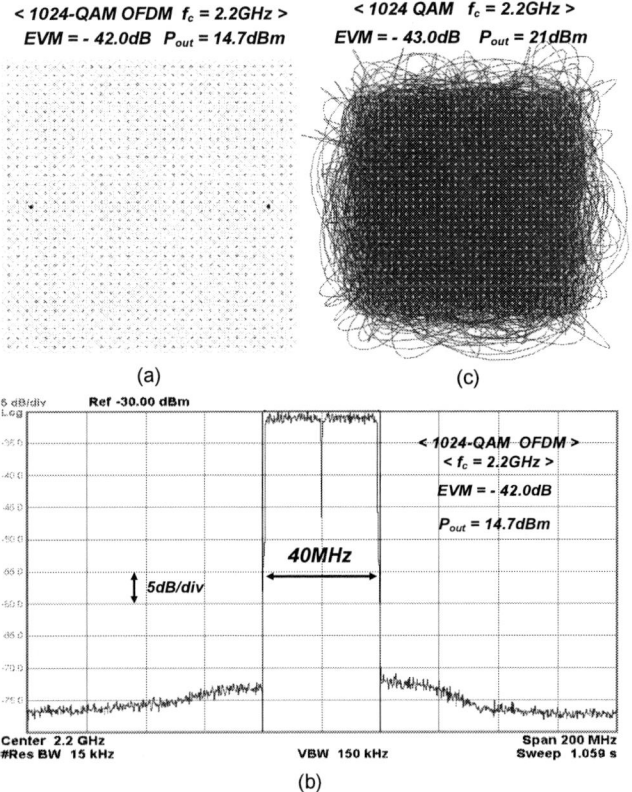

Fig. 7. (a) Constellation and (b) spectrum at average P_{out} of 14.7 dBm (after DPD) for an 802.11ax 40-MHz 1024-QAM OFDM signal with 13.1-dB PAPR. (c) Constellation with a 20-MHz single-carrier 1024-QAM signal at 6.8-dB PBO (after DPD).

allocated to the unary and binary cells of each PA, respectively.

For the linearity in the digital transmitters or DPAs, it is important to match the unary and binary cells. For a better matching, two sets of binary cells are allocated to main and peak PAs even though only one set is required to represent the complete I/Q domain. In addition, multiple delay cells are utilized to compensate for the unwanted delay mismatch between the main and peak PAs and between the unary and binary cells.

III. MEASUREMENT RESULTS

The prototype of the quadrature IQ-cell-shared Class-G CDD DPA, fabricated in a 65-nm RF CMOS process, occupies a chip area of 1.07×0.845 mm² including main and peak SCPAs, a four-phase signal generator, a LVDS receiver,

a power combining transformer, and pads as shown in the chip micrograph (Fig. 5). The prototype chip is mounted, and wire bonded on PCB.

Fig. 6 shows the measured EVM of the proposed quadrature DPA with an 802.11ax 40-MHz (20-MHz) 1024-

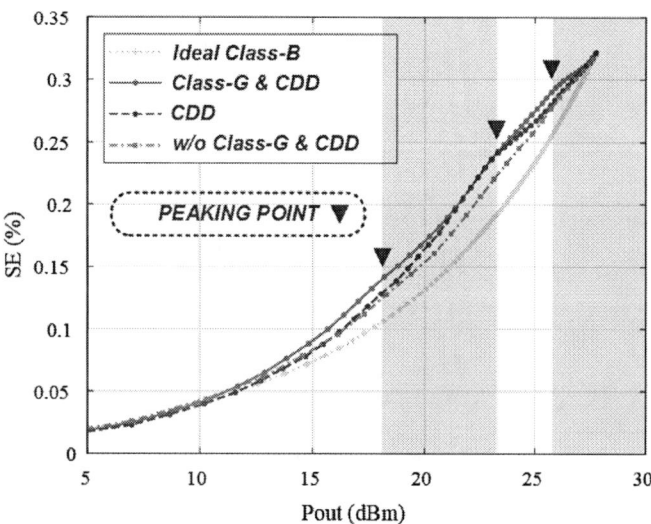

Fig. 8. SE vs. P_{out} for continuous-wave (CW) signal.

Table 1. Performance comparison with the state-of-the-art

	ISSCC 2017 [3]	JSSC 2016 [4]	RFIC 2018 [7]	**This work**
Architecture	Polar Class-G VMD	Quadrature Class-G	Quadrature Class-G	Quadrature Class-G CDD
Process	45 nm	65 nm	65 nm	65 nm
Supply	2.4/1.2V	2.4/1.2V	2.5/1.2V	2.55/1.25V
Resolution	9b (5+4)	7b (6+1)	11b (6+5)	12b (1+5+6)
frequency	3.5GHz	2GHz	2.2GHz	2.2GHz
Peak power	25.3 dBm	20.5 dBm	30.1 dBm	27.8 dBm
Peak PAE	30.4%	20%	37.0%	32.1%
Modulation	10-MHz 32-Carrier 1024 QAM	LTE 10-MHz 64 QAM	20-MHz Single-Carrier 256 QAM	20-MHz Single-Carrier 1024 QAM
Avg. power	14.8 dBm	14.5 dBm	22.5 dBm	21 dBm
PAPR	10.5 dB	6 dB**	7.6 dB	6.8 dB
System efficiency	18.0%*	12.2%	18.3%	18.4%
EVM (DPD)	-40.3 dB	-28.8 dB	-40.3 dB	-43.0 dB

* Power consumption of PA only excluding CORDIC and phase modulator
** Estimated from the peak/average power.

QAM OFDM signal with a 13.1-dB (12.4-dB) PAPR at 2.2GHz. The measured EVM is -42.0 dB (-43.3 dB) after digital predistortion (DPD) at 14.7-dBm (15.4-dBm) average P_{out}. The EVM floor is -42.7 dB (-43.5 dB) at an average P_{out} of 10.3 dBm (11.1 dBm), and the PA achieves EVM of <–40 dB for a more than 10-dB (12-dB) dynamic range after DPD. Fig. 7(a) and (b) depict the measured constellation and spectrum for a 40-MHz 802.11ax signal at 14.7-dBm average P_{out}. The system efficiency (SE) measured with a 20-MHz single-carrier 1024-QAM signal with 6.8-dB PAPR at 21-dBm P_{out} is 18.4%. The measured constellation is depicted in Fig. 7(c) and demonstrates −43.0-dB EVM.

Fig. 8 shows the measured SE vs. P_{out} at 2.2GHz with a continuous-wave (CW) signal. The SE of the proposed IQ-cell-shared Class-G CDD DPA clearly demonstrates multiple efficiency peaks associated with Doherty and Class-G in comparison to the ideal efficiency of Class-B PA and other operation modes with/without Doherty and/or Class-G operation. The maximum SE is 32.1% at 27.8-dBm P_{out}. Three additional peaks at 2.5-dB, 6-dB, and 12-dB PBO improve the overall efficiency. The SE depicted in this figure reflects the total power consumption in the complete quadrature transmitter chain. It is noted that the prototype quadrature DPA demonstrates a large digital power dissipation in logic circuits in a 65-nm CMOS process, which degrades the efficiency at the PBO region. However, the power dissipation in digital circuits scales down significantly in the advanced nanometer CMOS technology, and the efficiency improvement at PBO will be more noticeable in the advanced CMOS nodes. A comparison to the state-of-the-art PAs is shown in Table 1.

IV. CONCLUSION

A linear highly efficient transmitter based on a quadrature CDD SCPA is implemented in a 65-nm CMOS. The combination of quadrature IQ-cell-shared Class-G and CDD techniques improves the SE at the PBO region by providing multiple efficiency peaks, and the linearization techniques for unary and binary cells enhance the linearity of the quadrature DPA.

ACKNOWLEDGMENT

The author would like to thank Integrand Software for providing an electromagnetics simulation tool (EMX).

REFERENCES

[1] S. Hu, S. Kousai and H. Wang, "A Broadband CMOS Digital Power Amplifier with," in *IEEE ISSCC Dig. Tech. Papers*, 2015.

[2] S.-M. Yoo, J. S. Walling, O. Degani, B. Jann, R. Sadhwani, J. C. Rudell, and D. J. Allstot, "A Class-G switched-capacitor RF power amplifier," *IEEE J. Solid-State Circuits*, vol. 48, no. 5, pp. 1212–1224, May 2013.

[3] V. Vorapipat, C. Levy and P. Asbeck, "A Class-G Voltage-Mode Doherty Power Amplifier," in *IEEE ISSCC Dig. Tech. Papers*, 2017.

[4] W. Yuan, V. Aparin, J. Dunworth, L. Seward, and J. S. Walling, "A quadrature switched capacitor power amplifier," *IEEE J. Solid-State Circuits*, vol. 51, no. 5, pp. 1200–1209, May 2016.

[5] D. Chowdhury, L. Ye, E. Alon, and A. M. Niknejad, "An efficient mixed-signal 2.4-GHz polar power amplifier in 65-nm CMOS technology," *IEEE J. Solid-State Circuits*, vol. 46, no. 8, pp. 1796–1809, Aug. 2011.

[6] H. Jin, D. Kim, and B. Kim, "Efficient digital quadrature transmitter based on IQ cell sharing," *IEEE J. Solid-State Circuits*, vol. 52, no. 5, pp. 1345–1357, May 2017.

[7] S.-W. Yoo, S.-C. Hung and S.-M. Yoo, "A 1W Quadrature Class-G Switched-Capacitor Power Amplifier with Merged Cell Switching and Linearization Techniques," in *IEEE Radio Frequency Integrated (RFIC) Symp. Dig. Papers*, 2018.

A Frequency Tuneable Switched-Capacitor PA in 65nm CMOS

Zhidong Bai[#1], Ali Azam[#2], Jeffrey S. Walling[#$3]

[#]PERFIC lab, University of Utah, USA

[$]MCCI, Tyndall National Institute, Ireland

[1]zhidong.bai@utah.edu,[2]ali.azam@utah.edu,[3]jeffrey.s.walling@utah.edu

Abstract— A frequency tuneable, switched capacitor power amplifier (SCPA) is introduced that allows it to operate in relatively narrow channel bandwidths (10's of MHz) while enabling operation across a much wider frequency band (>1GHz). An SCPA is connected in series with a fixed inductor and a digitally programmable capacitor (DPC) that allows frequency tuning of the SCPA resonance. A prototype is fabricated in a 65nm CMOS process and embedded with the DPC on a printed circuit board for validation. The measured prototype operates from supply voltages of 1.2 and 2.4 V and achieves linear operation over 58.8% fractional bandwidth (1.4-2.5 GHz) with less than 1.7 dB output power variation. Due to its linearity, no digital pre-distortion is required to achieve measured ACLR<-30 dBc and EVM<3.9%-rms over all operating frequencies when transmitting an LTE 64-QAM, 20 MHz OFDM symbol at >14.1 dBm.

Keywords—Tuneable RF Circuits, RF Power Amplifier, RF PA, Switched Capacitor PA, SCPA, Polar PA.

I. INTRODUCTION

Demand for wireless spectral access is increasing at a relentless pace, owing to individual users of wireless devices demanding higher data throughputs to an ever-increasing world of connected devices. The increasing demand for wireless traffic has spurred the next generation of wireless networks to offer up to 1000× increases in throughput with proposed 5G services. These services will also support increased access for connected devices via the Internet-of-Things (IoT).

To provide improved data throughput, wireless systems primarily rely on improvements in signalling, flexibility in accessing the available spectrum and using multiple-input, multiple-output techniques (MIMO). Hardware solutions for the transmitter (TX) that can simultaneously allow RF systems to take advantage of these network level enhancements are a primary bottleneck in deploying future 5G communications.

In terms of signalling, spectrally efficient modulation techniques such as orthogonal frequency division multiplexing (OFDM) have been introduced that allow higher density of information in a given amount bandwidth. Modulation techniques such as OFDM provide higher spectral efficiency at the expense of operating with a high peak-to-average power ratio (PAPR). An increase in PAPR increases the output power backoff that an amplifier must operate in and hence, reduces the energy efficiency of the power amplifier (PA) in the TX. This is because PAs are most efficient when outputting their saturated output power and the efficiency degrades when operating in output power backoff.

There are many techniques that can improve the efficiency of a PA when operating in output power backoff. Recently, the switched-capacitor power amplifier (SCPA) has provided a

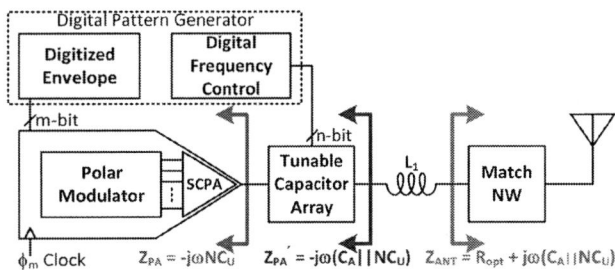

Fig. 1. Conceptual block diagram schematic of a frequency tuneable SCPA.

relatively elegant solution for increasing efficiency in output power backoff by reducing input power consumption as the required output power is reduced [1].

To address flexibility in accessing available spectrum, there are two common hardware solutions. The first option is to use PAs with wide instantaneous bandwidths that can operate across large segments of the wireless spectrum. This approach typically sacrifices the peak performance of the PA in favour of a flat, broadband frequency response in the amplitude domain and typically results in reduced efficiency when operating away from the center of the band. Nevertheless, there are several recent wideband approaches that have achieved outstanding results [2]–[4]. A second technique is to use discrete multiband designs that utilize multiple PAs/TXs to cover specific bands of interest in a wide bandwidth. This approach requires either independent design and optimization for each TX [5], or requires a low-loss passive network implemented on an integrated passive device (IPD) [6]. Additionally, this approach removes some of the flexibility in frequency planning that a network designer would require for optimal spectral usage. Alternatively, a multi-chip discrete wideband technique using an interposer board has been proposed [7], but again requires design of multiple unique PAs and a unique package.

In this paper, a frequency tuneable SCPA (Fig. 1) is presented that leverages the energy efficiency enhancements of an SCPA, while allowing discrete frequency tuning for coverage of a wide range of spectrum. The technique leverages the fact that the output impedance of an SCPA has a series RC equivalent circuit that is resonated by a series inductor (L_1, Fig. 1). The resonance of the SCPA can be adjusted by adding a digitally programmable capacitor (DPC) in series with the SCPA. In theory, this capacitor (C_A, Fig. 1) does not impact the linearity or output power of the SCPA, since all the charge contained in the SCPA can redistribute across C_A.

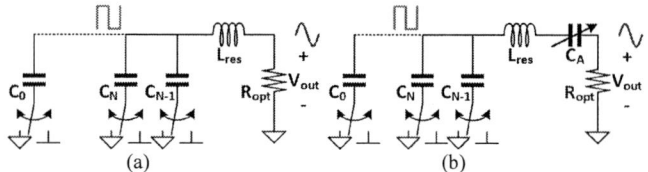

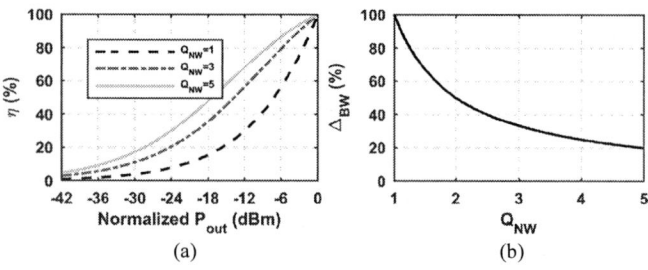

Fig. 2. Schematic of an (a) ideal SCPA and (b) a frequency tuneable SCPA

Fig. 3 (a) Ideal drain efficiency vs. output power and Q_{NW} and (b) Ideal fractional bandwidth vs. Q_{NW}.

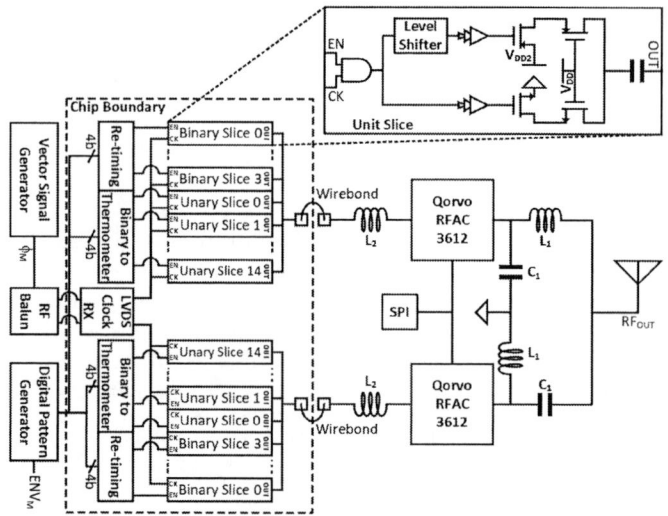

Fig. 4. Schematic of the proposed frequency tunable SCPA.

II. FREQUENCY TUNEABLE SCPA USING EMBEDDED DIGITALLY PROGRAMMABLE CAPACITORS

SCPAs are a special case of a class-D PA where the series resonant capacitor is divided into an array of unit capacitors that can be driven by independent switches. A summary of the standard SCPA operation is now provided, followed by a detailed description of the frequency tuning operation. Finally, details of a prototype frequency tuneable SCPA implementation are provided.

A. SCPA Operation

In an ideal SCPA (Fig. 2(a)), an array of capacitors is placed in series with an inductive reactance and an optimal termination resistance. The capacitors share a common top-plate, and each bottom plate is connected to a switch that can be connected to either ground or a fixed supply voltage. The switch is implemented as a cascoded CMOS inverter. The total capacitance in the array, C_{tot}, is given by the following:

$$C_{tot} = \sum_{i=0}^{N-1} C_i. \tag{1}$$

The capacitors can either be switched between V_{DD} and V_{GND} at the RF carrier frequency, f_0, or held at V_{GND}. The total number of capacitors being switched, C_{on}, can be changed by a digital input code, n, and is given by the following:

$$C_{on} = n \cdot C; 0 \le n \le N - 1, \tag{2}$$

The total capacitance not being switched, C_{off}, is given by:

$$C_{off} = (N - n) \cdot C$$

Due to charge re-distribution, the voltage amplitude at the top-plate of the capacitor array is given by the following:

$$V_{top} = V_{DD} \cdot \frac{C_{on}}{C_{tot}}. \tag{3}$$

An series inductor is resonant at the carrier frequency, f_0:

$$L_{res} = \frac{1}{C_{tot}(2\pi f_0)^2}. \tag{4}$$

The series resonant circuit selects the fundamental switching tone at the output and to produce a sinusoid at the load resistor, resulting in the following power, P_{out}, delivered to the load:

$$P_{out} = 2 \cdot \left(\frac{n}{\pi N}\right)^2 \cdot \frac{V_{DD}^2}{R_{opt}}. \tag{5}$$

Note that the SCPA is effectively a voltage mode RF-DAC. The output amplitude is controlled by n, which can be a time varying digital representation of the amplitude modulation applied to the SCPA. The series resonant circuit acts as a bandpass reconstruction filter to reconstruct the amplitude modulation as a time varying analog voltage across R_{opt}. Phase modulation is accomplished by modulating the carrier phase that drives the switching clocks. In this way, the SCPA acts as a polar PA, providing a polar vector-multiplication.

The ideal drain efficiency, η, of the SCPA can be found by first finding the input power required to drive the output stage. The switching power, P_{sw}, is due to charging and discharging the array capacitance. The array capacitance is the series combination of C_{on} and C_{off}; hence P_{sw} is given by:

$$P_{sw} = \frac{n(N-n)C}{N} \cdot V_{DD}^2 \cdot f_0 \tag{6}$$

The total input power to the output stage is given by the sum of P_{sw} and P_{out}. Note that the SCPA is a series resonant network and the quality factor of the network, $Q_{NW} = (2\pi f_0 C_{tot})^{-1}$. η, is the ratio of P_{out} to the total input power:

$$\eta = \frac{4n^2}{4n^2 + \frac{\pi n(N-n)}{Q_{NW}}} \tag{7}$$

η does not consider additional sources of loss [1], but some observations can be made regarding the effect of Q_{NW} on the performance of the SCPA. The efficiency at backoff is proportional to Q_{NW} (Fig. 3(a)), whereas the fractional bandwidth has the inverse relationship (Fig. 3(b)). This means that increases in efficiency can be traded off for reduction in bandwidth. This paradigm can be broken with a means to tune the SCPA over a broad range of frequencies.

One way to provide frequency tuning of the SCPA is now shown. Though it is demonstrated for a polar SCPA, note that it is possible to employ the same technique for quadrature and multiphase SCPA variants. Also note that it is possible to use a similar frequency tuning method in current-mode DPAs.

978-1-7281-1702-7/19 $31.00 © 2019 IEEE

B. Frequency Tuning

An SCPAs resonance can be adjusted by placing an additional series capacitor in the network, as shown in Fig. 2(b). The total capacitance, C'_{tot}, in the modified circuit is given as the series combination of C_{tot} and C_A:

$$C'_{tot} = C_{tot} || C_A. \tag{8}$$

The modified resonant frequency, f'_0, of the frequency tuneable SCPA is thus given by the following:

$$f'_0 = \frac{1}{2\pi\sqrt{L_{res}C'_{tot}}}. \tag{9}$$

This does not affect the linear operation of the SCPA, as the total charge held by the capacitor array is re-distributed across C_A. If C_A is implemented as a digitally programmable capacitor (DPC), the resonant frequency of the circuit can be tuned independent of the linear operation, allowing for narrow band operation centered around the programmed resonant frequency.

C. Frequency Tuneable SCPA Implementation

The schematic of the proposed frequency tuneable SCPA is shown in Fig. 4. An 8b SCPA is designed in a 65nm RF CMOS technology. The array is comprised of MIM capacitors with shields to minimize interaction with the substrate. The array is segmented and is comprised of 4-unary and 4-binary weighted bits and has a total array capacitance of 1.4 pF. The switch and driver stages are designed as custom cells that can be tiled with the adjacent cells to minimize parasitic capacitance. All digital logic is synthesized from Verilog code and is automatically placed and routed with timing constraints to ensure matched path delays for each control bit. The IC is chip-on-board bonded to a PCB where it is interfaced with the external components. The series resonant inductor is designed to include the bondwire parasitic and a high-Q wire-wound ceramic core inductor (L_2). The DPC is a Qorvo RFAC 3612 that is programmable via a serial peripheral interface (SPI). The DPC has 64 states and the capacitance has a capacitance range between 0.47-13 pF. The total capacitance of the array in series with the DPC can thus be tuned between 0.37-1.26 pF. An external LC balun, consisting of components L_1 and C_1, is used to transform the antenna impedance at the connector (e.g., 50 Ω) to the differential optimal termination impedance, R_{opt} = 12.5 Ω, at the center frequency of the band to be covered. LC baluns typically offer relatively narrow fractional bandwidths and if a wider bandwidth of operation is desired, can be replaced with a transmission line balun at the expense of board area. An additional advantage of the LC balun is that it allows easy swapping of surface mount components that can be interchanged to cover different frequency bands. The platform is completely digitally reconfigurable via a digital interface that can provide the digital modulation and program the DPC.

III. MEASUREMENT RESULTS

The frequency tuneable SCPA IC is fabricated in a 65nm RF CMOS process and chip-on-board bonded to a 4-layer FR4-06 PCB, as shown in Fig. 5. The area of the IC is 0.575mm², including all I/O pads. The core circuit area is 0.12mm². The system operates from two voltage supplies; all core logic, the

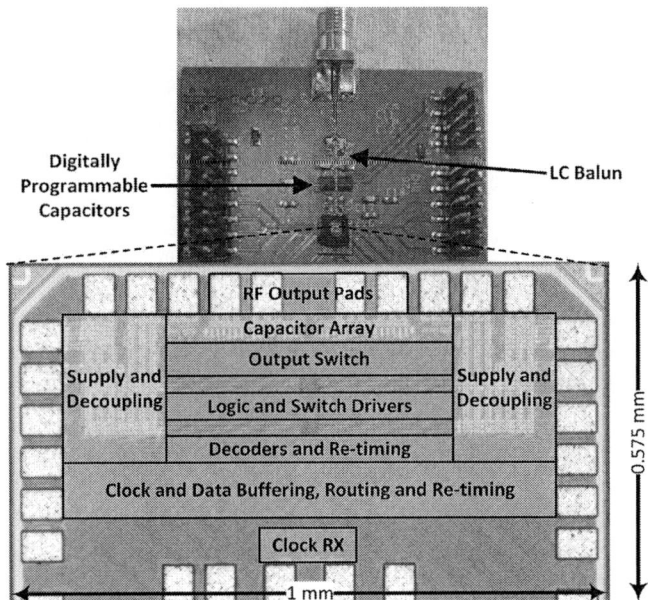

Fig. 5. (top) Fabricated PCB and (bottom) 65nm Chip microphotograph of the frequency tunable SCPA.

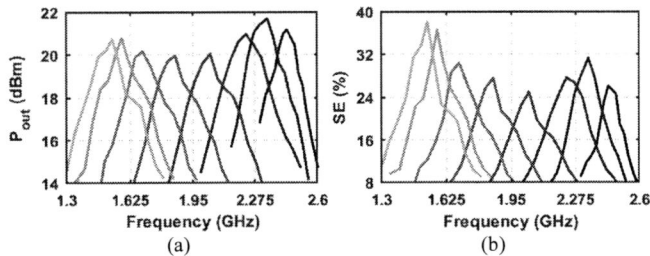

Fig. 6 Measured (a) output power and (b) system efficiency versus frequency for different tuned center frequencies.

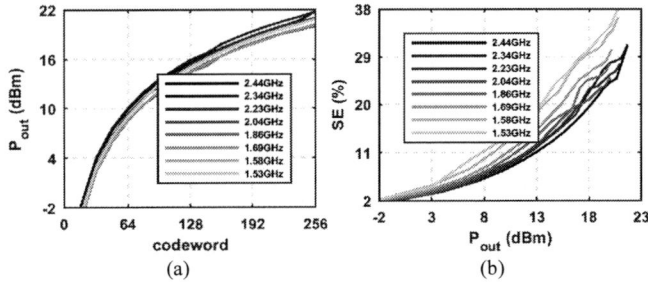

Fig. 7. Measured (a) output power versus input code and (b) system efficiency versus output power for different tuned center frequencies.

clock and driver stages are operated from 1.2V, while the output stage operates from 2.4 V. The output stage is cascoded so that no transistor operates with more than 1.2V across any two terminals. Unlike many CMOS PA measurements, it is noted that all measurement results presented include the losses from the PCB, the DPC, the LC balun and the SMD inductors.

The frequency tuneable SCPA is first measured for its static performance. The measured P_{out} is plotted vs. frequency in Fig. 6(a), while varying the DPC control code. In the plot, 8 different codes={63, 32, 16, 8, 4, 2, 1, 0} are plotted, where the highest code={63} corresponds to the lowest operation frequency. The

978-1-7281-1702-7/19 $31.00 © 2019 IEEE   297

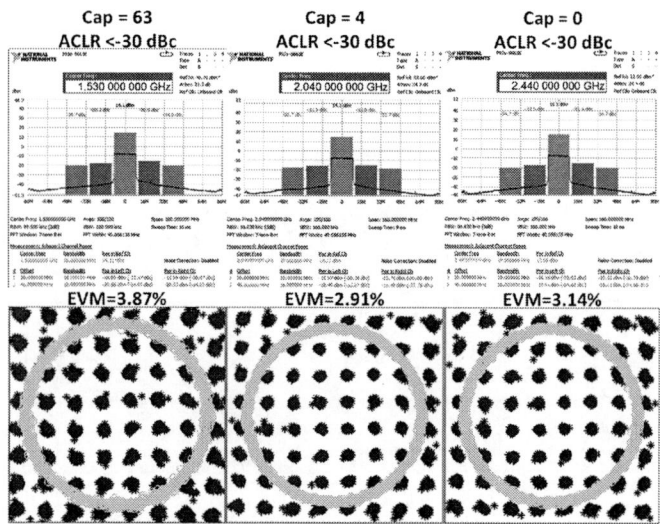

Fig. 8. Measured (top) PSD and (bottom) Signal constellations for LTE signals operating at different center frequencies.

PA can be tuned from 1.4-2.5 GHz and saturated P_{out} varies from 20-21.7 dBm across the frequency range. The peak system efficiency (SE), which includes all external losses and all power input to the board, ranges from 25-38.1% and is plotted in Fig. 6(b). The output power versus input envelope code and SE versus output power are plotted in Fig. 7(a) and (b), respectively for different tuned center frequencies. The linearity and backoff efficiency characteristics are similar for each tuned frequency.

Due to low-loss switches and precision capacitor matching, the prototype is linear with respect to the input code and can be operated without digital pre-distortion (DPD) across its frequency range. This is important because of the large number of operation states, DPD is undesirable, since it would require either a large lookup table and/or high power-consumption. An additional advantage of the proposed technique is that the high-Q filtering provided reduces the contribution of the out-of-band (OOB) noise more, which enhances the ability to co-exist with multiple wireless systems on the same platform.

To validate linearity of the proposed frequency tuneable SCPA across its range of operation, the dynamic performance is measured using an LTE signal (20 MHz, 64 QAM, data rate=100.8 Mbps) at different program center frequencies. Owing to the outstanding linearity, the measurement is made without use of DPD. The PSD estimate and signal constellation are plotted for codes ={63, 4, 0}, corresponding to center frequencies {1.53, 2.04, 2.44} GHz in Fig. 8. To enhance the

efficiency, the codes are de-troughed until the ACLR is at the limit for E-UTRA ACLR (-30 dBc) [8]. The measurement was performed for all programmable center frequencies across the band the ACLR and EVM are met without use of DPD at all available frequencies of operation. The average (linear) output power and SE across all frequencies is >14.1 dBm and 15.7%, respectively. The highest average output power occurs for code={0} and is 15.1 dBm at 2.44 GHz. The highest average SE occurs for code={63} and is 23.7% at 1.53 GHz.

IV. CONCLUSION

A frequency tuneable SCPA is presented that allows narrowband operation in discrete channels across a much wider band of operation. The core circuitry operates from 1.2 V, while the cascoded output stage operates from 2.4 V. The PA is compared to recent multiband and multi-standard PAs in Table 1. The proposed solution offers high average efficiency and output power across its range of frequencies and the measurements include all external measurement losses. It provides similar performance to the recent art, with enhanced flexibility in deployment, which is important for emerging 5G and IoT applications.

REFERENCES

[1] S. Yoo, J. Walling, E. Woo, and D. J. Allstot, "A switched-capacitor RF power amplifier," *IEEE J. Solid-State Circuits*, vol. 46, pp. 2977–2987, Dec. 2011.

[2] W. Ye, K. Ma, and K. S. Yeo, "A 2-to-6GHz Class-AB power amplifier with 28.4% PAE in 65nm CMOS supporting 256QAM," in *IEEE ISSCC Dig. Tech. Papers*, 2015, pp. 38–39.

[3] J.-K. Nai, Y.-H. Hsiao, Y.-S. Wang, Y.-H. Lin, and H. Wang, "A 2.8–6 GHz high-efficiency CMOS power amplifier with high-order harmonic matching network," in *Proc. of the IEEE International Microwave Symposium*, 2016, pp. 1–3.

[4] J. Park, Y. Wang, S. Pellerano, C. Hull, and H. Wang, "A 24dBm 2-to-4.3 GHz wideband digital Power Amplifier with built-in AM-PM distortion self-compensation," in *IEEE ISSCC Dig. Tech. Papers*, 2017, pp. 230–231.

[5] M. Fulde *et al.*, "A digital multimode polar transmitter supporting 40MHz LTE carrier aggregation in 28nm CMOS," in *IEEE ISSCC Dig. Tech. Papers*, 2017, pp. 218–219.

[6] J. Ko *et al.*, "A high-efficiency multiband Class-F power amplifier in 0.153 μm bulk CMOS for WCDMA/LTE applications," in *IEEE ISSCC Dig. Tech. Papers*, 2017, pp. 40–41.

[7] N. Kuo *et al.*, "A 0.4-to-4-GHz All-Digital RF Transmitter Package With a Band-Selecting Interposer Combining Three Wideband CMOS Transmitters," *IEEE Trans. Microw. Theory Tech.*, vol. 66, pp. 4967–4984, Nov. 2018.

[8] J. Jeong, D. F. Kimball, M. Kwak, C. Hsia, P. Draxler, and P. M. Asbeck, "Modeling and Design of RF Amplifiers for Envelope Tracking WCDMA Base-Station Applications," *IEEE Trans. Microw. Theory Tech.*, vol. 57, no. 9, pp. 2148–2159, Sep. 2009.

Table 1. Comparison to Recent Wideband/Multiband CMOS PAs

	Architecture	Freq. (GHz)	Supply (V)	Δ_{BW} (%)	P_{sat} (dBm)	Efficiency (%)	Modulation (Data Rate Mb/s)	P_{avg} (dBm)	Avg. Efficiency (%)	DPD	EVM (dB)	ACLR (dBc)
Ye ISSCC '15	Wideband w/ 2nd Harmonic Match	2-6	3.3	115.5	20.1-22.4	19-28.4$^&$	OFDM 20 MHz 64-QAM (54)	11.1-13.2	N/A	No	<4%	N/A
Park ISSCC '17	Wideband Multiresonant Match	2-4.3	1.4	78.4	23.9-24.9	27-42.7$^#$	20 MHz 64-QAM (120)	14.6$^$	15.6$^{#$}	PM-Free	2.95$^$	<-33.5$^$
Ko ISSCC '17	Discrete Multiband IPD	0.835, 0.898, 1.88, 1.95	3.4	N/A	31.7-32$^@$	46.3-58.5$^{&@}$	OFDM 20 MHz 16-QAM (67.2)	27.5-28.1$^@$	33.3-37.3$^{&@}$	No	3.5	<-32
This Work	Programmable Capacitor Array	1.4-2.5	1.2/2.4	58.8	20.0-21.7$^@$	25-38.1$^{*@}$	OFDM 20 MHz 64-QAM (100.8)	14.1-15.1$^@$	15.7-23.7$^{*@}$	No	2.9-3.9	<-30

$^$System Efficiency $^&$Power Added Efficiency $^#$Drain Efficiency $^$Modulation data @2.8GHz $^@$Includes External Losses

A Broadband High-Efficiency SOI-CMOS PA Module for LTE/LTE-A Handset Applications

A. Serhan[1], D. Parat[1], P. Reynier[1], R. Berro[1], R. Mourot[1], C. De Ranter[2], P. Indirayanti[2], M. Borremans[2],
E. Mercier[1], A. Giry[1]

[1]CEA, LETI, MINATEC Campus, F-38054 Grenoble, France
[2]Huawei Technologies Research & Development, Belgium NV

Abstract— **This paper presents a broadband high-efficiency linear Doherty power amplifier (DPA) for LTE/LTE-A handset applications. The proposed PA is implemented in a 130nm SOI technology and packaged using flip-chip on a laminate substrate. At 2.5GHz, the PA shows a PAE of 44%, a power gain of 27dB, and an E-UTRA ACLR of -35dBc at 28dBm output power using a 10MHz LTE uplink signal without DPD. Moreover, the PA reaches a maximum FOM (PAE+|ACLR|) of 80 and maintains a FOM greater than 70 over 31% of fractional bandwidth around 2.3GHz without using DPD. When using DPD, the ACLR is improved by 10dB leading to a maximum FOM of 90. To the best of our knowledge, these performances represent the best linearity-efficiency performances compared to recently published LTE PAs for handset applications.**

Keywords— **Doherty, LTE, LTE-A, Power Amplifier, SOI.**

I. INTRODUCTION

RF Front-end Module (FEM) design has significantly increased in complexity during the transition from 3G to 4G/5G standards due to the ever increasing signal bandwidth, PAPR, number of modes and frequency bands that need to be supported. Power amplifier (PA) is a key building block of the RF FEM which represents a large part of the cost and has a strong impact on battery life. Although CMOS PAs have been successfully demonstrated in commercial products for 2G/3G [1]-[2], GaAs-based power amplifiers remain the dominant choice for high efficiency linear PA. Many research activities and industrial efforts are currently engaged to demonstrate CMOS PAs with improved FOM [3] under high PAPR and wide bandwidth signals in order to take benefit from the increased integration capability and lower cost of CMOS technologies. The trend towards higher data rates leads to RF signals with high PAPR that makes the design of high efficiency linear PA very challenging. Linear amplification is traditionally achieved by operating PA in back-off (BO) where

conventional PA architectures show poor efficiency. PA architectures that use supply or/and load modulation represent attractive solutions for efficient amplification of high PAPR signals [4]. Moreover, DPA architecture require less complex interaction with the based-band to be linearized with DPD compared to Envelope Tracking (ET) architecture. In this paper, a SOI-CMOS DPA for LTE/LTE-A handset applications with broadband high-efficiency linear performances is presented.

II. CIRCUIT DESCRIPTION

Fig. 1 presents the simplified block diagram of the two-stage DPA. The output matching network (OMN) transforms the external load impedance $R_L = 50\Omega$ to the optimum load impedance Z_{opt} at the combining point. A two-section matching network is used to improve both the operating bandwidth and the out-of-band rejection. The impedance inverter (INV) is realized using a lumped π-network. The characteristic impedance of the impedance inverter sets the desired output back-off level. The main and auxiliary cells use common-source SOI LDMOS transistors with high breakdown voltage in order to ensure high reliability. The size of the power cells is selected as to reduce the knee voltage and hence increase efficiency. The inter-stage matching is optimized to provide the optimum load impedance to the driver stage while presenting the optimum source impedance to the main and auxiliary power cells. The driver stage uses a cascode topology involving a high speed NMOS transistor as a common source and a high voltage LDMOS device for the common gate transistor. Finally, a two-section input matching network (IMN) is added to provide the optimum source impedance for the driver stage while ensuring sufficient input matching level. The bias voltage of the auxiliary cell and the driver stage, and the phase difference between the main and auxiliary transistors are optimized as to improve the overall linearity-efficiency trade-off with emphasis on linearity.

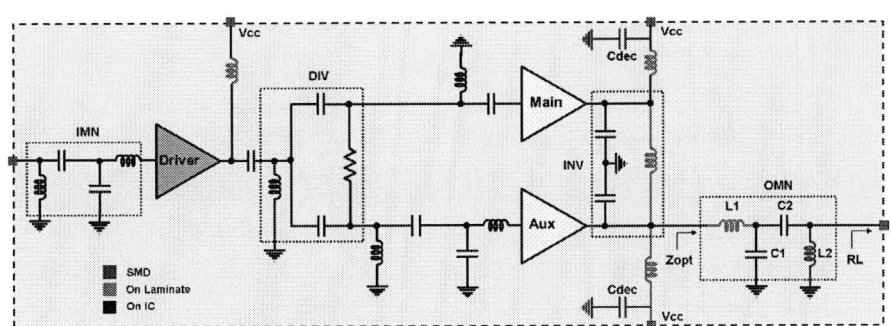

Fig. 1. Simplified schematic of the proposed Doherty power amplifier.

978-1-7281-1702-7/19 $31.00 © 2019 IEEE

III. CIRCUIT IMPLEMENTATION

The DPA is implemented in a 130nm SOI-CMOS industrial process from STMicroelectronics offering high breakdown voltage devices and a metal-stack with two thick metal layers allowing the integration of low-loss inductors. High quality factor MOM (Metal-Oxide-Metal) capacitors are used to support the high voltage swing at the output. Flip-chip technology is used for the assembly in order to minimize the source degeneration inductor of the power devices.

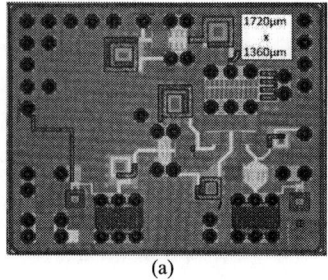 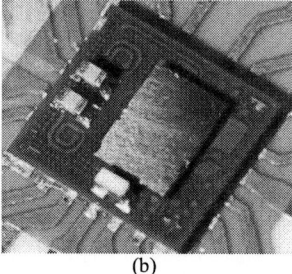

(a)   (b)

Fig. 2. (a) Die photo of the SOI IC, (b) photo of the laminate package including the SOI IC and the SMD components.

Fig. 2 (a) shows the die photo of the fabricated chip which includes all the passive and active components except the supply and the output matching inductors which are designed on the laminate package. Fig. 2 (b) shows the LGA laminate package mounted on the evaluation board and including the SMD components L2 and Cdec. Electromagnetic simulation of the integrated passive components, transistors accesses, and the laminate package are included to take into account the electromagnetic coupling and the interaction between the laminate package and the flip-chip IC.

IV. MEASUREMENT RESULTS

A. Small-signal and large signal CW measurements

Measurements are performed under 3.4V supply voltage for both driver and power stages. Fig. 3 shows the measured small-signal S-parameters of the proposed PA. The small-signal gain is higher than 26dB over a wide frequency range while the input and output VSWR are better than 3:1 over the frequency range from 1.9GHz to 2.7GHz.

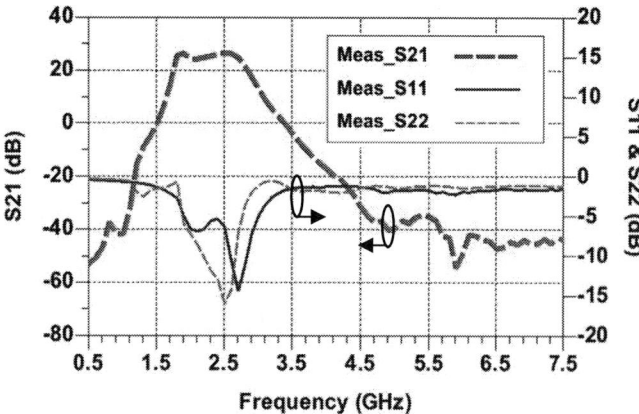

Fig. 3. Measured small-signal S-parameters of the proposed PA.

Fig. 4 shows the large signal CW performances measured at 2.5GHz for three different bias configurations. In linear Doherty (L-Doherty) mode, the auxiliary power cell is biased as to have a flat overall power gain response while preserving high efficiency in the back-off region. The PA achieves 57% of maximum PAE at a peak output power of 32.7dBm. At 28dBm output power, the PA has a PAE of 45% and a large signal gain of 27dB (blue line). In efficient Doherty (E-Doherty) mode, the auxiliary power cell is biased in deep class-C mode which enhances the PAE in the low-power region at the expense of linearity (red line). The linear and efficient Doherty modes can be combined by using adaptive bias for the auxiliary transistor. To demonstrate the advantage of load-modulation technique, the PA is measured with both main and auxiliary transistors biased in class-AB as to have the same small-signal power gain. Fig. 4 shows that the PAE in the Doherty modes is improved by about 13 points compared to that obtained in the class-AB mode at 27dBm output power.

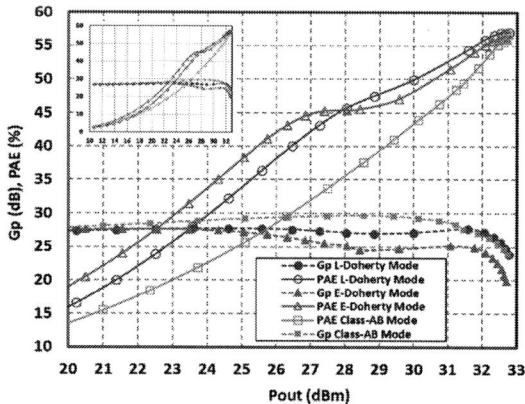

Fig. 4. Measured large signal PAE and power gain versus output power.

The second harmonic (H2) and third harmonic (H3) power levels are measured in E-Doherty and L-Doherty modes. Fig. 5 shows that the PA operating in L-Doherty mode has a significantly better linearity compared to the E-Doherty mode. Moreover, the H2 and H3 power levels measured in L- Doherty mode satisfy the harmonic rejection requirements.

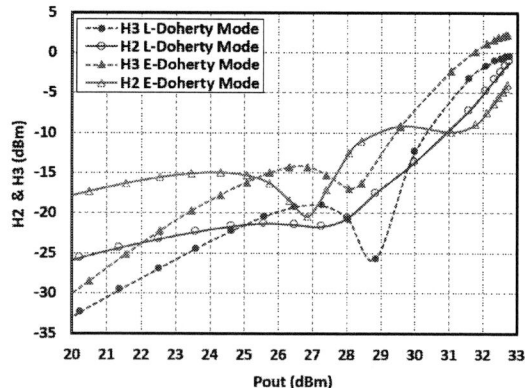

Fig. 5. Measured second and third harmonics power versus output power.

B. Measurements with modulated signal

LTE performances measured in L-Doherty mode with several uplink LTE signal configurations (10MHz 12RB,

10MHz 50RB, and 20MHZ 100RB) are shown in Fig. 6. At 28dBm average output power the PA achieves a PAE of 44%, a power gain of 27dB, and an E-UTRA ACLR lower than -33dBc without using DPD. Moreover, the ACLR is improved by using a memory-less polynomial DPD technique allowing an E-UTRA ACLR lower than -40dBc at 28dBm output power while maintaining similar PAE and Gp performances.

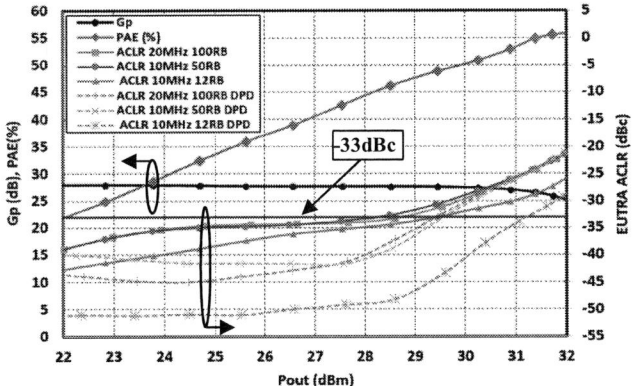

Fig. 6. PAE, power gain, and ACLR versus output power measured with a 10MHz 12RB, 10MHz 50RB, and 20MHZ 100RB QPSK uplink LTE signals.

The PA is characterized in L-Doherty mode over the frequency band from 1.9GHz to 2.7GHz, with a 12RB 10MHz QPSK LTE uplink signal. The PAE is maintained higher than 40% from 2GHz to 2.6GHz with an E-UTRA ACLR lower than -33dBc while delivering 28dBm of average output power, as shown on Fig. 7. The large signal power gain is higher than 27dB from 2.2GHz to 2.5GHz and is reduced to 26.6dB for the remaining part of the band to avoid efficiency-linearity degradation. To compare PA performances, the following Figure of Merit (FOM) is used as proposed in [3]:

$$\text{FOM= PAE+|ACLR|} \qquad (1)$$

As shown on Fig. 7, the proposed PA achieves a maximum FOM of 80 without using DPD.

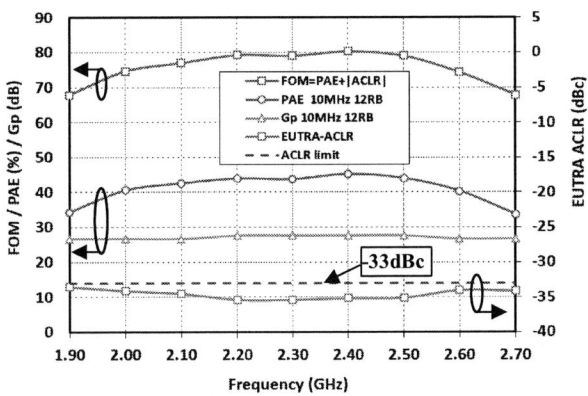

Fig. 7. PAE, power gain, E-UTRA ACLR, and FOM versus frequency at 28dBm measured with a 10MHz 12RB LTE signal without DPD.

Moreover, the PA is also characterized by applying polynomial memory-less DPD functions. As shown on Fig. 8, the E-UTRA ACLR is significantly enhanced, -48.2dBc at 2.3GHz, while maintaining similar PAE, Pout, and Gp performances. Hence, the PA achieves a maximum FOM of 90

from 2.2GHz to 2.5 GHz and maintains a FOM greater than 80 over 31% of fractional bandwidth around 2.3GHz.

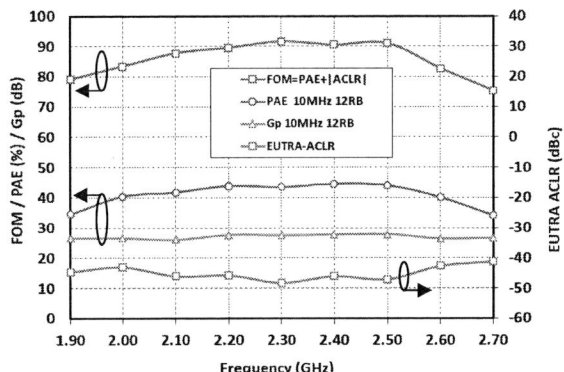

Fig. 8. PAE, power gain, E-UTRA ACLR, and FOM versus frequency at 28dBm measured with a 10MHz 12RB LTE signal with memory-less DPD.

The PA is also evaluated using 40MHz and 60MHz LTE-Advanced (LTE-A) intra-band carrier aggregated (CA) signals. As shown on Fig.9, the PA has an E-UTRA ACLR of -41.5dBc and -43dBc at 26dBm and 25dBm of average output power using 40MHz and 60MHz signals respectively. The corresponding PAEs are 35.5% and 29.8%. Note that the 3GPP specifications allow 1dB of maximum power reduction (MPR) for each additional carrier component (CC), as shown in Table 1, in order to compensate for the increase in PAPR and to maintain reasonable linearity performances.

Table 1. MPR and PAPR for different signal LTE-A CA signal bandwidths.

BW (MHz)	20MHz	40MHz	60MHz
MPR (dB)	1	2	3
PAPR (dB)	6.19	8.97	9.59

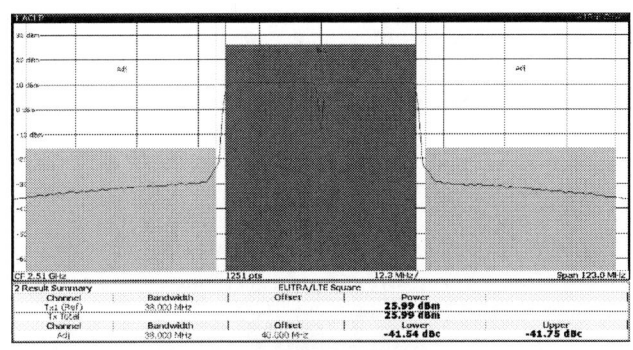

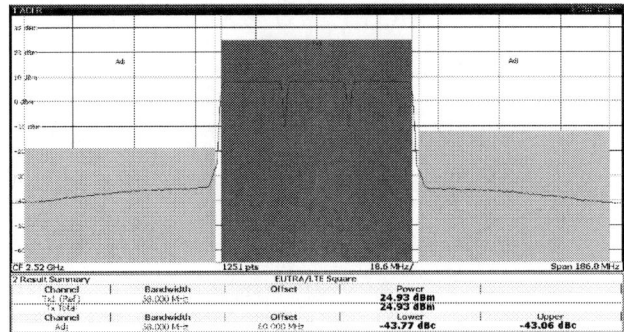

Fig. 9. Measured output spectrums for 40MHz (top) and 60MHz (bottom) intra-band carrier aggregated full-RB signals with DPD.

978-1-7281-1702-7/19 $31.00 © 2019 IEEE

Table 2. State-of-the-art of LTE and LTE-A integrated PA.

Reference	Freq. (GHz)	Pout (dBm)	PAE (%)	ACLR (dBc)	Vcc (V)	FOM	BW / N° of RB	DPD	Technology
[5]	1.88	27.9	33	-30	3.4	63	10MHz / FRB	No	0.11μm SOI CMOS
[6]	1.9	28.1	34.7	-32	3.4	66.7	20MHz / FRB	No	0.153μm SOI CMOS
[7]	1.85	27.2	37.7	-30	3.3	67.7	10MHz / FRB	No	0.18μm SOI CMOS
[8]	2.1	27.5	35.7	-33	4	68.7	10MHz / FRB	No	0.32μm SOI CMOS
[9]	1.85	27.5	38.5	-31	-	69.5	10MHz /FRB	No	0.18μm CMOS
[10]	1.9	26.2	34.3	-35.5	4	69.8	10MHz / FRB	No	0.32μm SOI CMOS
[11]	1.95	26	46	-30	APT	76	20MHz / FRB	Yes	0.18μm SOI CMOS
[12]	2.5	27.7	41.3	-36	3.4	77.3	10MHz / 12RB	Yes	InGap HBT
[13]	1.85	26	42	-37	3.5	79	10MHz / FRB	Yes	InGaP/GaAs HBT
[14]	1.9	27.5	45.8	-35	3.4	80.8	10MHz / FRB	Yes	InGaP/GaAs HBT
[15]	2.5	26.5	36.5	-39	ET	75.5	20MHz / FRB	No	0.18μm CMOS (TSV & HR)
[This work]	2.5	28	43.9	-35.1	3.4	79	10MHz / 12 RB	No	0.13μm SOI CMOS
		28	44	-47		91	10MHz / 12 RB	Yes	
		27	44	-40		84	20MHz / FRB		
		26	35.5	-41.5		77	40MHz (2x20MHz)/FRB		
		25	29.8	-43		62.8	60MHz (3x20MHz)/FRB		

ET: Envelope Tracking, APT: Average Power Tracking, FRB: Full Resource Block, TSV: Through Silicon Via, HR: High Resistivity

V. COMPARISON WITH THE STATE-OF-THE-ART

Table 2 summarizes the state-of the-art of recent LTE and LTE-A PA designs for handset applications. The PA proposed in this paper achieves state-of-the-art output power, efficiency and linearity with FOM reaching 80 for 28dBm linear output power at 2.4GHz without using any DPD or supply modulation technique. When using memory-less polynomial DPD, the proposed PA exceeds the state-of-the-art performances in terms of FOM.

VI. CONCLUSION

A broadband high-efficiency Doherty PA module fabricated in a 130nm SOI technology is presented. At 2.5GHz, the PA has a power gain of 27dB, a PAE of 44%, with an E-UTRA ACLR of -35dBc for 28dBm of average output power using 10MHz uplink signal. The proposed PA shows excellent broadband LTE performances, reaching a maximum FOM of 80 (90 with DPD) and maintaining a FOM higher than 70 (80 with DPD) over 31% of fractional bandwidth around 2.3GHz. Moreover, when measured with CA 40MHz and 60MHz LTE-A signals, the PA shows a PAE of 35.5% and 29.8% at 26dBm and 25dBm average output power respectively. These results validate the advantage of Doherty PA architectures and the strong potential of SOI-CMOS technologies for LTE/LTE-A handset applications.

ACKNOWLEDGMENT

The authors would like to thank Sylvie Gachon from STMicroelectronics for her helpful support and Christophe Zinck from ASE for module fabrication and assembly.

REFERENCES

[1] Skyworks/Axiom, "GSM/GPRS Quad Band Power Amplifier," AX508 datasheet, 2009.

[2] Broadcom/Avago, "WCDMA/HSPA Band I Power Amplifier," AJAV-6101-BLK datasheet, 2015.

[3] J. P. Young and N. Cheng, "Multimode multiband power amplifier optimization for mobile applications," in Proc. International Symposium on VLSI Technology, Systems, and Applications (VLSI-TSA), 2013.

[4] D. Jung, H. Zhao and H. Wang, "A Highly Linear Doherty Power Amplifier with Multigated Transistors Supporting 80MSymbol/s 256-QAM," in Proc. IEEE MTT-S International Microwave Symposium (IMS), 2018.

[5] K. Kim, J. Ko, S. Lee, and S. Nam, "A Two-Stage Broadband Fully Integrated CMOS Linear Power Amplifier for LTE Applications," IEEE Transactions on Circuits and Systems II: Express Briefs, vol. 63, no. 6, pp. 533–537, Jun. 2016.

[6] J. Ko et al., "A high-efficiency multiband Class-F power amplifier in 0.153μm bulk CMOS for WCDMA/LTE applications," in Proc. IEEE International Solid-State Circuits Conference (ISSCC), pp. 40-41, 2017.

[7] K. Kim, D.-H. Lee, and S. Hong, "A Quasi-Doherty SOI CMOS Power Amplifier With Folded Combining Transformer," IEEE Transactions on Microwave Theory and Techniques, vol. 64, no. 8, pp. 2605–2614, Aug. 2016.

[8] U. Kim, J. L. Woo, S. Park and Y. Kwon, "A single-chain multiband reconfigurable linear power amplifier in SOI CMOS," in Proc. IEEE MTT-S International Microwave Symposium (IMS), 2015.

[9] B. Park, J. Park, S. Jin, Y. Cho, J. Kim and B. Kim, "CMOS Power Amplifier on Top of Embedded Transformer for Compact Module," IEEE Microwave and Wireless Components Letters (MWCL), vol. 25, no. 10, pp. 678-680, Oct. 2015.

[10] U. Kim and Y. Kwon, "A High-Efficiency SOI CMOS Stacked-FET Power Amplifier Using Phase-Based Linearization," IEEE Microwave and Wireless Components Letters (MWCL), vol. 24, no. 12, pp. 875-877, Dec. 2014.

[11] P. Draxler, J. Hur, "A Multi-Band CMOS Doherty PA with Tunable Matching Network," in Proc. IEEE MTT-S International Microwave Symposium (IMS), 2017.

[12] K. Takenaka, T. Sato, H. Matsumoto, M. Kawashima, and N. Nakajima, "New compact Doherty power amplifier design for handset applications," in Proc. IEEE Topical Conference on RF/Microwave Power Amplifiers for Radio and Wireless Applications (PAWR), 2017.

[13] Y. Cho, D. Kang, J. Kim, K. Moon, B. Park and B. Kim, "Linear Doherty Power Amplifier With an Enhanced Back-Off Efficiency Mode for Handset Applications," IEEE Transactions on Microwave Theory and Techniques (TMTT), vol. 62, no. 3, pp. 567-578, 2014.

[14] B. Kim and Y. Cho, "Design of linear Doherty power amplifier for handset application," in Proc. IEEE MTT-S International Microwave Symposium (IMS), 2016.

[15] F. Balteanu, "Linear Front End Module for 4G/5G LTE Advanced Applications," in Proc. European Microwave Conference (EuMC), 2018.

2019 IEEE Radio Frequency Integrated Circuits Symposium

A 27 GHz Adaptive Bias Variable Gain Power Amplifier and T/R Switch in 22nm FD-SOI CMOS for 5G Antenna Arrays

Christian Elgaard[#*1], Stefan Andersson[#], Peter Caputa[#], Eric Westesson[#], Henrik Sjöland[#*2]

[#]Ericsson, Lund, Sweden
[*]Lund University, Lund, Sweden
[1]christian.elgaard@ericsson.com, [2]henrik.sjoland@ericsson.com

Abstract—A 27 GHz fully integrated, variable gain, two stage Power Amplifier (PA) and a Transmit/Receive (T/R) switch targeting 5G antenna array systems are presented. The PA uses adaptive bias, tracking the input signal amplitude, which improves saturated output power (P_{sat}) with 1.4 dB and 1 dB output compression (OP_{1dB}) by 3 dB. For a supply voltage of 1.2 V, the PA reaches a P_{sat} of 17.4 dBm and an OP_{1dB} of 16.5 dBm, with a power added efficiency of 19.5 % and 17.3 %, respectively. The power gain can be controlled with 5-bits from 5.2 to 34 dB. The T/R-switch has an insertion loss of 1.63/1.46 dB in TX/RX mode, and for reliability reasons all switch devices are on in TX-mode. The complete PA and T/R-switch only occupies 0.146 mm^2 in a 22 nm FD-SOI CMOS technology.

Keywords—power amplifier, adaptive bias, T/R-switch, millimeter wave integrated circuits.

I. INTRODUCTION

Fifth generation mobile communication (5G) will support millimeter-wave frequencies and beamforming, using large antenna array systems (AAS) with tens or even hundreds of antennas, where each antenna supports two polarizations [1]. Size and separation of the antennas in an AAS is typically about half a wave length, which at 30 GHz amounts to 5 mm. To generate transmit signals for such densely spaced antennas, full integration of the PA and the transmit/receive (T/R) switch into the CMOS transceiver chip is necessary. Effectively, many transceivers will be placed on one chip. In addition to the analog part, each chip needs to handle massive digital signal processing associated with beam-forming. To limit the size of the digital part, a short channel length advanced CMOS technology, must be used. A challenge with such technologies is the sensitivity to high voltage levels, which makes the implementation of the PA and T/R-switch particularly difficult. The T/R-switch must handle the voltage swing from the PA with low insertion loss, while protecting the low noise amplifier (LNA) from the strong PA output signal. Many published works focus either on the PA or the T/R-switch, thereby not addressing the issues arising when combining them.

Digital predistortion (DPD) is often applied to the PA input signal, which improves adjacent channel leakage ratio and error vector magnitude at the PA output. A DPD system can effectively combat amplitude to phase (AM-PM) variation introduced by the PA, and within limits also reduce the amplitude to amplitude (AM-AM) error. One limitation of AM-AM reduction for a PA using a DPD, however, is the increase of peak to average ratio of the input signal. This

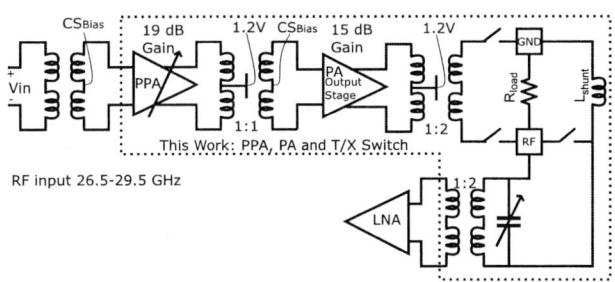

Fig. 1. The proposed power amplifier and T/R switch architecture, part of a fully integrated transceiver.

requires an increased dynamic range in the TX digital to analog converter (DAC), and an increased maximum signal level in the entire analog part of the transmitter, leading to more stringent requirements on linearity and output power capability.

To the authors knowledge, this paper presents the design and measurement of the first published fully integrated PA and T/R switch in 22 nm FD-SOI CMOS, see Fig. 1, targeting the 3GPP band n257, i.e. 26.5 - 29.5 GHz [2]. The PA, consisting of a variable gain pre-PA (PPA) and a PA output stage, utilizes analog linearization by dynamically adapting the PA bias to track the input envelope. Turning on the adaptive bias increases saturated output power (P_{sat}) with 1.4 dB and output referred 1 dB compression (OP_{1dB}) by 3 dB. Prior to the T/R-switch, the PA reaches a P_{sat} of 17.4 dBm and an OP_{1dB} of 16.5 dBm when the adaptive bias is active. The switch has an insertion loss of 1.63/1.46 dB in TX/RX-mode The results show that adaptive bias can reduce the dynamic range of a DPD required to combat AM-AM distortion, effectively reducing the required dynamic range for the TX DAC, TX-BB, and mixer. It is also shown that a low loss T/R-switch can be fully integrated in 22 nm FD-SOI CMOS in a reliable way.

II. CIRCUIT DESIGN

A. PPA

The PPA depicted in Fig. 2 is a common source (CS) common gate (CG) stage (i.e. a cascode stage) where both the CS and the CG transistors are implemented as 32 unit cells to control the gain. The gates of the CS transistors in all unit cells on each differential side, are all connected to the same bias voltage and input signal. The unit cells are switched on/off by controlling the CG transistor gate voltage bias, which will control the effective width of both the CG and

978-1-7281-1702-7/19 $31.00 © 2019 IEEE

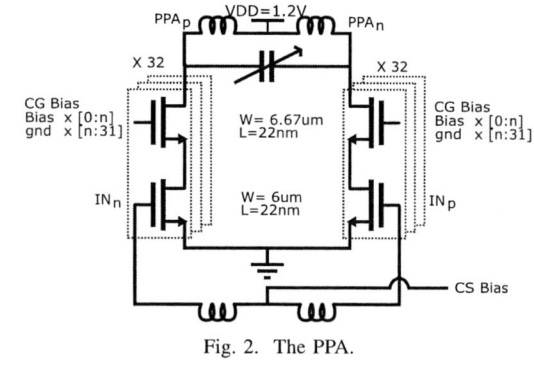

Fig. 2. The PPA.

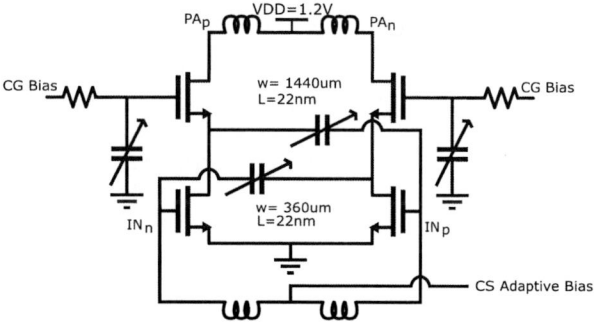

Fig. 3. The PA output stage.

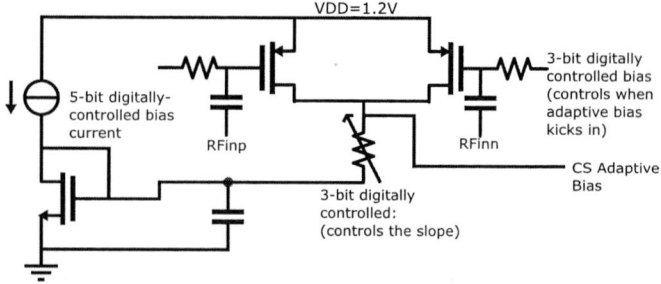

Fig. 4. Adaptive bias circuit for PA output stage CS transistor.

CS transistors. This results in a reduced bias current when the PPA gain is reduced. The input impedance is dominated by the CS transistor C_{gs} capacitance, which is resonated by the inductive part of the preceding transformer. A 5-bit digitally controlled capacitor is used to compensate for capacitance variations when changing the PPA gain. At resonance the input impedance becomes real and is simulated to $260\,\Omega$ for a Q-value of 20. The high input impedance is essential to reduce loading of the preceding mixer and to avoid the need for an additional gain stage between the mixer and the PPA. At the output transformer, between the PPA and PA output stage, a 3-bit digitally controlled capacitor is used to control the resonance frequency.

B. PA Output Stage

The PA output stage, shown in Fig. 3, uses a similar CSCG stage as the PPA, but with 2-bit programmable neutralization capacitors and digital control of CG gate voltage swing. Assuming high resistance from the bias source to the CG gate, the voltage swing at the CG gates is set by the output voltage swing and the capacitive voltage division between C_{dg} and the gate capacitor to ground. The swing of the CG gate voltage in turn sets that of the CS drain. To divide the output voltage swing evenly between the CG and CS transistors, a 2-bit digitally controllable capacitor between the CG gate and signal ground is added. The real part of the input impedance, determined mainly by the Q-value of the input transformer of the PA output stage, is simulated to $170\,\Omega$ at resonance.

C. Adaptive Bias

To boost OP_{1dB} and P_{sat}, an adaptive bias circuit, see Fig. 4, is used to increase the CS transistor gate bias voltage in the PA output stage as the input signal amplitude increases. The reference current to the adaptive bias circuit can be controlled with a 5-bit digital word. For the adaptive bias circuit to function correctly, it is important that it starts to increase the voltage bias at the input level where the PA output stage starts to compress, and that it also increases the bias voltage with the correct slope to compensate the gain compression at higher input signal levels. To a first approximation the input level starting point can be controlled by the 3-bit bias voltage level, effectively setting at which input amplitude the rectifying PMOS pair will start conduct current. The slope of the increase in gate bias voltage is set with the 3-bit tunable resistor, transforming the DC current from the PMOS pair into a DC voltage. In total 64 different adaptive bias settings can be selected (8 starting points and 8 slopes). Figure 5 shows measured results for output power and CS voltage bias as a function of the input signal level for all 64 combinations. For test reasons the CS bias voltage is also routed, so that it can be measured outside the chip, through an auxiliary path. The output signal of the adaptive bias circuit is designed to track the envelope of the input signal and the bandwidth is limited by the output impedance of the adaptive bias circuit, which is below $1\,k\Omega$ and the input capacitance of the PA output stage CS transistors, which is approximately $2 \times C_{gs} = 340\,fF$. The envelope of an OFDM-modulated signal has the features with the highest frequency components when the amplitude is low, but as the adaptive bias circuit does not change the bias level at low amplitudes, this is not a problem, and it is enough if medium to high signal levels can be tracked.

D. T/R Switch

Figure 6 shows the layout of the T/R switch, implemented in the thick upper copper and aluminum metal layers, including RF and supply pads, the TX 1:2 balun, and the RX 2:1 balun. The upper three pads show GND-Signal-GND connecting to antenna, and the lower two pads are VDD supplying the PA through the center tap of the balun. The architecture of the T/R switch is chosen such that all the three active switches, as shown in Fig. 1, are conducting when in TX-mode. This ensures low voltage across the switch transistors, thereby avoiding the risk of damaging them by

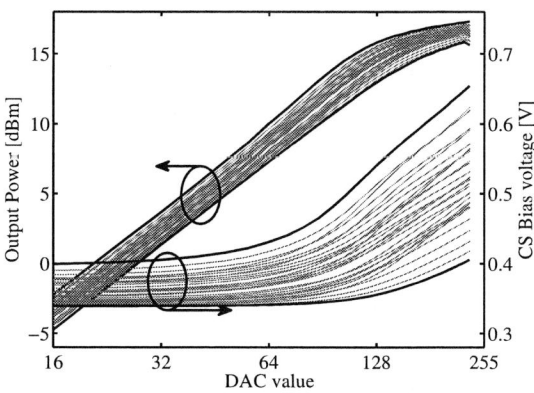

Fig. 5. Measured output power at 27 GHz and CS bias voltage from adaptive bias circuit vs. DAC signal level for all 8x8 adaptive bias settings. The two thick curves are the extremes for the adaptive bias when enabled.

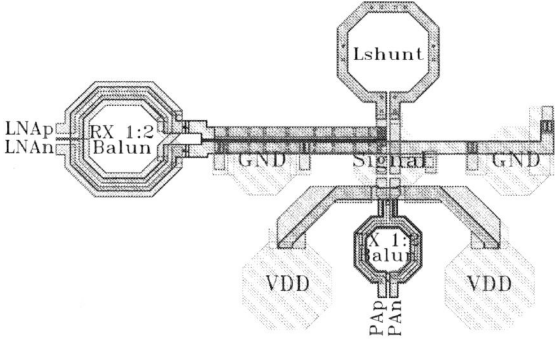

Fig. 6. The layout of the T/R switch and the TX/RX baluns, omitting active parts, vias, and lower metal layers.

high voltage stress in TX mode, without the need for transistor stacking, which has the drawback of increased switch on-resistance R_{on}. In addition, non-conducting transistor switches in TX mode would risk introducing additional non-linearities, as the switches could start to conduct current during voltage peaks. In TX mode the inductor L_{shunt} is used to resonate the parasitic capacitance of the RF pad whereas in RX mode it is used to conduct the received signal to the LNA together with the primary side of the RX 2:1 balun. Capacitive tuning is applied at the input of the LNA for final matching. In RX mode, parasitic capacitance of the switch between the RF pad and the TX balun will leak signal power into TX, which increases the loss in RX mode. To mitigate this, an extra switch is added between the TX balun and the GND pad, which places the parasitic capacitances of the two switches in series, effectively reducing the signal leakage. Figure 7 shows simulated losses in the T/R switch. At 27 GHz, TX and RX insertion loss (IL) are simulated to 1.63 and 1.46 dB, respectively. In addition, for the two 1:2 baluns TX and RX simulated losses are 1.22 and 1.41 dB respectively. Isolation from TX to RX is simulated to 22.1 dB, which at P_{sat} gives a 130 mV$_{rms}$ input signal to the LNA, which is acceptable since the circuit is targeted to operate in time division duplex.

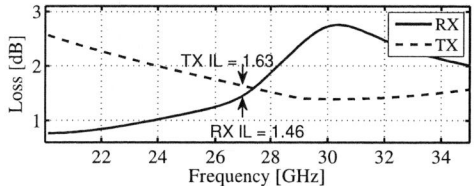

Fig. 7. Simulated T/R-switch losses in TX/RX mode.

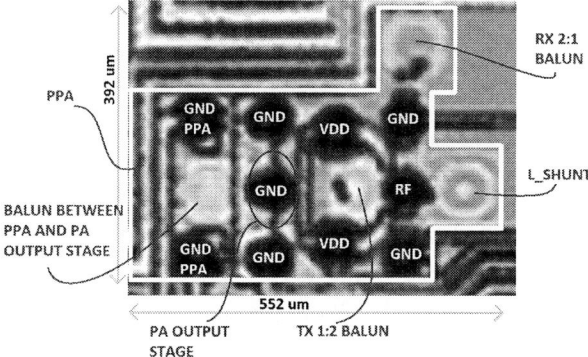

Fig. 8. Die photo showing the the PA and T/R-switch. The area inside the white polygon is 0.146 mm^2.

III. Measurement Results

The chip was fabricated in a 22 nm FD-SOI CMOS process and flip-chip mounted in a package that was mounted on a PCB. The PA, including decoupling capacitors, bias generation, and pads for both supply and RF occupies a die area of 0.129 mm^2. The complete PA and T/R-switch occupies a die area of 0.146 mm^2, see Fig. 8. The input signal to the PA is generated on-chip by an integrated DAC, fed through an on chip analog base band, and up-converted to carrier frequency by a mixer directly preceding the PPA. The PA was measured, using an R&S NRP2 Power Meter and an R&S FSM Signal & Spectrum Analyzer, with both adaptive and constant CS gate bias voltage, configured such that for low input amplitudes the CS-bias voltage was the same for both cases. Measurement results were obtained with identical settings for figures 9 - 11. Figure 9 shows the output power and the CS gate bias voltage as a function of TX-DAC value. The increased CS bias voltage at high input amplitudes when using adaptive bias increases P_{sat} by 1.4 dB. Measured drain efficiency for the PA output stage and normalized power gain for the complete PA is shown in Fig. 10. The gain is normalized to a simulated small signal power gain of 34 dB. When using adaptive bias, drain efficiency peaks at 20.1 %. Furthermore, OP$_{1dB}$ increases by 3 dB from 13.5 to 16.5 dBm, when using adaptive bias. A two tone test with 50 MHz separation around 27.2 GHz, was carried out to assess the linearity of the PA. Figure 11 shows the total power in the two IM3 tones, both with constant and adaptive bias, vs. output power for the fundamental tones. As can be seen, the power in the IM3 products when using adaptive bias is lower for all output power levels reachable when using constant bias. A 2.5 dB reduction in IM3 power can be seen at the highest constant bias two tone output power,

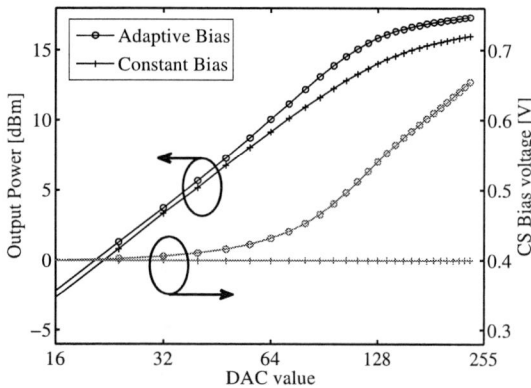

Fig. 9. Output power at 27 GHz vs. input signal to the TX-DAC.

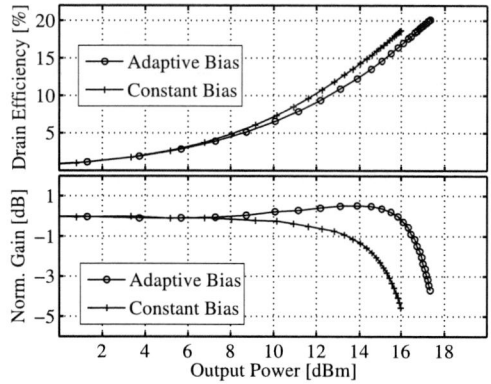

Fig. 10. Upper plot: Drain efficiency vs. output power for adaptive bias and constant bias at 27 GHz. Lower plot: Normalized gain vs. output power for adaptive bias and constant bias at 27 GHz.

i.e. 11.3 dBm. In addition, the adaptive bias case offers >2 dB extended fundamental output power. The 1 dB bandwidth was measured to more than 1.2 GHz, frequency ranging from 26.3 GHz to 27.5 GHz. Table 1 summarises the performance and compares with state-of-the-art PAs in 28 nm CMOS, since no published PAs in 22 nm CMOS process were found for similar frequencies. For fair comparison all PA performance are prior to the losses in the T/R-switch and the proposed PA

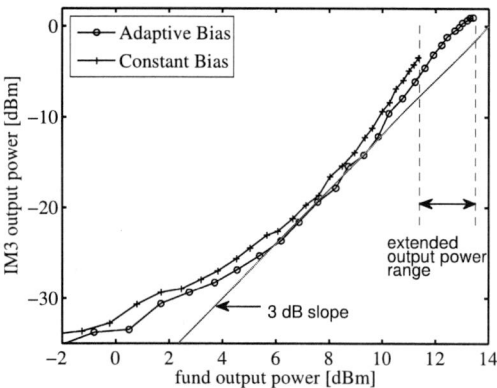

Fig. 11. IM3 tones output power vs. the fundamental tones output power for constant and adaptive bias. A 3 dB slope is added for reference.

Table 1. Performance Summary and Comparison

Parameter	This Work	APMC'18 [3]	RFIC'18 [4]	RFIC'17 [5]	RFIC'17 [6]
Technology [nm]	**22FDSOI CMOS**	28FDSOI CMOS	28FDSOI CMOS	28Bulk CMOS	28Bulk CMOS
Frequency [GHz]	**27**	26	31	27	32
Pwr.Combining	**1**	2	1	2	2
Gain Control [dB]	**28.8**	19	10	None	None
No. of Stages	**2**	2	2	2	2
Vdd [V]	**1.2**	1.5	0.7/1.98	1	1
P_{sat} [dBm]	**17.4**	20.6	17.9	18.1	19.8
P_{1dB} [dBm]	**16.5**	18.8	11.6	16.8	16
PAE_{max} [%]	**19.5**(1)	22.6(1)	25.5	41.5	21
PAE_{1dB} [%]	**17.3**(1)	16.6(1)	10	37.6	12.8
Power Gain [dB]	**34**(2)	33.1(2)	32.6	20.5	22
Area [mm^2]	**0.129**	0.144	0.508	0.361	0.59

(1)Calculated from measured drain efficiency and simulated power gain. Not including PPA power consumption.
(2)Simulated value since input signal is generated internally.

achieves the highest power gain, largest gain control range, and smallest die area.

IV. CONCLUSIONS

This paper introduces a variable gain PA at 27 GHz, utilizing adaptive bias that tracks the envelope of the modulated signal, and a T/R-switch. The adaptive bias increases the OP_{1dB} and P_{sat} from 13.5 to 16.5 dBm and from 16 to 17.4 dBm, respectively. In addition, for a two tone test with 50 MHz frequency difference, the adaptive bias reduces the power of the IM3-tones by 2.5 dB. The T/R-switch achieves a simulated TX/RX insertion loss of 1.63/1.46 dB. For reliability reasons, the transistors in the T/R-switch are all in highly conductive mode when transmitting. Both the PA and the T/R-switch are implemented in a 22 nm FD-SOI CMOS process and integrated on the same chip, in a transceiver targeted for 5G AAS.

REFERENCES

[1] B. Sadhu et al., "A 28-GHz 32-Element TRX Phased-Array IC With Concurrent Dual-Polarized Operation and Orthogonal Phase and Gain Control for 5G Communications," in IEEE Journal of Solid-State Circuits, vol. 52, no. 12, pp. 3373-3391, Dec. 2017. doi: 10.1109/JSSC.2017.2766211

[2] 3GPP TS 38.104 V15.0.0 (2017-12), Technical Specification, 3rd Generation Partnership Project; Technical Specification Group Radio Access Network; NR; Base Stations (BS) radio transmission and reception (Release 15). http://www.3gpp.org/ftp/Specs/archive/38_series/38.104/

[3] C. Elgaard, A. Axholt, E. Westesson and H. Sjöland, "A 26GHz 22.2DBM Variable Gain Power Amplifier in 28NM FD-SOI CMOS for 5G Antenna Arrays," 2018 Asia-Pacific Microwave Conference (APMC), Kyoto, 2018, pp. 965-967.

[4] F. Torres, M. De Matos, A. Cathelin and E. Kerhervé, "A 31 GHz 2-Stage Reconfigurable Balanced Power Amplifier with 32.6dB Power Gain, 25.5% PAEmaxand 17.9dBm Psatin 28nm FD-SOI CMOS," 2018 IEEE Radio Frequency Integrated Circuits Symposium (RFIC), Philadelphia, PA, 2018, pp. 236-239.

[5] Y. Zhang and P. Reynaert, "A high-efficiency linear power amplifier for 28GHz mobile communications in 40nm CMOS," 2017 IEEE Radio Frequency Integrated Circuits Symposium (RFIC), Honolulu, HI, 2017, pp. 33-36.

[6] P. Indirayanti and P. Reynaert, "A 32 GHz 20 dBm-PSAT transformer-based Doherty power amplifier for multi-Gb/s 5G applications in 28 nm bulk CMOS," 2017 IEEE Radio Frequency Integrated Circuits Symposium (RFIC), Honolulu, HI, 2017, pp. 45-48.

A 9 dB Noise Figure Fully Integrated 79 GHz Automotive Radar Receiver in 40 nm CMOS Technology

Tomotoshi Murakami[#1], Nobumasa Hasegawa[#2], Yoshiyuki Utagawa[#3], Tomoyuki Arai[#4], Shinji Yamaura[#5]

DENSO Corporation, Tokyo, Japan

{[1]tomotoshi_murakami, [2]nobumasa_hasegawa, [3]yoshiyuki_utagawa}@denso.co.jp,
{[4]tomoyuki_arai, [5]shinji_yamaura}@denso.co.jp

Abstract— This paper presents a low noise and fully integrated automotive radar receiver in 40 nm CMOS technology. The receiver adopts the direct conversion architecture, and it consists of a low noise amplifier (LNA), mixer, and analog baseband blocks. The three-stage LNA and the low noise mixer improve the whole receiver noise figure (NF). Integrated low-dropped out regulators (LDOs) generate 1.1 V for the LNA and the mixer core from 1.8 V supply. The receiver chip also includes a frequency doubler, phase-locked loop (PLL), bandgap reference bias circuit, and a serial peripheral interface (SPI). The die is packaged using the wafer level chip size package (WLCSP). The receiver achieves 9.0 dB NF at 81 GHz local (LO) and 76.8 dB maximum gain. Moreover, the receiver front-end shows -22.3 dBm IP1dB, and consumes 143 mW power dissipation, and 0.8 mm² chip area.

Keywords — automotive, mm-Wave, radar, CMOS, receiver, Low Noise Amplifier (LNA), mixer, NF

I. INTRODUCTION

The automotive radar systems have been developed over the past years to realize a more secure driving environment. To further expand automotive radar market, large volume production and highly integrated functions are required for cost reduction. For long-range radar (LRR) in the future, it must cover a range of more than 200 meters. To achieve this requirement, a lower noise receiver is necessary.

The conventional automotive radar has been developed in the latest SiGe technologies since decades ago, and several automotive radar transceivers realized in SiGe have been reported [1] - [5]. However, the requirement for higher performance and higher integrated functions of future radar is demanding, and it is a challenge to integrate digital circuits since SiGe BiCMOS technology is based on 130 nm / 180 nm CMOS technology.

On the other hand, the latest nano-scale CMOS technology becomes attractive for automotive radars, because low noise and low power consumption can be expected by improving the maximum unity current gain frequency f_T with technology scaling. CMOS transceivers also have been reported [6] - [8], but the noise or power dissipation performance is not sufficient for long-range radar.

In this paper, we present a highly integrated and low noise receiver for a 79 GHz automotive radar receiver in 40 nm CMOS technology. Section II briefly describes the receiver architecture. Section III discusses the building blocks design, and the experimental results are summarized in Section IV. Finally, Section V concludes this work.

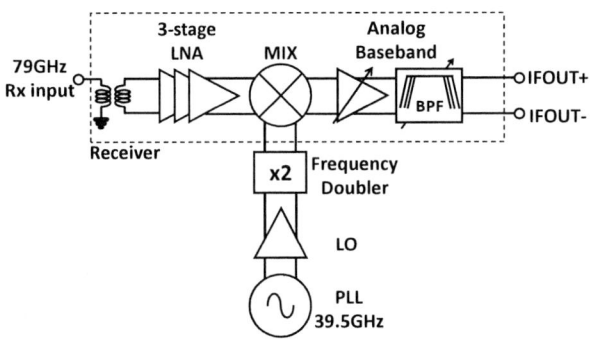

Fig. 1. Receiver block diagram

II. RECEIVER ARCHITECTURE

Recent automotive radar uses the Frequency Modulated Continuous Wave (FMCW). The operating principle of FMCW radar is to find the frequency difference of the reflected signal compared to a reference signal. The distance and the speed of the object are calculated from the frequency difference. Fig.1 shows the proposed receiver block diagram based on the Direct Conversion architecture. It consists of an LNA, a down conversion mixer, and the analog baseband blocks. The reflected signal from the object is converted from a single-ended signal to a differential signal with integrated balun. The three-stage LNA amplifies the signal and the mixer down converts the signal to intermediate frequency (IF). To suppress the noise due to the DC offset of the IF output level, the mixer has a double balanced configuration. The subsequent analog baseband blocks filter and amplify to a sufficient signal level for signal processing, and the output signal is sent to the following ADC and digital circuit. The analog baseband block includes a band-pass filter (BPF) and a variable gain amplifier (VGA) which provides a 64 dB maximum variable gain range. For a detection range of more than 200 meters, the receivers have a maximum gain over 70 dB. The reference LO signal, which is generated at around 40 GHz by integrated PLL, is amplified by a LO buffer and frequency doubler. The doubler output frequency is 80 GHz which is twice that of the LO signal. After that, the local signal is provided to the mixer in the receiver. A bandgap reference bias circuit, LDOs, and SPI are also integrated to provide reference bias, internal power supply, and digital control signals, respectively.

To realize a low noise receiver, it is necessary to minimize the noise contribution from each block. The receiver achieved

978-1-7281-1702-7/19 $31.00 © 2019 IEEE

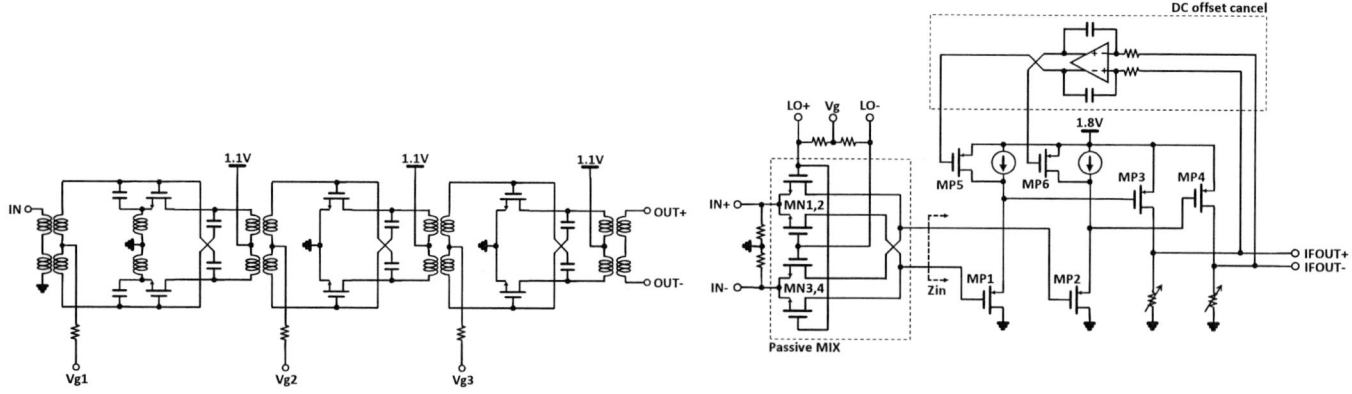

Fig. 2. Schematic of 3-stage differential LNA (left) and passive mixer with buffer amplifier (right)

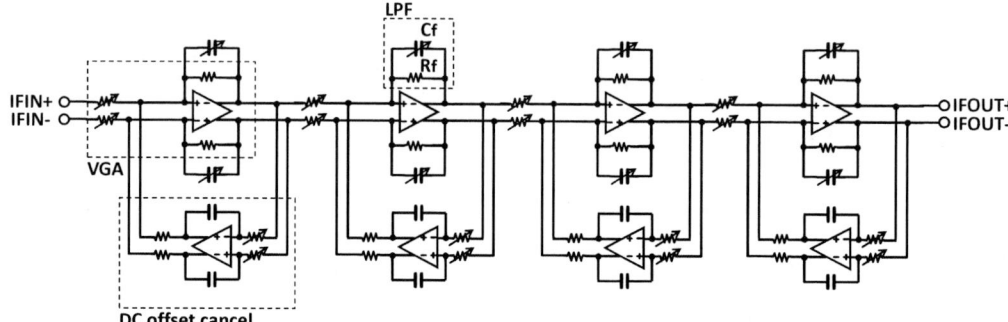

Fig. 3. Schematic of the analog baseband blocks

the low noise characteristics, due to the minimized NF not only of the LNA but also the mixer. Design details of the key blocks will be described in the following sections.

III. BUILDING BLOCKS

A. LNA

The simplified schematic of the three-stage differential LNA is shown in Fig. 2. The common-source (CS) amplifier is employed for a basic topology. It improves the 79 GHz performance compared to the cascode amplifier. Although the cascode amplifier has the advantage of high output impedance and good input-output isolation, the gain of the cascode amplifiers is not higher than that of the CS amplifier at 79 GHz, due to the parasitic capacitance on the node between the common source transistor and the common gate transistor. Furthermore, the thermal noise generated from the common gate transistor in the cascode amplifier has a large impact on the noise figure of the LNA, unlike sub-10 GHz operation.

However, the stability of the CS amplifier is not high, due to the poor input-output isolation in the CS amplifier. Therefore, the CS amplifier of each stage employs the cross-coupled capacitors to stabilize the amplifier by neutralizing gate-drain capacitance (C_{gd}). The gain of each amplifier is also enhanced by cancelling C_{gd}.

The LNA has a single input port and differential output port. The input balun converts the input signal from single-ended to differential. Due to the differential operation, the common-mode noise can be reduced and linearity is also

improved. Note that an inductive degeneration structure is adopted in the first stage to achieve a noise-free 50 Ω input matching. The target value of the NF is 7 dB for the LNA.

B. MIXER

The schematic of the mixer is shown in Fig. 2. It converts the frequency of the LNA output signal to IF frequency, and sends the signal to the following analog baseband blocks. The mixer consists of the double-balanced passive mixer (1.1 V operation) and a buffer amplifier (1.8 V operation). With this configuration, the voltage swing becomes large since the input impedance of the source followers (MP1, 2 in Fig. 2) is high. As a result, the NF of the mixer is improved.

The passive mixer has no DC current through the switches. It makes it possible to reduce the flicker noise. Moreover, the passive mixer also featuers high linearity. The width of the transistor should be large enough in order to provide a sufficiently low on-resistance. The gate-source voltage is chosen at the class-B bias point to satisfy both noise and linearity requirements.

The buffer amplifier is a cascade of source follower and common source (CS) amplifier with resistive load. The DC offset cancel loop is added to cancel the DC offset that is mainly caused by the feed-through from the RF to LO port, and to filter out strong reflections from nearby targets. Note that the source follower and the CS amplifier are mainly composed of pmos transistors (MP1-4 in Fig. 2). This is to reduce the generation of flicker noise compared to that of a nmos-type buffer.

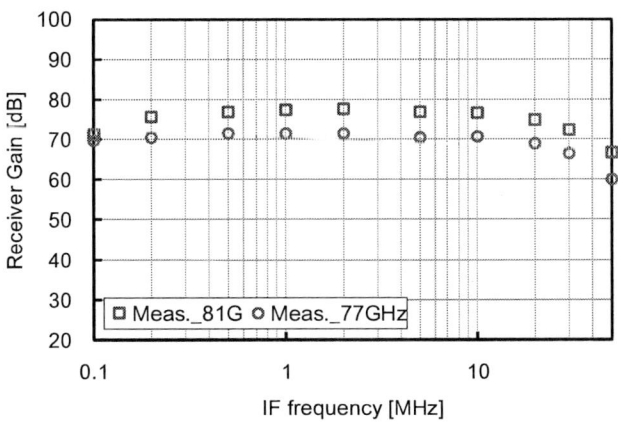

Fig. 4. Measured receiver gain versus IF frequency

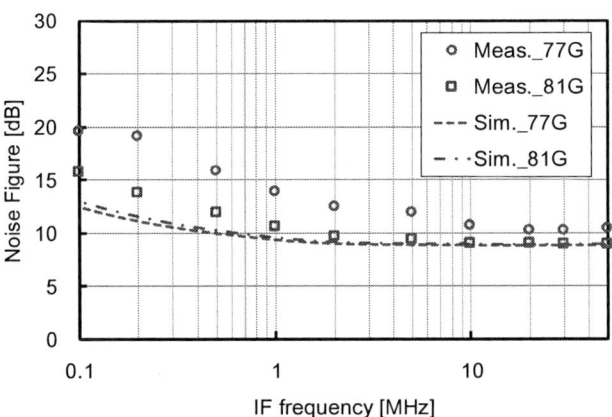

Fig. 5. Measured NF at 77G/81G versus IF frequency

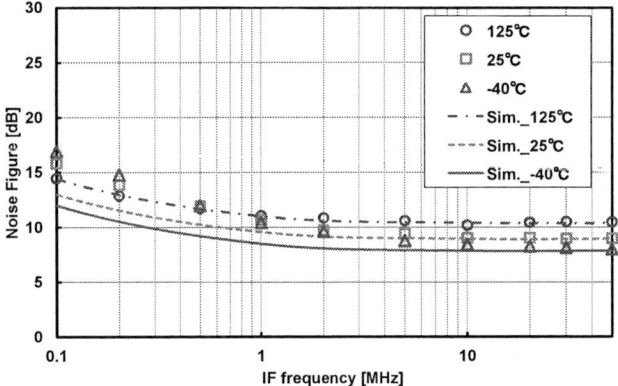

Fig. 6. Measured NF versus IF frequency over temperature

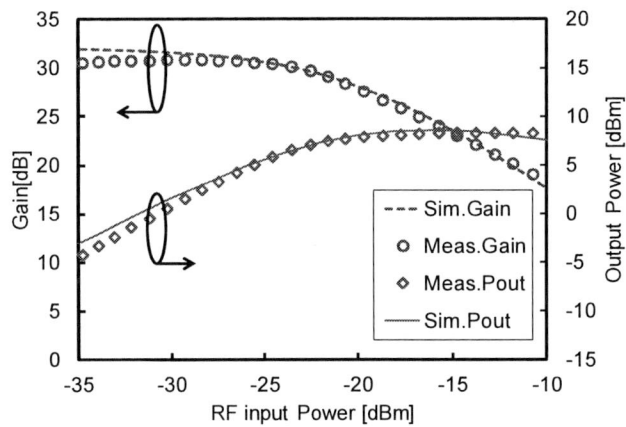

Fig. 7. Measured receiver front-end Pout versus RF input power at 81GHz

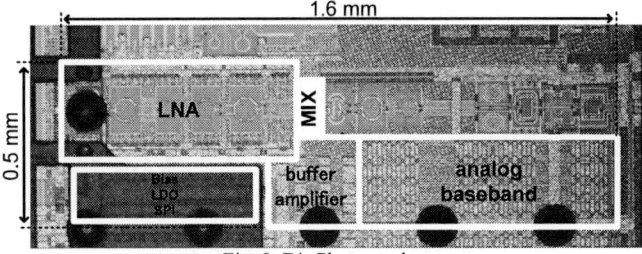

Fig. 8. Die Photograph

C. ANALOG BASEBAND BLOCKS

The analog baseband blocks are integrated on the same die to provide wide gain dynamic range and filtering characteristics. Fig. 3 shows the schematic of the analog baseband blocks. It consists of a four-stage VGA with a low-pass filter (LPF) and a DCOC loop. The VGA employs a resistive feedback fully differential amplifier. The gain is digitally programmed by configuring the feedback switched-resistor array. Each VGA stage provides a programmable gain from 0 to 16 dB with 1 dB step. The analog baseband provides a total dynamic range of 64 dB. The LPF consists of the resistor R_f and the capacitor C_f, and operates as a 4th order active RC Butterworth filter. The LPF bandwidth is from 3.125 MHz to 25 MHz. The DCOC loop also operates as a high-pass filter (HPF) to filter out the strong reflection from nearby targets. The reconfigurable HPF corner frequency is from 1 kHz to 100 kHz.

IV. EXPERIMENTAL RESULTS

The receiver chip was fabricated in a 40 nm CMOS process. The die is packaged using WLCSP and measured with an evaluation board. The receiver was measured at 1.8 V

power supply, while the LNA and mixer core were measured at 1.1 V power supply generated by internal LDOs. The receiver consumes DC total power of 143 mW, consisting of power dissipation of 22.8 mW in the LNA, 19.2 mW in the mixer, and 100.9 mW in the analog baseband blocks. The total DC power is 157 mW with LDOs. The receiver front-end consumes only 42 mW excluding the baseband blocks.

The gain and NF of the receiver have been measured versus IF frequency at 77/81 GHz, and the results are illustrated in Fig. 4, 5. The IF frequency range is from 100 kHz to 50 MHz. The NF was measured using the Y-factor method. The analog baseband blocks gain is set to 48 dB

Table 1. Performance comparison of receiver

	This work	[1]	[2]	[6]	[7]
Technology	40 nm CMOS	130 nm SiGe BiCMOS	130 nm SiGe BiCMOS	65 nm CMOS	65 nm CMOS
Integrated Blocks	LNA, mixer, baseband, doubler, PLL, LDO	LNA, mixer, TIA, VCO, div.	LNA, mixer, baseband, doubler, PLL	LNA, mixer, doubler	LNA, sub-Harmonic mixer, passive mixer
G_c (dB)	30.8	20-24	73	21	16
NF_SSB (dB)	9.0	12.5	5.9-7.0	12	13
IP1dB (dB)	-22.3	n/a	-21.6	n/a	-15 (*)
Power (mW)	143 (*)	91	594	400 (*)	28.2
Supply (V)	1.1 (RF), 1.8 (baseband)	1.8	1.6 (LNA), 2.7 (mixer)	n/a	1.2
Area (mm^2)	0.8 (*)	3.7 x 2.0 (*)	3.6 x 1.7	4.1 x 4.1 (*)	2.3
Notes	(*) LNA, mixer, baseband	(*) 4TX-4RX, VCO		(*) 4RX	(*) LNA only

in the measurement. In addition, the LO signal for the mixer is generated by the integrated PLL in the measurement. The receiver achieves a gain of 76.8 dB, and a single-sideband (SSB) NF of 9.0 dB (81 GHz LO, 20 MHz IF) at room temperature. The receiver also achieves an NF 10.3 dB at 77 GHz LO. The difference of an NF in 77 GHz and 81 GHz is caused by the LNA frequency characteristics. The NF of the receiver at 81 GHz was also measured versus IF frequency and temperature, and the results are shown in Fig. 6. The variation of NF is less than 2.1 dB when the ambient temperature changes from -40°C to +125°C.

Fig. 7 shows the gain compression characteristics of the receiver front-end at 0 dB VGA gain setting (81 GHz LO, 5 MHz IF). The receiver has a conversion gain (G_c) of 30.8 dB and an IP1dB of -22.3 dBm.

Fig. 8 shows the die micrograph of the receiver chip. The receiver and other peripheral circuits are integrated in the same die. The size of the receiver block is 1.6 mm x 0.5 mm. The performance of the receiver is compared with the other similar works in Table 1.

V. CONCLUSION

A fully integrated low noise receiver for a 79 GHz automotive radar receiver in 40 nm CMOS technology has been reported. The receiver noise figure is improved by reducing the noise contribution of the LNA and the mixer, and the receiver achieves an NF of 9.0 dB at 81 GHz. The receiver front-end has an IP1dB of -22.3 dBm. The power dissipation is 143 mW and the maximum gain is 76.8 dB. Low noise, high integrity, and low power dissipation performance are obtained compared with the reported transceivers. The demonstrated 79 GHz CMOS receiver is one of the best candidates for the automotive radar receiver of the future.

ACKNOWLEDGMENT

The authors would like to thank RF semiconductor development dept. members for valuable comments.

REFERENCES

[1] Andrew Townley, Paul Swirhun, and Ali Niknejad, "A 94GHz 4TX-4RX Phased-Array for FMCW Radar with Integrated LO and Flip-Chip Antenna Package," *IEEE Radio Frequency Integrated Circuit Symposium*, pp. 294-297, 2016.

[2] Rose Ben Yishay, Oded Katz, and Danny Elad, "High-Performance 81-86 GHz Transceiver Chipset for Point-to-Point Communication in SiGe BiCMOS Technology," *Radio Frequency Integrated Circuit Symposium*, pp. 417-420, 2015.

[3] Christoph Wagner, Josef Bock, and Lunus Maurer, "A 77GHz Automotive Radar Receiver in a Wafer Level Package," *Radio Frequency Integrated Circuit Symposium*, pp. 511-514, 2012.

[4] Scan T Nicolson, Kenneth H. K. Yau, and Sorin P. Voinigescu, "A Low-Voltage SiGe BiCMOS 77-GHz Automotive Radar Chipset," *IEEE TRANSACTIONS ON MICROWAVE THEORY AND TECHNIQUES*, vol. 56, pp. 1092–1104, May 2008.

[5] Li Wang, Srdjan Glisic, and J. Christoph Scheyu, "A Single-Ended Fully Integrated SiGe 77/79 GHz Receiver for Automotive Radar," *IEEE JOURNAL OF SOLID-STATE CIRCUITS*, vol. 43, pp. 1897-1908, Sep. 2008.

[6] Toshihiro Shimura, Hiroshi Matsumura, and Yoji Ohashi, "Multi-channel Low-noise Receiver and Transmitter for 76-81 GHz Automotive Radar Systems in 65 nm CMOS," *European Microwave Conference*, pp. 596-599, Oct. 2014.

[7] Hoang Viet Le, Hoa Thai Duong, and Efstrations Skafidas, "A CMOS 77 GHz Radar Receiver Front-end," *European Microwave Conference*, pp. 13-16, Oct. 2013.

[8] Jri Lee, Yi-An Li, and Shih-Jou Huang, "A Fully-Integrated 77-GHz FMCW Radar Transceiver in 65-nm CMOS Technology," *IEEE JOURNAL OF SOLID-STATE CIRCUITS*, vol. 45, pp. 2746-2756, Dec. 2010.

A Compact 76-81 GHz 3TX/4RX Transceiver for FMCW Radar Applications in 65-nm CMOS Technology

Liang Chen, Lei Zhang, Weiping Wu, Li Zhang, and Yan Wang

Institute of Microelectronics, Tsinghua University, China

zhang.lei@tsinghua.edu.cn

Abstract — This paper presents a compact 76-81 GHz 3TX/4RX transceiver for FMCW radar applications in 65-nm CMOS technology. Three individual transmitters (TXs) and four receivers (RXs) are integrated for MIMO operation to achieve higher resolution and sensitivity. A wideband frequency sextupler cascaded with a Wilkinson power divider is employed as the local oscillation (LO) chain to lower down the frequency of FMCW signal generator and simplify the design of LO distribution network. Each transmitter channel achieves above 12.2 dBm output power from 76-81 GHz and a peak 14.2 dBm output power is obtained at 77 GHz. The receiver conversion gain from low noise amplifier input to baseband output is programmable from 46.3 dB to 102.45 dB for adapting to different detection distances. The measured noise figure of the receiver chain is 11 dB at 5 MHz intermediate frequency (IF). The transceiver including all dc bonding pads occupies a silicon area of 1.8 mm × 4 mm, and the power dissipation of the whole chip is 721.8 mW.

Keywords — FMCW, CMOS, MIMO, radar, transmitter, receiver, frequency sextupler, compact layout

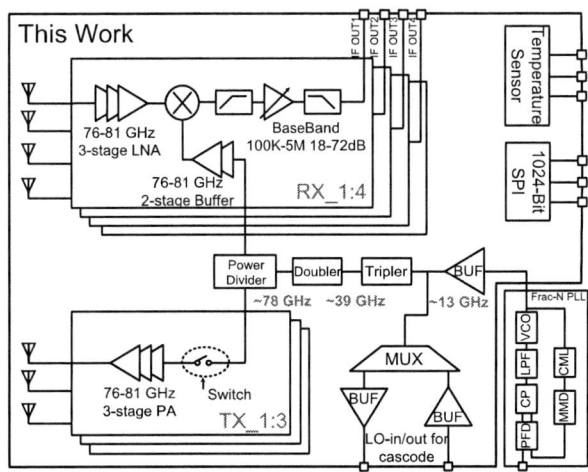

Fig. 1. Simplified block diagram of the proposed radar transceiver.

I. INTRODUCTION

In recent years, silicon based CMOS technology offers great opportunities for millimeter-wave applications such as automotive radar due to its rapid scaling down of transistor size. Intensive researches and development of millimeter-wave radar are underway due to its immunity to natural environment such as extreme temperature, bad light or weather conditions [1]. To overcome several challenges of CMOS technology compared with compound semiconductor process, some advanced transceiver architectures are proposed to improve the CMOS-based radar performance, such as frequency multiplier based transceiver to improve phase noise of the local oscillation (LO) signal [2][3]. In [2], a 77 GHz transceiver with injection locking frequency sextupler is proposed to lower down the local oscillation frequency and a 13.7 dBm output power with a 8.8 dB RX noise figure is achieved. An internal × 10 frequency multiplier based 77 GHz TRX is proposed in [3] and the radar system shows 9 dBm output power with a 13 dB conversion gain. Besides, many single channel radar transceivers with decent performance have been demonstrated [4]-[9]. In [9] , a frequency-modulated-continuous-wave (FMCW) radar transceiver with one transmitter and receiver path is presented and less than 1% ranging error from 1 m to 8 m distance is achieved. A fully-integrated FMCW radar system utilizing a fractional- synthesizer as the FMCW generator is proposed in [6], which provides a maximum detectable distance of 106 meters for a mid-size car.

However, besides the target velocity and distance information, angular identification is also required to recognize the traffic scenario around the vehicle for advanced automotive radar systems. Using more than one receive and transmit channels, additional angular information can be obtained. Therefore, multi-channel radar transceiver architecture is the trend for radar applications and higher resolution and sensitivity can be achieved further by the MIMO operation.

In this paper, a compact 76-81 GHz FMCW transceiver with three TX channels and four RX channels in 65-nm CMOS is presented to support angular identification and MIMO operation for automotive radar applications. A wideband frequency sextupler utilizing push-push topology is employed as the LO chain to improve phase noise of the FMCW signal generator and simplify the LO distribution network. Besides, long, medium and short-range transmitter patterns are achieved by the integrated individual three TX channels combined with the microstrip patch antenna with different gain and azimuth angle.

II. TRANSCEIVER ARCHITECTURE AND CIRCUIT DESIGN

Fig. 1 illustrates the simplified block diagram of the proposed radar transceiver. The transceiver incorporates three TXs and four RXs to support MIMO processing. On the receiver side, the received signals are amplified by the low noise amplifier (LNA) and down-converted to the intermediate frequency by a passive mixer and then be processed by the followed trans-impedance amplifier (TIA) and analog

978-1-7281-1702-7/19 $31.00 © 2019 IEEE

2019 IEEE Radio Frequency Integrated Circuits Symposium

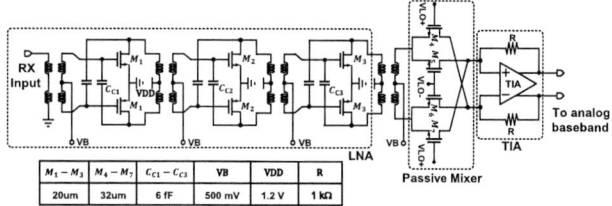

$M_1 - M_3$	$M_4 - M_7$	$C_{C1} - C_{C3}$	VB	VDD	R
20um	32um	6 fF	500 mV	1.2 V	1 kΩ

Fig. 2. Simplified schematic of the proposed receiver.

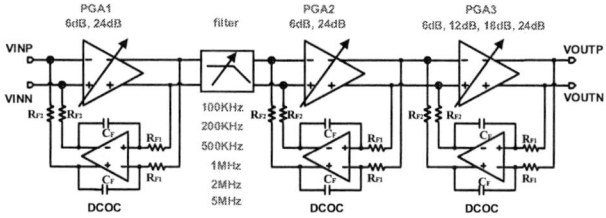

Fig. 3. Simplified block diagram of the proposed analog baseband.

baseband. Passive mixer is adopted here to reduce flick noise and improve the receiver linearity. The proposed analog baseband consists of a fifth-order Butterworth filter with a reconfigurable bandwidth and a 3-stages programmable gain amplifier (PGA) with dc-offset cancellation (DCOC) to eliminate the leakage-induced dc-offsets. On the transmitter sides, three individual transmit channels are integrated to achieve different transmitter patterns combined with the microstrip patch antenna. To lower down the frequency and phase noise of FMCW signal generator, a wideband frequency sextupler LO scheme is adopted in this paper. A pair of input / output LO buffers are designed for testing and multiple chip cascading. A temperature sensor is also integrated to monitor the chip temperature for automotive application.

A. Receiver Design

Fig. 2 presents the simplified schematic of the proposed receiver. Each receiver chain consists of a LNA, a passive mixer, a trans-impedance amplifier (TIA) to convert the down-converted current to voltage and a programmable analog baseband. The received signal is converted into differential by an on-chip multi-turn balun, which achieves power and noise matching at the same time, and then amplified by the three-stage common-source neutralized LNA. Neutralization capacitors are adopted to improve stability and power gain, thus the noise contribution of the following stages are compressed by the high gain LNA. To reduce receiver flicker

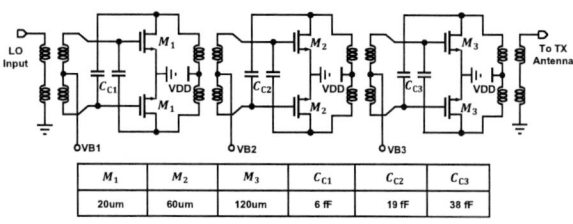

M_1	M_2	M_3	C_{C1}	C_{C2}	C_{C3}
20um	60um	120um	6 fF	19 fF	38 fF

Fig. 4. Simplified schematic of the proposed power amplifier.

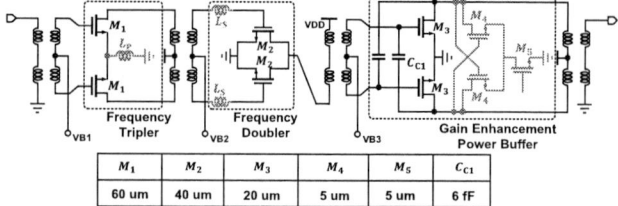

M_1	M_2	M_3	M_4	M_5	C_{C1}
60 um	40 um	20 um	5 um	5 um	6 fF

Fig. 5. Simplified schematic of the proposed frequency multiplier chain.

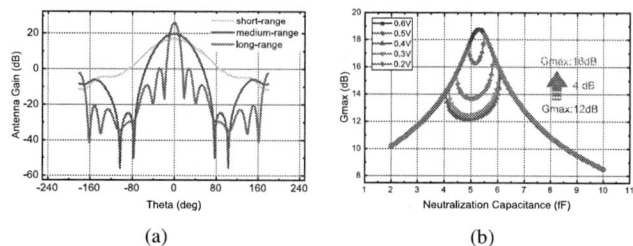

(a)   (b)

Fig. 6. Simulated (a) TX antenna radiation pattern under long, medium and short-range applications (b) gain enhancement of the proposed power buffer under different gain control voltage.

noise and improve linearity, passive mixer combined with a trans-impedance amplifier is adopted in this paper to convert the received high frequency signal to intermediate frequency.

To support different detection distance and accuracy in automotive radar transceiver, a programmable gain amplifier (PGA) to provide appropriate gain for different target distance and a low pass filter (LPF) with reconfigurable bandwidth for accuracy requirements are mandatory. Fig. 3 shows the block diagram of the proposed analog baseband. By combining the coarse and fine gain adjustment stage, which is achieved by the PGA1, PGA2 and PGA3, a gain adjustment range of 18 dB to 72 dB with 6 dB per step is obtained. To eliminate the leakage-induced dc-offsets, a DCOC loop is contained in each PAG stage. The fifth-order Butterworth filter, located between PGA1 and PGA2, provides a reconfigurable bandwidth including 100 KHz, 200 KHz, 500 KHz, 1 MHz, 2 MHz and 5 MHz, respectively.

B. Transmitter Design

A three-stage differential common source power amplifier is integrated in each transmit channel, whose schematic is shown in Fig. 4. Transistor width of the output stage M_3

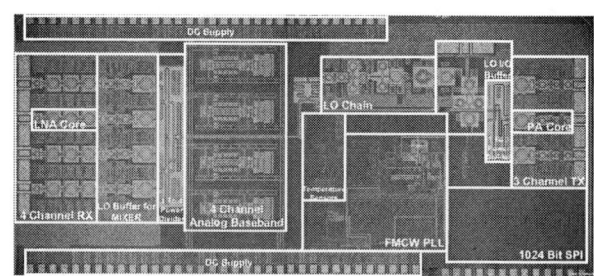

Fig. 7. Microphotograph of the chip. GSG ground pads are shared between adjacent channels to reduce die area.

978-1-7281-1702-7/19 $31.00 © 2019 IEEE   312

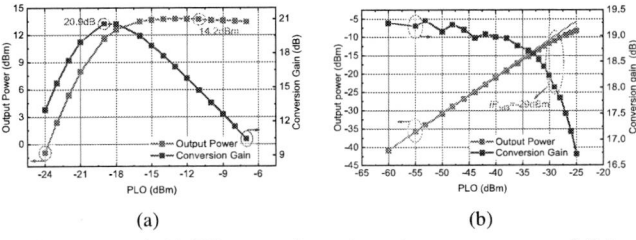

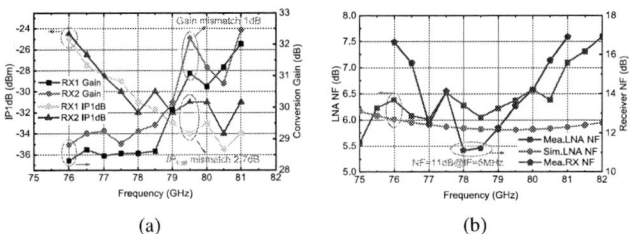

Fig. 8. Measured (a) TX conversion gain and output power vs. LO input power (b) RX input 1dB compression point at 77 GHz.

Fig. 10. Measured RX frequency response of (a) IP_{1dB} and conversion gain (b) noise figure at 5 MHz IF frequency.

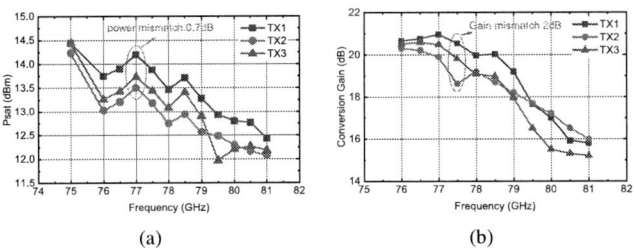

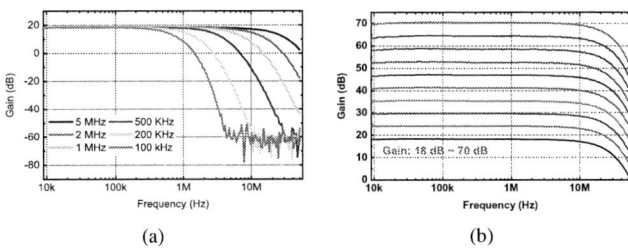

Fig. 9. Measured TX frequency response of (a) saturated output power and (b) conversion gain.

Fig. 11. Measured analog baseband (a) frequency response vs. bandwidth settings at 18 dB (b) frequency response vs. gain settings at 5MHz bandwidth.

is determined by the desired output power and the input and driver stages are sized to improve efficiency while not affecting linearity. Transformer based matching networks are employed for the benefit of compact footprint, convenient single-ended to differential conversion, simple separate DC biasing and wideband performance. Custom made parallel plate MOM capacitors are utilized for better accuracy and higher quality factor. Microstrip patch antenna with different gain and azimuth angle is co-designed with the integrated PA to achieve long, medium and short-range radar applications. The simulated TX antenna radiation pattern under different detection range applications is shown in Fig.6 (a) and a 66 degree azimuth angle is achieved at the short range mode.

C. LO Chain Design

As described in [3], a low frequency signal generator cascading with frequency multiplier is a good candidate to achieve better phase noise performance. In this paper, a frequency sextupler that consists of a frequency tripler and doubler is adopted to convert local FMCW signals to the desired 76-81 GHz range. Fig. 6 depicts the schematic of the proposed frequency sextupler. Transistor M_1 is biased at the pinch-off region to enhance harmonic components of the input signal. To improve output third harmonic components, a peaking inductor L_p is inserted at the tripler device M_1 common source terminal to resonate with parasitic capacitance C_P at the twice frequency of the input signal. The inter-stage matching network between frequency tripler and doubler keeps the third harmonic term and suppresses others. To achieve broadband performance, the push-push topology is employed in the design of frequency doubler to eliminate odd harmonic signals and double the frequency of input signal with the help of appropriate biasing. Series inductor L_s is added at

the gate terminal of the doubler device M_2 to lower down the input impedance quality factor for better power matching. G_m-boosted gain enhancement power buffer is adopted after the multiplier chain to enhance its output power and drive wideband power splitting network. A negative resistance is introduced to the load impedance network by the cross-coupled transistor pair M_4-M_5 and can be adjusted by the control voltage of transistor M_5. Fig. 6 (b) shows the simulated maximum power gain G_{max} of the proposed gain enhancement power buffer under different control voltage and a maximum 4 dB gain enhancement is achieved.

III. MEASUREMENT RESULTS

The proposed 76-81 GHz 3TX/4RX FMCW transceiver has been implemented in 65-nm CMOS general purpose (GP) 1P9M process, and the microphotograph of the whole chip is shown in Fig. 7. GSG probe ground pads are shared between adjacent channels to reduce die area and the chip area including all DC bonding pads is 1.8 mm × 4 mm.

The measured TX conversion gain and output power versus LO input power at 77 GHz is shown in Fig. 8 (a), where a 14.2 dBm saturated output power with a 20.9 dB conversion gain is achieved. Fig. 9 (a) depicts the TX frequency response of saturated output power P_{sat} and a peak 14.2 dBm output power is achieved at 77 GHz. The output power over 76-81 GHz varies from 12.2 to 14.2 dBm with a 0.7 dB power mismatch between three transmitter channels. A 21 dB TX peak conversion gain is achieved at 77 GHz with a 2 dB gain mismatch between TX1 and TX3, which is shown in Fig. 9 (b). The measured conversion gain at low frequency range is somewhat higher than that at high frequency due to inaccurate device modeling and electromagnetic simulation.

Fig.8 (b) reveals that the measured input 1dB compression point of the receiver path (LNA+Mixer+TIA) at 77 GHz is

Table 1. Performance summary and comparison of the 77 GHz CMOS automotive radar transceivers.

	[2]	[3]	[5]	[6]	[7]	[8]	[9]	This Work
Technology	65nm CMOS	65nm CMOS	90nm CMOS	65nm CMOS	65nm CMOS	65nm CMOS	90nm CMOS	65nm CMOS
Frequency [GHz]	75-82	76.8-77.4	73.5-77.1	75.6-76.3	75.2-79.2	76.0-76.7	78.1-78.8	76-81
Topology	Multiplier based TRX	Multiplier based TRX	VCO based TRX	PLL based TRX	VCO based TRX	PLL based TRX	VCO based TRX	Multiplier based TRX
Channel No. (TX/RX)	2/6	1/1	1/1	1/1	1/1	1/1	1/1	3/4
TX Output Power [dBm]	13.7	8.9	3.3-6.3	5.1	10	6.4	-2.8	14.2
TX Conv. Gain [dB]	-	7.4**	8.5**	13.7	14.3	15.18	14	20.9
RX Conv. Gain [dB]	31.6	13	8*	22.0	7	23	23.1	46.3-102.45
RX IP$_{1dB}$ [dBm]	-37	-27	-	-26.5	-28.2	-	-	-24
RX NF [dB]	8.8	41*	-	30*	-	14.8	15.6	11
LNA NF [dB]	6.6	7.95	6.8	7.4	8.7	6.0	6.8	6.0
DC Power [W]	1.43	0.467	0.92	0.243	0.275	0.52	0.517	0.72
Chip Area [mm^2]	3.63×2.91	1.82×2.82	2.4×1.2	0.95×1.1	3.7×2.93	1.03×0.94	3.5×1.95	1.8×4

* Calculated from measurement results   ** Power gain of a standalone PA

-29 dBm and the measured conversion gain ranges from 16.2 to 20 dB with a 1 dB gain mismatch between RX1 and RX2, which is shown in Fig.10 (a). The gain of analog baseband is not included in this figure due to testing limitation. The measured RX frequency response of IP_{1dB} and conversion gain is also presented in Fig.10 (a), where IP_{1dB} is higher at low frequency range due to higher conversion gain. Fig.10 (b) shows the measured noise figure of the receiver path and a standalone low noise amplifier. The minimum RX noise figure is 11 dB at 5 MHz IF frequency. For the standalone low noise amplifier, a 6 dB minimum noise figure is achieved over 76-81 GHz, which agrees well with the simulation results.

Fig. 11 presents the measured bandwidth and gain of the proposed analog baseband. The measured bandwidth can be reconfigured to 100 KHz, 200 KHz, 500 KHz, 1 MHz, 2 MHz and 5 MHz, while the gain can be programmable between 18 dB to 70 dB. Therefore, the RX conversion gain form LNA input to baseband output is 46.3 dB to 102.45 dB. Performance summaries of the proposed 3TX/4RX FMCW radar transceiver and comparison with other state-of-the-art works are listed in Table 1.

IV. CONCLUSION

This work has proposed a compact 76-81 GHz 3TX/4RX transceiver for automotive radar applications in 65-nm CMOS process. The transceiver employs a wideband frequency sextupler cascaded with a wilkinson power divider as the LO chain to lower down the frequency and phase noise of FMCW signal generator. The measured results show that the minimum output power is 12.2 dBm from 76-81 GHz and a peak 14.2 dBm output power is obtained at 77 GHz. The measured receiver conversion gain is programmable from 46.3 to 102.45 dB and the intermediate frequency bandwidth can be reconfigured as 100 KHz, 200 KHz, 500 KHz, 1 MHz, 2

MHz and 5 MHz, respectively. The measured receiver noise figure is 11 dB at 78 GHz with a fixed 5 MHz intermediate frequency. The overall chip size is 1.8× 4 mm^2 with a 721.8 mW dc power consumption.

ACKNOWLEDGMENT

This work is supported by the National High Technology Research and Development Program of China 2018YFB0105002, 2017YFB0102601, 2016YFB0101001, and Tsinghua University Initiative Scientific Research Program.

REFERENCES

[1] J. Park, H. Ryu, K. Ha, J. Kim and D. Baek, "76-81 GHz CMOS Transmitter With a Phase-Locked-Loop-Based Multichirp Modulator for Automotive Radar," *IEEE Trans. Microw. Theory Techn*, vol. 63, no. 4, pp. 1399-1408, Apr. 2015.

[2] Y. Hsiao et al., "A 77 GHz 2T6R Transceiver With Injection-Lock Frequency Sextupler Using 65-nm CMOS for Automotive Radar System Application," *IEEE Trans. Microw. Theory Techn*, vol. 64, no. 10, pp. 3031-3048, Oct. 2016.

[3] C. Cui, S. Kim, R. Song, J. Song, S. Nam and B. Kim, "A 77 GHz FMCW Radar System Using On-Chip Waveguide Feeders in 65-nm CMOS," *IEEE Trans. Microw. Theory Techn*, vol. 63, no. 11, pp. 3736-3746, Nov. 2015.

[4] V. H. Le et al., "A CMOS 77 GHz receiver front-end for automotive radar," *IEEE Trans. Microw. Theory Techn*, vol. 61, no. 10, pp. 3783-3793, Oct. 2013.

[5] Y. Kawano, T. Suzuki, M. Sato, T. Hirose and K. Joshin, "A 77 GHz transceiver in 90-nm CMOS," in *ISSCC Dig. Tech. Papers*, 2009, pp. 310-311.

[6] J. Lee, Y. Li, M. Hung and S. Huang, "A Fully-Integrated 77 GHz FMCW Radar Transceiver in 65-nm CMOS Technology," *IEEE J. Solid-State Circuits*, vol. 45, no. 12, pp. 2746-2756, Dec. 2010.

[7] C. Kim et al., "A CMOS centric 77 GHz automotive radar architecture," in *Proc. IEEE RFIC*, 2012, pp. 131-134.

[8] T. Luo, C. E. Wu and Y. E. Chen, "A 77 GHz CMOS Automotive Radar Transceiver With Anti-Interference Function," *IEEE Trans. Circuits Syst. I: Reg. Papers*, vol. 60, no. 12, pp. 3247-3255, Dec. 2013.

[9] T. Mitomo, N. Ono, H. Hoshino, Y. Yoshihara, O. Watanabe and I. Seto, "A 77 GHz 90-nm CMOS Transceiver for FMCW Radar Applications," *IEEE J. of Solid-State Circuits*, vol. 45, no. 4, pp. 928-937, Apr. 2010.

A Full-Band Multi-Standard Global Analog & Digital Car Radio SoC with a Single Fixed-Frequency PLL

Lucien J. Breems[1], Jan van Sinderen[1], Tom Fric[2], Hans Stoffels[2], Franco Fritschij[2], Hans Brekelmans[1],
Hendrik van der Ploeg[3], Ulrich Moehlmann[4], Robert Rutten[1], Muhammed Bolatkale[1], Shagun Bajoria[1],
Jan Niehof[1], Bert Oude-Essink[2], Gerard Lassche[2]

[1]NXP Semiconductors, Eindhoven, The Netherlands
[2]Catena, Delft, The Netherlands
[3]Catena, Son en Breugel, The Netherlands
[4]NXP Semiconductors, Hamburg, Germany

Abstract— **This paper presents a wideband car radio SoC for global multi-standard and multi-channel analog and digital broadcast radio. One of the major challenges of a wideband receiver is that the RF circuits and A/D converters need to have very high in-band IIP3 while the receiver NF should be low. These IIP3 and NF requirements are difficult to meet simultaneously. Key techniques to achieve low noise and very high linearity are high gain-bandwidth multi-inverter-based amplifiers supplied by PVT-tracking low-ohmic low dropout regulators, very linear high-speed 1-bit $\Delta\Sigma$ ADCs and a mixer-less wideband AM front-end architecture. In FM mode the measured in-band IIP3 is 17.5dBm at a NF of 6dB. At maximum gain the NF is 4.1dB in DAB mode.**

Keywords— **AM, FM, DRM30, DRM+, DAB, T-DMB, HD-radio, SoC, CMOS, receiver, wideband, fixed-oscillator, multi-channel, multi-band, tuner, LNA, inverter-based amplifier, harmonic reject mixer, ADPLL, delta-sigma ADC, SDR**

I. INTRODUCTION

A wideband RF receiver that receives and digitizes all channels in a complete band simultaneously, offers attractive advantages with respect to a traditional narrow-band receiver. It supports multi-channel reception with a single RF receiver and one PLL, saving power and area. Because channel selection is done in the digital domain, only a single fixed-

frequency PLL and simple frequency dividers are needed to derive the coarse mixer LO frequencies, ADC sampling clock and digital clocks. The simplicity of the clock generation and frequency planning mitigates interference issues like oscillator pulling and spurs, which is an important bottleneck in high-end automotive multi-band and multi-channel radio broadcast receivers.

On the down side, the major challenge of designing a wideband receiver is that, due to the lack of narrowband selectivity in the front-end, the receiver circuits (LNA, mixer, TIA, ADC) need to have large bandwidth, low noise and high linearity. In [1] a wideband receiver has been demonstrated for digital (HD-radio/DAB/T-DMB) car radio with state-of-the-art blocker performance and sensitivity. However, applying the wideband receiver concept for analog (AM/FM) radio reception aggravates block specifications with an order of magnitude which so far has not been shown feasible. The reason for the extreme noise and linearity requirements is that the (unconditioned) signal strengths of analog terrestrial radio channels can differ more than 100dB.

This paper presents a full-band multi-standard single-PLL radio SoC that meets automotive requirements for all global analog and digital radio standards. The receiver architecture is

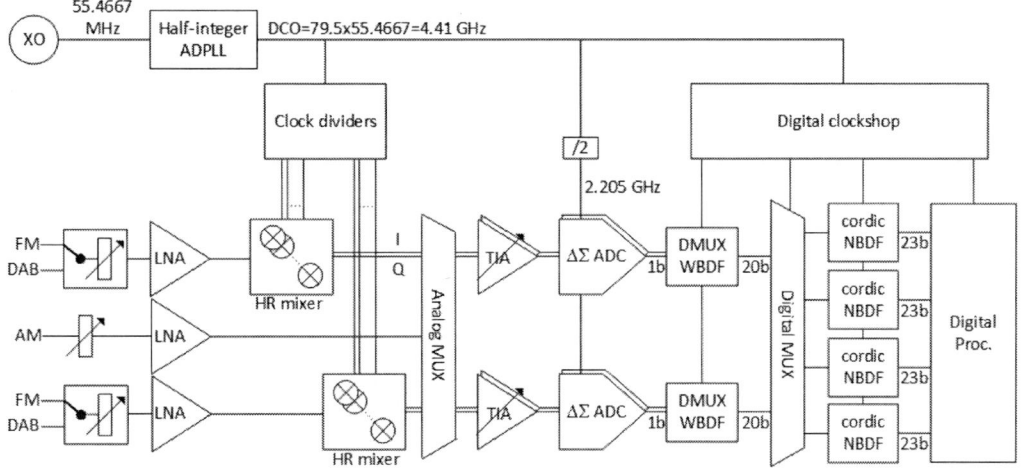

Fig. 1. Multi-band and multi-channel radio SoC block diagram with frequency planning (e.g. Europe, US).

presented in section II. In section III, the key circuits are described and in section IV measurements are shown. A benchmark and conclusions are given in section V.

II. RECEIVER ARCHITECTURE

A. Architecture

Figure 1 shows the block diagram of the multi-channel and multi-band radio SoC. The receiver front-end (RFE) consists of three RF pipes, an IF multiplexer (MUX) and two quadrature IF pipes with transimpedance amplifiers (TIA), ΔΣ ADCs and wideband decimation filters (WBDF). One RF pipe covers the LF, MF and HF bands for AM and DRM30 reception, while the two other RF pipes cover VHF-I, II and III for reception of DRM+, FM, HD-radio, DAB and T-DMB. The RFE is connected via a digital MUX to multiple digital pipes comprising a CORDIC and narrowband decimation filter (NBDF) for channel selection and channel filtering.

One wideband LNA is designed for covering the complete VHF frequency range. The RF signal is mixed down to baseband by a harmonic reject (HR) mixer with multiple LO phases that are derived from the PLL clock with frequency dividers (Figure 1). The AM RFE does not require a mixer as the ADC covers the complete band which avoids harmonic mixing. Usually, an AM receiver with a mixer requires a large amount of harmonic rejection and filtering because the relatively low AM LO frequency has a multitude of harmonics that appear in e.g. the AM, FM and DAB bands. The IF MUX can connect any RF channel to any ADC to guarantee uninterrupted operation of a receiver if another receiver is switching to a different band. The output of the ADC is filtered by the WBDF that suppresses out-of-band quantization noise. The output of the WBDF contains all channels in a band and is fed to the CORDIC/NBDF that select and filter up to four channels simultaneously. Further signal processing, such as I/Q mismatch correction, gain control, channel/source decoding, etc. is done in the digital signal processor.

B. Frequency plan

The frequency plan is shown in Figure 2. Because the ADC bandwidth is not sufficient to cover the full DAB band of 66MHz, the DAB band is split in two with LO frequencies of 184MHz and 221MHz. Therefore, two FM/DAB pipes support either full-band DAB (66MHz bandwidth), or simultaneously full-band FM and half-band DAB, or FM/DAB antenna diversity. The LO's at 45.9MHz, 78.7MHz and 110MHz are derived with simple integer dividers and can coarsely tune to DRM+ and FM bands.

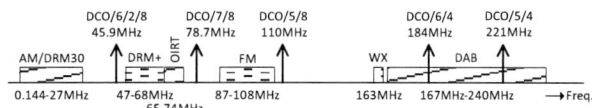

Fig. 2. Frequency plan (Europe, US example).

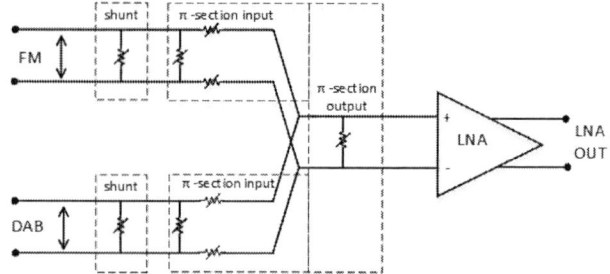

Fig. 3. DAB/FM attenuator.

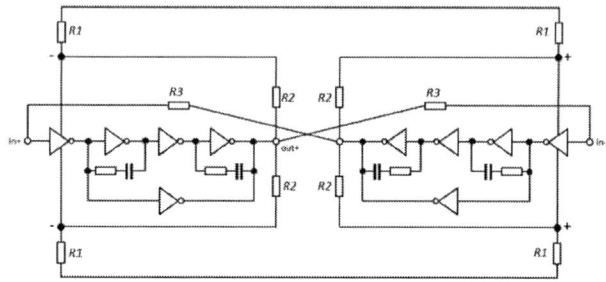

Fig. 4. DAB/FM LNA circuit concept.

The ADC bandwidth is large enough to accommodate for the coarse LO's. The LO frequencies are chosen outside the FM and DRM+ bands to mitigate interference (due to LO spurs) and de-sensitization (due to offset and 1/f noise). In case of DAB this is not an issue because of the larger channel bandwidth and robust modulation.

III. CIRCUIT DESIGN

The amplifiers in the LNA, TIA's and ADC's are all multi-inverter-based with the benefits of supply current re-use, low noise, large bandwidth, high linearity and rail-to-rail class AB output. The circuits are supplied from a 1.8V external supply through voltage regulators that generate 1.2V local supplies and provide forward and reverse isolation between the different supply domains.

A. FM/DAB RF circuits

The RF part of a FM/DAB pipe contains an attenuator (Figure 3), LNA (Figure 4) and mixer. The antenna signal is first attenuated with the RF automatic gain control (AGC) attenuator that comprises a shunt and pi network and provides constant input level and impedance for the LNA. Each FM/DAB pipe contains two switched differential attenuators with combined output that feeds the signal to the LNA. The pi-sections are scaled for 2dB attenuation steps and very large input signals are shunted with 6dB attenuation steps. The signal is amplified in the LNA which has a positive voltage gain dual feedback topology and provides constant gain. The nullor of the LNA is implemented with four CMOS inverters with a large gain-bandwidth product and therefore high IIP3 which is important for FM, and large gm/I_{drain} to realize a low noise figure (NF), important for DAB.

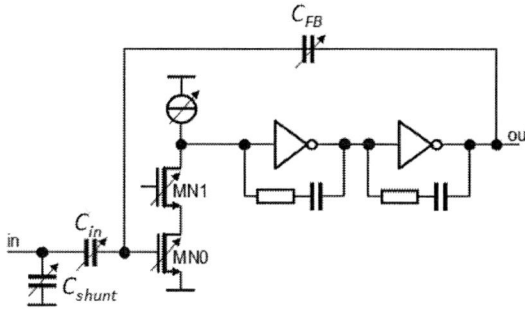

Fig. 5. AM LNA circuit concept.

The mixer consists of 6 or 14 balanced I/Q passive mixers, each driven with a phase shifted LO signal having 50% duty-cycle.

B. AM LNA

The AM LNA (Figure 5) has a capacitive feedback network to achieve high sensitivity and is suitable for reception with a passive capacitive antenna as well as an active antenna. The AGC is implemented by increasing C_{FB} with switches and then reducing C_{in} while keeping the load capacitance at the input constant. This AGC scheme realizes a high desensitization threshold. The AGC furthermore has a single shunt step to reduce the LNA input level. The LNA core contains a set of parallel cascoded first stage sections MN0-MN1. The number of stages can be varied for frequency compensation. The second and third stage are inverters.

C. TIA and continuous-time ΔΣ ADC

The TIA's are triple-inverter based with a Rauch lowpass filter (LPF) topology for anti-aliasing and removing the sum product from the mixers. The output of each TIA is connected to a continuous-time ΔΣ ADC (Figure 6). The ΔΣ ADC has a 4th order RC filter-based loop filter with local feedbacks (d_1, d_2) to create resonators. The modulator employs a 1-bit quantizer (ADC) and 1-bit feedback DACs (DAC1, DAC2, DAC3) which is inherently linear [2] but sensitive to far-off phase noise that is mainly contributed to by noise from the clock buffers in the clock distribution circuit. The clock drivers consume ~6mW to meet the clock jitter requirements.

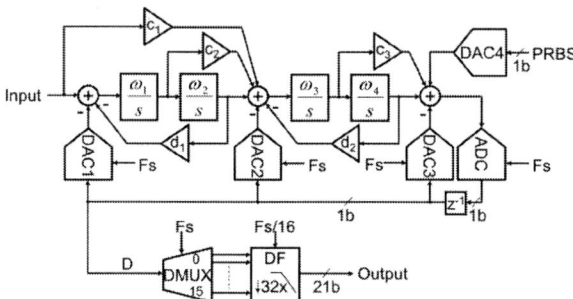

Fig. 6. Block diagram of 4th order continuous-time 1-bit ΔΣ ADC.

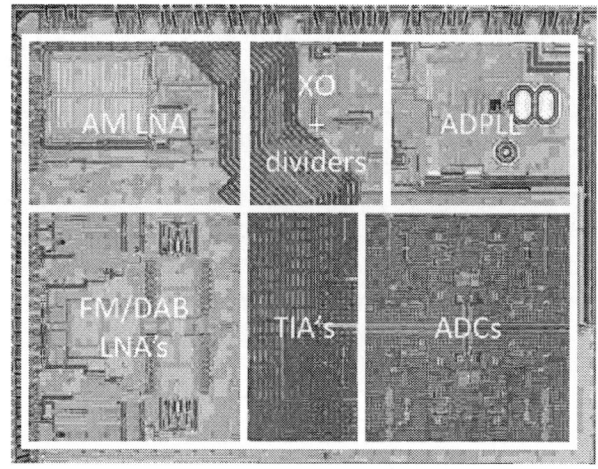

Fig. 7. Chip micrograph.

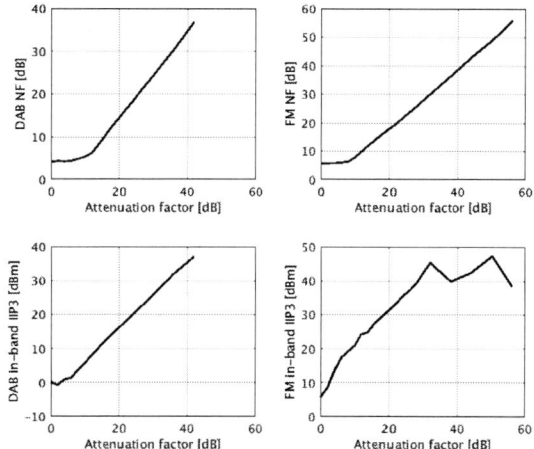

Fig. 8. Measured NF (top) and in-band IIP3 (bottom) (FM: Fin1=99.8MHz, Fin2=101.8MHz; DAB: Fin1=229.4MHz, Fin2=231.4MHz) versus attenuation factor for DAB (left) and FM (right).

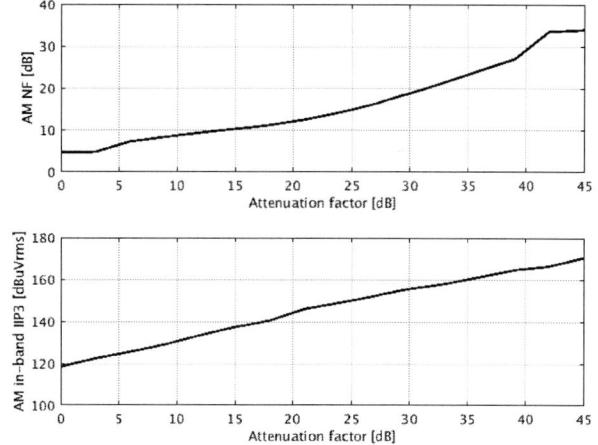

Fig. 9. Measured NF (top) and in-band IIP3 (bottom) (Fin1=1.4MHz, Fin2=1.79MHz) versus attenuation factor for AM.

Table 1. Performance summary.

Performance	Unit	[3]	[4]	[5]	[1]	This Work
Sensitivity DAB *	dBm	-102.5	-101.5	n.a.	-102	-103
Adj. Ch. Rej. DAB *	dB	48	37	n.a.	58	64
Far-off Select. DAB *	dB	?	53	n.a.	70	73
NF AM @ max. gain	dB	n.a.	n.a.	16.7	n.a.	4.6 **
NF FM @ max. gain	dB	3	n.a.	6.3	?	5.6 **
NF DAB @ max gain	dB	2.8	3	4.6	3.4	4.1 **
In-band IIP3 AM	dBuV$_{rms}$	n.a.	n.a.	95 (mid. gain)	n.a.	122.4 ** (3dB att. & 4.8dB NF)
In-band IIP3 FM	dBm	-18 (max. gain)	n.a	0 (mid. gain)	?	17.5 ** (6dB att. & 6dB NF)
In-Band IIP3 DAB	dBm	-21 (max. gain)	?	-1 (mid. gain)	-15 (max gain)	1.4 ** (6dB att. & 4.4dB NF)
Image Rejection	dB	61	?	40	100	100

* According to EN 62104:2008, chapter 7.3. ** Including ADC.

The ΔΣ ADC achieves <-120dBFS intermodulation distortion (for two -9dBFS tones) and the signal-to-noise and distortion ratio (SNDR) is ~100dB for any 200kHz FM channel within the 25MHz ADC bandwidth.

D. Half integer PLL

The half-integer sampler-based counter all-digital PLL (ADPLL) operates in bang-bang mode to achieve a low phase noise. The oscillator frequency can be programmed to operate between 76 and 84 times the reference frequency in half integer steps to support different crystal frequencies or optimize the multi-band coverage for different global regions.

IV. MEASUREMENT RESULTS

The receiver has been fabricated in TSMC 65nm CMOS technology (Figure 7). Figure 8 shows the measured NF and in-band IIP3 of the RFE (including ADCs) as function of receiver gain attenuation for FM and DAB reception. Different gain settings are used to achieve low NF of 4.1dB in DAB (VHF-III) mode and high IIP3 of 17.5dBm in FM mode (at 6dB attenuation resulting in a NF of 6dB).

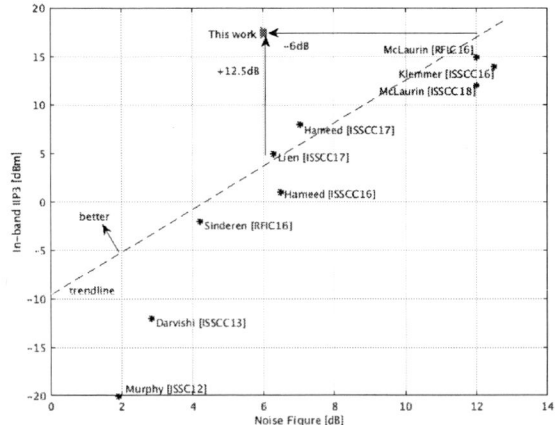

Fig. 10. NF versus in-band IIP3 benchmark with recent wireless receivers from ISSCC and RFIC.

Figure 9 shows the NF (50Ω source, unmatched) and IIP3 for the AM RF front-end. Current consumption of an FM/DAB pipe including ADCs is 166mA and an AM pipe consumes 137mA. The ADPLL current consumption is 20mA. Table 1 shows the performance benchmark. For DAB reception this work achieves highest sensitivity of -103dBm, adjacent channel rejection (ACR) of 64dB and far-off selectivity (FoS) of 73dB.

V. CONCLUSION

This paper presents a full-band multi-standard single-PLL radio SoC that meets automotive requirements for all global analog and digital radio standards. Key enabling techniques are high gain-bandwidth multi-inverter-based amplifiers supplied by PVT-tracking low-ohmic low dropout regulators (LDO), very linear high-speed 1-bit delta-sigma (ΔΣ) ADCs and a mixer-less wideband AM front-end.

Figure 10 gives a benchmark plot of in-band IIP3 versus NF of recent state-of-the-art published receivers. The dotted line shows the state-of-the art trendline. Typically, receivers with a low NF have low in-band IIP3 performance, while high NF receivers achieve high in-band IIP3. This work demonstrates an in-band IIP3 which is more than 12dB higher compared to receivers with similar NF, while the NF is 6dB lower compared to receivers with similar in-band IP3.

REFERENCES

[1] J. van Sinderen et al., "A Wideband Single-PLL RF Receiver for Simultaneous Multi-band and Multi-channel Digital Car Radio Reception," *Proc. RFIC*, 2016.

[2] L. Breems et al., "A 2.2 GHz Continuous-Time ΔΣ ADC with −102 dBc THD and 25 MHz BW," *ISSCC Dig. Tech. Papers*, Feb. 2016, pp. 272–273.

[3] M. Jeong et al., "A 65nm CMOS Low-Power Small-Size Multistandard, Multiband Mobile Broadcasting Receiver SoC", *ISSCC Dig. Tech. Papers*, pp. 460-461, Feb. 2010.

[4] J.-H. Chang et al., "A Multistandard Multiband Mobile TV RF SoC in 65nm CMOS", *ISSCC Dig. Tech. Papers*, pp. 462-463, Feb. 2010.

[5] K. Wang et al, "A SAW-Less First Folded-Conversion Second Down-Conversion Receiver for Multistandard Broadcasting Radio Application", *IEEE Trans. On Microw. Theory Techn.*, vol 61, no. 4, pp. 1674-1680, April 2013.

Laser Spectral Linewidth Reduction Using an Integrated Pound-Drever-Hall Stabilization System in 180 nm CMOS SOI

Mohamad Hossein Idjadi[#1], Firooz Aflatouni[#2]

[#]Dept. of Electrical and Systems Engineering, University of Pennsylvania, USA

[1]idjadi@seas.upenn.edu, [2]firooz@seas.upenn.edu

Abstract— An Integrated Pound-Drever-Hall laser stabilization system is demonstrated where the electronic and photonic components are monolithically integrated in the GF7RFSOI CMOS SOI process. An off-chip Fabry-Perot cavity with quality factor of 48000 is used as the frequency reference to reduce the linewidth of a commercially available distributed feedback laser from 246 kHz to 7 kHz. The electronic-photonic chip consumes 83 mW and occupies 1.6 mm^2 area.

Keywords— laser, silicon-photonics, frequency stabilization.

I. INTRODUCTION

Narrow linewidth lasers have far-reaching applications in communication [1], frequency metrology [2], and frequency synthesis [3], [4]. The Pound-Drever-Hall (PDH) method is one of the most widely used techniques for laser phase and frequency stabilization [5], where a feedback loop is utilized to lock the frequency of a laser to a reference cavity or resonator. Excellent frequency stabilization have been demonstrated using Bench-top PDH systems. However, these systems are typically expensive, bulky, and power hungry [7]-[9].

Recently, a novel integrated PDH system on a standard CMOS SOI platform has been demonstrated, where an on-chip reconfigurable Mach-Zehnder interferometer (MZI) serves as a frequency reference in the PDH loop [10]. The performance of such system can be improved if a photonic cavity with higher quality factor (Q) is utilized as the reference frequency instead of an optical frequency reference implemented on a standard CMOS process.

Here, we demonstrate the implementation of a PDH system, where an integrated electronic-photonic chip is used to lock a commercially available distributed feedback (DFB) laser to an off-chip photonic cavity. The electronic and photonic devices were monolithically co-integrated on the GlobalFoundries GF7RFSOI CMOS SOI process. Compared to the system reported in [10], use of power optimized low-noise electronic circuits and a higher Q optical reference frequency results in more than four times larger laser linewidth reduction at about one third of the power consumption for the implemented PDH system. This prototype system enables low-cost and compact laser frequency stabilization using different custom-made state-of-the-art photonic frequency references.

II. ARCHITECTURE AND PRINCIPLE OF OPERATION

Figure 1 shows the block diagram of the implemented Pound-Drever-Hall (PDH) stabilization system, where the laser output is modulated using a p-doped-intrinsic-n-doped (PIN) phase modulator creating two side-bands around the carrier frequency. The modulated optical signal is passed through an optical cavity serving as the optical reference frequency. The error signal, corresponding to the deviation of the laser frequency from the resonance frequency of the cavity, is generated and processed in the electrical domain by photo-detecting the optical signal at the cavity output. The photo-current is amplified using a trans-impedance amplifier (TIA) and down-converted using the same local oscillator that modulates the optical phase modulator. The down-converted signal is low-pass filtered, converted to a current, and injected to the gain section of the laser as the read-out (or error) current closing the feedback loop.

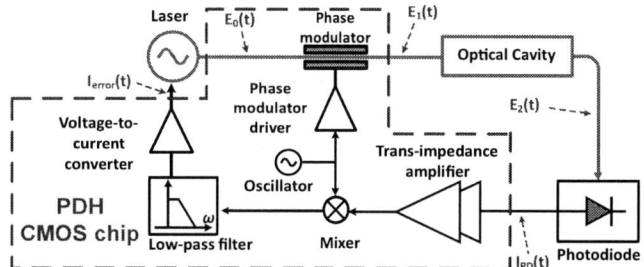

Fig. 1. The block diagram of the implemented Pound-Drever-Hall (PDH) system.

A. Theoretical analysis of the loop dynamics

Consider the case that the electric field at the output of the laser in Fig. 1 can be written as

$$E_0 = \sqrt{P_0}\, e^{i\omega t}, \tag{1}$$

where ω and P_0 are the laser frequency and power, respectively. As described earlier, the local sinusoidal oscillator at frequency of Ω modulates the laser output. The resulting electric field after the phase modulator is

$$E_1 = \sqrt{P_0}\, e^{i\omega t}\, e^{i\beta sin(\Omega t)}, \tag{2}$$

where β is the modulation index. Using Jacobi-Anger expansion, Eq. 2 can be simplified to

$$E_1 \approx \sqrt{P_0}\, e^{i\omega t}[J_0(\beta) + J_1(\beta)e^{i\Omega t} - J_1(\beta)e^{-i\Omega t}], \tag{3}$$

where $J_n(.)$ is the nth Bessel function of the first kind. The modulated light is filtered by the off-chip high Q cavity. The electric field of the optical signal at the cavity output is

$$
\begin{aligned}
E_2 = \sqrt{P_0}\, e^{i\omega t}[&J_0(\beta)T(\omega) + J_1(\beta)T(\omega + \Omega)e^{i\Omega t} \\
&- J_1(\beta)T(\omega - \Omega)e^{-i\Omega t}],
\end{aligned}
\tag{4}
$$

where $T(\omega)$ is the cavity frequency response. The cavity output is photo-detected and the resulting photo-current can be written as

$$i_{PD}(t) = R|E_2|^2, \qquad (5)$$

where R is the responsivity of the photodiode. Equations 4 and 5 show that the photo-current, $i_{PD}(t)$, has frequency components at DC, Ω, and 2Ω. This photo-current is amplified, down-converted by frequency Ω, and low-pass filtered to suppress the higher frequency component. In this case, the error signal under the slow-modulation assumption ($\frac{\Omega}{c} \ll 2\pi$) [6] can be written as

$$i_{error} = -I_0\Omega\frac{d}{d\omega}|T(\omega)|^2, \qquad (6)$$

where I_0 and c are the amplitude of the error signal and the full width at half maximum (FWHM) linewidth of the cavity power spectrum, respectively. Note that I_0 is a function of the loop gain, the laser power, and the optical loss.

If a Fabry-Perot cavity with a Lorentzian power spectrum profile (around its resonance) is used as the optical frequency reference, the magnitude squared of the cavity frequency response is written as

$$|T(\omega)|^2 = \frac{1}{1 + (\frac{\omega - \omega_{res}}{\pi c})^2}, \qquad (7)$$

where ω_{res} is the resonance frequency of the Fabry-Perot cavity. Figure 2 shows the frequency response of a Fabry-Perot cavity (blue) and the corresponding PDH read-out signal (red) for $\frac{\Omega}{c} = \frac{2\pi}{20}$. Note that the read-out (error) signal is asymmetric around the cavity resonance frequency. In other words, the read-out signal shows the difference between the laser frequency and the cavity resonance frequency and also whether the laser frequency is greater or less than the cavity resonance frequency.

As discussed in [10], the minimum achievable linewidth is limited by the total input referred current noise of the system and the opto-electronic loop gain. As a result, higher cavity Q-factor reduces the effect of the system noise on the PDH loop performance.

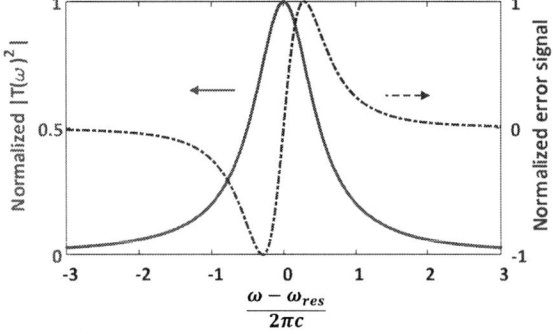

Fig. 2. Frequency response of the Fabry-Perot cavity around its resonance frequency (blue) and the normalized PDH read-out (error) signal (red).

B. Block diagram of the implemented PDH chip

Figure 3(a) shows the block diagram of the implemented PDH chip. The input photo-current is amplified and converted to a voltage using a TIA. The DC-offset compensation network corrects for any DC offsets due to the optical power fluctuations and/or process variation. An LC-VCO, serving as the local oscillator (LO), is used to drive the optical phase modulator and also down-convert the TIA output using a double-balanced mixer. The mixer output is low-pass filtered, amplified, and converted to a current using a voltage-to-current (VtoI) converter. The VtoI output is wire-bonded to the laser gain section input. The PIN phase modulator has a measured -3dB bandwidth of 150 MHz and an I_π of 64 mA at 150 MHz.

The microphotograph of the implemented PDH chip is shown in Fig. 3(b), where electronic and photonic blocks and devices are monolithically integrated in the GlobalFoundries GF7RFSOI 180 nm CMOS SOI process.

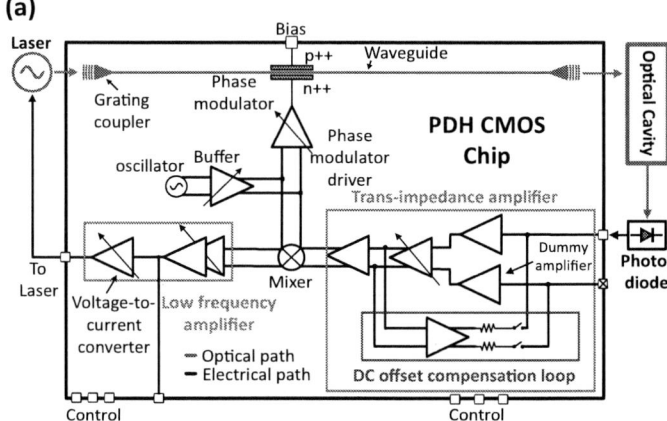

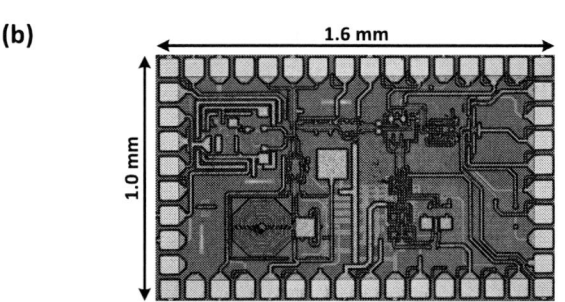

Fig. 3. (a) The block diagram of the PDH chip. (b) The microphotograph of the PDH chip fabricated in GF7RFSOI 180 nm CMOS SOI process.

III. MEASUREMENT RESULTS

A. Characterization of the electronic blocks

Figure 4(a) shows the schematic of the TIA which consists of a regulated cascode structure followed by a variable gain amplifier with total trans-impedance gain of -10 to 95 dBΩ. The schematic of the DC offset compensation network is shown in Fig. 4(b), where the DC mismatch is sensed at the TIA output and compensated by adjusting the input current of the TIA. As shown in Fig. 4(c), when the offset compensation

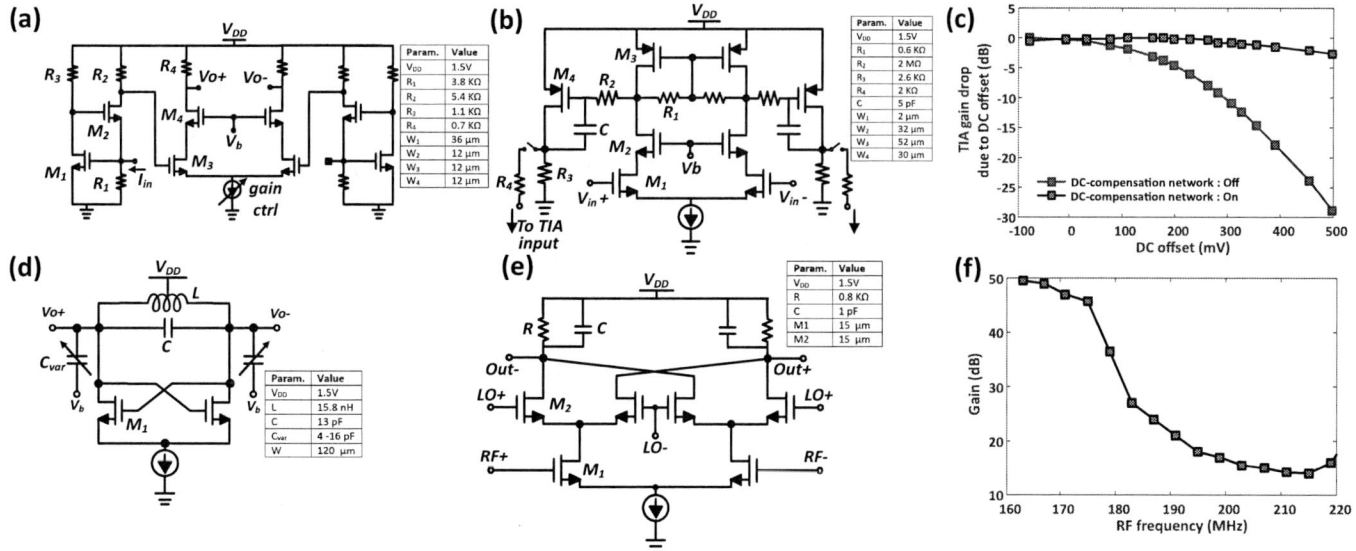

Fig. 4. (a) Schematic of the TIA, (b) schematic of the DC offset compensation network, (c) measured TIA gain drop due to introduced DC offset, (d) schematic of the cross-coupled LC VCO, (e) schematic of the mixer, and (f) measured conversion gain from the input of the TIA to the input of the VtoI (for fixed IF frequency of 5 MHz).

network is active, the TIA gain remains almost unchanged in the presence of large DC offsets.

Figure 4(d) shows the schematic of the on-chip cross-coupled LC VCO, which is designed to operate within the bandwidth of the on-chip PIN modulator. The VCO can be tuned from 150 MHz to 230 MHz and has a measured phase noise of -121.6 $\frac{dBc}{Hz}$ at 1 MHz offset. The input referred current noise of the CMOS chip (at the TIA input) is about 24 $\frac{pA}{\sqrt{HZ}}$ for LO frequency of 150 MHz. The schematic of the double-balanced mixer is shown in Fig. 4(e).

Figure 4(f) shows the conversion gain from the input of the TIA to the input of the VtoI. In this measurement, LO and RF frequencies are swept at the same time with a 5 MHz offset frequency.

B. Characterization of the Fabry-Perot cavity

To characterize the MICRON OPTICS-11150 tunable Fabry-Perot cavity, an Agilent 81642A narrow linewidth tunable laser is coupled into the cavity input and the cavity output is photo-detected and monitored on a spectrum analyzer. Figure 5 shows the measured transmission of the cavity, where a Lorentzian FWHM linewidth of c =32 pm at 1542.7 nm, corresponding to a quality factor of 48000, is measured.

C. Open loop measurement

Figure 6(a) shows the set up for PDH system measurements. To measure the open loop response, switches S1 and S2 are set to the open-loop mode. In this case, an Agilent 81642A tunable laser is coupled into the CMOS chip and the VtoI output current, the read-out signal, is converted to a voltage using a 50Ω resistor and monitored on an oscilloscope while the frequency of the tunable laser is

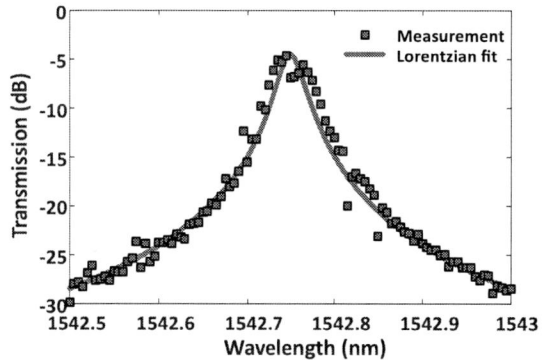

Fig. 5. Measured transmission of the Fabry-Perot cavity.

tuned. Figure 6(b) shows the measured read-out signal which is in agreement with the theory described by Eq. 6 (Fig. 2).

D. Closed-loop measurement

To measure the closed-loop performance of the implemented PDH system, switches S1 and S2 in Fig. 6(a) are set to the closed-loop mode. In this case, the output of an Emcore-CCJ5879 DFB laser is split into two branches using an 80/20 coupler, where 80% of the laser power is coupled into the PDH chip (using the on-chip grating coupler) and 20% of the laser power is used for spectral monitoring. The grating coupler has a measured efficiency of 28% for a coupling angle of 17°. The output of the PDH chip is coupled into the Fabry-Perot cavity. A Thorlabs FD80FC InGaAs photodiode is used to photo-detect the cavity output and the photo-current is wire-bonded to the PDH chip. The PDH chip output current is injected to the gain section of the DFB laser.

Table 1. Performance summary and comparison with other works.

Ref.	Frequency reference	Q-factor	Laser wavelength	Loop BW	Linewidth original-reduced	Power consumption	Area/Process
[7]	Si_3N_4 ring resonator	3×10^7	1577 nm	1 MHz	160 kHz-N/A	N/A	Bench-top / cavity: in-house fab.
[8]	Spiral cavity ring resonator	$\leq 10^8$	1551 nm	200 kHz	900 Hz-100 Hz	N/A	Bench-top / cavity: in-house fab.
[9]	TiO_2 athermal ring resonator	67×10^3	1549 nm	3 MHz	4.79 MHz-340 kHz	N/A	Bench-top / cavity: in-house fab.
[10]	Mach-Zehnder interferometer	38×10^3	1541 nm	<1 MHz	750 kHz-55 kHz	237 mW	2.38 mm^2 / 180 nm CMOS
This work	Fabry-Perot cavity	48×10^3	1550 nm	3 MHz	246 kHz-7 kHz	83 mW	1.6 mm^{2*} / 180 nm CMOS

*cavity is off-chip.

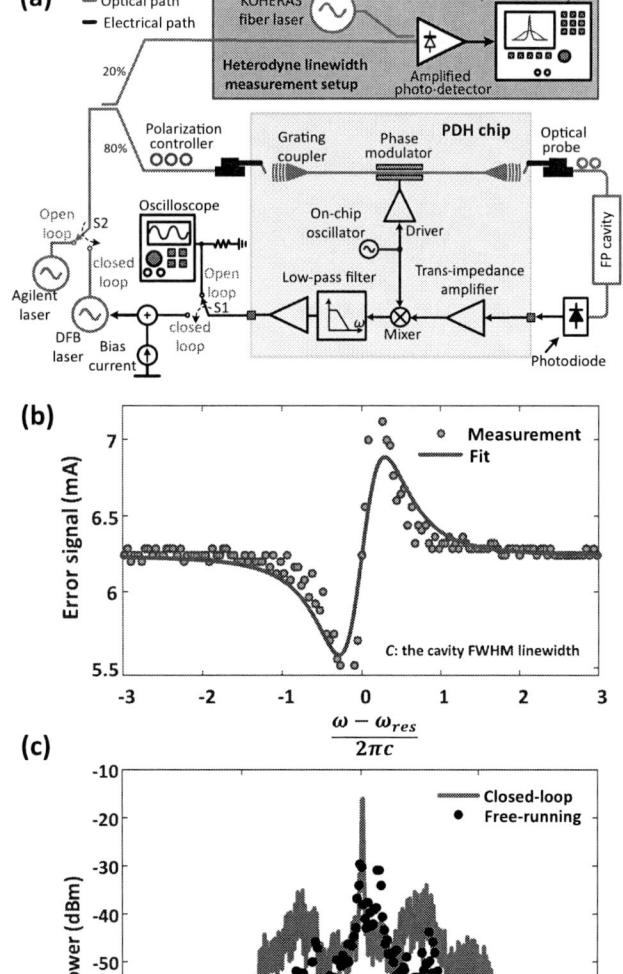

Fig. 6. (a) Open/closed loop measurement setup, (b) the measured PDH read-out (error) signal, and (c) the measured heterodyne spectrum of the DFB laser before and after PDH stabilization.

To measure the linewidth of the laser before and after PDH stabilization, the beat-note between the DFB laser (the 20% port of the power splitter) and a KOHERAS fiber laser (with linewidth of 9.3 kHz) is monitored on a spectrum analyzer.

Figure 6(c) shows the beat-note spectrum for free-running and PDH stabilized cases, where the FWHM linewidth of the free-running DFB laser (emitting 12 dBm) is reduced by a factor of 35 (from 246 kHz to 7 kHz). The resolution bandwidth for these measurements was set to 10 kHz. The performance of this work is compared with that of a few published works in Table 1.

IV. CONCLUSION

In this paper, we have demonstrated the realization of a PDH system, where the electronic and photonic components were monolithically co-integrated on a 180 nm standard CMOS SOI process and used to PDH lock a commercially available DFB laser to an off-chip Fabry-Perot cavity with quality factor of 48000. The measured closed-loop performance shows a linewidth reduction of more than 35 times. The chip occupies 1.6 mm^2 and consumes 83 mW. The implemented chip can be used to PDH lock different semiconductor lasers to custom-made optical cavities enabling realization of low-cost compact narrow linewidth light sources.

REFERENCES

[1] H. Al-Taiy et al., "Ultra-narrow linewidth, stable and tunable laser source for optical communication systems and spectroscopy," *Opt. Lett.* vol. 39, pp. 5826-5829, Oct. 2014.

[2] T. Udem et al., "Optical frequency metrology," *Nature*, vol. 416, pp. 233-237, Mar. 2002.

[3] F. Ashtiani et al., "Towards Integrated Wideband High Resolution Optical Synthesizers," in 2018 *IEEE/MTT-S International Microwave Symposium - IMS*, pp. 1316-1319.

[4] F. Ashtiani et al., "High-Resolution Optical Frequency Synthesis Using an Integrated Electro-Optical Phase-Locked Loop," *IEEE Transactions on Microwave Theory and Techniques*, vol. 66, no. 12, pp. 5922-5932, Dec. 2018.

[5] R. W. P. Drever et al., "Laser phase and frequency stabilization using an optical resonator," *Appl. Phys. B*, vol. 31, pp. 97-105, Jun. 1983.

[6] E. D. Black, "An introduction to Pound–Drever–Hall laser frequency stabilization," *Am. J. Phys.* vol. 69, pp. 79-87, 2001.

[7] D. T. Spencer et al. "Stabilization of heterogeneous silicon lasers using Pound-Drever-Hall locking to Si_3N_4 ring resonators," *Opt. Express*, vol. 24, pp. 13511-13517, Jun. 2016.

[8] H. Lee et al., "Spiral resonators for on-chip laser frequency stabilization," *Nat. Commun.* vol. 4, pp. 2468, Sep. 2013.

[9] E. S. Magden et al., "Laser Frequency Stabilization Using Pound-Drever-Hall Technique with an Integrated TiO_2 Athermal Resonator," in *Conference on Lasers and Electro-Optics*, 2016, paper STu1H.3.

[10] M. H. Idjadi, and F. Aflatouni, "Integrated Pound-Drever-Hall laser stabilization system in silicon," *Nat. Commun.* vol. 8, pp. 1209, Oct. 2017.

22nm FD-SOI Technology with Back-biasing Capability Offers Excellent Performance for Enabling Efficient, Ultra-low Power Analog and RF/Millimeter-Wave Designs

S.N. Ong[#1], L.H.K. Chan[#], K.W.J. Chew[#], C.K. Lim[#], W. L. Oo[#], A. Bellaouar[^], C. Zhang[^],
W.H. Chow[#], T. Chen[^], R. Rassel[^], J.S. Wong[#], C.W.F. Wan[#], J. Kim[#], W.H. Seet[#], D. Harame[%]

[#] GLOBALFOUNDRIES Singapore
[^] GLOBALFOUNDRIES USA
[%]now with Research Foundation State University Polytechnic at Albany, USA
[1]shihni.ong@globalfoundries.com

Abstract—This paper addresses the impact of back-gate biasing to DC, RF/millimeter-Wave (mmWave) and high frequency (HF) noise in 22nm FD-SOI technology (GLOBALFOUNDRIES' 22FDX® technology). The front-gate and the back-gate cut-off frequency fT, together with the maximum oscillation frequency fMAX, were extracted from the four-port S-parameters data. The maximum achieved front-gate/back-gate fT and fMAX for the NFET is 350/85 GHz and 370/23 GHz respectively. In addition, 22FDX® technology demonstrated a tuneable HF noise parameter by using the back-gate biasing to achieve best-in-class low noise level. Two front-end (FE) modules were presented, which exploit the unique feature of back-gate. This unique feature allows superior designs with excellent combination of performance, power consumption and development cost, for emerging applications such as IoT, Telecommunication UE, RF and mmWave circuits with high speed connectivity and networking.

Keywords— mmWave, RF, noise figure, 22FDX®, FDSOI, four-port, back-gate.

I. INTRODUCTION

FD-SOI technology exhibits excellent RF/mmWave performance with high cut-off frequency f_T and maximum oscillation frequency f_{MAX} [1]. The technology has low parasitic to the substrate and low gate resistance. This technology has a unique "second-gate" feature called the back-gate, due to an ultra-thin buried oxide (BOX) between the main gate and body substrate that allows designers to adjust the back-gate biasing for different kind of circuits optimization [1-3].

This paper evaluates the impacts of back-gate bias on the DC, RF/mmWave and HF noise figure-of-merit (FoM) using GLOBALFOUNDRIES' 22FDX® technology. It is presented for the first time that the noise figure and noise resistance is improved under forward body-bias (FBB), with positive back-gate bias for NFET, in contrary to [4]. In section IV, two front-end modules, i.e., 28GHz stacked Power Amplifier (PA) and SPST switch are presented to demonstrate the unique feature of the back-gate.

II. ON-WAFER CHARACTERIZATION

A. DC/RF Characterization

The impact of the back-gate bias to front-gate DC/RF characteristics has been investigated by measuring a 4-port transistor using a pair of GSGSG probes, as illustrated in Fig. 1 (a). DC current and S-parameters were characterized using the CASCADE RF probe station together with Keysight's PNA and B1500 DC parametric analyzer. Calibration was performed with the Short-Open-Load- Reciprocal thru (SOLR) method [5], to eliminate the parasitic down to probe tips.

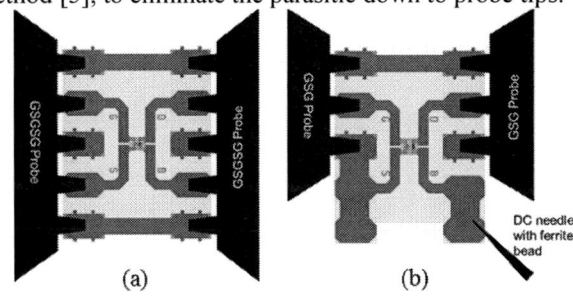

Fig. 1. Layout view of: (a) a 4-port transistor, with GSGSG probes; (b) a 3-port transistor, with GSG probes and DC needle with ferrite bead.

B. Four-Port De-embedding Methodology

Four-port Open-Short de-embedding method has been implemented to eliminate the parasitic of the GSGSG pad frame down to M1 reference plane. 4-port S-parameters of DUT (S_{DUT}), OPEN (S_{OP}) and SHORT (S_{SH}) structures were converted to Y-parameters (Y_{DUT}, Y_{OP}, Y_{SH}) using (1), with E is a 4-by-4 identity matrix.

$$Y = \left[Z_0 \left(\frac{E+S}{E-S} \right) \right]^{-1} \qquad (1)$$

The shunt admittance and the series impedance parasitic were removed using (2)-(3).

$$Y_{DUT\_noop} = Y_{DUT} - Y_{OP}; \; Y_{SH\_noop} = Y_{SH} - Y_{OP} \qquad (2)$$

$$Z_{deem} = \left(Y_{DUT\_noop} \right)^{-1} - \left(Y_{SH\_noop} \right)^{-1} \qquad (3)$$

Lastly, Z_{deem} was converted to S_{deem}, the de-embedded S-parameters of the transistor, using (4).

$$S_{deem} = \frac{Z_{deem} - Z_0 E}{Z_{deem} + Z_0 E} \qquad (4)$$

C. HF Noise Characterization

The impact of the back-gate bias to front-gate HF noise was evaluated by measuring a 3-port transistor using a pair of GSG probes with an additional DC needle with ferrite bead, for reason to prevent measurement oscillation, as illustrated in Fig. 1 (b). Four noise parameters were measured using MPI

978-1-7281-1702-7/19 $31.00 © 2019 IEEE

Prober TS3000, with Focus Microwave iCCMT-67100 tuner, Agilent PNA-X N5247A Noise Module, Agilent SMU B1500, Agilent Noise source 346CK01, and Miteq LNA JDM21-26505000-45-20P. Open-Short noise de-embedding was performed, followed by extraction of the drain/gate noise power spectral density (S_{id}, S_{ig}) and de-embedding of four-noise parameters [6].

III. BACK-GATE MODELING AND IMPACT TO FRONT-GATE AND BACK-GATE FOM

A. Transistor Cross Section View and Operation

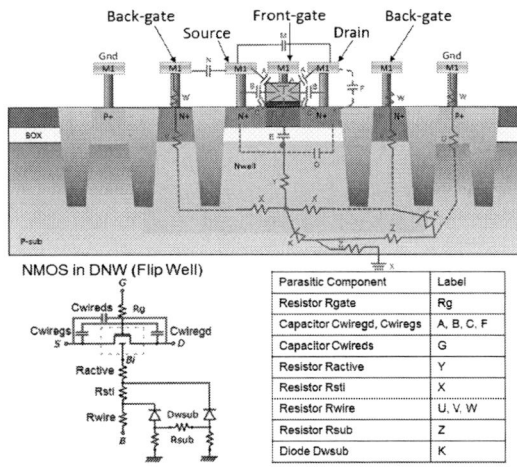

Fig. 2. Cross sectional view of the transistor with illustration of parasitic at the front-gate and back-gate [1].

The cross sectional view of the 22FDX® n-type flipped-well transistor with parasitic resistances, capacitances and diode are illustrated in Fig. 2 [1]. The back-gate bias is applied through the back-gate ring, which served as a "second-gate" for channel control.

B. DC Characteristic

The DC drain current I_d, transconductance G_m and output conductance G_{ds} at different back-gate biases of $|V_{bg}|$=0, 2, for NFET and PFET, are shown in Fig. 3. It shows that lower threshold voltage V_t can be achieved by simply applying a back-gate bias $|V_{bg}|$>0 to the transistor, under FBB condition. Higher drain saturation current I_{dsat}, higher output conductance G_{ds} and lower resistance R_{on} can be achieved as well with $|V_{bg}|$>0. Circuit designers are able to optimize their circuit to have either low power consumption at $|V_{bg}|$=0, or high speed operation with low R_{on} using $|V_{bg}|$>0. This is the unique feature of 22FDX®.

C. RF/mmWave Characteristic

By defining port 1 as front-gate, port 2 as drain, port 3 as back-gate and port 4 as source, the front-gate (FG) and back-gate (BG) cut-off frequency f_T and maximum oscillation frequency f_{MAX} can be extracted from Y-parameters, using equations (5) to (8).

$$FG\ f_T = \text{Mag}(H_{21}) \times \text{freq}; H_{21} = \frac{Y_{21}}{Y_{12}+Y_{13}+Y_{14}} @10\text{GHz} \quad (5)$$

$$BG\ f_T = \text{Mag}(H_{23}) \times \text{freq}; H_{23} = \frac{Y_{23}}{Y_{31}+Y_{32}+Y_{34}} @10\text{GHz} \quad (6)$$

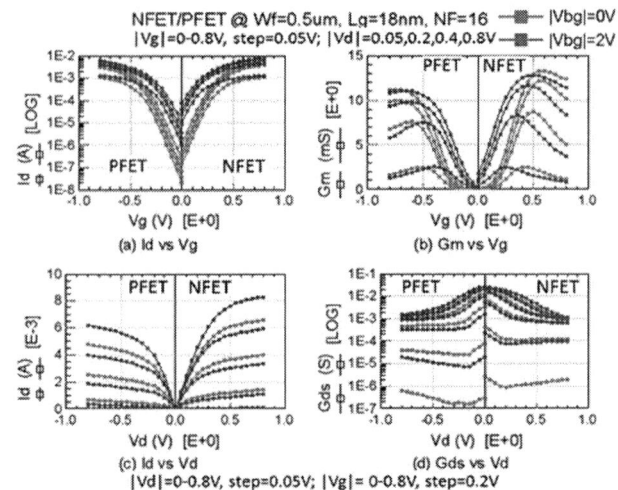

Fig. 3. DC characteristic versus gate and drain bias, of NFET and PFET, with (a) I_d-V_g, (b) G_m-V_g, (c) I_d-V_d, and (d) G_{ds}-V_d.

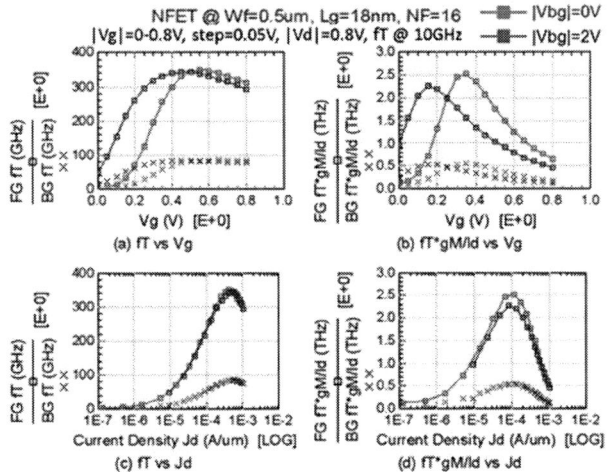

Fig. 4 FG and BG cut-off frequency of NFET, with (a) f_T-V_g, (b) f_T*g_M/I_d -V_g, (c) f_T-J_d, and (d) f_T*g_M/I_d -J_d.

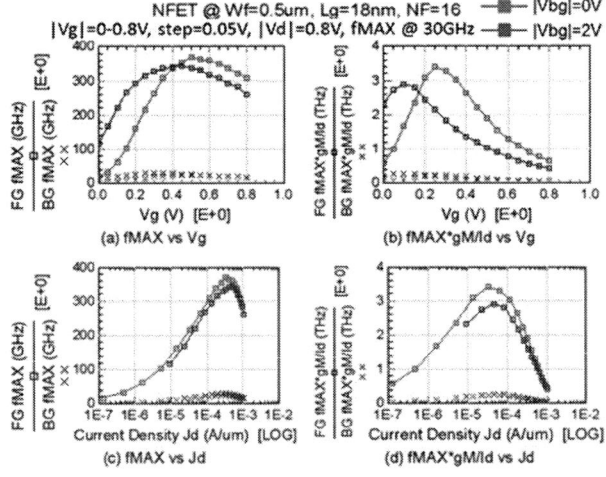

Fig. 5 FG and BG maximum oscillation frequency of NFET, with (a) f_{MAX}-V_g, (b) f_{MAX}*g_M/I_d -V_g, (c) f_{MAX}-J_d, and (d) f_{MAX}*g_M/I_d -J_d.

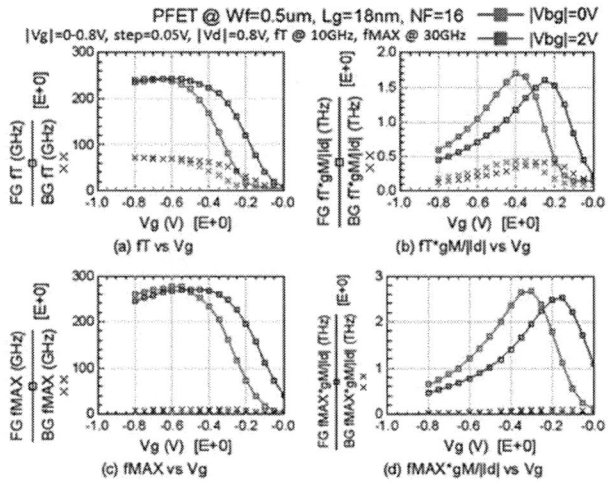

Fig. 6 FG and BG performance of PFET, with (a) f_T-V_g, (b) f_T*g_M/I_d -V_g, (c) f_{MAX}-V_g, and (d) f_{MAX}*g_M/I_d -V_g.

$$FG\ f_{MAX} = \sqrt{\mathrm{Mag}\left(G_{\max\_FG}\right)} \times \mathrm{freq};$$

$$G_{\max\_FG} = \frac{1}{4} \times \frac{|Y_{21}-Y_{12}|^2}{\mathrm{Re}(Y_{11})\mathrm{Re}(Y_{22})-\mathrm{Re}(Y_{21})\mathrm{Re}(Y_{12})}\ @30\mathrm{GHz} \quad (7)$$

$$BG\ f_{MAX} = \sqrt{\mathrm{Mag}\left(G_{\max\_BG}\right)} \times \mathrm{freq};$$

$$G_{\max\_BG} = \frac{1}{4} \times \frac{|Y_{23}-Y_{32}|^2}{\mathrm{Re}(Y_{33})\mathrm{Re}(Y_{22})-\mathrm{Re}(Y_{23})\mathrm{Re}(Y_{32})}\ @30\mathrm{GHz} \quad (8)$$

The FG and BG RF/mmW FoM, which are f_T, f_{MAX}, and their product with gain efficiency f_T*g_M/I_d and f_{MAX}*g_M/I_d, of NFET and PFET, at different back-gate biases of $|V_{bg}|$=0, 2, are illustrated in Figs. 4-6. It is observed that the peak of FG and BG f_T, f_{MAX} and their product with gain efficiency has shifted to lower V_g when $|V_{bg}|$ changes from 0 to 2V, for NFET and PFET. These FoMs peak at the same current density J_d, regardless of $|V_{bg}|$ being 0 or 2V. This indicates that, circuit designers are able to either obtain higher RF/mmW FoM at lower V_g and V_{bg}>0V for high speed design, or getting the similar RF/mmW FoM at same current density for low power design. The unique feature of FD-SOI provides another degree of freedom in circuit design, which enabled the designers to optimize their circuit in the desired operating region.

D. HF Noise Characteristic

HF noise FoM at 26 GHz versus Vg with different back-gate biases of $|V_{bg}|$=0, 2, of NFET, is shown in Fig. 7, with (a) drain current noise Sid, (b) normalized noise resistance Rn, (c) minimum noise figure NFmin, and (d) noise figure terminated at 50 Ohm source resistance NF50. With lower Vt and higher drain current Id for Vbg=2V, the intrinsic drain current noise Sid is higher. In spite of this, the higher Gm at lower Vg at Vbg=2V causing the Rn to be lower, and thus lower NFmin and NF50 as well. 22FDX® allows tuneable of HF noise parameter using back-gate bias Vbg. With Vbg>0V, lower noise figure and noise resistance can be achieved at low Vg regime, which is beneficial for circuit such as LNA.

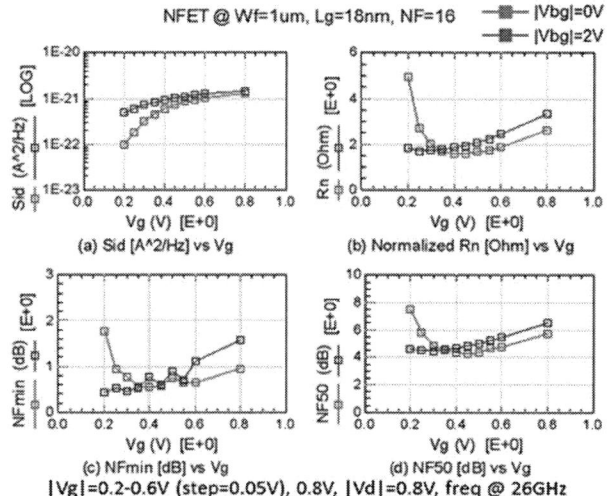

Fig. 7 HF noise FoM of NFET, with (a) S_{id}-V_g, (b) R_n -V_g, (c) NF_{min}-V_g, and (d) NF_{50} -V_g.

Table 1. NFET and PFET FoMs

	NFET @ V_{bg}=0V	NFET @ V_{bg}=2V	PFET @ V_{bg}=0V	PFET @ V_{bg}=2V
$|I_{dsat}|$ (uA/um)	819	1036	597	773
R_{on} (Ohm-mm)	0.356	0.313	0.369	0.306
Peak f_T (FG/BG) (GHz)	350 /85	343/83	244/73	244/73
Peak f_{MAX} (FG/BG) (GHz)	370/23	343/29	277/9	270/11
NF_{50} (dB) @26GHz	4.3	4.5	-	-

The NFET and PFET FoMs with different back-gate biases are summarized in Table 1. 22FDX® demonstrated better RF/mmWave performances as compared to other technology nodes such as bulk CMOS, PDSOI and FinFET, as shown in [1].

IV. MMWAVE CIRCUITS

The back-gate control adds a new dimension to mmWave circuit performance. There are two types of back-gate control in a circuit; one with low impedance to the back-gate bias and the other with high impedance. The former is also called static control where the threshold voltage of the device is controlled. The latter is also called AC floating back-gate bias (AFBG) in which the back-gate bias is done through a high-impedance (high resistor). In this case the drain/source to bulk capacitance is reduced.

Two examples are shown for the AFBG type. The first example is a power amplifier (PA), with the schematic using the AFBG concept is shown in Fig. 8. Performance comparison of the PA with and without back-gate is shown in Table 2. The results clearly show a boost in PAE by 4.7% and Pout improvement of 0.4 dBm.

Another example is the SPST switch insertion loss where the AFBG concept is used on the n-well of the super-low V_t (SLVT) transistors of 3 stacked devices [1, 7]. Fig. 9 shows the insertion loss (*IL*) versus frequency of the SPST switch. The floating BG exhibits a flatter frequency response especially with high BG bias (3V).

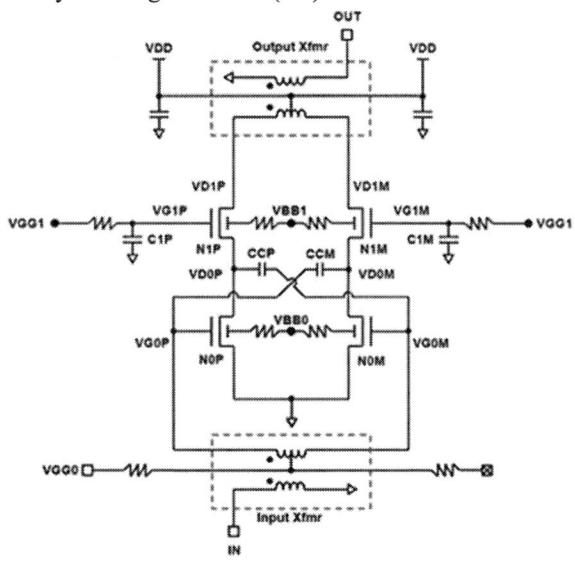

Fig. 8. Stacked PA Circuit Configuration (Common Sources are SG-types and the Cascode are EGU).

Table 2. PA Performance Comparison with/without Floating BG

2-Stack	SG-EGU (No AFBG) [#1]	SG-EGU (AFBG) [#2]
V_{DD} (V)	2.0	2.0
V_{BB} (V)	0	2 (VBB1: Cascode)
Gain (dB)	12.6	12.7
P_{1dB} (dBm)	18.3	18.8
P_{sat} (dBm)	18.8	19.2
Peak PAE (%)	38.2	42.9

[#1] Back-gates are shorted to ground on chip
[#2] Back-gates are biased with 10kOhm resister in series (AC floating bias) VBB0=0

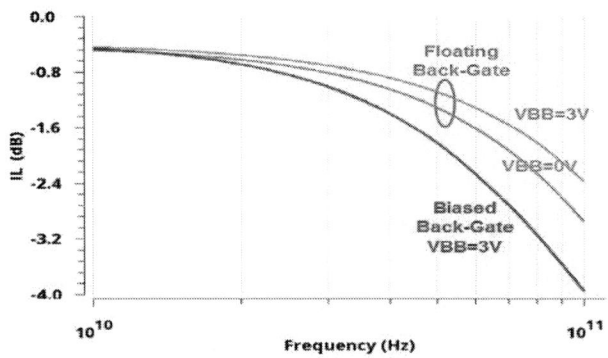

Fig. 9. SPST switch *IL*- Freq. with/without floating BG [1, 7].

V. CONCLUSION

The impact of back-gate biasing to DC, RF/mmW and HF noise of 22FDX® technology has been described. Applying forward back-gate bias has shown to reduce the threshold voltage, while increasing the drain saturation current I_{dsat} with higher output conductance G_{ds} thus lower resistance R_{on}. The peak f_T, f_{MAX} of FG and BG and their gain efficiency product has shifted to lower V_g with the peak values remain almost the same, when the back-gate bias increased. Besides, lower R_n, NF_{min} and NF_{50} can be achieved at low V_g with forward back-gate bias. The front-gate/back-gate achieved performance of the NFET for f_T is 350/85 GHz and for f_{MAX} to be 370/23 GHz. Dynamic back-gate biasing enables active trade-off of performance versus power. A 28GHz stacked Power Amplifier (PA) has been presented, achieving 19.2 dBm Psat, 12.7 dB Gain and 42.9% PAE. A SPST switch with improved *IL* versus frequency response has been demonstrated using AC floating back-gate with DC back-gate bias.

REFERENCES

[1] S. N. Ong, *et. al.,* "A 22nm FDSOI Technology Optimized for RF/mmWave Applications," in *Proc. IEEE RFIC Symposium*, Philadelphia, PA, June 2018, pp. 72-75.

[2] J. C. Barbé, *et. al.,* "4-port RF performance assessment and compact modeling of UTBB-FDSOI transistors," in *Proc. IEEE RFIC Symposium*, Phoenix, AZ, 2015, pp. 355-358.

[3] B. Kazemi, *et. al.,* "Back-gate bias effect on FDSOI MOSFET RF Figures of Merits and Parasitic Elements", *Proceeding ULIS*, 2017.

[4] H.Su, H.Wang, T.Xu R.Zeng, "Effects of Forward Body Bias on High-Frequency Noise in 0.18-um CMOS Transistors," *IEEE Trans. on Microwave Theory And Techniques*, Vol. 57, No. 4, April 2009.

[5] "Making Accurate and Reliable 4-Port On-Wafer Measurements", Cascade Microtech, Inc. App. Note 1-4.

[6] C. H. Chen, *et. al.,* "Extraction of the induced gate noise, channel thermal noise and their correlation in sub-micron MOSFETs from RF noise measurements," in *Proc. ICMTS.*, Kobe, Japan, 2001, pp. 131–135.

[7] A. Bellaouar (June 2018). *Millimeter-Wave Circuit Design and Techniques in FDSOI CMOS Technology*, Workshop presented at RFIC2018, Philadelphia, PA.

A Low Power Fully-Integrated 76-81 GHz ADPLL for Automotive Radar Applications with 150 MHz/us FMCW Chirp Rate and -95dBc/Hz Phase Noise at 1 MHz Offset in FDSOI

Ahmed R. Fridi[1], Chi Zhang[1], Abdellatif Bellaouar[1], Man Tran[2]

[1]GLOBALFOUNDRIES, RF and mmWave Design Team, Dallas TX, USA

[2]Mantric Technology, Toronto Ontario, Canada

Abstract—In this paper, a fully integrated 76-81 GHz All Digital PLL for FMCW automotive radar applications is presented. It features a 20 GHz digital PLL followed by a 4x multiplier and buffer. The proposed ADPLL is implemented in a 22nm fully depleted SOI CMOS technology. It achieves up to 150 MHz/us FMCW chirp rate over a 4 GHz bandwidth and dissipates 85mW only.

Keywords— Digital PLL, FMCW, radar, mmWave, 79 GHz.

I. INTRODUCTION

Millimetre-wave radars make it possible to quickly and precisely measure the radial velocity, range and azimuth angle of multiple objects. For this reason, the automobile industry is increasingly using this technology in advanced driver assistance systems (ADAS). These radars require high-quality frequency synthesizers that offer good frequency linearity, fast settling and low phase noise. Though analog PLLs implemented mainly in SiGe processes still dominate this field, **A**ll-**D**igital PLLs (ADPLLs) promise greater flexibility, area efficiency, lower power and better compatibility with advanced scaled CMOS technology nodes.

II. ARCHITECTURE DESCRIPTION

The block diagram of the proposed ADPLL is shown in Fig. 1. It consists of a 20GHz All Digital Fractional-N PLL followed by a multiply by 4 circuit followed by an amplifier to generate the linear FMCW frequency ramps in the 76-81 GHz band.

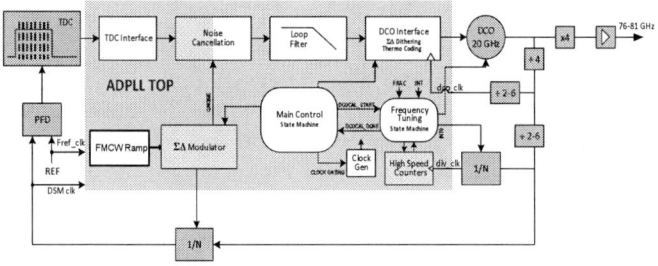

Fig. 1. ADPLL Architecture

The 20 GHz DCO frequency is divided by 4 by a full CMOS divider to around 5GHz where it can then clock a custom CMOS programmable divider. A DSM modulator is inserted in the feedback path to average to any fractional value. The output of the digital divider is then compared to a reference clock F_{ref}. The phase difference between these two clocks is digitized by the TDC and fed to a digital filter. The

locking process is divided into 2 phases: an open-loop phase where the TDC resolution is calibrated to a desired value and the DCO frequency is calibration through its coarse array to a frequency close enough to the ramp start frequency and a closed loop phase where the final frequency is reached.

A DSM quantization noise cancellation circuit is added to help attenuate the noise outside the loop bandwidth.

III. LOOP FILTER

The discrete time implementation of the loop filter is inspired from the conventional s-domain Type II single loop PLL with a single integrator and 2 poles where the zero and pole locations are given by (1):

$$\mathcal{H}_{zero}(s) = 1 + \frac{s}{2\pi f_z} \quad \text{and} \quad \mathcal{H}_{pole}(s) = \frac{1}{1 + \frac{s}{2\pi f_p}} \quad (1)$$

This approach has the benefit of using many of the s-domain analysis, tools and knowledge for early prototyping and also as a validation of the z-domain system simulations. The discrete time transfer functions are obtained by using a bilinear transform with pre-warping at f_p and f_z respectively.

$$H_{zero}(z) = 2\frac{(1+\alpha) - \alpha z^{-1}}{1 + z^{-1}} \quad \& \quad \alpha = \frac{1}{2}\left[\frac{1}{tan\left(\frac{\pi f_z}{F_s}\right)} - 1\right] \quad (2)$$

$$H_{pole}(z) = \frac{1}{2}\frac{\beta(1 + z^{-1})}{1 - (1-\beta)z^{-1}} \quad , \beta = 2\left[\frac{1}{tan\left(\frac{\pi f_p}{F_s}\right)} + 1\right]^{-1} \quad (3)$$

The overall loop filter z-domain transfer function (4) consists of a digital gain K_p, a zero (from 3KHz to 780KHz) and two digital poles to filter out the DSM excess quantization noise (from 1.6MHz to 33MHz). The wide range and flexibility of the loop parameter programming play a critical role in striking the right balance between phase margin, loop bandwidth, phase noise at 1 MHz offset and achieving chirp slopes as fast as 150 MHz/us.

$$H(z) = K_p \frac{\beta}{1 - (1-\beta)z^{-1}} \frac{\gamma}{1 - (1-\gamma)z^{-1}}\left(\alpha_1 + \frac{1}{1 - z^{-1}}\right) \quad (4)$$

Fig. 2 shows the loop filter implementation and Fig. 3 shows the perfect matching between the Matlab s-domain and z-domain models of the loop.

IV. DSM QUANTIZATION NOISE CANCELLATION

The DSM instantaneous quantization noise ε_Q is computed by subtracting the DSM output from its input after proper delay alignment. This quantization error causes an instantaneous jitter of the divided clock given by:

$$\delta T_k = \varepsilon_k\, T_{VCO} \qquad (5)$$

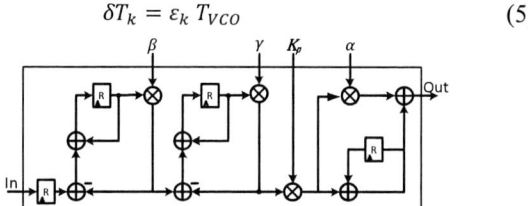

Fig. 2. Loop Filter Topology

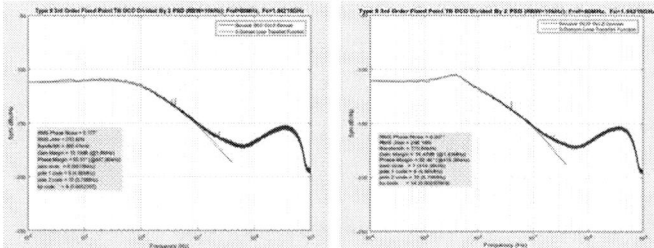

Fig. 3. The loop models agreement between the s-domain (red) and the z-domain (blue) Matlab models for different loop bandwidths and parameters.

This jitter is accumulated and scaled by the TDC gain. If K_{TDC} is the TDC code for half a reference clock cycle and N and M are the integer and fractional parts of the divider ratio then the scaled quantization noise is given by:

$$\varepsilon_{TDC} = \frac{K_{TDC}}{T_{REF}/2}\sum_k \delta T_k = 2\frac{K_{TDC}}{N+M}\sum_k \varepsilon_k \qquad (6)$$

The nominal value for the gain $\frac{2K_{TDC}}{N+M}$ is 105. It corresponds to a nominal TDC resolution δ_{res} of 4ps, an F_{vco} of 4.75GHz and a feedback divide ratio FB_{DIV} of 2. This value is rounded to 128 for simplified implementation.

$$gain = 2\frac{K_{TDC}}{N+M} = 2\frac{\frac{T_{REF}}{2}}{\delta_{RES}(N+M)} = \frac{FB_{DIV}}{F_{vco}\times\delta_{RES}} = 105 \qquad (7)$$

The DSM's quantization noise at the TDC output is removed by using a novel adaptive noise cancellation circuit based on an efficient implementation of the Normalized-**LMS** (NLMS) algorithm as shown in Fig. 4 and Fig 5. Hence the noise generated by the DSM modulator is accurately suppressed and real time tracking of operating condition changes (e.g. Temperature) is achieved.

The effectiveness of the adaptive noise cancellation circuit in suppressing the DSM noise is shown in Fig. 6 (a). Fig. 6 (b) shows its fast convergence (less than 1us) when enabled.

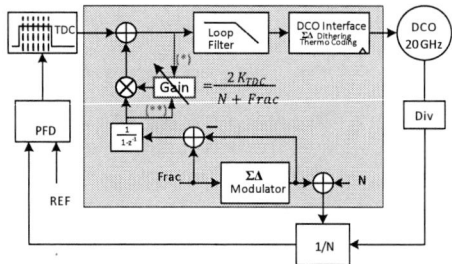

Fig. 4. The quantization noise cancellation concept.

V. DCO DIGITAL INTERFACE

The DCO interface connects the 25 bits digital loop filter output to the10 bits DCO fine array. The 2's complement of the digital loop filter is first converted to an offset binary code. The signal is then oversampled at $F_{DCO}/4$ and the 15 LSBs are dithered with a programmable 1st or 2nd order DSM modulator.

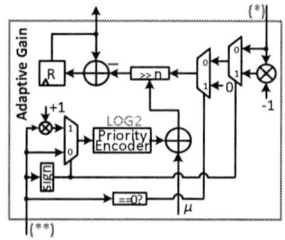

Fig. 5. Block diagram of the adaptive quantization noise cancellation.

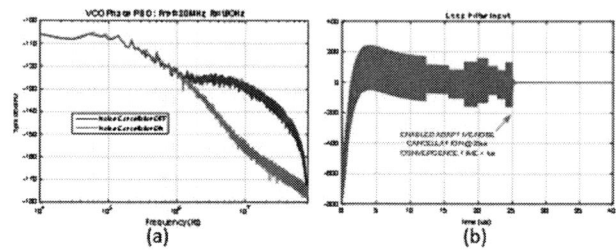

Fig. 6. (a) ADPLL PSD with and without quantization noise cancellation (b) ADPLL transient simulation of loop filter out after convergence of NLMS algorithm.

The Output of the DSM is then added to the remaining 10 MSB bits which are 2D thermo-coded and latched before they are sent to the DCO. The DCO interface is shown in Fig. 7.

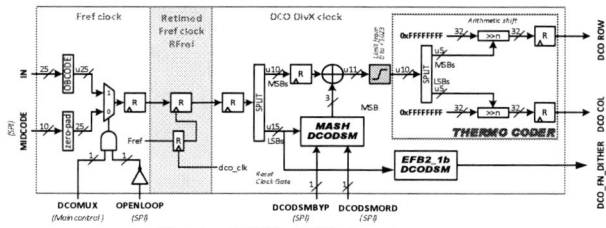

Fig. 7. ADPLL DCO Interface

VI. DCO CIRCUIT DESIGN

In order to meet the 76-81 GHz range requirement, the DCO has to cover a maximum frequency range of 19- 20.25 GHz. The DCO design is a tradeoff of four critical aspects: the frequency ramp bandwidth, the coarse and fine array resolutions, the phase noise and the power dissipation. The simplified schematic view of the DCO is shown in Fig. 8a.

A. Inductors

There are two important inductors in this design: One is the inductor of the main LC tank (L) which along with the capacitor arrays determines the correct centre and a second one (LS) to provide a good rejection of the DCO's 2nd harmonic (Fig. 8b). To achieve wide tuning range, a large capacitor array is required. Therefore the main tank's inductor L value needs to be small. In this design, the inductor was implemented using thick copper metal with a width of 10μm to achieve a Q of 18.

B. Coarse Capacitor Array

A 5-bits coarse capacitor array was designed to tune the DCO center frequency to compensate for frequency shifts due to process corners. It will also be used to set the starting frequency of each frequency ramp. The coarse capacitor array consists of 31 switched binary coded capacitor units (Fig. 9a).

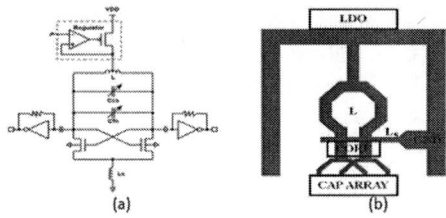

Fig. 8. (a) Simplified schematic of the DCO (b) Layout of the DCO inductors

The back-gate of the main switch N_0 is connected to its own front-gate and to the enable signal. The ON resistance (R_{ON}) of the switch is reduced when the switch is ON due to forward back-gate bias for better phase noise and low power dissipation. The coarse capacitor unit achieves an ON capacitance (C_{ON}) of 20.1fF and an OFF capacitance (C_{OFF}) of 7fF. The quality factors Q_{ON} and Q_{OFF} are 20 and 221.

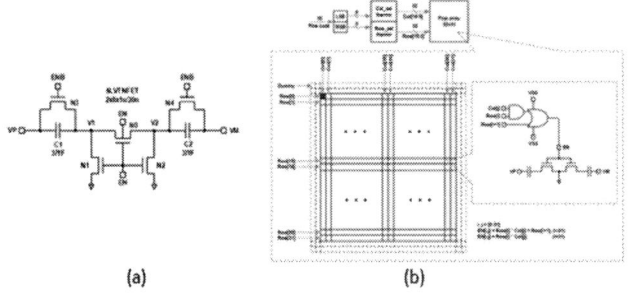

Fig. 9. (a) Schematic of switched capacitor unit in coarse array (b) Block diagram of fine capacitor array decoding

C. Fine Capacitor Array

The frequency ramping will solely rely on sweeping the fine capacitor array. This array has to cover the whole ramp range of 77–81GHz (19.25-20.25GHz). Another key DCO design parameter is its fine step resolution for lower quantization noise. As a trade-off between range and resolution, the fine capacitor array is a 10-bits thermometer coded array (Fig. 9b).

To ensure a very low parasitic loading the main LC tank, a novel MOM fine capacitor unit of 400aF with a tiny foot print is used (Fig. 9a). The unit achieves a C_{ON}/C_{OFF} of 346 / 234 aF and a Q_{ON} / Q_{OFF} of 28/57 for a resolution of 1.35MHz/code.

D. Cross-coupled Pair

The cross-coupled pairs' drains and common source are connected to the LS inductor directly to eliminate parasitic inductance which is critically important in mm-Wave design.

Only an NMOS pair is used in this design to improve the phase noise. This choice comes at the price of higher current to achieve the same gm as a CMOS pair. The DCO core consumes 36 mA while achieving 36mS of negative gm and 1.2V supply voltage.

VII. TDC DESIGN

The TDC is a key component in ADPLL design as it dominates the in-band noise and dictates the amount of flexibility the loop will have to accommodate fast chirps ramping times (need for larger loop bandwidths for fast frequency tracking) and meeting the 1MHz offset phase noise requirement for FMCW (need for smaller loop bandwidths for good attenuation of all excess in-band noise showing up at the 1MHz offset). The SSB quantization noise is:

$$L(f) = \frac{1}{12}\left(2\pi \frac{\delta_{res}}{T_{vco}}\right)^2 \frac{1}{F_{ref}} \tag{8}$$

Where δ_{res} is the TDC single stage resolution, F_{ref} is the reference frequency and T_{dco} is the DCO period after the divide by 4. In order to meet the -97 dBc/Hz in-band noise requirement, a reference frequency of 160MHz and a TDC resolution of 4 ps are needed. As shown in Fig. 10, a traditional single delay line ring oscillator TDC is selected. This architecture can only achieve one single gate delay resolution of 6ps. The access and control of the back-gate node in a forward bias manner helped reduce the V_{th} voltage of both NFETs and PFETs and boost the TDC stage resolution from 6ps to the 4ps needed.

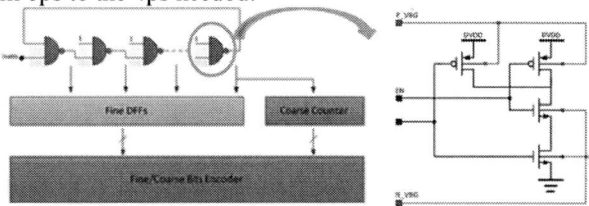

Fig. 10. Block diagram of the ring oscillator TDC with back-gate usage

VIII. FMCW MODULATOR

The FMCW linear frequency ramps are generated in the feedback path of the ADPLL. It can provide different ramp profiles with fully programmable slopes and frequency bandwidths. As shown in Fig. 11, the chirp is generated by continuously increasing the integer and fractional part of the divider ratio with fixed small frequency steps at a clock frequency of F_{ref}/R, where:

$$R = \frac{T_{ramp}}{N_{step}} F_{ref} \tag{9}$$

T_{ramp} and N_{step} are the ramp time duration and the number of frequency points respectively. In order to avoid glitches caused by clock domain crossing, a dedicated re-timing circuit is necessary. Another point of emphasis is to match the integer and fractional path delays to avoid frequency jumps when the fractional accumulator reaches its maximum and is ready to roll over.

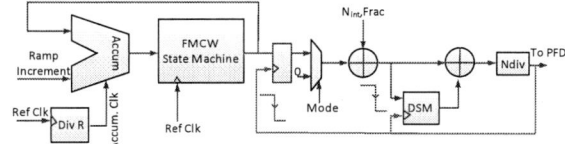

Fig. 11. Block diagram of FMCW modulator and timing.

IX. Measurement Results

The ADPLL was fabricated using GLOBALFOUNDRIES 22nm FDSOI CMOS technology. Fig. 13 shows the die photo, which has an area size of 0.74x1030 mm². The power consumption was 85mW where 52mW are dissipated by the 20 GHz ADPLL and 33mW by the 4X multiplier and buffer.

Due to measurement equipment limitation of a maximum frequency of 40 GHz, the phase noise of the ADPLL was measured at 20 GHz through a dedicated test port. The phase noise referred to 77 GHz, for a reference clock of 160MHz was -95 dBc/Hz at 1MHz offset for Chirp slopes of less than 50MHz/us and -92dBc/Hz for chip slopes of 100MHz/us or higher (Fig. 13a). The maximum Chirp slope measured was 150MHz/us for 4GHz FMCW bandwidth and 40us ramp time. The flyback time in saw tooth ramp mode is only 0.8us allowing for a very small idle time and a very efficient chirp sequence.

The different FMCW measurements were all scaled to 20GHz band too. Fig. 14 and Fig. 15 demonstrate the multimode capability of the chip with a saw-tooth ramp type with 100 MHz/us slope and 4GHz bandwidth and a triangular ramp with 150 MHz/us slope and 4GHz bandwidth respectively.

Fig. 12 Die Photo

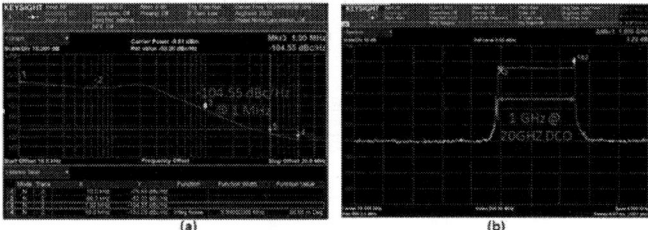

Fig. 13. (a) Phase noise at 19.25 GHz for 100 MHz/us chirp case (b) ADPLL FMCW spectrum with max hold for 4 GHz bandwidth case (1GHz x 4).

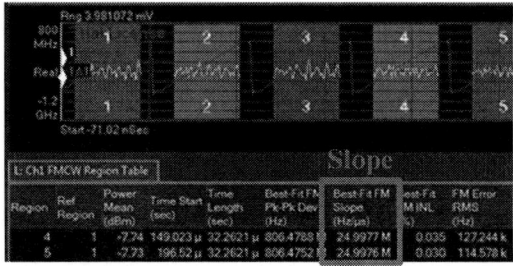

Fig. 14. A sawtooth FMCW Chirp with 100 MHz/us slope (25MHz x 4), 0.8us Fly-back and 4 GHz bandwidth.

A summary of the chip performance and comparison to prior art is shown in Table 1. The chip demonstrates state of the art FMCW performance at best in class 52mW power dissipation and 0.7mm² die area that includes all LDOs and pads.

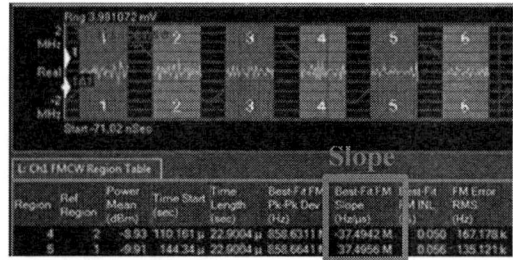

Fig. 15. A triangular FMCW Chirp with 150 MHz/us slope (37.5MHz x 4) and 4 GHz bandwidth.

Table 1 Summary and Comparison Table with Previous State of the Art Work

	This Work	**[2]**	**[5]**	**[6]**	**[1]**
PLL Type	Digital Frac-N	Analog Frac-N	Analog Frac-N	Analog Frac-N	Analog Frac-N
Technology	22nm CMOS FDSOI	130nm SiGe BiCMOS	65nm CMOS	65nm CMOS	45nm CMOS
Frequency Band (GHz)	76-81	75-83	76-81	76-81	76-81
FMCW Bandwidth (GHz)	4	8	4	4	4
Max Chirp Slope (MHz/us)	150	100	0.3	12.5	100
Frequency RMS Error (KHz)	480	3200	960(³)	-	-
Phase Noise @ 1MHz (dBc/Hz)	-95/-92(¹)	-97/-	-83/-	-91/-	-94/-91(¹)
Fly Back Time (us)	0.8	15	-	-	4
Power (mW)	52	240	320(²)	80	-
Area (mm²)	0.7	4.4	2.74	1.69	-

(¹) Second number for max chirp slopes (²) Synthesizer and 2X (³) Error at 0.3 MHz/us slope

X. Conclusion

This paper presents a very low power, fully integrated, 76-81 GHz ADPLL for FMCW automotive radars implemented in GLOBALFOUNDRIES 22nm FDSOI process. The ADPLL can generate chirps as fast as 150MHz/us with a best in class 52mW power dissipation for 0.7 mm² die size. It also features a very fast fly back time of less than 0.8us. The ADPLL achieves -95 dBc/Hz phase noise at 1MHz offset, for chirp rates of less than 50 MHz/us and -92 dBc/Hz for chirp rates of 100MHz/us or more.

References

[1] B. P. Ginsburg, and al., "A Multimode 76-to-81GHz Automotive Radar Transceiver with Autonomous Monitoring", IEEE ISSCC 2018.

[2] J. Vovnoboy, and al, "A fully integrated 75-83 GHz FMCW synthesizer for automotive radar applications with -97 dBc/Hz phase noise at 1 MHz offset and 100 GHz/ms max chirp rate", IEEE RFIC Symp., 2017.

[3] T. Fujibayashi, et al., "A 76- to 81-GHz Packaged Single-Chip Transceiver for Automotive Radar", IEEE BCTM, 2016.

[4] D. Weyer, and al, "A 36.3-to-38.2GHz -216dBc/Hz2 40nm CMOS Fractional-N FMCW Chirp Synthesizer PLL with a Continuous-Time Bandpass Delta-Sigma TDC Converter," IEEE ISSCC., 2018.

[5] J. Park, and al, "76-81-Ghz CMOS transmitter with a phase-locked-loop-based multi chirp modulator for automotive radar," IEEE Trans. Microw. Theory Techn, April 2015.

[6] H. Matsumura, and al, "Ultra-Low Phase Noise 76-81 GHz PLL Synthesizer for FMCW Radar in 65 nm CMOS", Proceedings of APMC 2012, Kaohsiung, Taiwan, Dec. 2012.

[7] D. Cherniak, and al, "A 23GHz Low-Phase-Noise Digital Bang-Bang PLL for Fast Triangular and Saw-Tooth Chirp Modulation." IEEE ISSCC, 2018.

An 82.2-to-89.3 GHz CMOS VCO with DC-to-RF Efficiency of 14.8%

A. Tarkeshdouz[#1], M. Haghi Kashani[#2], E. Hadizadeh Hafshejani[#3], S. Mirabbasi[#4], E. Afshari[*5]

[#]SoC Lab, The University of British Columbia, Canada

[*]UNIC Lab, The University of Michigan, Ann Arbor, USA

{[1]tarkeshdouz, [2]miladhk, [4]shahriar}@ece.ubc.ca, [3]ehsan_hadizade@yahoo.com, [5]afshari@umich.edu

Abstract—In this paper, we present a 90-GHz CMOS voltage-controlled oscillator (VCO) that operates from 82.2 to 89.3 GHz. The proposed VCO utilizes a device-centric method to achieve a high-efficiency oscillator with DC-to-RF efficiency closer to the efficiency limits of the employed transistors. Using the swing-independent bias condition for the CMOS transistors at the frequency of interest, we incorporate an active feedback network with minimal power overhead to provide the maximum achievable swing at the gate terminal of the core transistor, thereby increasing the power added efficiency. Utilizing this method, a high-efficiency VCO with center frequency of 85.75 GHz is designed and implemented in a 65-nm CMOS process. In this VCO, the trade-offs among efficiency and tuning range are also considered and the VCO achieves a peak DC-to-RF efficiency of 14.8% at 89.3 GHz and a wide tuning range of more than 8.3% while consuming only 8.5 mW of dc power from a 1.2-V supply. Compared to the prior art, the proposed VCO demonstrates a DC-to-RF efficiency of more than 14% at frequencies up to 100 GHz. The fabricated circuit occupies a silicon area of less than 0.25 mm² including the pads.

Keywords— DC-to-RF efficiency, power added efficiency (PAE), voltage-controlled oscillator (VCO).

I. INTRODUCTION

Due to availability of wider bandwidth and less populated frequency spectra, the lower frequency bands of the millimeter-wave (mm-wave) spectrum are being explored for a variety of communication, surveillance, and imaging applications [1]. This includes automotive radar at 77-GHz band with a great commercialized viability [2], passive radar imaging in W-band [3], and multi-gigabit wireless communication at 60 GHz [4]. For many of these applications, especially those that are portable, energy efficiency is a critical issue, requiring circuit and system solutions with minimal impact on battery life.

Recent examples of high-efficiency mm-wave oscillators have been demonstrated on CMOS in [5]–[8] to realize the above-mentioned systems. In [5], the fundamental VCO effectively manipulates the dc current of the drain of the transistors to minimize the time during which the transistor is on while maintaining the same fundamental generated power. The oscillator achieves 6.1% DC-to-RF efficiency in 0.13-μm CMOS with a limited tuning range. In [9], the theoretical limits on the drain efficiency for a single transistor are investigated and the major differences between theory and practice are explored to design a high-efficiency mm-wave oscillator utilizing a Class-E oscillator operating at 45 GHz. The design achieves a DC-to-RF efficiency of 15.6%. In this design, the underlying assumption that in an ideal Class-E

power amplifier the main switch is opened and closed with zero rise/fall time pulses is revisited. Using a CMOS transistor as a switch, the associated on-resistance and the parasitic drain capacitance are identified as the primary sources that cause departure from the ideal Class-E behavior. Thus, the maximum oscillation frequency (f_{osc}) for the minimal efficiency degradation is achieved. This approach, however, imposes a limit on f_{osc} in the design. In [10], an 88-GHz fundamental SiGe HBT oscillator is presented by utilizing a device-centric design technique to achieve a high efficiency. The oscillator benefits from a self-feeding oscillator topology, employing the technique presented in [11] to optimize the fundamental power generation and exhibit 19.4% DC-to-RF efficiency. The oscillator, however, does not fully exploit the device capacity as it requires a low supply voltage for the optimum condition.

In this work, we address the fundamental design challenges for efficient operations of CMOS transistors at very high frequencies and propose an oscillator architecture that offers the energy efficiency required for low-power applications. In Section II, the operation of a single-transistor CMOS transistor is investigated to develop effective mechanisms for enhancing the efficiency of mm-wave oscillators. In Section III, the design considerations for the proposed VCO are discussed. This section also describes the circuit implementation details. Section IV presents the measured results of the implemented oscillator, and finally Section V provides the concluding remarks.

II. A HIGH-EFFICIENCY SINGLE-ACTIVE-DEVICE-BASED OSCILLATOR

At very high frequencies, e.g., mm-wave frequencies, the share of passive devices in power loss become significant due to the limited quality factors (Q) offered by such elements. In [11], it is shown how the limited quality factor of passive components limits the maximum output power generated by the active transistors. Any oscillator operating at very high frequencies is highly impacted by such detrimental effects. The maximum achievable quality factor at high frequencies imposes a limit on the minimum size of the transistor required for oscillation buildup. This inherently increases the power dissipation by the transistor itself. In fact, the dominance of the parasitics in large transistors restricts power conversion capability in transistors, i.e., conversion of dc supply power to P_{out}. In CMOS, this forces the transistors to operate in current limited mode with a certain voltage swing on the gate, failing to generate large swing levels at the output. This significantly reduces the transistor efficiency when it is placed in a feedback loop to form an oscillator. As such, it is required to lower the supply voltage to enable close-to-efficient active operations. This is undesirable by itself as it restricts the generated power and may kill the oscillation.

To design a high-efficiency oscillator, it is necessary to obtain the power efficiency limits of a single transistor and find the ideal conditions for such an efficient operation. With the setting in Fig. 1(a), we obtain the maximum power added efficiency (PAE) for a single transistor without employing any wave-shaping techniques at the output. Key parameters for maximizing the efficiency, aside from the inductive load at the output, are the gate bias voltage, V_B, and the gate signal amplitude, S_G.

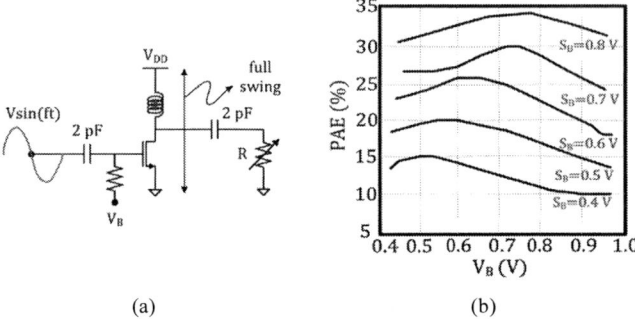

(a)   (b)

Fig. 1. (a) The setting utilized to obtain maximum PAE for a transistor; (b) PAE versus V_B for several voltage amplitudes on the gate.

As illustrated in Fig. 1(b), simulation results show that in CMOS transistors, PAE increases when the voltage amplitude on the gate increases. In this figure, PAE is plotted versus the gate bias voltage for different gate signal swing levels. In Fig. 2(a), the power consumption, P_{dc}, is shown for different gate bias voltage and gate signal swing levels. As observed, the peak PAE for larger swings in Fig. 1(b) occurs in the vicinity of the gate bias level (V_B) where P_{dc} becomes independent of the voltage swing level on the gate (S_G) as shown in Fig. 2(a). As can be seen, $V_B|_{optimum} = V_B \approx 0.8\,V$. This is an important finding in oscillator design where providing the start-up condition (which is correlated with V_B) does not degrade the transistor efficiency. This effectively allows to set the V_B required to provide the loop transfer when the transistor is placed in an oscillator architecture while accommodating a high peak PAE. In this context, V_B is a function of voltage swing and is larger for higher S_G. The inductive load on the drain side is the other parameter which controls V_B. Once the optimum V_B is set (Fig. 2(a)), the gate signal swing must be increased to achieve the best PAE performance (Fig. 1(b)). Simulations show that a voltage swing of 800 mV is required for $PAE_{max} \approx 33\%$. The pictorial waveforms in Fig. 2(b) explain the presence of such swing-independent point. The dc component (I_D) of the output drain current is a function of voltage swing on the gate as well as the gate bias level. In this context, V_B is the primary factor in determining I_D for bias voltages greater than $V_B|_{optimum}$ while the voltage swing on the gate (the shaded area below V_{th}) mainly controls this component for $V_B < V_B|_{optimum}$.

In single-transistor oscillator topologies, the output swing can be fed back to the input with the maximum feedback gain of 1. With limited maximum swing at very high frequencies, the returning swing may not be sufficient to reproduce the ideal efficient condition for the PAE of the transistor. Furthermore, to implement a single-transistor oscillator (for maximum DC-to-RF efficiency at mm-wave frequencies), the connection between the output and the input should provide the required power and phase condition for oscillation. The Colpitts oscillator topology is a structure where providing the desired start-up condition is more challenging. Additionally, the resulting voltage swing at the input side, due to a capacitive voltage division, is smaller than the output. In self-feeding oscillator topologies, a transmission line is employed with a proper loading condition to sustain oscillation [12]. However, the limited number of degrees of freedom in this configuration and limitations in usage of passive do not allow the transistor to capture the ideal condition for a high efficiency operation at both the input and output. Hence, a feedback network with a higher degrees-of-freedom is required. As pointed out, higher-order passive feedback circuits are not suitable options at high frequencies. This leads to utilization of active-based feedbacks where the degrees of freedom are higher while not consisting of passives. In this topology, the transistor in the active feedback network (AFN) does not assist with oscillation formation significantly and, as a result, consumes minimal power.

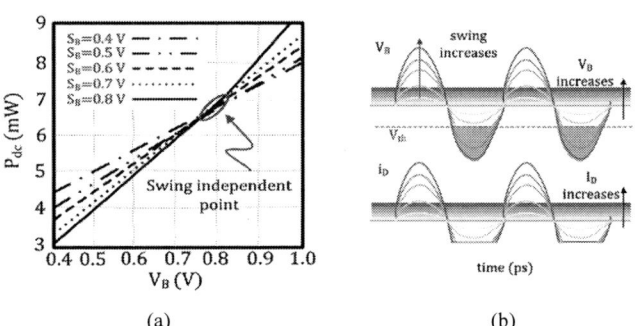

(a)   (b)

Fig. 2. (a) P_{dc} versus V_B for several swing levels; (b) Gate voltage and drain current waveforms for different levels of swing and bias voltage.

To increase the swing on the gate with $V_B = V_B|_{optimum}$, the AFN must source an in-phase current as shown by the red arrow in Fig. 3(a). To this end, we inspect the input impedance looking into the gate of the core transistor (Z_{in}). In Fig. 3(b), $|Z_{in}|$ is plotted versus frequency. By designing the circuit such that the peak in Z_{in} occurs within the desired frequency range, the amplitude of the gate signal is increased (remarks: in CMOS, the input power increases with increasing swing slightly even at higher frequencies and does not impact the efficiency significantly based on simulations). To accommodate the minimum power consumption in the AFN stage, the size of the transistor, the gate voltage bias, and the inductive load cannot be arbitrarily large and thus, the resulting S_G on the gate still might not be as large as desired. As $P_{dc}(total) = P_{dc}(core\ transistor) + P_{dc}(AFN)$, it turns out that that $S_G = S_{ideal}$ is achieved at the cost of excessive

978-1-7281-1702-7/19 $31.00 © 2019 IEEE

increase in $P_{dc}(AFN)$. This is one of the reasons why $V_B = V_B|_{optimum}$ (where PAE is approximately independent of the gate signal swing) and thus, the efficiency is minimally impacted if $S_G \neq S_{ideal}$.

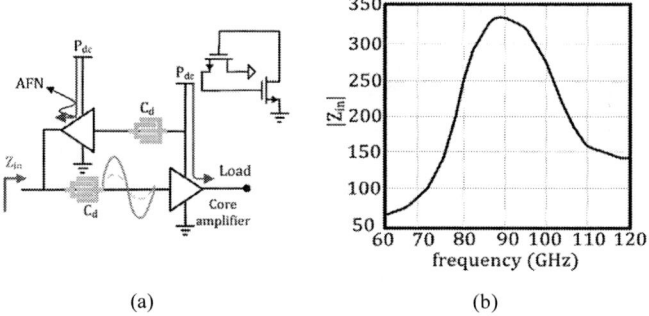

(a) (b)

Fig. 3. (a) Implementation of the proposed AFN; (b) Z_{in} vs. frequency.

III. CIRCUIT IMPLEMENTATION DETAILS

As a proof-of-concept for the presented circuit techniques, we have designed, implemented, and tested a fundamental VCO with the center frequency of 85.75 GHz in a 65-nm CMOS process. Fig. 4 displays the schematic view of the proposed VCO. The proposed VCO benefits from an included active-feedback single-transistor architecture. The core transistor has W = 15 μm with 10 fingers (of 1.5 μm each) and the AFN stage incorporates a smaller transistor with W = 8 μm with 8 fingers. To implement the decoupling capacitors (C_d), the interdigitated capacitors are used to improve the quality factor and provide a higher intrinsic self-resonance frequency (SRF). The VCO uses a varactor in the source of the core transistor to tune the frequency. To bias the transistors' sources and have freedom in selecting the value of the varactor's size and also lowering the silicon area, an LC tank is employed at the source of the transistor instead of an RF-chock. As such, the tuning range can potentially be extended as the value of C_{var} does not have to be close to C_{GS} necessarily. To achieve the maximum Q at the target frequency, the inductive loads are implemented as coplanar waveguide (CPW) transmission-lines with ground shielding.

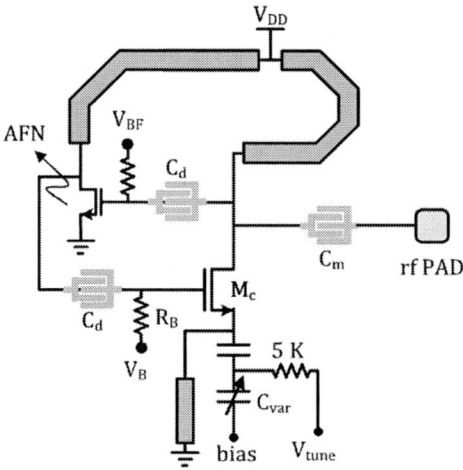

Fig. 4. the schematic view of the proposed VCO.

The equivalent inductance of the implemented transmission-lines are around 140 pH and 90 pH. The varactor in the core stage provides a reasonable tuning range (TR) and adding another varactor in AFN can extend the tuning range for an additional percent (~0.9%). At the output, the oscillator is connected to a small, low-capacitive RF pad (30 μm × 45 μm). The capacitance of the pad from the simulations is estimated to be around 10 fF. A small capacitor, $C_m = 20\,fF$, is placed at the output to form the required matching network with pad to transform the 50-Ω load to the optimum value of R_L required for maximum PAE. The inductance of supply (V_{DD}) traces and other interconnects are also carefully modeled and included in the simulation.

IV. MEASUREMENT SETUP AND EXPERIMENTAL RESULTS

For the purpose of testing, the chip is mounted on a printed circuit board (PCB) and all the dc biases are wire bonded. Fig. 5 shows the measurement setup for the VCO to determine the frequency of oscillation, phase noise and output power. A 75-to-110 GHz GSG (ground-signal-ground) Cascade Infinity Probe is used to probe the output of the VCO. The probe is directly connected to a harmonic mixer to mix down the signal. The local oscillator (LO) and intermediate frequency (IF) ports of the mixer are connected to the corresponding outputs on a spectrum analyzer. The output frequency is determined by sweeping LO frequency at various harmonics and measuring IF change.

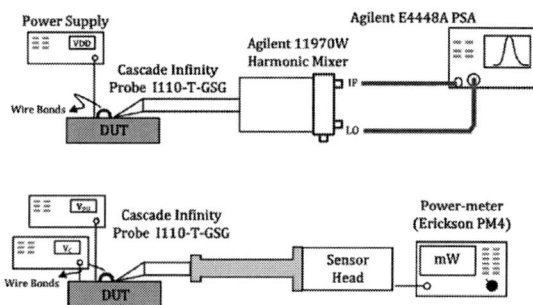

Fig. 5. The measurement setup for the proposed VCO.

The output power of the oscillator is measured using a VDI Erikson PM4 power meter. Figure 6(a) presents a chip photograph of the fabricated VCO. The measured peak output power and f_{osc} are plotted in Fig. 6(b). In Fig. 7 the simulated peak DC-to-RF efficiency and f_{osc} are compared with those from measurement. The measurement results are in good agreement with the simulations. The proposed VCO consumes ~8.5 mW of dc power and achieves a measured output power of 1 dBm for the record peak DC-to-RF efficiency of 14.8% when operating at 89.3 GHz. In Fig. 7(a), the output frequency can be continuously tuned from 82.2 GHz to 89.3 GHz which is translated to a frequency tuning range of more than 8.2%. Fig. 8 presents a phase noise plot measured at 89.3 GHz. The VCO attains the figure of merit (FoM) of −181.5 dBc/Hz and FoM$_T$ of −179.8 dBc/Hz at 1-MHz offset frequency. The performance of the chip is summarized in Table 1, along with a comparison with the prior art.

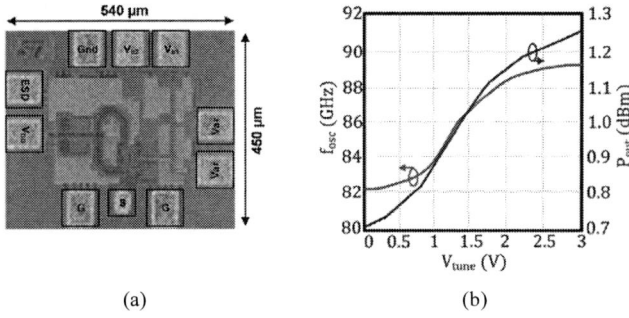

(a)   (b)

Fig. 6. (a) The chip photograph of the fabricated VCO; (b) The measured f_{osc} and peak output-power vs. V_{tune}.

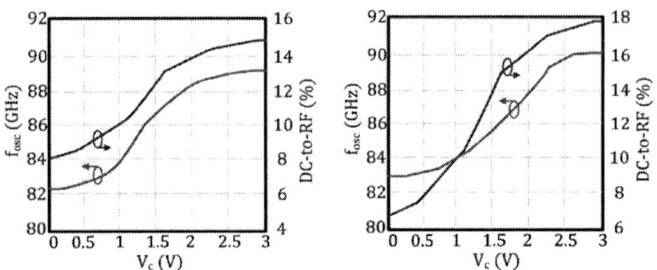

Fig. 7. f_{osc} and efficiency vs. V_{tune} (a) measured; (b) simulated.

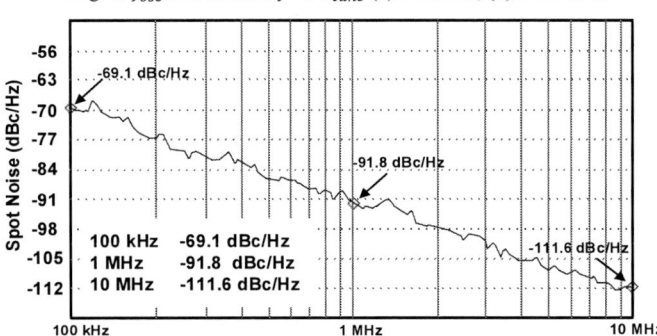

100 kHz	-69.1 dBc/Hz
1 MHz	-91.8 dBc/Hz
10 MHz	-111.6 dBc/Hz

Fig. 8. Sample phase noise plot measured at 89.3 GHz.

V. CONCLUSION

In this work, we propose an active-feedback-based single-transistor oscillator architecture to demonstrate a high-efficiency 85.75-GHz VCO that is implemented in a 65-nm CMOS process. A single-transistor active feedback network is used to provide the optimum signal swing condition at the gate of the core transistor for maximum PAE. This transistor offers additional degrees of freedom to realize the oscillator with the maximum PAE with minimal power overhead. The fabricated VCO achieves a record peak DC-to-RF efficiency of 14.8%, which to the best of authors' knowledge is the highest reported for mm-wave CMOS oscillators to date. The DC-to-RF efficiency number reported in this work is close to those from compound-semiconductor oscillators [13].

REFERENCES

[1] J. Laskar, et al., "The next wireless wave is millimeter wave," *Microw. J.*, vol. 8, iss. 8, pp. 22-36, Aug. 2007.

[2] J. Lee, et al., "A fully-integrated 77-GHz FMCW radar transceiver in 65-nm CMOS technology," *JSSC*, vol. 45, pp. 2746 –2756, Dec. 2010.

[3] A. Tomkins, et al., "A passive W-band imaging receiver in 65-nm bulk CMOS," *JSSC*, vol. 45, no. 10, pp. 1981 –1991, Oct. 2010.

[4] S. Emami, et al., "A 60 GHz CMOS phased-array transceiver pair for multi-Gb/s wireless communications," *ISSCC*, Feb. 2011, pp. 164–166.

[5] A. Tarkeshdouz, et al., "A 91-GHz fundamental VCO with 6.1% DC-to-RF efficiency and 4.5 dBm output power in 0.13-μm CMOS," *SSC-L*, vol. 1, no. 4, pp. 102-105, April 2018.

[6] M. Adnan, et al., "A 105-GHz VCO with 9.5% tuning range and 2.8-mW peak output power in a 65-nm bulk CMOS process," *T-MTT*, vol. 62, no. 4, pp. 753-762, April 2014.

[7] S. Kang, et al., "A 100GHz active-varactor VCO and a bi-directionally injection-locked loop in 65nm CMOS," *RFIC*, pp. 231-234, June 2013.

[8] S. Jameson, et al., "A 93.9 - 102.5 GHz Colpitts VCO utilizing magnetic coupling band switching in 65nm CMOS," *COMCAS*, 2015.

[9] E. Juntunen, et al., "High-power, high-efficiency CMOS millimeter-wave oscillators," *Journal of IET*, vol. 6, iss. 10, pp. 1158-1163, 2012.

[10] H. Aghasi, et al., "An 88-GHz Compact Fundamental Oscillator With 19.4% DC-to-RF Efficiency and 7.5-dBm Output Power in 130-nm SiGe BiCMOS," *SSC-L*, vol. 1, no. 5, pp. 106-109, May 2018.

[11] O. Momeni, et al., "High-power terahertz and mm-wave oscillator design: a systematic approach," *JSSC*, pp. 583-597, March 2011.

[12] R. Han, et al., "A CMOS high-power broadband 260-GHz radiator array for spectroscopy," *JSSC*, vol. 48, pp. 3090-3104, Dec. 2013.

[13] R. Weber, et al., "A 92 GHz GaN HEMT voltage-controlled oscillator MMIC," *IMS*, pp. 1-4, June 2014.

Table 1. Performance comparison of W-band VCOs with this work.

Ref.	Tech.	Freq. [GHz]	FTR [%]	P_{DC} [mW]	P_{out} [dBm]	DC-to-RF [%]	PN [dBc/Hz]	V_{DD} [V]	(3) FOM* [dBc/Hz]	(4) FOM$_T$* [dBc/Hz]
[5]	0.13-μm CMOS	91	0.5	46	4.5	6.1	−86.2 @ 1 MHz	1.8	−168.8 @ 1 MHz	−142.8 @ 1 MHz
[6]	65-nm CMOS	105	9.5	(1) 54	4.5	(1)5.2	−92.8 @ 1 MHz	1.2	−175.9 @ 1 MHz	−175.5 @ 1 MHz
[7]	65-nm CMOS	103.3	5.2	21	−2	3	−112.1 @ 10 MHz	1.2	−179.2 @ 10 MHz	−175.5 @ 10 MHz
[8]	65-nm CMOS	98	8.7	21.6	−1	3.7	−90 @ 1 MHz	1.8	−176.5 @ 1 MHz	−175.3 @ 1 MHz
[10]	0.13-μm BiCMOS	88	1.1	29	7.5	19.4	−107 @ 1 MHz	1.6	−191.1 @ 1 MHz	-
This Work	**65-nm CMOS**	**89.3**	**8.3**	**8.5**	**1**	**14.8**	**−91.8 @ 1 MHz**	**1.2**	**−181.5 @ 1 MHz**	**−179.8 @ 1 MHz**

(1) Buffer's P_{dc} not included. (3) $FOM = PN(f_{offset}) - 20log\left(\frac{f_{osc}}{f_{offset}}\right) + 10log(\frac{P_{DC}}{1mW})$. (4) $FOM_T = PN(f_{offset}) - 20log\left(\frac{f_{osc}}{f_{offset}} \cdot \frac{\Delta f}{10}\right) + 10log(\frac{P_{DC}}{1mW})$

978-1-7281-1702-7/19 $31.00 © 2019 IEEE

2019 IEEE Radio Frequency Integrated Circuits Symposium

A 62 GHz Tx/Rx 2x128-Element Dual-Polarized Dual-Beam Wafer-Scale Phased-Array Transceiver with Minimal Reticle-to-Reticle Stitching

Umut Kodak[#1], Bhaskara Rupakula[#2], Samet Zihir[$3], Gabriel M. Rebeiz[#4]

[#]University of California, San Diego, La Jolla, CA, USA

[$]Integrated Device Technology, San Diego, CA, USA

[1]ukodak@ucsd.edu, [2]bhaskara.rupakula@gmail.com, [3]Samet.Zihir@idt.com, [4]rebeiz@ece.ucsd.edu

Abstract — This paper presents a 62 GHz transmit/receive (TRX) dual-polarized/dual-beam wafer-scale phased-array with 2x128-elements spaced $\lambda/2$ apart in the x- and y-directions. Two reticles each containing 2x64-element arrays are stitched together in the IF/LO domain to form a super-reticle occupying an area of 21x42 mm^2. Each 64-element reticle-based phased-array includes RF-beam-forming TRX channels with 5-bit phase control, 9-bit gain control for each polarization, two 64:1 nested Wilkinson divider/combiners with bidirectional amplifiers, dual-transceivers with a common LO path for transmit and receive modes, and an active distribution network in the IF and LO domains to allow for tiling at the 64-element sub-array level. Also, high-efficiency quartz superstrate dipole antennas are used on the reticle to eliminate any RF transitions between the TRX channels and the antennas, resulting in a true wafer-scale array. The 2x128-element arrays scans to $\pm50°$ with side-lobe levels <-10 dB for both polarizations, achieving 43 dBm and 41 dBm saturated EIRP for V- and H-polarizations, respectively. To our knowledge, this paper presents the first dual-polarized dual-beam large-scale transmit/receive phased-array built using a reticle stitching technique.

Keywords — 60 GHz, dual-polarization, mm-wave systems, on-chip antenna, phased-array, stitching, transmit/receive, wafer-scale, wireless.

I. INTRODUCTION

Highly-complex and highly-integrated millimeter-wave phased-arrays in silicon technologies have become a reality over the past decade. The Wireless Gigabit Alliance (WiGig) and the new 5G standard have drawn attention the 60 GHz band, both for mobile and point-to-point applications, and there is a need for large number of elements to obtain high data-rates at large link distances ([1]-[3]).

The construction complexity of large-scale 60 GHz phased-arrays can be divided between the silicon chip and the printed-circuit board (PCB), one being more complex than the other depending on the method used. One way to build large-scale arrays is to use small beamformer chips, sometimes containing a transceiver, and these are attached to 2x2 or 4x4 antennas. These chips are then used to build scalable phased-arrays using expensive PCBs or interposer layers, and with 2-3 dB interconnect loss between the chip and antennas. Another way is to use a "wafer-scale" technique, where a scalable large beamformer chip containing 64-elements and on-chip high-efficiency antennas is used. This greatly decreases the PCB complexity and cost (especially

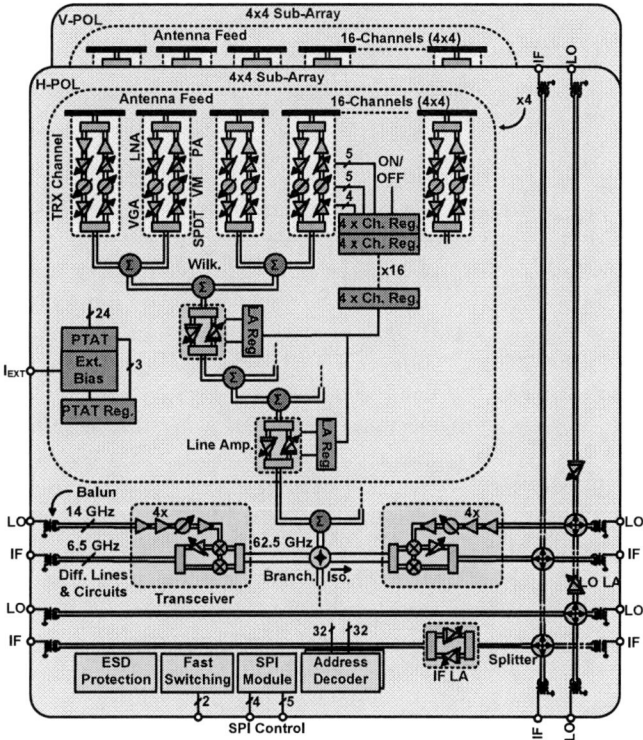

Fig. 1. Block diagram of the 64-element transmit/receive, dual-polarized, dual-beam phased-array with integrated dual-transceivers.

for dual-polarization, dual-beam designs), and improves the system NF and output power by 2-3 dB at 60 GHz ([2], [4]). However, the maximum reticle size allowed is 22x22 mm^2 which sets the upper limit in the scalable array design. It has been previously shown that the sub-reticle stitching technique breaks the reticle size limit and allows for large wafer-scale arrays, but this is achieved at a cost of a highly complex design and fabrication effort ([5]).

This work demonstrates the first reticle-to-reticle stitched wafer-scale phased-array with dual-polarization, dual-beam and up/down conversion on each reticle. The 64-element reticles are attached together using only LO and IF signals, thus allowing for bondwire-stitching between adjacent reticles on the same wafer and results in super-reticles and large millimeter-wave phased-arrays.

978-1-7281-1702-7/19 $31.00 © 2019 IEEE

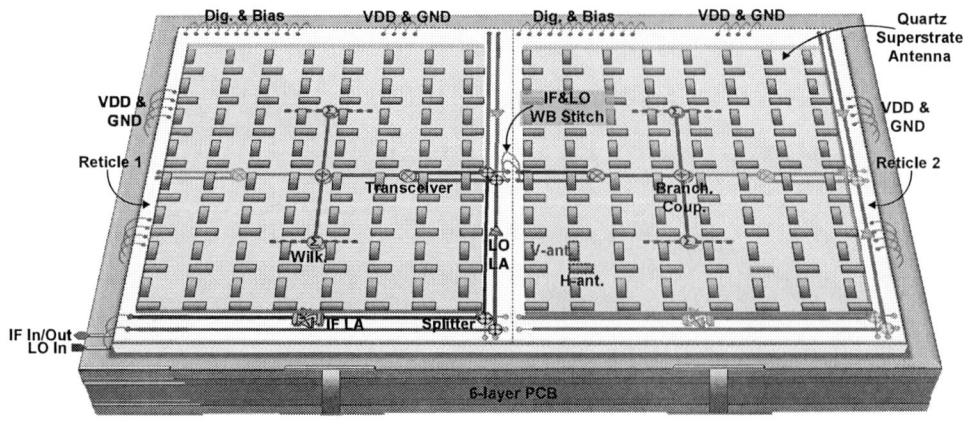

Fig. 2. Block diagram of the bondwire-stitched 128-element transmit/receive, dual-polarized, dual-beam phased-array with integrated dual-transceivers and quartz superstrate antennas. Shaded (gray) transceivers and IF/LO splitters are not used.

II. DESIGN

A. 2x128-Element Wafer-Scale Phased-Array Design

Fig. 1 presents the block diagram of the dual-polarized dual-beam 2x64-element wafer-scale phased-array reticle built in TowerJazz SBC18S3 process. The IF is at 6.5 GHz, and is up/down converted using bidirectional V and H transceivers using a 13-14 GHz LO. The transceivers share a common LO path with a x4 multiplier chain which allows for IF and LO distribution to multiple reticles and also outputs a 52-56 GHz LO signal at the transceiver with > 37 dBc harmonic rejection. Active mixers are used in the Tx and Rx path with 12 dB and 1 dB conversion gain, respectively. The radiated LO leakage is suppressed by > 30 dBc due to the tuned (bidirectional) line amplifiers and the TRX channels as well as using current DACs placed on each branch of the Tx mixer. The RF is distributed using 1:64 nested Wilkinson divider/combiner networks (for V- and H-polarizations) which employ bidirectional line amplifiers at each 2x2 and 4x4 sub-arrays to compensate the transmission line and division loss. The TRX channel contains two single-stage VGAs, a vector modulator and a PA with 3 dBm output P1dB in the Tx path. The Rx channel employs a differential LNA with an NF of 5 dB and similar VGAs and phase shifters. The Tx and Rx channels employ two differential SPDT switches with 2.5-3 dB insertion loss which leads to a 0 dBm output P1dB (in the transmit mode) and 8 dB NF (in the Rx mode) for each TRX channel.

The gain and phase states of each TRX channel as well as the operation mode (Tx or Rx) is digitally controlled using a serial-peripheral interface (SPI). The operation mode can also be controlled by fast switching circuitry (< 100 ns) and is independent from the SPI. The bias current is generated using a PTAT current generator with 3-bit reference current control. All digital and bias have ESD protection.

Each 64-element reticle employs two separate IF and LO distribution networks for infinite scalability (Fig. 1). The first network is located in the mid-west of the reticle, and is directly connected to the V and H transceivers. The second network is

fed from south-west of the reticle and contains a bidirectional two-stage IF line amplifier, a two-stage LO line amplifier, fully-configurable IF and LO splitters (9 dB division + ohmic loss per splitter) and a 35 mm transmission line (2 dB loss at 6 GHz). The IF line amplifier has a gain and output P1dB of 24 dB and 4 dBm at 6 GHz, respectively, and with 10 dB gain control. Two inductorless series-shunt SPDT switches are employed for bidirectionality and for compactness. The unidirectional LO line amplifier has a gain and output P1dB of 24 dB and 11 dBm at 14 GHz, respectively. The gain control is obtained using current-steering diode-connected BJTs at the cascode node with 5-bit 10 dB control. The IF and LO splitters are based on t-junction topology. Series-shunt BJT switches are used in four-branches and can be turned on/off independently. It can either divide/combine the IF signal or allow it to pass in any direction, which is critical for scalability.

The 2x128-element dual-polarized dual-beam wafer-scale phased-array is presented in Fig. 2. Each 64-element reticle has 6 sets of IF and LO pads and by configuring the distribution network, it is possible to build a scalable infinite-element phased-array in both x- and y-directions. In this work, a super-reticle containing two adjacent 64-element reticles is diced from the wafer and assembled on a 6-layer FR-4 based PCB. The IF and LO signals are distributed in an H-tree configuration on the super-tile. The south splitters are in the pass mode and the signal is directed from west to north (and vice versa for IF only). The middle splitters divide/combine the IF and LO signals, and feed the east and west transceivers in the first and second reticles, respectively. The 2x128-element array has only one stitching zone which is located at the super-reticle center, and where the 64-element reticles are stitched together for IF and LO routing using pad-to-pad bondwires. The VDD, ground, bias and SPI control are separate for each reticle and are directly connected to the PCB.

The dual-polarized differential dipole antennas are printed on a 100-μm thick quartz wafer and attached to each reticle (Fig. 2). A space of 1 mm is kept clear at the reticle edge

2019 IEEE Radio Frequency Integrated Circuits Symposium

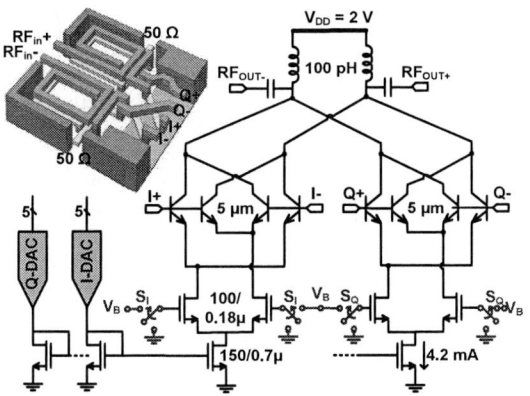

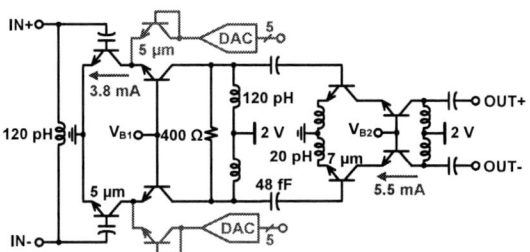

Fig. 3. Schematic of the vector modulator and layout of the I/Q generator.

Fig. 4. Schematic of the VGA with 5-bit gain control and PA.

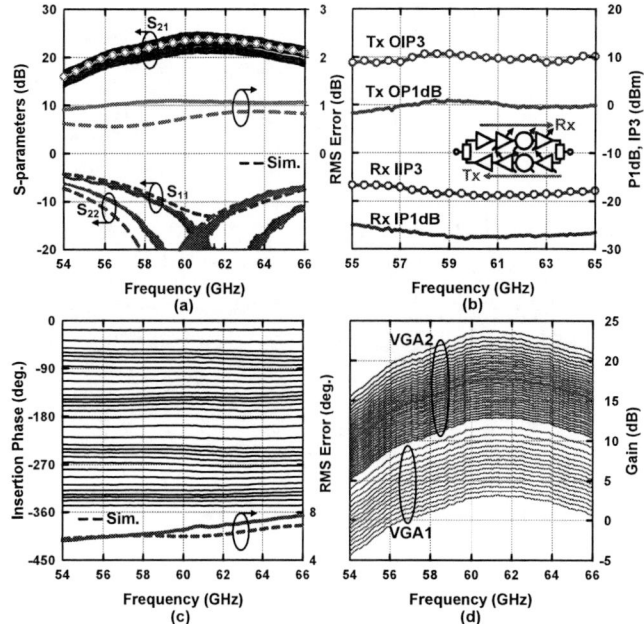

Fig. 5. Measured (a) Rx S-parameters for 32 phase-states, (b) Tx OP1dB and OIP3, and RX IP1dB and IIP3, (c) Rx phase-states and rms error and (d) Rx gain states for TRX channel breakout.

for wire-bonding (both for stitching and for VDD and SPI). The on-chip antenna feeds on M7 are electro-magnetically coupled to the dipole antennas on the quartz superstrate. The simulated antenna efficiency for V- and H-polarizations is 60% and 50%, respectively. The difference in efficiency is caused by the different transitions from the V- and H-Tx/Rx channels to the antenna feeds due to limited spacing in the dual-polarized unit-cell.

B. 60 GHz TRX Channel Design

The Rx channel consists of a single-stage LNA, two single-stage VGAs with 4-bit and 5-bit gain control and a 5-bit vector modulator placed between two VGAs. The vector modulator includes an I/Q generator based on $\lambda/4$-coupled transmission lines and followed by two variable gain stages (Fig. 3). The I/Q vectors are scaled using 5-bit current DACs and summed in current domain at the output.

The Tx channel consists of a single stage cascode PA, 4-bit and 5-bit single-stage VGAs and a similar VM as the Rx channel (Fig. 4). The 5-bit VGA employs a single-stage cascode amplifier and current steering method is used for gain control. The PA is a single-stage cascode design with 20 pH emitter degeneration for improved linearity. The Class A PA has a simulated output P1dB and Psat of 3 dBm and 5 dBm, with a PAE of 7% and 10%, respectively.

Fig. 5 presents the probe measurements of a single TRX channel. The Rx gain is 21-25 dB over 32 phase states with an rms gain error of 1 dB (Fig. 5a). In the Tx mode, the OP1dB and OIP3 are 0 dBm and 10 dBm, respectively, and in the Rx mode, the IP1dB and IIP3 are -27.5 dBm and -18.5 dBm

at 60 GHz, respectively (Fig. 5b). The phase states show no cross-overs at 54-66 GHz and the rms phase error is 6.5° at 60 GHz. The total gain control range is 20 dB including two VGAs (4-bit + 5-bit). The channel consumes 66 mW in the Rx mode and 80 mW in the Tx mode.

III. System Measurements

The 2x128-element dual-polarized dual-beam wafer-scale phased-array super-reticle is shown in Fig. 6. The array consumes 7.5 A and 8.6 A in the Tx and Rx modes, respectively, from a 2 V VDD. The chip temperature stabilizes at 50°C.

A standard gain WR-15 horn antenna is placed at R = 56 cm from the array in the far-field for pattern and EIRP measurements (between D^2/λ and $2D^2/\lambda$). The two reticle phases are equalized using a phase shifter in the LO paths at the transceiver level. The measured saturated EIRP is 43 dBm and 41 dBm for the V- and H-polarizations, respectively (Fig. 7). This agrees with simulations and can be calculated as EIRP = $P_T G_T$ = 44 dBm, where P_T = 0-1 dBm (P_{out}/element)+10log(N) = 21-22 dBm and G_T = 26 dB (directivity) - 3 dB to 4 dB (antenna efficiency) = 22-23 dB.

Fig. 8 presents the measured patterns for V-polarization and H-polarization in the H-plane (azimuth scan). The array scans to ±50° with low side-lobes (<-10 dB) and cross-polarization levels (<-30 dB) at 61.5 GHz. The 3-dB beamwidth for both polarizations is 6.5° at broadside and agrees with simulations for uniform illumination. The E-plane measurements (elevation scan) show side-lobe levels <-13 dB and 3-dB beamwidth of 12.6° for both polarizations (not shown due to brevity). The increased side-lobes in the H-plane

978-1-7281-1702-7/19 $31.00 © 2019 IEEE   337

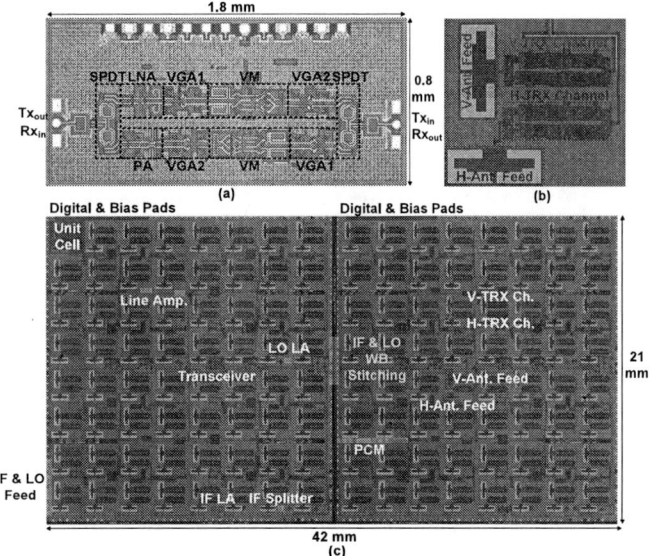

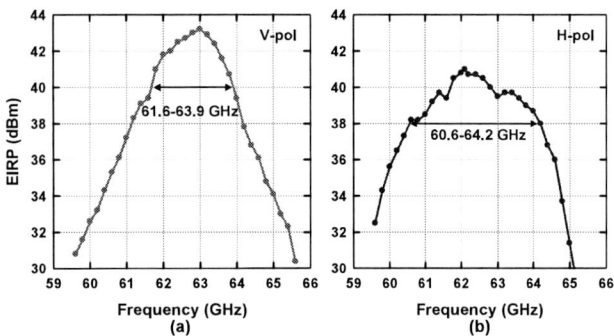

Fig. 6. Microphotograph of the (a) TRX channel, (b) unit cell, and (c) 2x128-element super-reticle wafer-scale dual-polarized array.

Fig. 7. Measured EIRP at Psat for (a) V-, and (b) H-polarization phased-array.

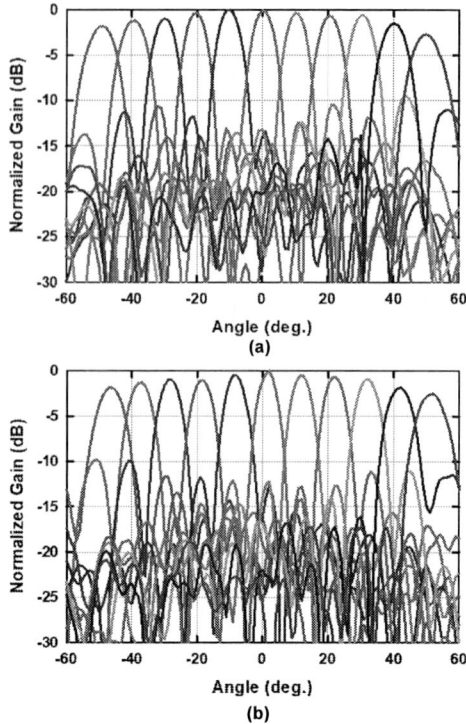

Fig. 8. Measured patterns at 61.5 GHz in azimuth scan: (a) V-polarization array in the Tx mode, and (b) H-polarization array in the Rx mode.

Table 1. Performance Summary

Frequency (GHz)	62
# of Elements	2x128
Polarization	Dual
Scan Angles E- /H- planes (°)	±50 / ±50
3-dB Beamwidth E- /H-planes (°)	12.6 / 6.5
Tx OP1dB / element (dBm)	0
Rx NF (dB)	8
Tx IRR & LO leakage (dBc)	< -60 / < -30
Power Diss. Tx/Rx per pol (W)*	20.5 / 18
EIRP at Psat (dBm)	41-43

* Chip VDD is 2 V, but a 2.4 V supply was used due to the IR drop on the PCB and on the chip.

scan is due to the 1 mm gap between the two 2x64-element arrays. Measurements in the Rx mode for the V-polarization array, and in the Tx mode for the H-polarization at 59 GHz and 63 GHz are nearly identical and not shown.

Table I summarizes the wafer-scale phased-array performance.

IV. CONCLUSION

This paper presents a 2x128-element dual-polarized dual-beam wafer-scale phased-array with integrated up/down converters and super-reticle tiling. It has been shown that it is possible to built phased-arrays larger than the standard reticle size (22x22 mm^2) using bondwire stitching. A saturated transmit EIRP of 41-43 dBm is achieved and the phased-array scans to ±50° with low side-lobes (<-10 dB). Stitching identical reticles into super-tiles proves the repeatability in wafer-scale designs and paves the way for the construction of near-infinite-element millimeter-wave phased-arrays.

ACKNOWLEDGMENT

This work was supported by DARPA MTO under the DAHI program. The authors thank TowerJazz for technical

discussions on wafer-scale arrays.

REFERENCES

[1] T. S. Rappaport et al., "State of the art in 60-GHz integrated circuits and systems for wireless communications," in Proc. IEEE, vol. 99, no. 8, pp. 1390–1436, Aug. 2011.

[2] M. Boers et al., "A 16TX/16RX 60 GHz 802.11ad chipset with single coaxial interface and polarization diversity ," IEEE J. Solid-State Circuits, vol. 49, no. 12, pp. 3031–3045, Dec. 2014.

[3] S. Onoe, "Evolution of 5G mobile technology toward 2020 and beyond," in IEEE Int. Solid-State Circuits Conf. (ISSCC) Dig. Tech. Papers, Feb. 2016, pp. 23–28.

[4] T. Kamgaing et al., "Ultra-thin dual-polarized millimeter-wave phased-array system-in-package with embedded transceiver chip," in IEEE MTT-S Int. Microw. Symp., pp. 1–4, May 2015.

[5] S. Zihir et al., "60-GHz 64- and 256-elements wafer-scale phased-array transmitters using full-reticle and subreticle stitching techniques," IEEE Trans. Microw. Theory Techn., vol. 64, no. 12, pp. 4701–4719, Dec. 2016.

A 1-4 GHz 4x4 MIMO Receiver with 4 Reconfigurable Orthogonal Beams for Analog Interference Rejection

Sajad Golabighezelahmad[1], Eric Klumperink[2], Bram Nauta[3]

IC Design Group, University of Twente, Enschede, The Netherlands

[1]s.golabighezelahmad@utwente.nl, [2]e.a.m.klumperink@utwente.nl, [3]b.nauta@utwente.nl

Abstract — A highly reconfigurable multi-beam MIMO receiver with 4 RF inputs and 4 outputs is proposed, allowing for digital MIMO but also analog interference rejection by spatial notch filtering through 4 reconfigurable orthogonal beams. A segmented constant-Gm vector modulator with improved interference tolerance and RF frequency range is proposed, allowing current-domain beamforming before I-V conversion by a transimpedance amplifier. A 1-4 GHz 22 nm FD-SOI prototype chip achieves >29 spatial notch suppression for in-band interference signals. In the notches, an IIP3 of +17 dBm and in-band B1dB of -12 dBm is achieved at 44.5 dB gain. Sub-3dB system noise figure is achievable in the corner points of vector modulator constellation. On the circle points, noise figure degrades about 2.5 dB. However, in-band, in-notch B1dB and IIP3 improve by 32dB and 43dB, respectively. The chip of 0.52 mm² active area consumes 75-115mW at an LO-frequency of 1-4 GHz from a 0.8V supply.

Keywords — Spatial filtering, Analog beamforming, Multiple-Input Multiple-Output (MIMO), Receiver, Vector Modulator, Interference rejection.

I. INTRODUCTION

The huge growth in wireless communication devices in scarce radio spectrum makes mutual interference increasingly a bottleneck. Spatial interference suppression by analog beamforming [1], [2], [3] can help but its Multiple-Input Single-Output (MISO) nature lacks the flexibility that Multiple-Input Multiple-Output (MIMO) systems can provide, e.g. to realize adaptive beamforming and support both spatial diversity and multiplexing gains. However, in digital MIMO, the ADCs are exposed to strong blocker signals. Adding analog spatial filtering before entering the ADCs can relax related dynamic range limits [4], [5], [6].

In this work, we present a new reconfigurable architecture supporting both digital MIMO and analog spatial filtering. Compared to prior art, performance in RF frequency range and linearity are improved in a power efficient way.

II. MULTI-BEAM RECEIVER ARCHITECTURE

Analog spatial notch filtering circuits can be added to a MIMO receiver array in different ways to reject interferers. In [4], [5], this is done by creating a beam in the direction of the interferer and subtracting a scaled version of the interferer from each of the individual antenna signals exploiting active gm blocks in baseband. Although frequency translated notch filtering is added to reduce RF gain somewhat, distortion in the active gm blocks poses a linearity limitation [5]. To improve interference tolerance, we propose to directly cancel the interferer current at the LNTA output in the current domain

before baseband I-V conversion and gain. Conceptually, we target multi-beam MIMO for interference rejection using multiple RF phased arrays. A traditional approach for generating multi-beam is a Butler matrix as exploited in [7]. However, the beams from a Butler matrix have fixed angular positions, lacking the flexibility for arbitrary spatial interference cancellation. While a passive Butler matrix can be highly linear, it consist of bulky passive devices that are dedicated for a rather limited and fixed frequency band and not very CMOS friendly. In this paper, a flexible and wideband multi-beam receiver is implemented using multiple RF phased arrays. We propose a novel architecture targeting orthogonal beams with a programmable spatial direction, exploiting a modified constant-Gm vector modulator [3].

Conceptually, separation of the interferer and desired signals in different beams is the key goal. As illustrated in Fig.1, this can be done by a set of flexible orthogonal beams for which each output beam is aligned with the nulls of other outputs. Thus steering one of them e.g. B_1 in Fig.1 towards the interferer results in its rejection at the remaining outputs. Note that in this scenario, recovery of the individual antenna signals is still possible in the digital domain. To do so, one could simply throw away the heavily distorted output beam containing the interference ($B_1 = 0$). By applying the inverse of the matrix representing the multi-beam RF front-end ($A_{N \times N}$), the input signals are recovered and a spatial notch is formed. In this paper we will just demonstrate a programmable orthogonal beam architecture. However, once flexibly programmable analog beamforming is implemented, more complex beam patterns with multiple deep notches are possible to cancel several in-band interferers simultaneously. For this purpose, both variable phase shift and gain are needed that we achieve here using constellation points of a vector modulator. The details of vector modulator are explained in the next section.

Fig.1 shows the block diagram of multi-beam RX. The LNTAs convert the input voltages to currents and the subsequent frequency down conversion, phase shift, and amplitude weighting all are performed in the current domain, allowing for interference cancelling by simple current summation at the input of Trans-Impedance Amplifiers (TIA). As this summing is done before I-V conversion linearity is improved. Moreover, frequency translated filtering at the output of the LNTA, which typically limits the RF bandwidth [4], [3], is not needed here.

978-1-7281-1702-7/19 $31.00 © 2019 IEEE

2019 IEEE Radio Frequency Integrated Circuits Symposium

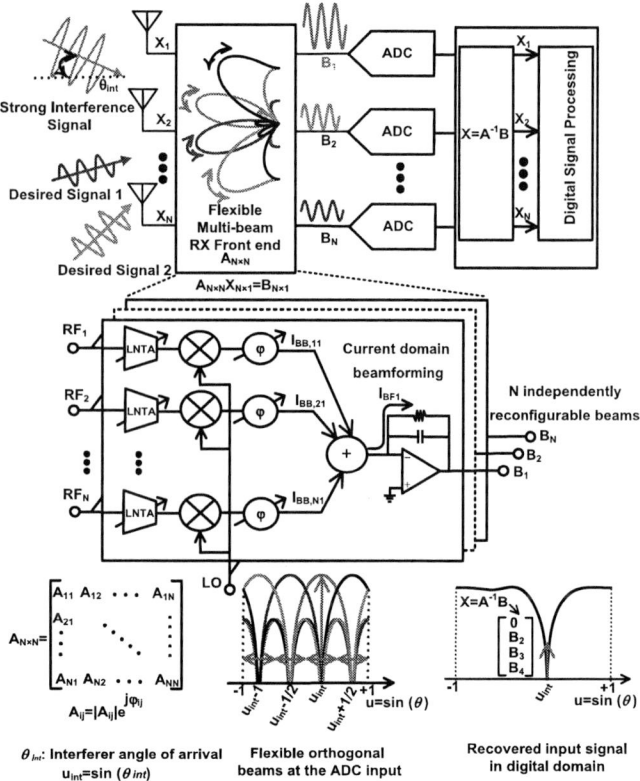

$A_{N \times N} X_{N \times 1} = B_{N \times 1}$

$$A_{N \times N} = \begin{bmatrix} A_{11} & A_{12} & \cdots & A_{1N} \\ A_{21} & & & \\ \vdots & & \ddots & \vdots \\ A_{N1} & A_{N2} & \cdots & A_{NN} \end{bmatrix}$$

$A_{ij} = |A_{ij}| e^{j\varphi_{ij}}$

θ_{int}: Interferer angle of arrival
$u_{int} = \sin(\theta_{int})$

Flexible orthogonal beams at the ADC input

Recovered input signal in digital domain

Fig. 1. Proposed multi-beam MIMO receiver using multiple RF phased arrays. Flexible orthogonal beams are created to form a notch at interference signal's angle-of-arrival.

III. CIRCUIT IMPLEMENTATION

A 4x4 multi-beam receiver is implemented as shown in Fig. 2, clearly showing the matrix operation. Each element of the matrix is implemented using a vector modulator, providing a complex weight for beamforming. This implementation is based on a low-power constant-gm vector modulator introduced in [3]. In this approach, a RF front-end including an LNTA followed by a 4-phase passive-mixer is sliced up into a number of binary weighted smaller cells. By static routing switches in the baseband part, each slice can contribute either to the I+, Q+, I- or Q- baseband output, producing a constellation square. A main limitation of the constant-gm vector modulator in [3] is its low gain and limited RF bandwidth. This is illustrated in Fig.3a, where the baseband signals are averaged using charge sharing at the output capacitors. This is power efficient but also results in voltage-mode operation of the passive mixer, hence the LNTA experiences a large voltage swing at its output node. High-frequency RF gain is limited by a pole at the LNTA output caused by the parasitic capacitance to ground (C_P in Fig. 3b) and a high output resistance.

To circumvent this bottleneck in frequency range we propose a vector modulator with current summation at the input of the TIA. Since the TIA provides low input impedance, the LNTA sees a low output impedance (as shown in Fig. 3b), leading to extended RF bandwidth of vector modulator.

Having less swing at the LNTA output also results in improved in-band output-referred linearity (OIP3). Moreover, in voltage mode operation, the amplification is limited to the intrinsic voltage gain of the LNTA [3], while a higher gain can be achieved exploiting a TIA with high feedback resistance.

Each vector modulator provides a 200 ohm input impedance employing self-biased inverter-based LNTAs, hence, 50 ohm low-noise input matching at each antenna interface is achieved since four vector modulators are connected to one RF input. The on-chip clock generation block produces 25% clock signals to drive the mixer switches. 15 slices are used to implement the vector modulator, which results in a 16 × 16 points square constellation. Using the points nearby the biggest circle fitting into the constellation square, a null depth of approximately 29 dB is achievable [3]. The TIA is implemented using simple inverter amplifiers using transistors with large channel lengths ($L = 500nm$) to achieve high open-loop gain and low flicker noise. Both the feedback resistor and capacitor of the TIAs are implemented in a programmable way with 3-bit digital control to provide a variable gain and baseband bandwidth for the receiver.

IV. MEASUREMENT RESULTS

A 1-4 GHz prototype chip with active area of 0.52 mm^2 (see Fig. 4a) was designed and fabricated in GlobalFoundries 22 nm FD-SOI technology and a QFN package with test fixture on PCB was used. The measured constellation is shown in Fig.4b, where points with equal weights but variable phase nearby the biggest circle fitting into the square constellation are used to create analog beams and notches at arbitrary direction. Fig.5a shows the measured beam patterns reconfigured to produce orthogonal beams targeting a notch at u=0.25 ($\theta = 14.5°$) with FLO=2.5 GHz at 1 MHz baseband. It is seen that a null depth of greater than 29 dB is achieved. To evaluate the notch suppression bandwidth, a notch is formed at broadside ($u = 0$) and RF frequency is swept at a fixed LO frequency of 2.5 GHz. As it is shown in Fig.5b, the achieved 20 dB notch suppression bandwidth is 2 GHz. The wideband suppression is due to immediate current-domain beamforming after voltage to current conversion by LNTA mainly because the current summation is not affected by parasitics of the circuit nodes. Fig.6a shows the total gain at 1 MHz baseband frequency across the swept LO frequency. Equivalent single element double sideband noise figure (NF$_{DSB,eq}$) versus LO frequency is shown in Fig.6b. NF$_{DSB,eq}$ is measured using single excitation method introduced by [4], [5]. Noise figure is measured both on the biggest circle fitting into the square constellation and on its corner points where all the vector modulator cells are in-phase and contribute to the signal gain, hence improving the NF by 2.5 dB. The best achievable system noise figure hence occurs in the corner points (providing analog beams and notches at u=0, u=1/2, u=-1/2, u=1 and digital beamforming can be performed later at arbitrary direction in the digital domain). Taking into account the SNR improvement of 6 dB, a sub-3dB system noise figure can be obtained. When exploiting analog beamforming

2019 IEEE Radio Frequency Integrated Circuits Symposium

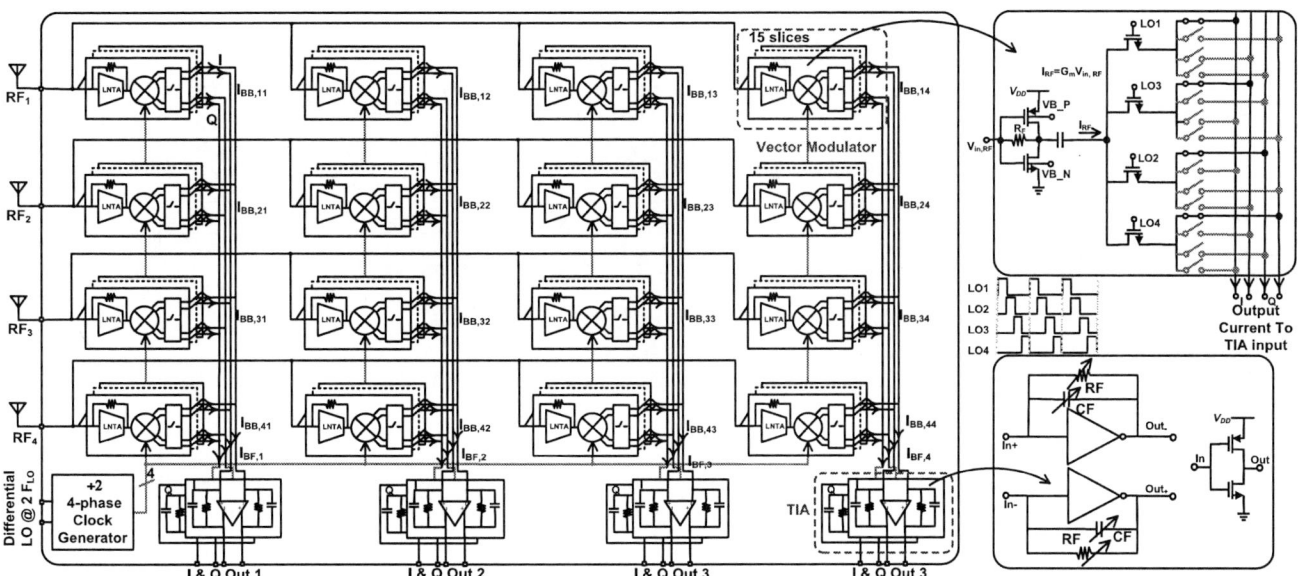

Fig. 2. Proposed multi-beam 4x4 MIMO receiver employing constant-gm vector modulators with current summing (interference rejection) at the input of the TIA.

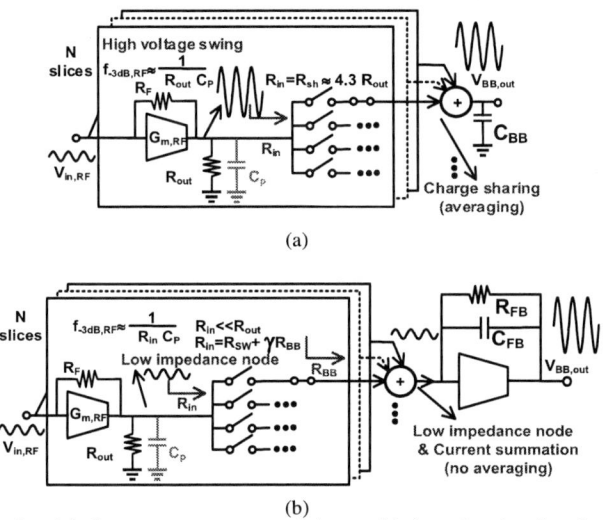

Fig. 3. (a) Constant-gm vector modulator with beamforming by charge sharing, i.e. I-V conversion with averaging [3]; (b) Proposed constant-gm vector modulator with current summing. This results in more achievable gain, better linearity, and RF frequency range.

with points nearby the circle in Fig.4b to realize notches in arbitrary directions, the noise figure degrades about 2.5 dB as mentioned above, but linearity improves much more, i.e. Spurious Free Dynamic Range benefits.

In-beam/in-notch IIP3 versus offset frequency of first tone are presented in Fig.7a. The results show +3 dBm out-of-band linearity in the main beam direction, limited by the linearity of the LNTA. In-band IIP3 of -26 dBm is achieved, corresponding to OIP3 of +18.5 dBm (44.5 dB gain). In measuring in-notch IIP3, it is assumed that a two-tone interfere is arriving at the receiver array from the spatial notch direction. Since the intermodulation product is suppressed by the notch filtering, the in-band IIP3 is improved to +17 dBm for in-notch incidence. This IIP3 remains approximately constant across the offset frequency, which fits to the wide

20 dB rejection bandwidth observed in Fig.5b. Fig.7b shows the conversion gain for the desired signal in terms of in-band, in-beam/in-notch blocker power level; putting a notch at the blocker's angle improves the blocker tolerance about 32 dB, achieving a high B1dB of -12 dBm.

Table1 summarizes the performance of the chip and benchmarks it with the state-of-the-art. The wideband RF bandwidth of the receiver covers the frequency range from 1 GHz to 4 GHz. The in-band, in-beam OIP3 as well as in-band, in-notch IIP3 values are the highest among MIMO receivers. Moreover, in-notch B1dB of -12dBm at 44.5 dB gain is the best achieved blocker tolerance for MIMO receivers. Taking into account 6 dB SNR improvement due to the analog beamforming, the receiver achieves sub-3 dB noise figure at the corner points. The noise performance is slightly worse than digital MIMO chips presented in [4], [5], but this is compensated by very significant improvements in linearity. The chip consumes 75-115 mW at LO frequency of 1-4 GHz, operating at faster clock speed compared to prior art, demonstrating power-efficient implementation considering the direct dependency of dynamic power consumption on LO clock frequency.

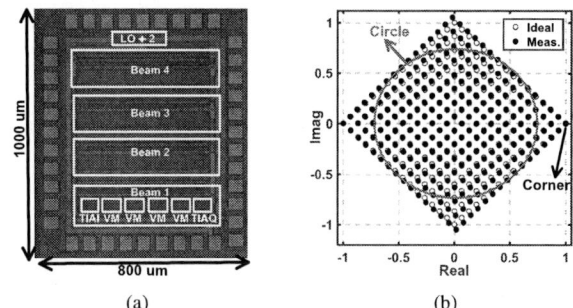

Fig. 4. (a) Die micrograph; (b) Constellation points.

978-1-7281-1702-7/19 $31.00 © 2019 IEEE   341

Table 1. Performance summary and comparison with state-of-the-art.

Architecture	ISSCC11 [1]	ISSCC13 [2]	JSSCC17 [3]	RFIC16 [6]	ISSCC16 [4]	ISSCC17 [5]	This work
Process	65 nm	65 nm	65 nm	65 nm	65 nm	65 nm	22 nm FD-SOI
Active area (mm^2)	0.44	0.97	0.2	3.8	1.69	1.44	0.52
Supply (V)	1.2	1.2	1	1.3-1.5	1.2	1.2	0.8
Array elements	4x1 MISO	4x1 MISO	4x1 MISO	4x4 MIMO	4x4 MIMO	4x4 MIMO	4x4 MIMO
RF Frequency (GHz)	1-4	0.6-3.6	1-2.5	10	0.1-1.7	0.1-3.1	1-4
Total Gain (dB)	16	-1	12	14	41	43	44.5
Single Element $NF_{DSB,eq}$ (dB)	10	5-8	6	9.5	2.2-4.6	3.4-5.8	10.9-11.6(Circle) 8.4-9.5(Corner)
Max.Spatial Suppression(dB)	>20	-	29	32	32	51-56	>29
20 dB Spat. Supp. Band. (MHz)	-	-	-	100 (1%)	15 (3%)	320 @ 500 MHz (64%)	2000 @ 2.5 GHz (80%)
In-band/in-beam OIP3(dBm)	+15	-	13	-	0	+11	+18.5
In-band/in-notch IIP3(dBm)	-	+9	+20	-	-7	+1	+17
Out-of-band IIP3 (dBm)	-	+20	+5	-	+11	-5	+3
In-notch B1dB (dBm)	-	-	-	-	-	-25	-12
Power (4 Elements, mW)	308@2.5 GHz	66-195 40 mW+43 mW/GHz	26-36 19mW+6.7 mW/GHz	145 (1 Elem.)	148@0.5 GHz	116-147 115mW+10 mW/GHz	75-115 62mW+13.2 mW/GHz

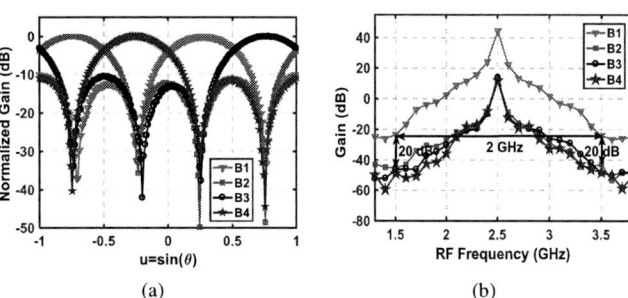

Fig. 5. (a) Measured beam patterns @ FLO=2.5 GHz with a notch at u=0.25; (b) 20 dB notch suppression bandwidth at broadside notch ($u = 0$).

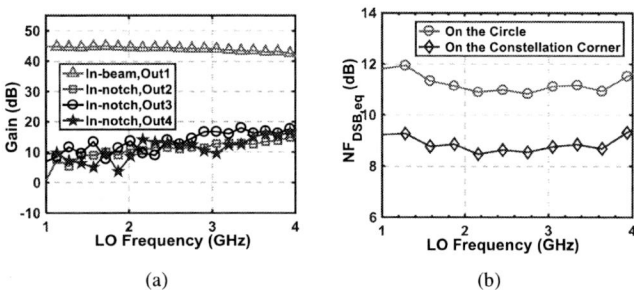

Fig. 6. (a) Total conversion Gain and (b) $NF_{DSB,eq}$ across LO frequency.

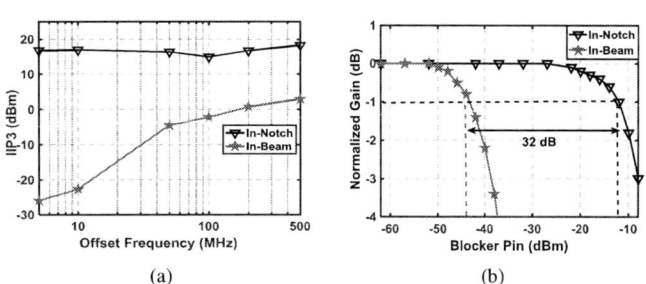

Fig. 7. (a) In-beam/in-notch IIP3 as function of offset frequency of first tone; (b) Conversion gain for desired signal versus in-band, in-beam/in-notch blocker input power.

V. Conclusion

In this paper, a highly flexible and reconfigurable multi-beam receiver was presented for spatial in-band blocker rejection. The maximum RF operating frequency is extended up to 4 GHz. Interference suppression early in the receiver chain before voltage amplification improves its linearity performance; in-band, in-notch IIP3 and B1dB of +17 dBm and -12 dBm, respectively, at 44.5 dB gain.

Acknowledgment

The authors would like to thank Gerard Wienk for CAD assistance and Henk de Vries for measurement setup. The authors also thank GlobalFoundries for supporting chip fabrication.

References

[1] M. C. M. Soer *et al.*, "A 1.0-to-4.0GHz 65nm CMOS four-element beamforming receiver using a switched-capacitor vector modulator with approximate sine weighting via charge redistribution," in *ISSCC*, Feb 2011, pp. 64–66.

[2] A. Ghaffari *et al.*, "Simultaneous spatial and frequency-domain filtering at the antenna inputs achieving up to +10dBm out-of-band/beam P1dB," in *ISSCC*, Feb 2013, pp. 84–85.

[3] M. C. M. Soer *et al.*, "Beamformer With Constant-Gm Vector Modulators and Its Spatial Intermodulation Distortion," *IEEE J. Solid-State Circuits*, vol. 52, no. 3, pp. 735–746, March 2017.

[4] L. Zhang *et al.*, "A scalable 0.1-to-1.7GHz spatio-spectral-filtering 4-element MIMO receiver array with spatial notch suppression enabling digital beamforming," in *ISSCC*, Jan 2016, pp. 166–167.

[5] ——, "A 0.1-to-3.1GHz 4-element MIMO receiver array supporting analog/RF arbitrary spatial filtering," in *ISSCC*, Feb 2017, pp. 410–411.

[6] S. Jain *et al.*, "A 10GHz CMOS RX frontend with spatial cancellation of co-channel interferers for MIMO/digital beamforming arrays," in *IEEE Radio Frequency Integrated Circuits Symp.*, May 2016, pp. 99–102.

[7] A. Tork *et al.*, "Reconfigurable X-Band 44 Butler array in 32nm CMOS SOI for angle-reject arrays," in *IEEE MTT-S International Microwave Symp.*, May 2016, pp. 1–4.

AUTHOR INDEX

Abughazaleh, Salah....................................251
Adam, G. C..111
Adepu, Naresh..243
Aflatouni, Firooz 319
Afshari, E...331
Ahasan, Sohail...103
Ahmed, Ahmed S. H. 135
Alakusu, U. 111
Alshammary, Hussam...................................143
Amakawa, Shuhei 175
An, Kyu Hwan............................179, 247, 287
Andersson, Stefan 303
Anderwald, M...283
Antonov, Yury 203
Arai, Tomoyuki...307
Arfaoui, W..27
Avenier, Grégory 23
Azam, Ali..295
Azizi, Shahrnaz 255
Babaie, Masoud..107
Bachmanek, Greg..251
Bai, Zhidong 295
Bajor, Matthew 95
Bajoria, Shagun..315
Baltus, Peter G. M...3
Barnett, K. 27
Bassi, M...167
Beevi K. T., Asma...255
Bellaouar, A..27, 31, 323
Bellaouar, Abdellatif.............................207, 327
Ben-Haim, D. 163
Berro, R. 299
Berroth, M..215
Bershansky, S. 163
Bevilacqua, A...167
Bhat, Anoop Narayan 155
Binaie, Ali 103
Blaauw, David 263
Blampey, Benjamin 19
Bolatkale, Muhammed 315
Bonen, S. 111
Bordelon, J. 27
Borremans, M...299
Bossu, G. 27
Boumaiza, Slim..63
Brandelstein, Ohad 251
Breems, Lucien J...315
Brekelmans, Hans 315
Brown, Thomas W. 251
Buchali, F..215
Buckwalter, James F...............................99, 143
Buczko, Michel...23
Burak, Abdurrahman.......................................75
Byrd, Justin M...115
Cali, Joseph D...115

Cao, Jinzhou 51
Cao, Yuhe...147
Caputa, Peter 303
Carlowitz, Christian 131
Carlton, Brent R. 255
Carta, Corrado 131
Cartwright, Justin A. 115
Caruso, M. 283
Casper, Bryan 235
Çelik, Umut 187
Chadwick, Steve A..115
Chan, L. H. K. 31, 323
Chang, Mau-Chung Frank 231
Charbon, Edoardo...107
Chen, Boshen 47
Chen, Huan-Neng 231
Chen, Liang 183, 311
Chen, T. 27, 31, 323
Chen, W. T. 111
Chen, Zhe 3
Chevalier, Pascal 23
Chew, K. W. J. 27, 31, 323
Chi, Baoyong 123
Chi, Taiyun 275
Chiang, Patrick Yin 219
Chiotellis, Nikolaos 263
Cho, Yunsung 179, 247
Choi, Sung Tae 287
Chow, W. H. 31, 323
Chung, Hyunchul 119
Chuo, Li-Xuan 263
Clavera, Berta Trullas 91
Clemente, Antonio 19
Cohen, Emanuel 87
Cressler, John D. 15
Cui, Bolun 211
Dadash, M. S...111
Dai, Yuefei...39, 67
Dalmia, Sidharth 251
Daneshgar, Saeid 235
Danneville, François 35
Dantoni, Francesco 155
Dasalukunte, Deepak 255
Dascurcu, Armagan 243
Dasgupta, Kaushik 235
Daughton, D. R. 111
Davis, Brandon 251
De Ranter, C. 299
Dehng, GK..267
Dehos, Cedric 19
Deng, Wei 279
Dinc, Tolga 43
Dong, Ruibing 175
Dorrance, Richard 255
Druml, M. 283

AUTHOR INDEX

Druml, R. 283
Du, Jieqiong....................................231
Du, X.-Q. 215
Duan, Zongming....................................39
Ducournau, Guillaume 23
El-Aassar, Omar 159
Elgaard, Christian 303
Ellinger, Frank 55, 131
Embabi, S. 27
Englund, Mikko 203
Entesari, Kamran 151
Ershadi, Ali....................................151
Ezri, Doron 87
Fadila, Ashbir Aviat....................................279
Farid, Ali A. 135
Farouk, Mahitab 63
Finocchiaro, Salvatore 155
Fisher, Iris 251
Flynn, Michael P. 223
Freiman, Amit 251
Fric, Tom 315
Fridi, Ahmed R. 327
Fritsche, David 131
Fritschij, Franco....................................315
Fu, Xi....................................279
Fujishima, Minoru 175
Galioglu, Arman....................................243
Gao, Hao 3
Gao, Li 239
Gertman, I. 163
Ghaleb, Hatem 131
Gianesello, Fréderic 23
Gidel, Vincent....................................23
Ginzberg, Nimrod 87
Giry, A. 299
Golabighezelahmad, Sajad....................................339
Gong, M. J. 111
Gong, Yunyi....................................15
Gonzalez-Jimenez, Jose Luis....................................19
Grbic, Anthony 263
Grözing, M. 215
Guitard, Nicolas 23
Gurbuz, Ozan 51
Gurbuz, Yasar....................................75
Gutierrez, L. E. 111
Hafshejani, E. Hadizadeh....................................331
Hagn, Josef 251
Hamza, Ahmed....................................143
Han, Guoxiang 95
Han, Jaeyeol 271
Hao, Yun 223
Haq, Faizan Ul....................................203
Haque, Tanbir 95
Hara, Shinsuke 175
Harame, D. 31, 111, 323

Harame, David L. 211
Harwood, Richard L. 115
Hasegawa, Nobumasa 307
Hay, Christopher....................................51
Hayden, Joseph S. 251
He, Jian 219
Helkey, Roger 99
Henderson, Greg 1
Heng, CL. 267
Hill, Cameron W. 143
Hofmann, Klaus....................................227
Hori, Shinichi....................................279
Horovitz, G. 163
Hosseinzadeh, Navid 99
Hsiao, CH. 267
Hu, Shang....................................219
Huang, Dong 39, 47
Huang, Min-Yu 275
Huang, Tzu-Yuan 275
Hung, Shih-Chang 291
Idjadi, Mohamad Hossein....................................319
Ilic, M. 283
Inac, Ozgur 251
Indirayanti, P. 299
Iordanescu, S. 111
Jain, Aditya 99
Jain, Sanket....................................243
Janardhanan, S. 27
Jang, Jae Sik 287
Jang, Seunghyun 7
Jang, Sunmin 223
Jaussi, James....................................235
Jensen, Jonathan 251
Jia, H. 111
Jin, LM. 267
Jo, Ikkyun 271
Johnson, Manoj 243
Joram, Niko 55
Jou, Chewn-Pu 231
Kadry, J....................................163
Kalyoncu, Ilker 75
Kane, Ousmane M. 35
Kaneko, Tomoya 279
Kang, Daehyun 287
Kang, Kai 59, 71
Kao, Te-Yu Jason 251
Kaps, Jonas 199
Kashani, M. Haghi 331
Kawabuchi, Masaru 279
Kawaminami, Masaru 263
Kaynak, Mehmet 75
Kim, Chulho 271
Kim, Hun-Seok 263
Kim, J. 31, 323
Kim, Jihoon....................................179, 247, 287

AUTHOR INDEX

Kim, Kihyun 287
Kim, Kwang-Seon 7
Kim, Meeran 287
Kim, Myoung-Gyun 271
Kim, Nam-Seog 271
Kim, Seokhyeon 247
Kim, Shinwoong 271
Kim, Yejoong 263
Kinget, Peter R. 95, 259
Klumperink, Eric 91, 339
Kodak, Umut 335
Koli, Kimmo 203
Kong, Sunwoo 7
Kosunen, Marko 203
Krishnamurthy, Sashank 139
Krishnaswamy, Harish 103, 243
Kristem, Vinod 255
Kubozoe, Ryo 279
Kunihiro, Kazuaki 279
Kuo, CM. 267
Kushner, Lawrence J. 115
Lakshminarayanan, Sreekesh 227
Lassche, Gerard 315
Le, S. T. 215
Lee, Hui-Dong 7
Lee, Jeong Ho 179, 247, 287
Lee, Jongwoo 271
Lee, Jooseok 247, 287
Lee, Kwang-Chun 7
Lee, Sangho 287
Lee, Sangyeop 175
Lee, Seokwon 271
Lee, Woojae 247
Lee, Wooram 43
Leenaerts, Domine M. W. 3
Legrand, Charles-Alex 23
Lepilliet, Sylvie 35
Levi, R. 163
Levinger, R. 163
Li, DP. 267
Li, Pei 39, 67
Li, Sensen 275
Li, Songhui 55
Li, Tong 83
Li, Yutian 123
Liao, Bingbing 39, 67
Liao, Qiwen 219
Lim, C. K. 31, 323
Lin, Fujiang 39
Lin, HT. 267
Lin, Jianfu 123
Liu, Bangan 279
Liu, Hanli 279
Liu, Huihua 59, 71
Liu, Jian 219

Liu, Renzhi 255
Liu, Yangzi 83
Liu, Yao 127
Lo, Chilun 271
Long, John R. 211
Lopez, Mario A. Santana 255
Low, EC. 267
Lu, Rundao 223
Lucci, L. 111
Lucci, Luca 35
Luo, Xun 79
Luxey, Cyril 23
Lv, Wei 39, 67
Ma, Qian 119
Ma, Ruichang 123
Madison, Gary M. 115
Maistro, D. Dal 283
Maksymova, I. 283
Malhotra, Harman 227
Mangal, Vivek 259
Mansour, R. R. 111
Mantravadi, M. 27
Manzillo, Francesco Foglia 19
Matsumoto, Daiki 279
Matta, John T. 115
Mehmood, Atif 199
Meier, Thomas 199
Mercier, E. 299
Meredith, James M. 115
Mertens, K. 283
Messaoudi, N. 111
Milon, Victor 23
Min, Byung-Wook 11
Min, Hao 83
Minn, Donggyu 287
Mirabbasi, S. 331
Miyoshi, Satoru 263
Moehlmann, Ulrich 315
Momeni, Omeed 171
Motoi, Keiichi 279
Mourot, R. 299
Mueller, M. 283
Mukatel, Tsvika 251
Müller, A. 111
Murakami, Tomotoshi 307
Muralidharan, Sriram 51
Nakamura, Takeshi 279
Natarajan, Arun 127, 243
Nauta, Bram 91, 155, 339
Navara, David 227
Nguyen, Huy Thong 195
Niehof, Jan 315
Niknejad, Ali M. 139
Ning, Kang 99
Oh, Michael 115

AUTHOR INDEX

Okada, Kenichi.................................279
Ong, S. N.............................27, 31, 323
Oo, W. L................................31, 323
Oshima, Naoki...............................279
Östman, Kim B..............................203
Otto, Michael.................................207
Oude-Essink, Bert...........................315
Padovan, F....................................167
Pan, Dongfang.................................39
Panazzolo, F..................................167
Pang, Jian....................................279
Parat, D......................................299
Park, Byungho................................251
Park, Byungjoon..........................179, 247
Park, Hyun-Chul..........................179, 247
Park, Jaehong............................179, 247
Park, Jeehoon...................................7
Park, Minyoung...............................255
Park, S......................................215
Park, Sangyong...............................247
Pasteanu, M..................................111
Peeters, Michael................................2
Peng, Yatao..................................107
Purushothaman, Vijaya Kumar 91
Qi, Nan......................................219
Qian, Huizhen Jenny..........................79
Qiu, Junjun..................................279
Quadrelli, F.................................167
Qunaj, Valdrin...............................191
Rassel, R.................................31, 323
Rebeiz, Gabriel M. 119, 159, 239, 335
Regev, Dror...................................87
Reiss, Norbert...............................227
Reynaert, Patrick........................187, 191
Reynier, P...................................299
Rodwell, Mark J. W...........................135
Roy, Ankur Guha..............................251
Rubino, C....................................283
Rueddenklau, U...............................283
Ruffino, Andrea..............................107
Rupakula, Bhaskara...........................335
Rutten, Robert...............................315
Ryynänen, Jussi..............................203
Saengchan, Rattanan..........................279
Saha, Prabir..................................51
Sakalas, Mantas...............................55
Sakalas, Paulius..............................55
Schatzmayr, E................................283
Scheiblin, Pascal.............................35
Schuh, K.....................................215
Sebastiano, Fabio............................107
Seet, W. H................................31, 323
Seol, Gyungseon..............................271
Serhan, A....................................299
Shen, SC.....................................267

Shilo, Shimi..................................87
Shirane, Atsushi.............................279
Shoham, Doron................................251
Shu, Yiyang...................................79
Siddabathula, M...............................27
Siligaris, Alexandre..........................19
Simsek, Arda.................................135
Singh, Amitoj................................251
Singh, Baljit................................251
Sjöland, Henrik..............................303
Son, Juho......................179, 247, 287
Song, Zhe......................................3
Song, Zheng..................................123
Sover, Raanan................................251
Stadius, Kari................................203
Stoffels, Hans...............................315
Stuenkel, Mark E.............................115
Suh, Bohee...................................287
Suh, Bosung...................................11
Syed, Shafiullah.............................207
Takano, Kyoya................................175
Tang, Yang....................................47
Tarkeshdouz, A...............................331
Taylor, R.....................................27
Teng, Jeffrey W...............................15
Thakkar, Chintan.............................235
Thurner, P...................................283
Tiebout, M...............................167, 283
Tran, Man....................................327
Tseng, PS....................................267
Tsodik, Genadiy...............................87
Tsvelykh, I..................................283
Turner, Steven Eugene........................115
Uhl, A.......................................215
Utagawa, Yoshiyuki...........................307
Valdes-Garcia, Alberto........................43
van Der Ploeg, Hendrik.......................315
van Der Zee, Ronan...........................155
van Sinderen, Jan............................315
Vehovc, S....................................283
Verma, Ashutosh..............................271
Voinigescu, S. P.............................111
Walling, Jeffrey S...........................295
Wan, C. W. F..............................31, 323
Wang, Fei....................................275
Wang, Hao....................................171
Wang, Hua................................195, 275
Wang, X. Shawn...............................231
Wang, Yan..............39, 47, 67, 183, 311
Wang, Yun....................................279
Wentzloff, David.............................263
Westesson, Eric..............................303
Wong, Chien-Heng.............................231
Wong, J. S................................31, 323
Wright, John..................................95

AUTHOR INDEX

Wu, Bowen 39
Wu, MJ. 267
Wu, Nanjian 219
Wu, Rui 279
Wu, Tianjun 59, 71
Wu, Weiping 311
Wu, Yanhui 39
Wu, Yunqiu 59, 71
Wuertele, J. 283
Xia, Jingjing 63
Xiao, Bin 251
Xiong, Liang 83
Xu, Daiguo 39
Xu, Hongtao 83
Xu, Hua 39
Yamaura, Shinji 307
Yan, Na 83
Yang, KH. 267
Yang, Sung-Gi 179, 247, 287
Yang, W. 267
Yao, Chih-Wei 271
Yasuda, Makoto 263
Ye, Jialiang 123
Yin, Bozhi 219
Yin, Yun 83
Yoo, Sang-Min 291
Yoo, Si-Wook 291
Yoon, Youngchang 287
Yoshida, Takeshi 175
You, Dongwon 279
Yu, Yiming 59, 71
Zaghi, M. 283
Zhan, Kai 243
Zhang, C. 27, 31, 323
Zhang, Chi 207, 327
Zhang, Frank 207
Zhang, Haosheng 279
Zhang, Lei 67, 183, 311
Zhang, Li 47, 183, 311
Zhao, Chenxi 59, 71
Zhou, Jia 231
Zhou, Jie 79
Zhou, Jin 147
Zhu, Huabing 47
Zhu, Wei 67
Zhu, Yanping 67
Zihir, Samet 335